21st International Mass Spectrometry Conference (IMSC 2016)

Toronto, Canada
20 – 26 August 2016

ISBN: 978-1-5108-4931-0

Printed from e-media with permission by:

Curran Associates, Inc.
57 Morehouse Lane
Red Hook, NY 12571

Some format issues inherent in the e-media version may also appear in this print version.

Printed by Curran Associates, Inc. (2018)

Additional copies of this publication are available from:

Curran Associates, Inc.
57 Morehouse Lane
Red Hook, NY 12571 USA
Phone: 845-758-0400
Fax: 845-758-2633
Email: curran@proceedings.com
Web: www.proceedings.com

imsc 2016 toronto

21st International Mass Spectrometry Conference

August 20-26, 2016
Metro Toronto Convention Centre
255 Front Street West, Toronto, ON, Canada

Book of Abstracts by Author

Hosted by
The Canadian Society for Mass Spectrometry

CSMS SCSM
Canadian Society for Mass Spectrometry
Société Canadienne de Spectrométrie de Masse

21ST INTERNATIONAL MASS SPECTROMETRY CONFERENCE
Toronto, Ontario, Canada
August 20 – 26, 2016

Abdel Rahman, Anas,
King Faisal Specialist Hospital and Research Center

Contributing Authors:
Minnie Jacob, Department of Genetics, King Faisal Specialist Hospital and Research Center (KFSHRC), Riyadh, Saudi Arabia; Abbas Hawwari, King Abdullah International Medical Research Center, Al-Ahsa, Saudi Arabia; Hamoud AlMousa Department of Genetics, King Faisal Specialist Hospital and Research Center (KFSHRC), Riyadh, Saudi Arabia; Ayodele A. Alaiya, Stem Cell & Tissue Re-Engineering Program King Faisal Specialist Hospital and Research Center (KFSHRC), Riyadh, Saudi Arabia; Majed Dasouki, Department of Genetics, King Faisal Specialist Hospital and Research Center (KFSHRC), Riyadh, Saudi Arabia.

Title:
Combined Metabolomic and Proteomic Studies of DOCK8-Hyper IgE Syndrome

Introduction:
Hyper-IgE syndrome is a serious inborn error of immunity caused by mutations in DOCK8. Symptomatic patients with DOCK8 present with severe complication (high serum IgE), which may cause high morbidity and mortality. The overlapping symptoms and the unspecific molecular testing are encouraging to find novel mass spectrometry biomarkers specific for DOCK8 deficiency with clinical diagnostic value.

Methods:
Lymphoblastoid cell lines were established from a family with DOCK8 deficiency. In solution LC-Mass Spectrometry (LC-MS/MS) based proteomics and metabolomics profiling was performed. Data was analyzed using Ingenuity Pathway Analysis (IPA) and MetaboAnalyst respectively. The statistically significant features were combined to limit the pathway searching and molecular mapping.

Results:
Significant perturbations in several metabolic pathways including the nucleotide pool (adenosine, ADP, AMP, GDP, dGTP, IDP), nitrogen metabolism, and amino acid pathways were found. Thioredoxin was upregulated as well, probably in response to an altered redox state. 99 of the 108 dataset were mapped in the IPA, only two of them were implicated in two network system, GDP and ATP. GDP is implicated in cell signaling, molecular transport, nucleic acid metabolism, while ATP in cell signaling and interaction, cellular movement, and function Combined proteomics and metabolomics analysis identified multiple novel biochemical abnormalities in DOCK8 deficiency which most likely contribute to its pathogenesis and associated clinical manifestations. Further studies in additional patients with DOCK8 deficiency are needed to validate these findings in order to identify a disease-specific biomarker

Novel Aspect:
Develop novel mass spectrometry-based biomarkers which are specific for DOCK8 deficiency

Abu Kassim, Nur Sofiah,
The University of Queensland

Contributing Authors:
Paul N. Shaw, School of Pharmacy, University of Queensland, Brisbane, QLD; Amitha K. Hewavitharana, School of Pharmacy, University of Queensland, Brisbane, QLD.

Title:
Effect of mobile phase pH: implications for the LC-MS/MS bioanalysis of vitamin D –sulfate compounds

Introduction:
Mobile phase composition and pH in LC-MS/MS analysis are known to impact analyte ionisation, thus affecting the amounts and types of the ions formed. Bioanalytical method development should address the effects of different mobile phase pH values on the MS response, in addition to assessing LC peak shape and resolution; four vitamin D-sulfate compounds were investigated in this regard.

Methods:
Mobile phases were prepared using two buffers (0.01M) in water/ methanol at pH 2.56 (formic acid) and 7.50 (4-Methylmorpholine adjusted with formic acid). Chromatographic separations were carried out using a Pentafluorophenyl column. Deprotonated species ([M–H]–) of vitamin D-sulfate (D3-S, D2-S, 25(OH)D3-S and 25(OH)D2-S) were detected in ESI negative ion mode.

Results:
Vitamin D-sulfates are acidic in nature due to the presence of the ionisable sulfate moiety, and capable of producing negatively charged ions, by deprotonation. Retention times in minutes (at pH 2.56, 7.50) are D2-S (19.23, 7.23), D3-S (19.25, 7.24), 25(OH)D3-S (10.21, 4.57) and 25(OH)D2-S (10.22, 4.57). Retention times were longer and peak areas were reduced at low pH. These results indicate that the buffer ions in the mobile phase are participating in the retention process, and that in gas phase (of ESI source) the ionisation is favoured at high pH.

Novel Aspect:
The dependence mass spectrometry on LC mobile phase conditions is important in vitamin D-sulfates assay, and potentially, for other similar compounds.

Abu Kassim, Nur Sofiah,
The University of Queensland

Contributing Authors:

N.S Abu Kassim, School of Pharmacy, University of Queensland, Brisbane, QLD.; Amitha K. Hewavitharana, School of Pharmacy, University of Queensland, Brisbane, QLD.; Paul N. Shaw School of Pharmacy, University of Queensland, Brisbane, QLD.

Title:

Comparison of two protein precipitation methods for vitamin D LC/MS assay of human serum

Introduction:

Protein precipitation is a common approach used to release vitamin D compounds bound to lipids and proteins in serum/ plasma. Acetonitrile (ACN) and trichloroacetic acid (TCA) are two very effective reagents used for protein precipitation. In this study these two reagents were compared in a vitamin D LC/MS assay application.

Methods:

Serum sample (250uL) was mixed with 500µL of ACN or TCA (2%; w/v). The mixture was incubated for 15 mins, and then centrifuged for 10 mins. The supernatant was evaporated to dryness using a freeze drier. The residue was reconstituted in mobile phase before LC-MS analysis. Triplicate measurements were performed and peak areas were used to assess the efficacy of each set of conditions.

Results:

In this study, consistently lower (< 50%) peak areas were observed for the following vitamin D compounds: 25(OH)D3-S, 25(OH)D2-S, D3-S, D2-S, D3, D2, 25(OH)D3, 1,25(OH)2D3 and 24,25(OH)D3, when precipitating human serum with trichloroacetic acid. We were unable to detect three minor compounds (25(OH)D2, 1,25(OH)2D2 and 24,25(OH)D2) using either extraction method due to their very low levels in sample. One or both of the reasons listed below may be responsible for the observed differences between ACN and TCA as precipitants: 1. ACN is more efficient in releasing the vitamin from the matrix, 2. The ACN precipitated extract has less of the other compounds that can co-elute with vitamin D compounds therefore the ion suppression of vitamin D compounds by co-eluting compounds in the MS detector are likely reduced.

Novel Aspect:

The results illustrate the importance of optimising the extraction of the vitamin D compounds from serum prior to their measurement using LC-MS.

Agar, Jeffrey,
Northeastern University

Contributing Authors:

Jeniffer Quijada, Northeastern University; Nicholas Schmitt, Northeastern University; Christopher Thompson Bruker Daltonics; Michael Easterling, Bruker Daltonics.

Title:

Metabolic Labeling/Ultra-high resolution FTICRMS for Quantitative Neutron Encoded Lipidomics and Proteomics of Any Organism or Feed on an LC Timescale

Introduction:

Many metabolic labeling methods have been developed and proven effective for proteins or lipids. Few metabolic labeling methods, however, are effective for both proteins and lipids. Moreover, we are aware of no combined technique that employs isotopic fine structure. Here isotopic labeling is accomplished with inexpensive deuterated water, and differences expression are measured on an LC-timescale

Methods:

Labeled 2H-water is spiked into standard Drosophila feed and incorporated into all biomolecules. This results in (arbitrary) changes of isotopologue distributions (ACID), creating "heavy" biomolecules for isotope dilution MS. Ultra-high res. MS were acquired on a FT-ICR with a dynamically harmonized ICR cell, employing both absorption mode processing (AMP) and quadrupolar overtone detection (2ω).

Results:

During metabolic labeling the change in a molecule's mass depends upon its size. Small molecule average masses change by only a fraction of a Da (at deuterium levels used here), resulting in overlap with the natural abundance isotopologue distributions. As a result, detecting labeled lipids, endogenous peptides, and small proteins is facilitated by resolving isotopic fine structure. Large molecules, on the other hand, change mass by many Da – enough to create distinct (non-overlaping) isotopologue distributions. Labeling was optimized so that when molecular mass increased to the point where isotopic fine structure could not be resolved, the isotopic distributions of natural abundance and labeled molecules were completely separated, and resolving isotopic fine structure was no longer necessary. We demonstrate that can routinely be achieved with lipids and small proteins on an LC-timescale

Novel Aspect:

A single method for metabolic labeling LC-MS quantitation of intact proteins and lipids. 2-omega quadruolar overtone detection (2ω) on a commercial MS

Ahmadi, Soha,
University of Toronto

Contributing Authors:

Heinz-Bernhard Kraatz, Department of Chemistry, University of Toronto, Canada; Shaolong Zhu, Department of Chemistry, York University. Canada; Soha Ahmadi Department of Chemistry, University of Toronto, Canada; Derek Wilson, Department of Chemistry, York University. Canada.

Title:

Implication of tau-metal interactions in Alzheimer's Diseases; investigation by ESI-MS and IMS-MS

Introduction:

Tau hyperphosphorylation and aggregation is linked to neurodegeneration in Alzheimer's diseases (AD), however the actual etiology remains unknown. There has been a growing interest in investigating the etiology of biometals accumulation and its role in AD pathogenesis. ESI and IMS-MS was implemented to provide fundamental insight in interactions of biometals with native and phosphorylated tau.

Methods:

Native Tau-441(n-tau) and phosphorylated Tau-441 (p-tau) were incubated with metal ions (Cu(II), Zn(II), Fe(II),and Fe(III)) and were analysed with ESI and IMS-MS in order to probe the metal-protein complexes and their effect on tau conformational and structural changes. Different tau microtubule binding domains were also analysed to investigate the metal binding sites in tau.

Results:

ESI-MS has shown the direct interactions of n-tau with Cu(II), Zn(II), Fe(II),and Fe(III). The stoichiometry of metal-tau complexes has been determined, which was different for each of these metal ions. The formation of metal-tau complexes led to structural changes in tau and this was confirmed by IMS-MS and circular dichroism spectroscopy. Interestingly, the binding affinity of tau to metal ions was not increased by phosphorylation using Src and GSK-3β, and it may be completely limited for some of the metals. Our finding suggested that R1 microtubule binding domain is the binding region for Cu(II), but not for other metal ions.

Novel Aspect:

This study provides a fundamental insight in metal accumulation, which may open up different direction for development of future therapeutic for AD.

Ahrends, Robert,
ISAS

Contributing Authors:
Rene Zahedi, ISAS; Albert Sickmann, ISAS; Andreas Hentschel ISAS; Cristina Coman1, ISAS; Fiorella Solari, ISAS.

Title:
SIMPLEX: a combinatorial multimolecular omics approach for systems biology

Introduction:
Interconnected molecular networks are at the heart of signaling that mediate adaptive plasticity of cells. To gain deeper insights into the underlying molecular mechanisms, a comprehensive and representative analysis demands a deep and parallel coverage of a broad spectrum of molecular species. Therefore, we introduce SIMPLEX SImultaneous Metabolite, Protein, Lipid EXtraction.

Methods:
SIMPLEX is a novel strategy for the quantitative investigation of lipids, metabolites and proteins. Compared to uni-molecular workflows, SIMPLEX offers a fundamental turn in study design, since multiple molecular classes can be accessed in parallel from one sample with equal efficiency and reproducibility.

Results:
Application of this method in mass spectrometry based workflows allowed the simultaneous quantification of 360 lipids, 75 metabolites and 3327 proteins. The versatility of this method is shown in a model system for adipogenesis - PPARG signaling in mesenchymal stem cells – where we utilized SIMPLEX to explore cross-talk within and between all three molecular classes and identified novel potential molecular entry points for interventions, indicating that SIMPLEX provides a superior strategy compared to conventional workflows. The identification of a close collaboration between the PPARG signaling on the protein level and TAG metabolism on the metabolite level proves our initial motivation to investigate the system within the cellular context rather than a restricted single-sided way and allowed for the elucidation of a further level of feedback based PPARG control.

Novel Aspect:
SIMPLEX: a combinatorial multi-molecula omics approach for systems biology allows for the simultaneous processing of lipids, metabolites and proteins.

AIT-BELKACEM, Rima,
Fondazione Pisana per la Scienza ONLUS

Contributing Authors:

Liam A. McDonnell, 1/ Center for Proteomics and Metabolomics, Leiden University Medical Center, Einthovenweg 20, 2333ZC Leiden, The Netherlands 2/ Department of Pathology, Leiden University Medical Center, Albinusdreef 2, 2333ZA Leiden, The Netherlands 3/ Fondazione Pisana per la Scienza ONLUS, 56125 Pisa, Italy; Rima AIT-BELKACEM, Fondazione Pisana per la Scienza ONLUS, 56125 Pisa, Italy.

Title:

In-Source Decay and Pseudo-MS3 Top-Down Analysis of Peptide and Protein Ions Using an AP-UV-MALDI Source Coupled to a Q Exactive Mass Spectrometer

Introduction:

Atmospheric-pressure MALDI is a soft ambient ionization technique that generates singly charged peptide and protein ions. At high laser fluence, the laser induces in-source decay (ISD) fragmentation in the hot MALDI plume. When attached to a Q-Exactive Orbitrap, these fragment ions can be detected in FTMS full scan mode, and then selected for a pseudo-MS3 experiment.

Methods:

1 µL of high ISD-efficient liquid matrices (α-CHCA, 2,5-DHB, super-DHB, 1,5-DAN and SA) was spotted with 1 µL of analyte solution (peptide or protein) directly onto glass microscope slides. After volatile solvent evaporation, the MALDI laser energy was increased and the system optimized to detect the ISD fragments. Specific fragments were then isolated and subject to HCD pseudo MS3 analysis.

Results:

We have optimized AP-MALDI to generate in source fragmentation of peptides and proteins. Liquid UV-MALDI matrices were used for long-life MALDI signal, and exhibited high fragment ion yield and a high signal stability. ISD fragment ions of peptides and proteins were obtained using a Q-Exactive Plus, extensive a-, b-, c-, y-, and z- type fragment series were observed. Complete sequence coverage of small peptides and N-terminal or C-terminal sequence tags of intact proteins (thymosin beta 4 fragment and cytochrome C), also larger than 50 kDa (bovine serum albumin) could be obtained and assigned. Additional sequence information was obtained by using HCD as a pseudo MS3 method.

Novel Aspect:

AP-MALDI ISD and pseudo-MS3 analysis achieved through liquid matrix preparation, with increase/extent of fragment ion yield and high signal stability

AIT-BELKACEM, Rima,
Fondazione Pisana per la Scienza ONLUS

Contributing Authors:
Liam A. McDonnell, 1/Center for Proteomics and Metabolomics, Leiden University Medical Center, Einthovenweg 20, 2333ZC Leiden, The Netherlands 2/ Department of Pathology, Leiden University Medical Center, Albinusdreef 2, 2333ZA Leiden, The Netherlands 3/ Fondazione Pisana per la Scienza ONLUS, 56125 Pisa, Italy; Rima AIT-BELKACEM, Fondazione Pisana per la Scienza ONLUS, Pisa, 56121 Italy.

Title:
MSI-directed LCM of the same tissue section: quantitative proteome analysis of regions with distinct MSI profiles

Introduction:
Mass spectrometry imaging (MSI) is able to resolve molecular distributions with cellular specificity but lacks the depth-of-coverage of LC-MS/MS. In this work we demonstrate how MSI can be combined with laser capture microdissection (LCM), of the same tissue section, to enable MSI directed isolation of cell populations, which can then be further characterized by quantitative LC-MS/MS.

Methods:
Mouse brain tissue sections were cut and mounted onto LCM-compatible polyethylene naphtalate (PEN) coated glass slides. MALDI MSI analysis of the PEN slides using AP-MALDI provided a molecular map of the tissue sections. The MSI data was then examined and small, 1mm2 areas were isolated from molecularly distinct regions (as defined by MSI) and the peptide extracts analyzed by MALDI MS and LC-MS/MS

Results:
PEN slides are used for the isolation of cells with laser capture microdissection. When mounted onto the PEN coating, a UV laser (in this case a frequency tripled Nd:YAG operating at 355 nm) can be used to isolate specific groups of cells. This same laser is also used for MALDI-MSI of tissues but typically at lower fluence. We have developed a method for AP-MALDI MSI of PEN mounted tissue sections. MALDI-MSI data quality is identical to that obtained from regular glass slides. Following the MALDI-MSI experiment LCM may be performed, also without removing excess MALDI matrix. 1mm2 areas were isolated from regions with distinct peptide profiles (as measured by MALDI MSI); MALDI MS analysis of tissue extracts confirmed the specificity of the LCM isolation, and which were then analyzed by LC-MS/MS.

Novel Aspect:
Development of high spatial and mass resolution MSI coupled to an automated LCM.

AKASHI, SATOKO,
Yokohama City University

Contributing Authors:

Satoko Akashi, Yokohama City Univ.; Akihisa Osakabe, Waseda Univ.; Kazumi Saikusa Hiroshima Univ.; Hitoshi Kurumizaka, Waseda Univ.; Yasushi Kawata, Tottori Univ.

Title:

Behavior of Intrinsically Disordered Regions in Protein Complexes

Introduction:

Intrinsically disordered regions (IDRs) of proteins, especially of eukaryote, are responsible for various critical functions. So far, several intrinsically disordered proteins (IDPs) consisting of a single polypeptide chain have been examined by ESI-ion mobility mass spectrometry (ESI-IM-MS) but little information is available on the gas-phase behavior of IDRs in protein complexes.

Methods:

Proteins with IDRs were prepared using E. coli overexpression system. ESI-IM and ESI-TOF mass spectra were obtained using Waters Synapt G2 HDMS or Q-tof2. Samples were prepared in solutions at various concentrations of ammonium acetate.

Results:

Single-polypeptide IDPs usually present wide charge and collision cross-section (CCS) distributions in ESI-IM-MS, reflecting the "disorder" characteristics. For ESI-MS of histone multimers, which have highly basic IDRs at N-terminus of each histone protein, a high concentration of ammonium acetate (>2 M) was required to avoid dissociation of the multimer structure in the gas phase. Contrary to single-polypeptide IDPs, histone multimers presented very narrow distributions of charge state and CCS, suggesting that IDR characteristics of the tail regions were almost lost. This could be due to reduction in electrostatic repulsion in the histone N-terminal regions by a high concentration of ammonium acetate. In this paper, ESI-IM-MS of NCP, a complex of a histone octamer and double-stranded DNA, will also be presented and electrostatic effects on the behavior of the IDRs will be discussed.

Novel Aspect:

It was experimentally proved that the behavior of IDRs in protein complexes is highly affected by electrostatic effects.

Alaee, Mehran,
ECCC

Contributing Authors:

Tommy Bisbicos, ECCC; Grazina Pacapvicius, ECCC.

Title:

A rapid and robust method for determination of 35 phthalates in influent, effluent and biosolids in wastewater treatment plants

Introduction:

Phthalate di-esters, commonly referred to as phthalates, are a wide group of compounds extensively used as plasticizers in polymers such as PVC, and as carrier agents in cosmetics. Phthalates are clear, odorless, flavorless, and stable compounds, that are easily blended into polymers without changing the appearance of the polymer, and can subsequently leach out into the environment.

Methods:

Phthalates were extracted from water and biosolid using a QuEChERS method. 10 mL of acetonitrile was added to a 10 mL water sample or 0.1 g of biosolid spiked with a mixture of D4 recovery phthalates. 3 grams of NaCl was added to separate the acetonitrile and water phases. Disperse solid phase extraction (dSPE) was used to remove interferences from biosolid samples.

Results:

The ubiquitous presence of phthalates in solvents used during the extraction, cleanup and chromatographic process and the laboratory environment is a major hindrance in phthalate analysis. In order to keep the background contamination from solvents to a minimum, the LC/MSMS was equipped with a scrubbing column before the injection port, and a QuEChERS method, which uses significantly less solvent than traditional extraction methods was used.The UPLC/MSMS determinations were accomplished in 4 minutes. The average recovery for phthalates in HPLC water, Lake Erie surface sediment and biosolids from a secondary wastewater treatment plant were 94%+/- 8%, 94% +/- 36% and 88% +/- 33% respectively. This method was applied to the analysis of more than 150 samples from wastewater treatment plants from across Canada. Results from this investigation will be discussed in this presentation.

Novel Aspect:

LC/MSMS analysis of 35 phthalates in 4 minutes, phthalates extraction using QuEChERS method, solid phase dispersion cleanup

Alagesan, Kathirvel,
Max Planck Institute of Colloids and Interfaces

Contributing Authors:
Daniel Kolarich, Department of Biomolecular Sciences, Max Planck Institute of Colloids and Interfaces, 14424 Potsdam, Germany; Daniel Varón Silva, Department of Biomolecular Sciences, Max Planck Institute of Colloids and Interfaces, 14424 Potsdam, Germany; Dana Michel Department of Biomolecular Sciences, Max Planck Institute of Colloids and Interfaces, 14424 Potsdam, Germany.

Title:
SYNTHETIC GLYCOPEPTIDES ARE UNIQUE TOOLS FOR FUNCTIONAL GLYCOPROTEOMICS

Introduction:
Functional glycoproteomics requires an insight into the protein and site specific glycan structure distribution to investigate structure-function relationships. Tandem MS (MS/MS) approaches are among the most powerful glycoproteomics tools, nevertheless reliable glycopeptide identification providing concomitant information on both, the glycan and peptide moieties still remains a challenging task.

Methods:
We have developed novel and simplified approaches to purify and produce a panel of glycosylated amino acids carrying N-linked glycans with various structures, which are subsequently used in standard solid phase peptide synthesis using previously reported Fmoc protocols. A well-defined glycopeptide library containing >100 glycopeptides plus their unglycosylated counterparts was synthesized.

Results:
A well-defined synthetic glycopeptide library represents a highly valuable tool to systematically investigate qualitative and quantitative functional glycoproteomics aspects. This glycopeptide library was the basis for a systematic investigation of glycopeptide fragmentation and ionisation behaviour in LC-MS-CID/HCD/ETD, which in turn allowed us to optimise analysis conditions for biologically relevant samples in glyco-biomarker research such as the entire panel of human Immunoglobulins. Limitations in glycopeptide ionisation were overcome by CaptiveSpray Nanobooster™ ionisation as systematically investigated using a commercially available Mass Spectrometry Retention Time calibration mix in addition to selected synthetic glycopeptides. Our glycopeptide library has yielded a comprehensive dataset allowing the systematic comparison of various search engines and fragmentation methods.

Novel Aspect:
This synthetic glycopeptide library presents a unique resource for developing novel qualitative and quantitative glycoproteomics applications.

Alagesan, Kathirvel,
Max Planck Institute of Colloids and Interfaces

Contributing Authors:
Udo Reichl, Max-Planck-Institute for Dynamics of Complex Technical Systems, 39106 Magdeburg, Germany; Otto von Guericke University Magdeburg, Inst. of Experimental Internal Medicine, 39120 Magdeburg, Germany; Matthew Campbell, Faculty of Science and Engineering, Macquarie University, NSW 2109, Australia; Erdmann Rapp Max-Planck-Institute for Dynamics of Complex Technical Systems, 39106 Magdeburg, Germany; glyXera GmbH, 39120 Magdeburg, Germany; Peter H Seeberger, Department of Biomolecular Sciences, Max Planck Institute of Colloids and Interfaces, Potsdam, Germany; Marcus Hoffmann, Max-Planck-Institute for Dynamics of Complex Technical Systems, 39106 Magdeburg, Germany; Daniel Kolarich, Department of Biomolecular Sciences, Max Planck Institute of Colloids and Interfaces, Potsdam, Germany; Nicolle H Packer, Faculty of Science and Engineering, Macquarie University, NSW 2109, Australia; Rene Hennig, Max-Planck-Institute for Dynamics of Complex Technical Systems, 39106 Magdeburg, Germany; glyXera GmbH, 39120 Magdeburg, Germany; Robyn Peterson, Faculty of Science and Engineering, Macquarie University, NSW 2109, Australia.

Title:
IgCarbKB: A GLYCOMICS AND GLYCOPROTEOMICS FOCUSED IMMUNOGLOBULIN KNOWLEDGE BASE

Introduction:
Immunoglobulins (Ig's) and Fc receptors are complex glycoproteins and key components in the innate and adaptive immune systems. Their specific glycosylation is well known to be crucial for maintaining and modulating effector functions. Stable, reliable & sensitive glycomics and glycoproteomics workflows are a prerequisite to investigate the functional role(s) of glycosylation in the immune system.

Methods:
Ig's heavy and light chains derived from healthy donors and myeloma patients were proteolytically digested, HILIC enriched and the resulting (glyco)peptides in-depth analysed by nano-RP-LC–ESI iontrap MS/MS and RP-LC-Orbitrap MS/MS. PGC nanoLC ESI-MS/MS & xCGE-LIF provided detailed glycomic profiles of the released N-glycans whereas the O-glycans were analysed by PGC nanoLC ESI.

Results:
We present a detailed glycomic and glycoproteomic map of the entire human Immunoglobulome (IgA, sIgA, IgD, IgE, IgG and IgM) acquired by a variety of analytical platforms and retrieved from the literature. Each individual Ig class showed distinct glycomic profiles, which furthermore differed on the individual sites of glycosylation of the respective Ig's. Not surprisingly, glycomic and glycoproteomic profiles differed significantly in all myeloma Ig's analysed. In order to make glycomic and glycoproteomic knowledge on Ig's accessible we have also developed an open source knowledge base IgCarbKB (www.IgCarbKB.org) where literature as well as acquired data on Ig's specific carbohydrate information is stored and made accessible. IgCarbKB presents an open access, curated and continuously extended resource for any Ig glycosylation related functional and structural data.

Novel Aspect:
As a first step towards understanding the functional role of Ig glycosylation, we have created a comprehensive Ig-carbohydrate knowledgebase.

Alberici, Rosana,
Thomson Mass Spectrometry Laboratory-UNICAMP

Contributing Authors:
Núbia C. A. Queiroz, Pharmacy School-UNICAMP; Rosana M. Alberici, Thomson Mass Spectrometry Laboratory-UNICAMP; Mary Ann Foglio Pharmacy School-UNICAMP; Gislaine C. Nogueira, School of Food Engineering-UNICAMP; Marcos N. Eberlin, Thomson Mass Spectrometry Laboratory-UNICAMP; Thais L. Silva, Thomson Mass Spectrometry Laboratory-UNICAMP; M. Angela A. Meireles, School of Food Engineering-UNICAMP; Ilza M. O. Souza, Pharmacy School-UNICAMP; Soraya El-Khatib, Thomson Mass Spectrometry Laboratory-UNICAMP.

Title:
Development of new sustainable cosmetic products using standardized Artemisia annua L. extract monitored by electrospray ionization mass spectrometry

Introduction:
There is a pressing need for the standardization of herbal extracts due to current consumer demands for safe, effective and high-quality products. It is often based on the measurement of marker compounds whose levels correlate with the activity in bioassay and clinical testing. Here, ESI-MS fingerprinting was used to standardize Artemisia annua L. extract for new sustainable cosmetic products.

Methods:
Artemisia annua L. extracts were obtained using both Soxhlet system and supercritical CO2 extraction. Samples were analyzed by direct infusion ESI using both positive and negative ion modes, in a high-resolution mass spectrometer, the Q Exactive™ Hybrid Quadrupole-Orbitrap. The antioxidant, antimicrobial and anti-inflammatory activities were tested by DPPH method, MIC and ELISA, respectively.

Results:
Artemisia annua is rich in sesquiterpene lactones that present a wide range of biological activities, exhibiting therapeutic potential as anti-inflammatory and antimicrobial. Mass spectra of conventional and supercritical CO2 extracts showed [M+Na]+ adduct of m/z 305.13 related to artemisinin (C15H20O5) as the base ion of the spectra, beyond its [M+H]+ and [M+K]+ ions. Deoxiartemisinin (C15H20O4) was also detected in both extracts as [M+Na]+ of m/z 289.14. Dihydroartemisinic acid and artemisinic acid were identified as [M-H]- of m/z 235.16 and m/z 233.15, respectively. The extract obtained from supercritical CO2 extraction showed absolute abundance of the ions related to artemisinin about hundred times higher than those in the conventional extract. Preliminary results show there is a possible correlation between the phytochemical profiles monitored by ESI-MS and therapeutic activities.

Novel Aspect:
ESI-MS fingerprinting is a liable tool to standardize and certify herbal extracts for use in new and effective natural cosmetic products.

Al-Dulaymi, Mays,
University of Saskatchewan

Contributing Authors:
Anas El-Aneed, College of Pharmacy and Nutrition, University of Saskatchewan, Saskatoon, SK, Canada.

Title:
Tandem Mass Spectrometric Analysis of Novel Positively Charged Peptide-Modified Gemini Surfactants.

Introduction:
Gemini Surfactants are cationic lipids which have been used as a gene delivery vectors. A novel series of peptide-modified gemini surfactants was synthesized and showed promising results in vitro. The purpose of this work is to elucidate the MS/MS disccoaication behaviour of these novel molecules. Such knowledge is essential for understanding their biological fate at cellular and tissue levels.

Methods:
Exact mass measurements are achieved using a hybrid quadrupole orthogonal time-of-flight mass spectrometer (QqToF-MS). A multi-stage tandem mass spectrometric analysis was conducted using a triple quadrupole linear ion trap mass spectrometer (QqQ-LIT MS). Both instruments were operated in the positive ionization mode and are equipped with electrospray ionization (ESI) source.

Results:
Single stage QqToF-ESI-MS data confirm the molecular structure of all tested gemini surfactants. The mass accuracies are below 10 ppm for all tested compounds. Tandem mass spectrometric (MS/MS) analysis showed common MS/MS fragmentation behaviour among all tested compounds, resulting in the production of a universal fragmentation pathway. In particular, it was observed that the loss of one of the tail regions was dominant within the evaluated structures. In addition, the MS/MS dissociation was also centered on the substituted amio acids in which peptide bonds were cleaved for the generation of diagnostic product ions. The genesis of the observed product ions are established using MS3 experiments.

Novel Aspect:
New universal MS/MS fragmentation pathway is established. It will be utilized for the development of qualitative and quantitative MS-based methods.

Alharbi, Hattan,
Toxicology Graduate Program at U of Saskatchewan

Title:
An accurate analysis of the internal concentration of pesticide in whole larvae tissue of Japanese medaka using LC-QqQ-MS without clean-up step.

Introduction:
Fish are used to determine effect concentrations of chemicals in the contexts of chemical safety, drug discovery and environmental risk assessment. Several studies are investigating pesticide residue in fish using tissue specific (liver) of whole adult fish, which are costly, time-consuming and labour intensive. Different chromatographic techniques have been used to determine internal concentratio

Methods:
larvae were homogenized in 1 mL of a 1:1 mixture of hexane:ethyl acetate, followed by ultrasonication for 1 min. After extraction, samples of larvae and exposure solutions were centrifuged at 10,000 rpm for 5 min, and the organic phase was removed and concentrated under a gentle stream of nitrogen until dry. Next, samples were reconstituted in of 100% of methanol for LCMSMS analysis

Results:
The recovery was between 94-98% in tissues sample. The matrix effect was negligible and ranged from 97-101% for chlorpyrifos and 99-102% for malathion. No difference was observed between the matrix matches calibration and solvent based calibration. LOD ranged from 0.19-1.67 ng/g and LOQ ranged from 0.73-6.2 ng/g. The method was successfully applied to study the internal concentration of chlorpyrifos and malathion in whole larvae tissue at high and environmentally relevant exposure scenarios Conclusion: A quick, easy and efficient method was developed for analysis of pesticide in whole larvae tissue of small fish. The method was validated and had the acceptable range of recovery and matrix effect. The method can be successfully applied for investigation of environmental contaminants in fish with low cost of analysis without affecting the accuracy.

Novel Aspect:
An accurate, easy and cheap method was developed to investigate chemicals residues in small fish with no clean up step and minimum matrix effect.

Allan, Laurie,
Bruker Ltd.

Contributing Authors:
Louis Maljers, Bruker Daltonics; Zicheng Yang, Bruker Daltonics.

Title:
A Targeted Quantitative LC-MS/MS Method for 431 Positive and Negative Ion Pesticides in a Single Analysis

Introduction:
This work describes the simultaneous analysis of 431 pesticides in different vegetable matrices by a LC-MS/MS TQ system. Calibration curves for all compounds were used to evaluate dynamic linear range, correlation coefficients, limits of detection, and signal/noise ratio for each compound. This method used several matrices: tomato, garlic, beans and celery.

Methods:
MS acquisition methodology was based on Compound Based Scanning. Retention time of each pesticide was acquired with windows of 0.5 minutes with at least two transitions over 15 min. Average peak width was 8 seconds and points per peak set to 10 The software calculates the scan time for each compound taking into account polarity, taking into account the number of co-eluting acquisition windows.

Results:
AOAC QuEChERS official method was used for sample extraction. 10 g of fresh vegetable was treated to get a final matrix content of 1 gram per milliliter. Blank matrices were spiked stock solution containing all pesticides under study to make calibration curves. Compounds had linear dynamic ranges with a correlation coefficient > 0.993 and response factor std. dev. below 20%. Thermo-labile compounds like Aldicarb or Procloraz were quantified at low concentration levels without degradation as a result of the new Vacuum Isolated Probe Heated ESI source. Moreover, due to high temperature in the source pesticides with "poor ionization" can be detected even at sub-ppb levels. Sensitivity of overlapping compounds, 50 ng/mL for the positive compound Fenbuconazole and 100 ng/mL for the negative compound Fipronil with a r2>0.99 were reached easily by the developed pesticide method.

Novel Aspect:
A fast LC-MS TQ method with great sensitivity containing an extreme number of transitions using fast polarity switching.

Alley, William,
Waters

Contributing Authors:
Ying Qing Yu, Waters Corportation.

Title:
Giving Ion Mobility to the Masses: What Collisional-cross Section Values can do for Glycan Analysis of Biotheraputic Monoclonal Antibodies

Introduction:
Glycan profiling is important in the biopharmaceutical industry as glycans are critical quality attributes. Structural characterization is difficult and requires multiple analytical techniques including mass spectrometry (MS) and tandem MS. Here, we discuss the use of ion mobility-MS data to improve glycan identification when performed with calibrated chromatographic retention time in Glucose Uni

Methods:
Monoclonal antibodies were denatured and glycans were released and labeled with RapiFluor-MS tag. Labeled glycans were analyzed by HILIC UPLC/FLR/IMS-MS. Data were acquired, processed, and analyzed using an IMS QTof system controlled by UNIFI. Data processing included a glycan GU scientific library search for identification. Traveling wave ion mobility drift times were converted to collisional c

Results:
We have established an automated glycan UPLC/FLR/MS workflow that includes a library search of GUs and accurate mass measurements. However, some ambiguity may remain. We used a GU window of ±0.1 GUs and a pair of glycans (F(6)A2[3]G(4)1 and F(6)A2[6]G(4)1) with GUs of 6.69 and 6.56, respectively, is difficult to assign without more data. We investigated the use of reduced CCS values to minimize structural ambiguities. We have determined that F(6)A2[3]G(4)1 has a reduced CCS of 1319 Å2 and F(6)A2[6]G(4)1 has a reduced CCS value of 1310 Å2. Additionally, F(6)A2G(4)2 is an abundant mAb glycan and F(6)A2G1Ga1 can be present. Correct identification based on GUs and accurate mass is difficult for this pair. IMS reduced CCS values can distinguish these two isomers, leading to correct identifications when performing library searches. Using reduced CCSs, structural determination is simplified

Novel Aspect:
IMS MS data is used to supplement calibrated retention time and accurate mass searching for glycan assignment without tandem MS data.

Almostafa, Mervt,
Memorial University

Contributing Authors:
Joseph Banoub, Chemistry Department, Memorial University.

Title:
Tandem mass spectrometry determination of the structure of a heterogeneous mixture of Lipid As isolated from the LPS of Aeromonas liquefaciens SJ-19a

Introduction:
Bacteria are well known to cause many serious human and animal diseases. There are various numbers of bacterial species associated with fish diseases, which leads to the losses in both marine and freshwater of farmed fish.2 Lipid A as a part of LPS is located in the outer membrane and often is the basis to distinguish and classify the bacteria. We report in this poster the electrospray ionizatio

Methods:
The heterogeneous mixture of lipid As was isolated by the aqueous phenol method from the LPS of the rough wild strain of Gram-negative bacteria Aeromonas liquefaciens (SJ-19a). Hydrolysis of the LPS was with 1% acetic acid, and purification was by chromatography using Sephadex G-50 and SephadexG-15. ESI-MS and low-energy CID-MS/MS analyses were performed with a triple-quadrupole and a FTICR-MS.

Results:
The Preliminary analysis of the lipid As mixture was conducted by ESI-MS (- ion mode) and the spectrum obtained suggested that the lipid A SJ-19a was composed of a heterogeneous mixture of different lipid A molecules.. Low-energy CID-MS/MS experiments confirmed the identities of the various mono-phosphorylated β-D-GlcpN-(1→6)-α-D-GlcpN disaccharide entities. This lipid As mixture was asymmetrically substituted with fatty acids such as ((R)-14:0(3-OH)), (14:0(3-(R)-(O-12:0)) and (14:0(3-(R)-O-(14:0)) located on the O-3, O-3', N-2 and N-2' positions, respectively. The low-energy CID-MS/MS and FTICR-MS/MS allowed the determination of the fatty acid acylation positions on the H2PO3→4-O'-β-D-GlcpN-(1→6)-α-D-GlcpN the disaccharide backbones of the different lipid A mixture.

Novel Aspect:
This heterogeneous mixture of lipid As is composed of seven different substituted lipid As, formed from the incomplete biosynthesis of their

Alonso, David,
LECO Corporation

Contributing Authors:
Joe Binkley, LECO Corporation; Lorne Fell, LECO Corporation.

Title:
A novel benchtop time of flight GC/MS system for high performance analysis of human urine

Introduction:
Effective profiling methods are critical for the investigation of environmental exposure. Metabolomics is particularly useful due to phenotype proximity and quick insight into system perturbations. A major problem continues to be incomplete sample characterization due to compound chemical diversity, wide concentration range of metabolites and the complexity of biological matrices.

Methods:
Samples were dried, treated with urease and derivatized. A novel benchtop GC-TOFMS and associated software were used for comprehensive data collection and processing. Compounds were characterized using retention index values and similarity searches against large, well-established databases. Different group samples were quickly compared retrospectively with Target Analyte Finding processing.

Results:
A main objective of this research investigation was the development and implementation of an effective workflow for comparison of urine samples. A major bottleneck, regardless of the applied technique, is quick and confident characterization of individual metabolites in samples. In this study, the combination of robust chromatography and the increased dynamic range and lower detection limits of the GC-TOFMS system resulted in rapid identification of xenobiotics (e.g., halogenated compounds, phenols) and metabolites including: Acids, diacids, amino acids, fatty acids, bases, monosaccharides, disaccharides, sugar phosphates, and sterols. Software tools were used to quickly interrogate the rich data retrospectively in a targeted manner after compounds of interest were identified.

Novel Aspect:
Increased numbers of compounds were identified with powerful software tools and a novel benchtop GC-TOFMS with improved dynamic range and sensitivity.

Alonso, David,
LECO Corporation

Contributing Authors:
Joe Binkley, LECO Corporation.

Title:
A novel benchtop time of flight GC/MS system for high throughput qualitative and quantitative analysis of drugs of abuse in human urine

Introduction:
Drug abuse is a major problem around the world. Forensic laboratories are experiencing increased workloads and resource reductions. There is an immediate need for high throughput methods resulting in conclusive identification of controlled substances. In this study, preparation and analysis times were reduced without sacrificing thorough characterization of samples.

Methods:
Standard extraction techniques were used to prepare non-derivatized urine samples analyzed by a novel benchtop GC-TOFMS. Data acquisition and processing times were optimized through use of the instrument acquisition parameters and powerful data processing software. Processing methods included automated peak finding, retention index calculations, quantitation and searches against large databases.

Results:
High performance GC-TOFMS analysis facilitated drug identification through the acquisition of full range, non-skewed mass spectral data, along with fast acquisition rates, extended dynamic ranges and lower detection limits. Comprehensive data processing resulted in the identification of heterocyclic compounds such as alkaloids including nicotine and cotinine, as well as, narcotics, steroids and pharmaceuticals in urine samples. Excellent calibration linearity with a concentration range from 10 to 1000 ng/mL for a variety of controlled substances. Targeted Analyte Finding and calibration methods were applied for the quantification of opiates, barbiturates and synthetic drugs of abuse. Detectability of non-derivatized drugs ranged in the very low ppb to ppt levels for drugs of abuse in urine.

Novel Aspect:
The improved dynamic range and detection limits of a novel GC-TOFMS resulted in rapid detection of low level, non-targeted compounds in urine.

Altelaar, Maarten,
Utrecht University

Contributing Authors:
Oscar Krijgsman, Netherlands Cancer Institute; Daniel S. Peeper, Netherlands Cancer Institute; Linsey Raaijmakers Utrecht University; Kristel Kemper, Netherlands Cancer Institute; Albert J.R. Heck, Utrecht University; Violette Gautier, Utrecht University; Thierry Schmidlin, Utrecht University; Gianluca Maddalo, Utrecht University.

Title:
Advanced (phospho)proteomics to uncover dynamics and pathway dependence in melanoma drug resistance uncovers novel companion targets

Introduction:
Malignant melanoma is mainly fuelled through the ERK pathway, mainly due to oncogenic BRAFV600E. Development of targeted therapy against BRAFV600E is associated with unprecedented clinical benefits, however, eventually all melanoma patients become resistant. Therefore, there is a dire need to find either alternative drug targets, which requires systems-level proteomics approaches

Methods:
Studies were performed on patient-derived matched treatment naïve, sensitive and resistant melanoma cells, the latter two treated with BRAF inhibitor. After lysis, proteins were digested and fractionated by HpH, followed by either LC-MS or enrichment of the phosphopeptides. Samples were analysed using nano-flow LC-MS. Besides proteomics we performed RNAseq and metabolomics.

Results:
We identified ~10,000 proteins and phosphosites in our phosphoproteomics screen. The power of our study is highlighted by the fact that we were able to identify different proteins previously reported to be involved in melanoma drug resistance. Importantly, integrating the proteomics and phosphoproteomics data, we could draw a clear picture of the proteome rewiring upon drug resistance. From this picture we could designate several potential combination targets whose interference generated an additive effect in targeted melanoma therapy, which we confirmed with cell viability and colony formation assays. Combining metabolomics experiments helped to explain the mechanisms of resistance and the effect of inhibition of the combination targets was followed by targeted phosphoproteomics. Finally we combined the RNAseq with the phosphoproteomics data and show mutation specific phosphorylation

Novel Aspect:
Comprehensive phosphoproteomics screening of BRAFV600E mutant cells reveals novel drug co-targets and phosphorylation controlled pathway dynamics

Amirav, Aviv,
Tel Aviv University

Contributing Authors:
Tal Alon, Tel Aviv University; Alexander Fialkov, Tel Aviv University.

Title:
Molecule Identifier Software for Improved Sample Identification by GC-MS

Introduction:
The major strength of GC-MS in sample identification is its ability to perform library search for identification with names and structures. However, library identification cannot always be trusted and many compounds are not included in the library. Thus, we wish to be able to confirm or reject library identification and provide sample elemental formula from unit resolution GC-MS data.

Methods:
We developed the Tal-Aviv Molecule Identifier (TAMI) software that automatically combines with the NIST library. TAMI inverts the molecular ion isotopomeric pattern and the measured mass (which it can improve its mass accuracy) into elemental formulae and automatically confirms or rejects library identification. In case of rejection it provides via a single click the sample elemental formula.

Results:
The TAMI software helps in: A) Improved mass accuracy of quadrupole mass analyzers to <50 ppm; B) Combines the improved mass accuracy of the molecular ion with isotope abundance analysis for the provision of elemental formulae. TAMI begins with automated confirmation or rejection of NIST identification via isotope abundance analysis. In case of a confirmation the sample identification is unambiguous since it is done by an independent set of data and method. Upon rejection, TAMI independently provides a table of elemental formulas with declining order of matching to the experimental data. Thus, TAMI upgrades unit resolution quadrupole based GC-MS to be similar in sample identification capability to costly accurate mass GC-TOF-MS. Furthermore, TAMI operates with standard centroid data files and is easy to work with. You are invited to bring your files for its demonstration.

Novel Aspect:
New software was developed for improved sample identification by GC-MS and for the provision of elemental formula from quadrupole GC-MS data files.

Amirav, Aviv,
Tel Aviv University

Contributing Authors:
Alexander Fialkov, Tel Aviv University; Tal Alon, Tel Aviv University; Uri Keshet Tel Aviv University.

Title:
GC-MS with Cold EI and its Unexpected Benefits

Introduction:
GC-MS are widely used analytical instruments. However, the vast majority of its applications are achieved with electron ionization, 30 m long capillary columns and 1 ml/min column flow rate. Consequently, despite GC-MS availability from several vendors most of the GC-MS performance parameters are similar and their improvement is not addressed.

Methods:
GC-MS with Cold EI is based on GC and MS interface with Supersonic Molecular Beams (SMB) and electron ionization (EI) of cold compounds in the SMB. The GC eluting compounds are mixed with helium, expand from a supersonic nozzle into a vacuum chamber, vibrationally cooled, skimmed, collimated into a SMB, are ionized in a contact-free EI ion source and are mass analyzed by a single quadruple MS.

Results:
Several unexpected benefits of GC-MS with Cold EI are demonstrated: A) We found by experiments and simulation that NIST library identification probabilities are improved with Cold EI as the molecular ions are the most sample characteristic ions. B) Sample identification is superior to that of any other GC-MS since bigger and more labile compounds elute, exhibit molecular ions and elemental formula are provided with isotope abundance analysis software. C) Significantly extended range of compounds are amenable for analysis via the use of short columns, high column flow rates and a contact-free fly-through ion source. D) Compound independent ionization yield is provided. E) Under 2 minutes analysis time is uniquely achieved with column flow programming. F) The selectivity against matrix interference on the Cold EI molecular ion is as effective as MS-MS on a fragment ion.

Novel Aspect:
Demonstration of unexpected GC-MS with Cold EI benefits in identification, range of compounds, selectivity, response uniformity and speed of analysis.

Anacleto, Joe,
Bruker

Contributing Authors:
CARSTEN BAESSMANN, Bruker Daltonik GmbH, Bremen; Javier Lopez, Bruker Espanola, S.A., Madrid, SPAIN; Miguel Perez Bruker Espanola, S.A., Madrid, SPAIN; Louis Maljers, Bruker, Fremont, CA.

Title:
Parts-per-trillion level high-throughput quantitation of glyphosate, aminomethylphosponic acid (AMPA) and glufosinate in water samples by LC-MS/MS

Introduction:
The global use of glyphosate, glufosinate and AMPA herbicides are strictly controlled by international regulations. We describe a method employing derivatization using 9-fluorenylmethyl chloroformate (FMOC) coupled with UHPLC-MS/MS tandem mass spectrometry to deliver a rapid and reliable method for the determination of glyphosate, AMPA and glufosinate in water samples.

Methods:
Standards were spiked with glyphosate, AMPA and/or glufosinate. Incubate for 2 hours in Sodium tetraborate buffer and FMOC solutions. Phosphoric acid was added to stop the derivatization. Chromatographic gradient was between 2mM ammonium acetate and methanol, respectively, flowing at 0.4 mL min-1 through a C18 column. MS detector worked in MRM mode with heated-ESI negative ionization mode.

Results:
European guidelines require LC retention time of the in-matrix analyte must be within 2.5% of the retention time in the standard, and relative abundance of the MRM (SRM) transitions signals must be within 30% of the ratio obtained for the standards. Therefore, identification quality controls were tested using our method. The obtained results (especially LOQ and LOD) were low enough to meet the regulation about glyphosate and derivatives compounds in water analysis. The linearity and RSDRF of the calibration curves were between 5 and 20% and R2 0.999 respectively for the three compounds analyzed. S/N ratios of 10ppt points for all compounds were >= 10 and no significant carry over were detected following the 500ppt spiked sample. In summary, the method showed excellent linearity and sensitivity without exhibiting carryover whilst meeting the European Union regulations.

Novel Aspect:
Single method for multiple polar pesticides that is rapid, sensitive, low-cost and quantitative. Meets European MRL Guidelines.

Anacleto, Joe,
Bruker

Contributing Authors:
Petra Decker, Bruker Daltonik GmbH, Fahrenheitstr. 4, 28359 Bremen, Germany; Karin Wendt, Bruker Daltonik GmbH, Fahrenheitstr. 4, 28359 Bremen, Germany; Carsten Baessmann Bruker Daltonik GmbH, Fahrenheitstr. 4, 28359 Bremen, Germany.

Title:
Simultaneous quantitation and confirmation of about 500 pesticide residues in food extracts using LC-QTOF accurate mass spectrometry

Introduction:
Fast and comprehensive full scan accurate mass screening and quantitation became an excellent tool in food control in particular if hundreds of pesticides have to be measured in a short time frame. Additionally to the high number of targets the technique takes advantage of unknown evaluation and retrospective analysis.

Methods:
A set of 481 pesticides was selected. Matrix based dilution series (10 µg/kg – 2000 µg/kg, QuEChERS-extracts: tomato, summer squash, potato, orange) of the pesticide-mix were analyzed by an UHPLC connected to an impact II QTOF (Bruker) mass spectrometer using a 15 min gradient. Data acquisition is performed in alternating between full scan and bbCID all ion fragmentation mode.

Results:
Automatic data evaluation is performed using TASQ 1.1 (Bruker Daltonik GmbH), a dedicated software for target analysis screening and quantitation. For automated confident identification we use RT, accurate mass, isotopic pattern and up to 3 qualifier ions in full scan and 7 qualifier ions in bbCID. For tomato extract 454 of 481 pesticides or 94.3 % of the pesticides could be could be identified at 10 µg/kg and 98.9% at 50 µg/kg. For orange extract the numbers were 95.0% at 10 µg/kg and 98.9 at 50 µg/kg; for summer squash extract 91.3% at 10 µg/kg and 95.2% at 50 µg/kg; for potato 98.9% at 10 µg/kg and 99.6% at 50µg/kg.

Novel Aspect:
This study evaluates the performance achieved with a high performance QTOF, new software developed for this purpose and a very high quality data base.

Anders, Ulrike,
ETH Zurich

Contributing Authors:
Fatima Hibti, Horiba Jobin Yvon, Palaiseau, France; Jonas Schaefer, University of Zurich, Dep. of Biochem., Zurich, Switzerland; Renato Zenobi ETH Zurich, Dep. Chem. and Appl. Biosciences, Zurich, Switzerland; Detlev Suckau, Bruker Daltonics, Bremen, Germany; Chiraz Frydman, Horiba Jobin Yvon, Palaiseau, France; Andreas Plückthun, University of Zurich, Dep. of Biochem., Zurich, Switzerland.

Title:
Direct interfacing of imaging SPR with MALDI MS

Introduction:
The coupling of SPRi and MALDI MS technologies meet the increasing demand for high-throughput analysis in pharma and life science research. In addition, it provides information on binding kinetics, binding affinity in real time and identifies the interaction partners with no need of purification, separation or labeling in real time.

Methods:
SPR imaging measurements were performed on the SPR Plex II (Horiba, France) working with POE and NHS-esters functionalized gold slides. MS detection was done with a commercial MALDI TOF mass spectrometer (Ultraflex II TOF, Bruker Daltonics, Germany). The gold slide was mounted in an adapter. MS measurements were performed in the linear or reflected positive ion mode with standard settings.

Results:
We will show an investigation of specific as well as off-target binding of various DARPins against ribosomal protein S6 kinase 2 (RPS6KA2), a potential drug target. Designed ankyrin repeat proteins (DARPins) are a very promising class of non-immunoglobulin proteins that rival antibodies for target binding, but are genetically engineered. They are more robust than antibodies and can be tagged for an aligned chemical coupling to solid supports. DARPins were immobilized, dissolved pure protein was injected, and the KD was determined based on the measured SPR data. When injecting a complex mixture of proteins containing RPS6KA2 (like a cell lysate) a MALDI MS measurement on the SPRi-chip directly after the SPRi experiment verifies the specific interaction between DARPins and RPS6KA2 even in the low fmol range.

Novel Aspect:
SPRi-MALDI MS enables the multiplexed investigation of specific interactions out of a crude mixture of cells even in the low fmol range.

Anderson, Lissa,
National High Magnetic Field Laboratory

Contributing Authors:

Greg T. Blakney, National High Magnetic Field Laboratory; Lissa C. Anderson, National High Magnetic Field Laboratory; Alan G. Marshall Florida State University; Nathan K. Kaiser, National High Magnetic Field Laboratory; Donald F. Smith, National High Magnetic Field Laboratory; John P. Quinn, National High Magnetic Field Laboratory.

Title:

21 Tesla Fourier Transform Ion Cyclotron Resonance Mass Spectrometer for Top-Down Proteomics

Introduction:

FT-ICR mass spectrometry offers the highest achievable mass resolving power and mass accuracy of any mass analyzer, which is especially important for characterization of large biopolymers. We report here top-down protein analysis LC- 21 tesla FT-ICR MS/MS, which is the highest field ICR system to date. The instrument is a part of the National High Field FT-ICR User Facility.

Methods:

The instrument features a dual linear quadrupole ion trap for sensitive and accurate ion accumulation, isolation, and dissociation, and a third higher capacity linear quadrupole trap that efficiently transfers ions to a novel dynamically harmonized ICR cell. Ion fragmentation methods include CID, ETD, and UVPD. Reverse-phase liquid chromatography is performed with columns packed in-house.

Results:

High ICR frequency at 21 tesla facilitates 100,000 mass resolving power at m/z 1200 at 1 Hz scan rate, and ~1 ppm mass accuracy for proteins as large as 30 kDa (and a few ppm at 50 kDa). Multiple (3-20) ion fragmentation events are accumulated in the high ion capacity linear trap prior to injection into the ICR cell to maximize spectral S/N and scan rate. Targeted antibody analysis results in sub-ppm mass accuracy and 80% sequence coverage for light chain, and 3 ppm accuracy and 30% sequence coverage for heavy chain, including glycosylation site localization. Analysis of human colorectal cancer cell lysate results in identification of more than 400 unique proteins corresponding to ~1500 unique proteoforms, with <1% false discovery rate, from a total of 8 90-minute LC-MS/MS experiments, including proteins as large as 54 kDa. Work supported by NSF DMR-1157490 and the State of Florida.

Novel Aspect:

High mass protein characterization on a chromatographic time scale by LC-21T-FT-ICR MS.

Anderton, Christopher,
Pacific Northwest National Laboratory

Contributing Authors:
Jessica K. Cole, Pacific Northwest National Laboratory, Richland, WA, USA; James J. Moran, Pacific Northwest National Laboratory, Richland, WA, USA; Alexandra Cory Pacific Northwest National Laboratory, Richland, WA, USA; William B. Chrisler, Pacific Northwest National Laboratory, Richland, WA, USA; Beau Morton, Pacific Northwest National Laboratory, Richland, WA, USA; Ljiljana Paša-Tolić, Pacific Northwest National Laboratory, Richland, WA, USA; Ryan S. Renslow, Pacific Northwest National Laboratory, Richland, WA, USA; Jennifer M. Mobberley, Pacific Northwest National Laboratory, Richland, WA, USA; Yasemin Yesiltepe, Pacific Northwest National Laboratory, Richland, WA, USA; Amy A. Boaro, Pacific Northwest National Laboratory, Richland, WA, USA; Hayley Cardamone, Pacific Northwest National Laboratory, Richland, WA, USA; Stephen R. Lindemann, Pacific Northwest National Laboratory, Richland, WA, USA; Mary S. Lipton, Pacific Northwest National Laboratory, Richland, WA, USA.

Title:
The role of mass spectrometry in a multimodal investigation into how nitrogen sources govern carbon partitioning within a phototrophic consortium

Introduction:
Interactions between microbial photoautotrophs and associated heterotrophic organisms are ubiquitous in nature, and their prevalence enables them to exert significant impacts on global biogeochemical cycling. Using a multimodal approach, we investigated a model phototrophic consortium to elucidate the control of nitrogen source on carbon uptake into the community.

Methods:
Bulk isotope incorporation, protein stable isotope probing, and high lateral resolution secondary ion mass spectrometry analyses were performed in conjunction with quantitative PCR and fluorescence microscopy on a model consortium containing one autotrophic cyanobacterium and 18 associated heterotrophic species to measure the uptake of H13CO3- with 15NO3- and/or 15NH4+ in the biofilm.

Results:
We find that uptake of H13CO3-, 15NO3-, and/or 15NH4+ reveal that, though NH4+ was acquired approximately twice as rapidly as NO3-, there was no significant difference in community H13CO3- incorporation. Genome counts of community members in the consortial biofilm show significant alterations based on the available nitrogen source, though not in ways easily predictable from members' ability to assimilate NO3-. Examination of the biofilms with high lateral resolution secondary ion mass spectrometry reveal significant differences in the rate at which heterotrophs acquired cyanobacterially-fixed carbon, where carbon was much more rapidly transferred to heterotrophs when NO3- was the sole nitrogen source than when amended with NH4+. These results match proteomic data where a possible mechanism of rapid carbon transfer is due to altered iron acquisition and pyruvate metabolism.

Novel Aspect:
Stable isotope probes can bridge bulk and spatial mass spectrometry analyses and reveal significant biogeochemical processes.

Anderton, Christopher,
Pacific Northwest National Laboratory

Contributing Authors:
Gary Stacey, Divisions of Plant Sciences and Biochemistry, University of Missouri, Columbia, MO, USA; Akos Vertes, Department of Chemistry, The George Washington University, Washington, DC, USA; Beverly J. Agtuca Divisions of Plant Sciences and Biochemistry, University of Missouri, Columbia, MO, USA; David W. Koppenaal, Environmental Molecular Sciences Laboratory, Pacific Northwest National Laboratory, Richland, WA, USA; Sylwia A. Stopka, Department of Chemistry, The George Washington University, Washington, DC, USA; Ljiljana Paša-Tolić, Environmental Molecular Sciences Laboratory, Pacific Northwest National Laboratory, Richland, WA, USA.

Title:
Spatially resolved analysis of symbiotic soybean-rhizobia interactions by laser ablation-electrospray ionization mass spectrometry

Introduction:
Symbiotic interactions between bacteria and plants are of increasing scientific significance for reasons ranging from understanding global carbon and nitrogen cycling to developing advanced crops (e.g., drought resistance). To explore the mutualism between legumes (e.g., soybeans) and soil bacteria (i.e., rhizobia), we spatially analyze intact root nodules in their native state by LAESI-IMS-MS.

Methods:
Using a LAESI setup coupled to a Waters Synapt G2S, spectra from free-living rhizobia and uninfected soybean roots where compared to spectra of intact nodules, which contain both rhizobia and plant tissue. Putative assignments were based on the PlantCyc database, but the identity of many biochemical constituents were confirmed by accurate m/z, collision cross sections and tandem MS analysis.

Results:
We found that glutathione and arginine, among others, were abundant in free-living rhizobia. Simple carbohydrates were detected in all samples, but were much more abundant in the root and nodule than in the rhizobia. Soyasaponins, a diverse class of oleanane triterpenoid glycosides, were detected only in soybean tissue. IMS enhanced the detection of low intensity ions and separated isobaric species. For example, IMS resolved monosaccharide and theophylline, which were found to be more prevalent in the root and nodule, respectively. These results provide insight into the molecules specifically related to the symbiosis (e.g., nitrogen fixation). Lastly, through consecutive laser pulsing, we were able to controllably ablate though the nodule and collect spatially resolved MS. These results indicate that specific metabolites are linked to the different anatomical layers of the nodules.

Novel Aspect:
These results establish the feasibility of screening plant tissue directly to elucidate phenotypes with superior biological nitrogen fixation ability.

Anwar, Ahdia,
University of Waterloo

Contributing Authors:

Ahdia Anwar, University of Waterloo; Jarrod Psutka, University of Waterloo; Stephen W.C. Walker University of Waterloo; J. Larry Campbell, AB Sciex; W. Scott Hopkins, University of Waterloo.

Title:

Differential Mobility Spectrometry of Protonated Nucleobases

Introduction:

Through the study of DNA and RNA constituents, a greater fundamental understanding of the physical and chemical properties of these nucleic acids is obtained. Protonation and ionization of nucleobases directly influence higher-order structure, dynamics, and energetics. Ultimately, the function of DNA and RNA, as well as their interactions with proteins and drug molecules, is affected.

Methods:

Here we explore nucleobase tautomer separation in the gas phase. Preferred protonation sites are determined and subsequently controlled through the introduction of various chemical modifiers. Experimental outcomes provided by differential mobility spectrometry (DMS) studies are supported by quantum chemical calculations.

Results:

DFT calculations have identified preferred protonation sites. Separation of these nucleobase tautomers has been observed in the DMS cell. Varying chemical modifiers within the DMS cell have demonstrated the ability to further separate these isomers.

Novel Aspect:

A method of nucleobase tautomer separation and preferential formation is described.

Aoki, Jun,
Osaka University

Contributing Authors:
Yosuke Kawai, Osaka University; Michisato Toyoda, Osaka University; Kentaro Terada Osaka University.

Title:
Surface Analysis of Murchison Meteorite with MALDI Stigmatic Imaging Mass Spectrometer

Introduction:
The Murchison meteorite have been intensively studied for their prebiotic organic compounds. The elemental composition of this meteorite has been investigated by combination of Gas or Liquid chromatography and mass spectrometry analysis in bulk. In the next step, the spatial distribution of these organic compounds in Murchison meteorite becomes concerned information.

Methods:
We use Matrix Assisted Laser Desorption/Ionization (MALDI) method for soft ionization of organic matter in the Murchison Meteorite. Conventionally, the spatial resolution of imaging mass spectrometry (IMS) with MALDI is limited to about 10 - 100 μm. Therefore, we developed new MALDI-IMS (MULTUM-IMG2) with spatial resolution of 1 μm, high enough for observation of Murchison meteorite.

Results:
Ion distributions at the sample plate are magnified and projected with the ion optical lens system onto the detector. MULTUM which has four toroidal sector electric fields constitute a figure-eight trajectory is inserted into this system for extending ion flight path. We applied this new apparatus to observation of Murchison meteorite. Small piece of Murchison meteorite was mounted on a sample plate, and then polished to flat surface same with the surface of sample plate. This sample was measured by MULTUM-IMG2 and the distribution information of some elements and organic matters were obtained with high spatial resolution. We try to investigate relation between the distribution of organic compounds and the mineral structure of meteorite to clarify the creation process of organic matters in such meteorite.

Novel Aspect:
This analysis technique is useful for observing distribution of elements and organic substances in Murchison meteorite with high spatial resolution.

Aristizabal Henao, Juan,
University of Waterloo

Contributing Authors:

Ken D. Stark, University of Waterloo; Richard W. Smith, University of Waterloo.

Title:

Semi-quantitative lipidomic profiling of docosahexaenoic acid (DHA)-containing complex lipids in human breast milk and dried milk spots.

Introduction:

Breast feeding and the lipids in milk, particularly DHA content, have been linked to health benefits. Adapting a lipidomic approach would aid in understanding DHA metabolism in the mammary gland, while the use of dried milk spot (DMS) would enable research in developing countries and remote locations to better understand the impact of malnutrition on breast milk lipids.

Methods:

Breast milk was collected for technical triplicates with DMS prepared by applying 100uL onto chromatography paper. Lipids were extracted using chloroform/methanol. Samples were spiked with di-C17:0 phosphatidylcholine and were analyzed using a Thermo Q-Exactive mass spectrometer. Instrument conditions included reversed-phase UHPLC, positive ESI-MS, and top-5 data-dependent MS/MS acquisition.

Results:

Thirty-eight highly abundant acyl-specific lipids (from top 5 acquisition) were identified in wet milk with the majority being triacylglycerols with some diacylglycerols and phosphatidylcholines. Acyl chains of these abundant species ranged mainly from C10:0 to C18:2. Analysis of DMS samples revealed a similar profile of abundant lipids. DHA (C22:6n-3)-containing species were not observed in the list of 38, but neutral loss scanning for a mass of 345.2668 (loss of DHA and NH3 from ammoniated parent ion [M+NH4]+) identified 21 lipids that contain DHA. These DHA lipids were found in all of the wet and dried samples and there were no significant differences in their recoveries. These results suggest that valid lipidomic profiles can be obtained from DMS, but analyses that account for biological variation and a further examination of minor lipid species needs to be completed.

Novel Aspect:

Lipidomic profiles of abundant lipids plus docosahexaenoic acid-containing lipids were similar in both wet milk samples and dried milk spots.

Artaev, Viatcheslav,
LECO Corporation

Contributing Authors:

Georgy Tikhonov, LECO Corporation, USA; Yury Khasin, MSC-CG, Monenegro; Anatoly Verenchikov MSC-CG, Monenegro.

Title:

Advancements in Increasing Resolution in Multi Reflecting TOF Mass Spectrometry with Folded Flight Path

Introduction:

High resolution multi-reflecting time-of-flight mass spectrometry (HRMR-TOFMS) with Folded Flight Path (FFP) is successfully applied to the analysis of complex mixtures. High Speed HRMR-TOFMS with mass resolution up to 50,000 is compatible with GC and GCxGC and dramatically improves throughput and quality of the results. However many applications demand even higher mass resolution.

Methods:

Data were obtained using MR-TOFMS with FFP prototypes using a Electron Impact source. Several methods of improving resolving power of the HRMR-TOFMS were evaluated, including increase of the number of reflections and thus increase of the ion flight path within the ion optics with 3rd order energy focusing, as well as implementing novel ion mirror configuration to obtain higher order focusing.

Results:

The number of ion beam reflections (and the resolving power) is controlled by applying appropriate voltages to the multi-slot lens block, positioned in the space between the two planar gridless ion mirrors. We replaced production 16-slit lens block with narrow-spaced 22-slit one on a prototype GC-HRT unit. The instrument achieved resolving power up to 50K for full mass range, without any changes to the mass analyzer size. It was also found that higher resolving power can be achieved by optimizing ion mirror electrode configuration and voltages. In such a new configuration we were able to achieve 4th order TOF focusing with respect to energy spread combined with the full third order TOF focusing with respect to the spatial, angular and energy spread. The experimental resolution was demonstrated reaching >100K for the full mass range while detecting trace analytes in complex matrices.

Novel Aspect:

Ion mirrors with 4th order energy and full 3rd order spatial TOF focusing, experimental resolution >100,000 in the novel GC-HRMR-TOFMS.

Austin, Daniel,
Brigham Young University

Contributing Authors:
Sandra Osburn, Brigham Young University; Dimitri Deserranno, Creare, LLC.

Title:
Probing Surface-induced Dissociation of Exospheric Neutrals using an Ultra fast Rotor Reactor and GC-MS Analysis

Introduction:
High-speed neutral-surface impact dissociation obscures chemical analysis of planetary exospheres by flyby mass spectrometers. Molecules likely dissociate in ways similar to ion dissociation induced by surfaces, collisions, or photons, but neutrals have different electronic structure, so unique fragmentation pathways are expected.

Methods:
An ultra-fast rotor (4000 m/s tip speed) made of carbon fiber is being developed to study these reactions. A slower rotor (titanium, 1300 m/s) has been built for preliminary experiments using lower-energy reactions (racemization, isomerization). Volatile and semi-volatile compounds introduced into the vacuum chamber and impacted by the rotor are cryogenically trapped and then analyzed by GC-MS.

Results:
Preliminary results on the slower rotor demonstrate the overall process, and will be presented. Theoretical analysis shows the effects of molecular weight and number of bonds per molecule on the lifetime of vibrationally excited species. The design of the faster rotor will also be described, including technical challenges in running a rotor at speeds approaching the theoretical limits of tensile strength. Further, a novel inlet is being developed for future flight instruments that eliminates unwanted impact fragmentation. This inlet uses an array of microchannels. Although the same number and energy of thermalization impacts occur, the timescale of thermalization is fast by 2 orders of magnitude compared with conventional thermalization chambers. The neutrals are cooled on a timescale shorter than their dissociation lifetime, allowing MS analysis of intact, native compounds.

Novel Aspect:
First instance of hyperthermal impact-induced dissociation and reactions of cold neutrals

Ayala Cabrera, Juan Francisco,
University of Barcelona

Contributing Authors:
E. Moyano, Department of Analytical Chemistry, Faculty of Chemistry, University of Barcelona, Martí i Franquès 1-11, 08028, Barcelona; J. F. Ayala-Cabrera, Department of Analytical Chemistry, Faculty of Chemistry, University of Barcelona, Martí i Franquès 1-11, 08028, Barcelona; F. J. Santos Department of Analytical Chemistry, Faculty of Chemistry, University of Barcelona, Martí i Franquès 1-11, 08028, Barcelona.

Title:
Evaluation of atmospheric pressure ionization techniques for the determination of fluorotelomer alcohols and perfluorinated sulfonamides

Introduction:
Interest and concern in fluorotelomer alcohols and perfluorinated sulfonamides and sulfonamidoethanols is increasing. They might cause estrogenic effects in the living organisms and once introduced into the environment, they are transported to long-distance and are degraded to persistent and toxic perfluorinated carboxylic acids (PFOA) and sulfonates (PFOS).

Methods:
Fluorotelomer alcohols and perfluorinated sulfonamides are generally analyzed by GC-MS using classical ionization techniques (EI and CI), but sensitivity problems are observed. In this work, the applicability of alternative atmospheric pressure ionization sources (ESI, APCI, APPI) is investigated to propose a sensitive and selective UHPLC-MS/MS method for their determination in selected samples.

Results:
Target compounds are commonly analyzed by GC-MS using CI mode because it provides characteristic ions at high mass, although poor ionization efficiency and chromatographic peak shape hindered sensitivity. In contrast, EI is not frequently used due to the low abundance of molecular ions and the lack of specific fragments. Regarding atmospheric pressure ionization sources, ESI showed the worst performance due to inefficient deprotonation of fluorotelomers, while APCI and APPI significantly favor deprotonated ions, which are used as precursor ions for MS/MS. APCI shows the best performance using methanol as organic modifier while acetonitrile provides the best ionization efficiency for APPI. After characterizing the full-scan and MS/MS spectra and establishing fragmentation pathways, a selective and sensitive UHPLC-APPI-MS/MS method is proposed for the analysis of real samples.

Novel Aspect:
Critical evaluation of API sources (ESI, APCI, APPI) for the determination of volatile polyfluorinated compounds by UHPLC-MS/MS at low levels.

Azargun, Mohammad,
Memorial University

Contributing Authors:
Travis Fridgen, Memorial University.

Title:
Mass spectrometry fragmentation methods (BIRD, SORI-CID, IRMPD): powerful techniques to study physical chemistry of guanine quadruplexes

Introduction:
Guanine nucleobases can self-assemble to form stable tetrameric structures, called G-tetrads. These structures form in guanine rich parts of nucleic acids such as DNA. Tetrads can sit on top of each other with metal cations such as K+ between them to form G-quadruplexes. G-quadruplexes contribute to significant biological processes, and have found many applications for for therapeutic purposes.

Methods:
Blackbody infrared radiative dissociation (BIRD) was used to obtain temperature dependant rate constants and Arrhenius plots. Master equation modeling was employed to model Experimental BIRD rate constants to extract threshold dissociation energies, E0. Infrared multiphoton dissociation (IRMPD) spectroscopy along with computational chemistry was used to elucidate complex structures.

Results:
Preliminary electrospray mass spectrum of a solution composed of 9-ethylguanine and alkali metals revealed that complexes with multiples of four 9-ethylguanine, G-tetrad units, form dominantly and other complexes such as dimeric and trimeric complexes exist only in a relatively small amount. G-quadruplexes containing alkali metal cations were detected, except Li+ which is incapable of forming G-quadruplex due to its great intrinsic tendency to form tetrahedral structures. IRMPD spectra reveal that all alkali metal cation-containing G-quadruplexes share almost identical spectra. This showed that these clusters have similar sandwich-type structures. BIRD experiments along with master equation modeling divulge that K(9eG)8+ has the highest activation energy and stability which explains why K+ is the ion of choice for G-quadruplexes in DNA.

Novel Aspect:
G-quadruplexes intrinsic structures, and binding energies were studied.

Baba, Takashi,
Sciex

Contributing Authors:
Paul R. S. Baker, Sciex; J. Larry Campbell, Sciex; James W. Hager Sciex; Bruce A. Thomson, Sciex; J. C. Yves Le Blanc, Sciex.

Title:
Obtaining complete structural identification of lipids by using low-energy electrons in a novel branched radio-frequency ion trap

Introduction:
For over a decade, low-energy electron beams have been used as a dissociation tool in mass spectrometry. When analyte ions are exposed to these electrons, they undergo non-ergodic processes of electron capture or excitation, resulting in fragmentation patterns that are complementary to conventional collisional activation. a great deal of efforts has been made to maximize the overall efficiencies.

Methods:
We have developed a new type of RF ion trap for injecting analyte ions and electrons simultaneously. The trap has two axes, one of which is used as the primary analyte ion path, with an orthogonal axis used to guide electrons to those analyte ions. In addition, a magnetic field is applied along the electron beam path to confine the low-energy electrons from any perturbations from RF field.

Results:
With these advances in overall efficiency, a new application in lipidomics was demonstrated recently. Singly charged lipid precursor ions were excited by interactions with the electron beam, resulting in "near complete" structure analysis of phospholipids, sphingolipids andacylglycerols, where lipid class or head group, acyl's double bond number and positions, and regioisomerism were clearly displayed in a single spectrum. To make this method "complete", we expanded our analyses to cis/trans isoforms of intact phosphatidylcholines (PCs) and triacylglycerols (TG). As a result, we have added the ability to distinguish some cis and trans isoforms to the capabilities of this technology. This finding allows not only the qualitative analysis of lipids containing trans double bonds in biological lipids, but also the relative quantitative analysis of such double-bond isomers in a mixture.

Novel Aspect:
Development of a novel branched radio-frequency ion trap and its new application on lipidomics

Bachman, Martin,
AstraZeneca

Contributing Authors:
Eric Hall, Labcyte, Sunnyvale, CA; Brian Musselman, IonSense, Saugus, MA; Jonathan Wingfield AstraZeneca, UK; Lars Majlof, Labcyte, Sunnyvale, CA; Luke Ghislain, Labcyte, Sunnyvale, CA; Ian Sinclair, AstraZeneca, UK; Sammy Datwani, Labcyte, Sunnyvale, CA.

Title.
Seconds per sample with Echo-DART-MS: acoustic loading of a mass spectrometer for high-throughput label-free screening

Introduction:
MS-based detection is a promising approach to label free high throughput screening. We complement the fast and precise Labcyte Echo® technology to eject nanolitres of sample from a 384-well plate and the IonSense Direct Analysis in Real Time (DART®) ionization to generate analyte ions that are detected by a Waters mass spectrometer. This integrated platform can analyse 1,200 samples per hour.

Methods:
2.5 nL droplets are acoustically ejected at rates up to 200 Hz (droplets per second) onto a substrate where heated gas containing metastable helium atoms flows from the DART source and ionizes the analyte. An elongated vacuum assisted sample interface (VAPUR®) collects the thermally desorbed sample ions and permits efficient transfer to a Waters SQD2 mass spectrometer.

Results:
A variety of small-molecule compounds were used to optimize the components and experimental parameters of the Echo-DART-MS. Thermal desorption kinetics and degree of carryover were evaluated for different substrate types including a gold mesh and a heated filament. An interleaved 384-well plate of water and caffeine executed at 3 seconds per sample produced 100% true positive and 0% false negative hits. Analysis of 1,000 aqueous samples, each containing a unique compound, showed that the technique is general across the small-molecule compound class. Evaluation of spectral purity (molecular ion peak detected in the correct well) was carried out using custom software and correctly identified >70% of compounds. DART ionisation is tolerant to high salt concentrations needed for biochemical assays, which makes Echo-DART-MS a promising platform for high-throughput screening applications.

Novel Aspect:
An acoustic droplet ejection source integrated with ambient ionization, Echo-DART-MS, for high throughput label free screening.

Bachman, Martin,
AstraZeneca

Contributing Authors:
Eric Hall, Labcyte, Sunnyvale, CA; Rich Ellson, Labcyte, Sunnyvale, CA; Rhys Jones Waters, Wilmslow, UK; Mike Morris, Waters, Wilmslow, UK; Daniel Addison, AstraZeneca, UK; Luke Ghislain, Labcyte, Sunnyvale, CA; Lars Majlof, Labcyte, Sunnyvale, CA; Jon Wingfield, AstraZeneca, UK; Ian Sinclair, AstraZeneca, UK; Sammy Datwani, Labcyte, Sunnyvale, CA; Steven Pringle, Waters, Wilmslow, UK; Rick Stearns, Labcyte, Sunnyvale, CA; Joe Olechno, Labcyte, Sunnyvale, CA.

Title:
Echo-MS: A versatile platform for ultrafast mass spectrometry analysis in drug discovery

Introduction:
Biochemical high-throughput screening assays are central to drug discovery efforts and are typically developed around expensive label-based technologies. We introduce a label-free platform combining acoustic droplet ejection and mass spectrometry capable of analyzing 10,000 samples per hour while dramatically reducing sample consumption and overall cost.

Methods:
We combined a Labcyte Echo550 acoustic liquid dispenser with a Waters Xevo G2-S qToF mass spectrometer. The system features an XY-stage and is able to sample from a 384-well plate directly to a heated ion transfer system at 1,500 Hz. Electrospray ionization is achieved by charging the sample at the well surface. A data processing algorithm was designed to deal with large sample batches.

Results:
We will show how label-free screening by mass spectrometry reduces assay development time and cost compared to traditional methods, whilst increasing the quality of the produced data. The low sample volumes required for analysis allow for a single well to be interrogated reproducibly multiple times without impacting the well contents, enabling live kinetic data to be generated from up to 384 samples at a time. Buffer and surfactant conditions were optimized for various biochemical reactions to allow efficient enzyme-mediated substrate turnover while minimizing ion suppression. We will present data from the first high-throughput biochemical screen of 250,000 compounds, identifying new and novel inhibitors of a relevant drug target. We will discuss further anticipated applications of this technology within drug discovery and basic science, and its future developments.

Novel Aspect:
Novel hyphenation of acoustic droplet ejection and mass spectrometry for high throughput label free screening and other applications in drug discovery

Bader, Tobias,
Zweckverband Landeswasserversorgung (and Leuphana University Lüneburg)

Contributing Authors:
Wolfgang Schulz, Zweckverband Landeswasserversorgung Langenau (Germany); Klaus Kümmerer, Leuphana University of Lüneburg (Germany); Rudi Winzenbacher Zweckverband Landeswasserversorgung Langenau (Germany).

Title:
General strategies to increase the repeatability in environmental non-target screening by liquid chromatography-high resolution mass spectrometry

Introduction:
Despite the continuous developments in LC-HRMS, non-target approaches still suffer from the lack of reliable processing tools. Results obtained by peak finding algorithms comprise an unknown amount of false positives while real peaks of interest might be missed. Here, we present a novel data evaluation strategy based on a combinatorial approach, which allows addressing the above mentioned issues.

Methods:
Different models of data processing were evaluated, particularly with regard to the repeatability. The evaluation included different water matrices which were analyzed repeatedly leading to more than 250 direct injections that were incorporated in the assessment. As proof of concept, different matrix types were spiked with 130 isotopically labeled standards at different concentration levels.

Results:
Based on meaningful characteristics, different processing approaches for LC-HRMS data were compared. We found that the analysis of multiple replicates is the key factor as, on the one hand, it provides the option of integrating filters to minimize the false positive rate and, on the other hand, allows correcting false negative results occurring during the peak finding. The developed processing strategies clearly point to an enhanced data quality since both the repeatability as well as the peak recognition could be considerably improved. Despite the stringent filter criteria, recovery rates of up to 93% were reached by direct injection of 10 pg on-column. The proposed model, comprising three replicates, filters less than 5% of the standards recognized at 100 ng/L and thus indicates the general applicability of the presented strategies, even if considering more complex matrices.

Novel Aspect:
-Combinatorial approach for validation of data processing -Replicates decrease false positives/negatives -Signal fluctuation used as a powerful filter

Bader, Theresa,
Helmholtz Zentrum München Research Unit Analytical BioCheoChemistry

Contributing Authors:
Basem Kanawati, Helmholtz Zentrum Munich, Research Unit Analytical BioGeoChemistry; Constanze Müller, Helmholtz Zentrum Munich, Research Unit Analytical BioGeoChemistry; Daniel Hemmler Helmholtz Zentrum Munich, Research Unit Analytical BioGeoChemistry; Technical University Munich, Chair of Analytical Food Chemistry; Franco Moritz, Helmholtz Zentrum Munich, Research Unit Analytical BioGeoChemistry; Stephanie Rebensburg, Helmholtz Zentrum Munich, Institut of Virologie; Johanna Götz, Helmholtz Zentrum Munich, Institut of Virologie; Ruth Brack-Werner, Helmholtz Zentrum Munich, Institut of Virologie; Philippe Schmitt-Kopplin, Helmholtz Zentrum Munich, Research Unit Analytical BioGeoChemistry; Technical University Munich, Chair of Analytical Food Chemistry.

Title:
Characterization of anti-HIV-1 active compound classes from Cistus incanus via LC-fractionation and ultrahigh resolution mass spectrometry (FT-ICR-MS)

Introduction:
HIV is still a global health issue with 1.2 million HIV-related deaths in 2014. The development of new anti-HIV-1 drugs is crucial as the virus develops resistances against the existing 27 drugs. Since natural products are a promising source for new drugs, the anti-HIV-1 active water extract and polyphenol enriched extract of Cistus incanus were analyzed with mass spectrometry based metabolomics.

Methods:
Fractionation of both extracts, water and polyphenol enriched, where established to reduce the chemical complexity. The extracts were fractionated by C2-SPE and subsequently by HPLC-PDA via different selectivity (RP C18). The resulting fractions were measured by ultrahigh resolution mass spectrometry (FT-ICR-MS) and the anti-HIV-1 activities were tested in a full HIV replication system (EASY-HIT).

Results:
The polyphenol enriched extract of Cistus incanus shows a better anti-HIV-1 inhibition than the aqueous extract with an EC50-value of 0.7 µg/ml compared to the EC50-value of 7.8 µg/ml, respectively. Furthermore, several fractions of these extracts show true anti-HIV-1 activity and the ultrahigh resolution of the FT-ICR-MS allows the prediction of the molecular formulas of metabolites in these fractions. The spearman correlation of the anti-HIV-1 activity distribution and the intensity distribution of the masses with known elemental composition in all fractions lead to the characterization of significant compound classes. Polyphenols were identified as major players in the mined elementary composition datasets. These results lead to the conclusion that among others, polyphenol compounds are responsible for the anti-HIV-1 activity of Cistus incanus.

Novel Aspect:
The anti-HIV-1 activity of Cistus incanus was investigated and a characteristic polyphenol compound class could be identified with mass spectrometry.

Bai, Yu,
Peking University

Contributing Authors:

Huwei Liu, Peking University; Sensen Shen, Peking University.

Title:

Metabolomic Analysis of Mouse Embryonic Fibroblast Cells in Response to Acute Starvation with and without Atg7

Introduction:

Autophagy-related protein 7 (Atg7) is essential in the formation of the autophagophore and is indispensable for autophagy induction. Autophagy will exist in lower level or even be blocked in cells without Atg7. Even though the possible signaling pathways of Atg7 have been proposed, the metabolomic responses under acute starvation in cells with and without Atg7 has not been elucidated.

Methods:

An ultra-performance liquid chromatography-Q-Exactive mass spectrometry (UPLC-MS) method was employed to conduct metabolomic analysis of Atg7-deficient (Atg7-/-) mouse embryonic fibroblast cells (MEFs) vs wild-type MEFs along with the starvation time.

Results:

Significantly altered metabolites were identified in response to nutrient stress, which were mainly associated with amino acid, energy, carbohydrate, and lipid metabolism. For the wild-type MEFs, the induction of autophagy protected cell survival with some up-regulated lipids during the first two hours' starvation, while the subsequent apoptosis resulted in the decrease of cell viability after four hours' starvation. For the Atg7-/-MEFs, apoptosis perhaps led to the deactivation of tricarboxylic acid (TCA) cycle due to the lack of autophagy, which resulted in the immediate drop of cellular viability under starvation. These results contributed to the metabolomic study and provided new insights into the mechanism associated with Atg7-dependent autophagy.

Novel Aspect:

This is the exploratory research of the metabolite responses to the acute starvation in cells with and without Atg7 using metabolomics.

Bain, Ryan,
Purdue University

Contributing Authors:
Stephen T. Ayrton, Purdue University; Christopher J. Pulliam, Purdue University; R. Graham Cooks Purdue University.

Title:
Reactions in gas-phase, solution-phase and electrospray ionization

Introduction:
Microdroplets generated by spray-based ionization techniques including desorption electrospray ionization (DESI), electrosonic spray ionization (ESSI), electrospray ionization (ESI) and paper spray (PS) are characterized by reaction acceleration that produces rates that are multiple orders of magnitude higher than bulk-phase rates.

Methods:
We chose well-characterized solution- and gas-phase chemistry in order to determine the conditions under which a chemical reaction can be simply monitored without accelerated reaction in the course of analysis. Furthermore, we have explored accelerated reaction conditions to see if ion-molecule reactions can be promoted in ESI.

Results:
Reaction acceleration in ESI is observed at greater distances between the source and the MS due to increased concentrations and surface/volume ratios. The accelerated reactions occur in droplets where the conditions are between ion/molecule reactions and solution-phase. A reaction that progresses differently in the gas-phase than in bulk was developed to see if ESI at greater distances could promote both pathways. This experiment was performed using acetone and phenylhydrazine which forms acetone phenylhydrazone instantly. The gas-phase hydrazone undergoes rearrangement and ammonia loss by a Fischer rearrangement. In solution-phase the excess acetone reacts with the intermediate to form an imine. This can be accelerated by increasing the source-MS distance, while the gas-phase product cannot be accelerated, indicating that reaction acceleration follows solution-phase mechanisms.

Novel Aspect:
Solution-phase reactions can be accelerated in charged microdroplets with no interference for gas-phase processes.

Bajrami, Besnik,
Broad Institute of MIT and Harvard

Title:
RED-LCMS as an orthogonal technique for high throughput fragment based drug discovery

Introduction:
Binding affinity of a small molecule drug candidate to a therapeutically relevant target is considered as the first basis of the candidate's efficacy. In the fragment based screening, a cascade of multiple orthogonal technologies is required for reliable detection and characterization of fragment binding affinity to a target.

Methods:
Given the limitations of the current screening techniques, especially for weak binding ligands, we have developed a rapid equilibrium dialysis mass spectrometry based (RED-LCMS) approach which enables efficient, rapid and sensitive low-affinity ligand discovery. Furthermore RED-LCMS can be utilized to determine the KD values of individual ligands from the equilibrium mixture.

Results:
We demonstrate this system by using carbonic anhydrase as the target protein to illustrate the ability of the method to identify and rank non-covalent binding ligands with a wide range of affinity(low nM to high µM). A pilot study with randomly selected 100 fragments was spiked with carbonic anhydrase ligands (weak and strong) was incubated with protein with each fragment present at a P:L ratio of 1:1. All the spiked carbonic anhydrase ligands were identified from the pools. Accurate Kd values for the ligands were obtained by varying the ligand concentrations while keeping the protein concentration constant.

Novel Aspect:
This approach requires little sample preparation and provides a fast, robust, high throughput easy to use system for ligand-protein binding assays

Baker, Andrew,
Waters Corporation

Contributing Authors:
Jayne Kirk, Waters Corporation; Jason Wildgoose, Waters Corporation; Keith Richardson Waters Corporation; Martin Green, Waters Corporation; Mark Wrona, Waters Corporation.

Title:
Lipid analysis via a novel data independent technique on a QTof combining a scanning quadrupole mass filter and ultra fast data acquisition system

Introduction:
QTof mass spectrometry is an established tool for discovery and quantitative applications. A mode of operation where the resolving quadrupole mass filter is scanned, concomitantly with MS/MS data acquisition is described. This method combines the specificity of MS/MS fragmentation while maintaining a powerful, unbiased, data-independent approach for analysis of phospholipids in a complex matrix.

Methods:
The quadrupole mass filter of a Xevo G2-XS QTof was modified to operate in a continuous, repetitively scanning mode. The collision cell alternates between low and high voltage with each scan to provide fragment ion profiles that can be assigned to the precursor ion profiles resulting from scanning the quadrupole transmission window.

Results:
A rat brain phospholipid extract was injected onto a LC system equipped with a C18 column. This complex sample contains many closely eluting lipids, with precursor ions that fragment to generate the same (or similar) product ions, are shown to be clearly resolved with clean and unambiguous MS/MS spectra using 2D-MSMS. The main difference between data independent acquisition (DIA) and 2D-MSMS is the behaviour of the quadrupole. Instead of remaining open and transmitting all ions it slides over the selected mass range during both low and high energy scans. Filtering the of the precursor ions by the quadrupole increases the selectivity of the method. This set up produces two-dimensional datasets resembling nested ion mobility (IMS)-MS data. This opens up the possibility of precursor and fragment alignment with a tolerance much tighter than the quadrupole mass isolation window

Novel Aspect:
2D-MSMS is a valuable tool in the analysis of complex lipid sample sets by providing confidence in fragment ion assignment and lipid identification.

Baker, Andrew,
Waters Corporation

Title:
Targeted and Untargeted Analysis of Pesticides in Fish Oil Using an Ion Mobility Enabled Q-TOF Mass Spectrometer

Introduction:
Nontargeted pesticide screening requires high resolution chromatographic separations coupled with high mass spectral resolution, especially in complex matrices. Incorporation of orthogonal ion mobility separation gives better specificity for compound identification (cleaner MS and MS/MS spectra) even in the presence of isobaric contaminants. In targeted MRM experiments, IMS increases S/N ratios.

Methods:
Fish oil (0.5 g) or fish oil samples spiked with pesticide mixtures (down to 10 mg/kg) were extracted using a QUECHERS method (5 mL Water, 30 mL Acetonitrile, 1.5 g NH4OAc, 6 g Mg2SO4 30 uL HOAc). The supernatant was analyzed by LC/IMS/MS on a VION IMS Q-Tof mass spectrometer. Pesticides were identified on the basis of retention time, mass measurement and CCS from a library of known pesticides.

Results:
For untargeted screening, the Collisional Cross Section (CCS) is used as another defining property, along with retention time, mass measurement accuracy, fragment ion presence and chromatographic correlation. Pesticides in fish oil were identified using these criteria down to the 10 mg/kg level (ppB). In addition to using the CCS as a point of identification, the low and high energy spectra obtained during data independent acquisitions are filtered using the ion arrival time distribution. This dramatically cleans up the resulting spectra, removing confounding fragment ions. For targeted screening, mobility filtered MRM experiments were used to lower detection limits, primarily by reducing background chemical noise.

Novel Aspect:
Ion mobility filtering of spectra for increased confidence in assignment and increased specificity in targeted assays

Bararpour, Nasim,
Centre universitaire romand de médecine légale, Lausanne - Genève (CURML)

Contributing Authors:
Beryl Royer-Bertrand, Department of Computational Biology, University of Lausanne, Lausanne, Switzerland; Timothée Joye, Unit of Toxicology, University Centre of Legal Medicine, Lausanne-Geneva, Switzerland; Marc Augsburger Unit of Toxicology, University Centre of Legal Medicine, Lausanne-Geneva, Switzerland; Alexandre Moulin, Department of Ophthalmology, University of Lausanne, Jules-Gonin Eye Hospital, Lausanne, Switzerland; Aurélien Thomas, Faculty of Biology and Medicine, Lausanne University Hospital, Lausanne University, Lausanne, Switzerland & Unit of Toxicology, University Centre of Legal Medicine, Lausanne-Geneva, Switzerland; Francine Behar-Cohen, Department of Ophthalmology, University of Lausanne, Jules-Gonin Eye Hospital, Lausanne, Switzerland; Carlo Rivolta, Department of Computational Biology, University of Lausanne, Lausanne, Switzerland.

Title:
Metabolomic imaging mass spectrometry of high-risk metastasizing uveal tumors classified upon BAP1 mutational status

Introduction:
Uveal melanoma (UM) is the most common primary intraocular tumor of adult characterized by high metastatic risk. Mutations in BAP1gene are observed in more than 80% of metastasizing tumors. Here, UM tissue sections were classified based on BAP1 mutational status. We then investigated in situ metabolomic changes by imaging mass spectrometry (IMS) comparing patients with low and high metastatic risk

Methods:
IMS analyses were carried on UM tissue sections of 15 patients classified as low and high metastatic risk regarding BAP1genotype. Acquisitions were performed by MALDI LTQ-Orbitrap MS in negative mode with 9-aminoacridine (9-AA) matrix. Region selections were based on in situ classification and pathologist annotations.Metabolic patterns assessment was then driven by using supervised learning method

Results:
Multivariate analysis was performed to classify samples and determine whether discriminant metabolites were up-or down regulated in high risk metastatic group. Metabolite signature including species from amino acid, phospholipid and fatty acid classes allows unambiguous distinction between both phenotypes. Integration and comparison of the candidates into cellular and functional pathway were performed using over-representation analysis into the Kyoto Encyclopedia Genes and Genomes (KEGG) pathway database (www.genome.jp). Over-representation analysis displayed some metabolomic pathways were meaningfully perturbed namely, taurine, glycerophospholipids and glutathione metabolism pathways. Moreover, pathway analysis was performed to interconnect the metabolites to proteins activity and gene regulation highlighting the link between the detected metabolites and BAP-1 gene.

Novel Aspect:
In combination with genotyping, in situ mapping of hundred metabolic traits might open new insights into in situ biology and diagnosis of high risk UM

Bärenfänger, Melissa,
Institute of Organic Chemistry, University of Hamburg

Contributing Authors:
Prof. Dr. Bernd Meyer, Organic Chemistry, University of Hamburg.

Title:
Analysis of human Glycoproteins with different LC-MS/MS and NMR techniques

Introduction:
Glycosylation is an important posttranslational modification of proteins since their presence influences many biological activities. Altered glycosylation can lead to many diseases and is found in patients with e.g. cancer, inflammatory diseases and neuronal diseases. Therefore, their precise analysis plays an important role to get a better understanding about structure and related function.

Methods:
Here we show different techniques for glycan analysis of human glycoproteins, involving sample enrichment and separation with different LC methods and analysis with online LC-MS as well as offline MALDI MS and additional NMR Data. Combination of these techniques leads to a complete characterization with information not only about glycan composition but also about branching and anomericity.

Results:
The bottom up approach involved enzymatic digest with trypsin. Analysis of peptides and glycopeptides was carried out on an ESI-q/TOF. Release of glycans was performed by treatment with PNGase F, followed by further LC-MS/MS experiments. Incorporation of NMR-Data was archived by 3D cross correlation, resolving chromatographic overlap and generating pure NMR spectra of single components. This allowed an accurate determination of glycan structures. A middle down approach was tested with an incomplete digest with trypsin as well as with a digest with endopeptidase Arg-C, generating larger glycopeptides. Increasing the peptide residue led to better ionization in ESI-MS experiments, helping to detect less abundant glycoforms. With increased peptide residue, chromatographic behavior of glycopeptides became more similar and isolation of glycopeptides depending on their glycosite was possible.

Novel Aspect:
With the combination of LC, MS and NMR techniques, a complete characterization of complex glycan structures from human glycoproteins was achieved.

Barrow, Mark,
University of Warwick

Title:
Petroleomics and environmental monitoring: a more detailed view using FTICR MS

Introduction:
Petroleum-related components can enter the environment due to natural seepage or human activity. There is a need, therefore, to understand the transport and transformation of petroleum components in the environment. Mass spectrometry has been playing a vital role in advancing understanding of the composition of petroleum samples, giving rise to "petroleomics."

Methods:
Petroleomics and environmental research at the University of Warwick make use of a 12 T solariX Fourier transform ion cyclotron resonance (FTICR) mass spectrometer (Bruker Daltonik GmbH, Bremen, Germany). Electrospray ionization (ESI), atmospheric pressure chemical ionization (APCI) and atmospheric pressure photoionization (APPI) have been used, and gas chromatography (GC) has been coupled.

Results:
Extraction of bitumen from the oil sands (Alberta, Canada) produces approximately three barrels of water for every barrel of synthetic oil. It is necessary to improve methods for the analysis of process waters and to monitor the potential impact upon the environment. There is crossover between petroleomics and environmental monitoring for the oil sands samples. Naphthenic acids were traditionally targeted for many years, making negative-ion ESI the ionization method of choice. FTICR MS studies have revealed a much greater range of components. In addition to direct infusion of samples, coupling with GC highlights the presence of isomers. Recent work has compared sample profiles after different sample extraction methods were used. Thus, FTICR MS has been playing a vital role at the forefront of not only petroleomics, but also oil sands environmental monitoring.

Novel Aspect:
Ultrahigh resolution data of samples after different extraction methods, with both positive-/negative-ion ESI and positive-/negative-ion APPI

Barsch, Aiko,
Bruker Daltonics

Contributing Authors:
Sven Meyer, Bruker Daltonik GmbH, Bremen, Germany.

Title:
Bioactive compounds from Bees. Identification of characteristic metabolites in propolis extracts via UHPLC-MS and –MS/MS based metabolite profiling

Introduction:
Propolis is a natural resin produced by bees to protect their hives. Its ingredients show a large variety of biological activities, for instance against microorganisms or cancer cell lines. Here, samples from different bee yards were analyzed by UHPLC-QTOF-MS in order to identify propolis metabolites. Using metabolic profiling several compounds characteristic for specific extracts were identified.

Methods:
European poplar propolis samples from seven bee yards were extracted with 70% EtOH and analyzed in triplicate by high resolution UHPLC-QTOF-MS (Bruker Impact II and Thermo Scientific Dionex RSLC). The data evaluation was performed using MetaboScape, PathwayScreener (Bruker Daltonics) and MetFrag [1] software. [1] S.Wolf, S.Schmidt, M.Müller-Hannemann, S.Neumann, BMC Bioinformatics 2010, 11:148.

Results:
LC-MS chromatograms from propolis samples showed a general chemical similarity. Further investigation by PCA revealed compounds with significantly different abundance between samples. Of these, four characteristic compounds as well as six compounds of major abundance in all propolis extracts were further investigated. Their identification revealed that these are known ingredients in poplar type propolis and are related to flavonoid biosynthesis. Following this non-targeted data mining, data was used in a targeted workflow. Metabolites contained in the KEGG pathway for flavonoid biosynthesis served as basis to create a list of approx. 150 related compounds. LC-MS data files were screened for their occurrence by generating high resolution extracted ion chromatograms. Many positive hits were found and will be investigated with the aim to correlate bioactivity with identified compounds.

Novel Aspect:
Unique identification of propolis ingredients and their chemical pathways using UHPLC-QTOF-MS and –MS/MS data.

Barsch, Aiko,
Bruker Daltonics

Contributing Authors:
Matthias Szesny, Bruker Daltonik GmbH, Bremen, Germany; Jörn Kalinowski, Bielefeld University, Bielefeld, Germany; Marcus Persicke Bielefeld University, Bielefeld, Germany; Heiko Neuweger, Bruker Daltonik GmbH, Bremen, Germany; Frederik Walter, Bielefeld University, Bielefeld, Germany; Nikolas Kessler, Bruker Daltonik GmbH, Bremen, Germany; Stephanie Kaspar-Schoenefeld, Bruker Daltonik GmbH, Bremen, Germany.

Title:
Increasing arginine production in C. glutamicum by rational strain design using a combination of metabolomics and proteomics

Introduction:
C. glutamicum is a bacterium used for biotechnological production of amino acids and other metabolites. Arginine is of commercial importance in cosmetic and pharmaceutical industries and as food additive. Here we highlight that combining rational strain design with metabolomics and proteomics is a powerful tool to increase production of desired metabolites in a biotechnological workhorse.

Methods:
Three mutant strains were compared to wildtype C. glutamicum extracts. Metabolomics and proteomics data have been acquired on an impact II QToF-MS (Bruker Daltonics). The MetaboScape software was used for processing of metabolomics data, whereas MaxQuant was used to process proteomics data. Mapping of detected changes to the arginine pathway enable data interpretation in a biological context.

Results:
Metabolomics and proteomics data were acquired to gain insights into changes introduced by rational strain design to increase arginine production in C. glutamicum. Metabolomics data analysis resulted in the tentative identification of an unknown compound — more abundant in the mutant strains — as glutamylvaline. Several known compounds in the arginine biosynthetic pathway could automatically be identified. Proteomics data revealed significant changes of proteins involved in the arginine biosynthesis pathway. Mapping alterations detected by both OMICS approaches on biochemical pathway maps enabled quick formulation of hypotheses for the observed changes in the biological context. Our results demonstrate that combination of non-targeted omics techniques enables in-depth investigation of changes in C. glutamicum caused by rational strain design to increase production of desired metabolites.

Novel Aspect:
Combination of non-targeted metabolomics and proteomics workflows link data to biology by pathway mapping

Bartels, Benjamin,
IMPRS

Contributing Authors:
Aleš Svatoš, MPI-CE, Jena; Purva Kulkarni, IMPRS.

Title:
Mapping rough terrain - laser ablation electrospray ionization (LAESI) of non-flat, biological samples

Introduction:
To guarantee optimal laser focus during laser-based ionization experiments the sample is required to be as flat as tissue cross-sections. When analyzing surface of insects, bacteria or plants, the requirement of flatness is seldom met. So far effects of the sample's surface topology on signal intensity are largely ignored. Our method allows for the proper laser focus during imaging experiments.

Methods:
Topological maps of sample surfaces were acquired with a confocal distance sensor (µ-Epsilon). Metabolic profiling of the samples was performed subsequently with a custom-built laser ablation electrospray ionization source connected to a Synapt HDMS Q-ToF instrument (Waters). Laser focus was maintained by adjusting the sample's position in the 3rd dimension based on the acquired topological maps.

Results:
Topological maps of cotton and savoy cabbage leafs were acquired with lateral resolutions from 1 mm down to 10 µm using the confocal sensor. Telescope infrared-laser optics are employed in a custom-built laser ablation electrospray ionization source to focus an OPO-pumped pulse laser operating at 2940 nm wavelength. This renders the system susceptible to non-flat surface topology. Acquired topological maps were applied to perform metabolic profiling/imaging on cotton and savoy cabbage leaf surfaces with topological features differing in height up to 2 mm. Ablation marks were found to be ellipses with major and minor axes of roughly 90µm and 40µm, respectively. All processes are integrated in LabVIEW software and the measurements are executed automatically. Using a lab-developed software package, the resulting data files were integrated into a data cube for visualization and analysis.

Novel Aspect:
Laser ablation of non-flat samples is guided in the 3rd dimension for perfect laser focus allowing for beam shaping optics with short focal lengths.

Batey, Michael,
Waters Corporation

Contributing Authors:
Emmanuelle Claude, Waters Corporation; MicShazia Khan, Queen's Medical Research Institute, University of Edinburgh; Mark Towers Waters Corporation; Logan Mackay, SIRCAMS, University of Edinburgh; Phillippa Hart, Waters Corporation; Ruth Andrew, Queen's Medical Research Institute, University of Edinburgh.

Title:
Demonstration of enhanced ion mobility separation of derivatised isobaric steroids by DESI and MALDI TOF mass spectrometry

Introduction:
Steroids are poorly ionised by DESI and MALDI so derivatization is required to enhance ionisation. Isomeric steroids can be separated by LC or with limited efficiency by ion mobility separation (IMS). In this work we present the use of several agents to derivatise isobaric steroids, enabling enhanced IMS for direct sample analysis techniques like MALDI-MS and DESI-MS.

Methods:
Steroids were derivatised using p-tolenesulfonyl (PTSI). Experiments were carried out on a SYNAPT G2-Si mass spectrometer. For MALDI analyses, the steroid solutions were mixed 1:1 with a solution of 5mg/mL 9-aminoacridine 4:1 EtOH:H20. DESI spray conditions were set at 1.5µl/min, 98:2 MeOH:H2O with 5bar pressure.

Results:
PTSI derivatised steroid standards were firstly analyzed by DESI. The isobaric steroids 17a-estradiol-PTSI and 17b-estradiol-PTSI showed clear separation with a drift time difference of 0.38ms. When the two isobaric compounds were mixed 1:1, there was clear separation with a Rs=1. Androsterone-PTSI and trans-androsterone-PTSI demonstrated a greater drift time difference of 0.55ms. When mixed 1:1, separation was almost baseline with a Rs=1.2. The remaining three isobaric steroids were more challenging. Epitestosterone-PTSI had the shorter drift time of 3.53ms, followed by Testosterone-PTSI with a drift time of 4.01ms. DHEA-PTSI had the longest drift time of 4.29ms. Similar drift times were observed when the standards were spotted onto liver tissue sections and subjected to imaging mass spectrometry by DESI. For MALDI analyses, the drift time differences were comparable to DESI datasets.

Novel Aspect:
Investigation of steroid derivatisation to improve IMS of isobaric species for direct ionisation mass spectrometric techniques, DESI and MALDI.

Batey, Michael,
Waters Corporation

Contributing Authors:
Phillippa Hart, Waters Corporation; Emmanuelle Claude, Waters Corporation; Mark Towers Waters Corporation.

Title:
An Investigation of DESI Imaging at varying acquisition rates with real time display for optimized tissue imaging

Introduction:
Mass spectrometry imaging is an established tool in biological research. For routine clinical research adoption, workflows need to be optimized with faster throughput and improved informatics to allow imaging and mass spectral information in relevant timeframes. Here we investigate improvement of DESI imaging workflow with variation of acquisition speed and real time display of the ion images.

Methods:
MS imaging experiments were performed with and without ion mobility separation using Waters Xevo G2-XS and SYNAPT G2-Si mass spectrometers. DESI spray conditions were set at 1.5µl/min, 98:2 MeOH: water with nebulizing gas pressure of 7bar. During acquisition, DESI ion images were displayed in real time and imaging datasets were subsequently processed with High Definition Imaging 1.4 software

Results:
Proof-of-principal experiments have been carried out using a thin section of saggital mouse brain. An optical image was taken, co-registered with HDI and used to define the area to be imaged. Three rectangular regions of tissue were defined longitudinally. The surface of the tissue section was rastered line-by-line with mass spectra collected at predefined X,Y coordinates. Pixel size in the X direction was defined by the speed of the stage movement and acquisition rate of mass spectra. The Y-direction was defined by the distance between two lines of acquisition. All regions were imaged at a pixel size of 100 µm. As expected the signal intensities of the endogenous molecules decreased with the increase of the acquisition rate. Spectral resolution and mass accuracy were not affected with the variation of the speed of acquisition due to the decoupled nature of the ion source.

Novel Aspect:
Evaluation of a various speed continuous raster mode of DESI imaging on an ion mobility enabled Q-TOF

Baykut, Gökhan,
Bruker Daltonik GmbH

Contributing Authors:

Claudia Kriete, Bruker Daltonik GmbH, Fahrenheitstrasse 4, 28359 Bremen, Germany; Gökhan Baykut, Bruker Daltonik GmbH, Fahrenheitstrasse 4, 28359 Bremen, Germany; Christian Berg Bruker Daltonics Inc., 40 Manning Road, Manning Park, Billerica, MA 01821, USA; Jochen Friedrich, Bruker Daltonik GmbH, Fahrenheitstrasse 4, 28359 Bremen, Germany; Michael L. Easterling, Bruker Daltonics Inc., 40 Manning Road, Manning Park, Billerica, MA 01821, USA; Matthias Witt, Bruker Daltonik GmbH, Fahrenheitstrasse 4, 28359 Bremen, Germany; Walter Roeck, Bruker BioSpin AG, Industriestrasse 26, 8117 Faellanden, Switzerland; Christopher J. Thompson, Bruker Daltonics Inc., 40 Manning Road, Manning Park, Billerica, MA 01821, USA; Steve L. VanOrden, Bruker Daltonics Inc., 40 Manning Road, Manning Park, Billerica, MA 01821, USA; Roland Jertz, Bruker Daltonik GmbH, Fahrenheitstrasse 4, 28359 Bremen, Germany; Christoph Gosteli, Bruker BioSpin AG, Industriestrasse 26, 8117 Faellanden, Switzerland; Eugene N. Nikolaev, The Institute for Energy Problems of Chemical Physics Russian Academy of Sciences, Leninskij pr. 38, k.2 Moscow Russia 119334.

Title:

2ω-FT-ICR Mass Spectrometry with Quadrupolar Detection

Introduction:

The advantage of the double-frequency (2ω) detection in FT-ICR over single-frequency (1ω) detection is that it provides twice the resolving power (RP), or allows measurements with 1ω-equivalent RP in half of the time. Multi electrode detections were proposed decades ago but never successfully implemented because of the dominant presence of abundant 1ω-harmonics and weak multi-frequency signals.

Methods:

Our techniques to control the magnetron motion in the ICR cell applied to 1ω-detections were also essential to significantly reduce the 1ω-harmonics in 2ω-detections. New detection electronics improve the 2ω signal intensities and allow fast switching between excitation and detection modes using the same electrodes. Furthermore, it allows 1ω as well as 2ω-detections switchable by software-control.

Results:

The 2ω detection measurements were performed at 7T in a dynamically harmonized cell. Using 2ω-detection in narrowband the 1ω resolving power of 30Mio. at m/z 702 was increased to 60Mio. Oil deposits were studied by applying APPI, ESI and LDI also using absorption mode processing (AMP). RP: 800,000 with 1ω; 1.6Mio. with 2ω; >2Mio. with 2ω+AMP. More than 17,000 molecular formulae were identified in an oil deposit sample with an RMS mass error <150ppb. Over 50,000 molecular formulae could be identified with a similar RMS mass error using continuous accumulation of selected ions (CASI) in a mass window of 55Da. For isotopic fine structure (IFS) in real-time, a mixture of pharmaceuticals was analyzed with LC/MS (duty cycle of 1Hz) leading to a broadband RP of 450,000 at m/z 400. Isotopic patterns reveal IFS with excellent correlation to relative ratios of individual atoms and accurate masses.

Novel Aspect:

Routine 2ω-detection of fundamental resonance frequency doubling the resolving power or reducing the measurement time.

Beach, Daniel,
National Research Council Canada

Contributing Authors:
Michael Quilliam, National Reseach Council Canada; Elliott S. Kerrin, National Reseach Council Canada.

Title:
Differential Mobility Spectrometry of Polar Algal Biotoxins for Enhanced Selectivity in LC-MS Analysis

Introduction:
LC-MS is widely used for analysis of algal biotoxin, but challenges remain with small polar toxins including paralytic shellfish toxins (PSTs) and β-N-methylamino-L-alanine (BMAA). The orthogonality of differential mobility spectrometry (DMS) to LC and MS make it promising for use as a filter to separate toxin isomers and remove chemical interferences after LC separation before MS detection.

Methods:
A Selexion DMS was modified for external metering of carrier gas modifier. Comprehensive studies of the DMS separation of BMAA isomers and PSTs found optimization of carrier gas modifier and dispersion voltage (DV) to be most important. Existing HILIC methods for each analyte class were used to develop HILIC-DMS-MS/MS methods, which were applied to the analysis of toxins in shellfish and algae.

Results:
DMS conditions were identified for separation of all isomeric analytes from one another. In all cases, the accurate metering of low concentrations of solvent vapor modifiers in the DMS carrier gas was essential for achieving sensitive and selective DMS analysis. Carrier gas modifier, DV and analyzer temperature all had an impact on dissociation of labile PSTs during DMS, and different settings were required than those used for more stable analytes. The utility of these methods for eliminating matrix interference, increasing selectivity and lowering toxin LODs is demonstrated through analysis of PSTs and BMAA in algal and shellfish samples and reference materials. Robust quantitative analysis using matrix matched (for PSTs) and isotope dilution (for BMAA) quantitation is also demonstrated.

Novel Aspect:
Practical improvements to commercial DMS Quantitative HILIC-DMS-MS/MS methods for regulated biotoxins Insight into dissociation of labile ions in DMS

Beard, Adam,
Merck Research Laboratories

Contributing Authors:
Rong-Sheng Yang, Merck & Co. Inc.; Huaming Sheng, Merck & Co. Inc.; Li-Kang Zhang Merck & Co. Inc.; Roy Helmy, Merck & Co. Inc.

Title:
Application of Titanium(III) Chloride as a Reagent for the Detection of Nitro and N-oxide-Containing Potentially Mutagenic Impurities Using UHPLC/HRMS

Introduction:
Due to ICH guidelines, it is important to characterize all PMIs that could arise during the process of synthesizing and developing a drug. Two prevalent examples of PMIs are compounds containing N-oxide and nitro functional groups. We present a novel application of titanium(III) chloride reduction combined with UHPLC/HRMS to analyze PMIs for the presence or absence of these functional groups.

Methods:
A number of model N-oxide and nitro-containing compounds were treated with a commercially available TiCl3/HCl reagent and then directly analyzed by UHPLC/HRMS to monitor for reduction to the corresponding amine-containing compounds. The model compounds cover functionalities including aliphatic and aromatic N-Oxides, quinoline N-Oxide, Quinoxaline N,N'-dioxide, and aromatic nitro groups.

Results:
We present a broadly applicable platform method for the characterization of a variety nitro and N-oxide containing PMIs. We additionally demonstrate that in some cases minor changes to the reaction solvent can lead to significant improvements to the extent of conversion to the reduced product. This simple method has proven invaluable in the identification of PMIs since with it the presence or absence of N-oxide and nitro functional groups that are considered structural alerts can be quickly determined. Additionally, for poorly ionizing compounds containing aryl-nitro groups, the reduction of the nitro groups to their corresponding anilines results in an increase in ionization efficiency with ESI-MS thus aiding in the identification and quantitation of these potentially mutagenic impurities by UHPLC/HRMS.

Novel Aspect:
TiCl3 reduction of N-oxide groups is an established practice in metabolite identification, but is previously unexplored for PMI characterization.

Beck, Alain,
Centre d'Immunologie Pierre Fabre (CIPF)

Title:
State-of-the art orthogonal chromatographic, electrophoretic and MS methods for mAbs and ADCs structural assessment at multiple levels

Introduction:
Antibody Drug Conjugates (ADCs) are based on highly cytotoxic Small Molecular Drugs (SMDs) covalently attached via conditionally stable linkers to antibodies (mAbs) and are among the most promising next generation empowered biologics for cancer treatment. ADCs are more complex than naked mAbs, as the heterogeneity of the conjugates adds to the inherent microvariability of the biomolécules.

Methods:
The development and optimization of ADCs rely on improving their analytical and bioanalytical characterization by assessing multiple critical quality attributes (CQAs), such as the distribution and position of the drugs, the amount of naked antibody, the average drug to antibody ratio (DAR), and the residual drug-linker and related product proportions (SMDs) to name a few.

Results:
Progresses of multi-level (top, middle, bottom,SMDs) state-of the art MS methods (Native MS, Ion Mobility-MS, CE-MS, 2D-LC-MS, Extended Bottom Up, Top Down Sequencing) combined with chromatographic and electrophoretic techniques will be presented. They will be illustrated by extensive structural characterization of reference FDA and EMA approved therapeutic mAbs and ADCs. CE coupled to MS is now established as a valuable method, orthogonal to LC, for peptide mapping at the bottom and middle levels, and progress toward top-level analysis of ADCs under native conditions. Top-down MS and sequencing, middle-down MS with ETD fragmentation, and extended bottom-up proteomics with enzymes will be presented as well as alternative middle down workflow toward a full sequence coverage of IdeS digested and reduced fragments of IgG including LCMALDI with in-source decay.

Novel Aspect:
Novel native MS, Ion Mobility-MS, CE-MS and 2D LC-MS methods for ADC analysis and bioanalysis.

Becker, Chris,
Protein Metrics Inc.

Contributing Authors:

Yong Kil, Protein Metrics Inc.; Doron Kletter, Protein Metrics Inc.; Marshall Bern Protein Metrics Inc.; Wilfred Tang, Protein Metrics Inc.; Christopher Becker, Protein Metrics Inc.; Eric Carlson, Protein Metrics Inc.

Title:

New Developments in Intact Mass Analysis of Proteins and Protein Complexes

Introduction:

Software for intact mass analysis inverts or "deconvolves" the charging and measurement processes to reconstruct a neutral mass spectrum from a mass over charge (m/z) spectrum. We present a new deconvolution algorithm that models peptide charging and mass analysis imprecision, applies no regularization bias, and performs well for the full range of MS instruments.

Methods:

We obtained data from a variety of proteins, including intact and reduced monoclonal antibodies (mAbs) and antibody drug conjugates (ADCs), from all the instrument vendors. We tested our Intact Mass software by comparing measured and calculated masses and mass deltas (for example, 162 Da peak spacing from glycosylation), and checking the intensities of half-mass and double-mass artifacts.

Results:

We find that the model based algorithm gives greater sensitivity than MaxEnt to low abundance proteoforms. The model based algorithm also offers greater control of visual appearance of the deconvolved mass spectra, because the algorithm can, for example, preserve or discard isotope resolution. We also find that slicing the chromatogram by elution time can computationally enrich for low abundance species, for example, ADCs with unusually high or low drug payload, or mAbs with missing, truncated, or sialylated glycans. We will report on experiments applying model based deconvolution to various biopharma samples, including ADCs, mAbs with both Fc and Fab glycosylation, and mAbs with closely spaced variants such as oxidations and sulfur gains and losses.

Novel Aspect:

New charge deconvolution algorithm

Beltrao, Pedro,
EMBL-EBI

Title:
Evolution and function of protein phosphorylation networks

Introduction:
Cells need to constantly adapt to changes in conditions and use phospho-signalling different to transfer information from sensors to the effectors of cellular responses. Advances in mass-spectrometry now allow us to identify phosphosites in large scale and quantify their changes across different conditions.

Methods:
However, little is know about how the thousands of recently discovered phosphosites evolve and how they modulate protein function and coordinate cellular responses. I will introduce some of the computational methods our group has been working on to address these questions. I will introduce some of the computational methods our group has been working on to address these questions.

Results:
We are developing methods to infer the most likely age of phosphosites by combining phosphorylation data from multiple species, protein sequences and phylogenetic trees. We have observed that only a small fraction of phosphosites are ancient in origin and that these are more likely to cause deleterious consequences when mutated. In parallel we have set a computational pipeline to analyse changes in phosphosite abundances under different conditions. We have compiled a dataset of 150 perturbations in human cell lines where we can estimate the activities of approximately 200 kinases.

Novel Aspect:
These data allows to characterize cell signalling states across conditions and to infer associations between kinase activities and the phosphorylation

Beltrao, Pedro,
EMBL-EBI

Title:
Evolution and function of protein phosphorylation networks

Introduction:
Cells need to constantly adapt to changes in conditions and use phospho-signalling different to transfer information from sensors to the effectors of cellular responses. Advances in mass-spectrometry now allow us to identify phosphosites in large scale and quantify their changes across different conditions.

Methods:
However, little is know about how the thousands of recently discovered phosphosites evolve and how they modulate protein function and coordinate cellular responses. I will introduce some of the computational methods our group has been working on to address these questions. I will introduce some of the computational methods our group has been working on to address these questions.

Results:
We are developing methods to infer the most likely age of phosphosites by combining phosphorylation data from multiple species, protein sequences and phylogenetic trees. We have observed that only a small fraction of phosphosites are ancient in origin and that these are more likely to cause deleterious consequences when mutated. In parallel we have set a computational pipeline to analyse changes in phosphosite abundances under different conditions. We have compiled a dataset of 150 perturbations in human cell lines where we can estimate the activities of approximately 200 kinases.

Novel Aspect:
These data allows to characterize cell signalling states across conditions and to infer associations between kinase activities and the phosphorylation

Benesch, Justin,
University of Oxford

Title:
Quantitative IM-MS for Structural Biology

Introduction:
Ion mobility mass spectrometry (IM- MS) is attractive for structural biology due to the ability to interrogate individual states of proteins and the assemblies they form. However, its full potential is far from realized, with most studies relying on qualitative comparisons, which hampers direct structural insight and integration of the data with that obtained via other structural biology means.

Methods:
We perform nanoelectrospray IM-MS experiments on Waters and Thermo instrumentation customised for high-m/z operation. Software is home built, and freely available from benesch.chem.ox.ac.uk.

Results:
We have developed IM-MS experiments and computational tools to allow their quantitative application to structural biology. We use MS and tandem MS to identify and quantify the relative abundances of different protein stoichiometries present in solution, while real-time experiments allow the extraction of quaternary fluctuations. This leads us to obtain equilibrium and rate constants for the underlying protein-protein and protein-ligand interactions. Concurrently, IM-MS measurements give us information as to the physical size of the proteins and provide powerful restraints in modelling the structures of protein assemblies. We will illustrate our approaches by application to molecular chaperone proteins, and emphasise integration with complementary MS methods (cross-linking and hydrogen/deuterium exchange) and other structural biology techniques.

Novel Aspect:
Experimental and computational strategies for capitalising on IM-MS data for structural biology.

Benigni, Paolo,
Florida International University

Contributing Authors:
Francisco Fernandez-Lima, Florida International University.

Title:
Optimizing analytical tools for the structural characterization of molecules in complex samples by using High resolution IMS and FT-ICR MS

Introduction:
The analytical platform developed in our laboratory enables high resolution ion mobility measurements in tandem with ultra-high mass analysis. Because FT-ICR MS has long acquisition rates IMS analysis must be performed utilizing discrete sampling. The challenge is accurately sampling the mobility of an ion. This work highlights a novel methodology for the characterization of complex mixtures.

Methods:
In trapped ion mobility spectrometry analysis, the mobility range that is injected into the mass analyzer is defined by a voltage ramp. The subsequent mobility range is defined by the step increment in the voltage across the ramp. In this work the effect of overlapping voltage ramps is explored as a way to increase sensitivity, separate isomers, and perform accurate mobility measurements.

Results:
Results show that there are significant advantages to using oversampling selected accumulation TIMS such as, increased in peak definition, described by the number of points across a peak, increased resolving power in the analysis of isomers, and the characterization of different states for biological molecules. The use of OSA-TIMS also increases the signal to noise ratio of the MS analysis 2-6 fold for Agilent's tunemix. When oversampling with a 1V ramp higher S/N is observed while retaining the high resolution of narrower ramp sizes. This methodology was successfully applied to the analysis of complex mixtures, such as the PAHs in a coal tar as well as illustrates unsupervised detection of molecular features in complex mixtures. Automated peak detection requires highly defined peaks, from oversampling, in order to accurately identify the ion mobility for each chemical formula.

Novel Aspect:
This work provides the workflow for OSA-TIMS FT-ICR MS analysis and the advantages in, resolving power, signal to noise and feature detection.

Benkaci-Ali, Farid,
University of Sciences and Technology Houari Boumediene, Faculty of Chemistry, Laboratory of Functional Organic Analysis,

Contributing Authors:
Eppe Gauthier, University of Liège, Laboratoire de Spectrométrie de Masse L.S.M, Allée du 6 Août, Bât B6c, 4000 Liège (Sart-Tilman), Belgium; Mékaoui Radja, Université des Sciences et Technologies Houari Boumediene, Faculté de Chimie, Laboratoire d'Analyse Organique Fonctionnelle, U.S.T.H.B, B.P. 32 El Alia, Bâb Ezzouar, Algiers, Algeria.

Title:
GCxGC-MS analysis of the effect of the extraction time and cryogenic grinding (N2 at -196°C) on the cumin seeds volatile oil

Introduction:
Steam distillation assisted by microwave (SDM) and hydrodistillation assisted by microwave (HDM) are respectively advanced steam distillation (SD) and hydrodistillation techniques (HD).

Methods:
They are considered as accelerated techniques extraction performed at atmospheric pressure. The effect of the cryogrinding (CG) on the yields and composition of volatile oils was studied.

Results:
There were differences in the major compounds cumin aldehyde (HDAM-CG: 20.50–53.35%, SDAM–CG: 21.68–66.6%), p-cymene (HDAM-CG: 14.1–37.54%, SDAM-CG: 6.06–20.18%), β-pinene (HDAM-CG: 4.53-14.74%, SDAM-CG: 4.3–32.65%) and γ-terpinene (HDAM-CG: 1.26-14.70%, SDAM–CG: 5.13%–18.23%).

Novel Aspect:
A high number of components eluted were separated and identified, constituting compounds of differentiation between the techniques used

Benkessou, Fadwa,
Karolinska Institutet

Contributing Authors:
Ibrahim El Serafi, Experimental Cancer Medicine (ECM), Clinical Research Centre (KFC), Department of Laboratory Medicine, Karolinska Institutet Huddinge, Novum, Stockholm, Sweden; Moustapha Hassan, Experimental Cancer Medicine (ECM), Clinical Research Centre (KFC), Department of Laboratory Medicine, Karolinska Institutet Huddinge, Novum, Stockholm, Sweden; Brigitte Twelkmeyer Anesthesiology and Intensive Care, Clinical Research Centre (KFC), Department of Clinical Science, Karolinska Institutet Huddinge, Novum, Stockholm, Sweden; Manuchehr Abedi-Valugerdi, Experimental Cancer Medicine (ECM), Clinical Research Centre (KFC), Department of Laboratory Medicine, Karolinska Institutet Huddinge, Novum, Stockholm, Sweden; Fadwa Benkessou, Experimental Cancer Medicine (ECM), Clinical Research Centre (KFC), Department of Laboratory Medicine, Karolinska Institutet Huddinge, Novum, Stockholm, Sweden.

Title:
Sensitive and robust method for the quantification of dimethylacetamide and its primary metabolite monomethylacetamide in human plasma

Introduction:
Dimethylacetamide (DMA) is a solvent that causes hepato-neurotoxicity. DMA is metabolized to monomethylacetamide (MMA) in the liver. Mechanisms underlying DMA/MMA toxicities are not understood.

Methods:
LC-MS and C18 ODS-AQ column were used to analyze DMA and MMA in plasma where SIM mode with an optimized flow rate and gradient were used. The method was validated in accordance with FDA guidelines

Results:
Calibration curve was linear was within the range (1 µM-4000µM; n=12). LOQs for DMA and MMA were 1.8 µM and 8.6 µM, respectively. The method has achieved selectivity, accuracy, reproducibility and stability (CV < 10%). DMA and MMA recovery in plasma was [97% - 100%] and [76% - 100%], respectively.

Novel Aspect:
This is the first validated method for a simultaneous detection and quantification of DMA and MMA that can be used in clinic and environmental studies

Benter, Thorsten,
University of Wuppertal

Title:
Ion boosters, superchargers, chemical modifiers – ion chemistry at work. But how?

Introduction:
The ever increasing performance of mass spectrometers is often gauged in terms of the mass resolution or sensitivity of the "analyzer". In parallel, over 50 new acronyms for ion sources have been coined in the past two decades suggesting a comparable development boost. Chemical modifiers added to the high pressure region allegedly boost the performance of many AP methods. How does that work?

Methods:
A Bruker micrOTOF, esquire6000 ion trap (both equipped with a capillary/skimmer/ion guide stage), amaZon ion trap with a capillary/two-stage funnel/ion guide stage, and a SCIEX 5500 triple quadrupole with SelexION DMS cell were employed. Kinetics simulations were done with a customized version of the RS Monte Carlo kinetics code and Cantera 2.1.1., and ab initio calculations with Gaussian 03

Results:
When delivering considerable amounts of e.g. CH3CN to the spray region in nESI, peptide ions are often generated with favorably shifted higher charge state distributions, sometimes called super-charging. In contrast, CH3OH is adversely affecting this process. In DMS though, CH3OH strongly promotes the DMS separation efficiency. The experimental and theoretical results presented clearly demonstrate the pivotal role of ion cluster chemistry, which is governing the complex sequential and parallel processes involved in the collision-dominated regions in the mass spectrometer. It is further discussed, why hardly any cluster ions are detected in API mass spectra recorded with modern analyzers. On one hand, de-clustering is essential for unequivocal analyte ion identification; on the other hand such processes render mechanistic studies of cluster ion-molecule chemistry virtually impossib

Novel Aspect:
Studies of ion-molecule chemistry for the interpretation of results observed with differential ion mobility stages and supercharging agents

BERDEAUX, Olivier,
ChemoSens - INRA (FRANCE)

Contributing Authors:
Elodie A.Y. MASSON, UMR CSGA - 1324 INRA - 6265 CNRS – University of Burgundy Franche-Comté, Dijon, France; David ROPARTZ, UR 1268 INRA – BIA - Nantes, France; Olivier BERDEAUX UMR CSGA - 1324 INRA - 6265 CNRS – University of Burgundy Franche-Comté, Dijon, France; Estelle SIBILLE, UMR CSGA - 1324 INRA - 6265 CNRS – University of Burgundy Franche-Comté, Dijon, France.

Title:
Ganglioside characterization by high resolution mass spectrometry and ion mobility

Introduction:
Gangliosides (GG) are sialic acid-containing glycosphingolipids particularly abundant in the nervous system. They exhibit a wide variety of structures depending on both the oligosaccharide chain and the ceramide moiety. About 200 GG species have been described so far, but their diversity and biological roles are far from being completely elucidated.

Methods:
An efficient method of identification and quantification is crucial to apprehend this huge heterogeneity. In this study, we focused on the characterization of the ceramide portion of GG, identifying the long chain base (LCB) and the fatty acid. A commercial standard of GD3 was analysed with both a Thermo LTQ-Orbitrap XLTM and a Waters Synapt G2-Si equipped with ion mobility.

Results:
Whatever the equipment used, fragmentation of GD3 40:1 molecular species revealed the presence of four LCBs (d16:1, d17:1, d17:0 and d18:1) for one ceramide type, suggesting the presence of four GG isomers. This observation was confirmed by the MS3 analyses conducted with the Orbitrap, which allowed isolation and fragmentation of the ion corresponding to the ceramide. The technique of ion mobility offered the possibility to separate the four detected LCBs, after MS/MS fragmentation of GD3 40:1. We were thus able to detect and separate the LCBs, according to their differences of steric hindrance, for each isomer of GG. In this way we deduced the combination LCB/fatty acid (ex: GD3 d16:1/24:0) for each GD3 molecular species. Moreover, this separation presents the advantage to individualize each molecular isobaric species of GG for their ulterior quantification.

Novel Aspect:
Ion mobility applied to GG analysis allowed separation and further characterization and quantification of the various ceramide isomers.

Berezovski, Maxim,
University of Ottawa

Contributing Authors:

Jeffrey W. Keillor, Department of Chemistry and Biomolecular Sciences, University of Ottawa; Gleb G. Mironov, Department of Chemistry and Biomolecular Sciences, University of Ottawa; Christopher M. Clouthier Department of Chemistry and Biomolecular Sciences, University of Ottawa; Abdullah Akbar, Department of Chemistry and Biomolecular Sciences, University of Ottawa.

Title:

Simultaneous Analysis of Enzyme Structure and Activity by Kinetic Capillary Electrophoresis and Mass Spectrometry

Introduction:

The development of biophysical and bioanalytical methods that can simultaneously monitor protein activity and conformation is an important goal and a significant challenge. In particular, it is of paramount importance to observe activity and conformation in real-time and in solution, for the investigation of biomolecular interactions relevant to the modulation and regulation of cellular function.

Methods:

We developed a method based on kinetic capillary electrophoresis, coupled on-line with UV detection and ion mobility mass spectrometry (CE-UV-IM-MS). Kinetic CE-UV separated protein conformers and monitored their interconversion dynamics in solution. Ion mobility mass spectrometry analysed conformer size, exact molecular weights and structures of an enzyme, its substrates, inhibitors and products.

Results:

To observe protein conformations simultaneously with their enzymatic activities, we chose to study human tissue transglutaminase (TG2). The biological transamidase activity of TG2 is regulated through a dramatic conformational change between the open and closed forms, which is allosterically controlled by binding either calcium (II) ion or GTP/GDP, respectively. In the current study we separated two TG2 conformers and measured their enzymatic activity with two peptide substates. In addition, we were able to determine values for the apparent potency and efficacy of several inhibitors with respect to their ability to shift the conformational equilibrium.

Novel Aspect:

This is the first report of a method able to simultaneously monitor protein conformation and activity in a single experiment.

Berezovski, Maxim,
University of Ottawa

Contributing Authors:
Nasrin Khan, Department of Chemistry and Biomolecular Sciences, University of Ottawa; Gleb G. Mironov, Department of Chemistry and Biomolecular Sciences, University of Ottawa.

Title:
Direct Detection of Endogenous MicroRNAs and Their Post-transcriptional Modifications in Cancer Serum by Capillary Electrophoresis-Mass Spectrometry

Introduction:
The miRNA sequences and they quantitative levels are used as unique signatures associated with cancer diagnosis and prognosis after anticancer treatment. The problem is that existing bioanalytical methods such as microarrays and a quantitative polymerase chain reaction are sensitive, but not capable of identifying the posttranscriptional modifications of miRNA.

Methods:
Here we report a Capillary Electrophoresis-Mass Spectrometry (CE-MS) method, which performs a multiplex, direct analysis of miRNAs from biological samples. The CE separation and following MS analysis provides label-free quantitation and reveals modifications of miRNAs.

Results:
Coupling CE with mass spectrometry makes it more powerful for analysis of post-transcriptional modifications of miRNAs since it fuses CE separation with multiplex detection, precise mass identification and fragmentation in tandem mass spectrometry. We isolated small RNAs from human blood serum using a commercially available kit, then directly analyzed by CE-MS. Using the CE-MS method, we detected two endogenous human circulating miRNAs, a 23-nucleotide long 5′-phosporylated miRNA with 3′-uridylation (iso-miR-16-5p) and a 22-nucleotide long 5′-phosporylated miRNA (miR-21-5p) isolated from B-cell chronic lymphocytic leukemia serum. The CE separation and following MS analysis provides label-free quantitation and reveals modifications of miRNAs.

Novel Aspect:
This is the first report of mass spectrometric detection of endogenous miRNAs from cancer serum.

Betgovargez, Edna,
Sciex

Contributing Authors:

Bryan Fonslow, SCIEX, Brea, CA; Ying Zhang, Pfizer, Chesterfield, MO; John R. Yates III The Scripps Research Institute, La Jolla, CA; Olga V. Friese, Pfizer, Chesterfield, MO; K. Steven Cook, Pfizer, Chesterfield, MO; Jason Rouse, Pfizer, Andover, MA.

Title:

Intact immunoglobulin gamma analysis by CESI-MS

Introduction:

IgGs are the most commonly used therapeutic mAbs. The ability to do comprehensive characterization of these molecules with high sensitivity is essential to their development & for regulation of their efficacy, bioavailability & biosafety. CESI-MS helps improve the sensitivity, speed & comprehensiveness of mAb characterization. Here, CESI-MS is used under different separation conditions for the

Methods:

IgGs were desalted & diluted in either water or ammonium acetate for sample stacking. IgGs were also reduced with DTT or digested with IdeS protease. ~25 ng of protein was analyzed using a Sciex CESI 8000 and Thermo Fusion MS using the CESI neutral capillary cartridge. Acetic acid and ammonium acetate-based background electrolytes were used for separations from 20-60 minutes.

Results:

Triplicate analysis of CHO- and NS0-derived IgG2 molecules analyzed on CESI-MS yielded similar separation profiles and good migration time and MS spectra reproducibility. Further analysis of separations revealed MS spectral differences between the different IgG2 peaks. NS0-derived IgG2 deconvoluted spectra have higher masses detected, consistent with minor amounts of C-terminal lysine, in comparison to the trace levels in the CHO-derived IgG2 deconvoluted spectra.

Novel Aspect:

Exploiting CE separation and ultralow nanoflow of CESI to improve comprehensiveness of intact therapeutic mAb characterization

Betgovargez, Edna,
Sciex

Contributing Authors:

Stephen J. Lock, SCIEX, Warrington, Cheshire.

Title:

The application of Capillary Electrospray Ionization to the detection of Neuropeptides

Introduction:

Vasoactive intestinal peptide (VIP) & Pituitary adenylate cyclase-activating polypeptide (PACAP) are very basic peptides (PI's > 10) making them difficult to analyze by LC-MS methods. In this work we describe the development of a CESI-MS method for the analysis of these types of peptides at biologically relevant levels in serum as an alternative approach.

Methods:

Peptides were extracted from serum using a basic SPE protocol. The reconstituted extracts were then injected by pressure onto a 30 µm ID x 91 cm polyethylenimine coated capillary. The CE separation used reverse polarity with a mixture background electrolyte of 5% Acetic acid:30% Acetonitrile:65% Water. Mass spectrometer was set in MRM mode at an ionspray voltage of 1600-1800V.

Results:

A CESI-MS method has been developed which has been shown to be able to detect VIP and PACAP below 1 pg/uL (biological level) in serum. The signal to noise currently observed for both peptides exceeded 25:1 for a 1 pg/uL serum spike. As the SPE extraction can also be further optimized, there is the potential to achieve lower detection levels for both peptides in biological samples. This should enable this approach to be applied to other sample volume limited matrices, such as mouse models, and opens up this methodology for other peptide assays which are challenging for standard LC-MS approaches.

Novel Aspect:

CESI-MS analysis and detection of basic neuropeptides at physiological levels in biological samples

Bhattacharya, Nivedita,
CSIR-National Chemical Laboratory, India

Contributing Authors:

Dharmesh Parmar, CSIR-National Chemical Laboratory, India; Ajeet Singh, Center for Applications in Mass Spectrometry, Pune, India; Anil Kishen University of Toronto, Canada; Dipali Nagda, Sinhgad Dental College and Hospital, Pune, India; Venkat Panchagnula, CSIR-National Chemical Laboratory, India.

Title:

Mass spectrometry based proteomics and metabolomics of early dental plaque from caries-free Indian subjects

Introduction:

Dental caries is a public health issue. As dental caries infects teeth, plaque developed on teeth surface can provide useful molecular information for earlier intervention that may help in better management of disease. In this work, proteomics and metabolomics are carried out from early dental plaque under cariogenic and carioprotective insults from caries-free subjects with mass spectrometry.

Methods:

Same subjects (n=10) were exposed to glucose (cariogenic), fluoride (carioprotective) challenges and without any exposure (control) on different days and dental plaques were collected at the end of 2 hours. Proteomics (Agilent 6540, nanoLC-MS/MS) and metabolomics (Thermo Q-Exactive Orbitrap, C-18 LC-MS) analysis were performed on pooled samples for different treatment groups.

Results:

Proteomics data was analyzed with MassWiz. Proteins such as histatins, statherins, cystatin responsible for remineralization were found in across all treatment groups. Proteins responsible for innate immunity such as immunoglobulins and bactericidal permeating increasing (BPI) proteins were found in glucose challenged group. Bacterial proteins from non-cariogenic bacteria were observed. Metabolomics data analysis revealed presence of organic acids, amino acids and amines. Relative profiles for organic acids decreases in fluoride group than the other groups. Citrulline increased in glucose group compared to other groups and is known to increase pH of oral microenvironment. Proteomics and metabolomics data suggests that remineralization events, proliferation of innate immunity, non-cariogenic bacteria and increased pH are some of the caries evading mechanisms in caries-free subjects.

Novel Aspect:

First work reporting metabolomics and bacterial proteins from early dental plaque in Indian subjects with cariogenic and carioprotective challenges.

Bieler, Andre,
University of Bern

Contributing Authors:
K. Altwegg, University of Bern; H. Balsiger, University of Bern; J.-J. Berthelier LATMOS; J. De Keyser, Belgian Institute for Space Aeronomy; B. Fiethe, TU Braunschweig; S. A. Fuselier, Southwest Research Institute; T. I. Gombosi, University of Michigan; A. Korth, Max-Planck-Institut für Sonnensystemforschung; B. Marty, Université de Lorraine; T. Owen, University of Hawaii; H. Rème, Université de Toulouse / CNRS; IRAP; M. Rubin, University of Bern; J. H. Waite, Southwest Research Institute; P. Wurz, University of Bern.

Title:
Mass Spectrometry in Space: Highlights of the Rosetta Mission

Introduction:
Since August 2014, after a 10 year journey through our Solar System the European SpaceAgency's (ESA) spacecraft Rosetta is in orbit around comet 67P/Churyumov-Gerasimenko. One of the key instruments on this space mission is a suite of mass spectrometers, called ROSINA (ROSETTA ORBITER SPECTROMETER FOR ION AND NEUTRAL ANALYSIS).

Methods:
ROSINA consists of one Double Focusing Mass Spectrometer (DFMS) with Nier-Johnson configuration and one Reflectron Time of Flight mass spectrometer (RTOF). Both instruments use electron impact ionization and are designed to measure absolute and relative abundances as well as isotopic ratios in the tenuous neutral and ion gas cloud (the so called coma) surrounding comet 67P.

Results:
By escorting comet 67P through the Solar System for more than two years we have the opportunity to examine the characteristics of this dynamicly evolving system. ROSINA detected changes in the relative and absolute abundances due to seasonal effects and large scale heterogeneities on the comet nucleus. The measured D/H ratio in water contradicts the popular idea that a large fraction of Earth's water is due to impacts from comets. On the other hand ROSINA measurements indicate that comets might be responsible for significant amounts of noble gases in Earth's atmosphere. Several species were detected for the first time in comets. The biggest surprise was the large abundance of the reactive molecular oxygen, a highly elusive molecule in interstellar objects. In many ways the results obtained with ROSINA impact our current understanding of our Solar System.

Novel Aspect:
First extended in situ examination of a cometary coma with mass spectrometers in space.

Bier, Mark,
Carnegie Mellon University

Contributing Authors:
Fan Wang, Carnegie Mellon University; Urs Schlunegger, Carnegie Mellon University; Stephan C. Tjader Carnegie Mellon University; Sanford D. K. Roman, Carnegie Mellon University; Alexander A. Aksenov, Carnegie Mellon University; Logan Plath, Carnegie Mellon University; Jie Jiang, Carnegie Mellon University; Abdil Ozdemir, Carnegie Mellon University.

Title:
Development of a Heavy Ion Mass Spectrometer Incorporating a Linear Ion Trap – Part I

Introduction:
We developed a linear ion trap (LIT) incorporating radial ejection with an extended m/z range to >300kTh. The goal was to extend the mass range into the MegaDa's for routine analysis of complexes such as viruses and synthetic nanoparticles. We describe the testing and results based on dynode/channel electron multiplier (CEM) detectors. We intend to show high-m/z cryodetector results in Part II.

Methods:
The new modular and flexible mass spectrometer consists of a MALDI source, two octopoles, a LIT with radial ejection, and two CEMs. To trap high-m/z ions, a pulsed gas is supplied. System control and acquisition is facilitated by a National Instruments computer driven by LabVIEW software. Both frequency scanning and AC amplitude scanning is possible. Samples were prepared using standard methods.

Results:
The upper-m/z limit of ion traps is based on the geometry, AC amplitudes, AC frequencies and the q-eject point. To obtain high-m/z results using a standard LIT geometry, we used frequency scanning between 400kHz to 5kHz, Vpp < 2kV and higher than normal trapping pressures. Although frequency scanning requires maintaining constant amplitude of the AC signal, a broad m/z range is easily achieved. We have implemented a custom transformer with broadband amplifier and applied the signal to the LIT in bipolar fashion. MALDI results for apomyoglobin (16.9kDa), GFP-poly-oligo ethylene oxide methacrylate hybrid (27+kDa), albumin (66.7kDa), and IgG (148.5 kDa) with multimers will be shown. The system has the flexibility of using different detection schemes such as a STJ high mass detector (Part II). In the future, we intend to fragment trapped ions not readily possible with TOF analyzers.

Novel Aspect:
A new LIT mass spectrometer has been developed for analysis of high mass ions.

Bierstedt, Andreas,
Federal Institute for Materials Research and Testing

Contributing Authors:
Jens Riedel, Federal Institute for Materials Research and Testing; Steffen Bräkling, University of Wuppertal; Reto Glaus Federal Institute for Materials Research and Testing; Hendrik Kersten, University of Wuppertal; Joschua Rieger, University of Wuppertal; Igor Gornushkin, Federal Institute for Materials Research and Testing.

Title:
Development and characterization of an airborne laser-spark ion source for ambient desorption / ionization mass spectrometry

Introduction:
A novel direct sampling ionization scheme for ambient mass spectrometry is presented. Desorption and ionization is achieved by a quasi-continuous laser induced plasma in air. Since there are no solid or liquid electrodes involved the ion source does not suffer from chemical interferences or fatigue originating from erosive burning or from electrode consumption.

Methods:
A laser plasma was ignited under ambient conditions in front of a modified TOF MS atmospheric pressure interface, using a high repetition rate DPSS laser operating at 532 nm and 26 kHz and an aspherical lens with a focal length of 8 mm. Emission spectroscopy (40-1100 nm) and time resolved studies on specific plasma parameters revealed insight into the physical and chemical plasma properties.

Results:
Plasma ignition can be performed in rare gases and under ambient conditions. The hot plasma zone was kept at a certain distance from the sample region. Thus, effective collisional cooling seemed to prevent thermal fragmentation. Every single spark generates a shockwave, providing new reactive species, which expands concentrically from the hot region. Under ambient conditions primary charge carriers (ions and electrons) as well as VUV radiation initialize reaction cascades equivalent to other ambient ionization methods, such as DART or DBD. Mass spectra of polar/nonpolar hydrocarbons, sugars, pharmaceuticals and natural biomolecules in food were observed. Comprehensive emission spectroscopic measurements and time resolved electron current studies revealed insight into some plasma properties, such as the emitted high energetic radiation and the time evolution of the expanding plume.

Novel Aspect:
A high repetition rate DPSS laser is used for ignition of a highly sensitive quasi-continuous airborne laser induced plasma ion source for ambient MS.

Bieske, Evan,
University of Melbourne

Title:
Photoisomerization of Molecular Ions in a Tandem Ion Mobility Spectrometer

Introduction:
Molecular photoisomerization plays a role in biological systems (cis-trans isomerization of retinal in animal vision) and also in molecular technologies (in light activated molecular machines). Recently we have shown that it is possible to study photoisomerization of molecular ions such retinal protonated Schiff base in a tandem ion mobility spectrometer.

Methods:
Electrosprayed molecular ions are introduced into a custom-built tandem ion mobility spectrometer in which the target isomer is selected in the first stage, followed by irradiation with a pulse of tunable light and separation of the photoisomers in the second stage. By plotting the photoisomer yield as a function of laser wavelength one can generate a photoisomerization action spectrum.

Results:
A variety of molecular ions have been examined using the new approach, including biological molecules (protonated derivatives of retinal and beta-ionone) and molecules used in molecular switches (protonated spiropyran/merocyanine, azoheteroarenes). For retinal protonated Schiff base (RPSB), there is evidence for at least 6 different isomers in the gas phase, distinguishable by their collision cross sections and photo-responses. Isomers with larger collision cross sections, which correspond to trans/cis isomers, are interconvertible through absorption of visible light and also collisions. The most compact isomers do not respond to visible light, but can be interconverted through collisions, and are assigned to isomers in which the polyene chain has cyclized. Other molecular ions, such as protonated azoheteroarenes, exhibit just two isomers interconvertible using light or collisions.

Novel Aspect:
Tandem ion mobility spectrometry combined with laser excitation forms the basis of a new form of action spectroscopy for molecular ions.

Binkley, Joe,
LECO Corporation

Contributing Authors:
Christina N. Kelly, Leco Corporation, Saint Joseph, MI, USA; Lorne M. Fell, Leco Corporation, Saint Joseph, MI, USA; Jonathan D. Byer Leco Corporation, Saint Joseph, MI, USA; David E. Alonso, Leco Corporation, Saint Joseph, MI, USA; Elizabeth Humston-Fulmer, Leco Corporation, Saint Joseph, MI, USA.

Title:
Comparison of "Weird and Wild" Jelly Bean Flavors Using HS SPME - Gas Chromatography - Mass Spectrometry

Introduction:
Bring your children to work day is a national event in the United States which often excludes children with parents in the sciences due to lab safety protocols. This paper details a safe experiment which was designed to create interest in the sciences for all ages, while also demonstrating the capabilities of a new benchtop Time of Flight GC/MS to effectively differentiate some very exotic flavor

Methods:
Single jelly beans were placed into respective 20mL HS vials along with 5mL of HPLC grade $H2O$ and 1g of NaCl. A 2 cm PDMS/Carboxen/DVB SPME fiber was used to extract aroma compounds from the headspace of each jelly bean and deliver them to the GC–MS for analysis. An Rxi-5SilMS column was used for the chromatographic separation. The TOFMS data were acquired from 30-500 m/z at 20 spectra/second.

Results:
The benchtop time of flight GC/MS data were used to differentiate several pairs of jelly beans that had identical appearances, but vastly different flavor profiles. Examples of these visually identical jelly bean pairs which were differentiated include the following: lime / lawn clippings, berry blue / toothpaste, buttered popcorn / rotten egg, chocolate pudding / canned dog food, peach / barf, caramel corn / moldy cheese, and tutti-fruity / stinky socks. The children were able to predict the good vs bad flavored jelly beans based on the published sensory descriptors for the analytes which were identified by GC/MS in each of the jelly bean varieties. The kids found the experiment exciting and were exposed to analytical chemistry in a fun and educational way.

Novel Aspect:
A new benchtop GC/TOF system offers improved sensitivity and spectral deconvolution to differentiate flavor profiles of visually identical jelly beans

Binkley, Joe,
LECO Corporation

Contributing Authors:
Jonathan D. Byer, Leco Corporation, Saint Joseph, MI USA; Lorne M. Fell, Leco Corporation, Saint Joseph, MI USA; David E. Alonso Leco Corporation, Saint Joseph, MI USA; Christina N. Kelly, Leco Corporation, Saint Joseph, MI USA.

Title:
Comprehensive Two-Dimensional Gas Chromatography (GC×GC) with Simultaneous TOFMS and Flame Ionization Detector Analysis of Petroleum Products

Introduction:
Combining the separation power and increased peak capacity of Comprehensive Two-Dimensional Gas Chromatography (GC×GC) with the ability of TOFMS to provide rich data to deconvolution algorithms allows for complete characterization of complex petroleum matrices, with the addition of quantitative results from flame ionization detectors (FID).

Methods:
Several possible configurations have been tested to develop a workflow which would allow either subsequent or simultaneous acquisitions of GC×GC TOFMS and FID data on the same GC×GC instrument. Multiple configurations were explored, including splits at various points. Specific peaks were quantitated by the FID and then identified by correlating relative retention times with the MS data.

Results:
Four main configurations were explored: 1. Installing independent GC×GC column sets in separate GC inlets (Front and Back) and terminating at respective detectors (TOFMS and FID). This option only allowed data collection via subsequent injections. 2. Connecting independent GC×GC column sets in a single inlet by way of a dual-hole ferrule and terminating at respective detectors (TOFMS and FID). 3. Installing a single primary column splitting to two second dimension columns terminating at respective detectors. 4. Installing a controlled-flow splitter after both the first and second dimension columns in an attempt to match flow rates. Options 2-4 allowed simultaneous acquisition of TOFMS and FID data.

Novel Aspect:
The combination of data from TOF and FID with GCxGC allow for both characterization and area percent quantitation of petroleum samples.

21ST INTERNATIONAL MASS SPECTROMETRY CONFERENCE
Toronto, Ontario, Canada
August 20 – 26, 2016

Birsan, Alex,
Phytronix Technologies Inc.

Contributing Authors:
Jean Lacoursiere, Phytronix Technologies Inc.; Pierre Picard, Phytronix Technologies Inc.; Serge Auger Phytronix Technologies Inc.

Title:
Generic sample extraction workflow for analysis of complete Drugs of Abuse panel by LDTD-MS/MS (screening) and LC-MS/MS (confirmation)

Introduction:
New regulations in the US are impacting the way Tox labs perform drug screening and confirmation in urine samples. Laboratories require a generic sample preparation for all drug classes. To increase the sample screening throughput, Laser Diode Thermal Desorption Mass Spectrometry (LDTD-MS/MS) was used. Using the same sample extract, positive sample were reanalyzed using LC-MS/MS for confirmation.

Methods:
LDTD-MS/MS offers specificity combined with an ultra-fast analysis for an unrivaled screening method. Generic extraction method for all drugs was developed to unify screening and confirmation methods. More than 100 drugs from different classes (opiates, benzodiazepines, etc.) are analyzed simultaneously in urine, with quantitative screening results obtained in less than 9 seconds/sample.

Results:
LDTD-MS/MS operated in MRM mode allows rapid measurement of all drugs desorbed simultaneously. Specific transitions are monitored for each drug to quantitate calibrator levels. Ionization is performed in positive mode for group one and negative mode for the second group. Analysis includes spiked drugs in urine and real samples. The '2 standard deviation' approach was used for the validation of the LDTD-MS/MS screening method. All compounds give good responses around the cutoff level. Drug concentrations in real samples were also evaluated in LC-MS/MS with a long gradient to separate each drug class. For each drug in real samples, correlation of the data generated by LDTD-MS/MS and LC-MS/MS, less than 10% of false positive and no false negative are obtained. The benzodiazepine, opiate and marijuana metabolite groups require B-glucuronide enzyme treatment for drug detection.

Novel Aspect:
Generic sample preparation allows robust drug screening of urine sample using the LDTD technology. Sample-to-sample run time of 9 seconds is achieved

Blank, Katrin,
Carleton University

Contributing Authors:

Hillary Weinert, Department of Chemistry, Carleton University, Ottawa, Ontario, Canada; Jeffrey C. Smith, Department of Chemistry, Carleton University, Ottawa, Ontario, Canada; Sam Williamson Department of Biology, Carleton University, Ottawa, Ontario, Canada; Kenneth Storey, Department of Biology, Carleton University, Ottawa, Ontario, Canada; Carlos R. Canez, Department of Chemistry, Carleton University, Ottawa, Ontario, Canada.

Title:

Investigating the lipidomic dynamics of torpor through examination of hibernating squirrel liver tissue and dehydrated frog leg tissue

Introduction:

Animals that live in northern climates have different ways of coping with cold long winters and many enter into a state of decreased physiological activity known as torpor. Low body temperature, slow breathing and heart rates, as well as a low metabolic rates are possible through radical biomolecular changes. To date, the lipid compositional dynamics of torpor have not been investigated.

Methods:

Hibernating/non-hibernating squirrel liver samples, hydrated/dehydrated frog leg muscle tissue samples were lysed via homogenization and sonication; lipids were extracted via a Bligh and Dyer extraction. Lipids were separated via reversed phase high performance liquid chromatography and analyzed via specific precursor ion or neutral loss scanning modes using a triple quadrupole mass spectrometer.

Results:

Initial results have indicated that the ground squirrel increases the fluidity of its membranes during hibernation by increasing the degree of lipid unsaturation, which decreases the density and viscosity of the membrane. Recently we have expanded our study to investigate whether similar biomolecular mechanisms exist in the membranes of freeze-tolerant wood frogs as they experience temperatures below 0oC. To pursue this, the leg tissue of a normal (hydrated) wood frog has been compared to the leg tissue of a cold (dehydrated) wood frog using the same methodologies as described above. Our results will summarize the lipid species that are present in these tissues as well as highlight common lipidomic dynamics that occur during torpor. The results will also provide insight into the biomolecular mechanisms at play that allow cells to survive during these extreme stresses.

Novel Aspect:

This is the first investigation into the lipidomic dynamics of torpor.

Blanksby, Stephen,
Queensland University of Technology

Contributing Authors:
Todd Mitchell, University of Wollongong; Berwyck Poad, Queensland University of Technology; Alan Maccarone University of Wollongong.

Title:
New ion-mobility and ion-activation strategies for lipid structure elucidation

Introduction:
Contemporary lipidomics relies heavily on workflows combining liquid chromatography with mass spectrometry. Such strategies typically provide evidence for many hundreds of lipid species in biological extracts but can also belie an even greater molecular complexity arising from the presence of isomeric lipids. This presentation describes new ion-mobility and ion-activation strategies to separate an

Methods:
Experiments were conducted on tandem linear ion-trap mass spectrometers (QTRAP5500 and 6500, SCIEX). Instruments were modified to deliver high concentration ozone to the collision cell and thus the acquisition of ozone-induced dissociation mass spectra. Each instrument was fitted with a DMS device (SelexION, SCIEX) to facilitate mobility separations of ions generated by electrospray ionisation

Results:
Preliminary experiments have shown the ability of differential-mobility mass spectrometry to separate a homologous series of ionised deoxysphingosine isomers that differ only in the position of a carbon-carbon double bond. The isomeric separation is found to be greater than that achieved using classical drift-mobility technologies and, moreover, differential-mobility is found to be an excellent companion to ozone-induced dissociation (OzID) mass spectrometry. The unique OzID dissociation patterns are predictable and facilitate unambiguous identification of the isomer, and – in the presence of suitable stable isotope labeled standards – could be used for quantification in complex mixtures. Finally, differences in OzID patterns related to the stereochemistry of the carbon-carbon double bond (i.e., cis versus trans) will be discussed.

Novel Aspect:
A novel combination of ion-mobility and ion-activation workflows is demonstrated for the identification of isomeric lipids within complex extracts

Blockhuys, Frank,
University of Antwerp

Contributing Authors:
Julie Cautereels, University of Antwerp.

Title:
A Quantum Chemical Tool for the Prediction of Mass Spectra and the Identification of Fragmentation Pathways: Inter-side-chain Interactions in Peptides

Introduction:
Due to the nature of the processes on which MS is based (chemical reactions rather than transitions between energy levels) MS is one of the few remaining techniques for which no quantum chemical methods are routinely available for the systematic reproduction or prediction of spectra. Methods such as RRKM-QET or QCEIMS suffer either from being an incomplete model or from a high computational cost.

Methods:
QCMS2 (Quantum Chemical Mass Spectrometry for Materials Science) is based on DFT/B3LYP/6-311+G* calculations and selects fragmentation pathways based on the relative values of bond orders, and reaction (for bond cleavages) and transition-state energies (for rearrangements). For work on peptides the code omits the breaking of irrelevant bonds and incorporates the Mobile Proton Model (MPM).

Results:
QCMS2 was developed using small organics with different functional groups. The main features of their EI spectra were correctly reproduced, but more importantly, unexpected fragmentation pathways were predicted and subsequently experimentally corroborated using MS/MS measurements. QCMS2 is currently being used to study the influence of inter-side-chain (ISC) interactions on the ESI CID MS fragmentation of non-cyclic tripeptides, as model systems for proteins. Calculations on Ser-His-X, Asn-His-X and Asp-His-X indicate that the influence of ISC interactions, which is hardly or not at all implemented in common mass spectral interpretation and assignment software, on the initial protonation and the subsequent fragmentations is considerable: the ISC interactions allow amino acids other than the most basic to be protonated and fragmentation of the side chain to occur more easily.

Novel Aspect:
The MPM was verified and extended since fragmentation pathways via six- or eight-membered-ring intermediates are also energetically favorable.

Bodai, Zsolt,
imperial college london

Contributing Authors:
Simon Cameron, Imperial College London; Frankie Bolt, Imperial College London; Julia Abda Imperial College London; Zoltan Takats, Imperial College London; Haixing Wang, Imperial College London.

Title:
Rapid Evaporative Ionization Mass Spectrometry – The ultimate tool for real-time analysis?

Introduction:
Rapid Evaporative Ionization Mass Spectrometry was introduced as method for real-time, intrasurgical tissue identification method, based on the direct combination of electrosurgery and mass spectrometry. The technology was soon picked up by other application areas ranging from microbiology to food analysis. Recent results relevant to environmental analysis are demonstrated in the present paper.

Methods:
Environmental water, soil and aerosol samples were collected in London and Cambridge. Samples were analysed at Imperial College, using a commercially available REIMS source mounted on a Xevo G2-XS quadrupole-TOF mass spectrometer (Waters). Model aerosols were created using a laboratory-built ultrasonic nebulizer system.

Results:
Soil and water samples were spiked with various pesticides and PPCP components and subjected to direct REIMS analysis as well as pellets of filtered or centrifuged residue of the surface water samples. LODs in the low-to-middle ng/ml range were obtained using the REIMS approach. Similar results were obtained for cyanobacteria/cyanobacterial toxins. Feasibility of detecting and speciating water-borne pathogens was also tested, both via direct analysis of natural and municipal water samples and following pre-concentration. A novel REIMS interface was also constructed featuring matrix introduction capability and a heated collision surface for the declustering and further ionization of aerosol particles. The novel interface was successfully used for the analysis of environmental samples.

Novel Aspect:
Applicability of the REIMS technology for sample preparation-free environmental monitoring was successfully demonstrated

Boersema, Paul,
ETH Zurich

Contributing Authors:
Johan Bijzet, University of Groningen; Paola Picotti, ETH Zurich; Christian E. Zimmerli ETH Zurich; Bouke P.C. Hazenberg, University of Groningen.

Title:
A novel mass spectrometry based assay for the diagnosis and typing of systemic amyloidosis

Introduction:
In systemic amyloidoses (SA) different precursor proteins misfold, aggregate as amyloid fibrils and are deposited in various tissues, thereby causing severe functional damage of the affected organs. Here, we developed an LC-MS based method that assays in one analysis the major specific systemic amyloidosis proteins, their eventual mutation and additional proteins as generic indicators of SA.

Methods:
We developed an SRM-MS assay targeting 13 major SA proteins and often occurring mutations plus 20 additional proteins to evaluate their diagnostic potential for generic SA (289 precursors, 1134 transitions). Heavy isotope labeled synthetic peptide analogs for all targeted peptides were spiked in. Fat tissue aspirates of patients were digested and measured during the 35 min SRM assay.

Results:
Fat tissue aspirates can be processed for direct analysis by LC-SRM MS. 58 samples consisting of control samples and patient samples with varying levels of ATTR, AA, AL-L and AL-K were measured with the SRM-assay. The amyloidogenic precursor protein was correctly detected with >80% specificity and sensitivity. The correctness of diagnosis increased with increasing disease severity. Possibly present mutations were detected simultaneously with the assay. An amyloid signature of 20 proteins was developed to diagnose the presence of amyloid fibrils in fat tissue aspirates from systemic amyloidosis patients regardless of type and to gauge SA severity. We further validated the assay with a set of 84 blinded samples. The high specificity and sensitivity of the assay combined with its relative straightforwardness make it an interesting alternative to the contemporary SA diagnosis pipeline.

Novel Aspect:
A one-stop-shop LC-SRM assay that in a 35 minute run simultaneously diagnoses, types and grades 13 different SA types and their potential mutations.

Boice, Aaron,
Agilent Technologies

Contributing Authors:

Crystal Cody, Agilent Technologies; Stephen Madden, Agilent Technologies; Alex Apffel Agilent Technologies; Yinghang Yang, Agilent Technologies; Xinning Jiang, Agilent Technologies; Xiangdong Li, Agilent Technologies; Ed Darland, Agilent Technologies; Norton Kitagawa, Agilent Technologies; Yuqin Dai, Agilent Technologies.

Title:

An Integrated Workflow for Qualitative Flux Analysis by Accurate Mass LC/MS

Introduction:

The field of metabolic flux analysis suffers from data analysis challenges; difficult-to-use software, manual data processing, and a lack of visualization tools to communicate results. We propose a new workflow using accurate mass LC/MS of stable isotope labels. This workflow integrates rapid data mining, simplified review of data analysis, and visualization of results on metabolic pathways.

Methods:

The targeted qualitative flux workflow starts by automatic target list creation based on metabolite pathways and a database. Stable isotope labelled feedstock is introduced in a biological system, from which samples are analyzed by accurate mass LC/MS. Batch isotopologue extraction information is then graphically reviewed, and then results are visualized directly on metabolic pathway diagrams.

Results:

JJ01 chondrosarcoma cells with a known isocitrate dehydrogenase (IDH1) mutation were incubated with U-13C-Glutamine and U-13C-Glutamate and extracted at various time points after label introduction. High resolution LC/MS analysis using ion pair C18 reverse phase chromatography was performed in negative ion mode. A target list was generated by software based on 17 resolved TCA metabolites. High resolution profile data was mined for the targets and their isotopologues, and results displayed for review in a histogram by sample group. Results were then exported directly into a program to visualize information rich pathways (position info, atom mapping, structures) with content downloaded from BioCyc. Results are overlaid on the pathway, and the visualization program allowed for observation of time-dependent label incorporation throughout the TCA cycle.

Novel Aspect:

Rapid and automated mining of stable label isotopologue data can then be quickly reviewed and visualized on metabolic pathways.

Bonneil, Eric,
IRIC-Université de Montréal

Contributing Authors:
Sibylle Pfammatter, IRIC-Université de Montréal; Pierre Thibault, IRIC-Université de Montréal.

Title:
Improved isotopic labelling quantitation for large scale proteomic analyses using High Field Asymmetric Waveform Ion Mobility Spectrometry (FAIMS)

Introduction:
Misalignment of isotopic clusters due to conflicting peptide or background peaks for SILAC quantitation as well as co-fragmentation of precursor ions for TMT quantitation can lead to skewed isotopic ratios. The addition of FAIMS has been successful in diminishing the occurrence of co-fragmentation and background level. Here we evaluate the benefit of FAIMS for isotopic labeling based quantitation.

Methods:
The FAIMS device has an analytical gap of 1.5 mm enabling higher resolution with 100% N2 and lower gas flow rate. A cylindrical stainless tube was welded to a 1/16" OD nut acting as a tip holder and connected to the column outlet. Samples were analyzed on a set-up comprising a C18 precolumn (5 mm*360 µm id) and a C18 analytical column (15 cm * 150 µm id) coupled to an Orbitrap Elite instrument.

Results:
With a two-proteome model (yeast and human), we evaluated the extent of distorted TMT ratios to be 62% with nano-LC-MS. FAIMS substantially decreased the extent of ratio distortion down to 18%. We used FAIMS to profile HEK293 proteome changes upon a time resolved heat shock treatment. FAIMS increased the number of identified peptides by 93% (7047 identified peptides with FAIMS compare to 4189 without FAIMS). Fuzzy c-means clustering of regulated proteins were performed on the 1383 identified proteins with FAIMS enabling the grouping of 502 proteins in 4 clusters. All identified heat shock proteins were found in the cluster grouping proteins whose abundance increased over heat shock time. For SILAC quantitation, we obtained a 3-fold increase for identified and quantified SILAC pairs with FAIMS. We probed the HEK293 phosphoproteome changes upon heat stress with SILAC and LC-FAIMS-MS.

Novel Aspect:
NanoLC-FAIMS-MS provides increased peptide identifications and fold change accuracy in isotopic labelling proteomics compare to multidimensional LC-MS

Borton, David,
LECO Corporation

Contributing Authors:
David E. Alonso, LECO Corporation; Viatcheslav Artaev, LECO Corporation.

Title:
Direct Insertion Probe Coupled to a High Resolution Time of Flight Mass Spectrometer for Rapid Targeted and Non-targeted Analysis of Drugs of Abuse.

Introduction:
Forensic analysis of confiscated drugs of abuse often involves lengthy sample prep prior to GC-MS or LC- MS analysis. Direct insertion probes (DIP) allow for the analysis of solid and solution phase samples with no sample prep, resulting in high throughput. EI provides a universal ionization method and the mass spectra are comparable to the extensive EI mass spectral libraries available.

Methods:
Solid samples of drugs of abuse were analyzed by a prototype DIP coupled to a multi-reflecting (MR) LECO Pegasus® GC-HRT with no sample preparation. The probe with sample was inserted into the ion source and heated. Evaporated analyte was ionized by EI. The resulting data was subjected to a mass spectral library search, with accurate mass providing additional confirmation of the compounds present.

Results:
Several drugs of abuse were analyzed. Bath salts were sampled using a syringe plunger pressed into the salt to pick up a small amount of solid which was transferred to a DIP sample vial with no sample prep. Synthetic cannabinoids in a botanical matrix were sampled in similar fashion. The probe was introduced into the ion source and heated to 400 °C to evaporate analyte into the gas phase where it was ionized by 70eV EI ionization. Analysis using library matching with spectral databases returned matches of high similarity for illegal drugs. The bath salts were comprised of ethylone while a known cannabinoid sample contained JWH-018 and synthetic cannabinoid 5-fluoro ADB was found in an unknown botanical sample. The high resolution and accurate mass of the mass spectrometer added further confirmation for the cannabinoids in spite of other interfering compounds from the botanical matrix.

Novel Aspect:
DIP analysis of drugs of abuse requires no sample prep. MR High res ToF yields accurate mass data for screening targeted and non-targeted compounds.

Bottaro, Christina,
Department of Chemistry, Memorial University

Contributing Authors:
Stefana Egli, Department of Chemistry, Memorial University.

Title:
Fresh Approaches to the Analysis of Small Polar N-Nitrosamines in Water using MIP-DESI-MS

Introduction:
N-nitrosamines, e.g., N-nitrosodimethylamine (NDMA), are likely human carcinogens found in many beverages, wastewater, and drinking water, etc. Many nitrosamines show a range of polarities, some with negative logP (octanol/water) values, and low environmentally-relevant concentrations (< 1 ng/mL). These properties features make nitrosamines a challenging target for analytical chemists.

Methods:
The US EPA Method 521 (2004) for nitrosamines in drinking water uses solid phase extraction followed by GC-MS/MS. This is a multi-step procedure with a high tolerance for poor recoveries. We have replaced traditional solid phase extraction with molecularly imprinted polymers coupled to desorption electrospray ionization-mass spectrometry (MIP-DESI-MS).

Results:
These MIPs use simple molecular recognition to adsorb nitrosamines from aqueous matrices. Detection can then be carried out under ambient conditions using DESI-MS/MS directly on the polymer surface. The performance of the MIPs in the uptake of nine nitrosamines was evaluated through binding tests in distilled water, drinking water and wastewater effluent in terms of sensitivity, selectivity and other analytical figures of merit. Various factors influencing sensitivity in DESI-MS, including matrix effects, spray composition, capillary voltage, and geometric relationships (e.g. capillary to surface angles and distances) were optimized. The method is suitable for semi-quantitative screening for nitrosamines, and the technique is fast and consumes only small amounts of solvent, making it environmentally friendly.

Novel Aspect:
MIPs for water-borne nitrosamines are new and by extension so is the MIP-DESI-MS approach.

Botting, Catherine,
University of St Andrews

Contributing Authors:
Swati Arya, Biomedical Sciences Research Complex, University of St Andrews, St Andrews, Fife, KY16 9ST; Imre Lengyel, University College London Institute of Ophthalmology, 11-43 Bath Street, London, EC1V 9EL; Alan Stewart Biomedical Sciences Research Complex, University of St Andrews, St Andrews, Fife, KY16 9ST; Sally Shirran, Biomedical Sciences Research Complex, University of St Andrews, St Andrews, Fife, KY16 9ST; Silvia Synowsky, Biomedical Sciences Research Complex, University of St Andrews, St Andrews, Fife, KY16 9ST.

Title:
SWATH Analysis of Plasma from Age-Related Macular Degeneration (AMD) Patients

Introduction:
AMD is the leading cause of visual impairment in the elderly. Focal (drusen) or diffuse (basal linear laminar deposits) sub-retinal pigment epithelial deposits in the macula are an AMD hallmark. We apply SWATH analysis to patient plasma samples, comparing the sub-proteome of patients who are homozygous wild type, heterozygous or homozygous mutant in an implicated protein.

Methods:
Albumin and IgG depleted plasma samples were, after sample clean up and digestion, subjected to SWATH analysis on a Sciex TripleToF 5600+ instrument, with Biognosys iRT peptides spiked in. A rigorous regime of repeat injection and complete technical replicates was employed. Data were searched against a library of pooled samples using Sciex software.

Results:
Our depleted and fractionated plasma protein library yielded 245 protein identifications. Five homozygous wild type patient samples, seven heterozygous and eight homozygous mutant patient samples were analysed by variable window SWATH. The data was processed using ProteinPilot v5.0 and PeakView v2.2. It was visualized as volcano and PCA plots. The data for proteins showing significant changes in expression level between the two homozygous (wild type and mutant) datasets was manually verified at the XIC level.

Novel Aspect:
Quantitative changes in expression levels have been found in several proteins which may play a role in AMD.

Bouledroua, Moncef,
Faculté de Médecine & LPR

Contributing Authors:
Fatima Talhi, Physics Department, Badji Mokhtar University, Annaba, Algeria.

Title:
Radiative association of 36Ar+ and 38Ar+ with hydrogen

Introduction:
In a recent paper [Barlow et al., Science 342, 1343 (2013)], the ArH+ ionic system has been detected in the Crab Nebula.

Methods:
Theoretical

Results:
In a recent paper [Barlow et al., Science 342, 1343 (2013)], the ArH+ ionic system has been detected in the Crab Nebula. We accordingly propose to examine in this work the radiative association of the argon ions 36Ar+ and 38Ar+ with atomic hydrogen and calculate the related temperature-dependent rate coefficients. To do so, we have to construct the transition dipole moments and the potential-energy curves via which both Ar+ and H species approach each other. The corresponding data points are borrowed from the most recent and reliable calculations of Stolyarov and Child [PCCP 7, 2259 (2005)]. Once the curves are constructed, the rate coefficients are computed quantum mechanically and analyzed in the temperature range 1-10000 K.

Novel Aspect:
ArH+ radiative association calculations

Bowman, Andrew,
Wichita State University

Contributing Authors:

Alexandre A. Shvartsburg, Wichita State University; Rinat A. Abzalimov, City University of New York; Gordon A. Anderson GAA Custom Engineering.

Title:

Broad Separation of Lipid Isomers Using High-Definition Differential Ion Mobility Spectrometry and a Range of Ionizing Species

Introduction:

Lipids exhibit isomers of multiple types, e.g., transacylation, chain-length, double bond migration, and cis/trans. Resolving them is crucial for metabolomics, but chromatography and tandem MS succeeded only in part. Differential IMS or FAIMS was proven capable of such separations, in particular upon silver cationization. We now explore the utility of FAIMS and metal cationization in breadth.

Methods:

A planar-gap FAIMS system with stabilized bisinusoidal waveform is coupled to LTQ XL ion trap via a slit aperture/ion funnel interface. The He/N2 and H2/N2 buffers are formulated by digital flow meters. Lipid standards and their mixtures were dissolved in CHCl3/MeOH, doped with salts of ammonia or metals (Na, Ag, Cu, Ni), and infused to an ESI source in front of curtainplate/orifice FAIMS inlet.

Results:

We focused on glycero- and phospholipids, covering over 20 isomer pairs of all four types. First, the FAIMS and CID spectra were obtained for individual samples, and the former aligned using non-isobaric lipids as internal calibrants. We then varied the FAIMS conditions to optimize the resolution based on those spectra and confirmed the separation employing mixtures in MS and MS/MS modes. Best performance was usually achieved in He/N2 with ~60% He. Isomers of all types were resolved for protonated or ammoniated species, with transacylation appearing the easiest, followed by chain-length, cis/trans, and double bond migration. We substantially separated ~75% of all pairs by FAIMS/MS, and adding the MS/MS step would likely help. Argentination has improved the resolution in some but not all cases. Preliminary results suggest cationization by Cu and Ni to be of benefit for selected isomers.

Novel Aspect:

Broad exploration of high-resolution FAIMS separations for diverse lipid isomers cationized by protons, ammonia, and main group and transition metals.

BOYACI, Ezel,
University of Waterloo

Contributing Authors:

Barbara Bojko, University of Waterloo; Nathaly Reyes-Garcés, University of Waterloo; Janusz Pawliszyn University of Waterloo; Germán Augusto Gómez-Ríos, University of Waterloo.

Title:

High throughput non-depletive SPME method for LC-MS/MS determination of free and total concentrations of multi-residue in small sample volumes

Introduction:

Non-depletive determination of compounds in biological matrices has significant importance in bioanalyses. The main constraints in such studies are the small sample volume, the necessity for rapid sampling and possible lack of sensitivity. Here, solid phase microextraction (SPME) was explored to address the listed limitations for high throughput determination of multi-residue in complex matrices.

Methods:

Triple quadrupole TSQ Vantage equipped with heated electrospray ionization ion source was used under selected reaction monitoring mode. Chromatographic separation of nicotine was achieved in an Ascentis Express HILIC column while diclofenac and DEET were separated in an Ascentis Express F5 column. HLB particle coated TF-SPME blades and fibers were made in laboratory and used in the study.

Results:

The probed analytes were selected based on their divergent plasma protein binding characteristics. Initial evaluations showed that extraction from such small system (100 µl) with thin film-SPME can disturb the existing protein-drug binding equilibrium. Therefore, to conduct bioanalytical studies in such microsystem at non-depleting conditions miniaturized SPME fibres with coating length of 2 mm and coating thickness of 6 µm were developed. The small volume of extraction phase on the fibre ensured non-depletive extraction while the fast equilibrium is secured by the thin coating. The final method allows high through put analysis of compounds possessing distinct physicochemical properties in small volumes of complex matrices, enables fast and reliable protein-drug binding studies, as well as repetitive extractions from the same sample without disturbing the system equilibrium.

Novel Aspect:

New SPME-based high throughput assay for multi-residue non-depletive bioanalysis in small sample volumes for toxico- and pharmacokinetics

Boyarkine, Oleg,
EPFL

Contributing Authors:
Vladimir Kopysov, LCPM, ISIC, EPFL.

Title:
New developments in 2D UV-MS Cold Ion Spectroscopy of Biomolecules

Introduction:
We recently integrated UV photofragmentation spectroscopy of cold ions with high-resolution Orbitrap mass spectrometry (MS) to measure two-dimensional (number of fragment ions vs m/z and UV wavelength) UV-MS fingerprints of ions. In addition to UV and mass spectra alone such fingerprints may contain systematic correlations between m/z of photofragments and UV wavelength.

Methods:
A use of appropriate mathematical methods reveals such correlations in 2D UV-MS data matrices and allows for identification of chemical composition of the probed ions. We earlier demonstrated the synergy of this instrumentation for quantitative analytical identifications of isomers of biomolecules, as well as for recognition of conformers of protonated peptides cooled to cryogenic temperatures.

Results:
(1) We use 2D UV-MS for detecting conformer-selective resonance energy transfer (RET) between the aromatic rings of Tyr and Phe residues in pain-relief drugs enkephalins. The distance between these aromatic rings is one of the key structural parameters that govern the pharmacological activity of these opioid peptides. In contrast to the known condensed-phase structures, these distances, evaluated within the Forster's RET model for the gas-phase ions, appear to be substantially different in the stereoisomers of enkephalin with very different pharmacological efficiencies. This observation suggests that gas-phase structures might be a better pharmacophoric metric for ligand peptides that bind to hydrophobic pockets of membrane proteins. (2) We now extend 2D UV-MS method to the deep UV spectral region to cover other chromophores and the absorption by peptide bonds.

Novel Aspect:
Scanning wavelength from UV to DUV gradually increases the complexity of fragmentation mass spectra, simplifying their interpretation.

Bregy, Lukas,
ETH Zurich

Contributing Authors:
Pablo Martinez-Lozano Sinues, Department of Chemistry and Applied Biosciences, ETH Zurich, Zurich, Switzerland; Malcolm Kohler, Department of Pulmonology, University Hospital Zurich, Zurich, Switzerland; Nina Stebler Department of Pulmonology, University Hospital Zurich, Zurich, Switzerland; Yvonne Nussbaumer-Ochsner, Department of Pulmonology, University Hospital Zurich, Zurich, Switzerland; Renato Zenobi, Department of Chemistry and Applied Biosciences, ETH Zurich, Zurich, Switzerland; Thomas Gaisl, Department of Pulmonology, University Hospital Zurich, Zurich, Switzerland; Diego García-Gómez, Department of Chemistry and Applied Biosciences, ETH Zurich, Zurich, Switzerland; Martin Gaugg, Department of Chemistry and Applied Biosciences, ETH Zurich, Zurich, Switzerland; Yannick Suter, Department of Chemistry and Applied Biosciences, ETH Zurich, Zurich, Switzerland.

Title:
Real time SESI-MS analysis of exhaled breath identifies altered metabolic signature in chronic obstructive pulmonary disease

Introduction:
Diagnosis of chronic obstructive pulmonary disease is currently limited to spirometry. Therefore we explored VOC profiles in exhaled breath from patients suffering from COPD using real-time SESI-MS.

Methods:
Exhaled breath analysis on COPD patients and healthy controls by untargeted, high-resolution mass spectrometry was performed using secondary electrospray ionization-mass spectrometry (SESI-MS). A sensitivity in the low ppt range can be reached.

Results:
14 metabolites, including oxidative stress indicators and amino acid related compounds were found to be highly specific to differentiate COPD patients from healthy controls. The most discriminating features showed a prediction accuracy of 88.9% with a sensitivity of 92.9% and a specificity of 86.4%.

Novel Aspect:
Real-time SESI-MS analysis of breath is a powerful and easily applicable method to get new insights into the pathophysiology of a lung disease.

Brenner, Florian,
Vienna University of Technology

Contributing Authors:
Szilvia Török, National Korányi Institute of Pulmonology, Budapest, Hungary; Olga Kelemen, National Korányi Institute of Pulmonology, Budapest, Hungary; Balazs Döme Medical University of Vienna, Vienna, Austria; Herbert Hutter, TU Wien, Vienna, Austria; Balazs Hegedüs, Medical University of Vienna, Vienna, Austria.

Title:
Elemental and Molecular Imaging of Tumour Tissues treated with a Combination of Sunitinib and Evofosfamide with ToF-SIMS

Introduction:
Time of Flight Secondary Ion Mass Spectrometry (ToF-SIMS) represents an attractive technique for the analysis of intracellular distbutions of anti-cancer therapeutics in biological specimen. The mass spectra not only reveal the localisation of exogenous compounds within the sample, but also can distinguish different histological domains of the tissue due to their molecular composition.

Methods:
For the experiments distribution of two interacting modern anti-cancer drugs was studied. MIce were used as model organisms. Treated tumour and control tissues were analysed with a ToF-SIMS V mass spectrometer, equipped with a 25 keV bismuth liquid metal ion gun. The measurements were performed in the High Current Bunched Mode (higher mass and lower spatial resolution).

Results:
The distribution of the anti-cancer drugs sunitinib and evofosfamide was analysed with a spatial resolution in the single-digit μm scale by the use of ToF SIMS. The combined treatment is assumed to improve the efficacy, because sunitinib tends to provoke hypoxic conditions within the tumour, thus activating it for binding of evofosfamide. Appropriate target elements that are contained in the drug (e.g., fluorine, bromine) and remain with a low natural abundance were used for detection. The fluorine intensity (sunitinib) was distinctly reduced in the necrotic parts of the tumour compared to the non-necrotic areas of the tissue. Furthermore, the hypoxic areas in the tissue (shown by staining) were found to be related to increased fluorine intensities. In these areas also the bromine intensities (evofosfamide) were found to be significantly increased, thus confirming the primary hypothesis.

Novel Aspect:
Distribution analysis of modern anti-cancer drugs in cancer tissues with a lateral resolution in the single-digit μm scale with ToF-SIMS.

Britz-McKibbin, Philip,
McMaster University

Contributing Authors:
Amy Hector, Department of Kinesiology, McMaster University; Michelle Saoi, Department of Chemistry and Chemical Biology, McMaster University; Martin Gibala Department of Kinesiology, McMaster University.

Title:
Characterization of the Muscle Metabolome: Does Oral Bicarbonate Function as an Ergogenic Aide in Strenuous Interval Exercise?

Introduction:
Contracting skeletal muscle requires significant energetic demands during acute high-intensity exercise that results in metabolic acidosis with potential deleterious health outcomes. Alkalinizing agents (i.e. sodium bicarbonate) have been shown to delay the onset of fatigue by buffering against decline in blood pH, however the exact mechanisms of action within muscle tissue remain unclear.

Methods:
A placebo-controlled "cross-over" study design was performed on untrained male subjects (n=7) who ingested sodium bicarbonate (2 x 0.2 g/kg) or sodium chloride (2 x 0.2 g/kg) 1.5 hr prior to high intensity interval exercise on a ergometer cycle. Muscle tissue biopsies and plasma specimens were collected from each subject at three time intervals with metabolomic studies performed by CE/MS.

Results:
Over a hundred polar/ionic metabolites from freeze-dried muscle tissue extracts were consistently measured by CE-MS under positive and negative ion mode detection, whereas CE with indirect UV detection was used for analysis of major electrolytes. A dilution trend filter was used to characterize reproducible yet authentic metabolites from pooled muscle and plasma specimens. Complementary statistical methods were then applied to classify metabolites associated with strenuous exercise (time: rest, post-exercise, recovery) and treatment effect (bicarbonate ingestion), including multilevel PLS-DA. O-acetylcarnitine, lactate, sodium and glutamic acid were significantly changed in both plasma and muscle following exercise, whereas muscle glycine and plasma potassium, 4-hydroxyproline and 3-methylhistidine were selectively modulated by bicarbonate pretreatment, including several unknown ions.

Novel Aspect:
The first "paired" analysis of time-resolved metabolome changes in muscle tissue and plasma from a well-controlled exercise/dietary human intervention

Brown, Kerene,
NSERC CREATE York University and University of Toronto

Contributing Authors:
Mark Bayfield, York University; Derek Wilson, York University.

Title:
Structure and Dynamics of RNA binding to the Non-Canonical RNA Recognition Motif (RRM2) in the Human La Protein

Introduction:
La is a RNA chaperone associated with viral pathogenesis and translation regulation. Its main role is binding RNA pol III transcripts protecting them from exonucleases. This binding mode is typical of the N-terminal domain but RNA binding to the C-terminal is not well understood. Here we study the RNA binding mechanism of the CTD, to a ssRNA and IRES domain IV of the Hepatitis C viral mRNA.

Methods:
His-Tagged hLa-CTD was over-expressed in E. coli cells and purified using affinity chromatography. Native and ion mobility mass spectrometry (MS) were done in order to observe and compare the free protein and the protein/RNA complexes. Time-resolved electrospray ionization (TRESI) hydrogen deuterium exchange (HDX) mass spectrometry was done in order to localize RNA binding in the CTD of hLa.

Results:
Native and ion mobility MS reveal a more compact species upon RNA binding to the CTD of hLa. TRESI-HDX-MS reveals the structural differences of hLa-CTD when it binds to the ssRNA vs the HCV RNA. Binding to the ssRNA reveals an unusual mode of binding. The binding was expected to be similar to a La homolog p65, in which there was a helical extension upon RNA binding. This extension was of the helix that lies across the typical RNA binding surface of a RNA recognition motif (RRM). However, our data suggest a helical unwinding exposing the RNA binding site, which it is thought to bind. On the other hand, when HCV RNA binds to the CTD the same helix becomes less solvent accessible and the peptide following this helix (in the unstructured region) becomes more structured which is evident by a large decrease in deuterium uptake. This suggest 2 distinct RNA binding modes in the CTD of hLa.

Novel Aspect:
Structural insights on hLa-CTD and the conformational changes that occur when it binds single-stranded non-poly(U) vs structured hairpin HCV RNA.

Bruinen, Anne,
M4I - Maastricht University

Contributing Authors:
Nino Iakobachvili, M4I, Maastricht University, Maastricht, the Netherlands; Ron M.A. Heeren, M4I, Maastricht University, Maastricht, the Netherlands; Shane R. Ellis M4I, Maastricht University, Maastricht, the Netherlands; Olga J.G.M. Goor, Institute for Complex Molecular Systems, Eindhoven University of Technology, Eindhoven, the Netherlands ;Laboratory of Chemical Biology, Department of Biomedical Engineering, Eindhoven, the Netherlands; Patricia Y.W. Dankers, Institute for Complex Molecular Systems, Eindhoven University of Technology, Eindhoven, the Netherlands ;Laboratory of Chemical Biology, Department of Biomedical Engineering, Eindhoven, the Netherlands; Laboratory for Cell and Tissue Engineering, Departme.

Title:
3D Imaging ToF-SIMS Tandem Mass Spectrometry of Cellular Structures

Introduction:
ToF-SIMS enables sub-micron resolution molecular imaging of complex samples. In this study we applied 3D imaging to both biological and synthetic polymer systems. To perform complete characterization and enable interpretation of imaging results unambiguous identification is required. Here we present 3D imaging data acquired using a new instrument enabling simultaneous MS and MS/MS acquisition.

Methods:
A PHI NanoTOF II modified for simultaneous acquisition of MS1 and MS2 data via orthogonal deflection of ion packets into a collision cell and second ToF analyser was used. The system is equipped with a Bi (cluster) LMIG, a C60+ ion beam and an argon GCIB for soft sputtering. Cellular structures as well as supramolecular systems as an extracellular matrix model were subjected to 3D MS/MS imaging.

Results:
3D analyses of supramolecular biomaterials reveal differences in surface and bulk composition. The material consists of Upy-polycaprolactone polymers in which Upy-tetrazine guest molecules are mixed. Distinct fragements were formed during ionization and identified by MS/MS. A model click compound (TCO) was used as a protein substitute to monitor the chemical reaction that will be used to functionalize the polymer in a later stage. We found TCO mainly at the surface, which indicates that it is possible to only functionalize the surface of the biomaterial with this type of chemistry. The analysis of single cells and small cellular systems benefit from the combination of the LMIG for high 2D spatial resolution and a cluster ion beam to achieve high depth resolution to visualize subcellular structures. We were able to identify different lipid molecules and visualize them in a 3D model.

Novel Aspect:
The new capabilities offered by parallel MS and MS/MS enabling both high resolution 3D imaging and molecular identification are presented.

Bueno Duarte, Gustavo Henrique,
ThoMSon Mass Spectrometry Laboratory

Contributing Authors:
Gilberto Carlos Franchi Junior, University of Campinas; Vinicius Veri Hernandes, University of Campinas; Eduardo César Meurer Federal University of Paraná; Alexandre Eduardo Nowill, University of Campinas; Rosineide Costa Simas, University of Campinas; Marcos Nogueira Eberlin, University of Campinas.

Title:
METABOLOMICS BY LC-ESI-QTOF-MS IN NOD/SCID MICE UNDER CHEMOTHERAPY TREATMENT: POTENTIAL BIOMARKERS OF LEUKEMIA

Introduction:
Acute Lymphoid Leukemia (ALL) is the most common pediatric cancer. Despite advances in the treatment of ALL, protocols are based on a combination of vincristine, dexamethasone and a asparaginase (VXL). The NOD/SCID mice is the animal model to study ALL. In this work, we applied LC-ESI-QTOF metabolic fingerprinting to study the effect of VXL treatment.

Methods:
In order to improve the amount of information obtained, we compared different methodological approaches (sample preparation and chromatographic columns). The optimal approach was used to evaluate mice plasma samples from healthy, leukemia and leukemia treated with VXL groups. In data processing, we performed Chemometrics analysis using Metaboanalyst 3.0 software.

Results:
Protein precipitation followed by phase separation using a blend of organic solvents proved to be the best approach for sample preparation. The use of Poroshell C18 column for organic compounds and Synergy Hydro column for the aqueous compounds allowed the increase of the detected features. Chemometrics analysis was performed by means of Principal Component Analysis, Hierarquical Cluster Analysis, Partial Least Squares Discriminant Analisys. For aqueous extract, spectra from positive ion mode showed clear differentiation between the studied groups. The PLS-DA model built showed R2 0.9763 and Q2 0.9306 values. These results showed the possibility of diagnosis and prognosis in NOD/SCID mice using metabolic fingerprinting by LC-ESI-TOF-MS.

Novel Aspect:
Use of metabolomics approaches for studying the effect of administration of Vincristine, Dexamethasone and Asparaginase in NOD/SCID mice.

Burt, Michael,
University of Oxford

Contributing Authors:
Alexandra Lauer, University of Oxford; Kasra Amini, University of Oxford; Robert Burleigh University of Oxford; Ang Guo, University of Oxford; Edward Halford, University of Oxford; James Pickering, Aarhus University; Benjamin Winter, University of Oxford; Mark Brouard, University of Oxford.

Title:
Nanosecond velocity- and spatial-map imaging using the Pixel Imaging Mass Spectrometry (PImMS) camera.

Introduction:
The Pixel Imaging Mass Spectrometry (PImMS) camera is an event-triggered sensor that records the (x,y,t) coordinates of incident photons to a precision of 12.5 ns over 50 µs. When combined with a position-sensitive ion detector, PImMS enables multi-mass imaging in time-of-flight mass spectrometry and, consequently, for new information to be extracted from velocity- or spatial-map ion images.

Methods:
Time-of-flight imaging MS is used to demonstrate three applications of PImMS: microscope imaging, where a defocused laser is used to simultaneously analyse a 1 x 1 cm sample; Coulomb explosion imaging, which in combination with laser spectroscopy reveals the time evolution of an ion's structure; and 3D slice velocity-map imaging, which directly obtains dynamical information from ion images.

Results:
Multi-mass spatial imaging is demonstrated using a linear ToF-MS with mass and spatial resolutions of 2000 and 20 µm. This instrument also broadens the resolvable mass range by compensating for the spread in surface ion velocities using post-extraction delayed acceleration (PEDA). Velocity-map slice imaging and Coulomb explosion imaging are demonstrated using ToF-MS and laser spectroscopy. In the former, fragments spherically expand following photodissociation and are "sliced" by PImMS into a series of images. Dynamical properties are typically extracted from the simulated central slice of a flattened sphere, so this approach allows their direct determination. The latter technique correlates Coulomb explosion fragment momenta to establish the parent molecule's structure; this can be used to establish chirality or, using pump-probe schemes, to monitor time-dependent structural changes.

Novel Aspect:
The PImMS fast-imaging sensor is combined with time-of-flight mass spectrometry to create new applications for spatial- and velocity-map ion imaging.

Burton, Lyle,
Sciex

Contributing Authors:
Lijuan Kang, Janssen Research and Development, Johnson & Johnson, 1400 McKean Road, Spring House, PA, USA 19477; Naidong Weng, Janssen Research and Development, Johnson & Johnson, 1400 McKean Road, Spring House, PA, USA 19477; Wenying Jian Janssen Research and Development, Johnson & Johnson, 1400 McKean Road, Spring House, PA, USA 19477.

Title:
An automated workflow for absolute quantitation of large therapeutic proteins in biological samples using intact LC-HRMS

Introduction:
One common workflow for the quantitation of protein therapeutics utilizes LC/MS in a 'bottom-up' fashion. One limitation of this signature peptide approach is its inability to measure the protein as a whole entity. To date, intact analysis has been mostly used for qualitative characterization; here we present an automated workflow for the absolute quantitation of intact protein therapeutics.

Methods:
A human mAb was immuno-affinity purified from mouse plasma followed by Q-TOF LC-HRMS. Prototype software was created which automates deconvolution and quantitation of the multiply charged spectra. Tryptic digestion was also performed and a signature peptide monitored on a triple quadrupole. Quantitation using the intact spectral peak was compared to those obtained using the LC peak of the peptide.

Results:
The performance of the assay was evaluated by three runs conducted on different days. The calibration curves for both the intact (top-down) and signature peptide (bottom-up) approaches all had linearity with r-squared > 0.99. Quality control samples showed acceptable accuracy and precision. The overall assay performance of intact workflow was largely similar to those for the bottom-up approach. Current workflow of affinity-capture-LC-HRMS-automated deconvolution and quantitation is an effective approach to quantify large therapeutic proteins such as mAb in biological samples at intact level. The prototype software streamlined the process of peak deconvolution, peak integration, construction of calibration curve, and calculation of concentrations. In comparison to bottom-up approach, the major limitation of current workflow is sensitivity, which needs further improvement.

Novel Aspect:
Absolute intact protein bioanalysis – automated and compared to signature peptide approach.

Byer, Jonathan,
LECO Corporation

Contributing Authors:
Jack Cochran, Restek Corporation; Joe Binkley, LECO Corporation.

Title:
Fast Pesticide Residue Analysis Using a Novel Benchtop Time-of-Flight Mass Spectrometer

Introduction:
A method was developed to analyze 203 pesticides in less than 9 mins using a new benchtop gas chromatography/mass spectrometer (GC/MS). Pesticide residue analysis remains an important requirement for food and feed, and sample throughput is of high importance for laboratories looking to lower the cost of sample analysis.

Methods:
A mixture of 203 pesticides (GC Multiresidue Pesticide Standard #1 to #9; Restek) was analyzed on a novel benchtop GC/MS capable of 50 spectra/s with a mass range from m/z 10 to 1500. An analysis typically developed on a 30m 5MS GC column was translated to a 15m x 0.25mm i.d x 0.25 µm d.f (Restek) for high-throughput analysis.

Results:
A mixture containing 203 pesticides was analyzed in less than 9 mins. A majority of the components were resolved by chromatography and spectral deconvolution. A food commodity was spiked with a subset of 34 components (GC Multiresidue Pesticide Standard #5) relevant to the commodity to generate a calibration curve from 0.100 to 2000 ppb in matrix. The samples were also fortified with PCB-18, PCB-28, and PCB-52 for use as internal standards. All of the components were detected at the 1.0 ppm level in matrix, with the majority of components detected at 0.100 ppm, with excellent linearity to 2000 ppm. An unfortified sample was used to measure concentrations of these incurred pesticides in the food commodity.

Novel Aspect:
New benchtop GC/MS capable of high-throughput and low level pesticide residue analysis.

21ST INTERNATIONAL MASS SPECTROMETRY CONFERENCE
Toronto, Ontario, Canada
August 20 – 26, 2016

Byer, Jonathan,
LECO Corporation

Contributing Authors:
Karl Jobst, Ontario Ministry of the Environment and Climate Change; Kevin Siek, LECO Corporation.

Title:
Advantages of Comprehensive Two-Dimensional Gas Chromatography-High Resolution Time-of-Flight Mass Spectrometry in Petroleomics

Introduction:
Comprehensive two-dimensional gas chromatography high resolution time-of-flight mass spectrometry (GC×GC-HRToFMS) at 25 000 resolving power FWHH can be leveraged to distinguish features like heteroatom class, type (degree of unsaturation), and carbon number in crude oil. In addition, this technique was also used to distinguish key mass splits such as C3 vs SH4 (0.0034 Da).

Methods:
Crude oil samples diluted in dichloromethane were analyzed using GC×GC-HRToFMS. Chromatographic separation was achieved using an Rxi-5MS primary column and an Rxi-35SilMS secondary column. Mass Spectra were acquired with a mass range from 15 to 650 m/z in EI mode, and from 50 to 650 m/z in CI mode (CH4), with an acquisition rate of 150 spectra/s. Data were processed using ChromaTOF-HRT v1.91.77.

Results:
The added benefit of a second dimension of chromatographic separation and the resultant increased peak capacity compared to 1D GC, combined with high resolution (25K) MS, dramatically enhanced the capacity to distinguish and identify additional sample constituents in the crude oil sample. A workflow was developed integrating Automatic Peak Finding with High Resolution Deconvolution™, and User Defined Mass Defect Plots. The mass defect plot display was configured to parallel the structured nature of a GC×GC chromatogram, which when used in combination resulted in a more comprehensive chemical profile and characterization of a sample than previously achievable. A number of key mass splits in petroleomics were distinguished in the sample including CH4 vs O (0.0364 Da), CH2 vs N (0.0126 Da), and C3 vs SH4 (0.0034 Da).

Novel Aspect:
The application of GC×GC to resolve isobaric compound classes that would otherwise require ultrahigh resolution MS (i.e. FT-ICR and Orbitrap).

C. Correra, Thiago,
IQ-USP - Brazil

Contributing Authors:
André Santos Fernandes, Instituto de Química - Universidade de São Paulo - Brazil; Philippe Maitre, Laboratoire de Chimie Physique, Université Paris Sud, UMR8000 CNRS; Vincent Steinmetz Laboratoire de Chimie Physique, Université Paris Sud, UMR8000 CNRS.

Title:
Evaluation of the Sharpless epoxidation by ESI-MS and IRMPD spectroscopy

Introduction:
Catalytic processes are highly important for industrial activities. Nevertheless, some mechanisms and active catalytic species are not yet clearly comprehended, which is the case for Sharpless epoxidation, the first widespread reaction to offer stereochemical control and independence from chiral substrates. Although there have been some studies aiming this catalytic system, doubts still persist.

Methods:
To evaluate the actual mechanism and the role of conformational equilibrium in enantiomeric selectivity, mixtures of Ti(IV) alkoxides and different alkyl tartrates are analyzed through ESI-MS/MS. Relevant complexes are isolated and have their IRMPD spectra acquired from 600 to 1800 cm-1 at CLIO facility. Experimental results are supported by calculations at the B3LYP/6-31+G(d,p) level of theory.

Results:
These systems show a complex equilibrium between several structures, including monomers, dimers, trimers, with varying number of tartrates. There is also a fast exchange of alkoxide groups between the complexes and the alcohol solution. The active catalyst, previously depicted as a dimer ([Ti2(tartrate)2(OR)4]), was successfully observed in the carboxylate form. CID and IRMPD fragmentation show a loss of Ti(OR)4 that could be an evidence for an asymmetric species, with both tartrates coordinated to the same metallic atom, which seems to be corroborated by comparison between vibrational spectra and theoretical calculations. Furthermore, nucleophilic substitution in the tartrates are also observed during fragmentation, probably mediated by alkoxide attacks to the carbonyl groups, which could explain the low intensity of the C=O stretch band in the dimer's IRMPD spectra observed for DMT.

Novel Aspect:
Active catalytic structures may be different from previous proposals. The techniques used provide relevant new information about the reaction system.

C. Correra, Thiago,
IQ-USP - Brazil

Contributing Authors:
Laís Castrignano Tavares, Universidade de São Paulo - IQ-USP - Brazil.

Title:
Methyl-methoxy metathesis reactions of Si and Ge cations in the gas-phase

Introduction:
Organosilicon and organogermanium compounds are widely used in industrial processes and it is important to study their reactivity and proprieties, specially under PVD and CVD conditions. For this reason, the reactivity of (CH3)4X (X =Si, Ge) and (OCH3)4Si species and their respective trivalent cations were studied and a methyl-methoxy exchange reaction was observed.

Methods:
The target reactions between alkyl and alkoxy groups in Si and Ge cations were probed in the gas-phase by following the ion-molecule reaction kinetics in a 1 T FT-ICR coupled with a tunable CO2 laser used for IRMPD experiments. Experimental observations were supported by theoretical calculations of the potential energy surface of the reaction carried out at M06-2X/6-31G(2df,p) level of theory.

Results:
A series of organosilicon and organogermanium ions and neutrals were studied. Two of them reacted by the alkyl/alkoxy exchange pathway. By reacting Me4Si and (OMe)3Si+, a methyl migration from the neutral species to the cation was observed. When Me3Ge+ and (OMe)4Si were allowed to react, a methyl-methoxy exchange took place. For all systems, the main reaction is the adduct formation. However, none of the adducts formed took any role in the metathesis reaction pathway since adduct fragmentations induced by the CO2 laser led only to formation of the reactant species. These results suggest that the observed reactions occur in a concerted step or a multiple-step mechanism that takes place avoiding the adduct formation. Preliminary theoretical calculation results support this hypothesis and show that only a distinguished pathways can lead to the formation of the observed products.

Novel Aspect:
This novel alkyl/alkoxy exchange reactions have potential use in the gas-phase synthesis of functionalized organosilicon and organogermanium species.

Cahill, John,
Oak Ridge National Laboratory

Contributing Authors:
Vilmos Kertesz, Oak Ridge National Laboratory; Gary J. Van Berkel, Oak Ridge National Laboratory.

Title:
On-line, absolute quantitation of analytes from 20-40 µm microdissections by laser microdissection-liquid vortex capture-mass spectrometry

Introduction:
In most all mass spectrometry imaging (MSI) techniques quantitative surface sampling is difficult, requiring the use of arduous calibration protocols to mitigate matrix effects. In conventional ESI-MS/MS practice analyte/internal standard ratios are used to quantify. The ability to easily quantitate MSI using conventional internal standard approaches would be of great interest to the community.

Methods:
A recently developed hybrid laser capture microdissection-liquid vortex capture/ESI-MS system (LMD-LVC/ESI-MS) was used for surface sampling of dosed-polystyrene thin films and tissue sections. A Leica LMD7000 with an AB Sciex 5500 or 5600+ mass spectrometer was used for all experiments. Deuterated versions of the targeted analytes were used as internal standards for quantitation.

Results:
In the laser 'cut and drop' sampling mode whole sample microdissections are cut and dropped directly into the LVC probe and are dissolved in its entirety into solvent during transport to ESI-MS for analysis. This sampling mode ensures 100% capture efficiency and with the addition of an internal standard to the solvent of the system and an external calibration curve, absolute quantitation is possible. Measured concentrations of analytes in doped polymer films linearly increased with increased dosing concentration over several orders of magnitude. Quantitative propranolol concentrations measured in tissue from 20 and 40 µm microdissections closely matched (within 1σ) with quantitative 2.3 mm tissue punch-HPLC/ESI-MS measurements. This method did not require matrix-matched protocols or require any additional sample preparation, making it a relatively easy method for absolute quantitation.

Novel Aspect:
On-line laser capture microdissection-liquid vortex capture-ESI-MS enables easy and absolute quantitation of analytes with high spatial resolution

Campbell, J. Larry,
SCIEX

Contributing Authors:
J. C. Yves Le Blanc, SCIEX; Chang Liu, SCIEX.

Title:
Using Differential Mobility Spectrometry to Study Chiral Compounds

Introduction:
Separating and identifying chiral compounds is very challenging using MS. Often, derivatization and orthogonal separation methods (gas or liquid chromatography) are required. Several chiral separation techniques have been promoted using ion mobility (IMS, FAIMS, DMS), to varying degrees of success. Here, we discuss our recent efforts using differential mobility spectrometry (DMS) in this area.

Methods:
Experiments were conducted using a hybrid quadruople linear ion trap equipped with a DMS cell between the orifice and ESI source. Chemical modifiers could be added to the curtain gas at 1.5 % (v/v). In some cases, chiral analytes were derivatized by reactions with various acyl chlorides.

Results:
The racemic mixtures studied here (amphetamine, methamphetamine, warfarin) showed no separation using conventional DMS conditions (SV <= 4000 V; chiral chemical modifier – (S)-2-butanol). The lack of separation when using the chiral modifier can be explained by the fact that the hydroxyl group of the butanol will be involved in an intermolecular hydrogen bond with the protonated analytes at the site of protonation. For chiral separation to be enhanced by chiral solvent, the site of interaction must be at the chiral position. However, when SV=4000V and resolving gas was used to extend the residence time of the ions in the DMS cell, separation of the diastereomers ephedrine and pseudoephedrine were realized. Based upon this knowledge, we examined the separation of various enantiomeric and diastereomeric isomers using DMS, including transforming enantiomers to diastereomers.

Novel Aspect:
Rapid and sensitive method for separating and quantifying chiral compounds using differential mobility spectrometry

Campbell, J. Larry,
SCIEX

Contributing Authors:
J. C. Yves Le Blanc, SCIEX; Cathy Lane, SCIEX.

Title:
Analysis of Derivatized Glycans using Differential Mobility Spectrometry

Introduction:
Several factors impact ion behavior in differential mobility spectrometry analyses, including steric shielding or electronic delocalization of charge and intramolecular hydrogen bonds. Here, we describe the derivatization of glycans with isomeric tags to introduce these aforementioned factors to separate populations of analytes with implications to structure interrogation and multiplexed analysis.

Methods:
Experiments were conducted using a hybrid quadruople linear ion trap equipped with a DMS cell between the orifice and ESI source. Chemical modifiers could be added to the curtain gas at 1.5 % (v/v). Glycan analytes were subjected to derivatization using standard reductive amination protocols using isomeric aminopyridine, aminobenzoic acids, or aminobenzamides as reagents.

Results:
Our initial test set of derivatizing reagents was the isomeric set of 2-, 3-, and 4-aminopyridine (2-, 3-, and 4-AP). Our hypothesis was that glycans modified by different AP isomers would behave differently in the DMS cell. The closer the ring nitrogen to the glycan structure (i.e., hydroxyl groups that are potential sites of intramolecular hydrogen bonds), the weaker the interactions will be with chemical modifiers (e.g., IPA). Indeed, the 2-AP glycans were shifted to more positive CVs than the 3-AP glycans, revealing the greater clustering of the 3-AP glycans with IPA (i.e., more exposed charge site). In turn, the 4-AP glycans were shifted to more negative CVs than the 3-AP glycans, for the same theorized reason. Further experiments will discover if similar trends hold for other aminated derivatives of glycan analytes.

Novel Aspect:
Using derivatization chemistry to add DMS-active functional groups to analyte ions, increasing multiplexing capabilities

Canez, Carlos,
Carleton University

Contributing Authors:

Samuel W. J. Shields, Department of Chemistry, Carleton University, Ottawa, Ontario, Canada; Hillary P. Weinert, Department of Chemistry, Carleton University, Ottawa, Ontario, Canada; Jeffrey M. Manthorpe Department of Chemistry, Carleton University, Ottawa, Ontario, Canada; William G. Willmore, Institute of Biochemistry, Carleton University, Ottawa, Ontario, Canada; Magdalena Bugno, Department of Biology, Carleton University, Ottawa, Ontario, Canada; Jeffrey C. Smith, Department of Chemistry, Carleton University, Ottawa, Ontario, Canada; Karl V. Wasslen, Department of Chemistry, Carleton University, Ottawa, Ontario, Canada.

Title:

TrEnDi using isotopically-labelled diazomethane to increase sensitivity and selectivity of PE, PC and PS derived from complex biological samples

Introduction:

TrEnDi using isotopically-labelled diazomethane results in the complete methylation of phosphate moieties, carboxylic acids and primary amines. Significant sensitivity enhancements in tandem mass spectrometry-based analysis of phosphatidylethanolamine (PE), phosphatidylserine (PS) and phosphatidylcholine (PC) species from whole cell lipidome extracts were achieved after modification.

Methods:

Lipid extracts from HeLa cells were mixed with 1 uL of tetrafluoroboric acid and 300 uL of isotopically-labeled diazomethane for 5 minutes before solvent removal via evaporation. Derivatized lipids were resuspended in ethanol and separated using reversed phase HPLC and analyzed using specific precursor ion and neutral loss scanning modes on a triple quadrupole mass spectrometer.

Results:

Phosphate moieties, carboxylic acids and primary amines were methylated, providing a net fixed, permanent positive charge. Sensitivity gains for PE and PS lipids were achieved through the consolidation of ion fragmentation into one or two channels. The use of isotopically-labelled diazomethane enabled the distinction of modified PE and modified PC species by providing distinctive and exclusive precursor ion scans for both classes, which cannot be achieved using unlabelled diazomethane. Modest sensitivity gains for PC were observed. Significant sensitivity enhancements were observed for PE and PS. A total of 8 PE species and 11 PS species became observable following isotopically-labelled TrEnDi, which otherwise were below the limit of detection prior to modification. An additional 24 PE species and 6 PC species surpassed the limit of quantitation after derivatization.

Novel Aspect:

This is an important modification to recently reported chemistry that derivatizes phospholipids and permits unprecedented levels of sensitivity in MS.

Cao, Jingpei,
China University of Mining & Technology

Contributing Authors:
Xin Huang, China University of Mining & Technology; Xianyong Wei, China University of Mining & Technology; Jingxian Wang China University of Mining & Technology; Tianlong Liu, China University of Mining & Technology; Xiaoyan Zhao, China University of Mining & Technology; Xing Fan, China University of Mining & Technology.

Title:
GC/MS analysis of bio-oils obtained from fast pyrolysis of sawdust under different temperatures and atmospheres

Introduction:
It is more important and necessary to make full understanding of molecular composition of the bio-oil in order to upgrade the bio-oil as fuels and also as chemical feedstocks. The most widely used analytical method for pyrolysis bio-oil is Gas Chromatography/Mass Spectrometry (GC/MS) which can show different complex compounds present in bio-oil.

Methods:
Fast pyrolysis was performed at 400-600 oC in a drop tube quartz reactor heated by electrical furnace under Ar or H2 atmosphere. For each experiment, the reactor was heated up to the selected temperature and then 2 g sawdust was fed to the frit at a rate of approximate 0.1 g/min. The condensable vapor was collected in a tar trap series with methanol cooled with a chiller at −10 oC.

Results:
More than 100 chemical species were positively identified with GC/MS in the liquids and most of them are oxygen compounds. The content of phenols significantly increased, while the ethers content sharply decreased with raising pyrolysis temperature under both Ar and H2 atmospheres. The ketones content decreased slightly when raising the temperature from 400 to 600 oC, while the content of esters had a little change. Much higher relative contents of phenols and aldehydes in bio-oils under H2 atmosphere compared with Ar atmosphere at the range of reaction temperatures. The sugars content increased slightly when raising the temperature from 400 to 500 oC and then decreased when raising the temperature to700 oC under H2 atmosphere. In this case, levoglucosan concentration in bio-oils under Ar atmosphere was much higher than under H2 atmosphere.

Novel Aspect:
H2 was found to be helpful for decomposition of ethers and sugars, while Ar atmosphere showed good effect on removal of esters.

Cardenas, Johnny,
SCIEX

Contributing Authors:

Johnny Cardenas[1], Lei Xiong[2], Remco Van Soest[2] [1]SCIEX, Framingham, MA, USA, [2]SCIEX, Redwood City, CA, USA.

Title:

Improving the Sensitivity for an Immunocapture LC-MS Assay of Infliximab in Rat Plasma Using Trap-and-Elute MicroLC-MS

Introduction:

Quantitation of monoclonal antibodies (mAbs) in biological fluids is important during all stages of antibody drug development. While traditionally immunoassays are used, more recently LC-MS has been adopted because of its high selectivity, accuracy, and precision. We have studied the increase in sensitivity that can be made to an immunocapture assay for infliximab in rat plasma by using microLC-MS

Methods:

A magnetic bead immunoaffinity capture method was used to extract infliximab from rat plasma. The extracted antibody is digested using trypsin, and analyzed using a SCIEX QTRAP® 6500 LC-MS/MS system. MultiQuant 3.0.2 software was used for data analysis. For the High Flow LC-MS experiments we used a Prominence HPLC (Shimadzu), while for the MicroLC experiments a M3 MicroLC-TE system (SCIEX) was use

Results:

MRM transitions and parameters were optimized by injection/infusion of neat digestion protein standards. Chromatographic resolution was similar between the direct inject High Flow LC-MS method, and the trap-elute microLC-MS method, and the same throughput was realized for both methods. An improvement in S/N between High Flow LC-MS and MicroLC-MS with a factor of 6 was seen. Sensitivity improvement was studied by determining the LLOQ from calibration curves measured with both the direct inject High Flow LC-MS method and Trap-Elute MicroLC-MS. 5 Replicates of each concentration were injected. Using neat digested infliximab samples, the LLOQ was improved with a factor 5, while for plasma extracted samples the improvement was 4x. Robustness of the method was studied by injecting 1,000 samples with the trap and elute MicroLC-MS method. No clogging or significant deterioration in performance

Novel Aspect:

Improved sensitivity by using MicroLC-MS for an LC-MS immunocapture antibody assay in plasma for sample limited applications.

Cariou, Ronan,
LABERCA-Oniris, France

Contributing Authors:
Bruno Le Bizec, LUNAM Université, ONIRIS, Laboratoire d'Etude des Résidus et Contaminants dans les Aliments (LABERCA), Nantes, F-44307, France; Alexis Léon, LUNAM Université, ONIRIS, Laboratoire d'Etude des Résidus et Contaminants dans les Aliments (LABERCA), Nantes, F-44307, France; Gaud Dervilly-Pinel LUNAM Université, ONIRIS, Laboratoire d'Etude des Résidus et Contaminants dans les Aliments (LABERCA), Nantes, F-44307, France; Elsa Omer, LUNAM Université, ONIRIS, Laboratoire d'Etude des Résidus et Contaminants dans les Aliments (LABERCA), Nantes, F-44307, France.

Title:
Fishing halogenated environmental contaminants in biota based on isotopic pattern and mass defect provided by high resolution MS profiling

Introduction:
In the present work, we addressed the question of global seeking/screening organohalogenated compounds in a large panel of complex biological matrices, with a particular focus on unknown chemicals that may be considered as potential emerging hazards. A fishing strategy was developed based on untargeted profiling among full scan acquisition datasets provided by high resolution mass spectrometry.

Methods:
For filtering useful information, we developed a workflow involving a Visual Basic for Applications script aiming at pairing features according to the exact mass differences between Cl and Br isotopes, in order to point out potential organohalogenated clusters, preceded by an automated peak picking step based on the centWave function (xcms package of open access R programming environment).

Results:
The filtering script was successfully applied to a dataset generated from eel muscle upon liquid chromatography coupled to ESI-HRMS measurement from eel muscle extract, allowing for realistic manual investigations of filtered clusters. H/Cl-scale mass defect plots were used to visualize the datasets before and after filtering. Starting from 9,789 initial obtained features, 1,994 features were paired in 589 clusters containing at least an M+2 contribution. Hexabromocyclododecane, chlorinated paraffin series and various other compounds have been identified or hypothesized. Thus, the proposed methodology proved to be powerful in filtering a broad range of polyhalogenated organic compounds from HRMS datasets for fishing a range of contaminants in this fish extract at environmentally relevant levels, with a profitably limited sample preparation.

Novel Aspect:
The present work is a first step toward an automated and open access workflow dedicated to broad screening emerging halogenated contaminants in biota

Cariou, Ronan,
LABERCA-Oniris, France

Contributing Authors:

Ronan Cariou, LABERCA-Oniris; Gaud Dervilly-Pinel, LABERCA-Oniris; Farouk Jaber Lebanese University; Inas Abdel Malak, LABERCA-Oniris and Lebanese University; Bruno Le Bizec, LABERCA-Oniris.

Title:

Identification of Dechloranes by gas chromatography coupled to tandem mass spectrometry with atmospheric pressure chemical ionization (GC-APCI-MS/MS)

Introduction:

Dechloranes are a group of polychlorinated flame retardants raising concern, since they were detected in various environmental and biota matrices. As regard to analytical aspects, most authors use GC-MS couplings with EI or NCI. In the present work, we investigated the suitability of Atmospheric Pressure Chemical Ionization (APCI) as ionization technique for their identification.

Methods:

Native anti- and syn-Dechlorane Plus (DP), Dechlorane Dec-601, -602, -603, -604, -604CB, Chlordene Plus were from Wellington Laboratories.13C10-labelled anti- and syn-DP and Dec-602 were from Cambridge Isotope Laboratories. A 7880 series gas chromatography (Agilent) was coupled with to a triple quadruple mass analyzer fitted with an Atmospheric Pressure Chemical Ionization (Waters).

Results:

We first investigated the fragmentation patterns Dechloranes by APCI. Limited in-source fragmentation has been observed, [M]+• being a major peak for all compounds, with quite similar mass spectra in protic (H2O/MeOH 1:1, v/v) and dry conditions. Dechlorane 602 also showed major [M+H]+• ion, as well as [M+MeOH]+ adduct in protic conditions. Other dominant ions were, for example, [M-Cl]+ and [C5Cl6]+. Thus, highly specific precursor ions were selected and transitions were optimized (selection of fragment, collision energy) on a triple quadruple mass spectrometer. Finally a highly specific Selected Reaction Monitoring (SRM) method was defined. As a perspective, after extraction/purification developments, this method will be evaluated and applied to food and Human samples in order to produce original prevalence data for the French population in a risk assessment context.

Novel Aspect:

The present work is the first to report fragmentation pattern of dechloranes obtained with GC-APCI-MS. Intense molecular ions were observed.

Carlson, Erin,
University of Minnesota

Title:
Activity-Based Probes For Selective Penicillin-Binding Protein Visualization

Introduction:
Most bacteria are surrounded by a cell wall predominantly composed of a polymeric structure called peptidoglycan (PG) that is essential for cell survival. Consequently, the biosynthetic pathway for production of PG and the proteins required for its assembly are common targets for antibiotic therapy. Nonetheless, the structure and assembly of PG are only partially understood.

Methods:
Production of PG requires a suite of penicillin-binding proteins (PBPs), the individual roles of which are difficult to determine because each enzyme is often dispensable for growth, possibly due to functional redundancy. To address this challenge, we generated tools to enable selective examination of a single homolog or subsets of the PBPs.

Results:
We evaluated the PBP selectivity of a library of β lactam antibiotics in live cells using an unencapsulated derivative of D39 Streptococcus pneumoniae and Escherichia coli. The most selective compounds are being pursued for generating PBP-specific probes. Furthermore, we designed and synthesized a fluorescent cephalosporin C derivative (β lactam) to visualize PBP activity in live Bacillus subtilis and S. pneumoniae. Super-resolution fluorescence microscopy revealed that even PBPs that are located at a particular site (e.g., division septum) are not all intermixed. Rather, PBP subpopulations are discretely localized in bacterial cells.

Novel Aspect:
These new fluorescent probes have become important tools for correlating the activity and physiology of individual PBPs in their native environment.

Carr, Patrick,
University of Waterloo

Contributing Authors:
C. Zhan, University of Waterloo; R. A. Marta, University of Waterloo; M. Burt University of Waterloo; T. B. McMahon, University of Waterloo; E. Fillion, University of Waterloo; W. S. Hopkins, University of Waterloo; M. J. Lecours, University of Waterloo.

Title:
Exploring Mode-Selective Ligand Scrambling of Palladium Catalysts

Introduction:
An oxidative addition within a palladium catalytic cycle generates a square planar complex containing an organic residue intended for a new carbon-carbon bond. This complex is a "weak point" in the otherwise robust catalytic cycle - ligand scrambling can occur, which introduces side reactions removing the catalyst from the cycle.

Methods:
Mass isolated variants of a palladium catalyst were interrogated with a free electron laser (FEL) at the Centre Laser Infrarouge d'Orsay (CLIO) facility in Orsay, France. Monitoring catalyst fragmentation as a function of FEL wavelength yielded vibrational spectra for the catalysts. The computational study used a basin-hopping algorithm followed by high-level DFT optimizations.

Results:
This approach identified agostic interactions that stabilize catalyst geometries, which were confirmed via IRMPD. Excitation of all vibrational modes resulted in loss of an acetonitrile tag while excitation of the phosphine ligands accessed ligand scrambling pathways. This mode-selective behaviour seems to arise from non-statistical intramolecular vibrational energy redistribution and was not observed when arsenic was substituted in place of phosphorus. Following the first experiments, a time resolved electrospray ionization (TRESI) source was introduced. This enables isolation and IRMPD probing of the catalytic cycles' transient intermediates which can elucidate the catalytic mechanism.

Novel Aspect:
Mode selective ligand scrambling in palladium catalysts and a transient reaction intermediate observed by use of a TRESI source and probed by IRMPD.

Carroll, Frances,
Restek Corporation

Title:
HILIC, Polar, and Shape Selectivity of a FluoroPhenyl Phase

Introduction:
The FluoroPhenyl stationary phase has long been marketed as a phase that offers alternative, or orthogonal, selectivity to a C18. In addition to the traditional reversed-phase interactions, this phase also exhibits shape selective, polar, cation-exchange and even HILIC retention mechanisms which aid in selectivity of specific analytes.

Methods:
We aim to demonstrate the useful and alternate retention of the FluoroPhenyl stationary phase. We chose several relevant target analytes which we plan to use to exemplify the unique retention characteristics of the FluoroPhenyl phase when used in either HILIC or reversed-phase mode. All of these analytes have been pursued due to either poor retention, poor resolution, or both on a C18.

Results:
Shape selectivity properties of the FluoroPhenyl stationary phase allow for the separation of epimers not easily accomplished on a C18. The FluoroPhenyl phase allows retention based on hydrophobicity as well as the cation-exchange mechanism. This results in a mixed-mode column which supports reversed-phase and HILIC analyses. By utilizing the cation-exchange mechanism and high organic mobile phases, increased retention and sensitivity may be seen for compounds such as 4-MEI.

Novel Aspect:
Alternate retention of the FluoroPhenyl stationary phase

Carroll, Frances,
Restek Corporation

Title:
Achieving Fast, Efficient Separations in environmental applications using Superficially Porous Particle Column Technology

Introduction:
Superficially porous particles (SPP) have been proven to provide fast and efficient separations. These column allow you to more easily achieve peak separation and faster analysis times without expensive UHPLC instrumentation. The versatility of these columns makes them ideal for separations for a variety of environmental applications.

Methods:
The performance of SPP particle columns will be demonstrated on a variety of applications, including bisphenol A, pesticides, and perfluorinated alkyl acids using simple mobile phases and gradient methods.

Results:
In order to analyze large lists of compounds, especially across multiple classes, your column must be capable of spreading analytes out over the gradient to ensure accurate detector response and quantitation. In designing the Raptor™ ARC-18 column, we adjusted our bonding procedures to form an ideal ligand density that offers balanced retention for the rapid analysis of large, multiclass assays. With its balanced retention profile, the Raptor™ ARC-18 column is ideally suited to analyze a 204-compound pesticide screen in just 9.5 minutes. Restek LC columns offer the speed of superficially porous particles with the resolution of the highly selective stationary phases, allowing peak separation and faster analysis times to be achieved without expensive UHPLC instrumentation.

Novel Aspect:
Speed of superficially porous particles with the resolution of the highly selective stationary phases.

Caruso, Donatella,
Università degli Studi di Milano

Contributing Authors:
Gaia Cermenati, DiSFeB, Dipartimento di Scienze Farmacologiche e Biomolecolari, Università degli Studi di Milano, Milano, Italy.; Matteo Audano, DiSFeB, Dipartimento di Scienze Farmacologiche e Biomolecolari, Università degli Studi di Milano, Milano, Italy.; Silvia Giatti DiSFeB, Dipartimento di Scienze Farmacologiche e Biomolecolari, Università degli Studi di Milano, Milano, Italy.; Maurizio D'Antonio, Division of Genetics and Cell Biology, IRCCS San Raffaele Scientific Institute, Milano, Italy.; Emma De Fabiani, DiSFeB, Dipartimento di Scienze Farmacologiche e Biomolecolari, Università degli Studi di Milano, Milano, Italy.; Maurizio Crestani, DiSFeB, Dipartimento di Scienze Farmacologiche e Biomolecolari, Università degli Studi di Milano, Milano, Italy.; Enrique Saez, Department of Chemical Physiology, The Scripps Research Institute, La Jolla, USA.; Guido Cavaletti, Experimental Neurology Unit, Department of Surgery and Translational Medicine, Università degli Studi Milano-Bicocca, Monza, Italy; Iñigo Azcoitia, Departamento de Biología Celular, Facultad de Biología, Universidad Complutense de Madrid, Madrid, Spain; Luis-Miguel Garcia-Segura, Instituto Cajal, C.S.I.C., Madrid, Spain; Roberto C. Melcangi, DiSFeB, Dipartimento di Scienze Farmacologiche e Biomolecolari, Università degli Studi di Milano, Milano, Italy.; Nico Mitro, DiSFeB, Dipartimento di Scienze Farmacologiche e Biomolecolari, Università degli Studi di Milano, Milano, Italy.

Title:
Peripheral nerve structure and function: role of endogenous fatty acid synthesis

Introduction:
Changes in myelin fatty acid (FA) composition have been associated with peripheral neuropathy (PN), but the importance of nerve FA synthesis in maintaining myelin functionality is poorly understood. In our study, we explored the extent to which lack of the key regulator of FA synthesis, Sterol Regulatory Element Binding Factor-1c (Srebf1c), could result in the development of PN.

Methods:
The degree of peripheral neuropathy in Srebf1c null mice was investigated by behavioral tests. Morphological and morphometric analyses along with a transcriptomic approach were used to address the mechanism responsible for the development of PN. In addition, metabolomics analysis was performed by means of liquid Chromatography-tandem mass spectrometry (LC-MS/MS).

Results:
We demonstrated that the neuropathic phenotype observed in Srebf1c KO mice was due to Remak bundle alterations, hyper-myelinated small-caliber fibers and changes in myelin periodicity. Moreover, these animals displayed, in the peripheral nerves, decreased FA synthesis and glycolytic flux, but increased FA catabolism and mitochondrial function. The mechanism responsible for enhanced FA catabolism in Srebf1c knockout mice restricted to peripheral nerves is due to the accumulation of PC-C16:0/C18:1 and PC-C18:0/C18:1 specifically in the sciatic nerve detected by LC-MS/MS. Increased levels of these two endogenous Pparα ligands activate this transcription factor in vivo primarily in Schwann cells, resulting in increased FA catabolism and in the development of peripheral neuropathy. As evidence of this, treatment of Srebf1c null mice with a Pparα antagonist rescues their neuropathic phenotype.

Novel Aspect:
These data provide new insight into the contribution of endogenous FA synthesis to nerve homeostasis and function.

Castellanos, Anthony,
Florida International University

Contributing Authors:
Iru Paudel, Florida International University; Jeremy Chambers, Florida International University; Quentin Vanbellingen Florida International University; Monica Rodriguez Silva, Florida International University; Francisco Fernandez-Lima, Florida International University.

Title:
Drug delivery quantitation in single cells with sub micrometric spatial resolution TOF-SIMS

Introduction:
Time-of-Flight Secondary Ion Mass Spectrometry (TOF-SIMS) is the gold standard emerging imaging tool for biomedical studies with sub-micrometer resolution. Recent advances in the technology have enabled the acquisition of a larger amount of relevant chemical information of low concentration chemical species from small regions of interest with high spatial resolution.

Methods:
Cell culture samples were grown on Indium tin oxide coated conductive glass slides. Samples were treated with a gradient of drug concentrations. Washing and freeze-drying for TOF-SIMS analysis was carried out. High mass and high spatial resolution analysis were performed on single cells. Drug standards were analyzed in order to determine characteristic secondary ion species for semi-quantitation.

Results:
Characteristic secondary ions of the anti-tumor agents used to treat cell samples were identified using high mass resolution analysis. Signal from such ions were found to scale with treatment concentration. High spatial resolution analyses revealed sub-micrometer morphological features and allowed for the localization of drug molecular and fragment ions. Preliminary data suggests that the TOF-SIMS technique can offer valuable information regarding the specificity of drugs towards intracellular targets. In addition, it may be possible to semi-quantify introduced chemical species based on signal intensity of characteristic ions.

Novel Aspect:
For the first time, TOF-SIMS imaging is used for both localization as well as semi-quantitation to asses drug delivery on the single cell level.

Cautereels, Julie,
University of Antwerp

Contributing Authors:
Frank Blockhuys, University of Antwerp.

Title:
A Quantum Chemical Tool for the Prediction of Mass Spectra and the Identification of Fragmentation Pathways: Unexpected Fragmentations in Tripeptides

Introduction:
The existing methods for the prediction of the mass spectra of peptides can predict them only partially because they are, for the most part, based on empirical rules. Furthermore, the bulk of these methods only takes fragmentation of the backbone into account, while it is known that a large number of the non-predicted peaks are due to fragmentations involving the side chains.

Methods:
QCMS2 (Quantum Chemical Mass Spectrometry for Materials Science) is based on DFT/B3LYP/6-311+G* calculations and selects fragmentation pathways based on the relative values of bond orders, and reaction (for bond cleavages) and transition-state energies (for rearrangements). For work on peptides the code omits the breaking of irrelevant bonds and incorporates the Mobile Proton Model (MPM).

Results:
QCMS2 was developed using small organics with different functional groups: the main features in the mass spectra were correctly reproduced. In addition, a number of new fragmentation routes and mechanisms proposed by the calculations were experimentally observed and confirmed using MS/MS experiments. QCMS2 is currently being applied to predict the fragmentation patterns of tripeptides in ESI CID MS and focuses on the fragmentation of the side chain and the influence of hydrogen bonding in these side chains on the fragmentation. Calculations on Ser-His-X, Asn-His-X and Asp-His-X indicate that besides the traditional a-, b-, x- and y-ions, fragmentation of the side chains also occurs: at least one of the peripheral amino acids fragments but the central His never does. In addition, the inter-side-chain interactions allow small molecules to be split off more easily from the side chain.

Novel Aspect:
A number of the non-predicted peaks are assigned to fragmentations involving the side chains and the influence of inter-side-chain interactions.

Cha, Byungchul,
Thermo Fisher Scientific

Contributing Authors:
John Brann, Thermo Fisher Scientific; Terry Olney, Thermo Fisher Scientific; Pengxiang Yang Thermo Fisher Scientific; Christopher Elicone, Thermo Fisher Scientific.

Title:
Long Term Test on a New Four-Channel HPLC

Introduction:
We present a workflow on testing a new multi-channel LC-MS system. Total of ~125,000 samples were injected to test three systems with the same protocol for 9 months. Many data sets from LCMS were collected and analyzed statistically to improve the system performance and stability. By using a four-channel HPLC, we were able to trouble-shoot the LCMS with high efficiency.

Methods:
A test sample of AWLI (Atenolol, Warfarin, Lidocaine, Imipramine in 2 ng/mL) were prepared in 70/30 (v/v) MeOH/H2O. An HPLC with PFP columns was connected to a triple-quadrupole mass spectrometer. A single test consists of 8 batches of 48 injections (2 blanks and 10 AWLI injections for each channel) within 11.5 hours.

Results:
The new multi-channel HPLC reduces solvent consumption by 60% or better using positive-displacement pumps. It maximized MS instrument time by connecting up to four HPLC channels to a single mass spectrometer. High efficiency is the results, with 40 - 60 samples per hour and 960 - 1440 samples per day. The set of 40,000 injections from a single LC/MS system over 9 months, less than 2% variation on RT within and cross channels was observed during which columns were replaced four times. Since HPLC showed no significant cross talk, results led us quickly to the root cause of bad data. LC issues are tied to specific channel(s) while MS issues show on every channel. The four-channel HPLC improved the working efficiency dramatically to test a whole LC/MS system with respect to four times faster on collecting data than regular HPLC and more straightforward on trouble-shooting.

Novel Aspect:
A large LCMS data set (125,000 injections) for statistic studies of multi-channel HPLC-MS. Identification of unexpected workflow problems.

Cha, Min Ho,
Korea Institute of Oriental Medicine

Contributing Authors:
Jeeyoun Jung, Korean Institute of Oriental Medicine; Min Jung Kim, Korean Food Research Institute; Myung-Sunny Kim Korean Food Research Institute.

Title:
Metabolite profiling related with dysmenorrhea in Korean female subjects

Introduction:
Dysmenorrhea (menstrual pain) is one of menstrual disorders and many women are suffering from dysmenorrhea. In this study, we investigated the differences in the plasma metabolic profiles of patients with menstrual pain

Methods:
Seventy nine Korean female subjects were recruited from 7 traditional Korean medical hospitals. The pain degree of dysmenorrhea were checked five-scale by two expert and anthropometric and blood parameters were measured. Serum metabolic profiling was performed using ultrahigh-performance liquid chromatography-quadrupole time-of flight mass spectrometer. Statistical analysis was performed by SPSS

Results:
Average age and BMI of subjects were 37.21 years and 22.59 kg/m2, respectively. Among blood parameters, monocyte and D-dimer showed positive correlation with dysmenorrhea pain degree ($r=0.243$ and 0.294, $p=0.031$, 0.009, respectively). Controversially, level of BUN was negatively correlated with dysmenorrhea. Among 4,968 serum metabolites, 79 metabolites showed the relation with dysmenorrhea. Ethinylestradiol, a derivative of endogenous estrogen and management of menopausal symptoms, was significantly showed negative correlation with dysmenorrhea. Some of amino acid such as histidine, N-acetyl-L-histidine and L-cystathionine were also negatively correlated with dysmenorrhea.

Novel Aspect:
Our results demonstrated that several metabolites involved in hormone biosynthesis and amino acid metabolism were related with dysmenorrhea.

Chamot-Rooke, Julia,
Institut Pasteur, CNRS

Contributing Authors:
Martial Rey, Institut Pasteur, CNRS, Paris, France; Eric Durand, Institut de Microbiologie de la Méditerranée, CNRS, Marseille, France; Remi Fronzes Institut Pasteur, CNRS, Paris, France; Eric Cascales, Institut de Microbiologie de la Méditerranée, CNRS, Marseille, France.

Title:
XL-FASP: A New Integrated Cross-Linking Workflow to Study Extra Large Membrane Protein Complexes

Introduction:
Cross-Linking Mass Spectrometry (XL-MS) has proved useful to map structural details of functional protein complexes but the analysis of large membrane protein complexes remains very difficult. We propose here a new gel-free strategy based on a trifunctional cross-linker and Filter Aided Sample Preparation (XL-FASP), and apply it to the Type 6 secretion membrane core complex (T6SSJLM).

Methods:
Protein complexes (creatine kinase as a model and T6SSJLM) were cross-linked with NNP9 [1,2]. Click-chemistry was used to introduce a biotin in cross-linked proteins that were further trypsin digested on an Amicon filter (FASP) [3]. Peptides were eluted, cross-linked ones enriched with avidin beads and analyzed in LC-MS/MSMS on a Orbitrap Fusion. Data analysis was performed with Mass Spec Studio.

Results:
All steps of the workflow (cross-linking, click-chemistry, enrichment, MS analysis) were optimized on creatine kinase (CK) solubilized in detergent. The use of a filter, adapted from the FASP protocol [3], allowed us to achieve an excellent cross-linked peptide recovery. EThcD was chosen for fragmentation as it led to high sequence coverage and reporter ions for the cross-linked peptides. For CK, we have more than doubled the number of unique cross-links found, in comparison to all previous studies, showing the great sensitivity of our XL-FASP strategy. For the Type 6 secretion membrane core complex T6SSJLM (2 MDa), the large number of cross-linked peptides obtained could be used to improve the structural model proposed by electron microscopy. [1] Nury et al. Anal.Chem. 2015 [2] Durand et al. Nature 2015 [3] Erde et al. J. Proteome Res. 2014

Novel Aspect:
First gel-free integrated cross-linking FASP strategy allowing the analysis of extra-large functional membrane protein complexes.

Chao, Hsi-Chun,
National Taiwan University

Contributing Authors:
Sung-Chun Tang, Stroke Center and Department of Neurology, National Taiwan University Hospital, Taipei, Taiwan; Yufeng Jane Tseng, Graduate Institute of Biomedical Electronics and Bioinformatics, National Taiwan University, Taipei, Taiwan; Sin-Yu Yang Stroke Center and Department of Neurology, National Taiwan University Hospital, Taipei, Taiwan; Hsi-Chun Chao, School of Pharmacy, College of Medicine, National Taiwan University, Taipei, Taiwan; Guan-yuan Chen, School of Pharmacy, College of Medicine, National Taiwan University, Taipei, Taiwan; San-Yuan Wang, Graduate Institute of Biomedical Electronics and Bioinformatics, National Taiwan University, Taipei, Taiwan; Ching-Hua Kuo, School of Pharmacy, College of Medicine, National Taiwan University, Taipei, Taiwan; Hsiao-Wei Liao, School of Pharmacy, College of Medicine, National Taiwan University, Taipei, Taiwan.

Title:
Development of the concentration normalization method for cellular lipidomics studies by FIA-ESI-MS/MS

Introduction:
Lipidomics focuses on studying the metabolisms and functions of lipids. Cellular lipidomic studies have been widely used for investigating different biomarkers. Concentration normalization for comparing cell samples is essential to ensure the unbiased comparisons. We developed a new concentration normalization method based on the membrane phospholipid contents with high-throughput analysis.

Methods:
A flow injection analysis coupled with electrospray ionization tandem mass spectrometry (FIA-ESI-MS/MS) method is developed for measuring total phosphatidylcholine (PC) and sphingomyelin (SM) contents from the cell membrane through the precursor ion scan of ion 184 (PIS of 184). The cell extraction method was modified from Bligh and Dyer and different cell types were tested in this study.

Results:
Good linearity was observed between the lipid extract dilution factor and the reciprocal of the peak area of the total ion current in PIS of 184 within the dilution range of 1- to 16-fold (R^2= 0.998). The intra-day and intermediate precisions were below 10%. The accuracy ranged from 93.0% to 105.6%. The PIS of 184 method was evaluated using different numbers of HCT-116 cells and different lipid species were selected for test. The results showed that the peak areas of each species obtained from different cell numbers were within a 20% variation after normalization. Finally, the PIS of 184 method was applied to study ischemia-induced neuron injury using oxygen and glucose deprivation (OGD) neuron model. Changes in lipid profiles after OGD treatment could be observed which showed the applicability of the PIS of 184 method.

Novel Aspect:
A new concentration normalization method is developed for cellular lipidomic studies with more convenient, accurate, and precise than other methods.

Chaurand, Pierre,
Université de Montréal

Contributing Authors:
Martin Dufresne, Université de Montréal.

Title:
Metal-assisted LDI for high resolution imaging MS of neutral lipids from thin tissue sections.

Introduction:
Lipids play key roles in cellular membranes, signaling and energy storage. In particular, cholesterol, fatty acids and triglycerides play significant roles in the development of life-style associated diseases. We have explored the analysis of these compounds by high resolution laser desorption ionization imaging MS using metal layers directly deposited on tissue sections.

Methods:
Different tissue sections including brain, kidney, and liver were investigated. For the analysis of cholesterol and fatty acids, a ~20 nm-thick layer of sputtered silver is directly deposited on thin sections. For the analysis of triglycerides, a layer of sodium salt is first deposited followed by ~28 nm-thick layer of sputtered gold is used. LDI imaging MS is performed on a TOF MS system.

Results:
From the investigation of various tissue specimens, using silver-assisted LDI IMS we have thus far mapped and identified with certainty several compounds including, cholesterol, arachidonic acid and docosahexaenoic acid. While both fatty acids and cholesterol show great affinity for silver, other molecules such as triglycerides have strong affinities for other selected cations. For this purpose, the coupling of sodium salts and sputtered gold has been developed to enhance both desorption and ionization yielding a 30-fold increase of triglyceride signals compared to standard MALDI MS approaches. Although this method targets TAGs, we have also detected and imaged several other compounds including cholesterol esters, which are hard to detect by MALDI MS. These novel imaging MS strategies are being used to study atherosclerosis, Alzheimer's and non-alcoholic fatty liver diseases.

Novel Aspect:
Matrix-free metal assisted LDI methods are developed for the imaging MS of targeted lipid compounds with high specificity and high sensitivity.

Chen, Chao-Jung,
Graduate institute of integrated medicine, China Medical University

Contributing Authors:
Chung Y. Hsu, Graduate Institute of Clinical Medical Science, China Medical University, Taichung, Taiwan; Hsin-Yi Liao, Proteomics Core Laboratory, Department of Medical Research, China Medical University Hospital, Taichung 40402, Taiwan;; Chiz-Tzung Chang College of Medicine, China Medical University, Taichung, Taiwan; Ming-HsiuWu, Division of Neurology, Department of Internal Medicine, Chi Mei Medical Center, Tainan, Taiwan.

Title:
The discovery of HDL-associated protein biomarkers to predict stroke

Introduction:
High-density lipoprotein (HDL) is a lipid and protein complex that consists of apolipoproteins and lower level HDL-associated enzymes. HDL can protect against atherosclerosis via its cholesterol acceptor and effects in antioxidation, anti-inflammation, and antiapoptosis. ApoA1, a major lipoprotein on HDL, has antioxidant and anti-inflammation properties to prevent atherosclerotic vascular disease

Methods:
Plasma samples from healthy (n=50), diabetes (n=50), hyperlipidemia (n=50), uremia (n=50), chronic stroke and acute stroke subjects (n=100) were collected. HDL was purified with salt density ultracentrifugation and analyzed with MALDI-TOF for protein profiling. The modified ApoA1 was further separated with Bis-Tris gradient gel and identified with MALDI-TOF and nanoLC-MS/MS.

Results:
The expression levels of major lipoproteins on HDL were analyzed with intact protein profiling. The results showed that the expression level of major lipoproteins are similar among all disease groups (diabetes, uremia and hyperlipidemia). Compared to healthy, diabetes, and hyperlipidemia subjects, uremia and chronic/acute stroke subjects displayed more intense modified-ApoA1 bands. These modified Apoa1 bands are highly correlated with oxidation, glycation and carbamylaton

Novel Aspect:
Modified ApoA1 on gels can be used to evaluate HDL quality and predict risks of stroke.

Chen, Ping,
Dalian Institute of Chemical Physics, Chinese Academy of Sciences

Contributing Authors:
Jichun Jiang, Key Laboratory of Separation Science for Analytical Chemistry, Dalian Institute of Chemical Physics, Chinese Academy of Sciences, Dalian, 116023, People's Republic of China.; Haiyang Li, Key Laboratory of Separation Science for Analytical Chemistry, Dalian Institute of Chemical Physics, Chinese Academy of Sciences, Dalian, 116023, People's Republic of China.; Keyong Hou Key Laboratory of Separation Science for Analytical Chemistry, Dalian Institute of Chemical Physics, Chinese Academy of Sciences, Dalian, 116023, People's Republic of China.; Wei Liu, Key Laboratory of Separation Science for Analytical Chemistry, Dalian Institute of Chemical Physics, Chinese Academy of Sciences, Dalian, 116023, People's Republic of China.

Title:
Multi-reflection Time-of-Flight Mass Analyzer for Rare Earth Isotopic Geochemistry

Introduction:
Multi-reflection time-of-flight (MR-TOF) mass analyzer has achieved an ultrahigh mass resolution by extending flight path1-3. MR-TOF mass spectrometry, Combined with laser ablation or ion beam sputtering, could acquire accurate chemical and isotopic composition on micro-zone surface for rare earth isotopic geochemistry.

Methods:
Ion beam extraction from surface is introduced into reflection region, then repeatedly reflected between two sets of planar gridless mirrors arranged symmetrically. Five sets of periodic focusing lenses in free-field region are used to keep ion beam stable transmission. One pair of right-angled bend deflection lenses are used to achieve ion beam injection/ extraction.

Results:
We have already designed and constructed the MR-TOF-SIMS mass analyzer .The size is about 900×600×300 mm, and the vacuum kept at 1.2×10-5 Pa. A nanosecond laser is adopted as a sputtering source for copper surface. The 532 nm laser pulse energy is 30 μJ and 1 ns duration with repetition rate 1~25 KHz. With this system the total flight time of 63Cu+ can be several milliseconds and the total flight path over 600 m. As laps N grows, the mass resolution increase according. While ions flight laps N=1, the analyzer have full mass range and the resolution is ~2000. In experiment the optimum resolution is ~50,000 at laps N=48 and flight path over 200 m. The total ion transmission efficiency appears to decline with the laps increase until N>20, then remains stable at 25%. However, if N is larger enough, the transmission would continue to decrease because of scattering events with residual gas.

Novel Aspect:
To prevent influence of fringe field, one pair of right-angled bend deflection lenses was designed for ion beam injection/ extraction respectively.

Chen, Yi-Ting,
Chang Gung University

Title:
Integrated proteomic datasets for biomarker discovery and verification of bladder cancer

Introduction:
Bladder cancer is potentially lethal and is the most costly urological malignancy to manage. Base on the morbidity and mortality, diagnosis and treatment of the cancer are important issues for the better clinical practice.

Methods:
To discover potential biomarkers of bladder cancer, we employed a strategy combining isotopic labeling and LC-MS/MS analysis to profile proteomic changes in secretome, clinical urine, urine microparticles, and fresh-frozen bladder tumor specimens.

Results:
Seven differentially-expressed proteins have been selected as potential biomarker candidates for verification by immuno-assays or multiple-reaction-monitoring MS in additional clinical specimens. Overall, the urinary concentrations of the classical plasma proteins or acute phase proteins show the best AUC values for discrimination between age-matched control and bladder cancer patients. Proteomic analysis of urinary microparticles reveals strong association of TACSTD2 with bladder cancer. TAGLN2 shows the most significant overexpression in bladder cancer tissues and might be a useful molecular tumor marker for evaluating bladder cancer lymph node metastasis. Urinary TAGLN2 also represents a potential biomarker for non-invasive screening of bladder cancer.

Novel Aspect:
Integration of multiple clinical proteomes provide valuable information for protein origin, specificity and application for future validation studies.

Chen, Peter,
ETH Zürich

Title:
Threshold CID measurements probe organic and organometallic reaction mechanisms

Introduction:
Quantitative thermochemical measurements provide hard tests for chemical bonding models and proposed reaction mechanisms. Energy-resolved collision-induced dissociation experiments provide activation energies, and in many cases, bond strengths, for medium-to-large molecular ions that are not only models, but even the actual chemical species involved in catalysis.

Methods:
ESI-MS/MS, threshold CID

Results:
Examples are shown for non-covalent intermolecular interactions (dispersion), biological chemistry of radical SAM (S-adenosylmethionine) enzymes, and homogeneous catalysis (transmetalation via heterodinuclear metal complexes in the Sonogashira coupling), for which the quantitative data validate or invalidate chemical models proposed for each case.

Novel Aspect:
combination of synthesis, computation, and MS-based analytical methods

Chen, Peter,
ETH Zürich

Title:
Threshold CID measurements probe organic and organometallic reaction mechanisms

Introduction:
Quantitative thermochemical measurements provide hard tests for chemical bonding models and proposed reaction mechanisms. Energy-resolved collision-induced dissociation experiments provide activation energies, and in many cases, bond strengths, for medium-to-large molecular ions that are not only models, but even the actual chemical species involved in catalysis.

Methods:
ESI-MS/MS, threshold CID

Results:
Examples are shown for non-covalent intermolecular interactions (dispersion), biological chemistry of radical SAM (S-adenosylmethionine) enzymes, and homogeneous catalysis (transmetalation via heterodinuclear metal complexes in the Sonogashira coupling), for which the quantitative data validate or invalidate chemical models proposed for each case.

Novel Aspect:
combination of synthesis, computation, and MS-based analytical methods

Chendo, Christophe,
Aix-Marseille Université – CNRS, UMR 7273, Institut de Chimie Radicalaire, Marseille – France

Contributing Authors:

Laurence Charles, Aix-Marseille Université - CNRS, UMR 7273, Institut de Chimie Radicalaire, Marseille - France; Trang N. T. Phan, Aix-Marseille Université - CNRS, UMR 7273, Institut de Chimie Radicalaire, Marseille - France; Didier Gigmes Aix-Marseille Université - CNRS, UMR 7273, Institut de Chimie Radicalaire, Marseille - France; Dao Le, Aix-Marseille Université - CNRS, UMR 7273, Institut de Chimie Radicalaire, Marseille - France.

Title:

Multidimensional mass spectrometry to characterize by-products generated during MALDI of polystyrenes synthesized by controlled radical polymerization

Introduction:

While being popular for mass spectrometry of synthetic polymers, MALDI is not soft enough to form intact ions from macromolecules with fragile end-groups, as produced by controlled radical polymerization (CRP). Although reasonable, structures proposed for the variety of by-products formed in the MALDI source have never been validated via an in-depth characterization approach.

Methods:

Polystyrene (PS) prepared by three CRP methods were ionized in the MALDI source of a Synapt G2 HDMS mass spectrometer which allows orthogonal injection of ions into a reflectron TOF for improved resolution and accurate mass measurements, collision-induced dissociation (CID) experiments for individual end-group analysis, and ion mobility separation to CID for improved selection of precursor ions.

Results:

Although targeted PS with original end-groups were clearly evidenced by ESI-MS, no intact species could be observed in MALDI. Instead, various by-product distributions were detected depending on the polymerization technique. For each sample, high resolution mass measurements allowed elemental composition to be proposed for pairs of end-groups, then terminal moieties were individually characterized in MS/MS. Beside chains terminated with an endo- or exo-double bond, high resolution MS analysis revealed additional distributions that could be separated in the mobility cell before CID. These were not new by-products but the endo- or exo-terminated species with their alpha end-group having experienced H/Ag exchange. This study hence validated structural assumptions reported in the literature but also revealed that PS fragmentation behavior was strongly influenced by the nature of end-groups.

Novel Aspect:

Accurate characterization of end-groups in by-products generated upon MALDI of polystyrenes holding fragile terminations.

Chevallier, Olivier,
Queen's University Belfast

Contributing Authors:
Connor Black, Queen's University Belfast, Belfast, UK; Julia Balog, Waters Research Centre, Budapest, Hungary; Steven Pringle Waters, Wilmslow, UK; Sara Stead, Waters, Wilmslow, UK; Zoltan Takats, Imperial College London, London, UK; Christopher Elliott, Queen's University Belfast, Belfast, UK; Rachael Hill, Queen's University Belfast, Belfast, UK.

Title:
Real time detection of fish fraud using rapid evaporative ionisation mass spectrometry (REIMS)

Introduction:
The increasing number of reports regarding global food fraud scandals has brought food authenticity and safety to the attention of regulators, industry and consumers worldwide. In this study, we present for the first time an effective, near real time method to identify fish product speciation methods using rapid evaporative ionisation mass spectrometry (REIMS).

Methods:
Data was acquired using a Waters Medimass REIMS source coupled with a Waters Xevo G2-XS QTof mass spectrometer. All specimens were sampled using a monopolar handpiece that was equipped with a smoke evacuation line connected to the atmospheric interface of the mass spectrometer. Extracted data was subjected to multivariate analysis such as PCA followed by LDA.

Results:
Over 3000 spectra were acquired from five different authenticated species of fish; cod, coley, haddock, pollock and whiting. Spectral data were acquired between m/z 200-1200 resulting in the spectra being dominated by fatty acids in the 200-450 m/z region, glycerophospholipids in the 600-900 m/z region and triglycerides in the 900-1000 m/z region. Both PCA and LDA score plots, built using m/z 600-1200, identified clear signs of fish speciation. Leave 20% out cross validation of the PCA- LDA models resulted in a 95.90% correct classification rate. The multivariate model created from the fish database was then tested real-time on novel fish samples at two different sites, Belfast and Budapest. Authentic samples from all different species were sampled and subjected to real-time analysis during sampling using the same prototype software. All samples were correctly classified by our method.

Novel Aspect:
REIMS could provide a paradygm shift across authenticity applications by providing real-time and reliable method for the analysis of food products.

Chevolleau, Sylvie,
Axiom platform - UMR 1331 Toxalim INRA Toulouse - FRANCE

Contributing Authors:

M.H Meireles, Toxalim, Université de Toulouse, INRA, INP-ENVT, INP-EI-Purpan, Université de Toulouse 3 Paul Sabatier, F-31027 Toulouse, France; N. Naud, Toxalim, Université de Toulouse, INRA, INP-ENVT, INP-EI-Purpan, Université de Toulouse 3 Paul Sabatier, F-31027 Toulouse, France; S. Chevolleau Toxalim, Université de Toulouse, INRA, INP-ENVT, INP-EI-Purpan, Université de Toulouse 3 Paul Sabatier, F-31027 Toulouse, France; F. Pierre, Toxalim, Université de Toulouse, INRA, INP-ENVT, INP-EI-Purpan, Université de Toulouse 3 Paul Sabatier, F-31027 Toulouse, France; F. Guéraud, Toxalim, Université de Toulouse, INRA, INP-ENVT, INP-EI-Purpan, Université de Toulouse 3 Paul Sabatier, F-31027 Toulouse, France; L. Debrauwer, Toxalim, Université de Toulouse, INRA, INP-ENVT, INP-EI-Purpan, Université de Toulouse 3 Paul Sabatier, F-31027 Toulouse, France; I. Jouanin, Toxalim, Université de Toulouse, INRA, INP-ENVT, INP-EI-Purpan, Université de Toulouse 3 Paul Sabatier, F-31027 Toulouse, France.

Title:

Validation of a U-HPLC-ESI-MS/MS method for the quantification of reactive aldehydes produced by lipid peroxidation in the intestinal lumen.

Introduction:

Red or processed meat rich diets have been shown to be associated with an elevated risk of colorectal cancer (CRC). One major hypothesis involves dietary heme iron which induces lipid peroxidation. The quantification of the resulting reactive aldehydes (e.g. HNE and HHE) in the colon lumen is therefore of great concern since they may play a promoting role in CRC, by selecting precancerous cells.

Methods:

U-HPLC-ESI-MS/MS method has been developed and validated for HNE and HHE quantification in rat intestinal contents or faeces. Samples were derivatized in-situ using a brominated reagent (BBHA) in presence of pre-synthesized deuterated internal standards (HNE-D11/HHE-D5), extracted by solid phase extraction, and then analyzed by LC-positive ESI-MS/MS (MRM) on a TSQ Vantage mass spectrometer.

Results:

The use of BBHA allowed the efficient stabilization of the unstable and reactive hydroxy-alkenals HNE and HHE. The MRM method allowed selective detection of HNE and HHE on the basis of characteristic transitions monitored from both the 79 and 81 bromine isotopic peaks. This method was validated according to the European Medicines Agency (EMEA) guidelines, by determining selectivity, sensitivity, linearity, carry-over effect, recovery, matrix effect, repeatability, trueness and intermediate precision. The performance of the method enabled the quantification of HNE and HHE in concentrations 0.10-0.15 µM in fecal water. Results will be presented on the application of the validated method to the quantification of HNE and HHE in different fecal waters obtained from faecal of rats fed diets with various fatty acid compositions thus corresponding to different pro-oxidative features.

Novel Aspect:

Validation of an assay for the quantitative determination of two hydroxy-alkenals (HNE/HHE) by U-HPLC-ESI-MS/MS with internal calibration.

CHOI, INHEE,
KOREA INSTITUTE OF NUCLEAR SAFETY

Contributing Authors:
Cheol-Su Kim, Korea Institute of Nuclear Safety.

Title:
High precision Plutonium (Pu) isotope analysis of environmental samples with a double-focusing Multi Collector Inductively Coupled Plasma Mass Spectro

Introduction:
A method is presented for the preparation of environmental materials (seawater) and their accurate analysis, by multiple collector magnetic sector ICP-MS, for the isotopic composition of Pu.

Methods:
Pu isotopes concentration is extremly low and U(uranium)/Pu ratio is quite high, analysis of Pu in the environmental seawater sample demands highly sensitive instrument and an efficient U separation process. A multi collector inductively coupled mass spectrometry (MC-ICPMS) is an ideal instrument in the quantitative and qualitative analysis of Pu isotopes (239Pu, 240Pu, 241Pu) in the seawater.

Results:
Mixed HF (0.01M) and HCl (0.6M) as an eluent showed high recovery for Pu in the 1st and 2nd TEVA separation and also contributed high signal intensity by washing out effect in the sample inlet pathway before plasma. Additionally X-cone in the skimmer and Jet sampler cone combination dramatically increased Pu signal by a factor of 2~3. Detection limit and precision for Pu isotopes determination were compared at 3 different seawater amounts (5 kg, 15 kg and 25 kg) in the 3 depth groups (surface, 200m+500m, 750m+ 1000 m) to determine practical sampling amounts for surface and deep seawater.

Novel Aspect:
Accurate and precise determination of trace amount of Pu isotopes in seawater samples with minimizing the isobaric effect of U via chemial separation

Choi, Sam,
George Washington University

Contributing Authors:
Dr. Marta Zamarbide, George Washington University; Dr. Peter Nemes, George Washington University; Dr. M. Chiara Manzini George Washington University.

Title:
CE-nanoESI-MS for Untargeted Proteomic Characterization of Small Neuron Populations

Introduction:
Molecular characterization of small neuronal populations presents transformative potentials for studying neurodegenerative diseases, but requires detection technologies with high sensitivity, preferably mass spectrometry (MS). Here we develop a capillary electrophoresis nanoelectrospray ionization (CE-nanoESI) platform for MS to identify 300+ protein groups in <500 pg protein digest from neurons.

Methods:
Proteins were extracted from small populations of hippocampal neurons, trypsin-digested, and reconstituted to 3 microL. 500 pg digest (1 nL) was analyzed with a custom-built CE-nanoESI-MS system. Neuronal peptide ions were sequenced using quadrupole-Orbitrap MS (QE+) to identify proteins against the NCBInr database using SEQUEST with FDR < 1% against a reversed-sequence database.

Results:
We custom-designed a microscale CE and a CE-nanoESI interface, and developed specialized MS methodology to enable trace-level bottom-up proteomics. The CE platform was able to analyze 1 nL, or ~500 pg protein digest. The CE-nanoESI interface featured a co-axial sheath-flow design, a tapered-tip metal emitter, and the stable cone-jet spraying regime. 2 fg angiotensin II yielded signal/noise = 17, suggesting an ~350 zmol lower limit of detection. Quantification was linear over 5 log-order concentration with a ~2% RDS in migration time and <20% RSD in peak area. Furthermore, measurement of a 500 pg of protein digest from hippocampal neurons uncovered ~300 different protein groups (<1% FDR), including known neuronal-specific gene products (e.g., synapsin 1 and 2 and the neural cell adhesion molecule). High-sensitivity MS raises exciting new investigative opportunities for neuroscience.

Novel Aspect:
A CE-nanoESI-MS platform is developed with trace-level sensitivity for limited neuron populations.

Chong, Patrick,
Public Health Agency of Canada

Contributing Authors:
Garrett Westmacott, Public Health Agency of Canada; Michelle Shuel, Public Health Agency of Canada; Raymond Tsang Public Health Agency of Canada.

Title:
Detection of Bordetella pertussis Protein Pertactin using Parallel-Reaction Monitoring Compared to Western Blot Analysis

Introduction:
Bordetella pertussis is the causative agent of whooping cough. B. pertussis produces the virulence factor pertactin (Prn), which is used in acellular pertussis vaccines. Reports of Prn-deficient B. pertussis has been increasing, possibly leading to increased cases of whooping cough. In this study, we report the use of parallel-reaction monitoring to detect Prn-deficient strains of B. pertussis.

Methods:
In a blind study consisting of 22 Prn-positive and –negative isolates, B. pertussis pellets were lysed, quantified, and the protein extract was digested overnight with trypsin. Parallel-reaction monitoring was done on four tryptic peptides unique to Prn using an LTQ Orbitrap XL, and data was analyzed using Skyline software. The data was compared against the results of Western blots.

Results:
Compared to the results of the Western blot analysis, 10/10 of the expected Prn-positive strains were detected as positive by PRM with all four target peptides. Prn was not detected in 11/12 of the expected Western blot-Prn-negative B. pertussis strains. However, one of the four target peptides was detected in one of the Prn-negative strains. The remaining 3 peptides were not detected in this Prn-negative strain. Interestingly, the Prn-encoding gene was found to be wild-type in this strain, indicating that some or part of the Prn protein may be expressed in this strain. Further analysis will be required to ascertain the Prn status of this strain, and a larger blind panel of B. pertussis isolates will also be tested.

Novel Aspect:
To our knowledge, this is the first report of a targeted proteomics-based approach used to detect B. pertussis strains deficient in pertactin.

Chu, Ivan Keung,
Department of Chemistry, The University of Hong Kong

Contributing Authors:
Cheuk Kuen Lai, The University of Hong Kong.

Title:
Radical-Mediated Peptide Tyrosine Nitration

Introduction:
PTN is regioselective at the ortho-position of the tyrosine side chain and site-specific—only particular tyrosine residues within a protein are susceptible to nitration. This presentation describes the mechanism of radical-mediated tyrosine nitration elucidated at the molecular level.

Methods:
The mechanism of radical-mediated peptide tyrosine nitration (PTN) has been elucidated in detail at the molecular level using an integrated approach combining gas phase synthesis of prototypical tyrosine-containing peptide radical cations, ion–molecule reactions with nitrogen dioxide (•NO2), and DFT calculations.

Results:
The intrinsic properties of the tyrosine ortho-nitration process were examined through a combination of isotopic labelling MS experiments and DFT calculations. These experimental and theoretical investigations into radical-mediated tyrosine nitration revealed the necessity for phenoxyl radical formation prior to the production of distinct 3-nitrotyrosine products, which are influenced by the location of the radical and the tautomerization barriers for radical and proton mobility. Key [GGYG]•+ intermediates were characterized to probe the individual radical migration and isomerization steps occurring prior to the PTN recombination reactions; a two-step mechanism, involving the generation of phenoxyl radical intermediates with viable isomerization barriers (<10 kcal/mol) and concerted proton rearrangement prior to the formation of stable closed-shell ortho-nitration product is determined.

Novel Aspect:
This is the first report of experimental and theoretical investigations into radical-mediated tyrosine nitration.

claude, emmanuelle,
Waters Corporation

Contributing Authors:
James Langridge, Waters Corporation; Fiona Henderson, Wolfson Molecular Imaging Centre, Manchester, UK; Adam McMahon Wolfson Molecular Imaging Centre, Manchester, UK; Emrys Jones, Waters Corporation; Steve Pringle, Waters Corporation.

Title:
Metabolic phenotyping of cell lines by Rapid Evaporative Ionization Mass spectrometry

Introduction:
Human cell lines are routinely used as models for diseases such as cancer or for in vitro models for drug screening. Common techniques like FACS are usually biased, and do not allow the biochemical nature of the cell to be interrogated. Rapid Evaporative Ionization Mass Spectrometry (REIMS) is an emerging technique allowing the rapid, in situ characterization of biological samples and tissues.

Methods:
HeLa cells were cultured for 24hrs. At this time point half of the culture flasks were placed in a hypoxic conditions overnight whilst the other half remained under standard conditions. Subsequently, cells were frozen down as viable and stored at -80°C until analysis. Cells were thawed and re-suspended, washed in 150mM ammonium acetate and pelleted prior to REIMS analysis.

Results:
REIMS uses bipolar forceps which allow the cell pellet to be analysed in situ within the sample tube. The REIMS sampling method is almost instantaneous, and as such multiple data points can be collected rapidly without any cell treatment. In the analysis of these samples, 3 different tubes of the same cell type were analysed with 5 technical replicate measurements per tube in less than one minute per sample. The mass range acquired was from m/z 50-1000 and the associated spectra obtained were rich in lipid and metabolite molecular information with a high degree of intra and inter sample reproducibility. Cells subjected to different environmental conditions (hypoxic vs. normoxic) were rapidly and reproducibly analysed by REIMS to compare their metabolic differences. Differences between the samples were clearly demonstrated along with the molecules contributing to the differentiation.

Novel Aspect:
Fast and rapid REIMS analysis of metabolic phenotypes of cell lines subjected to differing environmental conditions.

claude, emmanuelle,
Waters Corporation

Contributing Authors:
Emmy Hoyes, Waters Corporation, Wilmslow, United Kingdom; Paul Murray, Waters Corporation, Wilmslow, United Kingdom; Lukasz Migas University of Manchester, Manchester, UK; Richard Chapman, Waters Corporation, Wilmslow, United Kingdom; Mark Towers, Waters Corporation, Wilmslow, United Kingdom.

Title:
An investigation of MALDI imaging with higher speed sample stage "rastering" from an ion mobility enabled Q-TOF mass spectrometer.

Introduction:
MALDI imaging is an established technique to determine the localization of specific species within biological tissue sections. Here, we evaluate an alternative acquisition method, scaling the stage speed, laser beam diameter and desired pixel size to improve acquisition rate independently of image resolution.

Methods:
Investigations were performed using MALDI SYNAPT G2-Si mass spectrometer equipped with an ion mobility cell and controlled by prototype software. To achieve higher imaging rates, sample stage movement and laser were controlled using the Waters Research Enabled Software (WREnS). Tissue imaging experiments were performed in MS and MS/MS modes, with and without ion mobility separation.

Results:
WREnS allows the creation of custom scripts for acquisition in both a "typewriter" mode and a "serpentine" mode. Initial experiments were carried out using thin sections of mouse brain sprayed with α-cyano-4-hydroxycinnamic acid. Spectra have been generated at rates up to 20 pixels per second in MS and MS/MS mode, independent of image spatial resolution. In a similar manner IMS-MS spectra are obtained at a rate of 10 pixels per second allowing the additional dimension of mobility separation to be fully included. For different spatial resolutions the laser focus was adjusted to provide optimum coverage in each case. The image quality from both typewriter and serpentine modes and the effect of the acquisition direction) will be compared. The direction of acquisition is important as overall acquisition time maybe reduced by choosing the appropriate direction and thus reducing overhead.

Novel Aspect:
Application and assessment of a high speed continuous raster mode of MALDI imaging on an ion mobility enabled Q-TOF

Clemen, Martin,
Christian-Albrechts-Universität zu Kiel

Contributing Authors:

Martin Clemen, Christian-Albrechts-Universität zu Kiel; Jürgen Grotemeyer, Christian-Albrechts-Universität zu Kiel.

Title:

Role of Charge Position - Radical Fragmentation Reactions from Azobenzene Derivatives containing Diethylamino Groups

Introduction:

Several dye molecules show a loss of C3H8 from diethylamino groups. Still, the mechanism of this fragmentation reaction is not clearly understood. Therefore, some azo dyes like ethyl red have been examined. Their mass spectral behavior shows the loss of C3H8 as a combination of two radical species.

Methods:

All mass spectra are recorded on a APEX-IV-Qe FT-ICR MS equipped with a 9.4 T magnet. For positive ESI the samples are diluted to a final concentration of 20 pmol/µL. Mass resolution is about 100000 and mass accuracy better than 1 ppm. Fragmentation is achieved by using either CID with argon as collision gas or PD with a cw ar-ion-laser. It is possible to gain MS2 and MS3 spectra.

Results:

o-Ethyl red shows a significant loss of H2O. This indicates the position of the proton to be between carboxyl- and azo-group. Further fragmentations are sequential losses of a methyl radical and an ethyl radical but not the other way around. This is due to stabilization of the fragment ion after the loss of a methyl radical. No indications are found that point to a concerted fragmentation mechanism. p-Ethyl red and ethyl yellow behave differently. Loss of an ethyl radical indicates the protonation to occur on the amine nitrogen. Also, a fragment ion can be observed that has to result from a combined loss including a methyl radical. In this case, protonation is most likely on the azo group. In summary, the position of the proton and the charge leads to very different fragmentation reactions for the azo dyes. MS3 experiments and quantum chemical calculations support these findings.

Novel Aspect:

It's showed the first time that loss of C3H8 from a diethylamino group is a clean twostep radical fragmentation reaction for the azo dye o-ethyl red.

Clendinen, Chaevien,
Georgia Instiiute of Technology

Contributing Authors:
David A. Gaul, Georgia Instiiute of Technology; Arthur S. Edison, University of Georgia, Athens; Rebecca Arnold Emory University; John A. Petros, Emory University; Facundo M. Fernandez, Georgia Instiiute of Technology.

Title:
Multiplatform Metabolomics of Recurrent Prostate Cancer.

Introduction:
MS and NMR are two frequently employed technologies in metabolomic studies. Though often used independently, when these technologies are combined, much more information of the metabolome is gained. In this study, we merged the results from these two techniques to investigate recurrence in prostate cancer (PCa) patient by serum analysis from men with different clinical outcomes after prostatectomy.

Methods:
Serum samples were collected from PCa patients prior to prostatectomy. UPLC-MS data was acquired on a ThermoScientific Ulitmate 3000 coupled to a Q Exactive HF in both positive and negative modes. LC parameters was adapted from Damen et. al 2015. NMR data was acquired on a Bruker 600 AVIII-HD with 5-mm cryoprobe and cooled SampleJet. Pre-processed data were imported into Matlab for analysis.

Results:
UPLC-MS collected both positive and negative modes revealed over 600 features that separated serum from men with recurrence from those without. The majority of these species were detected in the more polar regions of the chromatographic gradient (0.5 to 2 minutes). OPLS-DA showed separation of the two patient classes. NMR revealed an increase in several resonances in men with recurrence. Some of these resonances have been tentatively identified as lactate, alanine, and creatinine. Resonances corresponding to sugars are decrease in men with recurrence as well as metabolites such as taurine. Data analysis is ongoing, however, these preliminary findings indicate a clear difference in the metabolomic profiles between males that responded differently to surgery. NMR and UPLC-MS data help us to better characterize the metabolites responsible for these differences.

Novel Aspect:
The fusion of NMR and UPLC-MS information is useful to gain a more comprehensive understanding of the metabolic signatures of PCa recurrence.

21ST INTERNATIONAL MASS SPECTROMETRY CONFERENCE
Toronto, Ontario, Canada
August 20 – 26, 2016

Cody, Robert (Chip),
JEOL USA, Inc.

Contributing Authors:
Chikako Takei, Biochromato, Inc.; Michael J. Churchill, Biochromato, Inc.; Shinji Azuma Biochromato, Inc.

Title:
Temperature Regulated Thermal Desorption and Pyrolysis Coupled with Direct Analysis in Real Time Mass Spectrometry (DART)

Introduction:
A new temperature-regulated thermal desorption and pyrolysis (TDP) device is combined with DART to analyze materials including polymers, single fibers, drugs, and seeds. Compounds are clearly separated by differential thermal desorption. Separating the pyrolysis permits the identification of polymers that might otherwise be obscured by strong additive signals.

Methods:
The Biochromato, Inc. "IonRocket" TDP device was installed on a JEOL AccuTOF-DART mass spectrometer. Samples were placed into small wells on copper stages. Desorption and pyrolysis products were sampled with a glass tee mounted between the DART and the atmospheric pressure interface, positioned 1 mm above the sample holder. Samples were heated from 50°C to 600°C at rates of up to 600°C min-1.

Results:
TDP analysis of duct tapes showed thermal separation of plasticizers, resins, natural rubber adhesive and pyrolysis of polyolefin backing. Chemometric analysis permitted identification of manufacturer and brand of each tape. TDP/DART permitted analysis of powders and single fibers, and separated active ingredients and excipients in OTC drugs. Excipients could be detected that not seen by DART alone, but thermal degradation was observed for some drug components as the temperature increased. The IonRocket also facilitated analysis of fullerenes and high-temperature compounds. The IonRocket was tested with the glass tee and also with the IonSense Vapur interface. The latter required a ceramic "chimney" to guide compounds into the DART gas stream. Pyrolysis products from polyolefins were suppressed when the Vapur interface was used, but could be detected with the glass tee interface.

Novel Aspect:
A thermal desorption and pyrolysis device coupled to a DART-TOF mass spectrometer is useful for materials and mixture analysis.

Cody, Robert (Chip),
JEOL USA, Inc.

Title:
An ambient ionization toolbox: Comparison of helium and argon DART with complementary ambient ionization methods.

Introduction:
A variety of ambient ionization methods are needed to analyze widely varying sample types. Dopant-assisted argon DART with charge transfer was found to complement helium DART and O2–• attachment for small-molecule analysis. Large and nonvolatile samples that could not be desorbed by DART were analyzed by inlet ionization and an integrated paper spray attachment.

Methods:
A JEOL AccuTOF-DART mass spectrometer was used for all measurements. A syringe pump introduced dopants (e.g. chlorobenzene) for argon DART. Samples were aspirated into Orifice 1 for inlet ionization or O2–• attachment. A new movable stage was mounted on the linear rail to permit paper spray ionization without requiring removal of the DART source.

Results:
Because metastable argon does not ionize water, dopant-assisted argon DART is analogous to atmospheric pressure photoionization. Dopants can favor charge exchange or proton transfer depending on the proton affinity and ionization energy of the analyte, providing a means to control the types of ions formed. Whereas DART is a plasma-based ambient ionization method, inlet ionization and paper spray more closely resemble electrospray ionization and were used to analyze proteins and peptides as well as inorganics and compounds with fixed charges such as proanthocyanidins. An interesting forensic example concerns peptides of abuse. Any or all of these ionization methods can be selected depending on the sample type without removing the DART ion source.

Novel Aspect:
Dopant-assisted argon DART complements helium DART and O2–• attachment. Inlet ionization and paper spray can be used without hardware changes

Cole, Richard B.,
Sorbonne Universités, UPMC Univ Paris 06

Contributing Authors:

Ulli Hohenester, Commissariat à énergie atomique et aux énergies alternatives, GIF SUR YVETTE CEDEX , Fr; Pierre Barbier Saint Hilaire, Commissariat à énergie atomique et aux énergies alternatives, GIF SUR YVETTE CEDEX , Fr; Benoit Colsch Commissariat à énergie atomique et aux énergies alternatives, GIF SUR YVETTE CEDEX , Fr; Francois Fenaille, Commissariat à énergie atomique et aux énergies alternatives, GIF SUR YVETTE CEDEX , Fr; Christophe Junot, Commissariat à énergie atomique et aux énergies alternatives, GIF SUR YVETTE CEDEX , Fr; Richard B. Cole, Sorbonne Universités, UPMC Univ Paris 06, Institut Parisien de Chimie Moléculaire, Paris, Fr.

Title:

Comparison of direct introduction on Orbitrap Fusion and FT-ICR at high-end resolution for global metabolite screening

Introduction:

Orbitrap technology (resolution 500,000) and FT-ICR cells with higher ion capacity (resolution two million for low m/z values in Broadband mode) have benefited metabolomics, lipidomics and proteomics fields. Typically, global sample screening on FT-ICR instruments is performed at resolutions of 200,000 to 800,000. However, recent instrument developments may give rise to improved performance.

Methods:

956 small molecules were dissolved and diluted in mixtures of acetonitrile, methanol or 2-propanol with H2O and 0.1% formic acid. These solutions were analyzed by Flow Injection Analysis (FIA) on an Orbitrap Fusion at resolution 500,000 and were compared with direct infusion results obtained on a Bruker SolariX FT-ICR equipped with ParacellTM at resolution 1,800,000.

Results:

A complex sample mixture (956 molecules) has been investigated on both Orbitrap and FT-ICR instruments with the goal of assessing the limitations of resolution and speed for global metabolite detection. In total, of the 956 molecules studied, corresponding to 620 different molecular formulas present simultaneously in this very complex mixture, more than 200 signals were detected on both instruments (positive ion mode). If overloaded with ions, space charge effects in the FT-ICR cell can lead to mass shifts greater than 10 ppm. As a consequence, large m/z range acquisitions can be performed in 2.5 min by direct infusion, but a real resolution of only around 500,000 is obtained. The use of SIM-stitching allows the attainment of the theoretical resolution of 1.8 million. Use of appropriate isolation parameters permits achievement of the theoretical FT-ICR resolution for each segment.

Novel Aspect:

Global metabolite detection is shown at high-end resolution and high speed on Orbitrap Fusion and SolariX FT-ICR.

Cole, Richard B.,
Sorbonne Universités, UPMC Univ Paris 06

Contributing Authors:
Sergey Sladkevich, Sorbonne Universités, UPMC Univ Paris 06; Anne-Laurence Dupont, CRCC, Sorbonne Universités, Muséum National d'Histoire Naturelle; Michel Sablier CRCC, Sorbonne Universités, Muséum National d'Histoire Naturelle; Dalila Seghouane, CRCC, Sorbonne Universités, Muséum National d'Histoire Naturelle; Richard B. Cole, Sorbonne Universités, UPMC Univ Paris 06.

Title:
Identification of products of cellulosic paper decomposition at the wet-dry "tideline" interface using GC-MS and LTQ Orbitrap

Introduction:
A deeper understanding of the nature of compounds responsible for paper discoloration is essential to elaborating strategies for preservation and remediation of valuable paper specimens. The aim of the present study is to investigate the low molecular weight compounds forming in the "so-called" tideline, and to put an effort into identification of key products responsible for paper discoloration.

Methods:
GC-MS experiments were conducted on an Agilent 7890B GC system interfaced to a 5977A MSD mass spectrometer. Several extraction protocols and analysis configurations were employed in order to detect a wide range of products forming in the tidelines on pure cotton paper. The extracts were prepared either by direct solvent extraction, or by a chromophore release and identification protocol.

Results:
Identified low-weight products of oxidative cellulose degradation were classified. First, compounds possessing conjugated double bond systems were identified. Various furan derivatives belonging to this group (e.g. furfural and hydroxymethyl furaldehyde), forming as a result of dehydration of glucose and pentose units may be responsible for some visible paper discoloration. Second, simple diols (ethylene and propylene glycol) and their condensation products were detected using GC-MS and Orbitrap LTQ. Third, detection of a series of carboxylic acids, from formic to nonanoic, confirms previously obtained results for tidelines and aged paper. Fourth, carbohydrate derivatives formed during scission and oxidation of cellulose chains were separated. Intense chromatographic peaks were detected for different anhydro- (i.e. levoglucosan) and dianhydro- (i.e. levoglucosenone) aldose derivatives.

Novel Aspect:
Two distinct mass spectrometry techniques were used to identify several groups of compounds forming in paper "tidelines" during cellulose degradation.

Compton, Philip,
Northwestern University

Contributing Authors:
Luca Fornelli, Northwestern University; Neil Kelleher, Northwestern University.

Title:
High Throughput Top Down Proteomics Utilizing Ultra Violet Photo Dissociation.

Introduction:
Over the past decade, top down proteomics has morphed from a technique capable of identifying a handful of proteins to one capable of identifying thousands of unique molecular species. However, the thorough characterization of these "proteoforms" remains challenging, especially at higher mass, with current fragmentation modalities.

Methods:
An Orbitrap Fusion Lumos mass spectrometer was equipped with an ultra violet laser aligned with the axis of the ion traps through a laser window custom machined into the vacuum flange. Triggering of the laser was integrated into the instrument control software to enable data dependent acquisition in a high-throughput mode.

Results:
Recently, the Kelleher laboratory has developed several metrics; one for protein identification and one for protein characterization. These metrics allow us to keep score when comparing fragmentation modalities. As an initial case study, yeast whole cell extracts were subjected to GELFrEE fractionation followed by reversed phase chromatography. Data dependent acquisition then proceeded using HCD, ETD and UVPD. Analyzing these data enable a clear look into the utility of UVPD for both identification and characterization when compared to current methods.

Novel Aspect:
High throughput top down analysis utilizing Ultra Violet Photodissociation on an Orbitrap Fusion Lumos mass spectrometer.

Consta, Styliani (Stella),
Department of Chemistry

Contributing Authors:
Myong In Oh, Department of Chemistry, University of Western Ontario.

Title:
Molecular simulation study of the dissociation rate and mechanisms of protein complexes in droplets

Introduction:
Computational studies of the stability of protein complexes are presented. We look into the stability of the complexes at variable conditions of pH as the droplet evaporates. We determine the rates of dissociation of weakly bound complexes and also factors that increase the stability of the complexes.

Methods:
Advanced molecular modelling methods are employed to represent the systems in a realistic manner. Atomistic molecular dynamics and Monte Carlo methods are employed to sample the molecular states of the systems. New order parameters are introduced to monitor the protein dissociation and the shape of the droplets where the proteins are found.

Results:
We introduce a method on how to deal computationally with droplets under variable pH conditions, which induce variable protonation states to the proteins as the droplet size changes. The mechanisms and the rates of protein dissociation in a droplet are discussed.

Novel Aspect:
The first computational study on the dissociation rate of protein complexes in variable pH is presented.

Consta, Styliani (Stella),
Department of Chemistry

Contributing Authors:
Mahmoud Sharawy, Department of Chemistry, University of Western Ontario; Myong In Oh, Department of Chemistry, University of Western Ontario.

Title:
Understanding the stability of non-covalent complexes of macromolecules in charged droplets by molecular simulations

Introduction:
Computational studies of the factors that affect the stability of non-covalent complexes in aqueous droplets are presented. These factors are: droplet evaporation rate, finite size of the droplet, increase in the concentration of ions in evaporating droplets and distinct droplet morphologies arising from the charge-induced instabilities. We classify the effects of the various factors.

Methods:
Advanced molecular modelling methods are employed to represent the systems in a realistic manner. We study complexes of proteins, double stranded DNA (dsDNA), G-quadruplex. The acidity of the droplet is taken into account and its variation with droplet size. Atomistic molecular dynamics and Monte Carlo methods are employed to sample the molecular states of the systems.

Results:
We find that charged droplets attain distinct ``spiky'' morphologies for a charge squared to volume ratio above the Rayleigh limit. These distinct morphologies affect the interactions within the complex. Consequently, a non-covalent complex may remain associated in an unstable droplet as long as there is enough solvent to accommodate the instability. In the presence of positive and/or negative ions, the ions may form adducts with the complex, which help to stabilise it. For instance, a DNA-containing droplet with a net charge that is less negative than 50% of the dsDNA charge, leads to the dsDNA maintaining a double-stranded state in the gas phase. However, the conformation of the dsDNA may change depending on the location of the ions on the strands. Dissociation rates of weakly bound protein complexes are computed and the effect of ions in the stability is analyzed.

Novel Aspect:
The first computational systematic study on the factors that determine the stability of weakly non-covalent complexes is presented.

Cooks, R. Graham,
Purdue University

Contributing Authors:
Xin Yan, Purdue University; Christopher Pulliam, Purdue University; Ryan Bain Purdue University; Yafeng Li, Purdue University.

Title:
Accelerated Organic Reactions in Microdroplets: Phenomena and Fundamentals

Introduction:
Microdroplets generated by spray ionization methods such as desorption electrospray ionization (DESI), electrosonic spray ionization (ESSI), and paper spray (PS) show reaction acceleration of C-N, C-C and other bond-forming reactions by several orders of magnitude {X.Yan, R. Bain and R. G. Cooks, Angew. Chem, 2016].

Methods:
A variety of spray-based ionization techniques are utilized in this work including electrosonic spray ionization (ESSI), desorption electrospray ionization (DESI), and paper spray (PS) ionization. Reactive ionization was performed with the Claisen-Schmidt base-catalyzed condensation, the Hantzsch dihydropyridine synthesis, and several more.

Results:
We show that i) the degree of reaction acceleration can be controlled by altering the extent of solvent evaporation ii) that discontinuities in solvent evaporation, either through ion ejection or sudden evaporation of all (remaining) solvent molecules when the droplet reaches vacuum, will truncate reaction it at whatever stage it has reached iii) that acceleration is associated with reaction at droplet surfaces as shown by competitive substituent effect studies on Claisen-Schmidt reactions of substituted benzaldehydes iv) that acceleration can be utilized to simultaneously derivatize and ionize analytes and so optimize chemical analysis by mass spectrometry v) that the use of multiple emitters allows scale-up and collection of product on surfaces at 100 mg/hour rates vi) that the use of accelerated reactions has considerable value in undergraduate organic teaching laboratories.

Novel Aspect:
We have shown that reactions in charged microdroplets are accelerated by both solvent desolvation and surface effects.

Cooks, R. Graham,
Purdue University

Contributing Authors:

Zane Baird, Purdue University; Valentina Pirro, Purdue University; Alan Jarmusch Purdue University; Eyas Hattab, University of Louisville; Clint Alfaro, Purdue University; Aaron Cohen-Gadol, Indiana University School of Medicine.

Title:

Intrasurgical Brain Tumor Diagnosis and Surgical Margin Definition by Desorption Electrospray Ionization-Mass Spectrometry

Introduction:

Surgical intervention is a primary treatment option for gliomas. The best patient outcomes are associated with complete tumor resection and minimal damage to adjacent normal tissue. We report an intraoperative method for the rapid analysis of tissue smears by DESI-MS which provides diagnostic information and can offer surgical guidance. Preliminary results of initial clinical cases are presented.

Methods:

Analysis was performed using a customized mass spectrometer modified for use in the operating room. Biopsies (with stereotactic positioning), were smeared using a 3D-printed device. DESI spectra had lipid and metabolite signals. Analysis in positive and negative ion modes took ~3 min per biopsy. Multiple biopsies were analyzed for each case. Data were processed by multivariate statistics.

Results:

Biopsies were taken from the tumor core and margin during surgery. Biopsies were smeared and analyzed intraoperatively by DESI-MS acquiring positive and negative ion mode data, including lipids and oncometabolites such as N-acetylaspartic acid (NAA) and 2-hydroxyglutaric acid. Glioma predictions matched pathology (true rates >97 %), confirming that DESI-MS molecular fingerprint can serve as an accurate cancer diagnostic tool. NAA was significantly suppressed in gliomas but decreased proportionally with the tumor cell percentage (TCP). We are exploring how lipid and oncometabolite changes detected by DESI-MS correlate with TCP, to assess the residual tumor at the surgical margin. This information is currently unavailable in surgery. We envision such measurements improving surgical intervention and outcomes, by more accurately assessing surgical margins and tumor heterogeneity.

Novel Aspect:

Intraoperative DESI-MS analysis of tissue smears allows brain cancer diagnosis and detection of tumor infiltration at surgical margins.

Cornett, Shannon,
Bruker

Contributing Authors:
Gökhan Baykut, Bruker Daltonik GmbH, Fahrenheitstrasse 4, 28359 Bremen, Germany; Eugene N. Nikolaev, The Institute for Energy Problems of Chemical Physics Russian Academy of Sciences, Leninskij pr. 38, k.e Moscow, Russia; Michael L. Easterling Bruker Daltonics Inc. 40 Manning Rd, Billerica, MS 01821, USA; Christopher J. Thompson, Bruker Daltonics Inc. 40 Manning Rd, Billerica, MS 01821, USA; Jochen Friedrich, Bruker Daltonik GmbH, Fahrenheitstrasse 4, 28359 Bremen, Germany; Matthias Witt, Bruker Daltonik GmbH, Fahrenheitstrasse 4, 28359 Bremen, Germany.

Title:
MALDI FT-ICR Imaging: Double the Performance or Double the Samples using Quadrupolar Detection

Introduction:
For FT-based MALDI imaging, reducing the time to acquire each spectrum by milliseconds can shorten the total measurement time by hours. Quadrupolar detection (2ω) promises to dramatically enhance FTICR imaging studies by providing high MS performance with reduced transient length. In this study we evaluate the speed benefit of imaging with 2ω vs traditional 1ω and compare their MS performance.

Methods:
Rat kidney and brain sections were coated with matrix using a pneumatic sprayer. 7T and 12T solarix systems were used, each fitted with a dynamically harmonized cell modified with new electronics that improves the 2ω signal intensities and allows fast switching between excitation and detection modes at the same electrodes. MALDI sensitivity was optimized for m/z 200-900 and images were collected.

Results:
A major bottleneck for MALDI FT-ICR imaging is low sample throughput due to a compromise one must make between acquisition speed and high MS performance. With a small electronic change to a commercial system, one can achieve the same mass resolving power in half the acquisition time or 2x the mass resolving power from the same acquisition time. For example, with a 2s transient at 7T the 1ω signal produces a resolution of 300k at m/z 800 while the 2ω signal produces resolution of 300k with a 1s transient. The benefit for imaging studies is significant in that each measurement can be done in half the time without loss of mass accuracy or resolution. Alternatively, for the same acquisition time 2ω detection yields 2x the resolving power, 600k at m/z 800. Although no speed benefit is realized in this mode of operation there is benefit to resolving isotopic fine structure at higher m/z.

Novel Aspect:
2ω detection with MALDI FT-ICR doubles the resolving power or reducing the measurement time

Cornett, Jack,
University of Ottawa

Contributing Authors:
Chris Charles, A.E. Lalonde AMS Laboratory, University of Ottawa; Zakir Kazi, Canadian Nuclear Laboratories; Liam Kieser A.E. Lalonde AMS Laboratory, University of Ottawa; Xiaolei Zhao, A.E. Lalonde AMS Laboratory, University of Ottawa.

Title:
Measuring Actinides by Accelerator Mass Spectrometry

Introduction:
Isotopes of the actinide elements (U, Pu, Am and Cm) are present in the environment as a result of anthropogenic activities during the past 60 years. These isotopes are usually present in ultra-trace concentrations ranging from atomgram to femtograms in most types of samples. We have tested new methods to measure these isotopes using Accelerator Mass Spectrometry (AMS).

Methods:
We tested the performance of both iron oxide and neodymium fluoride matrices as target materials for the measurement of actinide isotopes at the A.E. Lalonde AMS Laboratory at the University of Ottawa. Actinide anions were extracted as oxide or fluoride anions at 35 kev energies from the Cs sputter ion source (Model SO-110, HVE Europa).

Results:
Pu, Am, and Cm form strong beams of both fluoride anions and oxide anions. The performance of both target matrices is similar when the sensitivity, precision, interferences and detection limits of the two different types of sputter source targets are compared. U also forms both strong oxide and fluoride beam. However the formation of U-235OF- ions produced interferences in the measurement of the much rarer isotope U-236. The beam intensities of UO- ions decreased during the analysis of individual targets and the presence of higher concentrations of uranium in a target reduces the UO- beam currents. This poisoning of the UO- ion beam can be controlled, but not totally eliminated by adding silicon nanoparticles to the uranium oxide sputter targets.

Novel Aspect:
We have developed improved methods to measure the actinides using AMS by changing the matrix of the sputter target.

Costello, Catherine,
Boston University School of Medicine

Contributing Authors:
Deborah R. Leon, Center for Biomedical Mass Spectrometry, Boston University School of Medicine, Boston, MA 02118 USA; Mark E. McComb, Center for Biomedical Mass Spectrometry, Boston University School of Medicine, Boston, MA 02118 USA; Dharmendra B. Goswami United States Department of Veterans Affairs, VA Boston Healthcare System, Boston, MA 02130 USA; Amanda G. Balderrama, Center for Biometals & Metallomics, Boston University School of Medicine, Boston, MA 02118 USA; Andrew M. Fisher, Center for Biometals & Metallomics, Boston University School of Medicine, Boston, MA 02118 USA; Ann C. McKee, United States Department of Veterans Affairs, VA Boston Healthcare System, Boston, MA 02130 USA; Bo Yan, Center for Biomedical Mass Spectrometry, Boston University School of Medicine, Boston, MA 02118 USA; Lee E. Goldstein, Center for Biometals & Metallomics, Boston University School of Medicine, Boston, MA 02118 USA.

Title:
Mass Spectrometry Imaging of Proteins, Lipids and Metals in Human and Mouse Chronic Traumatic Encephalopathy Tissues

Introduction:
Chronic traumatic encephalopathy (CTE) is now recognized as a distinct neurodegenerative condition that affects people with a history of repetitive traumatic brain injuries. Molecular profile changes and disease progression are not well understood. We report here on MALDI-imaging MS and MS/MS, UPLC-ESI-MS/MS and ICP-MS method development to investigate molecular and metal signature alterations.

Methods:
Chronic traumatic encephalopathy (CTE) is now recognized as a distinct neurodegenerative condition that affects people with a history of repetitive traumatic brain injuries. Molecular profile changes and disease progression are not well understood. We report here on MALDI-imaging MS, ESI-MS/MS and ICP-MS method development to investigate molecular and metal signature alterations.

Results:
Adjacent brain sections from post-mortem human and mouse CTE and control tissues were analyzed. Lipids were imaged by +/- MALDI-FTICR MS and MS/MS; 2-D maps were based on MS signal intensities. Then tissues were H&E stained. Statistical analysis was performed in regions of interest, guided by pathological information. The discrimination capability for each m/z value was evaluated (SCiLS Lab 2015b). Lipid patterns in CTE tissues showed clear alterations. Protein assignments were based on MALDI-MSI profiles of [M+nH]n+ and MALDI-MS & MS/MS analysis of on-tissue tryptic peptides and LC/MS/MS of peptides generated in-solution. MALDI- and ICP-MS data indicated blood brain barrier (BBB) disruption in CTE. Acknowledgements. VA-BU-CLF Brain Bank, brain donors & families, NIH P41 GM104603, S10 RR025082, S10 OD010724, NINDS U01NS086659-01, NIA P30AG13846, VA I01 CX001038 & BU ADC Pilot Grant.

Novel Aspect:
Lipid and protein alterations and BBB disruption in human and mouse CTE tissues evaluated using MALDI-MSI and ICP-MS

Costello, Catherine,
Boston University School of Medicine

Contributing Authors:

Rebecca S Glaskin, Center for Biomedical Mass Spectrometry, Boston University School of Medicine, Boston, MA 02118 USA; Valery G. Voinov, Department of Chemistry, Oregon State University, Corvallis, OR 97331 USA; Cheng Lin Center for Biomedical Mass Spectrometry, Boston University School of Medicine, Boston, MA 02118 USA; Yi Pu, Center for Biomedical Mass Spectrometry, Boston University School of Medicine, Boston, MA 02118 USA; Joseph S. Beckman, Department of Chemistry, Oregon State University, Corvallis, OR 97331 USA; Yang Tang, Center for Biomedical Mass Spectrometry, Boston University School of Medicine, Boston, MA 02118 USA; Douglas F. Barofsky, Department of Chemistry, Oregon State University, Corvallis, OR 97331 USA.

Title:

Applications of Ion Mobility Separation with Electron-based Dissociation Tandem Mass Spectrometry

Introduction:

The combination of Ion Mobility with MS/MS can generate conformational, molecular weight and structural information. However, CID often results in only partial structure definition, whereas electron-based dissociation methods (ExD) generate peptide, protein and glycan sequence information while preserving labile modifications, and provide much more extensive glycan linkage information than CID.

Methods:

We installed modifications to a 6560 IMS-QTOF MS (Agilent Technologies) and a 12-T SolariX™ Qh- FT-ICR MS (Bruker Daltonics), to enable IMS-ExD-MS and MS/MS experiments. In the IMS-QTOF MS, a radio-frequency-free electromagnetostatic cell (e-MSion) is placed between the quadrupole and collision cell. The FTMS is equipped with a prototype Trapped Ion Mobility Spectrometry (TIMS) device.

Results:

IMS separations enable the resolution of analytes on the basis of their collision cross-sections; even individual charge states of a single analyte often exhibit ≥2 conformations. We are applying IMS-ExD-MS/MS to structural studies of native/modified peptides and proteins and to native/derivatized glycans and glycoconjugates. With these systems we are able to study conformation-specific ExD of linear oligosaccharides and branched milk sugars (e.g, LNT, LnNT), glycopeptides (e.g., RNase B glycoforms Man5-9), peptides (e.g., substance P, angiotensin) and proteins (e.g., ubiquitin). For charge states that exhibit multiple conformations, we obtain distinctive ExD spectra for each state. Acknowledgements: NIH grant P41 GM104603 and a gift from Agilent Technologies University Relations Program. We are grateful to many colleagues at Agilent and Bruker for their assistance in these projects.

Novel Aspect:

Combination of Ion Mobility Separation with ExD tandem high performance mass spectrometry

Cox, David,
SCIEX

Contributing Authors:
Romain Carriere, Institut des Sciences Analytiques - UMR CNRS, Villerbanne (Lyon); Lemoine Jerome, Institut des Sciences Analytiques - UMR CNRS, Villerbanne (Lyon); Blandine Rougemont Institut des Sciences Analytiques - UMR CNRS, Villerbanne (Lyon); J.C. Yves LeBlanc, SCIEX, Concord, ON, Canada.

Title:
Scout-MRM; a method to acquire large numbers of MRM without predefined retention times

Introduction:
In MRM mode, a triple quadrupole instrument offers the high duty cycle and can monitor a large number of analytes. With Scheduled-MRM, this may include 1000s of MRM per analysis, and requires reproducible LC conditions for ALL analytes. A daunting task. Here we propose the use of Scout-MRM, which are used to accommodate in real time any changes that are associated with LC parameters.

Methods:
Pesticide standards were obtained from SCIEX. LC was performed using a Shimadzu Nexera UFLC system operated at a flow rate of 600uL/min with a with a 2x100mm (1.7u) XBridge column. A water (5mM ammonium acetate) and methanol gradients was used. All samples were analysed using a QTRAP® 5500 system equipped with a Turbo V TM source using research grade software.

Results:
The concept of Scout-MRM is to rely on a subset of analytes that are used to assess any variation in absolute retention time and determine which other MRM should be monitored. In other words, the Scout-MRM is used to monitor any changes that could occur with the chromatography; once detected, a group of associated MRMs, eluting at a later time are then automatically monitored. The Scout-MRM detection is performed in real-time and can be detected even with dramatic changes in chromatography conditions. Therefore, it is possible to have a large number of MRMs that can be monitored, without having to manage their associated retention time for scheduling. In a typical application, each Scout-MRM can have as many as 50 additional MRMs that are associated with it, thus allowing management of large MRM sets with little need to adjust any retention time information.

Novel Aspect:
Scout-MRM to automatically manage retention time of scheduled MRM under chaging chromatographic conditions (gradient, system setup, dead volume...)

Crawford, Elizabeth,
Universität des Saarlandes

Contributing Authors:

Dietrich Volmer, Institute of Bioanalytical Chemistry, Saarland University, Campus B2.2, 66123 Saarbrücken, Germany; Cemal Esen, Department of Mechanical Engineering, Applied Laser Technology, Universitätsstraße 150, Ruhr-University Bochum, 44801 Bochum, Germany.

Title:

Snap-shot Monitoring of Acoustically Levitated Containerless Microreactions via Ambient Ionization Mass Spectrometry

Introduction:

Acoustic levitation for levitating small samples has emerged as the most practical method of object levitation. Levitating liquid and solid samples creates a containerless microenvironment ideal for real-time reaction monitoring. In this work, acid-catalyzed degradation of erythromycin A was directly initiated within single liquid droplets for in stillo degradation and monitored by ambient MS.

Methods:

A Direct Analysis in Real Time (DART) source was coupled to a Bruker HCT ion trap mass spectrometer for data-dependent MS/MS analysis of erythromycin and its major acid-catalyzed degradation product. Varying acid additions (0–5 % formic acid) directly to the levitated substrate droplets (in stillo) in the acoustic field were carried out with a total miniaturized reaction volume of 6 uL.

Results:

Real-time reaction screening is used for understanding mechanistic pathways and for determining compound stability and kinetics profiles. Acoustic levitation supported the concept of a contactless microreactor and, combined with DART ambient ionization, which permits samples in their native state to be directly ionized, instantaneous mass spectrometric detection for real-time reaction monitoring was achieved. The major acid-derived degradant of erythromycin A, anhydroerythromycin A was directly monitored from microreaction mixtures with total reaction volumes of 6 µL (in stillo), as well as from scaled-up reactions in conventional reaction vessels (in vitro) with 400 µL. Reaction miniaturization was successfully achieved and the acid-catalyzed reaction results, as completed in stillo after 2 minutes, were nearly identical to in vitro results achieved after 20 minutes.

Novel Aspect:

Reaction miniaturization was permitted by acoustic levitation and in stillo microreactions were immediately monitored via ambient tandem MS.

Curtis, Sharon,
John Holmes Mass Spec Facility Ottawa

Contributing Authors:
Paul Mayer, University of Ottawa; Yardley Cuthbert, University of Ottawa; Patrick Leal University of Ottawa.

Title:
Dual Polarity Mixing Source.

Introduction:
The dual spray source of a waters electrospray mass spectrometer provides an environment where ions can be generated, manipulated, collided, selected and ultimately detected. By independently controlling the voltage on one of the emitters we show the interaction of combining results from a variety of ESI polarities. Here we discuss preliminary results from just some of the reaction possibilities.

Methods:
The electrospray source of a Micromass Global mass spectrometer has been modified such that the lock mass probe has an independent voltage supply and gas lines. This allows various configurations of polarity mixing within the source. The analyte probe determines the polarity of the instrument while the lock mass probe can be selected to species in negative, neutral or positive modes.

Results:
Preliminary results show that the source can be used to create chemically interesting species. An example being the charge state reduction of a protein by combining the protein in positive mode and interacting with a negative spray of water, OH- ions, within the source. The charge states of the protein can be seen to shift towards lower mass as the OH- ions abstract the protons from the protein.

Novel Aspect:
We assess dual polarity mixing electrospray mass spectrometry as a technique for observation of unique chemical species and discuss its implementation

Cuyckens, Filip,
Janssen Pharmaceutical Companies

Title:
Sensitive quantification of drug metabolites with and without authentic standards

Introduction:
Metabolites are often present in relatively low concentrations, buried in a large background of endogenous compounds and their MS response can significantly differ from the parent molecule. However, authentic standards are often lacking.

Methods:
Different techniques can be applied and will be discussed, including radioactive detection, UV detection, organic Mass Spectrometry, Accelerator Mass Spectrometry and Inductively Coupled Plasma - Mass Spectrometry.

Results:
An overview will be given of different established and novel approaches for the quantification of metabolites in in vitro and in vivo matrices in the presence and absence of authentic standards. Also the implementation of 2D-LC approaches can have a significant impact on metabolite quantification by improving selectivity and/or allowing large volume injections to boost sensitivity. Depending on the circumstances (sample volume, sample matrix, compound structure, question to be answered, availability of a radiolabel, etc.) the right tool or combination of tools need to be selected since none of these techniques should be seen as the standard technique that suits all measurements.

Novel Aspect:
New approaches for the estimation of metabolite abundances when standards are lacking 2D-LC allowing large volume injections to enhance sensitivity

Cvačka, Josef,
Institute of Organic Chemistry and Biochemistry, Prague

Contributing Authors:
Barbora Rumlová, Department of Analytical Chemistry, Faculty of Science, Charles University in Prague, 128 43 Prague 2, Czech Republic; Vladimír Vrkoslav, Institute of Organic Chemistry and Biochemistry, Academy of Sciences of the Czech Republic, 166 10 Prague 6, Czech Republic; Risto Kostiainen Division of Pharmaceutical Chemistry and Technology, Faculty of Pharmacy, University of Helsinki, P.O. Box 56, 00014 Helsinki, Finland; Markus Haapala, Division of Pharmaceutical Chemistry and Technology, Faculty of Pharmacy, University of Helsinki, P.O. Box 56, 00014 Helsinki, Finland; Pavlína Nekvasilová, Department of Analytical Chemistry, Faculty of Science, Charles University in Prague, 128 43 Prague 2, Czech Republic; Josef Cvačka, Institute of Organic Chemistry and Biochemistry, Academy of Sciences of the Czech Republic, 166 10 Prague 6, Czech Republic.

Title:
Optimization and applications of low-flow APCI and APPI mass spectrometry with microfabricated heated glass nebulizer

Introduction:
Miniaturized separation systems used in bioanalytical chemistry require detectors compatible with micro- and nano-flows. Nanoelectrospray works well for polar compounds; however, it is rather inefficient at ionizing less polar species. Chip-based microfabricated sprayers seem to be a promising way for miniaturizing APCI and APPI sources (Anal. Chem. 2004, 76, 6659-6664 and 6797-6801).

Methods:
A borosilicate glass chip with a 50-µm i.d. fused silica capillary glued inside a rectangular channel (500 x 250 µm) and equipped with a platinum heater was used as a nebulizer for APPI and APCI. The chip was mounted on an x-y-z micromanipulator and attached to LCQ Fleet or LTQ Orbitrap XL (Thermo). The samples were delivered using a syringe pump or ACQUITY UPLC M-Class System (Waters).

Results:
The geometrical arrangement of the chip, APCI needle, and heated capillary orifice was optimized on Thermo mass spectrometers. The optimum position of chip outlet channel was in axis, 10 mm from the capillary orifice. The highest signals were observed for flow rates 10-12 µl/min (sample) and 30-40 ml/min (nebulizing gas). The response for reserpine continuously infused at 500 ng/min into the ion source (8 µl/min) was ca. 2 orders of magnitude higher than for commercial APCI source (100 µl/min). In APPI mode, the chip was positioned either in axis or with a tilt of 40 degrees, 10 mm from the capillary orifice. The highest signals were observed for high nebulizing gas flow rates (>100 ml/min). The device was successfully used for detection of non-polar lipids including FAMEs, triacylglycerols and lipids isolated from vernix caseosa. This work was supported by CSF project 16-01639S.

Novel Aspect:
Optimized geometry of an APCI/APPI source on Thermo instruments with a new generation of heated microfabricated chip. Novel applications for lipids.

Czar, Martin,
Eidgenössische Technische Hochschule Zürich

Contributing Authors:
Renato Zenobi, ETH Zurich.

Title:
Enzyme-substrate complexes in the gas phase: First steps towards gas-phase enzymology

Introduction:
Whether electrosprayed biomolecules resemble their native forms remains hotly debated. Here, we tackle the problem with a new, functional approach, using trypsin as a model enzyme. The aim is to isolate gaseous noncovalent trypsin-substrate complexes, and to see whether the acylenzyme (with the C-terminus of the substrate having dissociated after specific enzymatic cleavage) can be produced.

Methods:
Nano-ESI mass spectra were acquired using gold-coated glass capillaries with a Waters Synapt G2 (Q-ToF) mass spectrometer. ESI solutions contained trypsin and/or model substrate (N-benzoyl-capped synthetic hexapeptides of type Bz-XXRGGG) at equimolar (10 micromolar) concentrations. Ammonium acetate (500 mM) was used for "native" ESI. Measurements at low pH were done in 1% aqueous acetic acid.

Results:
ESI mass spectra of trypsin-substrate mixtures indicate that, for many peptides, the hydrolysis reaction goes to completion in the nano-ESI emitters prior to MS analysis. However, trypsin-bound N-terminal fragments are clearly visible in the high mass regime, in accordance with trypsin's N-terminal recognition properties. Hydrolysis rates can be slowed by lowering pH, though this appears to hinder substrate binding affinity. Trypsin complexes with peptides containing D amino acids (which react slowly in solution) survive the electrospray process, however no evidence for a gas-phase enzyme reaction is observed upon collisional activation. To enable studies of the short-lived complexes of trypsin with good substrates, we are currently working towards implementing a rapid-mixing device prior to MS analysis.

Novel Aspect:
First steps towards gas-phase enzymology. Also, the survival yield of trypsin-substrate complexes may be used to infer solution-phase affinity.

D'Agostino, Lisa,
University of Toronto

Contributing Authors:
Scott A. Mabury, University of Toronto.

Title:
Aqueous film forming foam (AFFF) components and related perfluoroalkyl and polyfluoroalkyl substances (PFASs) in Canadian surface waters

Introduction:
Dozens of novel PFASs have recently been identified in AFFFs, but environmental detection of these compounds has been limited to a few species near firefighting training areas. Over 40 AFFF-related PFASs were surveyed at AFFF-impacted, urban, and rural sites in Southern Ontario and Nunavut to determine whether these compounds may contribute to unidentified organofluorine in the water.

Methods:
Surface water grab samples (500 mL) were collected and extracted by mixed-mode WAX SPE in a clean lab. Extracts were analyzed by UPLC-ESI-MS/MS. Quantitation of selected PFASs was performed using authentic standards and isotopically-labelled internal standards, where available. Qualitative analysis was used to detect additional AFFF components with matching of retention times to AFFF extracts.

Results:
Some AFFF-impacted water had similar concentrations of perfluorohexane sulfonate (PFHxS) and perfluorooctane sulfonate (PFOS), perhaps due to degradation of perfluorohexane sulfonamide (FHxSA)-based AFFF surfactants detected in some samples. FHxSA was detected in all AFFF-impacted and urban samples (0.04-20 ng/L), indicating the prevalence of PFHxS precursors in surface waters. Fluorotelomer PFASs were also frequently detected in surface waters, including 6:2 fluorotelomer sulfonate (FTSA, up to 3.5 ng/L), 6:2 fluorotelomer sulfonamide alkylbetaine (FTAB, up to 25 ng/L), 6:2 fluorotelomermercaptoalkylamido sulfonate sulfone (FTSAS-SO2), and fluorotelomer betaines (FTBs). Sampling along the Welland River suggests that FTBs may be particularly persistent fluorotelomer PFASs as evidence indicates they do not decrease as rapidly as other fluorotelomer PFASs downstream.

Novel Aspect:
Several AFFF-related and novel PFASs were measured or detected in surface waters for the first time, including FHxSA, 6:2 FTAB, 6:2 FTSAS-SO2 and FTB.

Dai, Han,
GlaxoSmithKline

Contributing Authors:
Huizhen Zhao, GlaxoSmithKline; Sharon M. Sweitzer, GlaxoSmithKline; James L. Ellis GlaxoSmithKline; Yong Jiang, GlaxoSmithKline.

Title:
HDX-MS Studies of Human SIRT1 Activation by STACs

Introduction:
Sirtuins are a family of highly conserved NAD+-dependent deacylases that have been linked to a number of important biological processes across a broad span of diverse organisms. The validated, protective role of increased mammalian SIRT1 activity in metabolic diseases, inflammatory & autoimmune diseases, and neurodegeneration makes this enzyme an attractive therapeutic target.

Methods:
We have used hydrogen-deuterium exchange mass spectrometry (HDX-MS) to characterize the dynamic properties of full length human SIRT1 in the absence or presence of its substrates and/or ligands.

Results:
The HDX-MS results revealed the key structured regions of human SIRT1 required for its deacetylation activity and activation by sirtuin-activating compounds (STACs), which enabled the design of a mini-hSIRT1 construct and elucidation of the first crystal structure of a SIRT1/STAC complex. The key intramolecular interactions essential for the basal deacetylation activity observed in the crystal structure were confirmed using HDX-MS. HDX-MS profiling of full length SIRT1 in the absence or presence of substrate and STAC suggested an allosteric coupling between STAC and substrate binding, which is disrupted by a single point mutation E230K that impairs the SIRT1 activation by STACs.

Novel Aspect:
The combination of HDX-MS and structural biology provided important insights into the molecular mechanism of SIRT1 activation by STACs.

Dai, Yuqin,
Agilent Technologies

Title:
Using Ion Pair-Reverse Phase LC/Q-TOF Method to Maximize Coverage of Endogenous Metabolite Classes with Separation of Important Isomers

Introduction:
LC/Q-TOF MS is routinely used in metabolomics for discovery work. Here we present a high performance ion pair-reverse phase (IP-RP) LC/Q-TOF MS method for simultaneous detection of a diverse set of metabolites including amino acids, carboxylic acids, sugar phosphates, and nucleotides, etc. with separation of several pairs of important isomers, thus reducing the need for multiple LC/MS methods.

Methods:
A set of metabolites covering different chemical classes was selected for IP-RP LC/MS method development. The chromatographic separation was optimized on a C18 column using a gradient from water to methanol containing tributylamine and acetic acid. All analyses were performed using a negative ion electrospray ionization mode on a UHPLC system coupled to a Q-TOF MS.

Results:
A robust IP-RP LC/Q-TOF method was developed for simultaneous detection of a diverse set of endogenous metabolites and several pairs of biologically isomers such as citrate/isocitrate and D-glucose-6-phosphate/alpha-D-glucose-1-phosphate, etc. For the three pH conditions (5, 5.8 and 7.5), we found that the mobile phase at pH of 5.8 provides the overall best results in terms of the chromatographic separation and sensitivity among the metabolites. To further enhance signals of small, labile metabolites, we utilize an intelligent tuning algorithm to optimize the ion transmission of the Q-TOF. This method clearly demonstrated superior analytical performance in terms of the retention of ionic metabolites, reproducibility, and the coverage of metabolite classes compared with HILIC methods. An application of this method for analysis of real world biological samples is being studied.

Novel Aspect:
High performance IP-RP LC/Q-TOF method for profiling diverse classes of endogenous metabolites with separation of important isomers

Dane, John,
JEOL USA, Inc.

Contributing Authors:
Masaaki Ubukata, JEOL Ltd; Robert B. Cody, JEOL USA, Inc; Masahiro Ito JEOL Ltd; Toshiyuki Kato, JEOL Ltd; Jean-François Focant, University of Liège; Junichi Osuga, JEOL Europe S.A.S.; Bram van der Meer, JEOL Europe B.V.

Title:
Photoionization as a Soft Ionization Source for Comprehensive Two-dimensional GC (GCxGC) and High-Resolution TOFMS

Introduction:
Recently, we developed a PI ion source for a GC/HR-TOFMS system. GC/PI MS and GCxGC/PI MS are particularly useful for the analysis of hydrocarbons and aromatic compounds. Since the GC/HR-TOFMS system provides exact mass information for all of the ions generated by PI, target compound identification can be very selective, thus facilitating the identification of unknowns.

Methods:
Petroleum biomarkers including steranes, triterpenes and adamantanes in several crude oils dissolved in hexane were measured. The samples were analyzed using an AccuTOF GCv 4G (JEOL), GC/HR-TOFMS equipped with a Zoex ZX-2 thermal modulator for GCxGC separations. EI was used for hard ionization, and PI was used for soft ionization. A single-point drift compensation was used for mass calibration.

Results:
Hydrocarbons (i.e. n-Alkanes, PAHs, steranes, triterpenes and adamantanes) as well as other compound classes (chloroalkanes, ketones, FAMEs, phthalates and phenols) showed abundant molecular ion peaks with the lone exception being alcohols which predominantly produced [M-H2O]+. The HRMS was able to separate biomarkers from their interferences despite having identical integer masses but different exact masses. However, EI can produce homologue fragment ions that complicate the identification of sterane molecular ions. For example, the molecular ion for steranes (m/z 372.3756) and the isotopic fragment ion of methyl steranes (m/z 372.3712), require a resolving power of over 200,000 for peak separation. However, PI eliminates the presence of these interfering fragment ions, thus making it easier to interpret the GCxGC data for the crude oil samples.

Novel Aspect:
The combination of GCxGC with high-resolution mass spectrometry and photoionization is a powerful tool for biomarker identification in crude oils.

Dannhorn, Andreas,
Imperial College London

Contributing Authors:
Nicole Strittmatter, Drug Safety & Metabolism, AstraZeneca, Cambridge, CB4 0WG, UK; Richard J. A. Goodwin, Drug Safety & Metabolism, AstraZeneca, Cambridge, CB4 0WG, UK; Anna Mroz Department of Surgery and Cancer, Sir Alexander Fleming Building, Imperial College, London SW7 2AZ, UK; Jocelyn Tillner, Department of Surgery and Cancer, Sir Alexander Fleming Building, Imperial College, London SW7 2AZ, UK; Zoltán Takáts, Department of Surgery and Cancer, Sir Alexander Fleming Building, Imperial College, London SW7 2AZ, UK; Renata Soares, Department of Surgery and Cancer, Sir Alexander Fleming Building, Imperial College, London SW7 2AZ, UK; John G. Swales, Drug Safety & Metabolism, AstraZeneca, Cambridge, CB4 0WG, UK.

Title:
Mapping the spatial distribution of cassette-dosed drugs in tissue sections applying a targeted DESI-MSI approach

Introduction:
In preclinical development of drugs, cassette-dosing of compounds has become a powerful tool to help increase throughput whilst reducing the number of animals needed. In the presented study, we apply targeted desorption electrospray ionization-mass spectrometry imaging (DESI-MSI) to determine the spatial distribution of 4 orally dosed drugs and their main metabolites in rat brain, liver and kidney

Methods:
To determine the spatial distribution of the drugs (erlotinib, moxifloxacin, olanzapine and terfenadine) and their metabolites in different organs, DESI-MSI was performed on tissue sections. The drug metabolites, which were identified by DESI-MSI, within the different tissues, were confirmed by LC-MS analysis of extracted adjacent tissue sections and the data was correlated with the DESI images.

Results:
Using DESI-MS imaging, we were able to detect all drugs in the different tissue types, down to drug tissue concentrations in the low nmol/g range. The abundances for moxifloxacin and other chemically more complex molecules, such as phase 2 metabolites were rather low using DESI-MSI performed on an instrument using quadrupole-time-of-flight (Q-ToF) mass analyzer operated in full scan mode. Optimization of the desorption/ionization conditions and application of a targeted approach, operating a Q-ToF in targeted mode and transferring the method onto a triple quadrupol mass spectrometer, improved the signal significantly. Relative abundances for the detected drugs and their metabolites observed by targeted DESI-MSI showed good correlation with concentrations measured in LC-MS. The identified metabolites were phase one metabolites as well as phase two metabolites.

Novel Aspect:
This is the first study in which targeted DESI-MSI is used to map the spatial distribution of cassette-dosed drugs and correlate it with LC-MS data.

de Graaf, Erik,
Fondazione Pisana per la Scienza

Contributing Authors:
Sha Lou, Leiden University Medical Center, Leiden, the Netherlands; Liam McDonnell, Fondazione Pisana per la Scienza, Pisa, Italy; Judith Bovée Leiden University Medical Center, Leiden, the Netherlands; Davide Pellegrini, Fondazione Pisana per la Scienza, Pisa, Italy.

Title:
Novel automated quantitative microproteomics approaches for analyzing minute sample amounts in tumour heterogeneity

Introduction:
To study the interplay of tumor subpopulations laser capture microdissection (LCM) can be used to excise and isolate histopathologically different regions. While LC-MS has reached high sensitivities, sample preparation workflows for such small sample amounts remain challenging. In this work we developed protocols to quantitatively analyze ~1 µg of starting material, corresponding to 1 mm2 tissue.

Methods:
A modified microBCA assay was developed to reliably estimate protein content without sacrificing too much sample (10%, ~100 ng). Protein cleanup and digestion was performed using magnetic beads. Subsequent peptide processing was performed on a Bravo AssayMAP robot for automated sample cleanup using preconditioned labware, on-column TMT-labeling and off-line HpH fractionation.

Results:
A significant adsorbance of peptides was observed when desalting sample amounts below 7.5 µg, resulting in poor label free quantification. Quantitative reproducibility could be fully restored when using preconditioned labware. For relative quantification of 1 µg of sample a novel automated on-column TMT labeling protocol was developed with better labeling efficiency and a 2-fold increase in PSMs compared to conventional methods. Furthermore, automated off-line HpH fractionation using only 6 µg of sample resulted in a two fold increase of unique peptides and an increase in proteome coverage from 4000 to 6000 proteins when compared to a single 4h gradient using the same amount of sample. Finally, the entire integrated and automated sample clean-up, TMT labeling and HpH fractionation workflow was applied to study different parts of highly heterogeneous myxofibrosarcoma human tumor biopsies.

Novel Aspect:
This set of novel protocols enabled the analysis of 1 µg tissue with a significant improvement in proteome coverage and quantitative reproducibility.

de Graaf, Erik,
Fondazione Pisana per la Scienza

Contributing Authors:
Sha Lou, Leiden University Medical Center, Leiden, The Netherlands; Judith Bovée, Leiden University Medical Center, Leiden, The Netherlands; Liam McDonnell Fondazione Pisana per la Scienza, Pisa, Italy; Davide Pellegrini, NEST, Scuola Normale Superiore, Pisa, Italy.

Title:
Novel automated quantitative microproteomics approaches for analyzing minute sample amounts in tumour heterogeneity

Introduction:
To study the interplay of tumor subpopulations laser capture microdissection (LCM) can be used to excise and isolate histopathologically different regions. While LC-MS has reached high sensitivities, sample preparation workflows for such small sample amounts remain challenging. In this work we developed protocols to quantitatively analyze ~1 µg of starting material, corresponding to 1 mm2 tissue.

Methods:
A modified microBCA assay was developed to reliably estimate protein content without sacrificing too much sample (10%, ~100 ng). Protein cleanup and digestion was performed using magnetic beads. Subsequent peptide processing was performed on a Bravo AssayMAP robot for automated sample cleanup using preconditioned labware, on-column TMT-labeling and off-line HpH fractionation.

Results:
A significant adsorbance of peptides was observed when desalting sample amounts below 7.5 µg, resulting in poor label free quantification. Quantitative reproducibility could be fully restored when using preconditioned labware. For relative quantification of 1 µg of sample a novel automated on-column TMT labeling protocol was developed with better labeling efficiency and a 2-fold increase in PSMs compared to conventional methods. Furthermore, automated off-line HpH fractionation using only 6 µg of sample resulted in a two fold increase of unique peptides and an increase in proteome coverage from 4000 to 6000 proteins when compared to a single 4h gradient using the same amount of sample. Finally, the entire integrated and automated sample clean-up, TMT labeling and HpH fractionation workflow was applied to study different parts of highly heterogeneous myxofibrosarcoma human tumor biopsies.

Novel Aspect:
This set of novel protocols enabled the analysis of 1 µg tissue with a significant improvement in proteome coverage and quantitative reproducibility.

De Winter, Julien,
University of Mons

Contributing Authors:
Philippe Dugourd, Institut Lumière Matière, University Lyon 1; ChangMin Choi, Institut Lumière Matière, University Lyon 1; Pascal Gerbaux University of Mons; Quentin Duez, University of Mons; Thomas Josse, University of Mons; Fabien Chirot, Institut Lumière Matière, University Lyon 1.

Title:
Ion Mobility Mass Spectrometry for polymer ions : Arrival Time Distribution symmetry along the ion series

Introduction:
Ion Mobility Mass Spectrometry (IMMS) is an analytical tool able to separate ions in gas phase based on their 3D structure. This technique is certainly considered as one of the most powerful and versatile analytical tool currently available. Among all the research topics using IMMS most of them are related to biological applications but polymer characterization is also considered.

Methods:
For this study, linear polylactide (PLA) was synthesized in order to characterize them by Ion Mobility using two different ion mobility cells. The first part of this study was devoted to investigate PLA ions using a commercial IMMS (Waters® Synapt G2-Si) followed by a comparison with results recorded on a non-commercial IM-IMMS setup built by Prof Philippe Dugourd and Co-workers.

Results:
For the development of IMMS as a key structural tool, the ability of this method to access the secondary structure of polymer ions in the gas phase is starting to get explored in great details. In this context, we scrutinized the shape of the arrival time distribution (ATD) of polylactide ions of different molecular masses using a commercial and non-commercial Ion Mobility mass spectrometer equipped with 2 different mobility cells as well as 2 different electrospray sources. The high resolution and sensitivity of the experimental setups allow evidencing the occurrence of a loss of symmetry of the ATD signals when increasing the number of monomer units, which demonstrates the coexistence of different ion conformations. Interestingly, this loss of symmetry seems to be source dependent as the interconversion is not observed by any common collisional activation methods.

Novel Aspect:
The loss of symmetry for the ATD, for polymer ions, is not associated to a gas phase unfolding process (as proposed for proteins).

Dean, Natalie,
University of Victoria

Contributing Authors:

J. Scott McIndoe, University of Victoria.

Title:

A mechanistic investigation of the Hiyama Cross- Coupling reaction using ESI-MS

Introduction:

The palladium-catalyzed cross coupling of organosilicon reagents with organo halides is an efficient and attractive strategy for the formation of carbon-carbon bonds. The primary focus of ongoing research has been in expanding the scope of the reaction, while detailed studies of the mechanism of this reaction are limited.

Methods:

ESI-MS coupled with pressurized sample infusion (PSI) - a technique which allows for continuous reaction monitoring over time

Results:

Our research is focused in achieving mechanistic insight of the Hiyama reaction, in particular the role of the organosilicon compound in the transmetalation step. Our results thus far have revealed a complex and dynamic exchange reaction of anionic pentacoordinate silicon species - the identity and possible role of which will be presented.

Novel Aspect:

Continuous reaction monitoring of the Hiyama Cross-coupling Reaction by ESI-MS

Debrauwer, Laurent,
Axiom Platform - UMR 1331 INRA Toxalim - TOULOUSE - FRANCE

Contributing Authors:

DEBRAUWER L, Toxalim, Université de Toulouse, INRA, ENVT, INP-Purpan, Université de Toulouse 3 Paul Sabatier, F-31027 Toulouse, France; PLANCHE C, INRA, UR370 QuaPA, MASS Group, F-63122 Saint-Genès-Champanelle, France; CHEVOLLEAU S Toxalim, Université de Toulouse, INRA, ENVT, INP-Purpan, Université de Toulouse 3 Paul Sabatier, F-31027 Toulouse, France; ENGEL E, INRA, UR370 QuaPA, MASS Group, F-63122 Saint-Genès-Champanelle, France; NOGUER-MEIRELES MH, Toxalim, Université de Toulouse, INRA, ENVT, INP-Purpan, Université de Toulouse 3 Paul Sabatier, F-31027 Toulouse, France; RATEL J, INRA, UR370 QuaPA, MASS Group, F-63122 Saint-Genès-Champanelle, France.

Title:

Thermal degradation of sulfamethoxazole during food cooking

Introduction:

The occurrence of micro-pollutants in meat represents a public health risk. Parent toxicants are quite well monitored, but little is know about their fate during processes like cooking, which hinders comprehensive risk assessment. In this study LC-HRMS was used to investigate the fate of sulfamethoxazole, one antibiotic that can be found in meat, using various domestic cooking levels.

Methods:

Spiked meat was cooked according to 4 pan cooking levels (rare, medium, well-done, very well-done). Meat was extracted by accelerated solvent extraction and liquid extraction. All recoveries were checked using a radiolabelled molecule, allowing absolute quantification by radio-HPLC. Degradation products were identified by LC-high resolution MS and MS/MS carried out on a LTQ-Orbitrap XL instrument.

Results:

Radio-HPLC allowed for both highlighting and absolute quantification of 6 different SMX degradation compounds depending on the cooking conditions. These compounds were identified using reversed phase LC coupled to positive ESI-MS and MS/MS, and high-resolution exact mass measurements carried out on both molecular species and fragment ions. SMX was thus shown to undergo thermal degradation into an isomerization product (which complete structural elucidation with the help of NMR is in progress), acetylated derivatives, and compounds resulting from the opening or loss of the oxazole moiety of the parent SMX. All these compounds are described for the first time in cooked meat, although the oxazole modified compounds are known as SMX photodegradation products. The structural elucidation of each compound will be presented and a degradation scheme of SMX during meat cooking will be proposed.

Novel Aspect:

New data about identification of thermal degradation compounds of sulfamethoxazole formed during meat cooking.

Decroo, Corentin,
Umons

Contributing Authors:
Patrick Flammang, Umons; Julien De Winter, Umons; Guillaume Caulier Umons; Jérôme Cornil, Umons; Pascal Gerbaux, Umons; Vincent Lemaur, Umons.

Title:
Ion mobility mass spectrometry of saponins: from ion conformation to molecule structure.

Introduction:
For many decades, scientists have been concentrating their effort toward the structural characterization of natural molecules of pharmaceutical interest. The saponin family of molecules raises today a huge interest amongst the scientific community due to their great biodiversity and their involvement in biological phenomena with antibacterial and antifungal activities, for instance.

Methods:
LC-MS methods represent the most suitable way for saponin characterization. They are some difficulties to separate isomeric structures presenting minute structural differences. We propose now that the implementation of an additional separation dimension, say ion mobility, could allow deciphering even subtle structural differences allowing a better structural characterization of saponins.

Results:
Saponin molecules are secondary metabolite widespread in plants and marine animals such as Echinoderms. By definition, these molecules are all constituted of a glycone part (oligosaccharide) covalently attached to the aglycone that is based on a steroidic or triterpenoidic skeleton. Due to the huge variety in the nature of the monosaccharides and the steroids, the molecular diversity of saponins is extremely important. In the context of structural characterization of saponins on the basis of ion mobility, it is important to first understand the influence of the primary structure of the saponin congeners on the ion structures. In the present communication, we report preliminary results on the ion mobility analysis of saponins extracted from the common soy, namely Glycine max. The experimental data are generated on a Waters Synapt G2-Si mass spectrometer

Novel Aspect:
In this work, the strcutral characterization of saponin (ions) by ion mobility are investigated in first time

Delanghe, Bernard,
Thermo Fisher Scientific

Contributing Authors:
Torsten Ueckert, Thermo Fisher Scientific, Bremen; Carmen Paschke, Thermo Fisher Scientific, Bremen; Kai Frizemeijer Thermo Fisher Scientific, Bremen; Katja Tham, Thermo Fisher Scientific, Bremen; Hans Pfaff, Thermo Fisher Scientific, Bremen; David M. Horn, Thermo Fisher Scientific, San Jose; Daniel Lopez-Ferrer, Thermo Fisher Scientific, San Jose; Joseph Brown, Thermo Fisher Scientific, San Jose.

Title:
A new method for Label-free quantification in the Proteome Discoverer framework

Introduction:
Proteome Discoverer is a node-based workflow engine and platform for analysis of mass spectrometry-based proteomics data. Here we present a new workflow for label-free quantification that provides the full suite of quantitative capabilities only available previously to isotopically-labeled quantification. The workflow will be demonstrated on several datasets.

Methods:
A preliminary version of Proteome Discoverer 2.2 was used to interpret the data. Sequest HT was used as the primary search algorithm. Datasets were normalized on total peptide abundance and scaled abundances were subsequently used to compare the protein and peptides across the replicates and datasets.

Results:
The quantification values for proteins are calculated by simply summing the abundances of the peptide groups. The peptide groups abundances are also calculated by summing the abundances peptide features associated with the identified PSMs. For datasets with missing PSMs, the label-free workflow looks for those features in the remaining datasets using a “feature mapping” step. The “HeLa spike-in” dataset produced over 5000 protein identifications and over 34000 unique peptides, the majority of which were quantified. 80% of the proteins in the HeLa proteins were within 0.8 to 1.2 in the comparisons of the three different samples.More work will be presented on samples with biological significance as well as tools for statistical analysis within Proteome Discoverer as well as the ProteinCenter software.

Novel Aspect:
Scaled abundance of label-free results provides straightforward interpretation of label-free quantification results.

Delcourt, Vivian,
Université Lille 1 INSERM U-1192 PRISM - Université Sherbrooke

Contributing Authors:
Julien Franck, Université Lille 1 - INSERM U1192 - PRISM; Jusal Quanico, Université Lille 1 - INSERM U1192 - PRISM; Jean-Pascal Gimeno Université Lille 1 - INSERM U1192 - PRISM; Jean-François Jacques, Université de Sherbrooke; Xavier Roucou, Université de Sherbrooke; Michel Salzet, Université Lille 1 - INSERM U1192 - PRISM; Isabelle Fournier, Université Lille 1 - INSERM U1192 - PRISM.

Title:
Method for clinical application of MALDI Imaging driven Top-down microproteomics for discovery of tumor biomarkers

Introduction:
MALDI-MSI has been widely applied to clinical molecular histology to classify tumorous and healthy regions of a tissue. However, it's abilities of tracking intact proteins lacks a strong identification method. Recent advances in high resolution mass analyzer and fragmentation methods now allow the identification of a large number intact proteins out of complex samples with top-down proteomics.

Methods:
We developed a method combining MALDI-MSI for healthy and tumorous statistical classification of the tissue using spectral cluster . We then apply top-down analysis the different MALDI-MSI defined regions using two methods of microproteomics sample processing.

Results:
Results show that MALDI-MSI was able to effectively differentiate healthy, borderline and tumorous parts of a cancerous ovarian tissue. The three regions were then analyzed using top-down microproteomics and showed effectiveness to identify several potential ovarian cancer biomarkers in borderline and tumorous regions of the tissue. Alternative proteins which are small proteins translated from out-of-frame regions of a transcript were identified as well, showing impressive abilities to detect novel proteins. The method is showing strong capabilities to identify large number of intact proteins from low amounts of materials with minimal sample preparation and is expected to be used in the future for clinical characterization of diseases.

Novel Aspect:
Novel method combining MALDI-MSI and Top-down proteomics for clinical investigations

Deng, Bin,
York University

Contributing Authors:
Shaolong Zhu, The Centre for Research in Mass Spectrometry, Department of Chemistry, York University, Toronto, ON, Canada; Cristina Lento, The Centre for Research in Mass Spectrometry, Department of Chemistry, York University, Toronto, ON, Canada; Derek J. Wilson The Centre for Research in Mass Spectrometry, Department of Chemistry, York University, Toronto, ON, Canada.

Title:
Microfluidic-Enabled Time-Resolved Hydrogen Deuterium Exchange Mass Spectrometry (TRHDX-MS) for Equilibrated and Kinetic Epitope Mapping

Introduction:
TRHDX-MS is able to detect the higher order structure and millisecond conformational changes of proteins. The newly designed microfluidic chip enables ultra-fast HDX, acid quenching, and efficient protease digestion of proteins. Coupled with TOF-HRMS, peptide regions on myoglobin recognized by its antibody were identified for both pre-incubated and kinetically formed myoglobin-antibody complex.

Methods:
For equilibrated epitopes, myoglobin was pre-incubated with its antibody before TRHDX-MS. For kinetic epitopes, myoglobin was rapidly mixed with its deuterated antibody by TRHDX before MS. HDX profiles of the two types of myoglobin-antibody complexes were compared with native myoglobin. Peptide regions on myoglobin showing significantly less D2O uptake upon binding were identified as epitopes.

Results:
A sequence coverage of 98% was achieved by on-chip TRHDX-MS of native and antibody-bound myoglobin. Most of the peptides obtained from antibody-bound myoglobin reached a deuterium uptake plateau in less than 1 s, indicating the significance of sub-second detection by TRHDX-MS. In the equilibrium study, four epitopes along with one induced flexible region were identified. All epitopes were in agreement with previous literature data, showing the comparability of TRHDX-MS with conventional epitope methods. In the kinetic analysis, three epitopes were identified in the same regions. The allosteric conformational changes of myoglobin were not found because no time was given for the post complex formation. TRHDX-MS showed its ability to reveal structural epitopes with additional information about transient binding interactions, and complete elimination of allosteric artifacts.

Novel Aspect:
Kinetic epitope mapping eliminates allosteric effects and reveals millisecond binding dynamics of proteins by using microfluidic chip based TRHDX-MS.

DEPRETRE, Nicolas,
ChemoSens - INRA (FRANCE)

Contributing Authors:

Véronique DE SMEDT-PEYRUSSE, Nutrition et Neurobiologie Intégrée, INRA, UMR 1286 - University of Bordeaux- Bordeaux, France; Elodie A.Y. MASSON, UMR CSGA - 1324 INRA - 6265 CNRS – University of Burgundy Franche-Comté, Dijon, France; Pierre TRIFILIEFF Nutrition et Neurobiologie Intégrée, INRA, UMR 1286 - University of Bordeaux- Bordeaux, France; Olivier BERDEAUX, UMR CSGA - 1324 INRA - 6265 CNRS – University of Burgundy Franche-Comté, Dijon, France; Fabien DUCROCQ, Nutrition et Neurobiologie Intégrée, INRA, UMR 1286 - University of Bordeaux- Bordeaux, France; Nicolas DEPRETRE, UMR CSGA - 1324 INRA - 6265 CNRS – University of Burgundy Franche-Comté, Dijon, France; Estelle SIBILLE, UMR CSGA - 1324 INRA - 6265 CNRS – University of Burgundy Franche-Comté, Dijon, France.

Title:

Targeted lipidomic analysis of complex lipids in mouse synaptosomes by Hydrophilic Interaction Liquid Chromatography - Tandem Mass Spectrometry

Introduction:

The synapse, allowing communication between neurons, has been shown to display a specific lipid composition compared to other neuronal compartments, namely an enrichment in cholesterol and PUFA-containing phospholipids. To deeper explore the lipid composition of this unique compartment, we performed a lipidomic characterization of synaptosomes (isolated synaptic terminals) from mouse forebrains.

Methods:

Total lipids were extracted from synaptosomes. The main lipid classes, such as glycerophospholipids (GlyPL), gangliosides (GG) and neutral glycosphingolipids, ceramides (Cer) and sphingomyelins (SM) were separated using HPLC under HILIC conditions. Molecular species in each lipid class were structurally characterized and quantified using an Orbitrap and a QqQ mass spectrometers.

Results:

An effective separation of the GlyPL, SM, Cer, hexosylceramides (HexCer), lactosylceramides (LacCer) and GG was obtained. Individual species of GlyPL were fully characterized, providing information on the two fatty acids branched on the glycerol core. Similarly, the ceramide moieties (long chain base and fatty acid) of individual species of GG, SM, HexCer and LacCer were also characterized. As an example, our results show that synaptosomes are rich in GG, since we detected 16 GG classes, from the simplest (GM4) to the more complex GG (GQ1b). The GG profile contains mainly tetraosylGG (GD1b, GT1b, acetylated GT1b and GQ1b), but also significant amounts of GT3 and acetylated GT3. The ceramide moiety ranged from 34 to 42 carbons, with 0 to 4 unsaturations. Ceramides 36:1 and 38:1 appear to be the major ones and sphingosine (d18:1) is the most represented long chain base in this tissue.

Novel Aspect:

Our method offers a global and thorough characterization of complex lipids applied to mouse synaptosomes, highlighting a specific lipid composition.

Desai, Reena,
Anzac research Institute,Sydney,Australia

Contributing Authors:
Pekka Keski-Rahkonen, International Agency for Research on Cancer, F-69372 Lyon, France; David J. Handelsman, ANZAC Research Institute, University of Sydney, Sydney, NSW 2139, Australia; Reena Desai ANZAC Research Institute, University of Sydney, Sydney, NSW 2139, Australia; Mark Jimenez, ANZAC Research Institute, University of Sydney, Sydney, NSW 2139, Australia; D. Tim Harwood, Cawthron Institute, Nelson 7010, New Zealand.

Title:
Measurement of Estradiol in Human Serum by LC-MS/MS Using a Novel Estrogen-Specific Derivatization Reagent

Introduction:
Measurement of 17β-estradiol (E2) in the clinical setting requires high assay sensitivity and selectivity. Here, we describe a LC-MS/MS method that employs a novel derivatization reagent 1,2-dimethylimidazole-5-sulfonylchloride (DMIS) for high-sensitivity analysis of E2 and simultaneous analysis of underivatized testosterone(T) and dihydrotestosterone(DHT).

Methods:
Samples (200 µL) were extracted with MTBE. The evaporation residue was reconstituted in sodium bicarbonate and DMIS in acetone, and heated at 60 °C for 15 min before injection. Chromatographic separation used a Kinetex PhenylHexyl column (100×2.1 mm;1.7 µm) at 50 °C and a 9.5-min water-methanol gradient. Detection was performed with SCIEXAPI5000 system using photospray(APPI) and toluene as dopant.

Results:
Method validation was according to FDA guidelines. LLOQs in a 200-µL sample were 0.5 pg/mL for E2,25 pg/mL for T, and 0.10 ng/mL for DHT. Spiked samples at a concentration of 0.005, 0.08 and 0.4 ng/mL for E2; 0.05, 0.4, and 8 ng/mL for T and DHT, had mean accuracies of 96% (E2), 103% (T), and 101% (DHT). Extraction recoveries were 81–92% for all the analytes, with CVs between 4.1% and 10.3%.Carry over in blank injections after the highest calibrant was insignificant at 0.02–0.09% for all the analytes. None of the structurally related steroids studied interfered with the analytes of interest. The E2 serum reference samples BCR-576 (n = 8), BCR-577 (n = 5), and BCR-578 (n= 4) had mean bias of −1.9%, −7.6% and −8.2% , respectively, with CVs between 0.5% and 3.5%,demonstrating good accuracy .

Novel Aspect:
Derivatization of E2 with DMIS significantly improves assay sensitivity and enables monitoring of confirmatory mass transitions of the analyte.

Desbenoit, Nicolas,
Chair of Bioanalytical Science and Food Analysis, University of Bayreuth, Bayreuth, Germany

Contributing Authors:
A. Brunelle, CNRS, Institut de Chimie des Substances Naturelles, Gif-sur-Yvette, France.; A. Fatangare, Chair of Bioanalytical Sciences and Food Analysis, University of Bayreuth, Bayreuth, Germany.; B. Spengler Institute of Inorganic and Analytical Chemistry, Justus-Liebig-University, Giessen, Germany.; A. Walch, Institute of Pathology, Helmoltz Zentrum München, Neuherberg, Germany.; A. Roempp, Chair of Bioanalytical Sciences and Food Analysis, University of Bayreuth, Bayreuth, Germany.

Title:
imzML: Flexibility and exchangeability for mass spectrometry imaging data processing

Introduction:
imzML (www.imzml.org) is an open, vendor-neutral format for MSI data. It provides more flexibility in data processing as the user is no longer limited to proprietary software. Measurement results (full data sets) can be shared between collaborators without loss of information. Herein, we give an overview of recent software developments supporting imzML, and an application for multimodal MSI.

Methods:
Conversion to imzML was performed by converters from academic groups or vendors. MSI data were displayed using the following software tools: open-source (MSiReader, Cardinal...), freely available (BioMap...), and commercial (MALDIVision, SpectViewer...). Multimodal MSI experiments were performed combining TOF-SIMS & MALDI-Orbitrap using a single tissue section. Finally, tissue was stained with HE.

Results:
The data format imzML can be used to store MSI data in a public data repository (PRIDE database), & used as "open data" for metadata analysis or reprocessing. Moreover, imzML files can be used to analyze data from 1 instrument using the software tools mentioned above. Each of them offers unique features which can be of help for specific applications. We will give a brief overview of functionalities using the two different binary data types: Processed from PRIDE (Identifier: PXD001283); & Continuous. Finally, imzML can be used to analyze data from different instruments in 1 software. SIMS and MALDI have been used to study human tumor section. Thanks to the imzML flexibility, 16 adjacent SIMS images have been stitched giving over 3 million of pixels. The co-registration function from MSiReader allows to show a high correlation for the localization & the identification of lipid species.

Novel Aspect:
Overview & evaluation of imzML-based data processing strategies including applications such as public data repository, open data, & multicenter-study.

Dias, Herbert,
USP - Univesity of São Paulo

Contributing Authors:
Ricardo Vessecchi, Department of Chemistry, FFCLRP - USP, Ribeirão Preto, Brazil; Eduardo J. Crevelin, Department of Chemistry, FFCLRP - USP, Ribeirão Preto, Brazil; Antônio E. M. Crotti Department of Chemistry, FFCLRP - USP, Ribeirão Preto, Brazil; Talita C. T. Medeiros, Department of Chemistry, FFCLRP - USP, Ribeirão Preto, Brazil.

Title:
Fragmentation studies on deprotonated dihydrobenzofuran neolignan derivatives by accurate-mass electrospray tandem and sequential mass spectrometry.

Introduction:
Dihydrofuran neolignans (DNL) are a class of phenolic compounds that exhibit a diversity of biological activities, such as trypanocidal, leishmanicidal, among others. The aim of this work was to rationalize the fragmentation of some deprotonated synthetic DNL using ESI-MS/MS.

Methods:
Methyl esters of coumaric (a), caffeic (b) and ferulic (c) acids provided DNL 1a, 1b and 1c by oxidative coupling promoted by Ag2O. These compounds were converted into their acetylated (2a, 2c), hydrogenated (4a, 5a) and dehydrogenated (3a) derivatives and analyzed by accurate-mass (UltrOTOF-Q, Bruker) and MSn (IT Amazon-SL, Bruker) mass spectrometry.

Results:
Termochemical data estimated by Computational Chemistry at B3LYP/6-31+G(d,p) level indicated that deprotonation took place at the phenol hydroxyl for most of the compounds. Elimination of •C2H3O2 are characteristic for benzofuran-type neolignans (BNT) 3a and 5a, whereas loss of MeOH from deprotonated molecule (B) and CO from B (C) are common to all the dihydrobenzofuran-type neolignans 1a-1c, 2a, 2c and 4a. Diagnostic fragment ions were identified for all compounds (e.g. S[C-CO],U[C-C2H4O2] for 1a; R[B-CO2] for 1b; H[B-•CH3], I[H-CO], J[I-•CH3], K[J-•C2H3O2] for 1c; M[A-C2H2O], N[M-MeOH or B-C2H2O], O[N-CO] for 2a; G[A-•CH3], L[G-•CH3] for 2c; E[D-•C2H3O2] for 3a; P[B-MeOH] Q[P-CO], Z[Q-•C2HO] for 4a; and F[D-•C3H5O2] for 5a) and their formations were associated to specific structural features. Violations to the even-electron ion were eventually observed.

Novel Aspect:
Gas-phase fragmentation pathways of a series of deprotonated structurally-related DNL were rationalized and elucidated by the first time.

Díaz Moscoso, Danny Alejandro,
Los Andes University

Contributing Authors:
Chiara Carazzone, Laboratorio de Fitoquímica y Etnofarmacia. Universidad de los Andes; Jerson Felipe Benavides Cortés, Laboratorio de Fitoquímica y Etnofarmacia. Universidad de los Andes.

Title:
VENOM COMPOSITION'S STUDY FROM GENUS LONOMIA CATERPILLARS, USING SDS-PAGE AND HPLC-MS/MS

Introduction:
The genus Lonomia includes some venomous species of caterpillars, typical of Latin America, where they represent a public health problem because of the deadly reactions they can cause. The chemical composition of the venoms of some autochthonous Colombian species of Lonomia are virtually unexplored. Therefore, the aim of this study was to study the composition of these venoms through MS analysis.

Methods:
The specimens were collected in the Colombian Amazon. The venom was extracted with PBS, lyophilized, dissolved in 2%ACN, 0.2%TFA water, centrifuged and fractionated by RP-HPLC. Molecular masses of the components were estimated by SDS-PAGE. Fractions and gel protein bands were subjected to reduction, alkylation and digestion with trypsin, separated by RP-HPLC and analyzed by MS/MS for sequencing.

Results:
Molecular masses of the venom components were estimated to be below 300 kDa, using a 17% polyacrylamide gel. Lyophilized crude venom was loaded into a chromatographic column (Phenomenex LunaC18 100 Å, 150×3 mm, 3 µm) for fractionation, performed at 210 nm in a HPLC-DAD. Fractions were collected manually. Analysis by RP-HPLC-MS/MS of the fractions, showed that the venom is not only composed by peptides and proteins, but also by small molecules. The fractions were also subjected to SDS-PAGE under reducing conditions, and protein bands were excised from gels. Both fractions and bands were subjected to reduction with 10 mM DTT, alkylation with 50 mM iodoacetamide, followed by in-gel and in-solution digestion with trypsin respectively. The resulting tryptic mixtures were separated by RP-HPLC (Phenomenex KinetexC18 100 Å, 100×2.1 mm, 2.6 µm) and analyzed by ESI-MS/MS for peptide sequencing.

Novel Aspect:
In spite of the Colombian biodiversity, most species remain uninvestigated. This is the first study on the venom composition of a venomous caterpillar

Dikler, Sergei,
Bruker Daltonics Inc.

Contributing Authors:
Peter Marshall, GlaxoSmithKline, Stevenage, UK; Melanie Leveridge, GlaxoSmithKline, Stevenage, UK; Michelle Pemberton GlaxoSmithKline, Stevenage, UK; Carl Haslam, GlaxoSmithKline, Stevenage, UK; Neil Hardy, GlaxoSmithKline, Stevenage, UK; Gabriella Clarke, GlaxoSmithKline, Stevenage, UK; Adrian Dunn, GlaxoSmithKline, Stevenage, UK; John Hellicar, GlaxoSmithKline, Stevenage, UK.

Title:
Ultra High Throughput Screening by MALDI-TOF MS in Drug Discovery – Exceeding One Million Samples per Week

Introduction:
High-throughput screening is a key process in Drug Discovery for finding inhibitors against target enzymes. In contrast to fluorescent techniques, MS offers a label-free approach. MALDI-TOF provides an alternative for high-throughput detection using both 1536 and 6144 formats. Here we describe the development of Ultra High Throughput screening using MALDI coupled with nanoliter liquid handling.

Methods:
A Bruker RapifleX MALDI-TOF instrument with a laser repetition rate of 10 kHz was used in positive mode. Mass spectra were acquired with 200 laser shots per sample spot. MALDI targets were spotted with samples in either 1536 or 6144-format using a nanoliter liquid handling device. Sample deposition per spot was a "sandwich" of 25 or 100nL matrix/analytes/matrix spotted onto the MALDI target.

Results:
The development, validation and robust analysis of assays for both small molecule and peptide analytes using MALDI-TOF coupled to nanoliter liquid handling will be expounded in both 1536 and the higher density sample deposition 6144 formats. We will demonstrate that this technology can deliver an MS based readout at speeds of <0.25 sec per sample (including sample introduction/removal) and sample volumes of <1µL which are comparable with traditional fluorescent based Ultra High-Throughput screening approaches. Direct-detection of MALDI-TOF MS for screening assays should reduce the number of false positives, as fluorescent-based technologies are susceptible to interference from colored or fluorescent compounds which may be present in the compound collection. In addition, this technology will reduce the cost and assay development time commonly associated with fluorescent technologies.

Novel Aspect:
Perform Ultra High-Throughput Screening on over One Million biochemical samples per week using MALDI-TOF Mass Spectrometry.

Dikler, Sergei,
Bruker Daltonics Inc.

Title:
Enzymatic Assay Development Using Novel MALDI High-Throughput Screening System

Introduction:
Label-free detection methods based on mass spectrometry are gaining momentum in high-throughput screening (HTS) for drug discovery. Most of these MS methods are centered on multiplexed LC-MS and SPE-MS. MALDI-TOF has an inherent advantage over multiplexed LC-MS and SPE-MS techniques by being 10-100 times faster. This work focuses on developing biochemical assays using a novel MALDI-HTS system.

Methods:
Assay development was done in 384 well format using acetylcholinesterase or butyrylcholinesterase as target enzymes. The assay plates were processed on a MALDI PharmaPulse system, which includes a MALDI-TOF mass spectrometer, liquid handling robot, robotic arm, plate hotel and delidding device. All measurements were done in automated mode with 200-500 laser shots per spectrum.

Results:
This work focuses on assays where substrate (acetylcholine or butyrylcholine) and product (choline) are small molecules. Several buffers were compared including 40 mM TRIS at pH 7.5 (standard for this assay) and 40 mM ammonium bicarbonate at pH 8.6. Higher percent conversion was achieved for ammonium bicarbonate buffer comparing to TRIS in acetylcholinesterase reaction with a reaction time of 10 minutes using manual preparation or robotic preparation. More conditions will be explored using acetylcholinesterase and then adapted for butyrylcholinesterase assay. The next step is going to be quantitating the effect of known inhibitors including huperzine A and rivastigmine tartrate on both assays. Having a variety of optimized assays with different types of substrates and enzymes from different classes proves wide applicability of MALDI-HTS approach to drug discovery methodologies.

Novel Aspect:
Development of biochemical assays for novel ultrahigh-throughput screening system based on a MALDI-TOF mass spectrometer

Dilillo, Marialaura,
Fondazione Pisana per la Scienza

Contributing Authors:
Paolo Aretini, Fondazione Pisana per la Scienza - ONLUS; Avinash Yadav, Scuona Normale Superiore - Pisa ; Fondazione Pisana per la Scienza ONLUS - Pisa; Erik L. de Graaf Fondazione Pisana per la Scienza - ONLUS; Liam A. McDonnell, Center for Proteomics and Metabolomics and Department of Pathology, Leiden University Medical Center; Fondazione Pisana per la Scienza - ONLUS.

Title:
TARGETING CANCER-RELATED PROTEOFORMS UNCOVERED WITH PROTEOGENOMICS

Introduction:
Proteogenomics can overcome some deficiency that both proteomics and genomics alone can have in explaining the complex biology of a tumor. NGS is first used to point out the genetic mutations specifically associated with a patient's tumor. Custom protein database can then be defined and used to assign ID's and to target mutated peptides in bottom-up experiments.

Methods:
Exome Sequencing, RNA-Seq and Shotgun experiment were performed on MCF7 cell lines. A python (2.7) script (VarProt.py) was written to create the custom DB. Peptide fractions, collected with a Bravo Assay MAP, were analyzed with An OT Fusion with an Easy 1000. Targeted proteomics experiment were designed and analyzed using Skyline 3.5. PRM, DIA (10 m/z wind.), msxDIA (4x5m/z) were compared.

Results:
A specific protein database was created for the MCF7 cell line, containing 231 mutated proteins coming from 218 mutated genes. The presence of the mutant proteins was confirmed using bottom-up proteomics on a fractionated peptide digest. 115 mutant peptides (6-20aa, no missed cleavage) were selected and validated using Stable Isotope Standard (SIS – crude peptides from JPT). Six different transitions were selected for each mutant peptide, targeting the mutation identified with the proteogenomic pipeline. Relative areas for each transition were used to compare PRM, msxDIA and DIA. The PRM method gave the highest reproducibility and could identify the largest number of peptides; msxDIA gave the highest statistical similarity between endogenous and SIS.

Novel Aspect:
Our approach can uncover new gene mutation and thus used to identify mutant proteins specifically connected to disease development and progression.

Ding, Li,
SRL (Europe) Ltd.

Contributing Authors:
Li Ding, Shimadzu Research Laboratory (Europe) Ltd.; Aleksandr Rusinov, Shimadzu Research Laboratory (Europe) Ltd.; Hiroaki Nakanishi Shimadzu Research Laboratory (Europe) Ltd.

Title:
FTMS Based on the Ion Orbiting Frequency

Introduction:
Currently the Fourier transform mass spectrometer is mainly in forms of magnetic ICR trap or electrostatic Orbitrap. In the Orbitrap mass analyser, ions are orbiting around the central spindle, while at the same time oscillating in direction of the axis of the spindle. Only the axial oscillation frequency is converted to a mass spectrum because the orbital frequency has a strong energy dependence.

Methods:
We introduce a new electrostatic ion trap where ions also orbit around a central axis. The large ellipticity orbit of ion motion gives the chance to pick up the image charge signal using field forming ring electrodes. The orbital frequency in this device can be optimized to be independent on energy thus the image charge signal of the orbital motion is used to produce a mass spectrum.

Results:
The simulation of the orbital frequency trapping and mass analysis is reported. The simulation work bench includes the ion guide which passes through 2 stages of differentially pumped vacuum regions, a curved injection ion guide (CIG) situates in the ultra-high vacuum and surrounds the orbital frequency trap (OFT). Ions are cooled down in the first stage of the ion guide, and travel through to the CIG with very low transverse energy. Ions are injected into the OFT while they are travelling in the curved injection ion guide (CIG) and their travelling velocity are converted to the rotation momentum in the OFT. OFT voltage setting was optimised for different precession speed. It was found that a higher precession speed helps ion spread out in the whole space of the OFT thus increase the space charge tolerance. Image charge signal was recorded during the simulation and converted to MS by FFT

Novel Aspect:
Orbital frequency in electrostatic ion trap can be measured and transformed to a mass spectrum

Ding, Chuan-Fan,
Department of Chemistry, Fudan University

Contributing Authors:

Zhi C. Zuo, Fudan University; Wenning Wang, Fudan University; Yinjuan Chen Fudan University

Title:

Gas phase Complexation of α /β cyclodextrins with Amino Acids Studied by Ion Mobility-Mass Spectroscopy and Molecular Dynamics Simulations

Introduction:

There is a wider rim and a narrower rim in CDs and hydroxyl groups are located on these two rims, so the hydrophilic surface and hydrophobic cavity are formed. Based on such structural characteristics, CDs possess the ability to form inclusion complexes via embedding lipophilic groups or small molecules into the hydrophobic cavity to improve solubility, stability or bioavailability of drugs.

Methods:

Gas-phase structures of complexes of α-/β-cyclodextrin with Gly, Leu and Phe are investigated by ion mobility-mass spectrometry (IMMS) and theoretical calculation. IMMS analysis was performed on the Synapt G2 HDMS instrument. All the simulations were carried out in a vacuum with the package NAMD2.9,3 using the Amber Glycam06 force field4 for the CDs.

Results:

Gas-phase structures of complexes of α-/β-cyclodextrin with Gly, Leu and Phe are investigated by ion mobility-mass spectrometry (IMMS) and theoretical calculation. Complex ions with different sizes amino acids possess almost the same collision cross sections in β-CD system, while different results were observed in α-CD system. The experimental results indicated the exclusion structure of [α-CD+Leu]- and [α-CD+Phe]-ions and the other ions are of inclusion structure. To further identify our proposal, gas-phase complex structures are calculated via molecular dynamics simulations and the associated relative collision cross sections and free energy components of the most stable complex ions were also analyzed theoretically.

Novel Aspect:

The exclusion structure of [α-CD+Leu]- and [α-CD+Phe]-ions and the other ions are of inclusion structure.

Ding, Chuan-Fan,
Department of Chemistry, Fudan University

Contributing Authors:
Fuxing Xu, Fudan University; Haiyang Yang, Fudan University.

Title:
The improvement of mass resolution,collision induced dissociation efficiency and low mass cutoff by octopole field in linear quadrupole ion trap mass

Introduction:
Original three dimensional IT or LIT needed to be built with hyperbolic-shaped electrodes, and should be able to produce a pure quadrupole electric field. Higher-order electric field affect the performance of ion trap mass analyzers. Different higher-order fields will have different effects on the ion trap performance. The effect of higher field on the ion trap performance needs to be investigated

Methods:
The effect of positive and negative octopole fields on the performance of a linear quadrupole ion trap mass spectrometer was carefully investigated using both theoretical simulations and experiments. The linear ion trap with added octopole components was built by using one pair of larger semi-circular rods and one pair of smaller semi-circular rods.

Results:
The mass resolving power, ion signal intensity, and CID efficiency of LIT can be significantly increased by including positive octopole components, and the low mass cutoff effect can also be greatly recovered by positive octopole components. There is an optimum value of the positive octopole component, around A4/A2=2.6%, for improving the LIT mass resolving power and ion signal intensity, but higher A4 components can benefit the CID efficiency and improve the LMCO effect. Any amount of negative octopole component will certainly distort the mass resolving power and ion signal intensity of LIT. Some amount of octopole components can be easily produced besides quadrupole electric field by simply using one pair of larger semi-circular rods and one pair of smaller semi-circular rods to build a LIT.

Novel Aspect:
The mass resolving power, ion signal intensity, and CID efficiency of LIT can be significantly increased by including positive octopole components.

Donegan, Michael,
Waters Corp

Contributing Authors:
James Murphy, Waters Corp.; Angela Doneanu, Waters Corp.

Title:
High Throughput Microflow LC/MS: Sensitivity gains on a practical timescale

Introduction:
The ability to deliver high sensitivity at run times under 3 minutes compelled the need to develop a microflow device that can function on a UPLC timescale (<3 min). Presented within is a high throughput device that demonstrates improved sensitivity over UPLC methods on a timescale that is practical for everyday analysis.

Methods:
Standards were prepared in rat plasma and serially diluted down to the desired analytical range. Gradient conditions for both microflow and UPLC were identical and differed only in flow rate. The flow rate for UPLC is 600 µl/min while the flow for microflow is 12 µL/min. A 5 µl injection volume was used for both workflows.

Results:
A prototype BEH C18 iKey device (300 µm x 50 mm, 1.7 µm) was developed to allow higher flow rates during high throughput analysis. QC testing on the device consisted of running 1000 injections and evaluating reproducibility over the course of the run for the 4 analytes used in the study. Overall, the %CV across the run was less than 5% for all analytes. Using the 300 µm ikey, the linear range was 0.030-500 ng/mL for verapamil and propranolol, 0.122-1000 ng/mL for dextromethorphan, and 0.010-125 ng/mL for fluticasone. For all analytes, a blank injection after the high standard demonstrated carryover below the LOQ. Comparative analysis on a conventional flow UPLC/MS operated at 600 µL/min (2.1 x 50 mm BEH C18 column) showed decreased sensitivity with the limit of quantitation (LOQ) for verapamil and propranolol 2x higher than observed with the 300 µm ikey and fluticasone 4x higher.

Novel Aspect:
Microflow LC/MS using a 300 µm ikey resulted in gains of at least 2x in all compounds tested when compared to conventional UPLC with a 3 min runtime.

Donkuru, McDonald,
University of Saskatchewan

Contributing Authors:
Deborah Michel, Drug Design & Discovery Group, College of Pharmacy and Nutrition, University of Saskatchewan, 107 Wiggins Road, Saskatoon, SK S7N 5E5, Canada; Hanan Awad, Drug Design & Discovery Group, College of Pharmacy and Nutrition, University of Saskatchewan, 107 Wiggins Road, Saskatoon, SK S7N 5E5, Canada; Anas El-Aneed Drug Design & Discovery Group, College of Pharmacy and Nutrition, University of Saskatchewan, 107 Wiggins Road, Saskatoon, SK S7N 5E5, Canada; George Katselis, Canadian Centre for Health and Safety in Agriculture (CCHSA), and Department of Medicine, University of Saskatchewan, 104 Clinic Place, Saskatoon, SK S7N 2Z4, Canada.

Title:
HILIC-LC-MS/MS investigation of the intracellular fate of gemini surfactants designed for gene delivery

Introduction:
Dicationic gemini surfactants are promising lipids for nanoparticle-mediated gene delivery. They complex with DNA into lipoplexes capable of efficient non-viral transfection of cells. However, the fate of lipid nanoparticles after intracellular release of the DNA cargo is poorly studied. Therefore, we developed LC-HILIC-MS/MS methods to analyze gemini surfactants within nanoparticle-treated cells.

Methods:
Lipoplexes are formulated using two promising structurally-distinct gemini surfactants. Epidermal keratinocytes are treated with the lipoplexes, collected at different times, and extracted using Bligh/Dyer extraction. Subcellular organelles are also isolated using standard protocols. An isocratic, low-buffered "zwitterionc"-HILIC-LC-MS/MS method was developed, followed by analysis of the extracts.

Results:
The HILIC-LC-MS/MS method has 4-min run time, with a low limit of detection (3 ng/mL), and was validated as per the FDA guidelines. The analytes and their deuterated internal standards were monitored using multiple reaction monitoring on a triple quadrupole-linear ion trap instrument. The chromatographic method was applicable to 18 different gemini surfactant structures. Treated cells showed a steep rise in the analyte intracellular concentration to a maximum, followed by gradual decrease that terminates in a plateau. This can be attributed to the bio-transformation of gemini surfactants or their exocytosis from host cells. Further analysis showed the internalized gemini surfactants distributed to multiple subcellular compartments including plasma membrane, cytosol, mitochondria, nucleus. We are investigating the relationship between the cellular fate and the toxicity of tested compounds

Novel Aspect:
New simple HILIC-LC-MS/MS methods were developed assessing the cellular fate of lipid-based DNA delivery systems.

Dorman, Frank,
Penn State University

Contributing Authors:
Kari Organtini, Penn State University; Anne Myers, Ontario Ministry of Environment and Climate Change; Karl Jobst Ontario Ministry of Environment and Climate Change; Eric Reiner, Ontario Ministry of Environment and Climate Change; Jack Cochran, Restek Corporation; Adam Ladak, Waters Corporation; Douglas Stevens, Waters Corporation.

Title:
WHY IS FIREFIGHTING DANGEROUS? CHARACTERIZATION OF MIXED-HALOGEN DIOXINS AND FURANS IN FIRE DEBRIS USING GCXGC-TOFMS AND APGC-MS/MS

Introduction:
An increase in cancer rates among debris-exposed first responders have been associated with those present at the World Trade Center.1 The wide spread use of brominated flame retardants in common consumer products has extended the concern of health implications to long term exposure of first responders to fire debris, possibly due to exposure from combustion byproducts of these materials.

Methods:
In order to develop the analytical methodology needed to characterize these materials, this presentation will discuss the benefits of both GCxGC-TOFMS and Atmospheric-Pressure ionization GC (APGC)-MS/MS for the comprehensive characterization of the PXDD and PXDF's.

Results:
This presentation will focus on the investigation of the role mixed halogenated planar compounds, generated during the combustion of BFR-containing products, may play in the toxicity experienced by first responders. The emphasis of the project is placed on the generation of polyhalogenated dibenzo-p-dioxins and dibenzofurans in residential or commercial fire situations. These situations most accurately represent the conditions fire fighters are exposed to during their career. Additionally, data will be presented demonstrating that these compounds do, in fact, pose a human health risk, even though they are currently unmonitored across the globe.

Novel Aspect:
There is a lack of knowledge about these important compounds. This work may be the first report of a complete characterization of these materials.

Douix, Suzie,
Synchrotron Soleil

Contributing Authors:
Giuliani Alexandre, Synchrotron SOLEIL Paris; Dossmann Héloïse, Université Pierre et Marie Curie Paris.

Title:
Characterization of small molecules using Laser-induced Acoustic Desorption (LIAD)

Introduction:
Laser-induced acoustic desorption (LIAD) is vaporization technique to transfer nonvolatile, thermally labile compounds as neutral molecules into the gas phase. This method is be used to desorb the analyte, which can then be ionized by various means for further MS analysis. The benefit of this technic is the separation of desorption and ionization.

Methods:
Samples are be deposited on a uniform spot by a spray on a 12.6 µm aluminum foil. A Nd-YAG laser (Minilite I, continuum) operated at 532 nm is focused onto the metallic foil through the combination of a beam expander and a focusing lens.Power density of the order of 10 9 W/cm^2 is reached. We present two different homemade LIAD setups allowing ionization at the atmospheric pressure and under vacuum

Results:
In LIAD, a sample deposited onto a thin metal foil is irradiated from the back side with a series of high-energy laser pulses. It prevents from a direct exposure of the sample to the intense laser light, which makes it especially interesting for the analysis of light-sensitive molecules. High amplitude acoustic waves propagate through the foil and cause desorption of analytes species. It results in a plume of neutral molecules in gas phase. These molecules will then be ionized under vacuum by photon and electron impact. For those produced at the atmospheric pressure, ESI or APPI are used. The performances of both setups are compared and their applications will be presented. The final aim of the study is to be equally applicable to polar and nonpolar analytes. Some nonvolatile compounds such fluorotelomers will be investigated first.

Novel Aspect:
The benefits of LIAD operated at atmospheric pressure and under vacuum are compared along with different ionization methods.

Drake, Richard,
Medical University of South Carolina

Contributing Authors:
Evelyn Bruner, Medical University of South Carolina; Peggi Angel, Medical University of South Carolina; Shannon Cornett Bruker Daltonics; Anand Mehta, Medical University of South Carolina.

Title:
High resolution mapping of N-glycans in FFPE prostate cancer tissues using MALDI-FTICR and rapid MALDI-TOF mass spectrometry imaging

Introduction:
The prostate gland secretes hundreds of different abundant glycoproteins like prostate specific antigen (PSA), and understanding the role of glycosylation in relation to prostate cancer detection and progression remain critical questions. A new N-glycan MALDI imaging workflow has been applied to formalin-fixed paraffin-embedded (FFPE) tissues from pathology blocks, biopsy and tissue microarrays.

Methods:
FFPE human prostate cancer tissue blocks and tissue microarrays were cut at 5µm prior to deparaffinization and antigen retrieval. Peptide N-glycosidase F (PNGaseF) was applied by automated sprayer for glycan release, followed by CHCA matrix and detection on either a SolariX 7T FTICR-MALDI or rapifleX MALDI-TOF mass spectrometer. Immunostaining of PSA was done on adjacent tissue slices.

Results:
A recently developed MALDI imaging workflow for N-glycans has been applied to multiple FFPE prostate cancer tissues. Prior to imaging, each tissue was defined by a pathologist for tumor location and grade, location of stroma regions and areas with immune cell infiltrate. Immunostaining of PSA was performed to highlight the active glands that were producing it. Following N-glycan imaging using MALDI-FTICR for high mass accuracy, tumor associated glycans of either high mannose or multi-fucosylated complex glycan structures were identified. Non-sialylated glycans localized to regions of immune infiltrate and complex sialylated glycans were primarily localized in non-tumor stroma regions. The same tissues were analyzed by rapid MALDI-TOF at higher spatial resolution to further refine the differences in glycan co-localization with PSA, histopathology regions and clinical outcomes.

Novel Aspect:
MALDI imaging of N-glycans in FFPE tumor tissues can be used to identify functional and structural regions of the tumor microenvironment.

Du, Min,
Waters Corporation

Contributing Authors:
Weibin Chen, Waters Corporation; Andrew Borthwick, Waters Corporation; Ying Qin Yu Waters Corporation; Brad Williams, Waters Corporation; Ian Morns, Waters Corporation; Catalin Doneanu, Waters Corporation; Jackson Pope, Waters Corporation.

Title:
An LC/ESI-IM-MS/MS Assay for Identification and Quantification of Host Cell Proteins in Therapeutic Monoclonal Antibodies

Introduction:
Protein quantification in industrial settings is predominantly based on ligand binding assays (LBAs). Here we explored the capabilities of a 1D chromatographic separation, optimized for sample overloading conditions, coupled with ion mobility separation of peptide precursors and high-resolution (>25,000) MS-detection, for quantification of low-abundance HCPs present in protein biopharmaceuticals.

Methods:
A high-purity therapeutic mAb (Inflectra 10 mg/mL) was denatured, reduced, alkylated and digested overnight. Four protein digest standards were then spiked and the peptide mixture was separated on a 2.1 mm ID column specifically designed for chromatography with high sample loads (200-500 µg on-column). Peptide precursors were further separated in an ion mobility enabled QTOF mass spectrometer.

Results:
Several MS-based HCP assays have been recently developed [Anal Chem 2015, 87, p9186 and p10283], but they typically employ low-throughput, extensive 2DLC fractionation schemes. To address this shortcoming, we evaluated a single dimension chromatographic separation using a recently introduced CSH (charged-surface-hybrid) C18 stationary phase [Anal Chem, 2013, 85, p6936]. Peak capacity increased significantly up to 90 min gradients and sample amounts as high as 500 µg of mAb digest can be loaded without observing sample losses and with minimal carryover between injections. The 1DLC HCP assay was able to detect all 4 protein digest standards spiked in the Inflectra digest down to concentrations as low as 10 ppm. In a separate experiment, equimolar amounts of protein digests we spiked in Inflectra digest and protein changes in concentration were monitored using the Progenesis QIP software.

Novel Aspect:
A sensitive, robust LC/IM-MS/MS assay for identification and quantification of protein impurities (HCPs) in highly-purified therapeutic mAbs

Dubkov, Mikhail,
Ryazan State Radio Engineering University

Contributing Authors:
V.S. Gurov, Ryazan State Radio Engineering University; I.A. Kharlanov, Ryazan State Radio Engineering University; M.A. Burobin Ryazan State Radio Engineering University.

Title:
Multisectional monopolar mass-analyzer

Introduction:
Advantages of monopolar mass-analyzers are lower power consumption and simple design that determine their application in small-dimensional analytical devices for analysis of gaseous substances. Disadvantage of a monopolar mass-analyzer is lower analytical characteristics, in particular resolution. It is connected with limited time for the ion sorting and passage of unsorted ions to the detector.

Methods:
The main method to increase a flight system resolution is an increase of the ion sorting time due to increase of the mass-analyzer length. For the present task solution, extended design of the monopolar mass-analyzer electrode system is suggested to be assembled of several short similar sections containing field-forming electrodes.

Results:
Under design of small-dimensional systems, length of the electrode systems is limited, so the electrode system made of two short sections placed in relation to each other in a specific way can be used. Sections are turned around the longitudinal axis on 180º in relation to each other and there is a gap between sections. Researches have shown that unsorted ions in the area between electrode sections have higher transverse velocities, so they are scattered in this area and do not enter the second section of electrodes. Usage of such design of the monopolar mass-analyzer electrode system leads to increase of the resolution in 1.6 times in comparison with a monopolar mass-analyzer of ordinary geometry of the electrode system of the same length.

Novel Aspect:
Such design of the monopolar mass-analyzer electrode system allows improving a quality of the electric field in extended electrode systems.

Dubkov, Mikhail,
Ryazan State Radio Engineering University

Contributing Authors:
M.A. Burobin, Ryazan State Radio Engineering University; V.S. Gurov, Ryazan State Radio Engineering University; I.A. Kharlanov Ryazan State Radio Engineering University.

Title:
Two-channel monopolar mass analyzer

Introduction:
The design of the fourfold monopole mass spectrometer represented as a circular cylindrical electrode extending coaxially within a surrounding hollow-square cylindrical electrode (U.S. Patent) was suggested in 1975. However, there is an unsolved issue of interferences of mass analyzer adjacent channels and complexity to provide a coaxiality of electrodes along the whole length of a mass analyzer.

Methods:
The electrode system was previously suggested by us to be formed of short identical sections made according to the original technology (Patent RU 2393580). Such approach allows improving quality of the electric field in extended electrode systems. Two channel monopolar mass analyzer represented as two identical electrode systems located one above another and touched by rod electrodes is suggested.

Results:
Main advantage of the suggested design is a minimal cost of its production under presence of one sweep generator. New design allows saving advantages of the classical monopolar mass analyzer expanding area of its application. For example, the second channel of a mass analyzer can be a protection channel that is important under usage of a mass analyzer as a component of equipment for space researches. Also both channels can be used simultaneously, besides, value of the desired signal and consequently sensitivity of a mass spectrometer increase.

Novel Aspect:
The two-channel monopolar mass analyzer is represented as following development of monopole mass spectrometer design.

Duchoslav, Eva,
SCIEX

Contributing Authors:

Brad Patterson, SCIEX, Melbourne , Australia; J. Larry Campbell, SCIEX, Concord, ON, Canada; Yves LeBlanc SCIEX, Concord, ON, Canada.

Title:

Accelerating Lipid Profiling Acquisition Strategies with Differential Mobility Spectrometry and SWATH® Data Collection

Introduction:

Differential mobility spectrometry (DMS) provides effective separation of lipid species in shotgun lipidomics workflows. A potential coupling of this methodology with SWATH® acquisition was investigated in light of accelerating accurate mass lipid profiling strategies and development of dedicated data processing tools.

Methods:

Bovine heart extract (Avanti Polar Lipids, Inc.) was standardized for total amount of lipid of approximately 10µM in 45/45/10 DCM/MeOH/water. ESI/DMS/MS/MS data were collected with a hybrid TOF mass spectrometer equipped with a planar DMS system, in both positive and negative modes (DMS at 150ºC, SV = 4000V, DR off, 2-propanol). Multiple ESI data collection methods, both IDA and SWATH® were used.

Results:

In summary, 450 lipids were identified in either positive or negative acquisition modes. With generic IDA triggering criteria, TOF MS/MS spectra were collected for 65% of major lipid species. SWATH® MS/MS data captured fragmentation for several lipid class analogues in parallel. For example, 51 PC lipids (by total composition) detected in negative ion mode as formate adducts. They were easily confirmed based on accurate mass, isotope pattern and CV profile. Lipid data with isotope pattern failures pointed to the necessity to remove isotope contribution from less saturated PC analogues. With adjusting the SWATH® Q1 window width, CV range, CV step size of 1V and total cycle time of 12s the analysis time was less than 6 minutes per polarity.

Novel Aspect:

Integrating gas- phase separation and data- independent acquisition for comprehensive lipid profiling in higher throughput assays

Dumlao, Morphy,
School of Chemistry, University of New South Wales

Contributing Authors:
William A. Donald, School of Chemistry, University of New South Wales; Chun Hong Chan, School of Chemistry, University of New South Wales; Laura E. Jeffress School of Chemistry, University of New South Wales; J. Justin Gooding, School of Chemistry, University of New South Wales.

Title:
Solid-phase microextraction low temperature plasma mass spectrometry for the direct and rapid analysis of complex mixtures

Introduction:
Solid-phase microextraction is directly integrated with low temperature plasma ionisation mass spectrometry to rapidly detect target analytes in chemical mixtures, including urine. In this sampling and ionization method, the fibre serves: (i) to extract molecules from their native environment, and (ii) as the ionization electrode that is used to desorb and ionize molecules directly from the SPME.

Methods:
By use of a custom fabricated SPME fibre consisting of a stainless steel or graphene coated probe and low temperature plasma mass spectrometry, analytes can be detected at low ppb detection levels directly in water and urine. Organophosphates were not readily detected by this approach using an uncoated probe in negative control experiments.

Results:
For example, the use of the LTA coating significantly outperformed the use of a high alumina Zeolite Socony Mobil-5 (ZSM-5) coating of comparable thickness that is significantly more hydrophobic than LTA. The sampling and detection of chemical warfare agent analogues and hydrolysis products required less than 2 min per sample. A key advantage of this sampling and ionization method is that analyte ions can be directly and rapidly sampled from chemical mixtures, such as urine and seawater, without sample preparation or chromatography for sensitive detection by mass spectrometry. This ion source should prove beneficial for portable mass spectrometry applications because relatively low detection limits can be obtained without the use of compressed gases, fluid pumps, and lasers.

Novel Aspect:
Moreover, the ion source is compact, can be powered with a 10 V battery, and is tolerant of complex mixtures.

Dumlao, Morphy,
School of Chemistry, University of New South Wales

Contributing Authors:
J. Justin Gooding, School of Chemistry, University of New South Wales; William A. Donald, School of Chemistry, University of New South Wales; Laura E. Jeffress School of Chemistry, University of New South Wales; Chun Hong Chan, School of Chemistry, University of New South Wales.

Title:
Solid-phase microextraction low temperature plasma mass spectrometry for the direct and rapid analysis of complex mixtures

Introduction:
Solid-phase microextraction is directly integrated with low temperature plasma ionisation mass spectrometry to rapidly detect target analytes in chemical mixtures, including urine. In this sampling and ionization method, the fibre serves: (i) to extract molecules from their native environment, and (ii) as the ionization electrode that is used to desorb and ionize molecules directly from the SPME.

Methods:
By use of a custom fabricated SPME fibre consisting of a stainless steel or graphene coated probe and low temperature plasma mass spectrometry, analytes can be detected at low ppb detection levels directly in water and urine. Organophosphates were not readily detected by this approach using an uncoated probe in negative control experiments.

Results:
For example, the use of the LTA coating significantly outperformed the use of a high alumina Zeolite Socony Mobil-5 (ZSM-5) coating of comparable thickness that is significantly more hydrophobic than LTA. The sampling and detection of chemical warfare agent analogues and hydrolysis products required less than 2 min per sample.A key advantage of this sampling and ionization method is that analyte ions can be directly and rapidly sampled from chemical mixtures, such as urine and seawater, without sample preparation or chromatography for sensitive detection by mass spectrometry. This ion source should prove beneficial for portable mass spectrometry applications because relatively low detection limits can be obtained without the use of compressed gases, fluid pumps, and lasers.

Novel Aspect:
Moreover, the ion source is compact, can be powered with a 10 V battery, and is tolerant of complex mixtures.

Dupré, Mathieu,
Institut Pasteur

Contributing Authors:
Magalie Duchateau, Institut Pasteur, CNRS UMR 3528, Paris, France; Julia Chamot-Rooke, Institut Pasteur, CNRS UMR 3528, Paris, France; Valeria Calvaresi Institut Pasteur, CNRS UMR 3528, Paris, France; Diogo Borges, Institut Pasteur, CNRS UMR 3528, Paris, France; Christian Malosse, Institut Pasteur, CNRS UMR 3528, Paris, France.

Title:
An Optimized High-Throughput Top-Down Proteomics Platform for Improving Diagnostics in Clinical Microbiology.

Introduction:
The current technique used for microbial identification in hospitals is MALDI-TOF MS. However, it suffers important limitations, in particular for closely-related species or when the database used for the identification lacks the appropriate reference. We show here that top-down proteomics can go beyond these limitations and provides a new approach for an accurate bacterial characterization.

Methods:
Bacterial cell were harvested at exponential growth phase from cultures of clinical samples. Lysis was performed by mechanical disruption using glass beads and various buffers. After purification, intact proteins were analyzed in LC-MS/MS (monolithic column, Data Dependent Analysis in ETD or EThcD, dedicated top-down softwares) using an Orbitrap Lumos mass spectrometer.

Results:
We first set up a high throughput LC-MS/MS top-down proteomics platform dedicated to intact bacterial protein analysis. E. coli was chosen as a model. All steps of the workflow were optimized: protein extraction, on-line chromatographic separation, protein ion fragmentation and data analysis. Using optimized parameters, about 300 proteins, corresponding to more than 1,000 proteoforms could be obtained in a single run. We then used this platform for the characterization of bacterial pathogens (E. coli, Salmonella, Shigella) that are undistinguishable with MALDI-TOF although leading to very different clinical outcomes. For each pathogen, we identified specific proteoforms that could be further used as biomarkers. Several were found to differ only by a single amino acid. This highlights the great advantage of targeting proteoforms and not peptides for an accurate bacterial identification.

Novel Aspect:
The development of a top-down proteomics platform for bacterial pathogen identification is a major breakthrough in clinical microbiology.

Durr, Michael,
Justus Liebig University Giessen

Contributing Authors:
Andre Portz, Justus Liebig University Giessen; Christoph R. Gebhardt, Bruker Daltonics Bremen.

Title:
Real-time monitoring of surface reactions of biomolecules by means of cluster-induced desorption/ionization mass spectrometry

Introduction:
Reactions of peptides and proteins on surfaes are of interest in many fields of biochemistry and biotechnology. However, the chemical information obtained by standard surface analysis tools is limited. Here we show that desorption/ionization induced by neutral clusters (DINeC) can be successfully applied for studying (surface) reactions of oligopeptides in real-time by means of mass spectrometry.

Methods:
Oligopeptides were desorbed and ionized in high vacuum by neutral SO2 clusters. The impacting polar clusters both provide the energy for desorption as well as they serve as a transient matrix in which the analyte is dissolved during cluster surface impact. Thus desorption takes place at low cluster energy and without fragmentation. Mass spectra of the desorbed ions were recorded by ion trap MS.

Results:
The quantitative nature of the method was demonstrated using angiotensin II molecules individually adsorbed on gold substrates by means of electrospray ion-beam deposition. These adsorbates were desorbed and detected as intact molecules down to a coverage of 0.001 ML; a linear relationship between surface coverage and signal intensity was observed over three orders of magnitude. Real-time monitoring of surface reactions is demonstrated for isotope-exchange experiments with angiotensin II and different types of porphyrins. Dosing D2O led to a broadening of the isotopic pattern and a continuous shift towards higher m/z values. The results were modeled by means of Monte Carlo simulations taking into account reaction and back-reaction of H/D exchange. Several different rate constants could be extracted and were assigned to the different types of functional groups in the respective molecules.

Novel Aspect:
DINeC-MS allows for real-time investigation of surface reactions of biomolecules, H/D exchange rates for different functional groups were determined.

Durr, Michael,
Justus Liebig University Giessen

Contributing Authors:
Andre Portz, Justus Liebig University Giessen; Julian Heep, Justus Liebig University Giessen; Christoph R. Gebhardt Bruker Daltonics Bremen.

Title:
Mass spectrometry of peptides using a combination of thin-layer chromatography and desorption/ionization induced by neutral SO2 clusters

Introduction:
For identification of the separated components in thin-layer chromatography (TLC), TLC can be combined with different mass spectrometry techniques. Here we show the successful combination of TLC with desorption/ionization induced by neutral SO2 clusters (DINeC), which is an extremely soft and matrix free ionization method for mass spectrometry.

Methods:
Oligopeptides were mixed and drop cast on a thin-layer chromatography plate. After separation by means of TLC, the sample plate was moved stepwise through the SO2 cluster beam. The analytes were desorbed and inized by the impacting clusters without any fragmentation, mass analysis of the desorbed ions was performed in an ion trap mass spectrometer.

Results:
First experiments showed that oligopeptides such as angiotensin II, bradykinin, vancomycin, and bombesin, can be efficiently desorbed and ionized from the thin-layer chromatography plates by means of DINeC. The [M+H]+ signal of the respective substance is clearly detected. Stepwise scan of the thin-layer chromatography plate allows to measure a TLC chromatogram of the single components as identified by DINeC-MS. Close bands were separated demonstrating that thin-layer chromatography and DINeC can be successfully combined. As no further sample treatment, such as extracting the analytes for ESI-MS or application of a matrix for MALDI, is necessary, the analysis of oligopeptides with the combination of TLC and DINeC requires very little effort in sample preparation.

Novel Aspect:
Combination of thin-layer chromatography and DINeC-MS for analysis of oligopeptides. No additional sample preparation needed for MS from TLC plates.

Dusek, Martin,
Research Institute of Brewing and Malting

Contributing Authors:
Jana Olsovska, Research Institute of Brewing and Malting, Lipova 15, 120 44 Prague, Czech Republic; Michaela Scigelova, Thermo Fisher Scientific, Hanna-Kunath-Str. 11, 28199 Bremen, Germany; Martin Dusek Research Institute of Brewing and Malting, Lipova 15, 120 44 Prague, Czech Republic.

Title:
Proteomic and Metabolomic Analysis of Wort: Differentiation of Malting Barley Cultivars

Introduction:
Barley grain proteins can be classified into three groups according to their function: storage proteins, structural/metabolic proteins, and protective proteins. The profile of malt proteins in final worth preparation differ according to the barley cultivar and the malting process employed. Individual proteins critically impact the quality of the final product.

Methods:
We relied on the combination of SDS-page, high resolution accurate MS (HRMS), and advanced data processing sw. A tryptic digest of acetone-precipitated wort proteins was separated using reversed phase chromatography coupled to the Q Exactive with ESI source. Protein identification and differential expression were performed using Proteome Discoverer and SIEVE package, respectively.

Results:
We identified protein constituents of wort preparations made with malt coming from 8 different malting barley cultivars. Majority constituents corresponded to the prominent bands on SDS gel and included the family of hordeins and alpha-amylase-trypsin inhibitors. Other interesting proteins, e.g. lipid-transfer protein 1 implicated in foam retention, were also detected. We encountered several glycated species, likely the products of Maillard reactions during the malting process. We also performed the differential analysis of wort produced from the same malting barley variety, but grown at different locations. Our findings confirmed that multiple environmental factors (terroir) manifested themselves in the proteomic signature of wort preparations. Finally, we estimated the share of the relative contribution of malting barley and the brewing process to the final wort preparation.

Novel Aspect:
Differential analysis of wort proteome reliably classifies malt cultivars used in the brewing process.

Dusek, Martin,
Research Institute of Brewing and Malting

Contributing Authors:
Vladimíra Jandovská, Faculty of Science, Charles University in Prague, Prague, Czech Republic; Jana Olšovská, Research Institute of Brewing and Malting, Prague, Czech Republic; Alexandr Mikyška Research Institute of Brewing and Malting, Prague, Czech Republic; Pavel Čermák, Thomayer's Hospital, Prague, Czech Republic.

Title:
A novel approach for identification of biologically active phenolic compounds in Hops matrices using hybrid quadrupole-orbitrap mass spectrometer

Introduction:
Phenolic compounds, secondary metabolites of hops, represent a large family of compounds that could be subsequently divided into smaller groups based on the similarities between their chemical structures. The antibacterial, antifungal and antiviral properties of hops are well known, but there is a lack of information about antimicrobial activities of individual hop compounds.

Methods:
In our initial set of experiments we focused on the selective extraction of potential anti-microbially active compounds. The substances migrated and formed an inhibition zone surrounding a spot of minced hops applied into the inoculated solid growth medium. Active compounds were extracted by a newly developed extraction procedure based on a QuEChERS method and consequently analyzed using LC-HRMS.

Results:
The data were acquired by using a quadrupole-orbitrap hybrid mass spectrometer in targeted-MS2 experiment in both ionization modes. We attempted a differential analysis to sort out molecular features of compounds coming originally from hops, destined for further setting up of the targeted mass lists for confirmation and MSn experiments. Our attention then focused on a broad scale of phenolic compounds; a database, including more than 300 phenolic compounds, has been created for the purpose based on a literature survey. The unknown compounds extracted from inhibition zones were identified either by searching against the database or their structure was elucidated on the basis of their fragmentation spectra. On the basis of this experiment we extended the list of active compounds with potential anti-microbial activities.

Novel Aspect:
The method using of LC-HR/AM mass spectrometer was successfully used for identification of phenolic compounds with anti-microbial activity in hops.

Dutertre, Quentin,
Philip Morris

Contributing Authors:
Arno Knorr, Philip Morris; Mounir Rhouma, Philip Morris; Shoaib Majeed Philip Morris; Mark Bentley, Philip Morris.

Title:
A multi analyte method using liquid chromatography with high resolution accurate mass spectrometry for aerosol constituents in 3D tissue cultures

Introduction:
Aerosol constituents or their primary metabolites (e.g. glutathione adducts) can serve as markers for exposure normalization for 3D tissue cultures. Analysis of glutathione adducts is challenging using classical reverse phase chromatography, since adducted species are generally too polar to be adequately retained by this technique, therefore a novel chromatographic separation has been developed

Methods:
Aerosol deposition on tissues was performed using the Vitrocell® exposure system. Compounds were extracted from tissues using phosphate buffered saline and ethanol, using freeze/thaw cycles to lyse cells, and analyzed by liquid chromatography with high resolution accurate mass spectrometry. The separation used anion exchange (Biobasic AX®) and pentafluorophenyl (Kinetex® PFP) columns in series

Results:
An analytical method for quantifying multiple compounds using liquid chromatography with high resolution accurate mass spectrometry in full scan positive electrospray ionization mode (QExactive™, Thermo Fisher) was developed. Nicotine and six metabolites (nicotine-N-oxide, cotinine-N-oxide, cotinine, 3-hydroxycotinine, nicotine-N-glucuronide, cotinine-N-glucuronide) plus glutathione adducts of electrophilic aerosol constituents (acrolein, crotonaldehyde, 3-penten-2-one) were analyzed. All analytes of interest were separated within a single analytical run using two different analytical columns in series. Quantification was performed using stable isotope labeled standard compounds added prior to sample preparation of tissues and basolateral media. The method was designed for high throughput, with dynamic ranges covering five orders of magnitude between low and highly abundant species

Novel Aspect:
High throughput multi-compound analytical method using sequential analytical columns for the quantification of aerosol constituents and metabolites

21ST INTERNATIONAL MASS SPECTROMETRY CONFERENCE
Toronto, Ontario, Canada
August 20 – 26, 2016

Dziekonski, Eric,
Purdue University

Contributing Authors:
Ryan Hilger, Purdue University; Joshua Johnson, Purdue University; Scott McLuckey Purdue University; Robert Santini, Purdue University.

Title:
The Electrostatic Linear Ion Trap: A Compact, Low-Power, Fourier Transform Mass Spectrometer with Tandem MS Capabilities

Introduction:
The electrostatic linear ion trap (ELIT) is a DC-only mass analyzer that can be coupled to both continuous and pulsed ionization sources. The trap itself is roughly five inches long and consists of a field free region that is capped by two, axially opposing, reflectrons. As no RF supplies or cryogenic magnets are needed for its operation, it demands little power and can be used for field analysis.

Methods:
All experiments were performed on a home-built instrument where ions were generated by either nanoelectrospray or glow discharge ionization. Ions were accumulated, cooled, and bunched in a quadrupole equipped with LINAC electrodes after which they were pulsed into the ELIT. The image charge induced on a pickup electrode was digitized and Fourier transformed to generate a mass spectrum.

Results:
Our lab has previously shown that the ELIT is capable of performing ion isolation, recapture, surface induced dissociation, and absorption mode Fourier transform mass analysis. More recently, ion capture was effected by in-trap potential lift, as opposed to mirror switching, and the resulting image charge was detected with two off-center pickup electrodes. This configuration enabled us to record mass spectra of various carborane anions with pressure limited resolutions (FWHM) of up to 11,000 (CHB11Cl11-). In the interest of pushing the figures of merit of the ELIT, a new instrument has been constructed which includes differential pumping. With this instrument, we are currently exploring methods to electronically correct frequency shifts that arise from space-charge or perturbations in the output voltage of the power supplies when capturing ions via mirror switching.

Novel Aspect:
Correction of time-dependent frequency shifts observed in electrostatic linear ion traps that arise from pulsed loads and space-charge.

Easterling, Michael,
Bruker Daltonics

Contributing Authors:
Matthias Witt, Bruker Daltonik GmbH, Fahrenheitstrasse 4, 28359 Bremen, Germany; Philippe Schmitt-Kopplin, Helmholtz Zentrum Muenchen Deutsches Forschungszentrum fuer Gesundheit und Umwelt (GmbH) Ingolstaedter Landstr. 1 85764 Neuherberg; Christopher Thompson Bruker Daltonics Inc., 40 Manning Road, Manning Park, Billerica, MA 01821, USA; Basem Kanawati, Helmholtz Zentrum Muenchen Deutsches Forschungszentrum fuer Gesundheit und Umwelt (GmbH) Ingolstaedter Landstr. 1 85764 Neuherberg; Michael Easterling, Bruker Daltonics Inc., 40 Manning Road, Manning Park, Billerica, MA 01821, USA.

Title:
Evaluating Quadrupole Detection FTMS for Complex Mixture Profiling

Introduction:
The unique analytical power of FT-ICR has become a routine necessity in analysis of naturally occurring complex mixtures. Direct sample introduction into the ICR with minimal preparation allows a throughput advantage over LC-MS methods. Here we evaluate a novel harmonic detection scheme with substantially improved resolving power against benchmarked results for complex mixture characterization.

Methods:
Three naturally occurring benchmark complex mixture standards consisting of 1) Suwanee river fulvic acid 2) c elegans extract and 3) typical plant extract were all prepared for MeOH ESI. The samples were analyzed with a 12T "infinity cell" and statistically processed. Identical experiments were performed on 12T and 7T systems with a "ParaCell" and novel quadrupolar harmonic detection circuit.

Results:
The analytically challenging nature of complex mixtures has established high field ICR (>7T) as an essential tool for describing inherent chemical diversity on a molecular level. Recently a novel quadrupolar harmonic (2w) detection scheme has proven to double ICR resolving power within a given acquisition time for analyses of single components and complex mixtures. Specifically, application of 2w detection at 7T to direct infusion of crudes resulted in resolving powers similar to previous 12T experiments on legacy equipment. Mass accuracy was also conserved resulting in a similar number of peak determinations indicating that 2w detection potentially offers the requisite robustness needed for complex mixture workflows. Here we create a robust proof statement for these preliminary indications with a detailed analysis of essential spectral metrics for a variety of compounds.

Novel Aspect:
Application of a novel detection scheme based on quadrupolar harmonic ICR detection to the characterization of complex mixtures.

Egli, Stefana,
Memorial University of Newfoundland

Contributing Authors:

Christina S. Bottaro, Memorial University of Newfoundland.

Title:

Using MS with molecular imprinting to develop new environmental monitoring tools

Introduction:

Mass spectrometry is a powerful, multi-faceted technique used in the Bottaro group to assist in the development of environmental monitoring technology with the final purpose of integrating it with portable MS for easy direct on-site analysis. The technology is a thin-film molecularly imprinted polymer, MIP sorbent used to selectively adsorb compounds of interest in water analysis.

Methods:

MIPs are polymers prepared by polymerization in presence of an analyte,which upon removal leaves selective cavities.When exposed to an aqueous sample,the cavities re-bind the analyte based on a series of specific interactions.The analyte is analyzed directly from the surface of the MIP using desorption electrospray ionization mass spectrometry,DESI-MS or extracted and analyzed by GC-MS or LC-MS.

Results:

Thin-film MIPs targeting various applications have been developed, such as: - oil in water: light polycyclic aromatic hydrocarbons, thiophenes and phenols, -indicators of human impact, such as caffeine or cotinine, - emerging disinfection by-products, such as N-nitrosamines, - indicators of water treatment process efficiency, such as pharmaceuticals, - pesticides (neonicotinoids) and herbicides (2,4-D). This presentation aims to give the highlights of using thin-film MIPs coupled to MS in various applications. The MIPs were evaluated through binding tests in distilled water, fresh surface water, wastewater effluent, and seawater with no prior MIP conditioning or sample preparation. Method validation parameters such as sensitivity, selectivity, detection limits, response' linearity and other analytical figures of merit will be presented.

Novel Aspect:

MIPs coupled to MS are a rapid, selective and simple sample preparation technique that overcomes some of the challenges of traditional methods.

Eikel, Daniel,
Advion Inc.

Contributing Authors:
Christine Wang, Advion Inc.; Simon Prosser, Advion Inc.; Jamey Jones Advion Inc.

Title:
Ionization efficiency on silicon-based, micro-fabricated electrospray nozzles for micro flow LC-ESI-MS applications

Introduction:
ESI at nL/min flow provides enhanced MS sensitivity due to better ionization efficiency and reduced ion suppression. As flows increase to micro-flow (1-10 µL/min) the ionization efficiency drops significantly owing to larger sized ESI droplets and inefficient ionization. Here, we developed a silicon-based ESI emitter chip with various I.D. sizes (4-13µm) to improve µESI efficiency to nESI levels.

Methods:
ESI at nL/min flow provides enhanced MS sensitivity due to better ionization efficiency and reduced ion suppression. As flows increase to micro-flow (1-10 µL/min) the ionization efficiency drops significantly owing to larger sized ESI droplets and inefficient ionization. Here, we developed a silicon-based ESI emitter chip with various I.D. sizes (4-13µm) to improve µESI efficiency to nESI levels.

Results:
Emitter sizes obtained with MEMS covered 4.1 to 13.5 um ID and infusion experiments within the Schmidt/Karas model showed ideal ionization with no ion suppression up to 500 nL/min on the 5.5 um ID nozzle. Even at 8 uL/min on a 9.5 um ID nozzle the relative intensity of sugar to surfactant only dropped to a level of 80% – significantly better than a comparable pulled capillary emitter (value 1.8 of 10). Ideal ionization was also observed in the LC-MS model up to 8 uL/min suggesting that silicon-based ESI emitter nozzles can indeed significantly improve the range of near ideal ionization from low nL/min in nanoLC-MS to 8 uL/min and micro-flow LC-MS.

Novel Aspect:
Silicon-based ESI chip accommodates micro flow LC-MS applications at ionization efficiencies comparable to nano-ESI.

Eikel, Daniel,
Advion Inc.

Contributing Authors:
Simon Prosser, Advion Inc.

Title:
Liquid Extraction Surface Analysis (LESA) combined with nano-liquid chromatography (nLC) for analyte determination from biological surfaces.

Introduction:
The recent LESA analysis tool uses a direct solvent extraction from a surface of interest combined with Infusion-MS analysis. It is currently limited to 1 mm spatial resolution and can not distinguish between isobaric analytes due to the lack of HPLC separation. Here, we will evaluate a modified LESA system with added nLC chromatography after improved spatial resolution extraction of 400 um.

Methods:
A TriVersa-LESA system (Advion Inc.) was modified with a fused silica capillary mandrel (360/100 um OD/ID) and direct extract injection onto a 100 µm ID x 10 cm column prior to gradient nLC and on-board chip based nESI. Two samples were analyzed with an HCT ion trap (Bruker Corp.): Apple skin with a mixture of isobaric pesticides and Diclofenac dosed whole body mouse tissue sections.

Results:
The modified fused silica mandrel decreased the liquid extraction junction size to 400±20µm (determined by optical measurement of both removed material from a red marker as well as deposition of food color on tissue). LESA-nLC analysis from apple skin showed all five example pesticides detectable at the 20x below EPA tolerance level with separation of the isobaric pesticides Simazine and Carbaryl. Analyzed mouse sections (15 mg/kg PO Diclofenac) showed Diclofenac and its metabolites in heart/blood, stomach, kidney and liver. Isobaric 4'- and 5-Hydroxy-Diclofenac metabolites could be differentiated in liver and be confirmed against analytical standards. Adding nLC to a LESA experiment comes with a modest time penalty; however, with the set-up chosen here, it allows a 2 fold improvement in spatial resolution and enables the differentiation of isobaric analytes.

Novel Aspect:
Adding nLC separation to LESA results in higher spatial resolution and the ability to detect isobaric analytes from biological surfaces.

El Khoury, Maroun,
Thermo Fisher Scientific

Contributing Authors:
Daniel Lopez-Ferrer, Thermo Fisher Scientific, San Jose, CA; Chien-Hsun Chen, Thermo Fisher Scientific, San Jose, CA; Andreas FR Huhmer Thermo Fisher Scientific, San Jose, CA.

Title.
From The Ocean To The Table. An Integrated Mass Spectrometry Approach To Identify The Fish That Is In Your Plate.

Introduction:
The fishery market has growth in sales for the last 15 years. As a result, fish demand is producing a worldwide overexploitation of resources and fraudulent practices in the industry that account for 30% of the sales. In most cases, high priced fish species are substituted for lower value species. Here we described an integrated proteomic approach to authenticate fish species from muscle tissue.

Methods:
1 gr of fish muscle tissue was homogenized in water. The sample was then centrifuged to remove the insoluble material. Water soluble proteins were then heated at 95C for 5 min. After the heat treatment the sample was again centrifuged and the supernatant was aliquoted. One of the aliquots was submitted for bottom up and PRM proteomics analysis. A second aliquot was submitted for top down analysis.

Results:
A strategy for monitoring peptide biomarkers for fish authentication is described. This workflow consists in purifying thermos-stable proteins and analyzed them by mass spectrometry. First, proteins are analyzed using a fast shotgun proteomics approach. Due to the high protein sequence homology, it is quite difficult to assign the specific organism unless proteotypic peptides are detected. Further protein sequence alignment to will be used to build a parallel reaction monitoring assay using proteotypic peptides that will allow to detect the specie where the sample comes from. Further validation will be performed using top down mass spectrometry. This workflow was applied to frozen fish bought in a local grocery store that was simply labeled as hake. Using our strategy we certified that it was Merluccius paradoxus originally from South Africa.

Novel Aspect:
A fish authentication workflow achievable in less than 1 hour and represents an important advancement in the fight against fraud in the fish industry.

El Khoury, Maroun,
Thermo Fisher Scientific

Contributing Authors:
Gina Tan, Thermo Fisher Scientific, San Jose, CA; Daniel Lopez-Ferrer, Thermo Fisher Scientific, San Jose, CA; Andreas FR Huhmer Thermo Fisher Scientific, San Jose, CA; Barbara Dunn, Dept. of Genetics, Stanford University, Palo Alto, CA.

Title:
Systematic Integration of "Omics" Data to Improve Innovation in Beer Crafting

Introduction:
The beer industry has experienced a boom of microbrewing. Brewers seek to innovate by creating novel beers with unique organoleptic characteristics. We have acquired quantitative proteomics, and metabolomics data during fermentation. The integration of these datasets will allow for prototyping of new breeds of yeast to create create beers with distinct sensorial properties.

Methods:
Fermentations were carried out in triplicates over 48 hours. 10 different time points were collected. Cells were kept for proteomic and supernatants for metabolomic analysis. Peptide digests and metabolites were analyzed by LC-MS/MS analysis on a Q Exactive Plus mass spectrometer. Proteome Discoverer 2.1 and Compound Discoverer 2.0 were used to analyze proteomics metabolomics data respectively.

Results:
This work describes the process of explaining the differences among two popular beer strains, California ale and Hefe Weizen. As expected, ontology enrichment using ProteinCenter, the first time points of the fermentation for both strains showed that cell wall synthesis is over represented. After that, because glycolysis initiates, energy production and cell growth are more over represented. Metabolomics analyses revealed the absence of static metabolic phases. Instead, phases overlap allowing for the simultaneous growth and fermentation activity. The overlapping phases are a result of the high cell heterogeneity that makes individual cell to progress through fermentation at different rates which can be observed in the time-course progression of the individual metabolites. Metabolomics data has also revealed that the two strains have different growth rates.

Novel Aspect:
Systems biology in yeast to enhance the organoleptic properties of beer.

El-Aneed, Anas,
College of Pharmacy and Nutrition, University of Saskatchewan

Contributing Authors:
Jamie Willems, Department of Food and Bioproduct Sciences, College of Agriculture and Bioresources, University of Saskatchewan, 51 Campus Drive, Saskatoon, SK, S7N 1A8, Canada; George Katselis, Canadian Centre for Health and Safety in Agriculture (CCHSA), and Department of Medicine, University of Saskatchewan, 104 Clinic Place, Saskatoon, SK S7N 2Z4, Canada; Randy Purves Crop Development Centre, College of Agriculture and Bioresources, 51 Campus Drive, Saskatoon, S7N 5A8, Canada; Nicholas Low, Department of Food and Bioproduct Sciences, College of Agriculture and Bioresources, University of Saskatchewan, 51 Campus Drive, Saskatoon, SK, S7N 1A8, Canada; Waleed Mohammed Saeid, College of Pharmacy and Nutrition, University of Saskatchewan, 107 Wiggins Road, Saskatoon, SK S7N 5E5, Canada; Mona Khamis, College of Pharmacy and Nutrition, University of Saskatchewan, 107 Wiggins Road, Saskatoon, SK S7N 5E5, Canada.

Title:
Differential Ion Mobility and Tandem Mass Spectrometric Analysis of a Series of Chlorogenic Acid Isomers in Fruit Juice

Introduction:
Chlorogenic acids are among the most abundant phenolics found in the human diet. Of these, the mono-caffeoylquinic acids are the predominant phenolics found in fruits, such as apples and pears. In this research, the MS/MS behavior of the three most common mono-caffeoylquinic acids were determined. Differential ion mobility was employed to separate these isomers in juice without chromatography

Methods:
Three different mass spectrometric instruments equipped with ESI source were used for MS/MS analysis using both positive and negative ionization. The genesis of each product ion were conformed using MS3 experiments. ESI coupled with high-field asymmetric waveform ion mobility spectrometry-mass spectrometry (ESI-FAIMS-MS) was used to separate isomers within juice samples

Results:
MS/MS dissociation was primarily driven via the cleavage of the ester bond linking the quinic acid moiety to the caffeic acid moiety within tested molecules. Despite being structural isomers with the same m/z values and dissociation behaviors, the MS/MS data in the negative ion mode was able to differentiate the three isomers based on ion intensity for the major product ions, at m/z 191, 178 and 171. Subsequently, ESI-FAIMS-MS was employed for the separation of these compounds for the first time. By combining MS/MS data and FAIMS, a method for the separation and identification of mono-caffeoylquinic acids in apple/pear juice samples was developed with a run time of less than one minute. It is envisaged that this methodology could be used to identify pure juices based on their chlorogenic acid profile (i.e., metabolomics), and could also be used to detect juice-to-juice adulteration.

Novel Aspect:
Caffeoylquinic acid isomers were distinguished using MS/MS. ESI-FAIMS-MS separated the isomers within juice samples without chromatographic separation

Ellis, Shane,
M4I Institute

Contributing Authors:
Michael E. Belov, Spectroglyph, LLC, 101904 Wiser Parkway, Suite 104 Kennewick, WA 99338, USA; Shane R. Ellis, The Maastricht Multimodal Molecular Imaging Institute, Maastricht University, Maastricht, The Netherlands; Martin R. L. Paine The Maastricht Multimodal Molecular Imaging Institute, Maastricht University, Maastricht, The Netherlands; Ron M. A. Heeren, The Maastricht Multimodal Molecular Imaging Institute, Maastricht University, Maastricht, The Netherlands; Anthony D. Postle, Southampton Centre for Biomedical Research, NIHR Respiratory Biomedical Research Unit, University Hospitals Southampton, UK.

Title:
An Ion Funnel based MALDI-MSI Interface for Orbitrap Mass Spectrometers

Introduction:
MALDI-MSI has proven to be an immensely powerful for studying molecular distributions and local biochemical processes within complex surfaces. Its insight has been further improved via its coupling to high mass resolution analysers. Here we will present recent work describing the design, performance and application of a dual ion funnel-based MALDI source coupled to an Orbitrap analyser.

Methods:
The MALDI source contains the target plate at the entrance of a dual ion funnel (Spectroglyph) held at intermediate pressure (~1-10 mbar). Ions are generated via a 5 kHz 349 nm Nd:YLF laser focused to a ~25 um spot size. Ions are focused, trapped and efficiently transmitted through the ion funnels into the mass spectrometer. Tissue sections were prepared with standard MSI preparation protocols

Results:
First a performance evaluation of the MALDI system coupled to a high field Orbitrap analyser will be presented. The effect of ion funnel parameters and their influence on sensitivity, including entrance (sample) pressure, ion accumulation and voltages, along with MALDI laser fluence, ion cooling and clustering processes will be discussed. To demonstrate the performance for imaging studies we will present results acquired from lipid imaging in respiratory disease models where we seek to probe the underlying lipid metabolism of lung surfactant with high resolution. Mass accuracy throughout the imaging experiment can be maintained through the simultaneous orthogonal introduction of a lock mass ion by ESI. In addition, MALDI-MSI studies of small metabolites in tumor xenografts and drug imaging studies using will be presented.

Novel Aspect:
The design, evaluation and application of a dual ion funnel MALDI source for high sensitivity and high mass resolution MALDI imaging will be presented

ElNaggar, Mariam,
Prosolia

Title:
Emerging Applications in Biochemical Profiling and Imaging Atmospheric Pressure Mass Spectrometry

Introduction:
The continuous in situ microextraction provided by the flowprobe system enables analysis with few limitations to sample type. Applying flowprobe analysis to tissue, film, analytical surfaces, and cytological and microbial preparations, can result in characterization of drugs, metabolites, and other compounds of interest either as chemical images, time resolved intensity profiles, or mass spectra

Methods:
Samples were analyzed using the flowprobe extraction and ionization system mounted onto Thermo Exactives. Methanol and other solutions were used for analysis. Chemical images were acquired via automated sequential extractions covering a window of m/z ranges suitable for simultaneous acquisition of potential compounds of interest. Data were compiled using Firefly 2.0 and visualized with BioMap.

Results:
Preliminary data characterized profiling and imaging capabilities of the flowprobe. Monitoring for drugs, calibration curves could be linearly determined through tissue surfaces while lipid compounds associated with the tissue extracted from within the volume of the surface remained steady. Tissue samples with compounds of interest were characterized. Chemical images and profiles of dose response determined both disposition and the time resolved intensities of drugs and other diagnostic compounds in various tissue and sub-tissue types. These data correlated well to other techniques. Evaluation of mass spectral data and subsequent data from cells on cytospin slides revealed that fixation and analysis of the samples could be performed such that cells were preserved for subsequent morphological evaluation, even with spectra indicating the quantitative extraction of various phospholipids.

Novel Aspect:
Direct extraction, ionization, and identification of molecules from various preparations and workflows with samples preserved for subsequent analysis

Elsayed, Hanan,
College of Pharmacy, University of Saskatchewan

Contributing Authors:

Kevin Allen, College of Pharmacy and Nutrition, University of Saskatchewan, Saskatoon, SK, Canada; Darryl J. Adamko, Department of Pediatrics, College of Medicine, University of Saskatchewan, Saskatoon, SK, Canada; Anas El-Aneed College of Pharmacy and Nutrition, University of Saskatchewan, Saskatoon, SK, Canada.

Title:

Detection and Quantification of 16 Organic Acid Metabolites Excreted in The Urine of Respiratory Illness Patients Using a Novel LC-MS/MS Method

Introduction:

Asthma is a complex syndrome with symptoms that can overlap with other respiratory conditions such as chronic obstructive pulmonary disease (COPD). The goal of this study is to develop a sensitive and selective liquid chromatography-tandem mass spectrometric (LC-MS/MS) method to analyze 16 urinary organic acid metabolites that were differentially expressed in asthma patients.

Methods:

Light 12C and heavy 13C DmPA labeling agents are used for the analysis of the organic acid metabolites. The 13C labeled derivatives were used as an internal standard. Derivatized metabolites were separated on a C18 column using a gradient system. The analysis was conducted on an Agilent 1200 HPLC interfaced with an Qq-LIT-MS. The system was operated in the positive ion mode using MRM scans.

Results:

The fragmentation patterns of the 16 labeled metabolites show various diagnostic product ions that are selected for targeted MRM analysis. Chromatographic separation, MS/MS and the use of 13C labelled internal standards allows for the unambiguous identification of the target metabolites. A qualitative and quantitative targeted LC-MRM-MS/MS method is developed for the analysis of the 16 urinary organic acid metabolites in a run time of 30 minutes. The method is selective and showed the desired linearity within the range that will be applied for the analysis of patient urine samples. The method is currently being validated according to USFDA guidelines. Preliminary clinical data analysis showed promising results into the use of some the target metabolites as diagnostic biomarkers. Asthma patients were successfully separated form other respiratory illness patients.

Novel Aspect:

A novel LC-MS/MS method quantified 16 organic acid metabolites in urine, using a 13C labeled derivatives as internal standards

Engen, John R.,
Northeastern University

Contributing Authors:
F. Ulrich Hartl, Max Planck Institute of Biochemistry; Kristina Popova, Max Planck Institute of Biochemistry; Amit J. Gupta Max Planck Institute of Biochemistry; Florian Georgescauld, Max Planck Institute of Biochemistry; Manajit Hayer-Hartl, Max Planck Institute of Biochemistry; Andreas Bracher, Max Planck Institute of Biochemistry.

Title:
Chaperonin-assisted protein folding

Introduction:
Although long desired, the analysis of chaperonin-assisted protein folding pathways at high resolution has remained elusive. Due to advances in methodology, the folding of TIM-barrel proteins has now been observed at peptide-level resolution using hydrogen/deuterium exchange and mass spectrometry.

Methods:
Pulse-labeling deuterium exchange of free and GroEL/ES associated proteins was followed by pepsin digestion and UPLC MS. Mass spectrometry was performed with a Synapt QTof and data were processed with DynamX software.

Results:
In comparison to spontaneous folding, in which most elements of the DapA protein acquire structure simultaneously, GroEL/ES accelerated folding more than 30-fold by catalyzing segmental structure formation in the TIM-barrel. Analysis of folding by a similar protein from a bacterial species that lacks GroEL/ES revealed that segmental structure formation can occur in an chaperonin-independent manner for a different sequence that folds to a similar tertiary structure. Our results support the hypothesis that the chaperonin cage acts as a powerful folding catalyst for a set of proteins which otherwise fail to reach native state at a biologically relevant time scale.

Novel Aspect:
Peptide-level resolution of substrate protein folding in a GroEL/ES cage

Engst, Wolfram,
German Institute of Human Nutrition Potsdam- Rehbrücke (DIfE)

Contributing Authors:
Wolfram Engst, German Institute of Human Nutrition Potsdam-Rehbrücke; Stefanie Weiß, German Institute of Human Nutrition Potsdam-Rehbrücke; Delphine Saulnier German Institute of Human Nutrition Potsdam-Rehbrücke; Michael Blaut, German Institute of Human Nutrition Potsdam-Rehbrücke.

Title:
Determination of folate vitamers in feces and serum of Crohn´s disease patients using isotope-dilution UPLC-ESI-MS/MS

Introduction:
Crohn's Disease (CD) patients often show a serum folate deficiency. In order to find out whether serum folate status in CD patients reflects an altered folate metabolism in the gut, we determined folate vitamers in serum and feces of CD patients and controls. We developed a sensitive UPLC-ESI- MS/MS method for the quantification of different mono- and polyglutamylated folates in human samples.

Methods:
After solid phase extraction from fecal extract and serum, analytes were separated by RP-UPLC. MRM data were acquired in the positive ion mode using two MRM transition for determination of each analyte. We are able to analyze 8 monoglutamyl folate forms. Additionally, we developed a new hilic based chromatographic system which allows us to differentiate between intact polyglutamyl folate forms.

Results:
Using stable isotope labeled standards we quantified 3 predominant compounds (5-methyltetrahydrofolate (5CH3-THF), oxidized 5-methyltetrahydrofolate, folic acid) and their polyglutamylated forms (after cleavage with human conjugase) in 30 CD patients. Our results confirm 5CH3THF as the predominant folate in blood of all participants but with lower levels in the CD group compared to control group ($p < 0.05$). We found mainly polyglutamylated folates in feces. The concentration of the different polyglutamylated folate forms was higher in feces compared to their analogous monoglutamylated serum folate forms. We determined tetraglutamyl 5CH3THF as the dominant folate form within the fecal samples.The developed MS based method is a specific and sensitive procedure suitable for analysis of folate vitamers in human serum and feces which overcomes drawbacks of microbiological and immuno assays.

Novel Aspect:
New sensitive UPLC-ESI-MS/MS method for determination of folates in human feces and serum. Characterization of folate status of CD patients.

Enjalbal, Christine,
Université de Montpellier, Institut des Biomolécules Max Mousseron

Contributing Authors:
Didier Gagne, Institut des Biomolécules Max Mousseron (IBMM), UMR 5247, Université de Montpellier; Sonia Cantel, Institut des Biomolécules Max Mousseron (IBMM), UMR 5247, Université de Montpellier; Gilles Subra Institut des Biomolécules Max Mousseron (IBMM), UMR 5247, Université de Montpellier; Agnès Hagège, Institute of Analytical Sciences (ISA), UMR 5280, CNRS, Université Lyon 1; Jean Martinez, Institut des Biomolécules Max Mousseron (IBMM), UMR 5247, Université de Montpellier; Emmanuelle Cordeau, Institut des Biomolécules Max Mousseron (IBMM), UMR 5247, Université de Montpellier; Carine Arnaudguilhem, Laboratoire de Chimie Analytique Bio-inorganique et Environnement LCABIE-IPREM, UMR 5254; Brice Bouyssiere, Laboratoire de Chimie Analytique Bio-inorganique et Environnement LCABIE-IPREM, UMR 5254.

Title:
Selenium labelling combined with LC-ICP-MS, a new methodology in pharmacology to measure protein-ligand interactions

Introduction:
Measuring the affinity of a ligand for its target is an issue of central importance in life science. Due to the required sensitivity, radioactivity constitutes the method of choice for pharmacological studies of receptor/ligand system. We aim at replacing such hazardous technology by ICP-mass spectrometry for robust and sensitive quantification of peptide ligands in competitive binding assays.

Methods:
The capacity of inductively coupled plasma-mass spectrometry (ICP-MS) to provide quantitation of metallic and hetero elements, whatever the nature of the sample medium, prompted us to investigate this technique in combination with selenium chemical tagging through the creation of covalent bonds using conventional organic sulfur chemistry without disturbance of the affinity for the receptor.

Results:
To develop the analytical method and provide the proof of concept, membrane receptors were first chosen, and the strategy was validated on two GCPR models systems: AVPR and CCKR, involving peptides as native ligands. The analytical workflow based on RP-LC-ICP-MS analyses was optimized according to the biological matrices constraints (presence of many salts as well as alkaline metals) reaching remarkable LOQ at concentration of interest (50ppt of Se, 63 fmole of injected Se for vasopressin/AVPR system). The established method was first applied to the quantitative measurement of the affinity of selenium labeled ligands for their receptor (V1A or CCK2) through the determination of the dissociation constant (Kd). Inhibition constant (Ki) was deduced from competitive binding assay. These achieved results were in accordance to literature data recorded with radiolabeled reference ligands.

Novel Aspect:
LC-ICP-MS represents an attractive alternative to radioactivity to perform quantitative measurement of protein-ligand interactions in pharmacology.

Eremin, Dmitry,
Zelinsky Institute of Organic Chemistry

Contributing Authors:
Valentine Ananikov, Zelinsky Institute of Organic Chemistry, Russian Academy of Sciences, Leninsky Pr. 47, Moscow 119991, Russia.

Title:
A New Approach for Revealing Superatomic Cores in Transition Metal Complexes by ESI-MS/MS

Introduction:
Superatoms are now considered to be a promising base for developing new catalysts and for determination of a particular role of clusters in catalysis. The main problem is connected with the absence of a convenient method for revealing of new structural motives that could possess exceptional properties. We propose a practical procedure for detection of superatomic cores in metal acetylacetonates.

Methods:
High-resolution mass spectra were recorded on a Bruker maXis Q-TOF instrument equipped with an electrospray ionization (ESI) ion source. The measurements were performed in positive (+)MS ion mode. For the MS/MS mode for all samples nitrogen was used as a collision gas. Center of mass energy for survival yield 0.5 was chosen as a criterion of relative stability.

Results:
Electrospray ionization mass spectrometry (ESI-MS) with collision-induced dissociation (CID) provides a comparative determination of relative bond strength. We have investigated range of metal acetylacetonates by the means of ESI-MS/MS. Ions containing more than one transition metal underwent CID fragmentation. We have combined fragmentation analysis and survival yield analysis to reveal plausible superatomic cores. [Ni2O2] in Ni2(acac)3+ appeared to be the true superalkaline core which was confirmed by DFT calculations. Besides nickel, [Cu2O2] and [Pd2O2] also appeared to remain after fragmentation but being less stable. According to unique nature, we found the same core in nickel complexes with triacetylmethane and trifluoroacetylacetone. Abnormal stability was also revealed in mixed metal acetylacetonates (Ni/Ag, Ni/(VO), Ni/Cu, Ni/Pd).

Novel Aspect:
This approach confirmed the unique nature of M2O2 core. Thus we propose a new methodology for revealing superatoms in transition metal complexes.

ERRA-BALSELLS, Rosa,
Departamento de Quimica Organica, Facultad de Ciencias Exactas y Naturales, Universidad de Buenos Aires

Contributing Authors:
H. Nonami, Ehime University, Japan; T. Nakashima, Ehime University, Japan and Hokaido University, Japan; S. Morita NARO Kyushu Okinawa Agricultural Research Centre, Japan; H. Wada, NARO Kyushu Okinawa Agricultural Research Centre, Japan.

Title:
Intact trichomes single-cells analysis: metabolite profiling with internal electrode capillary pressure probe electrospray ionization MS

Introduction:
Plant trichomes are unicellular or multicellular epidermal outgrowths displaying various morphologies, functions, and distribution patterns across organs. For this reason, they are considered as an ideal model cellular system to study mechanisms underlying cellular differentiation, interaction and function at single cell resolution.

Methods:
picoPressure Probe; picoPressure Probe Electro Spray Ionization; Internal Electrode Capillary Pressure Probe Electrospray Ionization (IEC-PPESI); Analyzer: Orbitrap mass spectrometer (Exactive Plus, Thermo Fisher Scientific Inc., MA); nanoESI- Orbitrap mass spectrometer (Exactive Plus, Thermo Fisher Scientific Inc., MA)

Results:
In Solanum spp., trichomes have been classified into glandular or non-glandular types, depending on the presence or not of glandular cells atop the stalk cells. Since the metabolic ability of glandular cells to synthesize and accumulate metabolites, extensive omics studies have been conducted on isolated glandular cells. Metabolite analyses on stalk cells have been completely lacking thus far. This is due to difficulties in selective sampling or isolation of stalk cells without contamination of adjacent glandular or epidermal cells. Organic acids, carbohydrates, flavonoids and acyl sugars were detected from single stalk and glandular cells with sub-picoliter sap sample. The minimal cell sap removal from a stalk cell without severe disturbance allowed sequential analysis of adjacent glandular cell on the same type VI trichome, showing striking difference between two adjacent cell types.

Novel Aspect:
MS analysis of intact neighbor single cells with different metabolite composition

Fan, Xing,
China University of Mining and Technology

Contributing Authors:
Jingpei Cao, China University of Mining & Technology; Xing Fan, China University of Mining & Technology; Yunpeng Zhao China University of Mining & Technology; Xianyong Wei, China University of Mining & Technology; Miao Wang, China University of Mining & Technology.

Title:
Characterization of the oxidation products of Shengli lignite using mass spectrometers with "hard", "soft" and ambient ion sources

Introduction:
Although new ionization methods for MS have been developed in recent years, it remains a challenging task to obtain adequate molecular information from coal, a complex fossil fuel, through only one ionization mode. The combination of three MS ionization techniques contributes to detecting oxidation products, understanding the oxidation process of coal, and facilitating coal directional conversion.

Methods:
Coal sample was added to NaClO (6% active chlorine content) aqueous solution in a spherical flask to react for 4 h. Reaction mixture was separated to get filtrate and filter cake. The filtrate was sequentially extracted by ethyl ether and ethyl acetate to obtain extracts. Extracts were analyzed using GC/MS, TOF-MS coupled with ESI and DART ion sources.

Results:
Organooxygens, nitrogen-containing compounds, chloro-substituted species (CSSs) and sulfur-containing compounds in each extract were identified by MSs with different ion sources. According to GC/MS analysis, aliphatic acids and benzene polycarboxylic acids were the major species. Considerable CSSs were identified after oxidation, which proved the participation of chlorine through oxidation reaction. The kinds of species with molecular mass over 500 Da detected by ESI- and DART-MS were around 19.9% and 53.9%, respectively, which were not revealed by GC/MS. More species in Nn group were identified by DART-MS compared to the other two MS techniques, occupying 45.8% of the identified species. DART-MS obtained much more molecular information in the extracts and promoted better understanding of compositions and structures of coals and coal derivatives at molecular level.

Novel Aspect:
The combination of three MS methods provides comprehensive understanding of coal structures and facilitates coal directional conversion.

Fan, Xing,
China University of Mining and Technology

Contributing Authors:
Jingpei Cao, China University of Mining & Technology; Xianyong Wei, China University of Mining & Technology; Yunpeng Zhao China University of Mining & Technology; Junliu Xia, China University of Mining & Technology; Xing Fan, China University of Mining & Technology.

Title:
Sequential ultrasonic extraction of a Chinese coal and characterization of nitrogen-containing compounds in the extracts using HPLC/ESI-TOF-MS

Introduction:
Understanding the modes of occurrence and distribution of nitrogen and sulfur in coal is necessary for air pollution control. it is necessary to understand the composition and structure of organonitrogen compounds (ONCs) in coal at molecular level to develop effective approaches for nitrogen removal during combustion and value-added utilization as chemical feedstock.

Methods:
Dongming lignite (DL) was subjected to ultrasonic extraction sequentially through petroleum ether, carbon disulfide, methanol, acetone, and isometric carbon disulfide/acetone mixed solvent. Then enriched components in the extracts were separated by HPLC according to molecular polarity and ESI-TOF-MS selectively characterized heteroatom-containing compounds including isomers of ONCs in DL.

Results:
Sequential solvent extraction of DL followed by HPLC/ESI-TOF-MS analysis is proven to be an effective approach for separation and detailed characterization of ONCs. Sulfur-containing compounds was not found, which may be the ionization selectivity of ESI source. Acetone is an excellent solvent to dissolve ONCs in DL due to the hydrogen bonds of N–H···O between ONCs and acetone. Most of ONCs contain N-heterocyclic aromatic rings, including pyridine and quinoline for N1Oy class, pyridine, quinoline, and pyrrole for N2Oy class, and aliphatic amine group for N4 and N5 classes. Some isomers in the extracts were separated and identified by HPLC/ESI-TOF-MS, and the corresponding structural information was proposed. Characterization the chemical compositions of ONCs in coal should be beneficial to nitrogen removal and facilitate clean utilization of coal.

Novel Aspect:
Organonitrogen isomers in DL were separated by the HPLC and identified by the ESI-TOF-MS, and the corresponding structural information was proposed.

Fayad, Paul,
Thermo Fisher Scientific

Contributing Authors:

Kelly Flook, Thermo Fisher Scientific; Voktorija Vitkovske, Thermo Fisher Scientific; Vikrant Gohil Thermo Fisher Scientific; Rosa Viner, Thermo Fisher Scientific; John Rogers, Thermo Fisher Scientific; Agne Alminaite, Thermo Fisher Scientific; Paul Haney, Thermo Fisher Scientific; Kay Opperman, Thermo Fisher Scientific; Helene Cardasis, Thermo Fisher Scientific; Egle Capkauske, Thermo Fisher Scientific; Kestutis Bargaila, Thermo Fisher Scientific.

Title:

Development of an all-recombinant intact protein standard for LC MS application development and system suitability testing.

Introduction:

In recent years, interest in intact protein analysis by HPLC, LC-MS, and MSMS has increased significantly. Here, we describe the development of a multi-purpose recombinant intact protein standard for LC, LC-MS, and LC-MSMS quality control and application development.

Methods:

Multiple proteins were screened and seven were selected that 1) evenly covered a MW range of 12kD – 66kD, 2) presented clean, modification and adduct-free ESI spectra, and 3) whose ESI charge state distributions covered a wide m/z range from 500-2000. Mixing ratios were adjusted such that all seven proteins could be detected simultaneously in a single infusion MS experiment (R&D standard).

Results:

The initial R&D standard, which included cytochrome c, RNAse A, myoglobin , trypsin inhibitor, carbonic anhydrase, enolase, and BSA met our needs with respect to providing a standard sample that covered a large MW range, a wide m/z range, varying LC-retention characteristics, and consistent MSMS performance. To extend utility of the sample to the broader intact protein analysis community, improvements to long term stability were required, as well as the replacement of all animal origin proteins as per FDA regulations. As such, new recombinant protein candidates were synthesized, affinity captured, and purified. Selected candidates were fully characterized by UV HPLC and LC/MS/MS, and accelerated stability tests were performed on those which met purity, MW, and m/z distribution requirements. The new pre-commercial mixture will be presented here.

Novel Aspect:

A widely-applicable recombinant, affinity tag-free, high quality intact protein standard for LC-MS quality control and application development.

Fei, Fan,
McMaster University

Contributing Authors:
Michelle L. Mendonca, Department of Biochemistry and Biomedical Sciences, McMaster University; Brian E. McCarry, Department of Chemistry and Chemical Biology, McMaster University; Dawn M. E. Bowdish Department of Pathology and Molecular Medicine, Michael G. DeGroote Institute for Infectious Disease Research, McMaster University; Michael G. Surette, Department of Medicine, Farncombe Family Digestive Health Research Institute, McMaster University.

Title:
Comprehensive Metabolic and Transcriptomic Profiling of Streptococcus intermedius under Aerobic and Anaerobic Growth Environments

Introduction:
Streptococcus Anginosus/Milleri Group (SMG) are facultative anaerobic bacteria that asymptomatically colonize the upper respiratory, gastrointestinal and urogenital tracts. They are also common pathogens in pyogenic invasive infections, as well as pulmonary and urinary tract infections. Most SMG infections are polymicrobial and associated with co-infecting obligate anaerobic bacteria.

Methods:
To better understand the effect of oxygen on the growth and physiology of these organisms, we compared the global metabolomic and transcriptomic profiles of S. intermedius strain B196 under aerobic and anaerobic conditions. Intracellular and extracellular metabolites were measured using HILIC-LCMS.

Results:
The largest transcriptional changes were associated with induction of oxidative stress response genes under aerobic conditions. Modest changes in expression of genes associated with primary metabolism were observed under the two conditions. Differences in the abundance of specific metabolites were correlated with observed transcription changes in genes associated with their metabolism, implying that metabolism is primarily regulated at the transcriptional level. Rather than a large shift in primary metabolism under anaerobic conditions our results suggest a modest tuning of metabolism to support the accelerated growth rate of S. intermedius strain B196 in the absence of oxygen. For example, under anaerobic conditions, purine metabolism, pyrimidine de novo synthesis and pyrimidine salvage pathways were up-regulated at metabolic and transcriptional levels.

Novel Aspect:
This study provides a better understanding of differences between S. intermedius anaerobic and aerobic metabolism.

Fell, Lorne,
LECO Corporation

Contributing Authors:
Elizabeth Humston-Fulmer, LECO Corporation; Joe Binkley, LECO Corporation.

Title:
A Compendium of Food and Beverage Comparisons with a Novel Benchtop GC-TOFMS system

Introduction:
GC-TOFMS is an analytical technique that is broadly applicable for food and beverage samples. Among other capabilities, it offers the non-targeted characterization of samples and the ability to target specific analytes of interest in order to compare and contrast various samples. A novel benchtop GC-TOFMS system is demonstrated here for a variety of food and beverage products.

Methods:
A collection of food and beverage samples (fresh and dried herbs, fresh and frozen produce, flavored candies, flavored beverages, hops, and wines) were prepared for analysis with HS-SPME (using a DVB/CAR/PDMS fiber). The samples were subsequently analyzed with GC-TOFMS and characterized for both targeted and non-targeted.

Results:
A variety of food and beverage samples were successfully compared and contrasted using a novel benchtop GC-TOFMS system. The various volatile and semi-volatile analytes that contribute to the aroma and flavor of the products were collected with HS-SPME and separated with GC. Deconvolution offered an additional level of information based on the mathematical separation of the full m/z range TOFMS data. This aspect of the analytical approach allowed for identifying and quantifying individual analytes even when chromatographically co-eluting with other analytes in the matrix. In some cases, these co-elutions were important sample distinguishing features that would be hidden without this capability. Several examples are demonstrated.

Novel Aspect:
A novel GC-TOFMS system with improved sensitivity and linear dynamic range was used for these analyses.

Fell, Lorne,
LECO Corporation

Contributing Authors:
Jonathan Byer, Leco Corporation; David Alonso, Leco Corporation; Joe Binkley Leco Corporation; Elizabeth Humston-Fulmer, Leco Corporation.

Title:
Analysis of Urine SRMs with Comprehensive Two-Dimensional Gas Chromatography (GC×GC)-High Resolution Time-of-Flight Mass Spectrometry

Introduction:
In this study, various sample introduction techniques were used for the identification of volatile and semi-volatile compounds in two new standard reference materials (SRM, smoker's and non-smoker's urine). These reference materials are vital in the diagnosis and setup of standard operating procedures for metabolic profiling of urine in humans.

Methods:
Solid Phase Micro Extraction, Dynamic Headspace and Liquid Injection sample introduction techniques were used to analyze the SRMs. The applied methodology also included a combination of complementary hard and soft ionization techniques coupled to high resolution time of flight mass spectrometry, deconvolution and targeted processing methods which were used for characterization of samples.

Results:
Analyses resulted in confident identification of a wide variety of materials (e.g., polyaromatic hydrocarbons, phthalate metabolites, phenols and pain killers) and metabolite derivatives. GCxGC-TOFMS chromatograms were highly structured showing clustered classes of compounds and provided high quality spectral data that were searched against large, well-established databases. Comparison of the sample introduction techniques yielded expected results; dynamicheadspace and SPME having a larger volatile profile. High resolution time-of-flight mass spectrometry (HRT) resulted in additional benefits such as accurate formula determinations for fragment, molecular and adduct ions, as well as, increased selectivity that reduced background interferences. Comprehensive HRT data was probed multiple times via targeted and/or untargeted processing methods to identify important classes of compounds.

Novel Aspect:
Characterizing metabolites in smoker's urine standard reference materials via high resolution mass spectrometry and comprehensive GC (GCxGC).

Ferguson, Lee,
Duke University

Contributing Authors:
Gordon Getzinger, Duke University.

Title:
Exploring environmental chemical space through HR/AM mass spectrometry and cheminformatics: The example of wastewater-derived organic micropollutants

Introduction:
High-resolution accurate-mass mass spectrometry (HR/AM MS) provides a promising approach for comprehensive structural elucidation of numerous trace-level organic contaminants in aquatic media. However, comprehensive sample characterization is made difficult by inefficiencies in prioritization of detected features and postulated structures based on tandem MS analysis.

Methods:
We have developed a holistic, non-targeted screening workflow to address these challenges, which combines ultra-high resolution tandem mass spectrometry (Thermo Orbitrap Fusion), and computational mass spectrometry. We illustrate the utility of this approach to evaluate the occurrence of organic micropollutants in the effluent of a conventional domestic wastewater treatment facility.

Results:
Analysis of municipal wastewater effluent using an Orbitrap Fusion yielded 3,701 unique molecular features after data processing, for which >91% had associated tandem mass spectral data. Custom data processing scripts were deployed to pipe relevant feature data to computational mass spectrometry tools for molecular formula assignment based on isotope pattern and fragment spectrum decomposition (SIRIUS, formulas assigned to 90.2% of features), structure assignment from PubChem (postulated structures returned for >77% of features) and tandem mass spectral annotation and scoring (MetFrag, MAGMa, and CFM-ID). Finally, postulated structures were scored based on their structural similarity to possible environmental contaminants (i.e., compounds known to be produced in commerce) and to known wastewater pollutants. After rigorous data filtering, 607 micropollutants were tentatively identified.

Novel Aspect:
We have developed the first large-scale, automated workflow for non-targeted analysis of organic micropollutants by HRAM in environmental samples.

Fernandes, Anna Maria,
ThoMSon Mass Spectrometry Laboratory

Contributing Authors:
Luara Nicolai Piardi, Laboratório de Estudos da Dor, Institute of Biology, UNICAMP, Campinas; Elayne Vieira Dias, Laboratório de Estudos da Dor, Institute of Biology, UNICAMP, Campinas; Marcos Nogueira Eberlin Thomson Mass Spectrometry Laboratory, Institute of Cmemistry - UNICAMP; Cesar Renato Sartori, Laboratório de Estudos da Dor, Institute of Biology, UNICAMP, Campinas.

Title:
Dsorption Electrospray Ionization Mass Spectrometry Imaging (DESI-MSI) for Unravel of Biomarkers of Neuroplasticity in Pain and Stress

Introduction:
Neuroplasticity is the ability of the brain to change after external stimuli or injuries. How this process occurs at molecular level is still an important goal in neuroscience research. Time-consuming classical methods involve antibodies and/or radioactive materials. Herein, DESI-MSI was used for elicit chemical changes in brains of rats subjected to a mixed protocol of chronic pain and stress.

Methods:
C57BL/6JUniB rats were subjected to a combined experimental protocol of repeated social defeat stress and chronic inflammatory hyperalgesia that followed institutional and ethical guidelines (Protocol Number: 4019-1, IB-UNICAMP). DESI-MSI of brain slices were acquired by a Prosolia Omnispray 2D Ion Source coupled to a Q Exactive 2.2 SP1 mass spectrometer and produced via Biomap.

Results:
Rat brain slices from treated and control animals, i.e. subjected or not to the combined protocol of stress and chronic hyperalgesia, were analysed simultaneously in the same DESI-MSI experiment. Brain tissues were analysed over a range of 200 – 1000 Da. When comparing the treated group to the control group it was observed a decrease in the relative abundance of the ion of m/z 466.899 and an increase of the relative abundance of the ions of m/z 598.157, 600.110, 652.237, 654.190 in the negative ion mode. The accurate identification of these molecules is under investigation. For the first time, changes in the chemical brain tissue constitution after external stimuli were observed directly without the use of antibodies or radioactive materials. These findings indicate that this pioneer approach must be a powerful tool in neural plasticity studies.

Novel Aspect:
DESI-MSI is a powerfull toll to the direct visualization of biomarkers of neuroplasticity in rat brain slices.

Fernandez, Facundo,
School of Chemistry and Biochemistry

Contributing Authors:
Henrik Christensen, Institute for Robotics and Intelligent Machines, Georgia Institute of Technology; Nael McCarty, Emory+Children's Center for Cystic Fibrosis and Airways Disease Research and Department of Pediatrics, Emory University School of Medicine and Children's Healthcare of Atlanta; CHristina Jones School of Chemistry and Biochemistry, Georgia Institute of Technology; Brian Clowers, Department of Chemistry, Washington State University; Maria Eugenia Monge, CIBION, CONICET; Matthew Bernier, School of Chemistry and Biochemistry, Georgia Institute of Technology; Stephen Zambrzycki, School of Chemistry and Biochemistry, Georgia Institute of Technology; Jacob Huckaby, Energid Technologies; Anyin Li, School of Chemistry and Biochemistry, Georgia Institute of Technology; Jose Perez, School of Chemistry and Biochemistry, Georgia Institute of Technology; Xiaoling Zang, School of Chemistry and Biochemistry, Georgia Institute of Technology; Arlene Stecenko, Emory+Children's Center for Cystic Fibrosis and Airways Disease Research and Department of Pediatrics, Emory University School of Medicine and Children's Healthcare of Atlanta; Joel Keelor, School of Chemistry and Biochemistry, Georgia Institute of Technology; David Goodlett, School of Pharmacy, University of Maryland; Martin Paine, School of Chemistry and Biochemistry, Georgia Institute of Technology.

Title:
New Ambient MS and Ion Mobility MS Approaches for Multidimensional Complex Mixture Analysis

Introduction:
In this presentation we showcase the coupling of transmission mode Direct Analysis in Real Time (DART) to traveling wave ion mobility (IM) mass spectrometry for metabolomics, a new multimodal vacuum-assisted plasma ion (VaPI) source, a robotic 3D surface sampling approach for non-planar surfaces, and a new atmospheric pressure drift tube ion mobility-Orbitrap mass spectrometer.

Methods:
A Synapt G2, (Waters) equipped with a gas-ion separator tube interface (GIST, IonSense); a time-of-flight mass spectrometer (AccuTOF, JEOL); a quadrupole Orbitrap mass spectrometer (Q-Exactive Plus, Thermo); and a dual gate drift tube IM cell (MA3100, Excelims) were used for the experiments presented here. Robotic experiments used a second generation Universal Robots UR5 six axis arm

Results:
It was found that DART coupled to IM-MS offered high throughput and sufficient sensitivity for metabolomic analysis of biological samples of moderate complexity. It also offered complementary information to other types of ion sources such as APCI and ESI. VaPI was found to be as effective as DART, with the additional advantage of enabling several sampling modes simultaneously. Three dimensional objects were effectively investigated with a robotic thermal desorption sampling approach coupled to a 3D vision system for imaging acquisition, leading to the first automated ambient surface analysis tool for non-planar samples. Finally, initial results using a combination of high resolution drift tube ion mobility and ultrahigh resolution Orbitrap mass analysis demonstrated that exceedingly high peak capacities are possible with such system.

Novel Aspect:
The innovations presented in this talk involve the coupling of novel ion sources to IM-MS.

Fernandez, Marti,
Institut de Química Computacional i Catàlisi (IQCC) - Departament de Química - University of Girona

Contributing Authors:
Marti Fernandez, Institut de Química Computacional i Catàlisi, Departament de Química, University of Girona, Catalonia (Spain); Magda Parera, Institut de Química Computacional i Catàlisi, Departament de Química, University of Girona, Catalonia (Spain); Anna Dachs Institut de Química Computacional i Catàlisi, Departament de Química, University of Girona, Catalonia (Spain); Miquel Solà, Institut de Química Computacional i Catàlisi, Departament de Química, University of Girona, Catalonia (Spain); Anna Pla-Quintana, Institut de Química Computacional i Catàlisi, Departament de Química, University of Girona, Catalonia (Spain); Anna Roglans, Institut de Química Computacional i Catàlisi, Departament de Química, University of Girona, Catalonia (Spain).

Title:
The Rhodium-Catalyzed [2+2+2] Cycloaddition Reaction: A Mechanistic Study

Introduction:
The rhodium-catalyzed [2+2+2] cycloaddition reaction of three alkynes is a highly efficient method to obtain polysubstituted benzene derivatives in a single synthetic step. A deep knowledge of the reaction mechanism is fundamental in order to improve the process, but the low concentration and the transient nature of the intermediates involved complicate the identification of the species.

Methods:
Electrospray ionization mass spectrometry (ESI-MS) enables the gathering of detailed data by trapping and identifying short-lived intermediates in organometallic catalytic cycles. In addition to observing the molecular mass of ions for many reactive intermediates, this technique reveals structural information through characteristic fragmentation patterns by collision-induced dissociation (CID).

Results:
We present here an ESI-MS investigation of the [2+2+2] cycloaddition reaction between several diynes and phenylacetylene derivatives catalyzed by a cationic Rh(I) complex and BINAP-type ligands. The use of a cationic catalytic system makes it possible to detect all the intermediates proposed in the catalytic cycle. In addition, computer model studies based on density functional theory (DFT) support the structural proposal for the intermediates.

Novel Aspect:
Elusive intermediates of the Rh-catalyzed [2+2+2] cycloaddition have been detected for the first time, deepening our knowledge of the process.

Fialkov, Alexander,
Tel Aviv University

Contributing Authors:
Aviv Amirav, 1. School of Chemistry, Tel Aviv University, Tel Aviv 69978, Israel; Tal Alon, 1. School of Chemistry, Tel Aviv University, Tel Aviv 69978, Israel; Boaz Seemann 1. School of Chemistry, Tel Aviv University, Tel Aviv 69978, Israel; Svetlana Tsizin, 1. School of Chemistry, Tel Aviv University, Tel Aviv 69978, Israel.

Title:
LC-MS with Cold EI – The New System and Recent Applications

Introduction:
Electron Ionization (EI) LC-MS is highly desirable as it enables automated library based identification and is free of any ion suppression effects that plague ESI and/or APCI. Thus, bringing back EI to LC-MS is highly valuable if a robust, simple to operate and low cost EI-LC-MS system can be designed and built.

Methods:
We developed a new electron ionization (EI) LC-MS with Cold EI based on an Agilent 5977 quadruple GC-MS. Cold EI is EI of vibrationally cold molecules in supersonic molecular beams (SMB). The LC effluents are pneumatically sprayed followed by fast thermal vaporization. The vaporization chamber is connected to the supersonic nozzle via a capillary flow restrictor to suppress cluster formation.

Results:
A fly-through ion source was used and the ions were mass analyzed by the 5977 mass analyzer. Cold EI results in enhanced molecular ions which are often weak or missing in standard EI, yet it is fully compatible with library identification and it uniquely provides: 1. Library identification with compounds names and structures 2. Extensive fragment MS information 3. Our TAMI molecule identifier software enables the elucidation of elemental formula 4. The fly-through ion source eliminates any ion source degradation 5. Non-polar compounds analysis is enabled 6. Cold EI provides uniform, compound independent ionization efficiency 7. Cold EI does not suffer from any ion suppression or enhancement effects Our system was explored in the analysis of cannabinoids in Cannabis flowers, in the analysis of anti-oxidant compounds and cholesterol, free fatty acids and lipids in serum.

Novel Aspect:
Conversion of a commercially available MS of GC-MS into a novel electron ionization LC-MS is described along with its benefits and applications.

Fialkov, Alexander,
Tel Aviv University

Contributing Authors:
Uri Keshet, 1. School of Chemistry, Tel Aviv University, Tel Aviv 69978, Israel; Aviv Amirav, 1. School of Chemistry, Tel Aviv University, Tel Aviv 69978, Israel; Tal Alon 1. School of Chemistry, Tel Aviv University, Tel Aviv 69978, Israel.

Title:
Open Probe Fast GC-MS – Real Time Analysis with Separation

Introduction:
Ambient desorption ionization techniques such as DART, DESI, DSA, ASAP and LDTD are gaining attention and growing use. They enable real time sample on surfaces analysis, without sample preparation, while using the MS of LC-MS systems. However, they do not share the library identification strength of EI and the obtained mass spectra are often too complex hence require expensive HR-MS.

Methods:
We developed an Open Probe Fast GC-MS for real time analysis with separation, without sample preparation and with MS of standard GC-MS. It is based on a heated oven that is mounted on a low thermal mass fast GC which is open to room air with helium flow protection to prevent air penetration. Its operation is as simple as: touch the sample, insert it and get the GC-MS data in 30 s.

Results:
Several Open Probe Fast GC-MS forensic and other application were explored including: a) Synthetic cannabis on tobacco leaves analysis in 30 s; b) Several Cannabis flowers were analyzed for their content of active ingredients. c) Street drug heroin powder was analyzed in 30 s with all its main 6 ingredients separated and identified; d) Cocaine was analyzed on 50 Shekel bill; d) Trace TNT form human hand. e) Four pesticides were analyzed in 40 s on tomato that was sprayed with cockroach killer. Open Probe Fast GC-MS provides: A) Library based sample identification with names and structures B) Ultra-fast GC separation for improved mixtures analysis C) Uniform response is provided by EI D) Non-polar compounds are amenable for analysis E) Swabs can be used to bring samples from remote surfaces F) Low cost MS of GC-MS is used

Novel Aspect:
Open Probe Fast GC-MS was explored with several forensic applications. Real time analysis was achieved with separation and library identification.

Fioramonte, Mariana,
University of Campinas

Contributing Authors:
Fabio Cesar Gozzo, Dalton Mass Spectrometry Laboratory - University of Campinas.

Title:
New chemical cross-linking methodology enhancing zero length products

Introduction:
Chemical cross-linking (XL) is a very attractive technique to study protein structure, especially when no high resolution methods are possible. Zero length (ZL) cross-link using EDC is a very disseminated methodology on XL area, leading to the direct linkage between Lys and Glu/Asp. In this work we show a new methodology for ZL enhancing the product formation and targeting new amino acids.

Methods:
Myoglobin were diluted on appropriate buffer and reacted with the cross-linker for periods between 30 minutes and 2 hours. The reaction products were then analyzed by intact mass spectrometry (MS). The protein was digested and injected into a UPLC system coupled to a Synapt HDMS (Waters). The MS/MS spectra were acquired using data dependent acquisition. The XL species were identified using SIM-XL.

Results:
Intact protein analyses showed the formation of 22 ZL species, exactly the number of theoretical XL species when the tertiary structure is analyzed. Yet, the new methodology showed to be so efficient that even the HEME group was cross-linked to the protein. When compared with EDC at the same condition, the new approach presents purely ZL species as products, whereas the EDC methodology also produces species that contain the acid side chain modified with EDC, a non-desired subproduct. Moreover, the new approach reveals many cross-linked peptides between acid residues Glu/Asp and Ser, a new specificity for ZL cross-linking.

Novel Aspect:
New approach for zero length cross-linking to methodology to enhance reaction yield and specificity.

Fjeldsted, John,
Agilent Technologies

Contributing Authors:
Teresa Mairinger, BOKU University; Sarah Stow, Vanderbilt University; Jody May Vanderbilt University; John McLean, Vanderbilt University; Emma Rennie, Agilent Technologies; Erin Baker, Pacific Northwest National Labs; Stephan Hann, BOKU University; Tim Causon, BOKU University; Ruwan Kurulugama, Agilent Technologies.

Title:
Analytical Approaches in Support of Highly Accurate Collision Cross Section Measurements

Introduction:
The calculation of collision cross section using low drift field ion mobility-MS relies on the use of stepped field measurements to separate drift cell and non-drift cell time factors as applied to the Mason-Schamp equation. This work presents a theoretical and applied extension of the Mason-Schamp equation to single, static field operation applicable to various sample introduction configurations.

Methods:
Using stepped field methodology, CCS values were accurately determined for Agilent Tune Mix ions with a rigorous assessment of measurement uncertainty under various conditions. Under identical conditions panels of metabolites and proteins where analyzed. Using the CCS values determined for Tune mix ions, calibrations were made in support of single field operation and results compared.

Results:
A principle goal of this study is to establish traceable support of achieving standardized collision cross section (CCS) values for widely used MS tune and calibration ions based on the well-established Mason-Schamp equation. Preliminary data from multiple labs shows that CCS precision as low as 0.1% is achievable and with normalized experimental parameters, inter-lab agreement below 0.5%. This work presents traceable values, central to standardization and the curation of CCS values for publications and databases. We then transformed the Mason-Schamp equation to an expression which is analyte independent and can be used as the basis for calibration for single field operation in support of any chromatographic analytical system. Initial result show agreement to historical CCS data on peptide samples to within 2% and the promise of agreement to stepped field measurements of less than 1%.

Novel Aspect:
This work reports precise determination of CCS standard values and extends the Mason-Schamp equation to use in LC/MS applications for complex samples.

Flokova, Kristyna,
Laboratory of Plant Physiology, Wageningen University

Contributing Authors:

Harro J. Bouwmeester, Laboratory of Plant Physiology, Wageningen University, Droevendaalsesteeg 1, 6708 PB Wageningen, The Netherlands; Beatriz Andreo Jimenez, Laboratory of Plant Physiology, Wageningen University, Droevendaalsesteeg 1, 6708 PB Wageningen, The Netherlands; Mahdere Z. Shimels Laboratory of Plant Physiology, Wageningen University, Droevendaalsesteeg 1, 6708 PB Wageningen, The Netherlands.

Title:

New strategies to strigolactone determination in complex sample matrices

Introduction:

Strigolactones (SLs) are the most recently described group of plant hormones. Root-exuded compounds can be recognized by parasitic plants as a host plant signal resulting in enormous negative impact on yield of agriculturally important crops. Detailed insight to biosynthesis and metabolism of SLs is hampered by their limited stability and extremely low quantities in complicated plant/soil matrix.

Methods:

SLs were extracted with cold aqueous solution of acetonitrile or acetone and pre-concentrated using RP polymer-based sorbent with hydrophilic/lipophilic retention characteristics (Oasis® HLB columns, Waters). Ultra-high performance liquid chromatography-tandem mass spectrometry (UHPLC-MS/MS) method was employed to determine SLs in roots and root exudates of sorghum and rice plants.

Results:

We present highly sensitive and validated UHPLC-MS/MS method for simultaneous profiling of ten selected SL compounds. In order to minimize strong matrix effect, sample size of root exudate extracts and roots was reduced to 15 ml and 50 mg, respectively. Combination of rapid extraction with water miscible aprotic solvents and single-step high-throughput purification using macroporous polymer-based sorbents with both hydrophilic and lipophilic retention characteristics simplified current sample preparation procedures, notably improved stability and extraction recovery of selected SLs (≥ 85%). Individual SLs were separated in 10 minutes using an analytical column packed with sub-2-microne particles. Detection limits ranged from 0.1 to 5 fmol and the linearity for most of SLs was at least five orders of magnitude. Seven SLs were successfully determined in real samples.

Novel Aspect:

The advantages of approach to SL determination consist in possibilities to accommodate small sample size and highly selective and sensitive detection.

Focant, Jef,
University of Liege

Contributing Authors:

Benjamin L'Homme, Organic and Biological Analytical Chemistry Group, CART, University of Liège, Belgium; Chiara Calaprice, Organic and Biological Analytical Chemistry Group, CART, University of Liège, Belgium.

Title:

Cryogenic Zone Compression (CZC) and Comprehensive Two-dimensional Gas Chromatography (GC×GC) Coupled to (HR)MS for Measuring Hyper-Trace Compounds

Introduction:

The measurement of dioxins requires the utmost sensitive techniques. Gas chromatography (GC) coupled to sector HRMS (SIM mode) or QQQMS (MRM mode) for target analyses. HRTOFMS are used for non-target analyses, but suffers from limited sensitivity. Cryogenic Zone Compression (CZC) and Comprehensive Two-dimensional Gas Chromatography (GC×GC) can be used to enhance GC signals.

Methods:

CZC and GC×GC were implemented using cryogenic modulators. HRMS was performed on a Thermo Fisher DFS magnetic sector instrument (Bremen, Germany) and on a JEOL AccuTof T100GC (Tokyo, Japan). Sample preparation was carried out on a Power-Prep system (FMS Inc., Waltham, USA), or using MEPS (Trajan, Victoria, Australia) and VAMS (Phenomenex, Torrance, USA).

Results:

Methods have been developed to measure dioxins, PCBs, OCPs, and Dechloranes in low volume of human blood samples to estimate environmental exposure of selected populations. The use of MEPS and CZC-GC-HRTOFMS permitted the measurement of PCB-153 and DDE in 20uL serum samples. VAMS and GC-QQQMS were used to measure 24 OCPs and 6 PCBs in 40uL serum samples. Routine multistep automated clean-up and fractionation procedures were also optimized to afford lower solvent consumption for dioxin measurements at the sub-ag level based on CZC-sector HRMS. The various approaches illustrate some of the possibilities to achieve hyper-trace detection levels while reduce sample amounts to their minimum values. GC×GC-HRTOFMS further extend this to the inclusion of unknown analytes.

Novel Aspect:

Dedicated small size sample preparation approaches and cryogenic enhancement of GC signals allow HRMS instruments to perform hyper-trace measurements.

Focant, Jef,
University of Liege

Contributing Authors:

Edouard Louis, Gastroenterology Department, GIGA-R, Liège University Hospital, CHU-ULg, Belgium; Nicolas Di Giovanni, Organic and Biological Analytical Chemistry Group, CART, University of Liège, Belgium; Marie-Alice Meuwis Gastroenterology Department, GIGA-R, Liège University Hospital, CHU-ULg, Belgium.

Title:

Metabolomic Investigation of inflammatory bowel disease phenotypes using GC×GC-HRTOFMS

Introduction:

Inflammatory Bowel Disease (IBD) refers to chronic or recurring inflammation of the gastrointestinal tract, including ulcerative colitis (UC) and Crohn's disease (CD). Advanced metabolite profiling techniques are needed for biomarker discovery to better understand the metabolic imbalance, but also to possibly discriminate between CD patients and controls to contribute to diagnostic for CD.

Methods:

We developed, validated, and implemented a Comprehensive Two-dimensional Gas Chromatography High Resolution Time-of-flight Mass Spectrometry (GC×GC-HRTOFMS) method for metabolic profiling of human serum of CD patients under strict QA/QC guidelines. We developed a robust data processing approach based on templates relying on GC and MS quality features, and multiple statistical techniques.

Results:

First, the method was optimized using Design of Experiment (DoE) and further validated using a set of 20 representative analytes in QC serum pools. Second, validation was performed, using different phenotypes of inflammation, including high (n=12) and low inflammation (n=12), remission (n=11), and healthy patients (n=33). All 2D chromatograms were used to create a cumulative image used to define 524 good quality features. Raw feature volumes were normalized and corrected for analytical variation with LOESS protocol (locally weighted scatterplot smoothing). All features with RSD > 30% in QC were eliminated, which lead to 189 potential biomarkers further submitted to various statistical methods (uni and multivariate, feature selection, supervised learning) followed by re-sampling and test validation. Robust biomarkers were identified based on retention indices, and accurate MS data.

Novel Aspect:

We report sets of biomarkers capable to discriminate between various inflammation phenotypes representative of CD using GC×GC-HRTOFMS.

Forest, Anik,
Montreal Heart Institute

Contributing Authors:
Bertrand Bouchard, Montreal Heart Institute; Christine Des Rosiers, Montreal Heart Institute; Isabelle Robillard Frayne Montreal Heart Institute; Caroline Daneault, Montreal Heart Institute.

Title:
Comprehensive and comparative mass-spectrometry based untargeted lipidomics in plasma versus various tissues enhances coverage of the lipidome

Introduction:
The goal of a discovery-based untargeted approach in metabolomics is to analyze all detectable compounds, including unknowns, in a biological system. The application of this approach to lipidomics remains, however, challenging due in part to the presence of lipid isomers. We present a robust MS-based workflow for comprehensive lipidomics in a single sample using different biological matrices.

Methods:
Procedures for lipid extraction and analysis by high resolution HPLC-quadrupole-time-of-flight were adapted from Matyash et al. (2008) and Sandra et al. (2010) to enable wide coverage of compounds with diverse polarity and separation of isomers. Data analysis was achieved using Agilent Mass Hunter and MPP software, modified in-house, and lipid signal identity was manually confirmed by MS/MS.

Results:
Good recovery (>77%) was found for standard lipids with different polarities, ranging from lysophospholipids to triglycerides. For biological samples, the use of a long chromatographic gradient minimized ion suppression, while enabling detection of low abundance lipids and resolution of isomers. The latter is crucial for successful alignment and identification of lipid signals, including acyl chains on various head groups. Using 100 µL extracted plasma, we measured ~1,300 MS signals (CV<20%), which correspond to ~600 unique lipids, of which 450 have been characterized. This method was also applied to 3 different tissues. We were able to align MS signals from plasma and tissues samples and characterize an additional hundreds of lipid species, including cardiolipins. Data also show the diversity in the type, proportion and number of lipids species between plasma vs. tissues.

Novel Aspect:
The developed lipidomics workflow enables reproducible and semi-quantitative analysis of >450 lipid species, including isomers, in biological matrices

Fornarini, Simonetta,
Università degli Studi di Roma La Sapienza

Contributing Authors:
Davide Corinti, Dipartimento di Chimica e Tecnologie del Farmaco, Università degli Studi di Roma La Sapienza, I-00185 Roma, Italy; Nazzareno Re, Dipartimento di Farmacia, Università G. D'Annunzio, I-66100 Chieti, Italy; Cecilia Coletti Dipartimento di Farmacia, Università G. D'Annunzio, I-66100 Chieti, Italy; Barbara Chiavarino, Dipartimento di Chimica e Tecnologie del Farmaco, Università degli Studi di Roma La Sapienza, I-00185 Roma, Italy; Maria Elisa Crestoni, Dipartimento di Chimica e Tecnologie del Farmaco, Università degli Studi di Roma La Sapienza, I-00185 Roma, Italy.

Title:
Reactivity and Spectroscopic Features of Key Intermediates in Cisplatin Drug Chemistry

Introduction:
Cisplatin (cis-[PtCl2(NH3)2]) is one of the most potent anticancer drugs. Its activity is based on binding to the purine bases of DNA eventually leading to apoptosis of the cancer cell. However, how and why cisplatin reaches its target is currently a matter of debate. We have investigated the early events and primary complexes in the cisplatin reaction with biological targets.

Methods:
Electrospray ionization mass spectrometry (ESI-MS) is exploited to obtain cis-[PtCl(NH3)2(L)]+ complexes, isolating them in the gas phase, where they are characterized for their bimolecular reactivity, for their fragmentation behavior and for their structural features by Infra Red Multiple Photon Dissociation (IRMPD) spectroscopy combined with quantum chemical calculations.

Results:
IRMPD spectroscopy has disclosed the structure of [PtCl(NH3)2(L)(H2O)]+ clusters, delivered by ESI from the aqueous solution. These ions are found to conform to the encounter complex of cis-[PtCl(NH3)2(H2O)]+ with the incoming ligand L and represent the first direct evidence of a prototypical Eigen-Wilkins encounter complex in solution, lying on the reaction coordinate for ligand substitution and extracted by ESI for mass spectrometric analysis. Activated [PtCl(NH3)2(L)(H2O)]+ ions dissociate by loss of either H2O or L, the former process implying a ligand substitution event.

Novel Aspect:
IRMPD spectroscopy has revealed both structural details and reaction dynamics at the level of the isolated encounter complex.

Fornarini, Simonetta,
Università degli Studi di Roma La Sapienza

Contributing Authors:
Debora Scuderi, Laboratoire de Chimie Physique d'Orsay, Universitè Paris Sud, UMR8000 CNRS, Facultè des Sciences, 91405 Orsay Cedex, France; Barbara Chiavarino, Dipartimento di Chimica e Tecnologie del Farmaco, Università degli Studi di Roma La Sapienza, I-00185 Roma, Italy; Maria Elisa Crestoni Dipartimento di Chimica e Tecnologie del Farmaco, Università degli Studi di Roma La Sapienza, I-00185 Roma, Italy; Enrico Bodo, Dipartimento di Chimica, Università degli Studi di Roma La Sapienza, I-00185 Roma, Italy.

Title:
Sequential Cysteine Oxidation Probed by IRMPD Spectroscopy

Introduction:
Oxidation processes are ubiquitous events under both physiological and pathological conditions. The thiol side chain of cysteine residues in peptides and proteins is an active player in redox transformations, occurring in numerous oxidations states, including disulfides, sulfenamide (-SNR), sulfenic (-SOH), sulfinic (-SO2H) and sulfonic (-SO3H) forms.

Methods:
In this contribution, we report on a combined computational and experimental investigation of the intrinsic chemical properties and geometrical determinants of the conjugate bases of cysteine sulfenic, sulfinic and sulfonic acids, [cysSOx]- (x=1, 2, 3), obtained by electrospray ionization and assayed for IR activity.

Results:
Infrared multiple photon dissociation (IRMPD) spectra have been recorded in the highly informative mid-IR spectral range (750-1900 cm-1) employing a Paul ion-trap mass spectrometer coupled to the IR beamline of a tunable free-electron laser (FEL) at the CLIO (Centre Infrarouge d'Orsay) facility. The structural and spectral assignments are supported by ab initio static and DFT molecular dynamic simulations at finite temperature.

Novel Aspect:
This study has unveiled the deprotonation site, geometric determinants and spectroscopic signatures of the sampled oxoforms.

Fouquet, Thierry,
National Institute of Advanced Industrial Science and Technology (AIST), Japan

Contributing Authors:
Hiroaki Sato, National Institute of Advanced Industrial Science and Technology (AIST, Tsukuba, Japan).

Title:
The use of high resolution SpiralTOF MS and mass defect analysis for the characterization of coals and complex polymeric samples

Introduction:
The SpiralTOF analyzer recently introduced fills a gap between the time-of-flight analyzers (TOF) and the ultra high resolution ion cyclotron resonance devices. It provides a sufficiently high resolution to explore the composition of complex samples such as coal extracts or copolymeric samples and improve the visualization of data through the reliable use of Kendrick mass defect analysis (KMD).

Methods:
Deposits hydrothermally extracted from brown coals, coatings deposited by the plasma (co)polymerization of alkyl and perfluorinated acrylates as well as complex polymers produced from copolymers of ethylene and vinyl acetate (EVA) upon their hydrolytic and thermo-oxidative degradations have been characterized using a SpiralTOF with (matrix assisted) laser desorption ionization (MA)LDI.

Results:
The change in the composition of deposits from brown coal was readily displayed after a KMD analysis of the data recorded by LDI SpiralTOF MS. The plasma polymers deposited from the co-injection of two antagonist acrylates have not been described as "copolymers" for lack of clue. The MALDI SpiralTOF MS analysis of their soluble parts followed by a KMD analysis highlight the occurrence of co-oligomers i.e. their copolymeric nature. Mass spectra of EVA copolymers are complex owing to isobaric interferences between co-oligomers and isotopes. Their degradation products formed upon hydrolysis or thermo-oxidation make the spectra even more tedious to interpret. Relying on the high resolution of the SpiralTOF and combined with a KMD analysis using the proper base unit, the composition of EVA and aged EVA samples fractionated by size exclusion chromatography were assessed in a visual manner.

Novel Aspect:
The high resolution SpiralTOF is powerful enough for the mass defect analysis and visual display of complex spectra from coals and polymers.

Fouquet, Thierry,
National Institute of Advanced Industrial Science and Technology (AIST), Japan

Contributing Authors:
Hiroaki Sato, National Institute of Advanced Industrial Science and Technology (AIST, Tsukuba, Japan).

Title:
Convenient visualization of high resolution MS/MS spectra of synthetic polysiloxane ions using Kendrick mass defect analysis.

Introduction:
High resolution tandem mass spectrometry (MS/MS) is widely used for the characterization of polymer ions to get insights into the nature of the end-groups or architectures. However, MS/MS spectra could be tedious to describe owing to the number of product ions and inherent complexity of the fragmentation pathways. Their presentation to non-specialist audiences also remains an issue to overcome.

Methods:
Kendrick mass defect analysis (KMD) has been used to display the CID data from synthetic polysiloxane ions and highlight structural features, fragmentation routes and filiations. High resolution ESI-MS/MS spectra recorded from polydimethylsiloxane (PDMS), plasma polymerized hexamethyldisiloxane and polyhedral oligomeric silsesquioxane (POSS) have been re-analyzed using KMD as a user-friendly tool.

Results:
For an archetypal PDMS, a KMD plot readily discriminates two product ions series using dimethylsiloxane (DMS) as the base unit while their filiation is evidenced in the KMD plot using the expelled tetramethylsilane neutral as base unit. Structural features of product ions from a polycylosiloxane precursor ion found in a plasma-polymerized hexamethyldisiloxane – e.g. the number of cycles – are highlighted in KMD plots using DMS and silicon dioxide as base units. Finally, the fragmentation routes of a POSS bearing n-propylamine arms (the cage unfolding upon water release to yield a poly(azasiloxane) product ion and its depolymerization) are readily visualized on the KMD plots using water and azasiloxane as the base units. KMD plot is expected to accompany any high resolution MS/MS spectra of synthetic polymer ions and to become a convenient chart to present MS/MS data to wide audiences.

Novel Aspect:
The KMD analysis constitutes a versatile visualization tool for the CID data of polymer ions through intuitive point alignments.

Fournier, Isabelle,
University of Lille - France

Contributing Authors:
Philippe Saudemont, University of Lille, PRISM INSERM U1192; Benoit Fatou, University of Lille, PRISM INSERM U1192; Emilie Le Rhun University of Lille, PRISM INSERM U1192; Eric Leblanc, University of Lille, PRISM INSERM U1192; Michel Salzet, University of Lille, PRISM INSERM U1192; Marie Duhamel, University of Lille, PRISM INSERM U1192.

Title:
Bringing Mass Spectrometry Imaging from Pathologist Bench to Patient Bedside: From Diagnosis to Guided Surgery

Introduction:
Mass Spectrometry Imaging has now gained its letters of nobility demonstrating to be a robust technology to become the new pathologist tool. MSI coupled with Spatially-resolved Large Scale proteomics give access to tumor microenvironment studies and markers hunting for better diagnostic and treatment. The next step is to move from postoperative to intraoperative conditions for guided surgery.

Methods:
Unsupervised clustering on tryptic-peptides MSI data obtained from Human grade 3 gliomas define regions which serve for tissue microproteomics. Proteomics and MSI are then linked to the patient's clinical data. For real-time in-vivo, a novel instrument is developed based on laser ablation. The instrument capabilities are evaluated using a cohort of 100 patients with ovarian cancer.

Results:
Non supervised MALDI MSI analyses of glioma reveals after individual clustering a clear discordance with pathologist morphological annotation. Global clustering of the tissues together, on the other hand, gives 3 groups of molecular profiles. These groups were analyzed by tissue microproteomics leading to the identification of 2500 specific proteins. Each group demonstrates the upregulation of specific proteins involved in different mechanisms and signaling pathways. These data are then integrated with patient's clinical data e.g. IDH1 mutation, 1p19q codeletion allowing to improve glioma classification. For intra-operative studies, we demonstrate our new instrument to be operating in-vivo real-time conditions with minimum invasiveness. The instrument demonstrate its ability for accessing diagnosis of ovarian cancer using molecular profiles after building a classification model.

Novel Aspect:
Integration of MALDI MSI to clinical data and Development of a novel instrument for in-vivo real-time analysis for guided surgery

Franck, Julien,
University of Lille

Contributing Authors:
Xavier Roucou, Univ. Sherbrooke, Biochemistry department, Sherbrooke, Canada; Vivian Delcourt, Univ. Lille, INSERM, U1192, PRISM, Lille, France - Univ. Sherbrouke, Biochemistry department, Sherbrooke, Canada; Isabelle Fournier Univ. Lille, INSERM, U1192, PRISM, Lille, France; Michel Salzet, Univ. Lille, INSERM, U1192, PRISM, Lille, France.

Title:
Alternative proteins : the hidden world of potential biomarkers

Introduction:
A mature mRNA is a molecule presenting three different ORFs displaying several alternative ORFs. Recently, alternative proteins were detected in human cells although such a mechanism was considered as anecdotal in eukaryotes. These results suggest that the proteome is more complex and many proteins not considered so far remain to be identified and their molecular functions elucidated

Methods:
A phosphoproteome enrichment, a co-immunoprecipitation experiment and a full proteome extract available on the public data repository PRIDE (https://www.ebi.ac.uk/pride/archive/) were re-analyzed as well as a full proteome extract form SKOV3 cell lines. Tryptic peptides from SKOV3 were analyzed by nanoLC-HR-MS/MS and identified using Peptideshaker software and the human ORF/AltORF database.

Results:
To increase the number of potentially identified alternative proteins, a combination of multiple algorithms was used using the peptideshaker software. We were able to identify 10364 alternative proteins and 857 phosphorylated alternative proteins (1 peptide/protein). Re-analysis of data coming from co-immunoprecipitation of GFP tagged proteins from hela cells allows the identification of reference proteins as well as alternative proteins such as altSRRM2 translated from the SSRM2 gene in the CDS domain according to a +2 reading frame. Considering the SKOV3 cells, we identified the AKT2 protein as well as its associated alternative protein altAkt2. AKT has strong oncogenic function and is a key mediator of PI3K pathway function. AKT is highly activated in the late stage of ovarian cancer and it is possible to assume that AltAkt2 plays a major role in the progression of tumors.

Novel Aspect:
Alternative proteins not considered so far, can be now described as new potential biomarkers of pathologies.

Francois, Yannis-Nicolas,
LSMIS- University of Strasbourg

Contributing Authors:
Rabah Gahoual, Laboratoire de Spectrométrie de Masse des Interactions et des Systèmes (LSMIS), CNRS – UMR7140, University of Strasbourg; Lauriane Kuhn, Institut de Biologie Moléculaire et Cellulaire (IBMC), University of Strasbourg; Alain Beck Centre d'immunologie Pierre Fabre, Saint-Julien-en-Genevois; Emmanuelle Leize-Wagner, Laboratoire de Spectrométrie de Masse des Interactions et des Systèmes (LSMIS), CNRS – UMR7140, University of Strasbourg; Nassur Said, Laboratoire de Spectrométrie de Masse des Interactions et des Systèmes (LSMIS), CNRS – UMR7140, University of Strasbourg.

Title:
Structural characterization of antibody-drug conjugates by a combination of intact, middle-up and bottom-up techniques using sheathless CE-MS.

Introduction:
Antibody-drug conjugates (ADCs) represents a fast growing class of biotherapeutic products. Their production leads to a distribution of species exhibiting different number of conjugated drugs. We propose the development of complementary approaches implementing sheathless CE-MS to characterize the different aspects defining the structure of brentuximab vedotin.

Methods:
For intact and middle-up analysis, ADCs was buffer exchanged with 200 mM ammonium acetate buffer, pH 7.0 after or not IdeS digestion. For peptide mapping, sample was buffer exchanged against 50 mM ammonium bicarbonate buffer, pH 8.0. Before trypsin digestion, 10% of acetonitrile was added to the sample and 40% of isopropanol after digestion. Peptide mixtures were analyzed by CZE-ESI-MS/MS.

Results:
At the intact ADC level, we evaluated sheathless CE-MS system used as nanoESI infusion platform for native MS determination of the average DAR and ADC drug load distribution. Middle-up analysis using IdeS enzymatic digestion has been performed. Structural information of the expected Fc/2 and F(ab')2 subunits with the incorporation of 0 to 8 payloads have been obtained. Finally, we show the use of sheathless CZE-ESI-MS/MS for the bottom-up analysis of ADCs to enhance the level of characterization. We managed to perform the complete amino acid sequence characterization. Using the same set of data, glycosylations and drug-loaded-peptides were characterized. Regarding drug-loaded-peptides, we identified four drug-loaded-peptides linked to the inter-chain cysteines which gave precious information regarding localization of conjugated drugs and the assessment of the yield of drug incorporation.

Novel Aspect:
DAR determination. ADCs middle-up characterization. Complete ADCs primary structure characterization (sequence, glycosylations, drug-loaded peptide).

Francois, Yannis-Nicolas,
LSMIS- University of Strasbourg

Contributing Authors:
Charly Renard, Laboratoire de Spectrométrie de Masse des Interactions et des Systèmes (LSMIS), CNRS – UMR7140, University of Strasbourg; Alain Beck, Centre d'immunologie Pierre Fabre, Saint-Julien-en-Genevois; Rabah Gahoual Laboratoire de Spectrométrie de Masse des Interactions et des Systèmes (LSMIS), CNRS – UMR7140, University of Strasbourg; Emmanuelle Leize-Wagner, Laboratoire de Spectrométrie de Masse des Interactions et des Systèmes (LSMIS), CNRS – UMR7140, University of Strasbourg; Nassur Said, Laboratoire de Spectrométrie de Masse des Interactions et des Systèmes (LSMIS), CNRS – UMR7140, University of Strasbourg; Michael Biacchi, Laboratoire de Spectrométrie de Masse des Interactions et des Systèmes (LSMIS), CNRS – UMR7140, University of Strasbourg.

Title:
Glycoform Separation and Characterization of Cetuximab Variants by Middle-up Off-line CE-UV/ESI-MS

Introduction:
We report an original strategy to perform off-line CE-ESI-MS using CE-UV/fraction collection followed by ESI-MS infusion. Efficiency of our method was validated with the separation of cetuximab by middle-up approach. Two Fc/2 C-terminal lysine variants were separated. Regarding F(ab')2, 8 glycoform were identified in three different peaks following the presence of NGNA residues in glycan structure

Methods:
Sample was cleaved by IdeS (Genovis) to obtain two Fc/2 fragments and one F(ab')2 fragment. Sample was analyzed using a MDQ CE system (Beckman Coulter) coupled to a modified Proteineer FC (Bruker). Coated capillary with hydroxypropylcellulose was prepared in the laboratory. Infusion experiments were carried out with CESI 8000 system (Sciex) coupled to a Maxis 4G (Bruker)

Results:
We reported a strategy implementing off-line CZE-UV/fraction collection method followed by nanoESI-MS infusion for characterization of mAb. Assymetric CZE conditions involving different BGE at the ends of capillary have been described. Inlet BGE has been chosen as EACA 200 mM/ammonium acetate 25 mM, pH 5.70 and ammonium acetate 25 mM, pH 5.70 as outlet BGE. The three first peaks correspond to Fc/2 variants and the last three peaks correspond to F(ab')2 variants. Two C-terminal lysine variants of Fc/2 were identified and separated. Moreover separation of Fc/2 fragments allowed the glycoprofiling with the characterization of 7 glycoforms. Regarding F(ab')2 domain, deconvoluted mass spectra allowed identification of 8 glycoforms. This identification was supported by the decrease of effective mobility due to a reduction in net charge from the addition of negatively charged sialic acid.

Novel Aspect:
Middle-up characterization of mAbs, new off-line CE-ESI-MS methods, nanoESI, Glycoforms separation and characterization of mAb charge variants.

Fridgen, Travis,
Memorial University of Newfoundland

Contributing Authors:
Yanyang Chen, Memorial University; Graham J. Bodwell, Memorial University.

Title:
Innie or outie: Structures of M+[1,1,9,9-tetramethyl(2,11)terophyrenophane] (M=alkali metal) studied by mass spectrometric and computational methods

Introduction:
1,1,9,9-tetramethyl(2,11)teropyrenophane (TMTP or belt) presents a sizable cavity that molecules or ions can occupy. In this study, host-guest complexes composed of the TMTP and alkali metal cations (Rb+, Cs+) are studied by both experimental and computational methods. The purpose of these studies is to discern whether the ions might be inside or outside the TMTP in the ion-molecule complex.

Methods:
Complexes are trapped in a Fourier transform ion cyclotron resonance (FTICR) mass spectrometer and their structures are explored by blackbody infrared radiative dissociation (BIRD) kinetics and CID, and IRMPD spectroscopy. B3LYP/6-31+G(d,p) is used to perform geometry optimizations and frequency calculations in Gaussian 09.

Results:
BIRD studies indicate that there are two populations of ions that are electrosprayed. One is a very quickly-dissociating complex, while another is nonreactive under BIRD conditions over the temperature range studied. Computational methods were used to show that the weakly bound ion is likely the complex where the ion is outside the TMTP and the longer-living complex is the complex with the ion inside the TMTP, or wearing the belt. IRMPD spectroscopic studies in the C-H stretch region are underway. CID results on the TMTP radical cation will also be presented.

Novel Aspect:
Discerning whether an ion is wearing or holding the belt (innie or outie).

Fursey, Victor,
Bruker Daltonics Inc

Contributing Authors:

Sylvain Lehmann, 1-Laboratoire de Biochimie et Protéomique Clinique, Institut de Médecine Régénératrice et de Biothérapie, CHU de Montpellier-Hôpital St. Eloi, Montpellier, France; Audrey Gabelle, Centre Mémoire Ressources Recherche, CHU Montpellier, hôpital Gui de Chauliac, Montpellier. Université Montpellier I, Montpellier, F-34000 France ; 1- Laboratoire de Biochimie et Protéomique Clinique, Institut de Médecine Régénératrice et de Biothérapie,; Victor Fursey Bruker Daltonics Inc., 40 Manning Road, Billerica, MA, USA; Jerome Vialaret, 1- Laboratoire de Biochimie et Protéomique Clinique, Institut de Médecine Régénératrice et de Biothérapie, CHU de Montpellier - Hôpital St. Eloi, Montpellier, France; Pierre-Olivier Schmit, Bruker Daltonique S.A, 34, rue de l'industrie, 67160 Wissembourg, France.

Title:

Profiling of intact proteins in the CSF of Alzheimer's disease patients

Introduction:

Introduction: Proteomics investigations in biofluids have revealed that most proteins are present as a variety of proteoforms (combination of isoforms, proteolytic processing & PTM results). Measuring the intact mass of proteins by MS has the advantage to provide information on the relative amount of the different proteoforms. We have applied this approach for neurodegenerative diseases study.

Methods:

Methods: Cerebrospinal fluids (CSF) samples from 30 patients already diagnosed with Alzheimer's disease (AD), and controls were analyzed using an Impact II (Bruker Daltonics) with both a nanoflow and a normal-flow LC-MS analysis. After statistical analysis, Scheduled Precursor Lists have been used to fragment the proteoforms of interest and characterize them with BioTools (Bruker Daltonics).

Results:

Preliminary Data: The overlap of masses detected with the nanoflow and high-flow approaches was below 15%, and the mass accuracy obtained was below 2 ppm. Statistical analyses have been performed on more than 5000 compounds reproducibly detected in patient samples. MS results have been correlated with clinical data, to end up with a first list of more than 70 differentially regulated compounds. Those fragmented biomarker candidate have then been identified and characterized. The results confirmed that several isoforms of APP, clusterin or transthyretin were differentially expressed in the CSF of AD patients, when compared to the CSF of the other groups. Moreover, several isoforms of chromogranin/secretogranin (synaptic biomarker family) were detected, some showing a differential expression in the three clinical groups assayed.

Novel Aspect:

Neurological biomarkers identification and characterization based on TDCP approach allowed to identify differentially regulated proteoforms in CSF.

Furtos, Alexandra,
University of Montreal

Contributing Authors:
David H St-Pierre, Departement des sciences de l'activite physique de l'UQAM; Lilya Melbouci, Departement des sciences de l'activite physique de l'UQAM; Simon Comtois-Marotte Universite de Montreal.

Title:
LC-MS monitoring of lipid profile changes induced by a short term obesogenic diet in young rats

Introduction:
Increasing of obesity rates in children is alarming and understanding the role of dietis very important. Here young rats were exposed to a short term obesogenic diet and compared to control rats. We had observed that obesogenic diets induce changes in muscle contractility, gene expression and triglyceride composition. Here, LC-MS is used to monitor changes the diet has on lipid profile in muscle.

Methods:
Male Wistar rats were submitted to either a standard or an obesogenic diet and two muscles, EDL and SOL were used for comparisons. Lipids were extracted with MTBE and analyzed by LC-MS on a Synapt G2S instrument. Internal standards were added before and after the extraction. Mass spectra were acquired in positive and negative MSe mode between m/z 50-1200. MarkerLynx was used for data analysis

Results:
The method reproducibility at the lipid extraction level as well as the LC-MS analysis was evaluated. Both peak areas and retention times had good precisions. Several classes of lipids were observed and accurate mass analysis with data base search was used for identification. In negative mode, FA, PC and PE were detected as deprotonated species while ceramides were detected as formate adducts. In positive mode, PC, PE, SM and ceramides were present in their protonated form while TGs were detected mainly as ammonium adducts. Preliminary data on one SOL muscle sample compared between an individual under normal diet and an individual under an obesogenic diet showed differences in a number of different lipid classes. However, this result has to be confirmed and the differences validated on a statistically relevant number of samples

Novel Aspect:
Multidisciplinary project monitoring the dietary influence on the lipid infiltration process and muscle function alteration

Gabelica, Valérie,
Inserm, France

Contributing Authors:

Frédéric Rosu, IECB (CNRS UMS3033/Inserm US001/Univ. Bordeaux), France; Sandrine Livet, ARNA Laboratory (Inserm U1212/CNRS UMR5320/Univ. Bordeaux), France.

Title:

Native mass spectrometry and in-source collision-induced unfolding on a commercial helium drift tube ion mobility spectrometer

Introduction:

The meaningful comparison of collision cross section (CCS) values on different platforms is a prerequisite for using CCS for identification or structural assignment. Besides CCS calibration issues, the amount of internal energy imparted to the ions prior to the ion mobility cell is another source of experimental variation. Here we investigated how (DT)CCS(He) values change upon pre-IMS activation.

Methods:

The pumping system of an Agilent 6560 IMS-Q-TOF was modified, and the alternate gas kit was used to enable stable helium CCS measurements in the drift tube. The effect of source, transmission, and trapping parameters on the CCS distributions were tested on Ubiquitin7+, and on other collision-sensitive non-covalent complexes. A diverse set of potential CCS calibrants was then investigated.

Results:

The default tuning parameters are not the softest, and the entire voltage gradient was reoptimized. The source fragmentor and the trap entrance grid delta (TEGD) have the strongest influence on unfolding. Space charge in the trapping region also plays a role. Varying the TEGD allows reconstructing collision-induced unfolding (CIU) breakdown curves. We then investigated for a wide range of analytes (small peptides, proteins, nucleic acids, and rigid foldamers) whether and how the average CCS and width of the CCS distribution change as a function of pre-IMS activation conditions. Despite the absence of mass selection, CIU plots can be used to study gas-phase unfolding of biomolecules and noncovalent complexes produced from native solutions. We also propose a list of robust calibrants for the positive and the negative mode, which CCS values are insensitive to ion internal energy.

Novel Aspect:

CIU is controlled in the source or trapping regions of a DTIMS. Energy-dependent helium CCS are determined, and robust CCS calibrants are proposed.

Gabelica, Valérie,
Inserm, France

Contributing Authors:
Modesto Orozco, IRB Barcelona, BSC-IRB Research Program in Computational Biology, Barcelona, Spain; Joséphine Abi-Ghanem, ARNA Laboratory (Inserm U1212/CNRS UMR5320/Univ. Bordeaux), France; Hansel Gómez IRB Barcelona, BSC-IRB Research Program in Computational Biology, Barcelona, Spain; Clémence Rabin, ARNA Laboratory (Inserm U1212/CNRS UMR5320/Univ. Bordeaux), France; Frédéric Rosu, IECB (CNRS UMS3033/Inserm US001/Univ. Bordeaux), France; Leonardo R. Darre, IRB Barcelona, BSC-IRB Research Program in Computational Biology, Barcelona, Spain; Massimiliano Porrini, ARNA Laboratory (Inserm U1212/CNRS UMR5320/Univ. Bordeaux), France.

Title:
The structure of duplex nucleic acids in the gas phase

Introduction:
Native mass spectrometry aims at preserving the structure of biomolecular complexes upon gentle ionization and desolvation, in order to infer information about the starting solution structure from gas-phase mass, mobility or spectroscopy measurements. A crucial fundamental question is therefore the extent to which solution structural elements are preserved in gas-phase ions.

Methods:
Ion mobility MS of nucleic acids from NH4OAc solutions was recorded on an Agilent 6560 IMS-Q-TOF operated in helium. Experimental and calculated CCS were compared. Gas-phase structural models were generated by various types of MD simulations (unbiased, steered or replica exchange) in the gas phase. The stepwise desolvation of DNA in a charged droplet was also simulated.

Results:
For charge states typically produced from solution of physiological ionic strength, the experimental CCS values indicate a dramatic compaction compared to solution structures. Simulations indicate that double-stranded DNA or RNA undergo such compaction via new hydrogen bond formation between phosphate groups. At higher charge states, this effect is counterbalanced by Coulomb repulsion. MD in solution and in droplets suggest that ammonium ions relocate to the grooves, and mediate phosphate-phosphate interactions. Ammonia loss in the late stages of electrospray may then contribute to phosphate group self-solvation. Although not all base pair hydrogen bonds are lost, significant structural changes occur in the gas phase compared to the solution. However, the compaction pathway is now predictable and transposable to other DNA or RNA higher-order structures containing double-stranded regions.

Novel Aspect:
Nucleic acid higher-order structures are not as well conserved in the gas phase as initially thought, yet keep a memory of the solution.

Gaiffe, Gabriel,
Université Pierre et Marie Curie

Contributing Authors:
Maxime Bridoux, CEA, DAM, DIF, F-91297, Arpajon, France; Richard B. Cole, Sorbonne Universités, UPMC Univ Paris 06, CNRS, Institut Parisien de Chimie Moléculaire (IPCM), 4 place Jussieu 75252 Paris Cedex 05 France; Christine Costanza Laboratoire Central de la Préfecture de Police, 39 bis rue de Dantzig 75015 Paris.

Title:
Anion attachment mass spectrometry for improved analysis of explosive compounds

Introduction:
The development of rapid and efficient detection methods for the characterization of explosive compounds is of high importance to security forces concerned with terrorist threats. Using a MS approach, characteristic ions can be produced by attaching an anion to an analyte molecule in the negative mode of ESI-MS. Adducts can be detected with higher sensitivities as compared to the deprotonated.

Methods:
The formation of HMX, RDX, PETN, NG and R-salt adducts with different anions have been studied in ESI-MS . Afterwards, CID MS/MS experiments were performed to identify fragmentation pathways for these compounds. The formation of the deprotonated molecule resulting from dissociation of a series of anionic adducts employing different anions (A-) was measured to evaluate the strength of the H-A bond.

Results:
HMX and RDX, in the absence of an anion, were found to have similar fragmentation processes, each yielding the loss of neutral 74 Da fragments, whereas the other compounds produced mainly NO2- and NO3- as product ions. The fragmentation patterns of the adducts allowed the deduction that the gas-phase acidities of HMX, RDX and PETN are in the lower range of the tested anions, i.e., between fluoride and acetate, with RDX and PETN being less acidic than HMX in gas phase. Nitroglycerin was found to be more acidic. For all, it was particularly interesting to focus on the formation of chlorinated adducts. For HMX, we observed the formation of fragments that either contained or did not contain chloride. We explain this by postulating an unusual nucleophilic substitution reaction involving chloride anions. Although HMX and RDX are structural analogs, this behavior was not seen for RDX.

Novel Aspect:
Anion attachment mass spectrometry has been shown to present advantages for detection and characterization of explosives. The attachment of chloride w

Galba, Jaroslav,
AXON Neuroscience R&D Services SE

Contributing Authors:
Alena Michalicova, Institute of Neuroimmunology, Slovak Academy of Sciences; Vojtech Parrak, Axon Neuroscience R&D; Andrej Kovac Institute of Neuroimmunology, Slovak Academy of Sciences.

Title:
Liquid chromatography–tandem mass spectrometry method for determination of homocysteine and other biologically important thiols in plasma.

Introduction:
Thiols constitute the major portion of the total body antioxidants and they play a significant role in body defense against reactive oxygen species. The changes in thiols have been detected in various human disorders including neurodegeneration.

Methods:
Samples were reduced with tris-(2-carboxyethyl)-phosphine hydrochloride and deproteinized with trichloroacetic acid . Subsequently the thiols were derivatized with N-phenylmaleimide and the derivatization products were purified by solid phase extraction on mixed hydrophilic lipophilic balanced copolymer Oasis HLB. Separation of thiols was achieved by using HILIC chromatography.

Results:
Excellent linearity was observed for all analytes over their respective concentration ranges with correlation coefficients r2 > 0.99. The intra- and inter-day precision and accuracy were within ±20%. The analytical recovery for spiked plasma samples was in range from 92% to 108%. The limits of quantitation (LOQ) were 0.25 μmol/l for Cys-Gly and Hcys, 0.50 μmol/l for GSH and 5 μmol/l for Cys. Using this method, we analyzed plasma samples from transgenic rat model for tauopathy and detected significant changes of important antioxidant glutathione. This work was supported by APVV-14-0547 and VEGA 2/0159/15.

Novel Aspect:
We have developed short, simple and robust UPLC/tanden mass spectrometry method for analysis of biologically important thiols in plasma.

Galeano, Paula,
Universidad de los Andes

Contributing Authors:
Paula Galeano García, Laboratorio de Fitoquímica y Etnofarmacia, Universidad de los Andes. Bogotá. Bioprospección de los Productos Naturales Amazónicos, Universidad de la Amazonia. Florencia, Caquetá; Chiara Carazzone, Laboratorio de Fitoquímica y Etnofarmacia, Universidad de los Andes. Bogotá, Colombia.; Manuel Garavito Bioquímica y Biología Molecular de Parásitos (BBMP), Universidad de los Andes. Bogotá, Colombia; Barbara Zimmerman, Bioquímica y Biología Molecular de Parásitos (BBMP), Universidad de los Andes. Bogotá, Colombia.

Title:
LC-MS/MS analysis of compounds involved in De Novo pyrimidine biosynthesis in Solanum lycopersicum

Introduction:
Purine and pyrimidine nucleotides are the building blocks in the synthesis of nucleic acids, and they are precursors of primary and secondary metabolites. Pyrimidine nucleotides are synthetized by de novo pyrimidine biosynthesis, an ubiquitous metabolic pathway. The variation in levels of this metabolites offers information about the response to different stimuli that might occur in the organism.

Methods:
The extraction of S. lycopersicum metabolites was carried out by maceration with liquid N2 and subsequently SPE. The extracts were submitted to HILIC-HPLC-DAD-ESI-MS/MS. For the optimization of the MS/MS parameters, a solution of 5 mg/L of each metabolite was analyzed by direct infusion analysis (FIA) into the mass spectrometer at flow 1 µL/min.

Results:
The extraction with H2O:MeOH (80:20) and solid-phase extraction (SPE) on a Strata-XA cartridge, and posterior HPLC-ESI-MS/MS analysis of tomato leaves extracts allowed the determination of the glutamine, carbamoyl aspartate, dihydroorotate, orotate, and UMP which are intermediaries of de novo pyrimidine biosynthesis. Additionally, UTP, UDP-glu, CTP, GTP and ATP which are nucleotides produced by the metabolic pathway, and uridine and uracil that are metabolites used in mechanisms for pyrimidine recycling, can be identified.

Novel Aspect:
Development of a new method by HPLC-ESI-MS/MS for the quantification or level determination of the intermediates of the pyrimidine de novo synthesis.

Gates, Paul,
University of Bristol

Contributing Authors:

Norberto Lopes, School of Pharmaceutical Sciences of Ribeirão Preto, University São Paulo, Ribeirão Preto-SP, Brazil; John Walsby-Tickle, School of Chemistry, University of Bristol, Bristol, UK.; Christopher Arthur School of Chemistry, University of Bristol, Bristol, UK.; Paul Lawrence, School of Chemistry, University of Bristol, Bristol, UK.; Craig Butts, School of Chemistry, University of Bristol, Bristol, UK.

Title:

Probing ESI-MSn Fragmentation by 13C NMR and Computation

Introduction:

Structural elucidation of natural products is often hindered by complex fragmentations, rearrangements or by the absence of fragmentation in key parts of the molecule. Previous studies demonstrate that complexing natural products with metal cations can induce novel fragmentation routes that may extend the fragmentation into structural motifs of the analyte not previously accessible.

Methods:

A combination of sequential mass spectrometry, 13C NMR spectroscopy and computation have been used to probe the complexation of the natural product polyether ionophore antibiotic, lasalocid acid complexed with both sodium and calcium.

Results:

This investigation reveals a previously unrecognized fragmentation mechanism with potential applications to understanding metabolic and biosynthetic pathways. Differences in the binding sites of sodium and calcium within the polyether system are indicated by differences in the 13C spectrum and confirmed by computation. This difference in the binding enables alternative fragmentation routes to occur which in turn can lead to complementary structural data.

Novel Aspect:

A combined study by MSn, 13C NMR and computation leads to an increased understanding of complex formation and the different fragmentation observed.

Gaugg, Martin Thomas,
ETH Zürich

Contributing Authors:
Thomas Gaisl, Department of Pulmonology, University Hospital Zurich, Zurich, Switzerland; Yvonne Nussbaumer-Ochsner, Department of Pulmonology, University Hospital Zurich, Zurich, Switzerland; Lukas Bregy Department of Chemistry and Applied Biosciences, ETH Zurich, Zurich, Switzerland; Malcolm Kohler, Department of Pulmonology, University Hospital Zurich, Zurich, Switzerland; Yannick Suter, Department of Chemistry and Applied Biosciences, ETH Zurich, Zurich, Switzerland; Pablo Martinez-Lozano Sinues, Department of Chemistry and Applied Biosciences, ETH Zurich, Zurich, Switzerland; Nina Stebler, Department of Pulmonology, University Hospital Zurich, Zurich, Switzerland; Renato Zenobi, Department of Chemistry and Applied Biosciences, ETH Zurich, Zurich, Switzerland; Anna Engler, Department of Pulmonology, University Hospital Zurich, Zurich, Switzerland.

Title:
On-line breath analysis with secondary electrospray ionization reveals altered metabolic signatures between patients with different severities of COPD

Introduction:
Breath analysis of chronic obstructive pulmonary disease (COPD) patients with and without frequently reoccurring worsening of their symptoms ("exacerbations"), was performed on-line using ambient MS. The aim of the study was to discriminate between the two groups as well as getting an insight into underlying metabolic differences to better understand the pathophysiology of exacerbation events.

Methods:
44 COPD patients with and without frequent exacerbations were recruited and matched in terms of gender, age and smoking history. Breath analysis was performed in real-time using secondary electrospray ionization on a high-resolution time-of-flight mass spectrometer. Extensive data analysis was then applied to identify discriminative metabolic patterns.

Results:
We successfully identified 190 significant ($p < 0.05$) features in exhaled breath, which are significantly different between COPD patients with and without frequent exacerbations. Many of these molecules were found to form correlating chemical families, which speaks for a common underlying cause in metabolism. One of these groups is a series of nitro-aromatic compounds, which is increased in frequent exacerbators. Although unambiguous compound identification is still ongoing, it is reasonable to assume that these findings correlate with the increased oxidative stress caused by elevated nitric oxide production of the body's immune response to the pulmonary inflammation. Additionally, these compounds are highly carcinogenic, which is in agreement with the increased lung cancer risk in COPD patients. Other significant groups are short-chained fatty acids, as well as cysteine analogs.

Novel Aspect:
This study will help medical doctors to better understand the pathophysiology of COPD exacerbations, possibly facilitating diagnosis and treatment.

Gauthier, Jeremy,
Memorial University of Newfoundland

Contributing Authors:
Christina S Bottaro, Memorial University of Newfoundland; Stefana N Egli, Memorial University of Newfoundland.

Title:
Molecularly Imprinted Polymers with Desorption Electrospray Ionization Mass Spectrometry for Faster Environmental Monitoring of Neonicotinoids

Introduction:
Neonicotinoids are a class of insecticide developed as a less toxic alternative to organophosphates, methyl carbamates, and synthetic pyrethroids. Since their introduction, these pesticides have drawn public attention due to their implication in a decline of honeybee populations, as well as their environmental persistence.

Methods:
To increase the throughput of analysis of neonicotinoids in liquid matrices, we fabricate thin-film molecularly imprinted polymers (MIPs) for selective analyte uptake, which are also amenable to direct interrogation by desorption electrospray ionization (DESI)-MS.

Results:
Neonicotinoids are routinely analyzed by solid phase extraction with high performance liquid chromatography with MS (SPE-HPLC-MS), which can be time consuming. The application of our DESI-MS method has shown to improve the efficiency and throughput of analysis of neonicotinoids through the minimal sample preparation required. Optimizations of instrument parameters were performed to achieve optimal analyte detection. The developed method was then applied to thin-film MIPs, which through their simple mode of molecular recognition, increase method throughput by combining sampling, extraction, and analysis into one medium. Results from the development of an appropriate and applicable DESI-MS method will be shown, as well as work on the development of the MIPs.

Novel Aspect:
The combined use of novel thin-film MIPs with DESI-MS has significantly improved the throughput for analysis of neonicotinoids.

Gavriilidou, Agni Faviola Mika,
ETH Zurich

Contributing Authors:
Finn P. Holding, Astex Pharmaceuticals, 436 Cambridge Science Park, Milton Road, Cambridge, CB4 0QA, United Kingdom; Renato Zenobi, ETH Zurich, Department of Chemistry and Applied Biosciences, CH-8093 Zurich, Switzerland.

Title.
Studying non-covalent interactions with nano ESI-MS: An insight into the allosteric mechanism of M2 pyruvate kinase isoform (PKM2).

Introduction:
Allosteric regulation of multimeric proteins can occur through conformational changes induced/stabilized by binding to a ligand but it can be difficult to study the mechanism of allosteric regulation using ensemble measurements of ligand binding in protein solutions. In this study, native ESI-MS was applied to analyze the ligation states, oligomeric transitions and binding affinity of FBP to PKM2.

Methods:
MS experiments were carried out under native-like conditions in 200 mM ammonium acetate, pH 7.5. Prior to mass spectrometric analysis PKM2 protein and fructose-1,6-bisphosphate (FBP) solution stocks were prepared in the same ammonium acetate solution. Titration experiments of FBP against PKM2 were run using a native MS read-out on a Synapt G2-S HDMS (Waters, Manchester, UK) in positive ion mode.

Results:
It has previously been suggested that FBP activates PKM2 by inducing tetramerization.1 In this study, the oligomeric states of PKM2 were analysed in the presence and absence of FBP and the effect of increasing concentrations of FBP on the oligomeric states of PKM2 was tested. Our results indicate four molecules of FBP were bound to the active, tetrameric form of the enzyme. Binding of FBP to the dimer induced rapid tetramer formation and consequently FBP bound dimer was not observed in the mass spectra. The dynamics between unbound dimer, tetramer and ligand-bound tetramer were followed and the binding affinity of PKM2 for FBP was calculated by applying the Hill equation. The S437Y mutant of PKM2 which does not bind to FBP2 was used as a negative control and based on these results, positive cooperativity is proposed as a mechanism for activation of PKM2 by FBP binding.

Novel Aspect:
Native MS can significantly improve our understanding of complex enzyme mechanisms such as allostery.

Gelb, Abby,
University of Nebraska-Lincoln

Contributing Authors:
Eric D. Dodds, University of Nebraska-Lincoln.

Title:
Glycan Composition and Charge State Influence upon Collision Cross Sections of High Mannose N-Linked Glycopeptides

Introduction:
Mass spectrometry when coupled with ion mobility spectrometry (IMS) is a promising technique to study the challenging isomerism of biomolecules. Traveling wave ion mobility spectrometry (TWIMS) is an IMS technique, which requires calibration to measure collision cross sections (CCS). Interesting trends in the calculated CCSs of glycopeptides are presented based on glycan size and charge state.

Methods:
RNase B underwent trypsinolysis and glycopeptides were either analyzed directly or truncated (mannosidase digestion) and enriched by ZIC-HILIC tips from Protea. The glycopeptides were studied using a Waters Synapt G2-S HDMS instrument equipped with TWIMS. TWIMS measurements were calibrated to CCSs using charge-state matched peptide ions (polyalanine) with known CCSs.

Results:
Ion-neutral CCSs were measured for RNase B glycopeptides with varying amino acid sequences and varying glycan compositions, each in two charge states. When considering the measured CCS values of the glycopeptides, a systematic difference in the CCSs of glycopeptides with the same composition but different charge states was observed, with doubly charged ions being generally larger than their triply charged counterparts. Moreover, the CCS difference between charge states was found to increase as the size of the glycan increased (Man1-8). We speculate that these observations reflect a greater involvement of the glycan group in charge solvation for higher charge states and larger glycans; however, as the glycan size is reduced, its contribution to conformational differences between ions of different charge states is reduced.

Novel Aspect:
Systematic study on influence of glycan composition and charge state on the CCS of protonated high mannose N-linked glycopeptide ions.

Gelb, Abby,
University of Nebraska-Lincoln

Contributing Authors:
Eric D. Dodds, University of Nebraska-Lincoln; Christine Booth, University of Nebraska-Lincoln; Melanie A. Simpson University of Nebraska-Lincoln.

Title:
Glycomic Profiling of Human Hyaluronidase 1

Introduction:
Hyaluronidase 1 (Hyal1) is a key glycoprotein noted at elevated levels for both breast and prostate cancers. Hyal1 has three potential N-glycosylation sites (Asn 99, Asn 216, and Asn 350). These sites have been confirmed as occupied but the glycans and site heterogeneity have not been fully characterized. Therefore, a complete glycosylation profile of Hyal1 is of significant functional importance.

Methods:
FLAG tagged recombinant Hyal1 was transfected into human embryonic kidney 293F (HEK293F) suspension cells and expressed in conditioned media. Purified Hyal1 was then concentrated and deglycosylation using peptide N-glycosidase F (PNGaseF; Sigma). Released glycans were analyzed using HPLC (Waters Acquity UPLC with Agilent AdvanceBio Glycan Mapping Column) coupled to MS (Waters Synapt G2-S HDMS).

Results:
Hyal1 activity is seemingly lost when glycosylation is precluded by a point mutation at Asn 350 and has been shown to be involved in the metastatic process. When considering all types of known or suspected Hyal1 glycan compositions, the number of possible Hyal1 glycoforms totals 88. We have expressed Hyal1 successfully and are improving the purification protocol using ammonium sulfate cuts and ion exchange chromatography. These purification techniques have allowed us to enrich Hyal1 for maximum yield (roughly 2-fold higher) of the glycoprotein when compared to the anti-FLAG affinity column. Purified and deglycosylated Hyal1 will undergo separation of glycans using HPLC prior to MS; we will discern potential glycan targets, and these targets will undergo MS/MS experiments to characterize the glycan composition; this data will allow us to report a complete glycomic profile of Hyal1.

Novel Aspect:
Previously unreported glycomic profiling of human Hyal1 with potential implications in prostate cancer progression.

Genovesi, Luca,
Algorithme Pharma, an Altasciences Company

Contributing Authors:
Daniel Villeneuve, Algorithme Pharma Inc.; Georges Koudssi, Algorithme Pharma Inc.; Jean-Nicholas Mess Algorithme Pharma Inc.; Anahita Keyhani, Algorithme Pharma Inc.

Title:
Method Development of an Automated Hybrid LBA-LC/MS Assay for the Quantitative Bioanalysis of the Biotherapeutic Teriparatide in Human Plasma

Introduction:
Bioanalysis of large molecules by LC-MS entails multiple challenges, with sensitivity and selectivity at the forefront. Many issues have recently been circumvented using a hybrid LBA-LC/MS approach, where an affinity reagent is used to extract the analyte. The current investigation highlights the method development of a sensitive hybrid assay for the quantitation of intact teriparatide.

Methods:
Teriparatide was spiked in human plasma from 2.50 to 500.00 pg/mL and incubated with anti-teriparatide antibody and Streptavidin beads. The samples were processed on the KingFisherTM Flex. Teriparatide was eluted with 5% HCOOH in water. Chromatographic separation was achieved on a C18 column under gradient conditions using 0.1% HCOOH and ACN with 0.25% DMSO and analyzed with a Sciex QTrap 5500.

Results:
A hybrid LBA-LC/MS assay was developed for the quantitation of teriparatide in human plasma with a LLOQ of 2.5 pg/mL. In order to achieve this level of sensitivity, exhaustive optimization had to be performed. Extensive fine-tuning of the immunocapture process involved antibody screening and optimization of binding, washing and elution conditions. Since manual processing of the magnetic beads was time consuming and laborious, the extraction procedure was automated using a KingFisher™ Flex magnetic processor, which also improved assay reproducibility (%CV = 5.1, n = 96). This LBA-LC/MS assay for teriparatide demonstrated adherence to the strictest of industry standards for specificity (including hemolysed and lipemic), sensitivity, linearity, lot-to-lot recovery and stability. Inter-day (n=4) precision was within 14.0% and accuracy was between 96.4 and 102.6% through all QC levels.

Novel Aspect:
The first reported application that couples immunocapture with LC-MS/MS for the bioanalysis of teriparatide at low pg/mL concentration.

Gerbaux, Pascal,
University of Mons

Contributing Authors:
Lemaur Vincent, University of Mons; De Winter Julien, University of Mons; Cornil Jérôme University of Mons; Carroy Glenn, University of Mons.

Title:
Host-Guest Complexes Probed by Mass Spectrometry: bridging the gap between the solution and the gas phase

Introduction:
Host-guest complexes are now more and more investigated for their numerous applications. MS represents one of the tools to assess the exclusion vs inclusion nature of the complexes. The effect of the experimental conditions on the ion structures is often not considered. We then undergo a broad study to decipher the role of the solution and MS parameters on the nature of the detected ions.

Methods:
ESI is used to extract (or create) the non covalent complexes from aqueous solutions. Cucurbiturils are selected as receptors that are able to create strong complexes with selected amino guests. CID and ion mobility experiments are used to study the nature/decomposition of the ions. DFT calculations using the B97D functional and the 6-31g(d,p) basis set are performed to support the experiments.

Results:
In the first part of the communication, we will demonstrate that the ion topologies in gas phase can be affected by the equilibrium time in solution, especially for host-guest systems characterized by slow kinetics of complexation. As a model system, we selected, as the receptor, cucurbit[6]uril, whose complexation kinetics has been analyzed in great details by NMR spectroscopy. We used ion mobility and energy-resolved CID to follow the evolution of the ion topologies as a function of time in solution prior to the ESI analysis and highlight the structural time evolution. In the second part, we will monitor the CID reactions of multiply charged ternary complexes associating ns-CB[10] with amino guests. We will show that the CID reactions are both qualitatively and quantitatively dependent on the charged state and the nature of the guest molecule.

Novel Aspect:
Non covalent complexes are often analyzed by MS. Here, the influence of the experimental conditions on the topology of the complexes is demonstrated.

Gerislioglu, Selim,
The University of Akron

Contributing Authors:
Chrys Wesdemiotis, The University of Akron.

Title:
Chain-end and backbone analysis of poly(N-isopropylacrylamide)s by sequential electron transfer dissociation and collisionally activated dissociation

Introduction:
Full structural characterization of end group features and backbone sequences is essential for the development of thermoresponsive polymers that are used for drug delivary applications. Here, we applied collisionally activated dissociation (CAD) on select fragments generated by electron transfer dissociation (ETD) to characterize the precise chain connectivity of polyacrylamides.

Methods:
Experiments on carboxylic acid and triethoxysilane terminated pNIPAM samples were performed on multiply sodiated ions with a Bruker HCT Ultra II quadrapole ion trap mass spectrometer. For ETD, fluoranthane radical anions were produced in a negative chemical ionization (nCI) source. After accumulation of both species in the ion trap, ion/ion reaction time was arranged within 50-170 ms.

Results:
CAD on doubly sodiated ions resulted in rearrangements on the side chains and loss of the end group, which dominated the MS/MS spectra. The major fragments observed in ETD were side chain cleavages and hydrogen radical (-H•) loss, depending on the electron attachment site. Application of CAD on ETD product ions that contain a radical site suppressed the rearrangements by promoting homolytic C-C bond cleavages along the backbone. Application of CAD on an even electron product ion that lost the ω-end in the ETD stage resulted in backbone cleavages as well. Although the latter ETD product does not have any radical site, homolytic bond scissions were facilitated by the presence of a nearby salt bridge. Thus, not only the charge reduced precursor, but also other ETD fragments with unique structural elements can be used to obtain structurally diagnostic fragments via consecutive CAD.

Novel Aspect:
ETD fragments with either a radical site or a well stabillized charge site can be used to obtain structurally diognostic fragments via sequential CAD.

Ghiasikhou, Sahar,
ETH Zurich

Contributing Authors:
Prof. Renato Zenobi, Professor at ETHZ.

Title:
Capillary gap sampler: a new microfluidic platform directly coupled to ESI-MS for fast analysis of low sample amounts

Introduction:
Miniaturized sampling devices hyphenated with ESI-MS are very attractive analytical tools for early drug discovery. Reasons are that ESI-MS is compatible with microfluidics, and allows comprehensive, label-free sample analysis. This project aims at the development and implementation of a "capillary gap sampler"- a miniaturized sampling device that allows precise liquid handling or extraction.

Methods:
The capillary gap sampler is directly coupled to mass spectrometry. The core of the sampler consists of a liquid bridge of several nanoliters formed within a µm-sized gap between two capillaries, where one of the capillaries acts directly as the ESI-MS spray needle. Sample introduction in the range of a few nanoliters is performed via an open liquid bridge as a new microfluidic element.

Results:
We present a coupling of solid phase microextraction with the capillary gap sampler to address many challenges for small-volume sample cleanup prior to ESI-MS analysis. Recently we have shown the capability of our sampler to extract benzodiazepines from PBS.This would be interesting for less available or expensive biological samples within complicated matrices.

Novel Aspect:
Low sample volume (down to 10 nL), site specific sampling or extraction, online reaction monitoring coupled to MS (HTS).

Ghobadloo, Shahrokh,
University of Ottawa

Contributing Authors:
Gleb G. Mironov, Department of Chemistry, University of Ottawa, Canada; Shahrokh M. Ghobadloo, Department of Chemistry, University of Ottawa, Canada; Maxim V. Berezovski Department of Chemistry, University of Ottawa, Canada.

Title:
Protein Mass Spectrometry and DNA Aptamers Reveal Biomarkers of Human Monocyte

Introduction:
Monocytes are known as cells that express CD14. However, clinically significant subset of monocytes express excessively low levels of CD14. CD14 is not consistent monocyte marker because of its variable expression among monocytes. This study was designed to improve identification of human blood monocytes by marker that presents on all Monocytes, uniquely expresses and has known function.

Methods:
DNA aptamers were developed by Fluorescence Activated Cell Sorting (FACS) and Systematic evolution of ligands by exponential enrichment (SELEX). Monocyte specific aptamers were identified from created aptamers and used to isolate their targets of leukocyte membrane proteins. The isolated proteins sequenced by LC- Orbitrap Fusion Tribrid mass spectrometer and confirmed by flow cytometric methods.

Results:
The subsequent mass spectrometry investigation of the isolated proteins revealed CD91 as the aptamer target on monocytes. S100a9, Filamin A, Serpin and hsp90 were recognized as co-binding proteins with generated aptamer and CD91.

Novel Aspect:
Engaging mass spectrometry and FACS-SELEX created aptamers with specific to CD91, monocytes biomarker, and revealed co-binding proteins with CD91.

Gill, Chris,
Applied Environmental Research Laboratories - Vancouver Island University

Contributing Authors:
Dietrich Volmer, Institute for Bioanalytical Chemistry, Saarland University, Saarbrucken, Germany. ; Erik Krogh, Applied Environmental Research Laboratories - Vancouver Island University, Chemistry Department, Nanaimo, BC, Canada.; Achille Cappiello Chemistry, Urbino University, Urbino, Italy.; Kyle Duncan, Applied Environmental Research Laboratories - Vancouver Island University, Chemistry Department, Nanaimo, BC, Canada.; Gregory Vandergrift, Applied Environmental Research Laboratories - Vancouver Island University, Chemistry Department, Nanaimo, BC, Canada.

Title:
Condensed Phase Membrane Introduction Mass Spectrometry (CP-MIMS): Direct, Online, Measurements of Non-Volatile Analytes in Complex Aqueous Samples

Introduction:
Condensed phase membrane introduction mass spectrometry (CP-MIMS) is an on-line analytical method that directly interfaces mass spectrometry (MS) with complex samples. A membrane transfers permeating analyte mixtures to the MS via a liquid acceptor phase, rejecting the bulk sample matrix, obviating clean up & chromatographic separations, providing direct, real-time measurements in complex samples.

Methods:
Hollow fibre membranes, predominantly made from polydimethylsiloxane (PDMS) in both flow cell and several immersion probe geometries are coupled with a variety of atmospheric pressure ionization MS systems. Target analytes permeate the membrane and are transferred to the MS via a (condensed) methanol acceptor phase, whereupon they are resolved by selective ionization, m/z and/or MS/MS.

Results:
We have been developing CP-MIMS for direct, online measurements of environmental, pharmaceutical & bioanalytical targets in complex aqueous samples. Recent developments are presented, including the design and characterization of a wide range of interfaces for measurements in complex sample matrices, miniaturized interfaces for direct in-situ/in-vivo/small sample interrogation, thin/composite membranes for faster measurements, and the use of on-line internal standards in the acceptor phase to mitigate interference from co-permeants. Presented applications include continuous, online measurements of non-volatile analytes such as pharmaceuticals, hormones, vitamins, naphthenic acids and PAHs in complex samples such as live plant tissue, wastewater, seawater & artificial urine. Because CP-MIMS detects 'free' solution concentrations, its use to investigate analyte binding is also presented.

Novel Aspect:
Direct, on-line, trace level (pptr-ppb) measurements in complex aqueous sample matrices without sample cleanup or chromatographic separations.

Gill, Chris,
Applied Environmental Research Laboratories - Vancouver Island University

Contributing Authors:
Erik Krogh1,2, 1. Chemistry, University of Victoria, Victoria, BC, Canada. 2. Chemistry, Applied Environmental Research Laboratories - Vancouver Island University, Nanaimo, BC, Canada.; Gregory Vandergrift1,2*, 1. Chemistry, University of Victoria, Victoria, BC, Canada. 2. Chemistry, Applied Environmental Research Laboratories - Vancouver Island University, Nanaimo, BC, Canada.

Title:
Polymer Inclusion Membranes with Condensed Phase Membrane Introduction Mass Spectrometry (CP-MIMS): Improved Analytical Response Time and Sensitivity

Introduction:
CP-MIMS interfaces mass spectrometry (MS) with complex samples for direct, online measurements. Analytes permeate through a membrane and are then entrained to the MS via a liquid acceptor phase. The membrane may be doped with various co-solvents to increase analyte diffusivity and permeation, resulting in decreased response times and enhanced sensitivity at both the membrane and ionization source.

Methods:
Hollow fibre polydimethylsiloxane (PDMS) membrane immersion probes are coupled to electrospray ionization (ESI). Dopant is added to the acceptor phase and flows through the membrane lumen, forming a polymer inclusion membrane (PIM). The dopant choice is tailored to the target analytes for optimal analytical performance. Analytes permeate the PIM and are resolved by ionization, m/z, and/or MS/MS.

Results:
PIMs have been shown to increase analyte permeability, resulting largely from the greater diffusivity of analytes in solvent doped membranes. This yields faster analytical response times and a reduced duty cycle. In addition, increased permeability also leads to greater mass transport at the membrane level, resulting in improved analytical sensitivity. For some dopant/analyte combinations, we observe enhanced ion formation at the ESI source, leading to further sensitivity improvements. Preliminary results employing several hydrocarbon co-solvents in a methanol acceptor phase will be presented. Applications of this work for real-time, direct measurements of environmental contaminants in aqueous solution increase sample throughput, making this technique amenable to automation and pre-screening.

Novel Aspect:
Sensitive, fast, direct, inexpensive technique applicable to automation and pre-screening of complex samples with minimal sample preparation.

Gimenez-Cassina Lopez, Begoña,
UNICAMP

Contributing Authors:
Alexandra Christine Helena Frankland Sawaya, Pharmacy course. Institute of Biology, UNICAMP; Marcos Nogueira Eberlin, Thomson Laboratory, UNICAMP; Giovana Anceski Bataglion Thomson Laboratory, UNICAMP; Begoña Gimenez-Cassina Lopez, Pharmacy course. Institute of Biology, UNICAMP.

Title:
Phytosterols and phytoestrogens of red propolis determined by GC-MS

Introduction:
The use of natural products and phytotherapy has increased in the past decade.Red propolis was found a decade ago and has shown to have promising biological properties.Identification of substances in natural matrix is a tricky task and sometimes depends on the availability of standards.NIST GC-MS library together with Kovat`s index allows the identification of volatile and semivolatile substances.

Methods:
Analyses were carried out on a GC-MS Agilent Technologies system consisting of a gas chromatograph (7890A) coupled to a quadrupole mass spectrometer (MSD 5975 C) operating in the electron ionization (EI) mode (70 eV). The chromatographic separations were performed on a HP-5-MS fused silica capillary column. An alkane mixture (C8-C39) was injected for the Kovat`s index calculations.

Results:
Gas chromatography coupled to a triple quadrupole mass spectrometer allowed us to determine the chemical composition of red propolis, identifying 29 volatile and semi-volatile compounds. Among these substances, phytosterols and phytoestrogens stand out due to their important therapeutic properties. Phytosterols such as α-amyrin, 9,19(E) ciclolanost-3-en-3-ol, β-amyrin, 9,19(Z)ciclolanost-3-en-3-ol, cicloartenol, 20(29)-lupen-3-one, lupeol, β-amyrin acetate, α-amyrin acetate have structures analog to cholesterol. Flavonoids and chalcones are considered phytoestrogens. They are considered dietary estrogens and GC-MS analysis allowed the identification of homopterocarpin, 7-O-methylvestitol, vestitol, neovestitol, medicarpin and formononetin, whose structure is similar to estradiol.
Acknowledgments: FAPESP 2013/20313-3 for the financial support.

Novel Aspect:
Identification of phytosterols and phytoestrogens in propolis of different geographical locations and identification of its plant source.

Gionfriddo, Emanuela,
University Of Waterloo

Contributing Authors:

Justen Poole, University of Waterloo; Janusz Pawliszyn, University of Waterloo; Germán Augusto Gómez-Ríos University of Waterloo.

Title:

Solid Phase Microextraction-Transmission Mode and Direct Analysis in Real Time: an efficient tool for fast analysis of contaminants in complex samples

Introduction:

Recently, non chromatographic analytical methods allowed for the reduction of total analysis time. SPME enables rapid extraction of target analytes and minimizes matrix interferences thus representing a good option for direct coupling to ambient ionization methods. We report here the implementation of SPME-TM and DART for the multiresidue analysis of contaminants in food and environmental matrices

Methods:

The SPME-TM device was used for extraction of pesticides in surface water, milk, grape and orange juices without any sample treatment. One-minute extraction was followed by a quick (10s) rinsing in water that helped in minimizing matrix effects. Desorption/ionization (~20s) was carried out by positioning the SPME-TM device between the DART source and the inlet of the mass spectrometer

Results:

The simultaneous determination of 20 pesticides in environmental and food matrices was achieved. Three deuterated internal standards were evaluated to guarantee signal correction for all the analytes tested. The SPME-TM-DART-MS/MS method attained remarkable performance with only 1 minute of extraction. Satisfactory linearity (>0.99), from 0.1 to 100 ng/g, was obtained. Limits of detection ranged between 0.1 and 5 ng/g. Analytical accuracy ranged between 70 and 130%. The figures of merit obtained confirm the suitability of the method for fast and routine quantitative analysis at the MRL imposed by EU regulations for pesticide residues in food commodities (10 ng/g) with a total analysis time of only 3 minutes. As a proof of concept, SPME-TM-DART was also coupled to a high-resolution MS system (Orbitrap) to demonstrate the potential of this approach for retrospective non-targeted analysis.

Novel Aspect:

An ultrafast quantitative protocol for multi-residue analysis of pesticides in complex matrices was developed by SPME-TM-DART-MS/MS

Giorgi, Gianluca,
University of Siena

Contributing Authors:
Maura Marinozzi, University of Perugia, Department of Pharmaceutical Sciences, Via del Liceo,1 I-06123 Perugia (Italy).

Title:
STEREOCHEMICAL EFFECTS OF 2-(2'-CARBOXY-3'-PHENYLCYCLOPROPYL)GLYCINES

Introduction:
Stereochemical characterization and differentiation of enantiomers and diastereoisomers in the gas phase are fascinating topics in which mass spectrometry plays a key role. We wish to report here on the stereochemical effects of 2-(2'-Carboxy-3'-phenylcyclopropyl)glycines, designed as conformationally constrained glutamate analogs endowed with potential activity at metabotrobic glutamate receptor.

Methods:
Electrospray ionization in positive and negative mode has been carried out on a LCQ DECA mass spectrometer. MSn product ion experiments have been carried out by using helium as collision gas in the low energy regime. DFT calculations have been performed on different structures proposed for protonated and deprotonated molecules by using Gaussian 09.

Results:
Stereochemical effects of each protonated or deprotonated diastereoisomer owing to low energy collision induced dissociations in MSn experiments consist in the same fragmentation pathways but with different relative abundances of product ions or in specific pathways depending on the stereochemistry. Competitive losses of H_2O and NH_3, together with eliminations of CO and CO_2 are the main decomposition pathways for protonated species. CIDs of deprotonated species produce consecutive eliminations of H_2O, CO_2 and HNCO. In order to investigate properties and stability of different ion structures proposed owing to ionization and fragmentation, DFT calculations have been performed.

Novel Aspect:
Stereochemistry; gas phase ion chemistry; stereolibrary

Giuliani, Alexandre,
Synchrotron SOLEIL

Contributing Authors:
Aleksandar Milosavljevic, Synchrotron SOLEIL, France; Milos Rankovic, Institute of Physics, Belgrade, Serbia.

Title:
Electron induced activation: a mean for inner shell spectroscopy of ions

Introduction:
We report the coupling of an electron gun with a linear ion trap, for mass selected ion activation and action spectroscopy by electron impact. Electron tracing simulations of a 300 eV electron beam inside the ion trap are presented along with the design of the experimental setup. Examples of electron impact activation for peptides and proteins are presented in the 280 to 300 eV energy range.

Methods:
The experiments were performed by coupling a commercial linear ion trap mass spectrometer (LTQ XL, Thermo Scientific) to a custom-made electron gun assembly. The electron gun consists of an extraction part and a focusing part including XY deflectors and a thoriated-tungsten cathode. Ubiquitin, substance P and melittin were purchased from Sigma-Aldrich.

Results:
We present a system allowing energy resolved electron impact activation of m/z selected ions confined in a linear quadrupole ion trap. The instrument is based on an energy-tunable focused electron beam providing incident electron energies around Carbon K-shell excitation. We recorded MS2 spectra at selected electron activation energies and performed the action electron spectroscopy of trapped protein ions. Moreover, we report a comparative study of inner-shell protein ionization by electron impact and X-ray absorption. The electron impact ionization yield shows a strong incident energy dependence in a good agreement with previous soft X-rays studies. The atomic specificity of this method may find applications in tandem MS.

Novel Aspect:
The validity of energy-tunable electron activation of m/z selected trapped ions has been demonstrated.

Giuliani, Alexandre,
Synchrotron SOLEIL

Contributing Authors:
Laurent Nahon, Synchrotron SOLEIL; Francis Canon, CSGA, INRA.

Title:
Electron photodetachment, an efficient relaxation mechanism for electronically excited nucleic acids anions ?

Introduction:
Absorption of light in the visible or in the ultraviolet range may produce electronic excited species. Electron photodetachment (EPD) is known to be one of the major relaxation pathways of anions. EPD onsets have been observed lying at low photon energy, which is very different from cations. We have compared the relaxation mechanisms for nuclei acid ions in the UV using action spectroscopy.

Methods:
Action spectroscopy of electrosprayed ions has been carried out in a linear ion trap (LTQ XL, Thermo) coupled to a synchrotron radiation beamline in the 4 to 14 eV range (310 – 88 nm) at the SOLEIL facility. At each wavelength, selected cations or anions ions are stored in the trap and irradiated during 500 ms. Hexamers (dA6, A6 and T6) (Eurogentec) were placed in ammonium acetate buffer.

Results:
Tremendous amount of work has been produced to study the UV induced photochemistry and photophysics of individual nucleobases and nucleic acid strands in order to investigate their resistance towards radiations. Among all the relaxation pathways, MS based action spectroscopy offers the ability to observed fragmentation and ionization. Doubly and triply charged cations and anions of hexamers have been compared. EPD is observed from the absorption threshold of all anions and is the dominant product over the entire range investigated, fragmentations being much less abundant. For cations, the onset is found at higher energy. The only relaxation pathway below this threshold is fragmentation. We have thus compared fragmentation yields for anions and cations with identical charge state and found specific relaxation processes depending on the cation or anion nature of the species.

Novel Aspect:
We propose that the anionic nature of nucleic acids might enhance their UV radio-stability and that EPD may be an important source of slow electrons.

Godoy, Adriana,
University of Campinas

Contributing Authors:
Ana Valéria Colnaghi Simionato, LABi Tiselius and National Institute of Science and Technology in Bioanalytics (INCTBio), Institute of Chemistry – University of Campinas, SP-Brasil; Marcos Nogueira Eberlin, ThoMSon Mass Spectrometry Laboratory, Institute of Chemistry – University of Campinas, SP-Brazil; Marcelo Lancellotti Biochemistry Department, Institute of Biology – University of Campinas, SP-Brazil.

Title:
LC-MS/MS: Prospecting nucleosides as potential cancer biomarkers in in vitro tumor cell

Introduction:
During tumor's progress, the metabolism of nucleic acids increases leading to high metabolites secretion into the organism. Studies have shown a higher concentration of modified nucleosides in patients with cancer than in healthy individuals. Based on this fact, LC-MS/MS method was evaluated for further characterization of nucleosides as potential tumor biomarkers in cells in vitro.

Methods:
The LC-ESI(+)-MS/MS method on MRM mode was optimized for nine nucleosides standards. Analytes were extracted from culture medium by SPE. In order to find other nucleosides, a new strategy was explored using data-dependent acquisition with product ion scan and NL. Tests were performed in five tumor cell lines: prostate PC03, breast MCF7, glioblastoma U251, lung A549 and pancreas PANC-1.

Results:
All the targeted nucleoside and the internal standard were separated within c.a. 35 min. The relative amount of each analyte was obtained with the chromatographic peak areas; normalization was made according to the cell amount in culture medium. The preliminary results revealed a different profile of nucleosides synthesized by different tumor cells lineages, mainly for lung and pancreatic cancer. NL chromatograms indicated compounds with loss of -132 Da (ribose), some of these have already been identified by MRM mode, but other compounds can be detected at different retention times, and can be further identified. These compounds may reveal their potential as biomarkers as well as the ability to assign their specificity for different tumors. The next steps include identification of other nucleosides, method validation and nucleoside quantification.

Novel Aspect:
A sensitive and effective analysis on cell culture supernatant to prospecting nucleosides as potential cancer biomarkers in in vitro tumor cell lines.

Godoy, Adriana,
University of Campinas

Contributing Authors:
Matheus Henrique Petrarca, Faculty of Food Engineering, University of Campinas, Campinas, São Paulo, Brazil; Welligton S. Oliveira, Faculty of Food Engineering, University of Campinas, Campinas, São Paulo, Brazil; Helena Teixeira Godoy Faculty of Food Engineering, University of Campinas, Campinas, São Paulo, Brazil; Adriana Teixeira Godoy, Institute of Chemistry, University of Campinas, Campinas, São Paulo, Brazil.

Title:
Determination of phthalates in baby food by gas chromatography–mass spectrometry

Introduction:
Phthalates are esters of 1,2-benzenedicarboxylic acid, which are used in several areas (adhesives, cosmetics, packaging). In this study, a simplified sample preparation and GC-MS method was developed for the determination of seven phthalates: dimethyl (DMP), diethyl (DEP), dibutyl (DBP), diisobutyl (DIBP), di-n-octyl (DNOP), bis(2-ethylhexyl) (DEHP) and benzylbutyl (BBP) phthalates, in baby food.

Methods:
Phthalates are esters of 1,2-benzenedicarboxylic acid, which are used in several areas (adhesives, cosmetics, packaging). In this study, a simplified sample preparation and GC-MS method was developed for the determination of seven phthalates: dimethyl (DMP), diethyl (DEP), dibutyl (DBP), diisobutyl (DIBP), di-n-octyl (DNOP), bis(2-ethylhexyl) (DEHP) and benzylbutyl (BBP) phthalates, in baby food.

Results:
Limits of detection of 1 μg/kg were obtained for DMP, DBP, BBP and DNOP, 2 μg/kg for DEP, and 5 μg/kg for DIBP and DEHP, while limits of quantification of 2 μg/kg were obtained for DMP, DBP, BBP and DNOP, 5 μg/kg for DEP, and 10 μg/kg for DIBP and DEHP. Linearity was achieved for six-point matrix-matched calibration curves in the range from 10 to 100 μg/kg with determination coefficients higher than 0.99. Recoveries between 80 and 109%, 78 and 118%, and 75 and 114% were obtained for phthalate-free baby foods spiked at 10, 25 and 50 μg/kg, respectively, with relative standard deviations lower than 20% for all studied analytes. In order to evaluate the feasibility of proposed method, twenty samples of fruit-based baby food were investigated for the presence of the selected phthalates.

Novel Aspect:
Limits of detection of 1 μg/kg were obtained for DMP, DBP, BBP and DNOP, 2 μg/kg for DEP, and 5 μg/kg for DIBP and DEHP, while limits of quantificatio

Goguen, Robert,
IonSense

Contributing Authors:
Brian Musselman, IonSense.

Title:
Thermal Desorption Direct Analysis in Real-Time Mass Spectrometry (TD-DART-MS)

Introduction:
A novel off-axis Thermal Desorption (TD) device for use with DART ambient ionization source has been used for rapid compound analysis, The TD-DART enables use of sample introduction tools including a direct evaporation probe (DEP) and swab device for rapid analysis typically used to introduce samples into GC/MS units. Rapid analysis of polymers, chemicals and liquids will be presented

Methods:
An enclosed heated chamber has been produced to facilitate use of common thermal desorption devices with a DART source. The TD has two different sample introduction areas. The first is designed to permit use as a swab and detect device. The second port allows for the introduction of classical DEP probes, usually reserved for GCMS. A control unit with variable temperature control is also used.

Results:
The DEP probe was used to analyze different polymers. The rapid desorption of sample permitted detection of most substances detected in the GC/MS experiment in seconds. Pyrolysis of the polymer resulted in production of small components including monomers, which were then ionized by DART-MS. The DART was able to analyze the monomers from different polymers as well as some additives from the plastic. The sawb / detect sample introduction port of the TD was used for analysis liquid samples of common flavor compound pipeted onto the swab. Samples were also collected from different surfaces to assess efficiency of their transfer to the TD. Comparison of results from analysis of samples swabbed off a surface with results from samples directly analyzed by DART are provided in order to permit comparison of TD vs conventional direct analysis in real time for rapid sampling in open air

Novel Aspect:
An enclosed Thermal Desorption chamber equipped DART source capable of utilizing conventional thermal desorption GC/MS devices has been characterized

Goh, Byoungsook,
Gwangju Institute of Science and Technology

Contributing Authors:
Jinwoo Kim, Gwangju Institute of Science and Technology; Hyunesoo Kwon, Gwangju Institute of Science and Technology; Tae-Young Kim Gwangju Institute of Science and Technology.

Title:
High-throughput measurement of lipid turnover rate in HeLa cell based on metabolic heavy water labeling

Introduction:
Lipid metabolism is closely involved in vital biological processes. Disorders in lipid synthesis or degradation are known to cause various diseases like brain injury, heart failure, and obesity. The chemical diversity of lipid is a major barrier for studying quantitative lipid turnover on a global scale. Metabolic heavy water (2H2O) labeling was employed to examine lipidome dynamics in HeLa cell.

Methods:
Lipids were extracted from HeLa cells grown in 5% (mol/mol) 2H2O enriched media at 6 time points from 0 to 48 hours of labeling. The extracted lipids were analyzed with LC-MS/MS and normalized areas of the monoisotopic peak of identified lipids were calculated. The turnover rates were deduced by fitting the normalized peak areas at different labeling times into a first-order kinetics equation.

Results:
Using LipidBlast, a total of 371 lipid species including glycerophospholipids, glyceryl esters, and sphingolipids were identified from three biological replicates. In order to expedite the data processing from the peak quantification to the curve fitting, an in-house MS data analysis program was developed. During the first-order kinetics curve fitting process, three operational filters were exploited to rule out low-quality extracted ion chromatograms in the calculation of lipid turnover rate. After filtering, the turnover rates of 101 lipid species were determined. The turnover rates of 16 lipids that commonly analyzed in three biological replicates were in the range of 0.008 to 0.1 hr−1. Plasmenyl-PE 40:5 resulted in good consistency between the turnover rates measured in both positive and negative modes, which strongly supports the validity of our experimental platform.

Novel Aspect:
A technical platform to measure turnover rates in vivo of lipids on a global scale is developed and applied to a eukaryotic cell model.

Goshawk, Jeff,
Waters

Contributing Authors:
Michelle Wood, Waters; Nayan S. Mistry, Waters.

Title:
An Ion Mobility Screening Approach for the Detection of Toxicologically Relevant Substances

Introduction:
The utility and potential benefit of collision cross-section (CCS) data, as generated through use of ion mobility separation (IMS) techniques, was evaluated for the screening of more than 50 toxicologically-relevant substances.

Methods:
Solvent standards were analysed using an established UPLC-method in combination with a Vion IMS-QTof (Waters). Reference CCS values were measured and then incorporated into an existing toxicology library comprising retention time and exact mass data.

Results:
To assess the validity and reproducibility of the reference CCS values, blank urine and plasma samples, enriched with the same drug standards, were analysed using UPLC-HDMSE. In addition, a series of authentic biological samples were analysed. Data were processed and screened against the updated toxicology library. For all molecules investigated, observed CCS values were within 2% of reference values. The feasibility of using CCS as a potential substitute for RT was demonstrated by the successful application of an alternative chromatographic method with a shorter analytical run time.

Novel Aspect:
Collision cross-section as a screening parameter for toxicologically-relevant substances

Gozzo, Fabio,
Unicamp

Contributing Authors:

Hugo Cesar Ramos Jesus, Unicamp; Allan Jonathan Ferrari, Unicamp; Luciana Gonzaga de Oliveira Unicamp; Peter Leadlay, University of Cambridge; Mariana Fioramonte, Unicamp.

Title:

XPlex: A multi residue cross-linking strategy for protein structure determination

Introduction:

Chemical cross-linking has been used for protein-protein complexes structure determination but for tertiary structures a much higher number of cross-links must be obtained. That requires that other residues besides Lysine to be cross-linked. Here we describe Xplex, a cross-linking chemistry that can target several residues, increasing the number of cross-links that can be obtained for proteins

Methods:

Model proteins as well as targeted proteins were subjected to cross-linking using the XPlex chemistry. The digested cross-linked proteins were analyzed by LC-MS/MS using an Orbitrap HF mass spectrometer. The cross-links were identified using the SIM-XL software and manually validated using the software spectrum visualizer. The distance constrains were used in Rosetta input for structural modeling

Results:

The new cross-linking chemistry generates cross-links between Glu-Glu, Asp-Asp, Glu-Asp, Glu-Lys, Asp-Lys, Glu-Ser and Asp-Ser residues. Together, these residues corresponds to 25% of all residues in a protein on average, extending by a large margin the number of cross-links (and therefore the number of distance constrains) that can be obtained for a protein. Moreover, Lys poor proteins can now be efficiently analyzed by chemical cross-linking. A more even distribution of cross-links over the protein surface is also observed as acidic patches can now be cross-linked. The SIM-XL software was optimized to identify these new types of cross-links and the number of cross-links now obtained can reach 25% of the number of residues, enough to be used as distance constrains to model the tertiary structure of proteins

Novel Aspect:

A novel cross-linking chemistry for increased number of cross-links is developed targeting the determination of protein tertiary structure

Griffin, Patrick,
The Scripps Research Institute

Title:
Differential HDX to probe ligand mediated receptor signaling

Introduction:
The regulation of gene programs controlled by nuclear receptors (NRs) is driven by alterations in the conformational ensemble of the receptor upon ligand binding. Similar activation mechanisms are modulated by ligand binding to G protein coupled receptors (GPCRs). HDX analysis of both NRs and GPCRs has provide insight into activation of these critical mediators of cell signaling.

Methods:
Differential HDX studies were performed on the target protein in the presence and absence of ligands. Samples were incubated at 4°C for 10 s, 30 s, 60 s, 900 s, or 3600 s and processed using a fully automated system. HDX data was analyzed using HDX Workbench.

Results:
HDX studies probing the conformational dynamics of intact full length nuclear receptor complexes upon interaction with DNA and coactivator proteins demonstrate that DNA binding alters conformational dynamics of the nuclear receptor heterodimer in regions remote of the DBD. These alterations in conformational selection appear to be important for coactivator binding to the heterodimer suggesting that DNA acts as an allosteric ligand. In addition, HDX analysis of GPCRs in the presence of functionally selective modulators and effector molecules provides mechanistic insight into biased signaling.

Novel Aspect:
Mechanistic insight into the activation mechanisms of NRs and GPCRs has been gained by comprehensive differential HDX.

Gross, Jürgen H.,
Heidelberg University

Contributing Authors:
Martin Hähsler, Institute of Organic Chemistry, Heidelberg University, Im Neuenheimer Feld 270, 69120 Heidelberg, Germany; Sven Elbert, Institute of Organic Chemistry, Heidelberg University, Im Neuenheimer Feld 270, 69120 Heidelberg, Germany; Jürgen H. Gross Institute of Organic Chemistry, Heidelberg University, Im Neuenheimer Feld 270, 69120 Heidelberg, Germany; Gang Zhang, Institute of Organic Chemistry, Heidelberg University, Im Neuenheimer Feld 270, 69120 Heidelberg, Germany; Daniel Beaudoin, Institute of Organic Chemistry, Heidelberg University, Im Neuenheimer Feld 270, 69120 Heidelberg, Germany; Michael Mastalerz, Institute of Organic Chemistry, Heidelberg University, Im Neuenheimer Feld 270, 69120 Heidelberg, Germany.

Title:
Matrix-assisted laser desorption/ionization-time-of-flight-mass spectrometry of macromolecular cages in the 5–10 kDa range

Introduction:
Covalently weakly bound macromolecular cages with molecular masses in the 5–10 kDa range are demanding analytes for MS. MALDI presents the method of choice for these giant cages as MALDI is soft, ionizes neutrals, and yields spectra governed by singly charged ions. MALDI-MS allows monitoring of the synthetic progress, yields information on side products, and proves the purity of cage compounds.

Methods:
MALDI mass spectra were acquired using a Bruker Autoflex Speed time-of-flight mass spectrometer, typically in positive-ion mode. Solutions of the macromolecular cages were admixed to solutions of the matrix. Survey spectra were acquired in linear mode, covering 2–9 or 3–18 kDa. Reflector mode spectra were acquired to obtain isotopic resolution, which could be achieved for cages up to 8–10 kDa.

Results:
MALDI-TOF-MS yielded good mass spectra of the cages under study. The compounds were found to form molecular ions, M+•, or protonated molecules, [M+H]+. Boron ester cages with 24 borons resulted in complex isotopic distributions. Achieving high resolution was important for this class of cages. Optimization of the ion source lens voltage and the delay time of the pulsed ion extraction allowed a substantial increase of the resolving power. The use of monodisperse dendrimers provided valuable aid in tuning and mass calibration of the instrument. Small changes in molecular structure markedly influenced the intensity of the molecular ion peak and the occurrence of fragment ion peaks. Some compounds formed ions without fragmentation while others exhibited intensive fragment ion peaks, e.g., the position of alkyl substituents of triptycene building blocks could cause multiple alkene losses.

Novel Aspect:
Optimized experimental conditions for MALDI-TOF-MS of supramolecular cages deliver isotopic resolution of ions up to 8–10 kDa and 0.2 u mass accuracy.

Groves, Kate,
LGC and University of Leeds

Contributing Authors:

Alison Ashcroft, Astbury Centre for Structural Molecular Biology, University of Leeds; Milena Quaglia, LGC, Queens Road, Teddington, UK TW11 0LY; Adam Cryar LGC, Queens Road, Teddington, UK TW11 0LY.

Title:
Towards the measurement comparability of protein structural analysis

Introduction:
Higher order protein structural (HOS) analysis is of paramount importance in the biopharma area.While several analytical HOS methods exist, few data on platform comparability and sensitivity available to aid regulators and manufactures. A hydrogen deuterium exchange method was optimised, validated and an inter-platform comparison carried out on human growth hormone in the presence/absence of zinc.

Methods:
A HDX-MS method is here developed and validated through optimisation of a commercially available ACQUITY UPLC M-Class with HDX technology (Waters, Milford, USA). Mass spectrometry measurements are performed on a Synapt G2Si instrument (Waters) in MSE mode. IMS-MS and FPOP experiments were also perform alongside experiment on a number of different platforms including Raman, CD, NMR, CE.

Results:
Differential HDX-MS strategies developed here demonstrate that by understanding the variability of relative HDX-MS measurements, thresholds can be used to define significant HOS changes. Application of these methodologies to the model system revealed small HOS changes upon binding of rhGH to zinc ligands indicative of the zinc binding region and allosteric changes. Data suggested a concentration dependence relating to the ratio of rhGH:Zn, showing the sensitivity of the platform. Workflows have been here developed for assessing the sensitivity and comparability of different structural platforms using this model system. Varying sensitivity was observed across the platforms. This is an initial step towards the production guidelines to aid the bio-pharma area in the selection of the most appropriate structural platform.Data produced can also be used to benchmark novel analytical techniques.

Novel Aspect:
A model system & workflow to assess measurement comparability and platform sensitivity to protein HOS changes are here for the first time presented.

Gu, Binghe,
The Dow Chemical Company

Contributing Authors:
Bruce Bell, The Dow Chemical Company; Krishna Kuppannan, The Dow Chemical Company; Grant von Wald The Dow Chemical Company.

Title:
Non-volatile Mobile Phase Compatible LC/MS

Introduction:
One of the limitations of an LC-MS system is that non-volatile mobile phases cannot be used, while such mobile phases are sometimes required for LC-UV quantitation in order to get a low-noise baseline, attain symmetric peak shape, stabilize retention time, and offer alternative selectivity. The non-volatile mobile phases are particularly useful for quantitation of agricultural small molecules.

Methods:
To develop and evaluate three methodologies for converting non-volatile mobile phases into volatile ones.

Results:
This study reports the development of an interface/module which will be incorporated between LC and MS sections in order to achieve a) conversion of non-volatile mobile phase to MS-compatible one; b) retention of analytes so that it could be introduced into MS without loss of sensitivity. One approach was to use column switching to realize these aims. Preliminary data showed that not only were high-quality mass spectra acquired, it also achieved up to 15 times improvement in UV detection sensitivity for some analytes. Two other approaches with the use of either ion-depletion or ion-exchange column will also be explored. The advantage and disadvantage of each approach will be critically assessed.

Novel Aspect:
To develop a system so that LC/MS can run any mobile phases, including non-volatile ones.

Gueguen, Celine,
Trent University

Contributing Authors:
Sohidul Islam, Trent University; Vaughn Mangal, Trent University.

Title:
Evaluating changes in the molecular structures of photoirradiated aquatic dissolved organic matter using ultra high resolution mass spectrometry

Introduction:
High resolution Fourier transform ion cyclotron resonance mass spectrometry (FT-ICR MS) and Orbitrap are powerful tools to obtain detailed molecular information for complex aquatic dissolved organic matter (DOM). Knowledge of the structure can help understand the fate and reactivity of DOM in natural waters.

Methods:
Molecular composition of photoirradiated river and algal derived DOM, and aquatic fulvic acid standards were examined using QExactive Orbitrap and SolariX-XR FT-ICR-MS. Data were collected in negative mode in the mass range of 100–1500 m/z. Instrument settings and peak assignment procedure were as described in Mangal and Gueguen (2015).

Results:
Detectable differences in molecular structure were observed between FT-ICR-MS and Orbitrap at low molecular weight (< 400 Da), indicating that Orbitrap is better suited for low molecular weight DOM such as algal derived DOM. FT-ICR-MS analysis of photoirradiated DOM revealed the presence of three groups of molecular structures: (1) photo-resistant, (2) photo-labile, and (3) photo-produced. Molecular signatures consistent with refractory organic matter and black carbon were highly photo-labile, demonstrating that environmental solar irradiation levels are capable of removing refractory compounds from aquatic systems.

Novel Aspect:
These results demonstrate the applicability of ultra high resolution MS to monitor DOM molecular composition changes in response to photoirradiation.

Guerineau, Vincent,
CNRS ICSN

Contributing Authors:

Jean-Pierre Le Caer, Institut de Chimie des Substances Naturelles, CNRS, Gif-sur-Yvette, France; David Touboul, Institut de Chimie des Substances Naturelles, CNRS, Gif-sur-Yvette, France; Sylvie Auxilien Institute for Integrative Biology of the Cell, CEA, CNRS, Université Paris-Sud, Gif-sur-Yvette, France; Vincent Guérineau, Institut de Chimie des Substances Naturelles, CNRS, Gif-sur-Yvette, France; Satoko Yoshizawa, Institute for Integrative Biology of the Cell, CEA, CNRS, Université Paris-Sud, Gif-sur-Yvette, France; Laurent Laboureur, Institut de Chimie des Substances Naturelles, CNRS, Gif-sur-Yvette, France; Alain Brunelle, Institut de Chimie des Substances Naturelles, CNRS, Gif-sur-Yvette, France.

Title:

COMPARISON OF LC-MS AND SFC-MS ANALYTICAL STRATEGIES FOR THE CHARACTERIZATION OF MODIFIED TRANSFER RNA

Introduction:

Transfer RNAs (tRNAs) are involved in the protein translation leading to the protein synthesis. Chemical modifications of nucleobases, including methylations, were firstly described in the 60's. We have compared classical LC-MS approach with supercritical fluid chromatography SFC-MS, in terms of sensitivity, linearity, and specificity, for the quantification and characterization of modified tRNAs

Methods:

Three types of samples were successively analyzed: a nucleoside test mix (Sigma), well-characterized modified tRNA (E. coli and yeast), and uncharacterized tRNA species. LC-MS analyses were made with a LTQ-Orbitrap mass spectrometer (Thermo Scientific). SFC-MS experiments were performed with a modified LC 1260 system coupled with a Q-TOF 6540 (Agilent Technologies) mass spectrometer

Results:

LC-MS and SFC-MS strategies were firstly optimized. For LC-MS, classical approaches were employed, including an ammonium acetate buffer and a C18 column. For SFC-MS, a screening of different stationary phases was performed for improving base peak separation of modified nucleosides. Solvent make-up and ionization source conditions were adjusted, leading to the highest sensitivity using the ESI source in positive ion mode. Linearity and limits of detection and of quantification were measured for both techniques. Interestingly, SFC offers shortened analysis time whereas sensitivity was similar compared to LC-MS. Then total digests of E. coli and yeast tRNAs were investigated, leading to the detection of more than 20 modified nucleosides

Novel Aspect:

SFC-MS was compared to LC-MS for the detection of modified nucleosides

Guo, Hongbo,
University of Toronto

Contributing Authors:
Hui Peng, University of Toronto; Andrew Emili, University of Toronto.

Title:
High-Throughput Identification of the Protein Targets of Bioactive Natural Products

Introduction:
Determination of the physical interactions of bioactive natural products with cellular proteins is important for characterizing their mechanism of action, but unbiased screening approaches remain elusive. Here, we report a high-throughput mass spectrometry assay for systematic target identification based on ligand induced thermal stabilization.

Methods:
Here, we describe, validate and deploy an unbiased thermal shift assay – termed TILS (Target Identification by Ligand Stabilization) – based on quantitative mass spectrometric analysis of differential target precipitation, which shows considerably higher signal-to-noise than existing approaches measuring soluble protein levels.

Results:
We demonstrate the broad applicability of this approach by identifying both known and unexpected targets of diverse natural products (anti-cancer drugs, antibiotics, secondary metabolites), including a prominent, previously uncharacterized brominated compound that we establish is a selective inhibitor of bacterial fatty acid biosynthesis.

Novel Aspect:
The key innovation is to quantify changes in protein precipitation after sample heating.

Guo, Xinhua,
College of chemistry, Jilin University

Contributing Authors:
Shang Gao, College of Chemistry, Jilin University, P. R. China; Yanwei Cao, College of Chemistry, Jilin University, P.R. China.

Title:
Construction of accurate supramolecular DNA structures based on mass spectrometric analysis

Introduction:
Strand associations of DNA sequence d(TGmTC5T) and d(TC5TGnT) (m, n = 1, 2, 3) are studied by ESI-MS in acidic ammonium buffer solution, which provides information on the formation of i-motif structures affected by G base numbers in G-repeat and the linker between G- and C-repeat. Base on the results d(TG4AC7) was designed that formed the supramolecule containing both G-quadruplexes and i-motifs.

Methods:
The oligonucleotides (OPC grade) (40 ~ 100 μM) are annealed in 50 ~ 100 mM NH4OAc solution (pH 4.5) for ESI-MS analysis. Instrumentations ESI-MS (Bruker micro-Q-TOF) is used for DNA strand association studies. CD (JASCO-J-810), UV spectrophotometer (SHIMADZU-2550), native gel electrophoresis and Atomic force microscopy (AFM) are used for DNA structural characterization.

Results:
The tetramolecular ions formed by d(TGmTC5T) and d(TC5TGnT) (m, n = 1, 2, 3) were studied to investigate the effect of G-repeat on the formation of i-motif structures and the competition between WC duplex and i-motif. Furthermore, the affect of the linker between G- and C-repeats and terminal base was investigated. Significant tetramolecular ions were observed when m = 1, 2 and n = 1, and the sequences d(TGmAC5) (m < 5) was found that facilitated the formation of i-motif structures. Based on these results the sequence d(TG4AC7) was designed, which could form a stable i-motif core with single strand overhang as branched junctions to derive the formation of supramolecular structures including tetramolecular parallel G-quadruplexes and anti-parallel i-motifs. The construction of quadruplex DNA nanoarchitectures were characterized by native gel electrophoresis experiment and AFM imagines.

Novel Aspect:
A mass spectrometry based method was demonstrated for accurate construction of DNA supramolecules

Guo, Meiru,
Lanzhou Institute of Physics

Contributing Authors:
Detian Li, Lanzhou Institute of Physics; Yongjun Cheng, Lanzhou Institute of Physics.

Title:
Exploration and analysis of space atmosphere composition at 499 km sun-synchronous orbit

Introduction:
Mass spectrometer plays more and more important roles in space exploration. In order to solve the urgent requirement for space exploration, a miniature magnetic sector mass spectrometer is developed. It was sent to sun-synchronous orbit at 499 km on November 19, 2012. During more than three years exploration, we have gained a lot of date, and deduced important results.

Methods:
The mass spectrometer consists of three subsystems: physical, electric control unit and high voltage power. It's mass is 4.5kg, volume is 170×165×165mm, and operating power is 18 W. The mass spectrometer has two work modes that the one was designed to measure the neutral gases with the filament being on and the other was to measure ionosphere with the filament being off.

Results:
It not only has explored the neutral gases, but also detected the charged ions and abnormal ion current in space. The neutral gases include atomic oxygen, He, 12CO2, 13CO2, N2, O2, H2, CO, H2O etc. The results also show that O, He, N2, O2 come from orbit environment, H2O from satellite bay and/or mass spectrometer outgassing, H2 from orbit environment partially and outgassing, and CO, 12CO2, 13CO2 from products of C or hydrocarbon from satellite bay and/or mass spectrometer with O. The dependence of the quantity of H2O on time has been protracted. The variation of the compositions of space environment neutral gases against latitude was measured. The variation of the gases compositions in the same place against time was also measured. We think that the He comes from sun-wind. We also think that the change of the quantity of He is associated with the Earth's geology changing.

Novel Aspect:
A miniature mass spectrometer was developed, it has explored gas compositions in the satellite orbit and from the satellite in sun-synchronous orit.

Gurov, Victor,
Ryazan State Radio Engineering University

Contributing Authors:
D.V. Kirushin, Ryazan state radio engineering university, Gagarin st. 59/1, Ryazan, Russia 390005; A.A. Trubitsyn, Ryazan state radio engineering university, Gagarin st. 59/1, Ryazan, Russia 390005.

Title:
Simulation of transportation of ions by the electric field in gas in the case of gas flow and differential pressure

Introduction:
Development and introduction of new methods of the ionization under atmospheric pressure into practice has caused a sharp increase of mass spectrometry possibilities. Systems (interfaces) of ion transportation to the input of mass analyzers are important units of the sophisticated contemporary mass spectrometry equipment.

Methods:
The approach to simulation of the ion motion in gaseous medium and acceleration of them by the electric field in presence of the pressure gradient and gas flow is based on the Monte-Carlo method. Simulation of the electric field in volume of the interface is carried out by the boundary element method due to its ideal suitability for solution of potential theory tasks in electron-optical systems.

Results:
Calculation of the ion trajectory in the electromagnetic field is carried out according to numerical Runge-Kutta schemes as a result of which changes of position and velocity of the ion in a device between collisions are determined. When a current length of the ion trajectory part matches a free path length, the collusion "ion-molecule" is considered as happened. Following step of the simulation is a statistical sampling of molecule thermal velocity components along coordinate axes according to the Maxwellian distribution. Central limit theorem of the probability theory is used. Taking gas-dynamic phenomena into consideration supposes a shift of a zero mathematical expectation of the gas molecule thermal velocity axial component by the value of the gas flow velocity. Process of collisions of ions and molecules is described in approximations of the elasticity and isotropy.

Novel Aspect:
This approach is used for research of the ion transportation by capillaries from the area of atmospheric pressure to the mass spectrometer input.

Gurov, Victor,
Ryazan State Radio Engineering University

Contributing Authors:

A.A. Trubitsyn, Ryazan State Radio Engineering University; M.V. Dubkov, Ryazan State Radio Engineering University.

Title:

Device for simultaneous energy, angular and mass analysis of flows of ions emitted by the solid state surface

Introduction:

Process of the ion emission from the surface determines initial terms of its motion –energy and motion direction (polar angle and azimuth angle). The present research has suggested and simulated a device design for investigation of mass, energy and angular spectrums of secondary ions.

Methods:

Device for registration of energy and mass spectrums of ions consists in two stages. The first stage is an electrostatic energy analyzer with angular focusing not lower than the second order close by 90 degrees. The second stage is an isopath mass-analyzer. Usage of the position-sensitive detector as a collector of ions allows registering angular dependences.

Results:

Design of the energy analyzer is a combination of cylinders and rings. Suggested circuit provides an angular focusing of the forth order close by 90 degrees under significant energy dispersion. Within the range of input angles about two degrees, relative energy resolution of the device is 0.0005. Continuous ion flow focused by the energy analyzer field passes through the electrostatic interrupter forming a sequence of ion bunches getting in the isopath mass-analyzer input with electrodes of the cylindrical form. Design resolving power of such device is about 500. Registration of the ion flow by the polar angle is carried out by a position-sensitive detector. Successive rotation of the sample around the normal provides an analysis of spectrums in the whole solid angle.

Novel Aspect:

Research of the ion flow by mass, energy and angles allows specifying interaction mechanisms of outcoming ions with surface atoms.

Haberger, Markus,
Roche Diagnostics GmbH

Title:
Rapid characterization of biotherapeutic proteins by size-exclusion chromatography coupled to native mass spectrometry

Introduction:
In this study, an approach employing ultra-pressure liquid chromatography size-exclusion separation combined with native electrospray ionization mass spectrometry for the simultaneous formation, identification and quantification of size variants in recombinant bispecific antibodies (CrossMAb) was developed.

Methods:
native mass spectrometry; size-exclusion chromatography; ultra high performance liquid chromatography

Results:
SEC with UV detection facilitates fast and robust analysis, especially of non-covalent CrossMAb interactions. The coupling of SEC-UV to native ESI-MS not only allows the stepwise identification of abundant CrossMAb size variants (like dimers or free LC) by accurate mass determination, but also enables the enrichment and characterization of various low-abundant and non-covalent aggregate and fragment variants. Optimized native ESI-MS spray conditions and instrument settings represent a compromise between stabilization and artificial formation of protein complexes. Thus, the comparison of SEC-UV and SEC-MS data is needed to identify experimental artifact aggregate or fragment formation in the ion source of the applied MS system.

Novel Aspect:
size exclusion chromatography coupled to native ESI-MS for the characterization of bispecific antibody size variants

Hagino, Hiroyuki,
Japan Autmobile Research Institute

Contributing Authors:
Martin Tanner, Tofwerk; Olga Borovinskaya, Tofwerk; Hidetoshi Hikita Shoreline Science Research; Akihiro Shimono, Shoreline Science Research; Marc Gonin, Tofwerk; Katrin Fuhrer, Tofwerk.

Title:
Trace Multi-Elements Detection for Aerosol Particles by using Inductively Coupled Plasma High Resolution Time-of-Flight Mass Spectrometer

Introduction:
Inductively coupled plasma mass spectrometry is widely used for the monitoring of the aerosol particle in atmosphere. However, there are some issue of differentiate interferences (e.g. S+ and O2+). This study demonstrated the potential of measurement of trace multi-elements aerosol particles by using a Tofwerk inductively coupled plasma high resolution time-of-flight mass spectrometer (icpTOF).

Methods:
The aerosol inlet interface equipped with gas exchange device (GED) is used to introduce the aerosol particles to ICP ion source on icpTOF according to exchanging the air to argon gas using a gas exchange device. The aerosol particles samples were generated by the pneumatic dispersion of a dry environmental particles (NIST SRM 1648a or NIES CRM 8).

Results:
The icpTOF achieved quasi-simultaneous detection of the wide mass range (m/z 20-260), with a high temporal resolution of 30 us, a mass resolving power <3800. The determination of low levels of phosphorus (P) and sulfur (S) presents a challenge for conventional ICP-MS. The both of these isotopes suffer from major interferences resulting from nitrogen (N), oxygen (O) and hydrogen (H), reacting to form primarily 15N16O+ and 14N16O1H+ at m/z 31 and O2+ at m/z 32. To solve this problem, high mass resolution icpTOF allowed used to eliminate the effects of these ion interferences. Furthermore, the trace multi-elements (e.g. Na, Ca, Zn, Fe, Ba, Pb, U) allowed to detect for the environmental aerosol particles (i.e. NIST SRM 1648a or NIES CRM 8).

Novel Aspect:
In-suite analysis of aerosol particles with differentiate interferences and quasi-simultaneous detection of the wide mass range.

Haimovici, Liad,
Ontario Ministry of the Environment

Contributing Authors:
Eric Reiner, Ontario Ministry of the Environment and Climate Change; Karl Jobst, Ontario Ministry of the Environment and Climate Change.

Title:
Screening for dioxin-like compounds in sediment using modified QuEChERS and a GC-qTOF mass spectrometer with atmospheric pressure chemical ionization

Introduction:
Comprehensive sediment monitoring programs require a large number of samples and present a large cost and workload for the analytical laboratory. The purpose of this study is to combine QuEChERS extraction with detection by qTOF-MS to decrease the cost and turnaround time for targeted monitoring of polychlorinated dioxins and furans as well as non-targeted identification of halogenated organics.

Methods:
Samples from contaminated sites were extracted using a modified QuEChERS method, cleaned by carbon solid phase extraction, and analyzed using a Waters Xevo G2-XS QTOF (Wilmslow, UK) equipped with an atmospheric pressure gas chromatography (APGC) system (Wilmslow, UK) and a Zoex ZX2 thermal modulator (Zoex, Houston, USA) for comprehensive two-dimensional gas chromatography (GCxGC) experiments.

Results:
The modified QuEChERS method yields up to 30 samples/day compared to 10 samples in 3 or 4 days by the traditional Soxhlet-based method. The proposed method produces higher MDLs and lower recoveries, but the measured toxic equivalence quotients differed by only 16±10% compared to traditional methodology. The QTOF instrument used is capable of obtaining full-scan, high resolution (>20,000 FWHM) mass spectra in a time-frame compatible with GC and GCxGC experiments. The results show that the instrument performance is comparable to traditional magnetic sector instruments. It is also shown that GCxGC is capable of separating chlorinated DPE interferences that would otherwise necessitate a cleanup involving carbon SPE. Instrumental analysis using these approaches is compared to the classical method in terms of TEQs, precision, chromatographic separation, detection limits and workload capacity.

Novel Aspect:
This contribution represents the first screening method for dioxins and related compounds combining a modified QuEChERS procedure and GC(xGC)-QTOF.

Halada, Petr,
Institute of Microbiology AVCR

Contributing Authors:
Petr Halada, Institute of Microbiology, Prague; Daniel Kavan, Institute of Microbiology, Prague; Vit Dvorak Department of Parasitology, Charles University in Prague; Kristyna Hlavackova, Department of Parasitology, Charles University in Prague; Petr Volf, Department of Parasitology, Charles University in Prague.

Title:
Protein mass spectra database for rapid identification of field-caught phlebotomine sand flies

Introduction:
Phlebotomine sand flies are only proven vectors of leishmaniases. With the growing importance of phlebotomine-borne diseases precise sand fly identification is highly required to prevent the emergency of leishmaniasis and to ensure vector control and monitoring. MALDI-TOF MS protein profiling has been recently shown as a promising tool for rapid and simple identification of sand flies.

Methods:
Sand fly thoraxes were manually homogenized by a pestle and proteins were then acid-extracted for MALDI analysis in a mass range of 3-25 kDa. Reference database of sand fly protein spectra was built up under control of MALDI Biotyper software and successfully employed for the identification of field-collected specimens. The reference spectra were created from at least five specimens per species.

Results:
We have initiated the assembly of our homemade reference database, which currently consists of 16 different sand fly species representing three genera (Phlebotomus, Sergentomyia, Lutzomyia) and two Phlebotomus subgenera (Larroussius, Adlerius). The performance of the database was tested by a blind study showing more than 96% of correctly identified sand flies. We further analyzed and identified more than 800 field-caught specimens from two distinct geographical areas (Mediterranean and Caucasus regions). The enlargement of the database is in progress now to increase its performance and efficiency. We have proved that MALDI-TOF protein profiling is a suitable, time- and cost-effective method for species identification of large sets of field-collected sand flies.

Novel Aspect:
The reference database of MALDI-TOF protein mass spectra allowing realiable species identification of field-caught sand flies has been established.

Hall, Zoe,
University of Cambridge

Contributing Authors:
Julian L. Griffin, University of Cambridge; Catherine Wilson, University of Cambridge; Deborah Burkhart University of Cambridge; Gerard I. Evan, University of Cambridge.

Title:
Mass spectrometry imaging links Myc to aberrant lipid metabolism in lung cancer

Introduction:
Non-small lung cancer is a leading cause of cancer death. In adenocarcinoma, Myc has been shown to be frequently overexpressed, leading to uncontrolled cell proliferation. The detailed molecular mechanisms underpinning Myc-dependent tumorigenesis are not fully understood, in particular the role of Myc in driving lipid metabolism and biosynthesis.

Methods:
We use a novel transgenic mouse model of Kras-driven lung adenocarcinoma with reversible activation of Myc, to study the changes in the lipid profile of lung tumours with high Myc activity, and following deactivation of Myc. We characterised the spatial and temporal changes in lipid composition in lung tissue using advanced mass spectrometry imaging and surface analysis techniques.

Results:
We found that normal lung tissue was characterised predominantly by major lipid components of pulmonary surfactant. In contrast, tumour regions had increased signalling precursors. We further observed an increase in free arachidonic acid, which is released from membrane phospholipids by cytosolic phospholipase A2 (cPla2), in tumours with high Myc activity, and linked this to increased cPla2 activity. Moreover we report evidence at the metabolite and gene expression level, linking Myc to increased synthesis of arachidonic acid-derived eicosanoids via the lipoxygenase and cyclooxygenase pathways. Following Myc deactivation, specific metabolites and mRNAs associated with these pathways, and cPla2 activity, decreased with time. By inhibiting these pathways in vivo, a reduction in tumour burden and significantly reduced proliferation was observed.

Novel Aspect:
Lipidomics and surface analysis techniques are revealed as powerful analytical tools, highlighting potential new drug targets in lung cancer.

Hall, Zoe,
University of Cambridge

Contributing Authors:
Albert Koulman, University of Cambridge; Tom Ashmore, University of Cambridge; Julian L Griffin University of Cambridge; Nicholas Bond, University of Cambridge.

Title:
A molecular map of non-alcoholic fatty liver disease progression

Introduction:
Non-alcoholic fatty liver disease (NAFLD) is strongly associated with obesity and type 2 diabetes, and is the primary cause of liver disease in Western countries. NAFLD can progress from simple fat accumulation in the liver to cirrhosis and cancer. Currently the driver for this progression is not fully understood; in particular how NAFLD progression affects the distribution of lipids in the liver.

Methods:
Here, we use dietary and genetic mouse models of NAFLD, and translate results to clinical samples by correlating the spatial distribution of lipids in liver tissue with disease stage. We use matrix assisted laser desorption ionisation (MALDI) and high resolution mass spectrometry imaging (MSI) to directly sample lipids from tissue surfaces, maintaining spatial information.

Results:
By combining MSI with complementary molecular biology techniques, and sophisticated pattern recognition tools to analyse and interpret the vast quantity of data generated, we map molecular distributions and abundances across liver tissue with onset and progression of disease. Our results reveal several lipids change their zonal distribution during NAFLD, notably the arachidonic acid containing phospholipids, which become enriched in regions with increasing fat deposition and inflammation. We further provide a link between lipid composition and the regulation of enzymes involved in membrane remodelling, fatty acid release and eicosanoid formation. Overall this work delivers significant insights into the role of lipids and their metabolic pathways in NAFLD and demonstrates that NAFLD is associated not only with lipid enrichment, but also changes to the zonation of specific lipids.

Novel Aspect:
The ability to follow changes to lipid zonation in liver with disease progression paves the way for future studies using MSI for clinical research.

Hall, Chris,
Markes International Inc.

Contributing Authors:
Pete Grosshans, Markes International; Laura McGregor, Markes International; Matthew Edwards Markes International; Nick Bukowski, Markes International.

Title:
GCxGC with simultaneous mass spectrometry/flame ionization detection for routine petrochemical analyses

Introduction:
Precise characterization of petrochemical samples is crucial for quality control, and also to understand the reactions that take place during refining processes. Comprehensive two-dimensional gas chromatography (GCxGC) offers significant advantages over conventional chromatography for such analyses, with expanded separation space and the added benefit of highly structured groupings of compounds.

Methods:
Fast acquisition speeds make time-of-flight mass spectrometry (TOF MS) the ideal partner to GCxGC. However, best-practice in the petrochemical industry demands that flame ionization detection (FID) is used for target compound quantitation. This study explores the use of dual detection GCxGC to enable both targeted and non-targeted approaches to be performed simultaneously.

Results:
This approach will be shown to push the boundaries of hyphenated systems by providing three information-rich datasets per acquisition - namely robust quantitation by GCxGC-FID, untargeted screening by GCxGC-TOF MS (70 eV) and confirmatory identification by soft EI. The novel ion source technology employed here has been designed to overcome the inherent loss in sensitivity associated with lowering the ionization energy, meaning that both hard and soft ionization spectra can be acquired in tandem using the same on-column concentrations. Soft EI will be shown to allow speciation of isomers that would not have been possible using conventional 70 eV ionization – thus unlocking the fine detail of the sample composition. Simple yet effective data processing will also be demonstrated, proving that this technique can be a productive member of any high-throughput laboratory, by enabling sophi

Novel Aspect:
We introduce a novel, multi-hyphenated approach to provide unparalleled levels of detail on the composition of petrochemicals.

21ST INTERNATIONAL MASS SPECTROMETRY CONFERENCE
Toronto, Ontario, Canada
August 20 – 26, 2016

Hall, Chris,
Markes International Inc.

Contributing Authors:
Laura McGregor, Markes International; Pete Grosshans, Markes International; Matthew Edwards Markes International; Nick Bukowski, Markes International.

Title:
Comprehensive characterization of essential oils by flow-modulated GCxGC–TOF MS

Introduction:
Essential oils, due to their desirable aromatic qualities, are used as raw materials in the production of a wide range of cosmetic products, including fragrances and personal care products. The chemical composition of many essential oils exhibits a degree of natural variation, but even minor changes can adversely affect the quality of the final product.

Methods:
Rigorous quality control procedures are therefore needed to ensure that essential oils do not vary significantly from the initial reference batch, or contain any unwanted adulterants. Comprehensive two-dimensional gas chromatography with time-of-flight mass spectrometry (GC×GC–TOF MS) is the ideal choice for screening complex mixtures, such as essential oils.

Results:
The enhanced separation capacity offered by the coupling of two columns of different selectivity, combined with the highly sensitive mass spectral identification, provides a high-performance solution for rapid screening of essential oils – as evaluated in this study. This presentation shows the comparison of a range of essential oils by flow-modulated GCxGC-TOF MS combined with multivariate statistical analysis for fast evaluation of compositional differences.

Novel Aspect:
The use of a novel ion source design to allow tandem ionization, for collection of both hard and soft EI spectra in a single run.

Hamada, Mona,
College of Pharmacy and Nutrition, University of Saskatchewan

Contributing Authors:
Mona M Khamis, College of Pharmacy and Nutrition, University of Saskatchewan, Saskatoon, SK, Canada; Darryl J Adamko, Department of Pediatrics, College of Medicine, University of Saskatchewan, Saskatoon, SK, Canada.; Anas El-Aneed College of Pharmacy and Nutrition, University of Saskatchewan, Saskatoon, SK, Canada.

Title:
Determination of potential diagnostic urinary biomarkers of asthma and COPD using a novel DIL-HPLC-MS/MS method

Introduction:
Urine is rich in metabolites and is an ideal source for biomarker discovery. Obstructive airways inflammatory diseases sometimes show overlapping symptoms that hinder their early and correct diagnosis. Current clinical tests are tedious and cannot be used in young children. Therefore, we are developing tandem mass spectrometric (MS/MS) methods for targeted analysis of urine biomarkers.

Methods:
Recently, 1H-NMR analysis proposed 53 metabolites as potential diagnostic biomarkers among asthma and COPD patients. Metabolites are divided into 3 groups based on chemical nature. For group 1 (amines and phenols), we have developed an HPLC-MS/MS method using differential isotope labeling with dansyl chloride. An internal standard for every metabolite is created using C13-labeled reagent.

Results:
An HPLC-MS/MS method is established for the simultaneous quantification of 19 metabolites using C18 column. Linearity of the metabolites vary from 2 to 4800 ng/ml. The use of C13-labeled derivatives allows for the correction of any matrix effects as well as the unambiguous confirmation of the identity of each metabolite. Despite the challenges associated with method validation, we are able to fully validate the method as per the FDA recommendations. Non-blinded patients' samples are analyzed and the data are processed using principle component analysis (PCA). The obtained plots revealed differences among patients' groups suggesting the suitability of the metabolites as diagnostic biomarkers. Currently, we are analyzing samples collected at initial and subsequent clinical visits to confirm the validity of the selected diagnostic and prognostic metabolites.

Novel Aspect:
To the best of our knowledge, this is the first MS-based study investigating the validity of urine biomarkers for the diagnosis of asthma and COPD.

HAMADA, SATOSHI,
Nissan Chemical Industries,LTD.

Contributing Authors:
TAISUKE IHOZAKI, Kansai University; ISSEY OSAKA, Japan Advanced Institute of Science and Technology; RYUICHI ARAKAWA Kansai University; TOMOYUKI OZAWA, Nissan Chemical Industries, Ltd; HIDEYA KAWASAKI, Kansai University.

Title:
Imaging mass spectrometry of rat brain tissue using matrix-enhanced surface-assisted laser desorption/ionization with sputter-deposited platinum film

Introduction:
Imaging mass spectrometry (IMS) with MALDI allows the simultaneous detection and imaging of several molecules in biological tissue. However, the detection of many kinds of molecules is hindered by excessive ion yields of phosphatidylcholine (PC). In this study, we applied Surface-assisted laser desorption/ionization with sputter-deposited platinum film (Pt-SALDI) to IMS of biological tissue.

Methods:
Tissue sections of rat brain were analyzed using MALDI-IMS with DHB or Pt-SALDI. For the analysis of MALDI, DHB solution (30 mg/ml in 50% methanol aqueous) was sprayed on the tissue section. In the preparation of Pt-SALDI experiment, air-dried tissue section was coated by sputter-deposited Pt. FT-ICR MS was used for IMS of tissue sections to achieve complete characterization of the obtained ions.

Results:
MALDI-IMS detected mainly the dominant ion peaks of PC in the tissue section. In contrast, Pt-SALDI-IMS allowed the detection of glycerolipids such as diacylglycerol (DAG), but could not detect PC. Therefore, we used a hybrid method called matrix-enhanced (ME)-Pt-SALDI-IMS by combining MALDI and Pt-SALDI. In the experiment of ME-Pt-SALDI, the tissue section was sprayed by 5ml of DHB solution (30mg/ml in 50% methanol aqueous) and coated by sputter-deposited Pt. As a result, we achieved the simultaneous detection of PC and DAG in rat brain tissue sections. The sensitivity for the detection of these molecules was higher than that of MALDI-IMS or Pt-SALDI alone. The present simple ME-Pt-SALDI approach for the simultaneous detection of PC and DAG using two matrices (sputter-deposited Pt film and DHB matrix) would be useful in imaging analyses of biological tissue sections.

Novel Aspect:
We have developed a method combining the Pt-SALDI-IMS with MALDI (called ME-Pt-SALDI) for the simultaneous analysis of DAG and PC in the tissue.

Hammond, John,
Physical Electronics USA

Contributing Authors:

Scott R. Bryan, Physical Electronics; Ron M.A. Heeren, Maastricht Multi-Modal Molecular Imaging (M4I) Institute; Shin-Ichi Iida ULVAC-PHI; Nina Ogrinc Potočnik, Maastricht Multi-Modal Molecular Imaging (M4I) Institute; David M. Carr, Physical Electronics; Gregory L. Fisher, Physical Electronics; Anne L. Bruinen, Maastricht Multi-Modal Molecular Imaging (M4I) Institute; Takuya Miyayama, ULVAC-PHI.

Title:

Sub-Micron Imaging and Identification of Molecular Chemistry in Materials and Biological Research by TOF-SIMS Parallel Imaging MS/MS

Introduction:

A recently introduced TOF-TOF imaging mass spectrometer allows conventional TOF-SIMS (MS1) analysis and product ion (MS2) analysis to be achieved simultaneously and in parallel. Secondary ions for MS1 and MS2 analysis are produced from the same area of the surface by a pulsed and digitally raster-scanned 100 nm primary ion nanoprobe.

Methods:

Uncoated polymer samples and zebrafish tissue samples were analyzed in a PHI nanoTOF TOF-SIMS with parallel imaging MS/MS. Fragmentation of the molecular precursor ions, defined by a 1 Da precursor selection window, was accomplished by collision-induced dissociation (CID) at 1.5 keV in an activation cell of high pressure Ar gas.

Results:

Characteristic MS/MS spectra and parallel MS/MS images were obtained from several polymer-based specimens and polymer surface additives providing unambiguous molecular identification. Parallel imaging MS/MS was also used to probe the role of lipids and metabolites in disease states of zebrafish (D. rerio) infected with M. marinum. Elevated levels of α-tocopherol were observed in infected tissue as well as in the granuloma, but not in the necrotic cells. The role of phospholipids appears to differ, specific molecules being either elevated or depressed in the infected tissue.

Novel Aspect:

Sub-micron spatial resolution TOF-SIMS with parallel imaging MS/MS has been demonstrated on polymer and tissue samples.

Han, Kyoung-Moon,
NIFDS, MFDS, Republic of Korea

Contributing Authors:
Hye-Jin Cha, NIFDS, MFDS, Republic of Korea; Hyekyung Park, NIFDS, MFDS, Republic of Korea; Goeun Lee NIFDS, MFDS, Republic of Korea; Hyungsoo Kim, NIFDS, MFDS, Republic of Korea; Jisoon Shin, NIFDS, MFDS, Republic of Korea; Jaesuk Yun, NIFDS, MFDS, Republic of Korea; Young-Hoon Kim, NIFDS, MFDS, Republic of Korea; Hyewon Seo, NIFDS, MFDS, Republic of Korea.

Title:
In vitro metabolism of ketamine-derived designer drug methoxetamine by human liver microsomes using LC-HR-MS

Introduction:
Methoxetamine (MXE; 2-(3-methoxyphenyl)2-(N-ethylamino)-cyclohexanone) is a synthetic designer drug derived from ketamine and belongs to the arylcyclohexylamine class, which is used for its recreational and psychedelic effect.In this study, we introduce in vitro metabolism methoxetamine by human liver microsomes using LC-HR-MS

Methods:
In this study, the in vitro metabolism of MXE was incubated with human liver microsomes(HLM), and the reaction mixture was analyzed using liquid chromatography-high resolution mass spectrometry(LC-HR-MS) MXE was metabolized in HLM, yielding about 2 metabolites. These metabolites were structurally characterized on the basis of accurate mass analyses and MS/MS fragmentation paterns.

Results:
The retention times of MEX, M1 and M2 showed approximately 3.7, 3.2 and 3.5 min, respectively. The MS spectra of metabolites M1 and M2 showed a protonated molecular ion [M+H]+ at m/z 234.1488 and 220.1329. Also, the MS/MS spectra of protonated M1, M2 were observed at m/z 161.0954 and 175.1112. Based on these results, M1 and M2 were identified as O-demethyl- and N-deethyl-MXE, respectively. The results provide evidence for in vivo methoxetamine metabolism and they may be applied to the analysis of MXE and its relevant metabolites in biological samples.

Novel Aspect:
The metabolic pathways of MEX may aid in the development of analytical methods for monitoring the abuse of MXE in biological sample such as urine etc.

Hann, Stephan,
University of Natural Resources and Life Sciences - BOKU Vienna

Contributing Authors:
Michaela Schwaiger, Institute of Analytical Chemistry, University of Vienna, Vienna, Austria; Teresa Mairinger, Austrian Centre of Industrial Biotechnology (acib), Vienna, Austria and Department of Chemistry, University of Natural Resources and Life Sciences - BOKU Vienna, Vienna, Austria; Gunda Koellensperger Institute of Analytical Chemistry, University of Vienna, Vienna, Austria; Wolfhard Wegscheider, Department of General, Analytical and Physical Chemistry, Montanuniversitaet Leoben, Leoben, Austria.

Title:
Inter-comparison of modern methods for isotopologue ratio analysis in metabolomics: trueness and uncertainty

Introduction:
Flux analysis is considered a key technology addressing alteration of metabolism for optimization of cell factories. Flux modeling demands for accurate measurement of labeling patterns of metabolites with a high diversity of chemical properties and concentrations. We compare three novel MS-based strategies followed by error propagation for the quality assessment as proposed by EURACHEM.

Methods:
Monte Carlo simulation was used for the error propagation of influence factors for isotopologue ratios measured for metabolites of the primary carbon metabolism via GC-CI-QTOFMS [Mairinger T, et al., Anal.Chem.(2015),87:11792-802], LC-QTOFMS with data-dependent MS/MS and IC-orbitrapMS Furthermore, reproducibility and trueness were assessed with an in vivo 50% 13C-labeled cellular extract.

Results:
We present three complementary approaches capable of measuring isotopologue and tandem mass isotopomer ratios of the constituents of the primary metabolome such a sugars, sugar phosphates, amino acids and organic acids. The study showed that all of the developed methods are fit for purpose revealing reproducibilities in the low percent range for the analytes covered by all three approaches. Trueness of the determined isotopologue ratios was in agreement with the reference values of the in vivo labeled material with fully controlled labeling pattern. We show how to set-up a model equation for the propagation of influence factors in isotopologue ratio analysis enabling the assessment and impact of possible sources of error, e.g. peak integration and counting statistics.

Novel Aspect:
For the first time a cross validation of different platforms for measurement of isotopologue ratios of free metabolites is presented

Hao, Chunyan,
Ontario Ministry of the Environment and Climate Change

Contributing Authors:
Matthew Noestheden, SCIEX, Canada; Xiaoming Zhao, Ontario Ministry of the Environment and Climate Change; Holly Lee Ontario Ministry of the Environment and Climate Change; David Morse, Ontario Ministry of the Environment and Climate Change.

Title:
Applying Differential Ion-mobility Spectrometry to Improve LC-MS/MS Analysis of Emerging Organic Contaminants

Introduction:
Perfluoroalkyl acids and nonylphenol ethoxylates are widely used surfactants in various products that may cause potential adverse effects in the environment. Neonicotinoids have been investigated recently due to their possible link to bee colony collapse disorder. These compounds are organic contaminants of emerging concern and reliable analytical methods are critically needed for different enviro

Methods:
Samples were analyzed using a liquid chromatograph-tandem mass spectrometer (LC-MS/MS) modified with differential ion-mobility on a SCIEX QTRAP® 5500 or 6500 mass spectrometer system equipped with a SelexION® cell to remove or reduce the isobar interference and noise by permitting only ions with specific mobility through to the MS/MS detector.

Results:
False positive results for PFBS in mussels and PFPeA in fish samples were eliminated by ion-mobility. The interference for neonicotinoid pesticides from soil and pollen sample matrices was significantly reduced. An interference for nonylphenol monoethoxylate from the analytical instrument itself was also investigated. These results demonstrated that differential ion-mobility is a great alternative to reduce time-consuming sample cleanup and LC method re-optimisation efforts for complex matrices such as biota, soil and pollen.

Novel Aspect:
Differential ion-mobility is an effective tool to minimize chemical background and isobaric interference for emerging organic contaminants.

Healy, Robert,
Ontario Ministry of the Environment and Climate Change

Contributing Authors:
Nicholas Karellas, Ontario Ministry of the Environment and Climate Change; Julie Bennett, Ontario Ministry of the Environment and Climate Change; QingFeng Chen Ontario Ministry of the Environment and Climate Change; George Rioual, Ontario Ministry of the Environment and Climate Change.

Title:
A multi-year study of VOC emissions at a waste disposal facility using mobile LPCI-MS and APCI-MS instruments

Introduction:
Real-time analysis of volatile organic compounds (VOCs) in air is useful for the identification of their respective sources. On-line mass spectrometers can be deployed for the real-time detection and quantification of a wide range of VOCs in ambient environments.

Methods:
In this work, two complementary triple quadrupole mass spectrometers, fitted with atmospheric pressure chemical ionization (APCI) and low pressure chemical ionization (LPCI) sources, respectively, were deployed simultaneously to investigate emissions of VOCs associated with an Ontario-based waste disposal facility.

Results:
The facility contains a hazardous waste disposal and treatment site with a secure landfill and a liquid waste incinerator. Six separate field studies were undertaken between 2000 and 2015 to assess how emissions of VOCs have changed at the site as a function of time. Up to 29 VOCs were successfully identified and quantified using MS/MS in each study. Deployment of the mass spectrometers enabled the detection and quantification of polar VOCs including acetates, amines and ketones as well as non-polar aromatic VOCs including benzene and naphthalene. Concentrations of VOCs were found to decrease significantly at the site over the 15 year period, in particular since 2007. Concentration values are reported in the context of provincial guidelines and recommended odour thresholds for each study year. The implications of a recent expansion of the facility are also discussed.

Novel Aspect:
The two MS instruments were housed in vehicles to enable real-time "plume tracking" and the investigation of VOC concentrations in air spatially

Heck, Albert,
Utrecht University

Contributing Authors:
Fan Liu, Utrecht University; Richard Scheltema, Utrecht University; Philip Lossl Utrecht University.

Title:
Charting the cellular interactome by proteome-wide cross-linking mass spectrometry

Introduction:
Cross-linking combined with mass spectrometry (XL-MS) is a powerful approach to probe the structure of protein assemblies. However, currently XL-MS is mainly limited to the characterization of purified protein assemblies. However, ideallly XL-MS should probe protein complexes at the proteome level. Therefore, innovative methods are needed to overcome the hurdles in the field.

Methods:
Whole proteome samples were cross-linked with the MS-cleavable cross-linker disuccinimidyl sulfoxide (DSSO). The peptides were fractionated using SCX to enrich highly charged (presumably cross-linked) species and only the later SCX fractions were subjected to LC/MS analysis. The MS acquisition were performed on an ETD enabled LTQ Orbitrap Fusion mass spectrometer.

Results:
We established a XL-MS workflow that counteracts the two major difficulties in cross-linking data analysis, i.e., 1) the escalating size of the search space of the database, and 2) the inefficient fragmentation of either of the two linked peptides. The approach comprises three essential features, the use of MS-cleavable cross-linkers, a hybrid MS2-MS3 fragmentation strategy and a dedicated cross-link search engine for data analysis (XlinkX). To solve the database expansion problem, we applied the MS-cleavable cross-linker DSSO to be able to determine the precursor mass of each linked peptide moiety by the unique fragmentation pattern of the cross-link in CID MS2 acquisitions. To overcome the inefficient fragmentation, using data-dependent MS3 acquisition that is triggered by a specific mass difference. We applied this novel XL-MS strategy to E. coli lysate and HeLa proteome samples

Novel Aspect:
Developing a novel hybrid MS2-MS3 strategy to profile proteome-wide interactomes by cross-linking mass spectrometry

Hee, Daryl,
National University of Singapore

Contributing Authors:
Jun Liang Ong, National University of Singapore; Lawrence Soon-U Lee, National University of Singapore; Johannesen Yee Kiat Leaw National University of Singapore.

Title:
Development and validation of LC-MS/MS method for quantification of rifampicin, amoxicillin and faropenem in plasma and application in clinical study

Introduction:
Tuberculosis (TB) is the second leading cause of death from infectious disease worldwide. Combinations of rifampicin and beta-lactam antibiotics are being tested to optimize TB treatment. We have developed and validated a sensitive LC-MS/MS method for simultaneous quantification of rifampicin, amoxicillin and faropenem, to investigate the synergistic effect of these drugs in treating TB.

Methods:
Rifampicin, amoxicillin and faropenem were extracted from plasma by protein precipitation followed by chromatographic separation C18 column using gradient elution. The analytes were detected under positive ionization mode by multiple reaction monitoring in the mass spectrometer. Quantification of analytes was performed by using deuterium-labelled internal standards.

Results:
Excellent linearity ($r2 \geq 0.995$) was achieved for the analytes at different concentration ranges. The method was accurate (91 to 108%), precise (CV% < 7.8) and specific. Matrix effect was in the range of 92 to 112% while recovery was reproducible (CV% < 7.9) in the range of 64 to 105%. The analytes were found stable with more than 89% of recovered concentration for up to 4 and 24h at room temperature and 4˚C, respectively, as well as three freeze-thaw cycles and 48 h in 6 ˚C autosampler. The method was successfully applied to a clinical study of healthy subjects to evaluate the pharmacokinetics and anti-TB activity of amoxicillin and faropenem with rifampicin.

Novel Aspect:
First report on the simultaneous quantification of rifampicin and amoxicillin with faropenem for clinical application.

Heeren, Ron,
M4I, Maastricht University

Contributing Authors:

Ron M.A. Heeren, M4I, Maastricht University; Shane R. Ellis, M4I, Maastricht University; Tiffany Porta M4I, Maastricht University; Benjamin Balluff, M4I, Maastricht University; Berta Cillero-Pastor, M4I, Maastricht University.

Title:

Mass spectrometry in Systems Medicine: towards the molecular operating theatre

Introduction:

The modern approach to treating diseases is more and more based on Systems Medicine. Information about patient and disease is brought together systematically to help clinicians, to improve treatment accuracy, improve patient outcome and ultimately reduce the cost of healthcare. In systems medicine MS is an enabler to generate insights at the molecular, cellular, tissue and patient level combined.

Methods:

MS methods enable personalized medicine. MALDI-MSI is emerging as a novel tool that enables detailed molecular pathology of cells and tissue. Intraoperative diagnostics employs the detailed molecular information provided in real-time during a surgical procedure and takes full advantage of high throughput mass spectrometry. The two combined are the basis of the molecular operating theatre.

Results:

High throughput MALDI-Tof MSI has been employed for cellular phenotyping directly from human tissue sections. The MALDI-MSI study data was fused with H&E and IHC to increase the accuracy of the tissue typing experiments. Small molecules and metabolites were localized and identified using an ultrahigh resolution imaging FTICR-MS approach. This combined approached revealed various signalling pathways coupled to the tissue microenvironment. A common theme found through a variety of diseases is hypoxia. Studies in oncology and conditions related to tissue ischemia, such as heart failure, organ transplant and neurological stroke, revealed common molecular signatures. These molecular signatures can also be employed for intraoperative diagnostics using REIMS with electro-surgery. Combined, they provide the personalized diagnostic fundament of the molecular operating theatre.

Novel Aspect:

Translational mass spectrometry research towards the operating theatre of the future.

Heidelberger, Sibylle,
Sciex

Contributing Authors:
Neil R Kitteringham, MRC Centre for Drug Safety Science, Dept Pharmacology, University of Liverpool, UK; Rosalind E. Jenkins, MRC Centre for Drug Safety Science, Dept Pharmacology, University of Liverpool, UK; B Kevin Park MRC Centre for Drug Safety Science, Dept Pharmacology, University of Liverpool, UK; Francesco L Brancia, Sciex, Phoenix House, Warrington, UK; Thomas Knapman, Sciex, Phoenix House, Warrington, UK.

Title:
Comprehensive relative quantification of the cytochromes P450 by micro-LC and SWATH® acquisition and data processing using Cloud Computing.

Introduction:
Cytochrome P450 superfamily is very important due to their involvement with drug metabolism and safety, however their homology makes it difficult to identify the P450 involved in specific drug metabolim. The use of micro-LC and SWATH acquisition allows for rapid label-free quantification which enables closely-related proteins to be quantified retrospectively which is perfect for P450 profiling.

Methods:
Mice were exposed to inducers of the Cytochromes P450, and pooled microsomal fractions were prepared from the livers. Following protein extraction and digestion, a database of microsomal proteins and SWATH acquitision data was generated on a TripleTOF 6600® system (AB SCIEX, Framingham, USA). SWATH acquisition. Data was processed using the OneOmic platform.

Results:
OneOmics™ platform separated induced and non-induced mice based on their overall protein expression pattern, and that of the P450s. Relative quantification of uniquely discriminatory P450 peptides enabled the induction profile of each compound to be ascertained in unprecedented detail allowing for the possibility to identify and quantify peptides unique to Cyp2C50 and Cyp2C54 despite the fact that the proteins share 92% sequence identity.

Novel Aspect:
Micro-LC, data dependent and data independent acquisition (SWATH) to identify proteins with high homology

Heidelberger, Sibylle,
Sciex

Contributing Authors:
Daniel Blake, Sciex.

Title:
Characterization of sphingolipids from plasma by LC-Differential Mobility Spectrometry-MS/MS

Introduction:
Plasma sphingolipids are inherently similar in characteristics with some being isobaric compounds. Development of a rapid and facile method which allows for the identification and quantification of these similar analytes is of great interest. Differential Ion Mobility Mass Spectrometry allows for such separation and quantification.

Methods:
Plasma Sphingolipids (Glu-Sph, Gal-Sph, Lyso-Gb3, SPC and the internal standard C-17-SPC) were extracted by a simple protein precipitation. Separation was performed using a standard reversed phase column and DMS technology (SelexION® ION Mobility Technology, Sciex) using MRMs on a TripleQuad 6500 LC-MS/MS system (Sciex) operating in Low Mass Positive Turboionspray mode.

Results:
Simple chromatography was achieved on extracted samples with a runtime of 5 minutes. Retention times for all compounds were between 2.6 and 3 minutes. Injection of individual standards of all compounds following optimization of DMS parameters confirmed no interference between isobaric compounds despite no chromatographic separation. Sensitivity was determined by extraction of analyte free matrix and limits of detection were confirmed to be 0.1nmol/L for Lyso-Gb3 and Gal SPH, and 0.5nmol/L for SPC and Glu-SPH. Linearity of response was investigated and curves were shown to be linear up to and beyond 100nmol/L.

Novel Aspect:
Differential Mobility Spectrometry for plasma lipid analysis and direct analysis of similar sphingolipids.

Heijs, Bram,
Leiden University Medical Center

Contributing Authors:
Rob Tollenaar, Department of Surgery, Leiden University Medical Center, Leiden, The Netherlands; Manfred Wuhrer, Center for Proteomics and Metabolomics, Leiden University Medical Center, Leiden, The Netherlands; Peter van Veelen Center for Proteomics and Metabolomics, Leiden University Medical Center, Leiden, The Netherlands; Gabriella van Pelt, Department of Surgery, Leiden University Medical Center, Leiden, The Netherlands; Stephanie Holst, Center for Proteomics and Metabolomics, Leiden University Medical Center, Leiden, The Netherlands; Anand Mehta, Department of Microbiology and Immunology, College of Medicine, Drexel University, Philadelphia, PA, USA; Inge Briaire-de Bruijn, Department of Pathology, Leiden University Medical Center, Leiden, The Netherlands; Richard Drake, Department of Cell and Molecular Pharmacology and Experimental Therapeutics, Medical University of South Carolina, Charleston, SC, USA; Wilma Mesker, Department of Surgery, Leiden University Medical Center, Leiden, The Netherlands; Liam McDonnell, Fondazione Pisana per la Scienza ONLUS, Pisa, Italy; Arnoud de Ru, Center for Proteomics and Metabolomics, Leiden University Medical Center, Leiden, The Netherlands; Judith Bovée, Department of Pathology, Leiden University Medical Center, Leiden, The Netherlands.

Title:
Multimodal mass spectrometry imaging analysis of N-linked glycans and proteins from a single tissue section.

Introduction:
On-tissue digestion MALDI-MSI enables the spatially resolved analysis of proteolytic peptides directly from a tissue section. In this work we expand the on-tissue digestion MALDI-MSI methodology to analyze both N-glycans and proteolytic peptides, resulting in a novel, multimodal MALDI-MSI platform for the analysis of two highly related analyte classes from a single tissue section.

Methods:
In this study the repeatability and robustness of the multimodal on-tissue digestion approach was investigated by analyzing consecutive FFPE human leiomyosarcoma (LMS), myxoid liposarcoma (MLS), and colorectal carcinoma (CRC) tissue sections. Consecutive sections enabled the analysis of the digestions (PNGase F, sequential trypsin and control trypsin) by MALDI-MSI, and tissue extracts by LC-MS/MS.

Results:
N-glycan MSI was performed using a MALDI-TOF-MS platform and resulted in the detection of 59 unique N-glycan species. The identities of half of the glycan species were confirmed by MALDI-MS/MS of tissue extracts. The visualization of specific N-glycans revealed that their abundances were localized to specific morphological structures in LMS, MLS and CRC sections. Following the N-glycan analysis a trypsin digestion was performed, and the proteolytic peptides were analyzed by MALDI-FTICR-MSI. Compared to a control digestion (no deglycosylation), both the spectra and the visualizations for specific peptides were very similar, indicating similar measurement sensitivity and absence of sample preparation artefacts. After LC-MS/MS analysis of tissue extracts 183 unique peptides could be assigned to the MSI datasets from the control and sequential digestions of the LMS and CRC tissues.

Novel Aspect:
In this work we present the unprecedented and robust multimodal analysis of both N-glycans and proteolytic peptides from a single tissue section.

Heller, Manfred,
University of Bern

Contributing Authors:
Sophie Braga, CTC Analytics AG, 4222 Zwingen, Switzerland; Thomas Preiswerk, CTC Analytics AG, 4222 Zwingen, Switzerland; François Achermann University of Bern, 3010 Bern, Switzerland; Reto Bolliger, CTC Analytics AG, 4222 Zwingen, Switzerland; Natasha Buchs, University of Bern, 3010 Bern, Switzerland; Guenter Boehm, CTC Analytics AG, 4222 Zwingen, Switzerland.

Title:
Automated protein digestion on a RTC-PAL system directly coupled to nano-LC-MS2

Introduction:
Nanoflow reversed phase chromatography coupled to tandem mass spectrometry (nanoLC-MS2) has matured to an automated high-throughput technology in proteomics research. It is therefore beneficial to automate the protein digestion processes combined with the sample loading on the nano-LC-MS/MS instrument, in order to reduce mistakes and errors during sample handling.

Methods:
We adapted in-gel and in-solution protein digestion protocols, including reduction, alkylation and trypsin digestion, to be performed on a PAL RTC liquid handling – autosampler system (CTC Analytics AG, Switzerland). The current PAL RTC setup treats batches of six gel samples within about seven hours. Peptide extracts are loaded on a trap column mounted on a nano-UPLC system by an auxiliary pump.

Results:
Four batches of 1ng and 5ng MW standard proteins were processed by an experienced human operator or on the PAL RTC system during four consecutive days. Protein yields were determined by identification and quantification with MaxQuant software. No statistically significant differences in protein yield (non-normalized intensities) between batches and between procedures were found by one-way Anova test applying Tukey's honestly significant difference criterion to evaluate statistical significance (alpha = 0.05). The average coefficients of variation of the 12 standard proteins were 19.4% and 11.8% at 1ng protein, and 5.8% and 6.5% at 5ng for the manual and the automated procedure, respectively. Preliminary results with in-solution digests (HEK293 cells) indicate that median CVs of <2% (range 0.1-15%, N(proteins) = 1213) without, and <1% (range 0.008-5%) after normalization can be achieved.

Novel Aspect:
Uninterrupted, reproducible protein digestion on a commercial autosampler directly coupled to nano-LC-MS2 for automated proteomics research.

Hermannova, Martina,
Contipro a.s.

Contributing Authors:
Matej Simek, Contipro a.s.; Petra Liskova, Contipro a.s.; Dagmar Cozikova Contipro a.s.; Ludmila Grusova, Contipro a.s.; Zdislava Brunova, Contipro a.s.; Vladimir Velebny, Contipro a.s.

Title:
Novel UPLC-IMS-MS method for identification of hyaluronan oligosaccharides

Introduction:
Hyaluronidases and acidic or alkaline hydrolysis are extensively used for the preparation of HA oligomers with different chain length and the order of monosaccharide units. QC of prepared fragments is usually tested by AEC where the differentiation among reverse HA oligomers is impossible. LC-MS methodologies published so far are time-consuming and the usage of ion-pairing agents is necessary.

Methods:
HA oligosaccharides of different structures were provided by Contipro a.s.. UPLC-IMS-MS system from Waters containing UPLC H-class and Synapt G2-Si was employed for further analysis. CCS values were measured by direct infusion and in LC mode using different calibration standards by Synapt G2-Si and were also compared with the direct measurement of CCSs employing the QTOF 6560 from Agilent.

Results:
Extensive experiments were undertaken to set-up and optimize presented UPLC-IMS-MS on-line separation of a mixture of HA oligomers ranging in size from HA2 up to HA34, including two types of even- and two types of odd-numbered oligosaccharides. Unlike any other separation methods, MS-friendly mobile phase including volatile formic acid and acetonitrile was used. Furthermore, the choice of a small analytical UPLC column having particle size 1.7 um and pore size 130 Å and IM as the second separation dimension enabled us to resolve about 100 analytes in 16 minutes. Detected oligomers were identified based on their Rt, typical m/z and drift time or CCS which was found out to be the most important characteristic. We therefore evaluated the actual CCSs employing polyalanine and dextran as a calibrants and compared the results with the direct measurement of CCSs in the drift-tube instrument.

Novel Aspect:
Complete set-up of UPLC-IMS-MS method suitable for identification of HA oligomers with various chain length, including a number of positional isomers.

Herniman, Julie,
University of Southampton

Contributing Authors:
Philip Elves, University of Southampton, UK; G. John Langley, University of Southampton, UK; Christopher Hopley LGC, Teddington, UK.

Title:
Can ASAP Provide a Quantitative Screen for Rapid Detection of Drugs in Saliva?

Introduction:
Ambient ionisation offers a rapid route to analysis, with minimal sample preparation. Saliva provides non-invasive access to a readily accessible sample matrix. ASAP sampling conditions have been optimised and MRM methods developed using internal standards to afford ppb detection with R2> 0.99 for propanalol, cocaine and MDMA.

Methods:
ASAP sample loading, sample position and ASAP ionisation conditions were optimised using a Waters Xevo TQD mass spectrometer. 1 µL of sample, in methanol and/or synthetic saliva, was pipetted onto the glass capillary, five replicate samples were measured for each sample and calibrant concentration. Isotopomeric and analogue internal standards were used and MRM transition defined.

Results:
Initial experiments highlighted the importance of repeatable sample loading and orientation of the glass sample tune with respect to the ion source sampling cone orifice. Atenolol was chosen as an analogue and used as the internal standard for the analyte propranolol. Later the deuterium labelled isotopomer of cocaine was used as internal standard for the cocaine analysis. Sub ppm detection was achieved for all analytes, with good linear regressions observed throughout, irrespective of use of the analogue or isotopically labelled internal standard. Similar LODs and linearities were achieved with synthetic saliva sample matrix. These results suggest that this approach could be used for curb side testing for drugs in saliva. Further work is needed to determine whether these LODs can be achieved with miniaturised mass spectrometers.

Novel Aspect:
Quantitative ASAP for detection of drugs in saliva.

Herniman, Julie,
University of Southampton

Contributing Authors:
Robert Carr, Huntsman (Europe), Belgium; Tim J. Jenkins, Waters Corporation, UK; G. John Langley University of Southampton, UK.

Title:
Expanded Analytical Capability through the Application of Open Access UHPSFC-APPI-MS

Introduction:
Infusion analysis via an APPI source may be utilised for compounds that are too involatile for GC-MS or are incompatible with RP UHPLC-MS. This manual approach is time consuming and often requires the introduction of a dopant. A UHPSFC-APPI-MS method has been developed to produce rapid separation and extended ionisation capability that can easily be incorporated into an open access environment.

Methods:
UHPSFC separations were achieved using a Waters UPC2 TQD. Positive ion APPI mass spectra were acquired between m/z 150-1500. Sub 2 µm columns were used throughout, with organic modifiers and a gradient of 2-40 % over 3 minutes at a flow rate of 1.5 mL min-1. Toluene was infused via an external pump as both the APPI make-up solvent and dopant.

Results:
The APPI source conditions were initially optimised using a standard mixture of polycyclic aromatic hydrocarbons (PAHs). It was noted that in some cases both the molecular ion and the protonated molecule may be present in the same mass spectrum, therefore reducing sensitivity by attenuating the signal between the two ions. In subsequent analyses, improved sensitivity was obtained by introducing a toluene dopant externally via the make-up flow pump, forcing the production of predominantly molecular ion species. Successful assays have been developed for fullerenes, isocyanates, molecular wires, PAHs, other synthetic products incompatible with RP solvents and analytes in complex matrices. These methods have been incorporated into the existing open access provision provided via RemoteAnalyzer™ and have expanded the chemical space covered by automated analysis.

Novel Aspect:
The incorporation of automated UHPSFC-APPI-MS to expand chemical space in an open access MS environment.

Heywood, David,
Waters Corporation

Contributing Authors:
Keith Richardson, Waters Corporation, Wilmslow, United Kingdom; Praveen Harapanahalli, Waters Corporation, Wilmslow, United Kingdom; Christopher J Hughes Waters Corporation, Wilmslow, United Kingdom; Richard Chapman, Waters Corporation, Wilmslow, United Kingdom; Arkadiusz Grzyb, Waters Corporation, Wilmslow, United Kingdom; Kirsten Craven, Waters Corporation, Wilmslow, United Kingdom; Martin Green, Waters Corporation, Wilmslow, United Kingdom; Jason Wildgoose, Waters Corporation, Wilmslow, United Kingdom.

Title:
Advances in Targeted Omics Quantitation Using a Novel Scanning Quadrupole DIA Method

Introduction:
Context is driving current omics experiments, fuelling the development of "quan/qual" LC-MS acquisition methods. An alternative DIA mode of operation, whereby a resolving quad' is scanned during high-resolution precursor and product ion acquisitions, was applied for targeted lipidomics and proteomics quantitation experiments using transition extraction list and compound library based approaches.

Methods:
DIA data was acquired on a QToF mass spectrometer where the m/z isolation range of the quadrupole was continuously and repetitively scanned. Alternate MS scan data comprised of precursor and CID product ions. The resulting 2D data were processed, quantified and visualized using Skyline and OpenMS. Control, diabetic and obese plasma were analysed to determine quantitative lipid and protein changes.

Results:
Data were acquired by spiking, i.e. serially diluting, lipid standards in protein precipitated plasma and a four protein digest mixture into a proteolytic E.coli digest, respectively. For both experiments, transitions extraction lists were utilised to obtain quantitative information. Analysis of the data indicates that scanning quadrupole DIA enables over an order of magnitude more specificity than a static quadrupole operated with the same resolution, providing reduced FQR values and higher dynamic range. It was found that quadrupole transmission windows of 5-10 and 20-30 Da provided optimum lipid and protein identifications, respectively. Qualitative information from the same data sets was obtained by extracting lipid class information based on neutral loss or product ion extraction, and the use of libraries in the instance of the peptide-centric targeted proteomics experiment.

Novel Aspect:
Targeted quantitation of omics samples using a novel precursor quadrupole scanning based DIA method and open source informatics.

Heywood, David,
Waters Corporation

Contributing Authors:
Mark Towers, Waters Corporation, Wilmslow, United Kingdom; Lee A Gethings, Waters Corporation, Wilmslow, United Kingdom; Johannes PC Vissers Waters Corporation, Wilmslow, United Kingdom; Christopher J Hughes, Waters Corporation, Wilmslow, United Kingdom; David Eatough, Waters Corporation, Wilmslow, United Kingdom; Richard Tyldesley-Worster, Waters Corporation, Wilmslow, United Kingdom; Jonathan P Williams, Waters Corporation, Wilmslow, United Kingdom; Steven Lai, Waters Corporation, Beverly, MA, USA; Leanne Nye, Waters Corporation, London, United Kingdom.

Title:
Automatic CCS and MS/MS Library Creation and Application for Large Scale Metabolic Profiling

Introduction:
Metabolomics aims to identify and quantify changes in the chemical fingerprint of cellular processes and is a challenging analytical approach. Adding ion mobility increases peak capacity and selectivity and with the addition of a collision cross-section library together with MS/MS spectra of human metabolites will be shown to be an additional source of reference to aid metabolite identification.

Methods:
ESI and MALDI-MS were used to measure ion drift-times on Synapt G2-Si and ESI on Vion. In both instruments ions are accumulated in the trap T-Wave and periodically released into the T-Wave IM cell where they separate according to their mobility. Upon exiting the IMS cell, ions can be selected with the quadrupole and undergo CID for structural elucidation information prior to detection with ToFMS.

Results:
Using the IROA Mass Spectrometry Metabolite Library of Standards, comprising of, 605 non-redundant small molecules, IM was used to measure CCS values and fragmentation information. CCS measurements were performed in three differently laboratories and a mean CCS of each ionised metabolite provided the basis of a searchable CCS database. Preliminary data suggests the comparative results show excellent correlation (median <2%). A sub-set of the CCS values were verified on a Vion instrument to show transferability across different T-Wave IM instruments. Treated human plasma sample spiked with a number of metabolite standards, were analysed by LC-IMS-MS/MS. The results indicate that the addition of CCS measurements to searchable databases for Progenesis QI and UNIFI based workflows increases the specificity and selectivity of analysis, improving the confidence in metabolite identification.

Novel Aspect:
CCS and MS/MS library creation using novel bioinformatics approaches for metabolite identification.

Hieta, Juha-Pekka,
University of Helsinki

Contributing Authors:
Tiina J. Kauppila, University of Helsinki; Markus Haapala, University of Helsinki; Anu Vaikkinen University of Helsinki; Heikki Räikkönen, University of Helsinki.

Title:
Laser ablation atmospheric pressure photoionization (LAAPPI) for ambient mass spectrometry imaging of low polarity compounds in animal tissue

Introduction:
Laser ablation atmospheric pressure photoionization (LAAPPI) [1] is an ambient MS technique for the sensitive analysis of nonpolar and neutral compounds (<1000 amu) from water-rich samples. Previously, LAAPPI has been used for imaging of phytochemicals from Sage leaves [2]. Individual mass spectra from brain tissue have been collected using LAAPPI, but without imaging.

Methods:
The measurements were carried out using a Bruker micrOTOF mass spectrometer. The tissue was ablated using an OPOTEK IR Opolette 2940 nm laser. A heated nebulizer chip was used to evaporate the sample plume and to introduce dopant vapor. A krypton discharge lamp provided photons for ionization. The sample was cooled using a Peltier element and moved with a Thorlabs stepper motor stage.

Results:
LAAPPI-MS measurements of brain tissue samples yielded high intensities for cholesterol and glycerolipids. Significant improvements to the spatial resolution of the laser ablation were achieved, with even 60 µm laser spot size in tissue. The feasibility of LAAPPI MS imaging was tested using 12 µm thick slices of healthy rat brain and mouse brain gliomas. Compound profiles detected with LAAPPI and LAESI were compared to characterize the suitability ranges of the techniques. References: [1] A. Vaikkinen et al. Infrared Laser Ablation Atmospheric Pressure Photoionization Mass Spectrometry, Anal. Chem., 2012, 84 (3), 1630–1636. [2] A. Vaikkinen et al. Laser ablation atmospheric pressure photoionization mass spectrometry imaging of phytochemicals from sage leaves, Rapid Commun. Mass Spectrom., 2014, 28 (23), 2490–2496.

Novel Aspect:
MS imaging of neutral and nonpolar compounds from animal tissue by LAAPPI.

Hilliard, Mark,
NIBRT

Contributing Authors:
Ying Qing Yu, Waters Corporation; Pauline M Rudd, NIBRT; Sinead Hallinan NIBRT; William R Alley, Waters Corporation.

Title:
Development of N-and O- glycan workflows utilising a novel UPLC-mass detection system for biotherapeutic drug products.

Introduction:
N- and O- linked glycosylation is the most complex post-translational modification and can have major effects on the function of molecules used as biotherapeutics. As a critical quality attribute, glycosylation of biotherapeutic proteins is vital for therapeutic efficacy. Therefore, the ability to correctly monitor glycosylation, in a routine manner, is highly desirable.

Methods:
In collaboration with Waters Corporation we present an N-linked glycan monitoring method that combines an MS-friendly fluorescent label, RapiFluor-MS™ (RFMS), with a QDa, a novel mass detector for coupling with UPLC systems. For O-linked glycans, which still present several analytical challenges we present data for routine procainamide labelled glycan analysis using the QDa with UPLC separation.

Results:
We present a glycan monitoring method that combines an MS-enhancing fluorescent label, RapiFluor-MS™ (RFMS), with QDa detector. Using N-linked glycans derived from Herceptin as a test model, we demonstrate that our standard glycan workflows can be seamlessly integrated with the QDa. This set-up allows us to collect Glucose Unit (GU) values, perform exoglycosidase arrays for structural characterization, and collect mass information with the QDa. To address the current O-glycans challenges, we have developed a UPLC-HILIC-FLR-MS method using chemical release and procainamide labelling via reductive amination. The 1:1 stereospecific fluorescent labelling of the Bovine Fetuin glycans allows for their precise relative quantitation using fluorescence data. Our method thus compliments the QC environment in comparison to standard N, O-linked glycan LC-MS analysis.

Novel Aspect:
Routine N- and O-glycan separation and mass analysis using a small foot print UPLC/FLR/QDA analytical system

Hitzenberger, Jakob,
Friedrich-Alexander University Erlangen-Nuremberg

Contributing Authors:
Claudia Dammann, Friedrich-Alexander University Erlangen-Nuremberg; Nina Lang, Friedrich-Alexander University Erlangen-Nuremberg; Tomás Torres Universidad Autónoma de Madrid; Thomas Drewello, Friedrich-Alexander University Erlangen-Nuremberg; Norbert Jux, Friedrich-Alexander University Erlangen-Nuremberg; Dominik Lungerich, Friedrich-Alexander University Erlangen-Nuremberg; Giovanni Bottari, Universidad Autónoma de Madrid; Miguel García-Iglesias, Universidad Autónoma de Madrid.

Title:
Making the invisible visible: improved electrospray ion formation of metalloporphyrins/-phthalocyanines by attachment of the formate anion

Introduction:
We propose the axial coordination of the formate anion (HCOO–) to different porphyrins and phthalocyanines containing divalent metal centers (Zn, Mn, Mg, Co) as an effective and superior strategy for ion formation in electrospray ionization (ESI). This method is particularly useful when the oxidation of the neutral metallomacrocycle fails.

Methods:
The divalent metallo-porphyrins and –phthalocyanines (M(II)Pors/Pcs) were dissolved in DCM and diluted with DMF to a concentration of 5.0 ×10-5M. After addition of sodium formate (1.0 ×10-4M) and thorough mixing, the resulting solution was introduced to the ESI-MS by direct injection. ESI experiments were conducted with ESI-quadrupole time-of-flight and ESI-quadrupole ion trap mass spectrometers.

Results:
While focusing on Zn-porphyrins and –phthalocyanines, we show that formate is readily attached to a variety of other metals. A common reaction upon CID is CO2 loss and transfer of a hydride ion (H–) to the metal centre rather than electron transfer, as the metalloporphyrin possesses the lower electron affinity. In addition to single porphyrin units, multiporphyrin arrays were successfully analysed. In these, more than one formate anion attaches to the molecule, giving rise to multiply charged species, revealing that one formate anion can bind to two porphyrin units at a time. This latter motif additionally leads to the formation of dimers of the type: Por2HCOO–. CID allows to distinguish between isomeric ions, such as Por-HCOO–-Por vs. Por-Por-HCOO– formed by M(II)Pors/Pcs species bearing a peripheral heteroatom.

Novel Aspect:
Efficient ESI-ion formation of previously inaccessible M(II)Pors/Pcs allows molecular anion formation and elucidation of dimer-binding motifs.

Hoffmann, Waldemar,
Freie Universitaet Berlin

Contributing Authors:
Michael T. Bowers, University of California Santa Barbara; Wieland Schöllkopf, Fritz Haber Institute of the Max Planck Society; Jongcheol Seo Fritz Haber Institute of the Max Planck Society; Stephan Warnke, Fritz Haber Institute of the Max Planck Society; Xing Huang, Fritz Haber Institute of the Max Planck Society; Kevin Pagel, Freie Universitaet Berlin; Gert von Helden, Fritz Haber Institute of the Max Planck Society; Sandy Gewinner, Fritz Haber Institute of the Max Planck Society.

Title:
An Infrared Spectroscopy Approach to Follow beta-Sheet Formation in Peptide Amyloid Assemblies

Introduction:
Amyloidogenic peptides play a crucial role in disorders. The involved peptides undergo a spontaneous transition from a soluble, partially folded form into insoluble, β-sheet rich amyloid fibrils. The underlying assembly is not fully understood to date, but there has been increasing evidence that oligomeric folding intermediates that occur during the transformation process are the toxic species.

Methods:
IM-MS coupled to free electron laser-based IR spectroscopy was used to obtain direct structural information of individual oligomeric species. Absolute CCSs as well as IM-MS and m/z-selected IR spectra were obtained using an in-house constructured drift-tube IM-MS instrument. Condensed-phase FT-IR and TEM measurements further confirmed the characteristic amyloid morphology of the aggregates.

Results:
Our gas-phase data show clear structural differences for individual oligomers formed by three peptides. Tetrameric assemblies and higher oligomers of the fast-amyloid former VEALYL (wild type) show noticeable amide I features at 1625 cm-1, which are characteristic for beta-sheet structures. The IR spectra of VELYAL aggregates, however, indicate turn-like conformations, which confirms the non-amyloidogenic nature of this variant. For the last, slower amyloid-former YVEALL, highly polymorph pentameric assemblies are found. Compact YVEALL pentamers are shown to predominantly adopt an unfolded conformation while a considerably increased beta-sheet content is observed for more extended pentamers. The data represent the first spectroscopic study that successfully identifies specific pre-fibrillar oligomer states involved in the critical transition from unordered to beta-sheet rich structures.

Novel Aspect:
Our data reveal that already early oligomers consisting of four to nine peptide subunits can contain a significant amount of beta-sheet.

Hoffmann, Jan,
Group of Jun.-Prof. Morgner

Contributing Authors:
Heike Angerer, Group of Volker Zickermann.

Title:
Improving the resolution of LILBID mass-spectrometry

Introduction:
Laser-induced-liquid-bead-ion-desorption (LILBID) is an ion source, which has proven to be a versatile tool to analyze noncovalent interactions within biomolecules1. However, the completely home-built first generation LILBID2 instrument lacks the mass resolution and features of modern ones. Coupling the ion source to a modified Q-TOF enables both better resolving power and access to more features.

Methods:
For LILBID-MS droplets of 30µm size are transferred into vacuum and irradiated by IR laser pulses leading to the explosion of the droplet and thus release of analyte ions into gas phase. The ions are collected using an ion funnel and transferred into a quadrupole time-of-flight (Q-TOF) mass spectrometer. Different mechanisms of dissociation (laser power and collision cell) can be combined.

Results:
As a proof of principle experiment the 40 subunit protein complex I from Yarrowia lipolytica was measured on the LILBID Q-ToF and the individual proteins were analyzed. This second generation LILBID allows to distinguish proteins which are very similar in mass. It was even possible to detect post-translational modifications (PTMs). This resolution now allows to approach another interesting question: The possibility to investigate solution phase charge distributions of biological samples via their mass spectra, as the LILBID ionization process does not add additional charges to the solution complexes. Therefore we investigated native and chemically modified lysozyme. We could show that the peak distributions in the mass spectra for the proteins with different degrees of acetylation vary to mimic the changes of the solution charge states.

Novel Aspect:
LILBID has gained the possibility to study noncovalent biomolecules in a top-down manner by combining different mechanisms of dissociation.

Hogan, Chris,
University of Minnesota

Contributing Authors:
Carlos Larriba, IUPUI; Hui Ouyang, University of Minnesota.

Title:
Recent Developments in Gas Molecule Scattering Approaches for Collision Cross Section/Mobility Calculations in Polyatomic Gases

Introduction:
A critical step in applying ion mobility-mass spectrometry (IM-MS) for structural characterization of protein and multiprotein complex ions is comparison of measured mobilities/collision cross sections (CCSs) to predictions for candidate ion structures. Rigorous CCS calculation approaches for a variety of drift gases, including non-monoatomic gases, are therefore needed.

Methods:
CCS calculation algorithms largely depend upon monitoring gas molecule scattering from ion surfaces, as it is momentum transfer from gas molecules to ions which determines ion CCSs. We have developed a coarse-grained CCS calculation approach in which the extent of internal energy exchange from an ion to a gas molecule during collision can be prescribed.

Results:
In gas molecule scattering based CCS calculations, there are two limits for collision models: (1) Specular, elastic collisions, where the gas molecule translational energy is conserved and (2) diffuse, inelastic collisions, in which the gas molecule translational energy is independent of its incident energy. Via comparison of calculations to IM-MS measurements of metal iodide cluster ions and multiply charged ion liquid ions in Helium and molecular Nitrogen, we find that the specular-elastic collision model leads to excellent agreement with measurements when the gas molecule mass is significantly smaller than the mass of the atoms/functional groups within the ion. Conversely, the diffuse/inelastic collision model leads to better agreement when gas molecules are similar in mass or more massive than the atoms within the ion, as is the case for most organic ions with molecular nitrogen.

Novel Aspect:
New, computationally efficient algorithms have been developed for CCS calculation in polyatomic gases.

Holding, Finn,
Astex Pharmaceuticals

Contributing Authors:
Harpreet Saini, Astex Pharmaceuticals; Brent Graham, Astex Pharmaceuticals.

Title:
Comparison of the cellular effects two classes of Erk inhibitors in A375 mutant melanoma cells using a phosphoproteomic readout

Introduction:
Two potent inhibitors of Erk kinase have been described, GDC-0994 and SCH772984. They have differences in their mechanisms of action and it was unclear what the cellular ramifications of these similar, but varying, activities would be. In this study a phosphoproteomic approach was used to analyse the different cellular effects of the two compounds on Erk kinase itself and the kinome as a whole.

Methods:
A375 mutant melanoma cells were treated in triplicate with GDC-0994 and SCH772984 at five times EC50 and harvested after incubation for 2hr and 24hr. Non-drug treated cells (DMSO alone) and a Mek inhibitor were included as controls and were subjected to phosphoproteomic analysis using tryptic digestion, 10plexed TMT tagging and LCMSMS analysis.

Results:
Bioinformatic analysis was stratified into three stages: firstly to see the direct effects of the two drug treatments on the phosphorylation status of Erk itself compared to control samples. Secondly to detect changes in the family of Erk substrate proteins and then to see if there were any significant changes in the non-Erk related phosphoproteome - 'kinome remodelling'. The phosphorylation status of Erk itself showed significant differences between the two drug treatments at two phosphorylation sites and at 2hr and 24hr, corroborating data from traditional methods. Changes were also observed in the family of known Erk substrates and the wider phosphoproteome, and the technology has enabled detailed differentiation at the phosphopeptide level of the biological activity of two closely related inhibitors of the same enzyme.

Novel Aspect:
Differentiation of cellular activity of two potent inhibitors of Erk kinase using a phosphoproteomic readout

Hopkins, Scott,
University of Waterloo

Contributing Authors:

Luke Melo, University of Waterloo; J. Larry Campbell, SCIEX; Moaraj Hasan University of Waterloo; J.C. Yves Le Blanc, SCIEX; Hui Zhang, Pfizer; Mitch Verbuyst, University of Waterloo; Christian Ieritano, University of Waterloo; John Janiszewski, Pfizer; Chang Liu, SCIEX; Jefry Shields, Pfizer; Evan Shepherdson, University of Waterloo.

Title:

Combining Rapid Isomer Separations and Physicochemical Property Measurements with Differential Mobility Spectrometry

Introduction:

In vitro measurements of a candidate drug molecule's cell permeability routinely serves as a key step in a drug's overall development pathway. However, these measurements can be low-resolution, time-consuming, and labor-intensive. Here, we detail a technique using differential mobility spectrometry for determining drug physicochemical properties including passive cell permeability.

Methods:

A DMS system was mounted in the atmospheric region between a mass spectrometer's sampling orifice and its ESI source. The temperature of the transport gas in the DMS cell was maintained at 150 °C and methanol vapour was added to the curtain gas at 1.5 % (v/v). Analytes were dissolved in a 50:50 acetonitrile/water solution to a concentration of 100 ng/mL, and were infused into the ESI source.

Results:

Seven sets of substituted methylquinolin-8-ol derivatves were examined. The electronic resonance effect was found to be the dominant factor that influences DMS behaviour. The order of weakest to strongest ion/solvent interaction was 7-, 5-, and 6-substituted isomers, which could be explained by the delocalization of charge density when the substituent is at 5- or 7- positions, while this equivalent resonance stabilization does not exist for 6- substituted isomers. We also found support through strong correlations between DMS results and the Electrophilic Substituent Constant (σ+), originally developed by Brown and Okamoto. Experimental results also showed strong correlations with ion/solvent binding energies and drug physicochemical properties such as cell permeability and pKa. We propose that this DMS-MS technique can be used as a tool to probe the properties of drug candidates.

Novel Aspect:

Probing novel physicochemical properties of drug candidates by DMS-MS.

Hou, Keyong,
Dalian Institute of Chemical Physics, Chinese Academy of Sciences

Contributing Authors:
Shuang Wang, Key Laboratory of Separation Science for Analytical Chemistry, Dalian Institute of Chemical Physics, Chinese Academy of Sciences; Haiyang Li, Key Laboratory of Separation Science for Analytical Chemistry, Dalian Institute of Chemical Physics, Chinese Academy of Sciences.

Title:
VUV Lamp Based Chemical Ionization Source for On-situ Ion Trap Mass Spectrometry

Introduction:
Mass spectrometry should solve two tough issues for on-situ analysis: the miniaturization and analytical time. Ion trap coupled with discontinuous atmospheric pressure interface realized the miniaturization. Ambient ionization realizes on-line analysis. In this study, a new ambient chemical ionization source based on a vacuum ultraviolet lamp was designed and implemented in an on-situ ion trap.

Methods:
A commercial 10.6 eV VUV krypton (Kr) discharge lamp with dopant gas was used for positive chemical ionization source. Liquid and solid sample was collected on a thin film of PTFE, and a drift gas carry the gaseous samples generated from thermal desorption into the ionization chamber. A dopant gas is added into the ion source and assists the ionization of the analytes.

Results:
Acetone, butanone and water were chosen as the dopant gas. Twenty-seven different illegal drugs and ten pesticides were texted, the ionization source is soft and their characteristic ions are [M+H]+. The analysis time is less than 2 min, and the LODs of the drugs are at pg level for standard samples. The accuracy of the drug identification is improved by using tandem mass spectrometry analysis, and the pattern recognition of drugs was carried out using precursor ion and characteristic fragments. The humidity of gases promoted the detection of some compounds with higher ionization potentials that could not be detected well by single photon ionization, such as the detection of acids and alcohols. The results to confirm the on-situ performance, e.g. the stability and thesensitivity under under on-situ environment, will be presented.

Novel Aspect:
A VUV lamp based chemical ionization source ion trap mass spectrometry was designed and characterized for on-situ analysis of drugs and pesticides.

Hoyes, John,
Waters

Contributing Authors:
Chris Hughes, Waters; Keith Richardson, Waters; Jason Wildgoose Waters; Lee Gethings, Waters; Johannes PC Vissers, Waters; Jim Langridge, Waters.

Title:
Quantitation of Proteomics Samples Using a Novel Scanning Quadrupole DIA Method

Introduction:
Targeted LC-MS assays are increasingly applied in the biomarker area with an emphasis on validation and verification. A novel DIA mode of operation, which provides both qualitative and quantitative information, has been developed for a hybrid quadrupole/oa-time-of-flight (ToF) MS and applied to targeted proteomics experiments using transition extraction lists and compound library based approaches.

Methods:
NanoLC was performed with a Waters Nanoacquity UPLC equipped with a 75micron x 150mm column. The quadrupole was continuously scanned, isolating specific m/z regions prior to the oa-TOF. Data was acquired up to a frequency of 400 spectra/s. Alternate MS scans comprised precursor and CID product ions. The resulting 2D data, m/z (ToF) vs. m/z (quadrupole) were processed using Progenesis QIP and Skyli

Results:
Quantitative proof-of-principle data were acquired by serially diluting a four protein digest mixture into a proteolytic E.coli digest and quantitative information obtained by using transition extraction lists. Analysis of the data indicates that scanning quad DIA enables over an order of magnitude more specificity than a static quad operated with the same resolution and it was found that a quadrupole transmission window of approx 20 - 30 Da provided optimum protein identifications. Aliquots of control, diabetic and obese human plasma samples were digested with trypsin overnight. 2DMS data were collected and differentially expressed peptides and hence proteins across the three conditions were quantified. Results show that peptides from apolipoproteins exhibited the most significant changes in good agreement with expected changes in relation to disease and/or phenotype.

Novel Aspect:
Targeted quantitation in proteomics using a novel precursor quadrupole scanning based DIA method

HU, Jingyu,
NCS Testing Technology Co., Ltd.

Contributing Authors:
Rong QI, NCS Testing Technology Co., Ltd.; Guowu YANG, NCS Testing Technology Co., Ltd.; Mei HAN NCS Testing Technology Co., Ltd.; Yanxia HOU, NCS Testing Technology Co., Ltd.; Jingyu HU, NCS Testing Technology Co., Ltd.

Title:
Determination of Cd and Te in Metal Materials and Geochemical Samples by Dynamic Reaction Cell - Inductively Coupled Plasma Mass Spectrometry

Introduction:
As the most troublesome for analysis of mid-mass range isotopes, molybdenum oxide ions (MoO+ and MoO2+) cover form 108～132 m/z, in which none of the isotopes of Cd and Te is free to the interferences by ICP-MS. But Cd and Te should be strictly controlled down to ug/g level or even lower in metal materials and geochemical samples.

Methods:
In order to get the best resolution and signal-to-noise ratio for DRC, the reaction cell parameters have been optimized in detail. Samples were well dissolved by mixed nitric and hydrochloric acid and drops of hydrofluoric acid at appropriate temperature. Without matrix separation, the measures to eliminate and quantify the interferences of molybdenum oxide ions have been focused.

Results:
The data showed that the optimized collision reaction with He could sharply reduce the molybdenum oxide and dioxide ions. The interferences of MoO2+ to 126Te could be almost eliminated by collision cell technology, while the interferences of MoO+ to 114Cd could be eliminated by collision cell technology together with interference correction equation. The detection limits (3σ) of the established method were 0.8 ng□mL-1 for Te, and 0.3 ng□mL-1 for Cd respectively. The recovery rate of ppm level of Cd and Te in superalloy was about 87% to 116%. Good agreement has been achieved between the results by DRC-ICP-MS to that by AAS and AFS. By verification of superalloy CRMs, the method shows good accuracy and precision for determination of trace Cd and Te in superalloy. The method can also be applied to determination of trace Cd and Te in geochemical and environmental samples.

Novel Aspect:
The interferences of molybdenum oxide/dioxide ions to Cd and Te have been eliminated by collision cell technology together with correction equation.

Hu, Zhongming,
NSERC CREATE York University, University of Toronto

Contributing Authors:
Andrew Emili, 4 Department of Molecular Genetics, University of Toronto, Toronto, Ontario M5S 1A8, Canada; Edward M. Marcotte, Center for Systems and Synthetic Biology, Institute for Cellular and Molecular Biology, University of Texas at Austin, Austin, Texas 78712, USA; Gary D. Bader 4 Department of Molecular Genetics, University of Toronto, Toronto, Ontario M5S 1A8, Canada.

Title:
A platform to discover protein complexes in biology samples

Introduction:
Co-fractionation has been used to map protein complexes in several studies, but the detailed methodology has not been discussed and optimized. We developed a platform by standardizing and optimizing the parameters of both experiments and computations, and demonstrated its performance through applying it to C. elegans. We also developed an interface based plug-in (EPIC) to automate the platform.

Methods:
Co-fractionation experiments were performed on C. elegans lysate. Proteins present in each of the many thousands of fractions collected were identified and quantified by LC-MS/MS. Pair-wise protein elution profile similarities were calculated by distinct correlation metrics. To predict complexes membership, these features were integrated with functional evidences by machine learning algorithm.

Results:
We optimized the parameters used in the computational pipeline and selected the most informative correlation metrics for machine learning classification. To provide a more efficient solution to research community with least mass spectrometry machine time utilized, we provided a guideline of how to perform co-fractionation experiment to reach a satisfied level of prediction. Enrichment method is also used to deepen sampling coverage and improve prediction performance. Clustering algorithm was used to segment PPIs network to predict protein complexes. We also benchmarked our results against annotated co-complex interactions, which revealed good prediction performance at both protein-protein (PPI) interactions and complexes levels. To benefit the whole research community, we built an interface based Cytoscape plug-in, named EPIC, for automating the platform that can be implemented by others

Novel Aspect:
We generated WomMap, the largest protein complexes map for C. elegans, and developed a pug-in, named EPIC, for automating the whole platform.

Hudgens, Jeffrey,
National Institute of Standards and Technology

Contributing Authors:

Elyssia S. Gallagher, Biomolecular Measurement Div., National Institute of Standards and Technology; Ioannis Karageorgos, Biomolecular Measurement Div., National Institute of Standards and Technology; James J. Filliben Statistical Engineering Div., National Institute of Standards and Technology.

Title:

An Interlaboratory Determination of the Reproducibility of Hydrogen-Deuterium Exchange Mass Spectrometry (HDX-MS) Measurements

Introduction:

The term "precision" is represented in three components: Repeatability & Intermediate Measurement Precision, which are characteristics of specific instruments and operators within each laboratory; and Reproducibility, which is the precision observed when many laboratories measure the same object. NIST determined the reproducibility of proteolytic fragmentation HDX-MS via an interlaboratory study.

Methods:

To each lab NIST overnight shipped a kit of 26 vials on ice containing the test protein, NIST-Fab (47.5 kDa); perdeuterated Fab; D2O buffer; quench solution; and TCEP reducing agent. The kits assured uniform properties in all measurement steps up to the proteolysis step of the HDX-MS experiment. Participants returned spreadsheets of the measured peptide centroids for selected D2O immersion times.

Results:

Experienced investigators residing in an international cohort of commercial, university, foundation, and government laboratories, reported 10 datasets containing 61,380 measurements spanning immersion times of t = (30 to 14,400) s and a "fully deuterated" control. The datasets report 312 distinct peptides comprising the heavy and light chains of NIST-Fab; however, these proteomic datasets are very diverse with 40 % of peptide sequences appearing in only one or two datasets. Eight peptide sequences, comprising 3452 measurements, are reported by all laboratories. For these peptides measurement reproducibility ranges between 45% to 15% and a median of 25%. Most measurement discord is assigned to differences in back-exchange and to the diversity of equipment. These results provide a benchmark from which future improvements to HDX-MS methodology and harmonization can be measured.

Novel Aspect:

First interlaboratory comparison that has measured the reproducibility of Hydrogen-deuterium exchange mass spectrometry with a reference material.

Huguet, Romain,
Thermo Fisher Scientific

Contributing Authors:
Tim Stratton, Thermo Fisher Scientific; Vlad Zabrouskov, Thermo Fisher Scientific; Mark Berhow USDA; Jae Schwartz, Thermo Fisher Scientific; Chad Weisbrod, Thermo Fisher Scientific.

Title:
"The ETD-like fragmentation for small molecules"

Introduction:
Flavonoids are a class of secondary metabolites involved in many functions. An increasing number of studies are investigating their biological properties, focusing on the connection between antioxidant activity and chemical structures. Here we investigate their molecular structures using various fragmentation techniques (CID, HCD and UVPD) available on a modified Orbitrap Fusion Tribrid.

Methods:
We performed direct infusion experiments of various flavones and conjugated flavonoids with a modified Orbitrap Fusion Lumos. We performed CID, HCD and UVPD fragmentation experiments on these molecules for structural investigation. The data acquired from the different fragmentation techniques were used to reveal the molecular structures of each flavonoid using mzCloud and Xcalibur Qual Browser.

Results:
Common fragment ions using all three techniques were observed (331.1000, 617.1477, 471.0903, 308.0291 and 185.0420) but the UVPD option also provided unique fragmentation channels in both low and high mass range: 121.0283, 149.0231 and 763.204. These results present UVPD technique as a potential tool in the structural elucidation of conjugated flavonoid structures especially as CID and HCD fragmentation techniques are severely limited for flavonoid aglycones, O-glycosides, C-glycosides and acylated glycosides characterization. The diversity of chemical space in conjugated flavanoids is extremely large with the total number of combinations of conjugate sugars on different cores and their linking positions being unknown. UVPD surely offers the ability to gain unique fragmentation information which could lead to the position of conjugation that could be the key for identification.

Novel Aspect:
UVPD is an unmatched fragmentation process required for complete and accurate molecular structural elucidation of flavonoids and conjugated flavonoids

Huguet, Romain,
Thermo Fisher Scientific

Contributing Authors:
Michael Blank, Thermo Fisher Scientific; Shannon Eliuk, Thermo Fisher Scientific; Vlad Zabrouskov Thermo Fisher Scientific; Graeme McAlister, Thermo Fisher Scientific.

Title:
A simplified approach to fast and accurate, high throughput targeted MS2 quantitation using internal standards.

Introduction:
Quantifying peptides is often performed using selected reaction or parallel reaction monitoring. These techniques are more sensitive and reproducible than standard data-dependent analysis; yet, the breadth of these approaches is limited and methods are hard to develop and maintain. We present a novel and simplified approach to fast and accurate targeted MS2 quantitation using internal standards.

Methods:
All experiments were based around hybrid peptide samples, consisting of mixtures of unlabeled endogenous peptides combined with heavy internal standards. These samples were analyzed with an Orbitrap Fusion Lumos MS coupled to an EASY-nLC 1000 ultra-high pressure LC. The resulting LC-MS/MS data were searched using Proteome Discoverer 2.1 (1% FDR) and analyzed with Skyline.

Results:
Accurate quantification of high number of targets in a single analysis is obtained by reaching a balance between scan rate, sensitivity, spectral quality and scheduling. Generating an optimal method can be extremely time-consuming owing to intrinsic dependencies on matrix conditions, LC performance, and the need to balance the aforementioned properties. Herein we propose a very simple approach, which leverages the unique chemical properties of the heavy peptides and the high-resolution and accuracy of the Orbitrap analyzer to perform on-line¬ identification of the spiked-in standards without the need of a spectral library. We have compared the utility of this novel quantitative PRM-based method to traditional PRM analyses. We found that sensitivity and breadth of this novel PRM method was comparable to traditional approaches; yet, the method itself was much easier to setup and maintain.

Novel Aspect:
Development of a simplified approach to fast and accurate, high-throughput targeted MS2 quantitation using internal standards.

Huppertz, Laura Maria,
Institute of Forensic Medicine, Medical Center – University of Freiburg, Faculty of Medicine, University of Freiburg, Germany

Contributing Authors:
Ronja Peter, Institute of Forensic Medicine, Medical Center – University of Freiburg, Faculty of Medicine, University of Freiburg, Germany; Aurore Wicht, Institute of Forensic Medicine, Medical Center – University of Freiburg, Faculty of Medicine, University of Freiburg, Germany; Jürgen Kempf Institute of Forensic Medicine, Medical Center – University of Freiburg, Faculty of Medicine, University of Freiburg, Germany; Andrea Klehne, Bruker Daltonik GmbH, Bremen, Germany; Bjoern Moosmann, Institute of Forensic Medicine, Medical Center – University of Freiburg, Faculty of Medicine, University of Freiburg, Germany.

Title:
Detection and semi-quantitative determination of designer benzodiazepines in serum using LC-MSn

Introduction:
Benzodiazepines (BZP) play an important role in forensic toxicology as they are widely used for treatment of psychiatric disorders and as drugs of abuse. In 2012, the first designer BZPs were offered on the Internet providing an attractive alternative to prescription-only BZP. The presented method is an easy-to-use, semi-quantitative screening method for BZPs including newly emerged designer BZP.

Methods:
Chromatographic separation was achieved after liquid-liquid extraction on an Acclaim C18 column using formic acid/acetonitrile. The MS was operated in positive ESI generating data dependent MS2 and MS3 spectra. Compound identification was based on an in-house generated spectral library. Linearity of the method was evaluated from 5 to 500 ng/ml. A one-point calibration was used for quantitation.

Results:
The current spectra library contains 53 BZP including 11 designer BZP. The method can easily be extended once new designer BZP emerge on the drug market or according to specific needs of the user. The limit of detection was 5 ng/ml for the majority of the analytes. Nifoxipam - known for a poor ESI response and highly instable in serum - was the only compound not detectable serum. Concentrations within the linear range are reported as semi-quantitative result in the automatically generated report. Findings below or above this range are reported as '< cLow' or '> cHigh'. Semi-quantitative results were found to vary between ± 20 and ± 40 % in the lower calibration range and ± 10 to ± 25 % at medium and high concentrations. The assay allows identification of BZP even at sub-therapeutic concentrations for the majority of medical BZP, making the screening applicable to forensic analysis.

Novel Aspect:
The method allows automated semi-quantitative screening for BZP and designer BZP, and in contrast to IA unambiguous compound identification.

Huppertz, Laura Maria,
Institute of Forensic Medicine, Medical Center – University of Freiburg, Faculty of Medicine, University of Freiburg, Germany

Contributing Authors:
Volker Auwärter, Institute of Forensic Medicine, Forensic Toxicology, Medical Center – University of Freiburg, Faculty of Medicine, University of Freiburg, Germany; Sebastian Götz, Bruker Daltonik GmbH, Bremen, Germany; Florian Franz Institute of Forensic Medicine, Forensic Toxicology, Medical Center – University of Freiburg, Faculty of Medicine, University of Freiburg, Germany; Bjoern Moosmann, Institute of Forensic Medicine, Forensic Toxicology, Medical Center – University of Freiburg, Faculty of Medicine, University of Freiburg, Germany.

Title:
Rapid Method Updating for the Detection of Synthetic Cannabinoids in Human Urine by LC-QToF-MS using Software Assisted Metabolite Identification

Introduction:
New synthetic cannabinoids (SC) appear on the markets challenging toxicologists. For urine analysis metabolite identification is inevitable since most compounds are metabolized extensively. With software data-mining of pooled human liver microsome (pHLM) incubations the major in vitro metabolites of new SC are identified and used, along with their MS/MS info, to screen for SC uptake in urine.

Methods:
pHLM incubation was according to an in-house protocol. The MS was operated in positive ESI with data-dependent MSMS fragmentation. Datasets of the incubations were analyzed with MassMetaSite SW. Precursor and fragment information of the metabolites were added to a screening method. Authentic urine samples were analyzed using data independent broad-band-CID mode.

Results:
The highly potent and prevalent synthetic cannabinoid MDMB-CHMICA was chosen as amodel compound. Analysis of the pHLM incubations of MDMB-CHMICA with MassMetaSite revealed 10 metabolites with at least two fragment masses each. By manual data analysis 15 potential metabolites were identified. However, all main in vitro phase I metabolites were detected with the software. The metabolite and fragment information was used to generate a screening method with Bruker TASQ software. Additionally, the retention times were adapted and added to the database. Authentic forensic case samples were screened after enzymatic cleavage of glucuronides and processed with the updated TASQ method. Despite varying relative abundances of the detected metabolites, the in vitro and in vivo data showed good agreement with respect to the chosen MDMB-CHMICA metabolites.

Novel Aspect:
This workflow allows the rapid identification of metabolites and use of this information to update screening methods in routine forensic toxicology.

Hurst, Jerod,
University of Buffalo

Contributing Authors:
Mary A. Bisson, University of Buffalo, SUNY Department of Biological Sciences; Kaitlynn Donahue, University of Buffalo, SUNY Department of Biological Sciences; Diana S. Aga University of Buffalo, SUNY Department of Chemistry; Steven J. Ray, University of Buffalo, SUNY Department of Chemistry.

Title:
Phytochelatin Production and Sequestration of Heavy Metals as a Detoxification Pathway in Chara australis for use in Phytoremediation

Introduction:
Phytoremediation allows for a safe and environmentally friendly method for isolating and removing heavy metal contamination in soil and surface waters. Chara australis, a species of algae , has previously been shown to have redeeming qualities for use in phytoremediation techniques such as high biomass and hyperaccumulation of heavy metals. Phytochelatins are studied as a detoxification pathway.

Methods:
Advancements for developing analytical methods to quantify both characteristics will be reported utilizing liquid chromatography tandem- triple quadrupole mass spectrometry (LC-QqQ) and liquid chromatography-inductively coupled plasma mass spectrometry (LC-ICP/MS). Heavy metal treatment of these Chara algae were performed using cadmium and zinc at various time points.

Results:
Extraction recoveries for each phytochelatin studied reveal percent recoveries ranging from 81%-103%. Challenges in molecule stability for the quantification of phytochelatins were overcome by reducing the thiol bonds to prevent disulfide linkages for increased sample stability. Quantitation by matrix-matched calibration curves were performed leading to preliminary data showing glutathione levels in Chara at a concentration of 5 nmol/g and PC2-4 were present at concentrations near 1 nmol/g. Early advancement for studying metal-bound phytochelatins are reported.

Novel Aspect:
Detection of phytochelatins in their free and metal-bound state are studied using complimentary analytical mass spectrometry methods.

Husson, Steven,
University of Antwerp

Contributing Authors:
Kristien Van Camp, University of Antwerp; Geert Baggerman, University of Antwerp - VITO; Ronny Blust University of Antwerp.

Title:
Identification of endogenous neuropeptides of the zebrafish

Introduction:
(Neuro)peptides are small messenger molecules that are derived from larger, inactive precursor proteins by the action of processing enzymes. These biologically active peptides can be found in all metazoan species where they orchestrate a wide variety of physiological processes. Obviously, detailed knowledge on the actual peptide sequences, is of high importance when studying respective functions.

Methods:
A peptidomics approach therefore aims to identify and characterize the endogenously present peptide complement of a defined tissue or organism using liquid chromatography and mass spectrometry. Doing so, we set out to biochemically characterize endogenously present (neuro)peptides from the zebrafish brain.

Results:
The entire brain region of 7 adult zebrafishes were carefully dissected and (neuro)peptides were extracted using a specific extraction solvent that is extremely efficient in avoiding the presence of protein degradation products. The peptide samples were analyzed using a nanoLC instrument that is directly coupled with an LTQ-Orbitrap mass spectrometer to yield biochemical identifications of > 50 peptides that belong to 25 different precursor proteins. In addition, a remarkable amount of shortened forms (aminoterminally or carboxyterminally truncated) could be identified, yielding a total number of ~100 peptide identifications.

Novel Aspect:
Our peptidomics results might pave the way for functional studies and further aid ongoing genome annotations of an important model species.

Hyötyläinen, Tuulia,
University of Örebro

Contributing Authors:
Ismo Mattila, Steno Diabetes Center, Denmark; Jussi Posti, Department of Neurology, University of Turku, Finland; David K. Menon Division of Neurosurgery, Department of Clinical Neurosciences, University of Cambridge, Cambridge UK; Olli Tenovuo, Department of Neurology, University of Turku, Finland; Matej Oresic, Turku Centre for Biotechnology, University of Turku and Åbo Akademi University, FInland.

Title:
Mass spectrometric analyses in the identification of blood-based biomarkers of traumatic brain injury

Introduction:
Traumatic brain injury (TBI) is a major cause of death and disability worldwide, especially in children and young adults. TBI is an example of a medical condition where there are still major lacks in diagnostics and outcome prediction. Here we apply comprehensive metabolic profiling of serum samples from TBI patients and controls in two independent cohorts.

Methods:
Comprehensive two-dimensional gas chromatography combined with time-of-flight mass spectrometry (GCxGC-TOFMS) was applied for the analysis of serum samples after protein precipitation and two-step derivatization. Other mass spectrometric methodologies utilizing high resolution MS was utilized in the identification of candidate biomarkers.

Results:
A total of 465 metabolites were included in the statistical analysis. Among those were amino acids, sugar derivatives, hydroxyl acids, fatty acids as well as sterols and related metabolites. Almost 100 metabolites showed significant differences between the control patients and the three different types of TBI patients (mild, moderate and severe injury). Large fraction of the candidate biomarkers were unknowns. Thus, additional analyses using GCxGC-HRMS with both hard and soft ionization were done to identify the unknowns. In addition, UHPLC-QTOFMS was utilized for further verification of the identification. Based on metabolite profiles from TBI patients at the time of hospitalization, a model that accurately predicted the patient outcomes was developed. Addition of the metabolites to the established clinical model significantly improved prediction of patient outcomes.

Novel Aspect:
Our findings also suggest that the blood-based metabolic biomarkers may exhibit diagnostic potential in the full spectrum of TBI.

Iakubov, Adel,
Chemistry Department, Irkutsk State University

Title:
Observation of fractional charges in the mass spectroscopic experiments

Introduction:
In previous publications dealing with experimental mass spectrometry of tungsten hexacarbonyl, hexafluoroacetylacetone and its bidentate metal complexes M(hfac)2; M = Cu, Pd the obtained data have been not adequately systematized. In this paper, we analyse the previously published experimental data of the various bond dissociation energy.

Methods:
A modified Yukawa potential and a Lewis electron pair as the interaction transfer quantum, which is the exact solution of the problem dependence the chemical bond energy of its length, is used to analyse the experimental data.

Results:
Experimental results of the formation of ions can be interpreted only in terms of the formation of fractionally charged quasi-particles. The dissociative photoionisation processes obtained a series of bond W-CO dissociation energy of 4.03 eV, 3.46 eV, 2.13 eV, 1.31 eV, and 1.07 eV. The dissociative electron attachment process obtained a series of bond W-CO dissociation energy of 4.65 eV, 3.15 eV, 2.15 eV, 1.55 eV, 1.05 eV. This is a very good match with the discrepancies caused by the difference in the vibrational levels. If the bond dissociation energy values are converted to per-unit values by taking 4.65 eV as one then we see that the series of relative units corresponds to a series formed by a combination of two of the three numbers 1, 2/3, and 1/3.

Novel Aspect:
As mass spectrometry next to the fractional quantum Hall effect in which fractional values of the charge quantization are observed.

Ickert, Stefanie,
Federal Institute for Materials Research and Testing

Contributing Authors:

Jens Riedel, Bundesanstalt für Materialforschung und –pruefung, Richard-Willstaetter-Str. 11, 12489 Berlin; Sebastian Beck, Humboldt-Universitaet zu Berlin, Department of Chemistry, Brook-Taylor-Str. 2, 12489 Berlin; Kevin Pagel Fritz-Haber-Institut der Max-Planck-Gesellschaft, Faradayweg 4-6, 14195 Berlin; Michael W. Linscheid, Humboldt-Universitaet zu Berlin, Department of Chemistry, Brook-Taylor-Str. 2, 12489 Berlin; Johanna Hofmann, Fritz-Haber-Institut der Max-Planck-Gesellschaft, Faradayweg 4-6, 14195 Berlin

Title:

DNA Oligonucleotide Fragmentation in Tandem Mass Spectrometry Experiments

Introduction:

The exact structure imprinted by noncovalent bonds is of great importance to their biofunctionality albeit cannot be addressed directly by MS. In oligonucleotides a major intramolecular force are coulomb interactions between the charged ends in zwitterionic structures. The overall energy minimization of these structures should reflect in the stability of nucleotides in different charge states.

Methods:

A series of multiply negatively charged DNA oligonucleotides (n=15-40) has been subjected to a comprehensive collision induced dissociation (CID) mass spectrometry (MS) study to determine the energetic fragmentation threshold in dependence of the charge per base densities. The accompanying structural changes were studied in complementary ion mobility studies.

Results:

Incrementally increasing the CID energy resulted in sharp, well defined, energetic onsets above which the negatively charged oligonucleotides fragmented. Surprisingly the overall stability towards collision induced fragmentation (the exact energy of the fragmentation onset) was found to be independent of the sequence but to linearly scale with the charge-per-base (cpb) density. Along this linear increase in the parent ions stability towards lower charge density, a plateau around 0.45 cpbs could be observed. Ion mobility spectrometric determination of the collisional cross sections of the precursor ions corroborated that this plateau coincides with a transition from an elongated structure to a dense packing probably also preferred in native solution conditions. The findings suggest the overall geometrical stabilization being driven by charge distribution in zwitterionic molecules.

Novel Aspect:

This is the first report on nucleotide ions CID stability in correlation to their structure. Linearity along the cpb ratio suggests a general trend.

Ifa, Demian,
York University

Contributing Authors:
Chun Peng, York University; Alessandra Tata, York University; Michel Campos York University; Consuelo Perez, York University.

Title:
DESI-MS Imaging of AMMOENG 130 in Zebrafish.

Introduction:
Ambient mass spectrometry imaging (MSI) has become a powerful technique for the direct analysis of biological tissues in the open environment with minimal or no sample preparation. We introduce DESI-MSI as a novel, rapid and sensitive approach to localize the accumulation of a mildly toxic ionic liquid (IL), AMMOENG 130 in zebrafish (Danio rerio).

Methods:
Zebrafish were placed in 2 L aquaria and exposed to 1.25, 2.50 and 5.0 mg/L of AMMOENG 130 by adding the IL into the tank water. After 96h exposure, the surviving zebrafish were euthanized with tricaine mesylate. The fish were cryosectioned and the tissue sections thaw-mounted onto glass slides. The distributions of AMMOENG 130 and its metabolite in the tissues were mapped by DESI-MSI.

Results:
DESI-MS analysis of zebrafish gills demonstrated the appearance of a dealkylated AMMOENG 130 metabolite in the lowest concentration of exposure identified by high resolution mass spectrometry as the trimethylstearylammonium ion. The accumulation of AMMOENG 130 and its dealkylated metabolite in zebrafish tissue was found in the nervous and respiratory systems. AMMOENG 130 and the metabolite were capable of penetrating the blood brain barrier of the fish with significant accumulation in the brain. This ambient mass spectrometry imaging technique shows great promise for the direct analysis of biological tissues to monitor foreign, toxic and persistent compounds in aquatic organisms from the environment.

Novel Aspect:
The simultaneous characterization, distribution and metabolism of a toxic IL in whole body fish is reported by MSI for the first time.

Ihlenborg, Marvin,
Christian-Albrechts-Universität zu Kiel

Contributing Authors:
Jürgen Grotemeyer, Christian-Albrechts-Universität zu Kiel; Ann-Katrin Schuster, Georg-August-Universität Göttingen.

Title:
Peak Broadening Observations in APLI-IMS: Detailed Studies and Explanation by three Effects

Introduction:
In Ion Mobility Spectrometry only some remarks for the use of Atmospheric Pressure Laser Ionisation can be found. By using this method an unexpected peak broadening effect can be observed in comparison to the well-studied Atmospheric Pressure Chemical Ionisation when all other experimental parameters are kept constant. Herein we are able to explain this observation by three different effects.

Methods:
A commercial available IMS5000 from Dräger is equipped with a 3H-source for APCI and further modified to enable the use of a APLI technique using 266 nm of a Nd:YAG laser. Thus all measurements can be made by similar experimental conditions. The obtained spectra using APCI and APLI can be compared directly in drift time, FWHM and intensity.

Results:
Three different explanations can be given for the obtained peak broadening when APLI is used instead of APCI. First, the spectra reveal a wide range of different unstable ionic species with similar cross sections and thus no separation in the IMS can be obtained. With evaluated drift gas flows only one stable species can be detected. Secondly, with rising intensities of the laser a power broadening (spectroscopic term) can be obtained. At least one more stable ionic species can be observed to later drift times. Furthermore, the detected amount of ions does not change, thus the second stable species should be related to a cluster of several ionic fragments. Third, by delaying the entrance of the obtained ions into the drift region an ion-ion repulsion directly after the ionisation event can be proven. The resulting ion cloud rises up to the fourth fold of the used laser in diameter.

Novel Aspect:
Peak broadening in APLI-IMS can be explained by three effects: I) several unstable ionic species, II) power broadening and III) ion-ion repulsion.

Ingendoh, Arnd,
Bruker Daltonics

Contributing Authors:
Fernando Cava, Bruker Daltonics; Nazareth del Amo, Bruker Daltonics; Miguel Ángel Pérez Bruker Daltonics; Javier López, Bruker Daltonics.

Title:
Determination of Methotrexate (MTX) and its main metabolites by LC-UHR-QTOF in human serum of leukemia patients with high-dose administration

Introduction:
MTX treatment requires close drug level monitoring since renal failure as a side effect may result in toxic accumulation. Glucarpidase can reduce MTX levels in patients with renal failure by cleavage into non-active metabolites. But cross-reactivity of those metabolites in immunoassays causes overestimation of MTX. Here, we determinate MTX and its main metabolites in serum samples by LC-QTOF MS.

Methods:
Calibration standards, QC and serum samples were deproteinized with MeOH and centrifuged. 10 uL of supernatants were analyzed by LC-QTOF MS. MS analysis was done in FullScan and simultaneous broadband CID to verify MTX and is metabolites by product ions. Calibration curves and quantitation of MTX and its main metabolite 2,4-diamino-N10-methylpteroic acid (DAMPA) and other metabolites were done.

Results:
Acute case of a 47-years old woman diagnosed of acute lymphoblastic leukemia type B. High-dose of MTX during 24h was well tolerated. After the second MTX administration, serum creatinine levels increased dramatically indicating a renal failure. MTX clearance decreased, leading to a toxic concentration 36h after drug dose. Glucarpidase therapy was applied. A large discrepancy between the immunoassay and LC-QTOF results was observed, mainly during the first hours after glucarpidase administration. The DAMPA generation as main MTX metabolite caused cross-reactivity in the immunoassay and thus a huge overestimation of the MTX level. LC-QTOF MS analysis revealed this cause of overestimation and enabled as well to monitor the MTX and DAMPA elimination over some days after glucarpidase administration. It allows a reliable and fast quantitation of MTX and DAMPA within a few hours.

Novel Aspect:
Novel Aspect (150 characters): Accurate Methotrexate quantification by LC/MS/MS in human serum avoiding DAMPA metabolite interference.

Ishihama, Yasushi,
Kyoto University

Title:
Slicing Proteomics to Unveil the Human Proteome

Introduction:
Together with modern mass spectrometers with high resolving power and high mass accuracy, ultrahigh resolution LC based on meter-long monolithic silica capillary columns have been employed to expand the separation space for shotgun proteome samples. In this study, we combined this approach with a new strategy called slicing proteomics to increase the proteome coverage.

Methods:
Proteins were extracted and digested into peptides from Hela cell lysates with lysyl endopeptidase and trypsin. After StageTip desalting, peptides were further sliced by several methods such as HILIC separation using meter-long monolithic silica nanoLC columns for subsequent nanoLC-MS/MS analysis using meter-long C18 monolithic silica columns.

Results:
We developed new monolithic silica stationary phases for HILIC mode separation in a long column format. Using tryptic peptides from human HeLa cell lysate proteins, we obtained comparable numbers of identified peptides and proteins in HILIC with those in RPLC using a C18-modified monolithic silica column when shallow gradients were applied. Since HILIC mode LC-MS shows orthogonal selectivity to RPLC mode LC-MS, we introduced 'slice-shot proteomics', where only a single representative fraction in the first dimension separation is analyzed by the second dimension LC. Using HILIC-C18-MSMS, we successfully identified more proteins in slice-shot mode than one-shot mode due to the reduction of the sample complexity. We also introduced other slicing modes for protein N-terminal peptides and phosphopeptides with particular phosphorylation motif to increase the dynamic range of MS measurement.

Novel Aspect:
One-shot proteomics using ultrahigh resolution LC is combined with slicing proteomics approaches for analyzing sliced fractions at peptide level.

Islam, Sohidul,
Trent University

Contributing Authors:
Dr. Celine Gueguen, Trent University.

Title:
A Multi-Method Assessment of Photoirradiated Dissolved Organic Matter

Introduction:
Dissolved Organic Matter (DOM) largely influences the global carbon cycle. As one of the largest inland sea in the world, Hudson Bay receives large amount of DOM from more than 42 rivers. Previous studies showed an important impact of transported DOM from the surrounding river run off into marine ecosystem. But, little is known about DOM mineralization process at the fresh-marine interface.

Methods:
Photo irradiated DOM changes were assessed in fresh and marine water samples collected in the Churchill River estuary (Hudson Bay, Canada). The DOM molecular composition were analyzed using a 7.0 T FT-ICR-MS. Molecular weights and optical properties were determined using asymmetrical flow filed-flow fractionation (AF4), and UV/vis absorbance and excitation emission matrix fluorescence.

Results:
Our initial results of FT-ICR-MS analysis showed that the solar exposure induced significant molecular changes. The fluorescence and absorbance measurements revealed that the photodegradation rate is higher within the first few hours of solar irradiation. AF4 analysis illustrated that the higher molecular weight compounds shifted to lower molecular compounds with light exposure, congruent with the loss of large molecular structures using FT ICR-MS. This research work highlights that the multi analysis techniques can be used to identify photolabile, photoproduced and photoresistant compounds

Novel Aspect:
The molecular level characterization of photoirradiated DOM will be related to molecular weight and optical properties in Hudson Bay.

Jäger, Patrick,
Karlsruhe Institute of Technology

Contributing Authors:
Karin Fink, Institute of Nanotechnology, Karlsruhe Institute of Technology (KIT); Manfred M. Kappes, Insitute of Physical Chemistry and Institute of Nanotechnology, Karlsruhe Institute of Technology (KIT); Markus K. Armbruster Institute of Physical Chemistry, Karlsruhe Institute of Technology (KIT); Ulrike Schwarz, Institute of Physical Chemistry, Karlsruhe Institute of Technology (KIT); Patrick Weis, Institute of Physical Chemistry, Karlsruhe Institute of Technology (KIT); Miriam Himmelsbach, Institute of Physical Chemistry, Karlsruhe Institute of Technology (KIT); Katrina Brendle, Institute of Physical Chemistry, Karlsruhe Institute of Technology (KIT).

Title:
Gas phase Q and Soret Band Photoexcitation of Isolated Palladium Porphyrin Tetraanions

Introduction:
Pd(II) meso-tetra(4-sulfonatophenyl)porphyrin (PdTPPS) has attracted considerable interest due to its long triplet state lifetime and its high singlet oxygen sensitization efficiency for photodynamic therapy. In this respect, gas phase photodissociation spectroscopy was performed to get detailed insight into the electronic and optical properties of these kinds of systems.

Methods:
Action spectroscopy, combining photodissociation and photodetachment channels, was performed on a Bruker APEX II 7T FT-ICR mass spectrometer. The high mass resolution of this spectrometer, in combination with a ns pulsed OPO laser system allows spectroscopy of the metalloporhpyrins in defined charge and oxidation state of the metal center in absence of perturbations by the solvent environment.

Results:
We have used this experimental set up to study the response of isolated [Pd(II)TPPS]4- to electronic excitation in the spectral range of 410-700 nm. Electron emission, the primary decay channel observed after photoexcitation, was further investigated by photoelectron spectroscopy to gain a more detailed understanding of the corresponding detachment processes. Upon both Q and Soret band excitation, we find that electrons are emitted in three characteristic kinetic energy ranges. The corresponding detachment processes are identified as (delayed) tunneling emission from both excited singlet and triplet states (each of which is accessed by/after one photon absorption) as well as resonant two-photon detachment. The first triplet state lifetime of isolated [PdTPPS]4- is significantly longer than 10 μs, possibly on the 100 μs time scale.

Novel Aspect:
Our results yield new insight into the energy relaxation mechanism following electronic excitation of [Pd(II)TPPS]4-.

Jahn, Olaf,
Proteomics Group, Max-Planck-Institute of Experimental Medicine

Contributing Authors:
Hauke Werner, Dept. of Neurogenetics, MPI Experimental Medicine, Göttingen, Germany; Dörte Hesse, Proteomics Group, MPI Experimental Medicine, Göttingen, Germany; Stefan Tenzer Institute for Immunology, University Medical Center Mainz, Mainz, Germany.

Title:
Automated FASP digestion for quantitative myelin proteomics

Introduction:
Myelin, a plasma membrane specialization of glial cells, facilitates rapid nerve signaling and thus normal motor, sensory and cognitive functions. Proteomic analysis of myelin is challenging because of its high content of lipids and membrane proteins, and its unfavorable dynamic range. Thus, technical advances are required to understand how axon-glial interactions fail in neurological disorders.

Methods:
To improve the accessibility of myelin proteins and to minimize digestion variability, we developed a FASP-based in-solution digestion procedure and implemented it into a robotic liquid handling workstation (Tecan Freedom EVO 150). For label-free protein quantification, data were acquired on a QTOF LC-MS-system (Waters Synapt G2-S) and analyzed with the IsoQuant software (www.isoquant.net).

Results:
We used label-free protein quantification to compare CNS myelin from wildtype mice with myelin from mutant mice heterozygous for the CNP gene. In 3 independent experiments, each comprising 24 LC-MS runs (3 biological replicates per condition, duplicate digestion and injection), we consistently quantified ~930 proteins per experiment, resulting in a total of 1064 proteins of which 75% were present in all 3 experiments. The vast majority (~80%) of proteins reported to be associated with mouse CNS myelin (see compendium in Cell Mol Life Sci 2012, 69:2879-94) was readily covered. The CNP ratio between the two conditions, expected to be 2:1 due to heterozygosity, was exactly matched over all experiments (1.98 +/- 0.11). In conclusion, we have developed a robust workflow for accurate quantitative comparison of myelin proteomes, which will deepen our insights into myelin biology and pathology.

Novel Aspect:
development and automation of a special FASP procedure; improved label-free quantification of myelin proteomes

James, Andrew,
Sanofi Pasteur

Contributing Authors:
Manon Fradin, Sanofi Pateur Limited; Ben Messham, Sanofi Pateur Limited; Bruce Carpick Sanofi Pateur Limited.

Title:
Quantitation of residual proteins in an HSV-2 vaccine candidate using LC-MRM

Introduction:
This work focuses on the development of an MRM (multiple reaction monitoring) LC-MS method for quantitation of two HSV proteins: UL5 and UL29. The proteins were detected as residuals in an HSV-2 replication deficient, live-attenuated vaccine candidate, HSV529. The UL5 and UL29 proteins are expressed by a complementary cell line, AV529-19, used to produce the HSV529 candidate.

Methods:
After detection and characterization of UL5 and UL29 in HSV529 formulations by nano LC-MS, a novel MRM assay was developed using a high flow UPLC, on a Waters TQ-S system. An MRM method was developed for quantitation of the HSV transgene products using synthetic AQUA peptides derived from the UL5 and UL29 protein sequences.

Results:
The selection criteria for AQUA quantitation standards were optimized by analysis of the synthetic peptides using LC-MRM and subsequent confirmation of the complementary native sequences, detected in digested sample preparations. The MRM method for UL5 and UL29 had an LOD under 100 attomole, and LLOQ of approximately 200 attomole (on-column), for both proteins. Our results demonstrate the unique sensitivity of MRM methods for quantitation of peptides with high specificity for the residual UL5 and UL29 proteins, in a complex sample mixture. This method was designed around the selection of optimized synthetic standards for trace protein quantitation. Method development experiments were designed to establish method performance criteria, as well as to facilitate future assay qualification and validation.

Novel Aspect:
Trace quantitation of protein residuals in a complex viral vaccine candidate using peptide MRM analysis.

Jami Alahmadi, Yasaman,
Memorial University of NL

Contributing Authors:
Travis D. Fridgen, Supervisor, Memorial University.

Title:
Structure, thermochemistry, and reactivity studies of gas-phase alkali metal-coordinated monomer and trimers of ProLeu vs LeuPro dipeptide sequences

Introduction:
Protein and peptides play key role in regulating cellular activities. Nevertheless, misfolding of proteins may cause aggregation which is associated with disorders such as, neurodegeneration, and even death of cells. Here, mass spectrometric techniques are used to identify sequences and metal binding sites of ProLeu/LeuPro mono– and tri–dipeptides to help in understanding the behavior of proteins

Methods:
Experiments were done using a Bruker Apex Qe 7T FTICR. IRMPD spectra were recorded from 2700-3800 cm-1 at Memorial University and in the fingerprint region using the CLIO FEL. Structures were optimized and IR spectra were computed using B3LYPD3/6-31+G(d,p). The activation energies for loss of peptide from M(peptide)3+ complexes were determined using BIRD kinetics in the 293-390 K temperature range

Results:
IRMPD in the 2700–3800 cm−1 region have bands at 3580 corresponding to the O–H stretch of the free carboxylic acid group and a band at 3310 in M(ProLeu)+, when M = Rb and Cs, corresponding to hydrogen bonded amide stretches and a broad peak at 2990 corresponds with the C–H stretching. Interestingly, in all M(LeuPro)+ asymmetric NH2 stretches show a small band at 3300 that barely seen which is consistent with the calculated spectra. In the fingerprint region, bands between 1630 and 1750 cm-1 are assigned to the C=O stretching of two different carbonyl groups. Also set of pronounced features at 1170 correspond to COH bending and CH2 rocking. The spectrum between 1230-1550 correspond to C-COO stretching and HNC bending. BIRD results reveal the different structural isomers of the Na(ProLeu)3+ vs Na(LeuPro)3+complexes which proves that ProLeu isomer binds sodium ion stronger than the LeuPro

Novel Aspect:
Energetic/structure of alkali metal coordinated isomeric dipeptide; electrostatic interaction; influence of sequence on peptide-metal binding strength

Jamin, Emilien,
MetaboHUB-MetaToul-AXIOM

Contributing Authors:
Souria Tadrist, Toxalim, Toulouse University, INRA, INP, Toulouse, France; Pascal Bourdaudhui, Toxalim, Toulouse University, INRA, INP, Toulouse, France; Isabelle P. Oswald Toxalim, Toulouse University, INRA, INP, Toulouse, France; Laurent Debrauwer, Platform MetaboHUB-MetaToul-AXIOM, INRA, Toulouse, France; Thaïs Hautbergue, Toxalim, Toulouse University, INRA, INP, Toulouse, France; Cécile Canlet, Platform MetaboHUB-MetaToul-AXIOM, INRA, Toulouse, France; Olivier Puel, Toxalim, Toulouse University, INRA, INP, Toulouse, France.

Title:
Characterization of new secondary metabolites of Penicillium using isotopic labeling and HPLC-HRMS

Introduction:
Characterization of fungal secondary metabolomes has become a great challenge in the last decades. In view of this, we developed an analytical strategy for fungal secondary metabolomes characterization. This strategy has been applied to Penicillium verrucosum, a fungal contaminant of stored cereals, with the aim of identifying new metabolites, and potentially new mycotoxins.

Methods:
P. verrucosum was grown on wheat grains labeled with different stable isotopes: 97% 13C, or 53%/97%13C/15N. Secondary metabolites were extracted and analyzed by a non-targeted approach using HPLC-HRMS and electrospray ionization. Identifications were performed by comparison with standard compounds, or with theoretical spectra. Unknown secondary metabolites were annotated using molecular networks.

Results:
Substrate is the only source of carbon and nitrogen for fungal growth. Therefore, fungal secondary metabolites produced with a specific labeled substrate are labeled in the same way. The use of a 50% 13C labelling resulted in a characteristic isotopic pattern which enabled the specific detection of metabolites. Comparison of m/z detected in non-labeled, fully 13C, and 15N labeled conditions, led to the unambiguous determination of chemical formula. By this way, more than 90 secondary metabolites were annotated. Interestingly, 80% of these metabolites are totally unknown. Some known metabolites were confirmed by analyses of standard compounds and some identities were suspected using theoretical spectra (CFM-ID). Molecular network were applied on a curated MS/MS data set, showing a group of totally unknown metabolites, partially identified by interpretation of their MSn spectra.

Novel Aspect:
Specific detection and the identification of new fungal secondary metabolites produced by Penicillium.

21ST INTERNATIONAL MASS SPECTROMETRY CONFERENCE
Toronto, Ontario, Canada
August 20 – 26, 2016

Janiszewski, John,
Pfizer Inc.

Contributing Authors:

Germán Augusto Gómez-Ríos, University of Waterloo, Waterloo, ON Canada; Brendon Kapinos, Pfizer Inc., Worldwide Research and Development, Groton, CT USA; Chang Liu Sciex Inc. Concord, ON Canada; Janusz Pawliszyn, University of Waterloo, Waterloo, ON Canada; Tom Covey, Sciex Inc. Concord, ON Canada.

Title:

A continuous flow sampling interface for High-Throughput SPME-MS/MS analysis of in-vitro ADME samples.

Introduction:

We tested the applicability of solid-phase micro extraction (SPME) fibers for ADME (absorption-distribution-metabolism and excretion) sample preparation prior to ESI-MS analysis. The SPME fiber was dipped in the matrix (cell-culture media) to allow extraction/enrichment of the analyte of interest. After a quick rinsing in water, elution of the analyte was accomplished by insertion of the fiber i

Methods:

The open port sampling interface with flow rate 150 µL/min methanol, was plumbed directly into the electrospray emitter of the ion source (Turbo-V 5600+, Sciex). The nebulizing/sheath gas flow was used to optimize solvent stream at the port. Reverse-phase C18 SPME fibers with cylindrical coated surface dimension of Xby X were used. The fibers were dipped into sample for 5 min extraction period.

Results:

The sample matrices encountered in typical in vitro ADME assays (e.g. cell-based absorption, metabolic stability) are relatively well-defined and lower limits of quantification are ordinarily well within ESI-LC/MS/MS detection limits. Sample preparation for high-throughput LC/MS/MS typically occurs 'on-line' wherein analyte is loaded and trapped onto a cartridge column (~2X10mm) under aqueous conditions, the sample is desalted prior to elution with organic mobile phase and MS/MS detection. The online-SPE LC method has proven to be rapid and robust (~10sec/sample) at the expense of high solvent demand. As an alternative, we tested SPME fibers as an 'off-line' sample prep methodology. In this case the analyte is extracted/enriched from the matrices onto the SPME fiber coating. Then, elution of the analytes adsorbed on the fiber occurs in an open port sampling interface, where the analyte

Novel Aspect:

off-line sampler preparation (SPME), for High-throughput LC/MS/MS analysis

Janssen, Maurice,
MassSpecpecD BV

Title:
Direct enantiomer selective Mass Spectrometry of multi-component chiral mixtures by MS-PECD

Introduction:
Mass spectrometry is chirally blind, it cannot directly distinguish the two enantiomers of chiral molecules. Simultaneous, enantiomer-specific identification of chiral molecules in multi-component mixtures is extremely challenging. Many established techniques for single component analysis fail to provide selectivity in multi-component mixtures and lack sensitivity for dilute samples.

Methods:
Here we show how enantiomers may be differentiated by Mass-Selected PhotoElectron Circular Dichroism (MS-PECD) using an electron–ion coincidence imaging spectrometer [1,2]. Following an ionizing circular polarized laser pulse, ions and electrons are detected in coincidence on their respective time- and position sensitive detectors. The Mass-Selected PECD reveals that the compound is chiral.

Results:
As proof of concept, vapours containing ~1% of two chiral monoterpene molecules, limonene and camphor, are irradiated by a circularly polarized femtosecond laser, resulting in multiphoton near-threshold ionization with little molecular fragmentation. Large chiral asymmetries (2–4%) are observed in the mass-tagged photoelectron angular distributions. These Mass-Selected PECD asymmetries switch sign according to the handedness (R- or S-) of the enantiomer in the mixture and scale with enantiomeric excess of the chiral molecule. The results demonstrate that direct MS identification of mixtures of chiral molecules, without any prior enantiomeric separation or preparation is possible. Direct quantitative determination of enantiomeric excess in chiral mixtures is achieved in a table-top instrument using MS-PECD [3].

Novel Aspect:
[1] M.H.M. Janssen & I. Powis, PCCP, 16, 856 (2014) [2] M. M. Rafiee Fanood et al., Nature Communications 6, 7511 (2015) [3] www.massspecpecd.com

Janusson, Eric,
University of Victoria

Contributing Authors:
Dr. Scott McIndoe, University of Victoria.

Title:
Combined Spectroscopic-Mass Spectrometric Methodology for the Study of Homogeneous Catalytic Reactions

Introduction:
Catalyst activation is a complicated art, replete with arcane recommendations concerning reaction conditions. This work describes methods for the real-time investigation of homogeneous organometallic catalytic reactions through the use of combined, orthogonal analytical techniques: electrospray ionization mass spectrometry (ESI-MS) and ultraviolet-visible (UV/Vis) spectroscopy.

Methods:
We developed a technique known as Pressurized Sample Infusion (PSI). Much like a cannula transfer, PSI introduces a reaction mixture into both the flow UV-Vis and mass spectrometer by holding the flask at elevated pressure using an inert gas. The precatalyst, ligand and substrates may be injected directly into the flask from solutions prepared in the glovebox for air-free chemistry.

Results:
Tris(dibenzylideneacetone)dipalladium(0) (Pd2(dba)3) is one of the most popular palladium(0) sources and enjoys widespread use. We opted to study the activation of this precatalyst using a new and interesting method. The general trend of the activation was followed through both ESI-MS and UV/Vis. The non-innocence of the dibenzylideneacetone ligand was investigated in a variety of solvents including those less commonly used in coupling reactions. Some solvents appear to promote the release of the dibenzylideneacetone ligand from palladium prior to the addition of ligand. Additionally, several activation products were found to rapidly form in solution at room temperature without typical incubation strategies.

Novel Aspect:
A tandem, orthogonal technique was applied to mass spectrometry in order to investigate homogeneous catalytic activation methods.

Jarrold, Martin,
Indiana University

Title:
Charge Detection Mass Spectrometry: Instrumentation and Application to Virus Structure and Assembly

Introduction:
Charge detection mass spectrometry (CDMS) is a single particle technique where masses of individual ions are determined from simultaneous measurement of the m/z ratio and the charge of each ion. The technique has several favorable characteristics including the ability to determine masses of very large ions (beyond 100 MDa) and the ability to complex analyze mixtures.

Methods:
The mass resolution in CDMS is determined by the uncertainty in both the m/z and charge measurements. Historically, the mass resolution has been hindered by the low precision of the charge measurement. With recent advances it is now possible to measure the charge with almost perfect accuracy. Recent efforts to improve the m/z resolution will be described.

Results:
Several applications will be described including the use of CDMS to perform real time studies of hepatitis B virus (HBV) capsid assembly and the use of CDMS to analyze bacteriophages and enveloped viruses. HBV capsid assembly is a target of drug development. The real time CDMS studies show that the expected assembly pathway for formation of an icosahedral capsid is not followed. There are both fast and slow assembly pathways which occur in parallel. The ability of CDMS to measure large masses will be illustrated by measurements of the masses of infectious bacteriophage P22 and alphaviruses. Both phage P22 and the alphaviruses have masses beyond 50 MDa and both have mass distributions that display some heterogeneity.

Novel Aspect:
Charge detection mass spectrometry extends the mass range of MS and allows heterogeneous samples to be analyzed

Jaruga, Pawel,
National Institute of Standards and Technology

Contributing Authors:
Annie Jacob, Consolidated Safety Services, 10301 Democracy Lane, Suite 300 Fairfax, Virginia 22030; Miral Dizdaroglu, Biomolecular Measurement Division, National Institute of Standards and Technology, Gaithersburg, MD 20899, United States; W. Edward Johnson NOAA's Mussel Watch Program, National Oceanic & Atmospheric Administration, Silver Spring, MD 20910, United States; Kimani Kimbrough, NOAA's Mussel Watch Program, National Oceanic & Atmospheric Administration, Silver Spring, MD 20910, United States; Erdem Coskun, Biomolecular Measurement Division, National Institute of Standards and Technology, Gaithersburg, MD 20899, United States.

Title:
Measurements of biomarkers of oxidative DNA damage in zebra mussels by GC-MS/MS to evaluate the impact of environmental contaminants on aquatic life

Introduction:
Measuring legacy contaminants and chemicals of emerging concern in the environment is costly; lack of consideration of biological impacts can leave resource managers with insufficient information for decision making. Many contaminates can trigger DNA damage. We found elevated levels of mutagenic and/or lethal base and nucleoside modifications in the DNA of mussels from Ashtabula River estuary.

Methods:
Using GC-MS/MS with isotope dilution, we measured possible contaminant-induced changes in the DNA of zebra mussels. The biomarkers of oxidative DNA damage measured included 5-hydroxy-5-methylhydantoin, thymine glycol, 5,6-dihydroxyuracil, 4,6-diamino-5-formamidopyrimidine, 2,6-diamino-4-hydroxy-5-formamidopyrimidine, 8-hydroxyguanine and 8,5′-cyclopurine-2′-deoxynucleosides.

Results:
We identified and quantified elevated levels of oxidatively induced modifications in the DNA of mussels from Ashtabula River estuary, compared to those collected from the reference area. Results correspond to also higher concentrations of polyaromatic hydrocarbons and polychlorinated biphenyls and reflect the impact of the combination of pollutants on DNA of aquatic life. This type of biomarkers is needed to quantitatively identify the consequences of pollution that may lead to adverse health effects in animals and humans. They may also help determine the early changes at a subcellular level in animals and plants, as well as whether remedial action (e.g., removal of contaminated sediments) has proven reducing observed toxic effects. Similar assessment may be applied to monitor affected human populations by the measurement of levels of those oxidative DNA damage markers excreted in urine.

Novel Aspect:
Analyzed biomarkers are suitable for quantification of contaminant-induced damage to mussel DNA, enhancing their role as biomonitors.

Jaruga, Pawel,
National Institute of Standards and Technology

Contributing Authors:
Dorjbal Dorjsuren, National Center for Advancing Translational Sciences, National Institutes of Health, Maryland, USA; Nathan Donley, Oregon Institute of Occupational Health Sciences, Oregon Health and Science University, Portland, Oregon USA; Erdem Coskun Biomolecular Measurement Division, National Institute of Standards & Technology, Gaithersburg, Maryland, USA, andFaculty of Pharmacy, Gazi University, Ankara, Turkey; Miral Dizdaroglu, Biomolecular Measurement Division, National Institute of Standards & Technology, Gaithersburg, Maryland, USA; Aaron C. Jacobs, Center for Research on Occupational and Environmental Toxicology, Oregon Health and Science University, Portland, Oregon USA; David Maloney, National Center for Advancing Translational Sciences, National Institutes of Health, Maryland, USA; R. Stephen Lloyd, Center for Research on Occupational and Environmental Toxicology, and Department of Molecular and Medical Genetics, Oregon Health and Science University, Portland, Oregon, USA; Anton Simeonov, National Center for Advancing Translational Sciences, National Institutes of Health, Maryland, USA; Ajit Jadhav, National Center for Advancing Translational Sciences, National Institutes of Health, Maryland, USA; Amanda K. McCullough, Center for Research on Occupational and Environmental Toxicology, and Department of Molecular and Medical Genetics, Oregon Health and Science University, Portland, Oregon, USA; Marcus J. Calkins, Center for Research on Occupational and Environmental Toxicology, Oregon Health and Science University, Portland, Oregon USA.

Title:
Application of GC-MS/MS with isotope dilution for the discovery of inhibitors of DNA glycosylases as potential drugs in cancer therapy

Introduction:
DNA glycosylases have been identified as targets for cancer treatment. The therapeutic benefit for the inhibition of them has been proven by synthetic lethality with drugs that are used to limit DNA replication. We designed strategy to discover small molecule inhibitors (SMI) of human NEIL1, OGG1 and NTH1 as potential anticancer drugs.

Methods:
A fluorescence-based assay was used to screen SMI libraries for inhibitors of NEIL1, NTH1 and OGG1. Purine analogs, hydrazides and acyl hydrazones were found as top hits of these screens. Using gas chromatography/isotope-dilution tandem mass spectrometry (GC-MS/MS), we tested the release of substrates of NEIL1, NTH1 and OGG1 from DNA in the presence or absence of the inhibitors using damaged DNA.

Results:
Four purine analogs were found to be potent inhibitors of NEIL1 with the inhibitor potencies (IC50) varying from 4 μmol/L to 10 μmol/L. These compounds efficiently inhibited the excision of the main substrates of NEIL1, i.e., 4,6-Diamino-5-formamidopyrimidine (FapyAde) and 2,6-diamino-4-hydroxy-5-formamido-pyrimidine (FapyGua) from DNA containing multiple lesions. Three of the four potent NEIL1 inhibitors also inhibited the excision by NTH1 of FapyAde, FapyGua, thymine glycol, 5-hydroxy-5-methylhydantoin and 5-hydroxycytosine. For OGG1, five SMI with an IC50 value of less than 1 μmol/L were identified from a primary screen of ≈50000 molecules. These inhibitors displayed a >100-fold selectivity for OGG1 compared to other DNA glycosylases. Thus, they did not cause significant inhibition of NEIL1, NTH1 with IC50 values >50 μmol/L for all their substrates analyzed.

Novel Aspect:
This work forms the foundation for future discovery of cancer therapy drugs for the entire family of DNA glycosylases.

Jensen, Ole,
University of Southern Denmark

Title:
Deciphering PTM cross-talk by middle-down protein analysis using ETD MS/MS

Introduction:

Methods:
ETD MS/MS

Results:
Systems level analysis of histone H3 post-translational modifications reveals features of PTM crosstalk in chromatin regulation. Schwämmle V, Sidoli S, Ruminowicz C, Wu X, Lee CF, Helin K, Jensen ON. Mol Cell Proteomics. 2016 Jun 14. pii: mcp.M115.054460. Dynamic changes of histone H3 marks during Caenorhabditis elegans lifecycle revealed by middle-down proteomics. Sidoli S, Vandamme J, Salcini AE, Jensen ON. Proteomics. 2016 Feb;16(3):459-64. doi: 10.1002/pmic.201500285.

Novel Aspect:
Mapping and determination of PTM cross-talk and interplay.

Jhingree, Jacquelyn,
Manchester Institute of Biotechnology, University of Manchester

Contributing Authors:

Jacquelyn R. Jhingree, Manchester Institute of Biotechnology, University of Manchester, 131 Princess Street, Manchester, M1 7DN. UK.; Perdita E. Barran, Manchester Institute of Biotechnology, University of Manchester, 131 Princess Street, Manchester, M1 7DN. UK.; Bruno Bellina Manchester Institute of Biotechnology, University of Manchester, 131 Princess Street, Manchester, M1 7DN. UK.

Title:

Charge mediated compaction and extension of gas phase proteins studied by charge reduction reactions coupled to ion mobility mass spectrometry (IM-MS)

Introduction:

Gas phase studies show that the more charges a protein possess post electrospray ionization, the more unfolded it is. This is attributed to repulsive Coulombic effects caused by proximal charges. We examine conformations of charge reduced protein ions with n charge state generated from the electrospray process from ions with (n+x) charge state using charge reduction reactions coupled to IM-MS.

Methods:

Experiments were carried out on a Synapt G2 Si MS equipped with a glow discharge source for generating radical anions. ESI produced cations from a panel of proteins are charge reduced in the trap of the instrument. All species are mobility separated at constant wave amplitude and velocity then pulsed into a time-of-flight mass analyzer for arrival time measurement (ATD).

Results:

Structural changes in BPTI (structured), beta casein (disordered) and yeast enolase (dimer) are examined upon exposure to radical anions of 1,3-dicyanobenzene. Upon exposure, the [M+4H]^4+ ion of BTPI shows no change in the ATD. Charge reduced products, with the same number of charges, formed from charge reduction retain the same ATD. For BPTI this is a narrow distribution, indicative of a compact conformer, which is expected given the high number of disulfide bridges. A similar result is seen for the yeast enolase dimer, showing that it has a stable structure across its three charge states seen and that Coulombic effects do not play a significant role in altering the desolvated conformation. By contrast for the low and intermediate charges states of beta casein we see compaction when electrons are transferred. High charge states which are already extended do not compact significantly.

Novel Aspect:

Compaction is seen in a disordered protein of intermediate charge state; the structured protein retain its conformation even after two reductions.

Jiang, Jichun,
Dalian Institute of Chemical Physics, Chinese Academy of Sciences

Contributing Authors:
Yuanyuan Xie, Key Laboratory of Separation Science for Analytical Chemistry, Dalian Institute of Chemical Physics, Chinese Academy of Sciences, Dalian, 116023, People's Republic of China.; Ping Chen, Key Laboratory of Separation Science for Analytical Chemistry, Dalian Institute of Chemical Physics, Chinese Academy of Sciences, Dalian, 116023, People's Republic of China.; Yan Wang Key Laboratory of Separation Science for Analytical Chemistry, Dalian Institute of Chemical Physics, Chinese Academy of Sciences, Dalian, 116023, People's Republic of China.; Keyong Hou, Key Laboratory of Separation Science for Analytical Chemistry, Dalian Institute of Chemical Physics, Chinese Academy of Sciences, Dalian, 116023, People's Republic of China.; Wei Liu, Key Laboratory of Separation Science for Analytical Chemistry, Dalian Institute of Chemical Physics, Chinese Academy of Sciences, Dalian, 116023, People's Republic of China.; Haiyang Li, Key Laboratory of Separation Science for Analytical Chemistry, Dalian Institute of Chemical Physics, Chinese Academy of Sciences, Dalian, 116023, People's Republic of China.; Lei Hua, Key Laboratory of Separation Science for Analytical Chemistry, Dalian Institute of Chemical Physics, Chinese Academy of Sciences, Dalian, 116023, People's Republic of China.

Title:
Photoionization-generated dibromomethane cation chemical ionization TOFMS for rapid and sensitive detection of volatile sulfur compounds

Introduction:
Volatile sulfur compounds (VSCs) play significant roles in environmental related issues for human daily life. Most of the VSCs usually are highly toxic, and ppbv level VSCs can cause health problems to humans. Because of its high reactive and adsorptive properties, development of highly-sensitive and rapid-analysis techniques for detection of VSCs becomes necessary.

Methods:
A new photoionization-generated dibromomethane cation chemical ionization (PDCI) source has been developed for time-of-flight mass spectrometry (TOFMS). Reactant ions CH2Br2+ was first generated by photoionization, then ionized analytes via charge transfer and ion association. Besides, a nafion tube and membrane inlet were utilized for introduction of gas and liquid-phase analytes, respectively.

Results:
Five typical VSCs were chosen to evaluate the performance of the PDCI-TOFMS. The limits of detection (LODs) ranging from 0.01 to 0.2 ppbv were obtained in 50 seconds. Besides, H2S, CH3SH and CS2 could form [M+CH2Br2]+ ions, which are easy to distinguish from other ions in the mass spectrum by the apparent isotopic ratios of Br. In order to demonstrate its performance, the PDCI-TOFMS has been preliminarily utilized to analyze trace VSCs in sewer gases at various times during one day, the result shows that H2S and CH3SH were the major components of VSCs with concentrations of several hundreds and tens ppbv, respectively. Moreover, the rapid analytical capacity of this system was also demonstrated by analysis of trace VSCs in breath gases of healthy volunteers. Furthermore, this system will been applied to analyze odorous VSCs in refuse landfill and water samples in our following work.

Novel Aspect:
A new photoionization-generated dibromomethane cation was utilized to enhance the chemical ionization efficiency of the VSCs.

Jmaiff Blackstock, Lindsay,
Analytical and Environmental Toxicology Division, University of Alberta

Contributing Authors:
Ian Vander Meulen, Division of Analytical and Environmental Toxicology, Department of Laboratory Medicine and Pathology, University of Alberta, Edmonton, Canada, T6G 2G3; Xing-Fang Li, Division of Analytical and Environmental Toxicology, Department of Laboratory Medicine and Pathology, University of Alberta, Edmonton, Canada, T6G 2G3; Guang Huang Division of Analytical and Environmental Toxicology, Department of Laboratory Medicine and Pathology, University of Alberta, Edmonton, Canada, T6G 2G3; Yichao Qian, Division of Analytical and Environmental Toxicology, Department of Laboratory Medicine and Pathology, University of Alberta, Edmonton, Canada, T6G 2G3; Wei Wang, Division of Analytical and Environmental Toxicology, Department of Laboratory Medicine and Pathology, University of Alberta, Edmonton, Canada, T6G 2G3.

Title:
Aromatic Amino Acids: Newly Identified Halobenzoquinone Precursors

Introduction:
Drinking water disinfection is essential to prevent the spread of waterborne disease. Epidemiology studies show potential association between long-term disinfection byproduct (DBP) exposure and bladder cancer risk. Halobenzoquinones (HBQs) were identified in drinking water and are up to 1000x more cytotoxic than regulated DBPs. It is necessary to identify HBQ precursors to reduce their formation.

Methods:
Aromatic amino acid standards were prepared separately and exposed to free chlorine with or without Br - over time. Solid phase extraction (Oasis HLB) and nitrogen evaporation was employed to remove interferents and concentrate each sample. Analysis was performed using UHPLC-MS/MS. Positive HBQ identifications were made using retention time and two multiple-reaction-monitoring mode transitions.

Results:
Amino acids are ubiquitously present in natural water sources, highly soluble and not easily removed using current treatments. The inherent ring structure of aromatic amino acids (AAAs; phenylalanine, PHE; tyrosine, TYR; tryptophan, TRP) make them ideal HBQ precursor candidates for this study. 2,6-Dichlorobenzoquinone (2,6-DCBQ) the most frequently occurring and abundant HBQ, has been positively identified as a formation product resulting from reactions between free chlorine and AAAs. Each AAA: PHE, TYR and TRP was capable of producing 2,6-DCBQ. In vitro studies have identified a trend in the toxicity of di-halogenated DBPs: iodinated > brominated > chlorinated. Under high bromide conditions, the more toxic 2,6-dibromobenzoquinone was also produced from each AAA. Future work will systematically investigate the Cl-HBQ and Br-HBQ formation under various drinking water treatment conditions.

Novel Aspect:
This study provides new essential information to minimize HBQ formation during future water treatment processes, reducing risk to human health.

Joblin, Christine,
University Toulouse 3 / CNRS, IRAP

Contributing Authors:
Sarah Rodriguez Castillo, IRAP - Université de Toulouse 3 & CNRS, Toulouse, France; Dimitris Papanastasiou, Fasmatech Science + Technology, Athens, GR; Hassan Sabbah IRAP - Université de Toulouse 3 & CNRS, Toulouse, France.

Title:
Exploring Collision Induced Dissociation of coronene cation C24H12+ in a Linear Quadrupole Ion Trap driven by Rectangular Voltage RF Waveforms

Introduction:
Collision Induced Dissociation (CID) is performed in a segmented Linear Quadrupole Ion Trap (LQIT) driven by a pair of opposite phase rectangular RF waveforms. Excitation of ion motion in RF trapping fields produced by rectangular voltage pulses, rather than more classical sinusoidal waveforms, offers the ability to explore new regimes for dissociating ions in the presence of a buffer gas.

Methods:
CID experiments have been performed on the compact aromatic cation coronene C24H12+ formed by Laser Desorption Ionization (LDI). The dipolar excitation (DE) waveform is superimposed on the rectangular waveform to excite ion motion while pulsing argon gas. Ion optics simulations were performed to investigate ion motion and support the analysis of the results.

Results:
Using the AROMA setup (H. Sabbah et al., this conference), C24H12+ ions are formed by LDI, thermalized in an octopole cooler and stored in the LQIT. Off-resonance DE is applied with different amplitudes (1-5 Vpp), frequencies and excitation time (1-20 ms) while pulsing argon gas. Dissociation products are monitored on a oTOF mass analyzer. Over the range of frequencies examined secondary dissociation products can be formed. At low DE amplitudes consecutive loss of H atoms is observed consistent with previous studies performed in 3D ion traps and ICR cells using slow heating CID and sustained off-resonance excitation respectively. At higher amplitudes, fragments as small as C14+ were detected. Ion optical simulations are performed to monitor the levels of ion kinetic energy during excitation providing insights into the enhanced dissociation efficiency observed experimentally.

Novel Aspect:
Enhanced CID efficiency in a segmented linear quadrupole ion trap driven by rectangular RF waveforms is demonstrated for the aromatic species C24H12+.

Johnson, Mitch,
University of Missouri

Contributing Authors:
Valeri Mossine, Department of Biochemistry, University of Missouri, Columbia, MO, USA; Jiankun Cui, Department of Pathology & Anatomical Sciences, University of Missouri, Columbia, MO, USA; Zezong Gu Department of Pathology & Anatomical Sciences, University of Missouri, Columbia, MO, USA; C. Michael Greenlief, Department of Chemistry, University of Missouri. Columbia, MO, USA; Hailong Song, Interdisciplinary Neuroscience Program, University of Missouri, Columbia, MO, USA.

Title:
Method Development and Validation for the Quantitation of N-α-(1-deoxy-D-fructos-1-yl)-L-arginine (FruArg) using LC-MS/MS

Introduction:
Garlic is a prevalent plant botanical, and aged garlic extract (AGE) has been used as a nutritional supplement and implied to promote health benefits by exhibiting anti-oxidant and anti-inflammatory activity, as well as hypolipidemic and antiplatelet effects. It has recently been discovered that N-α-(1-deoxy-D-fructos-1-yl)-L-arginine (FruArg) is a major contributor to the bioactivity of AGE .

Methods:
A very sensitive analytical method was developed and optimized for quantitation of FruArg in mice plasma samples. Briefly, Phree® phospholipid solution was used to separate samples spiked with FruArg from potentially interfering compounds present in biological fluids. Eluates were collected and analyzed using liquid chromatography-tandem mass spectrometry (LC-MS/MS).

Results:
A full method validation was conducted including analysis of the lower limit of quantitation, linearity, range, recovery, selectivity, matrix effect, stability, accuracy, and inter/intra-day precision.

Novel Aspect:
This is a novel LC-MS/MS method for quantitation of a bioactive compound that is not well understood.

Josephs, Jonathan,
Thermo Fisher Scientific

Contributing Authors:
Eugen Damoc, Thermo Scientific; Stephane Houel, Thermo Scientific; Aaron Bailey Thermo Scientific; Kai Scheffler, Thermo Scientific.

Title:
Full Characterization of Heterogeneous Antibodies Under Denaturing and Native/Native-Like Conditions on a Hybrid Quadrupole-Orbitrap Mass Spectrometer

Introduction:
Therapeutic proteins are often complex due to various modifications. Intact proteins analyzed by MS are detected in higher charge states leading to spectral complexity. Protein analysis in native or native-like conditions with zero or minimal organic solvent and neutral or weakly acidic pH decreases charge state value resulting in mAb detection at higher m/z ranges with more spatial resolution.

Methods:
Monoclonal antibodies were desalted online for intact analysis using size exclusion chromatography (SEC) or reversed phase chromatography. Peptide mapping experiments were also conducted on the same commercially available hybrid quadrupole-Orbitrap mass spectrometer with modified instrument control software to allow for improved high mass transmission and mass detection up to 8000 m/z.

Results:
Significantly lower charge states observed under native conditions improves the spatial resolution of protein signals with significant benefits for the analysis of antibody mixtures. This requires the detection of masses beyond the standard mass range up to 6000 m/z requiring the modification to allow for mass detection up 8000 m/z. We observed that apparent native structures persist and ESI is benefitted by addition of low amounts of acetonitrile and formic acid in combination with the ammonium acetate-buffered mobile phase. For analyses on the peptide level we analyzed digests separated via reversed phase chromatography. All data acquired were submitted to a single software package for analysis aiming to obtain a complete picture of the molecules analyzed. Demonstrating the ability to fully characterize homogeneous and heterogeneous therapeutic proteins on one single platform.

Novel Aspect:
Full characterization of heterogeneous intact protein mixtures by improved mass separation on a modified quadrupole-Orbitrap mass spectrometer.

Josephs, Jonathan,
Thermo Fisher Scientific

Contributing Authors:
Aaron Bailey, Thermo Scientific; Eugen Damoc, Thermo Scientific; Kai Scheffler Thermo Scientific; Stephane Houel, Thermo Scientific.

Title:
Complete Characterization of a Cysteine-linked Antibody-Drug Conjugate Using a Hybrid Quadrupole-Orbitrap Mass Spectrometer with High Mass Range

Introduction:
Complete characterization of antibody-drug conjugates (ADCs) requires both peptide mapping and intact mass analysis. Native MS analysis preserves critical structural features, such as cysteine-linked ADCs. This produces decreased charge states (increased m/z). We have modified the instrument control software of a benchtop quadrupole-Orbitrap mass spectrometer to add native MS capability.

Methods:
Brentuximab vedotin was analyzed intact, as subunits or by peptide mapping by online chromatography coupled directly to a commercially-available Thermo Scientific Q Exactive Plus or Q Exactive HF Orbitrap mass spectrometers which were operated using a modified instrument control software to allow improved high mass transmission and scanning up to m/z 8000.

Results:
We measured the intact native average drug to antibody ratio (DAR) of Brentuximab vedotin using a native SEC-MS system. We observed excellent desolvation in 50 mM ammonium acetate. Intact analysis of the untreated ADC showed a distribution of 0, 2, 4, 6, and 8 DAR form; each form was observed to have a nested set of peaks corresponding to a distribution of N-glycans. For each DAR load we also found low level losses of a mass corresponding to 1-2 monomethyl auristatin E payloads and partial linker/spacer structure resulting from ether hydrolysis of a spacer moiety on the valine-citrulline-based linkers. We observed an average DAR of 3.9, with abundant species detected at < 10 ppm mass accuracy. Peptide mapping analysis resulted in 100% coverage of the amino acid sequence confirming the sites of linker-payload that are conjugated reduced interchain disulfide linkages.

Novel Aspect:
Peptide-mapping MS/MS and high resolution native-MS intact protein workflows performed for the first time on a single commercial Orbitrap MS.

Joudan, Shira,
University of Toronto

Contributing Authors:
Leo W.Y. Yeung, University of Toronto, Department of Chemistry; Scott A. Mabury, University of Toronto, Department of Chemistry.

Title:
Mammalian metabolism of the C-P bond in perfluorinated phosphinates (PFPiAs) and identification of resulting metabolites

Introduction:
Perfluoroalkyl acids (PFAAs) are a class of widespread contaminants that are completely recalcitrant under environmental conditions. A 2012 paper from our group showed the first biotransformation of any PFAA, where PFPiAs were metabolically hydrolyzed to perfluoroalkyl phosphonates (PFPAs) in fish. This work confirms the in vivo transformation in rats and identifies the remaining PFPiA metabolite.

Methods:
Rats were dosed with either C6/C8 PFPiA, C8/C8 PFPiA, or C8 PFPA via oral gavage. Blood, kidneys and liver were sampled to determine pharmacokinetic and partitioning properties. PFPAs and PFPiAs were quantified via UPLC-ESI-MS/MS using matrix matched calibration. 1H-perfluoroalkane metabolites were detected using a novel headspace solid phase microextraction method with negative CI-GC/MS.

Results:
All molecules were bioavailable, with blood absorption half-lives of 0.65, 1.1 and 1.3 hours, for C8 PFPA, C6/C8 PFPiA and C8/C8 PFPiA, respectively. C8 PFPA had a faster elimination half-life than the C6/C8 and C8/C8 PFPiAs, with values of 21, 52 and 66 hours, respectively. The faster absorption and elimination of the PFPA may be explained by its -2 charge or smaller size. Both PFPiAs were hydrolyzed in vivo to form PFPAs and 1H-perfluoroalkanes. 1H-perfluoroalkanes were not detected in rats dosed with C8 PFPA, suggesting the C-P bond cleavage requires the presence of a second electron withdrawing perfluoroalkyl group. Although metabolism decreases the concentration of PFPiAs, the products formed have the potential to be more persistent and toxic. Preferential formation of C6 or C8 PFPA from C6/C8 PFPiA could not be determined due to their rapid excretion upon formation.

Novel Aspect:
This is the first detected mammalian hydrolysis of C-P bonds in phosphinates and confirmation of 1H-perfluoroalkanes as metabolites of PFPiAs.

Julian, Ryan,
UC Riverside

Contributing Authors:
Chris Nellessen,

Title:
Photo-electron transfer dissociation of peptides for zwitterion identification

Introduction:
Interest in leveraging mass spectrometry (MS) for the study of three-dimensional biomolecular structure continues to increase, spurred by a growing body of results suggesting that native or native-lik

Methods:
UVPD experiments were performed in a ThermoFisher Scientific (San Jose, CA, U.S.A.) LTQ linear quadrupole ion-trap mass spectrometer modified to allow UV laser light from a 266 nm Nd:YAG laser (Contin

Results:
We report on a photo-induced electron transfer (PET) process that initiates electron transfer dissociation (ETD) in zwitterionic peptides, representing a new tool that can be used to detect gas-phase zwitterions. Characteristic c and z-type fragmentation is observed with extensive sequence coverage

Novel Aspect:
This is the first report of photo-electron transfer induced dissociation in peptides.

Jurneczko, Ewa,
University of Manchester

Contributing Authors:

Jan Commandeur, MS Vision; Perdita E. Barran, University of Manchester.

Title:

Integrating Drift Tube Mobility to existing QToF instruments for High Mass Native Mass Spectrometry

Introduction:

There has been an increase in the application of IM coupled with MS and a corresponding need to develop simple bolt on IM capabilities to existing MS instruments. Here a new IM device utilizing monolithic drift tubes as an alternative building material, is presented as a separation tool that can be retrofitted to existing QToF platforms, with a focus on its use for study of high mass native MS.

Methods:

The proposed prototype uses monolithic 3.9-inch long drift tube with 1.5-inch i.d. and 1.8-inch o.d. The resistive glass (Photonics) is lead silicate glass that has been reduced in hydrogen atmosphere to make its surface a semiconductor. The ends of tube were treated with thin-film metallized contact points and pressure measurement is enabled through a machined inlet

Results:

A fundamental component of the proposed monolithic IM device is the use of an efficient ion trap-and-release mechanism and RF ion funnel at the exit of the drift region, crucial for maintaining high ion transmission. Initial test results are shown from the first-principles calculations from kinetic theory, ion trajectory simulations (SIMION) and computer aided design (CAD) software which has been utilized to design the mobility device. The maximum theoretical resolving power at different drift voltages at ambient temperature is evaluated with a maximum operating voltage of 20 Vcm-1. This linear field drift cell IM device will be initially fitted into QToF-Ultima, by modifying the two original RF ion tunnels prior to the quadrupole region, and retaining the existing footprint of the instrument, as well as its vacuum capability.

Novel Aspect:

The novelty of such design is to provide a flexible bolt on mobility device that can be easily retrofitted in ToF instrumentation.

Kaftan, Filip,
Max Planck Institute for Chemical Ecology

Contributing Authors:
Johannes Pallua, Division of Developmental Immunology, University of Innsbruck, Innsbruck, Austria; Claudia Wöss, Division of Developmental Immunology, University of Innsbruck, Innsbruck, Austria; Ludger Beerhues Institute for Pharmaceutical Biology, Technical University Braunschweig, Braunschweig, Germany; Aleš Svatoš, Max Planck Institute for Chemical Ecology, Jena, Germany; Jerrit Weissflog, Max Planck Institute for Chemical Ecology, Jena, Germany.

Title:
Atmospheric pressure mass spectrometry imaging and its application in studying biological samples

Introduction:
Atmospheric pressure scanning microprobe matrix-assisted laser desorption/ionization source AP-SMALDI (TransMit, Germany) coupled to an Orbitrap mass spectrometer is a system with high sensitivity and spatial resolution suitable for analyzing biological structures. This powerful system was tested on plant, insect or forensic samples and includes also the application of a superbasic MAILD-matrices.

Methods:
Hypericum perforatum roots fixed in embedding medium were cut into sections (30 µm), covered with 9-aminoacridine (9-AA) and analyzed. In a similar way insect cross-sections were covered via sublimation with 1,8-di(piperidinyl)-naphthalene (DPN) and imaged. Another application of AP-SMALDI-FT MSI was dedicated to study illegal substances with focus on opioids in human hairs.

Results:
Based on AP-SMALDI-FT MSI experiments with root cross-section of H. perforatum was possible to visualize 2-D distribution of Toxyloxanthone B, 1,7-dihydroxyxanthone and other xanthone-like compounds in sub-epidermal root-layers of the root. The obtained results were in consensus with localization of transcript and protein of enzyme benzophenone synthase responsible for xanthones biosynthesis. AP-SMALDI-FT MSI experiments proved that it is possible to analyze longitudinally cut human hairs in order to study distribution of opioids, like a heroin, codeine, tramadol or methadone as well as their metabolites. On the basis of MSI of longitudinally cut hairs the life history of individuals was studied to follow more than one and a half years of drug abuse. Moreover, ambient MSI experiments with CHCA matrix shown that drugs can be detected even from uncut hairs.

Novel Aspect:
Opioid metabolites are detectable in uncut human hairs via AP-SMALDI-FT. DPN can be successfully applied in MSI without matrix signals below 1000 m/z.

Kakkoura, Maria,
The Cyprus Institute of Neurology and Genetics, Nicosia, Cyprus

Contributing Authors:
Christiana A. Demetriou, The Cyprus Institute of Neurology and Genetics, Nicosia, Cyprus; Kleitos Sokratous, The Cyprus Institute of Neurology and Genetics, Nicosia, Cyprus; Maria A. Loizidou The Cyprus Institute of Neurology and Genetics, Nicosia, Cyprus; Kyriacos Kyriacou, The Cyprus Institute of Neurology and Genetics, Nicosia, Cyprus; Andreas Hadjisavvas, The Cyprus Institute of Neurology and Genetics, Nicosia, Cyprus.

Title:
MASS SPECTROMETRY-BASED TARGETED METABOLOMICS OF SERUM IN GREEK-CYPRIOT WOMEN REVEALS FAVORABLE HEALTH EFFECTS OF GENE-MEDITERRANEAN DIET INTERACTIONS

Introduction:
The Mediterranean diet (MD) protects against chronic disease risk. The underlying MD protective mechanisms, however, remain unclear. Using a liquid chromatography-tandem mass spectrometry (LC-MS/MS) method, we measured the concentration of 14 metabolites in the serum of 564 healthy Greek-Cypriot women, in order to investigate whether SNP-MD interactions influence serum metabolite concentrations.

Methods:
The levels of 14 metabolites involved in the one-carbon metabolism and xenobiotic metabolism, were measured in the serum of 564 women, with a low or high MD adherence. Analysis was performed on an Acquity UPLC system coupled to a Xevo-TQD MS operated on multiple-reaction monitoring mode. Main effects of MD and its interactions with SNPs on metabolite concentrations were tested by linear regression

Results:
MD adherence was assessed by food frequency questionnaire and principal component analysis (PCA). Participants were previously genotyped for 9 SNPs in genes involved in one-carbon metabolism, oxidative stress and xenobiotic metabolism. Increasing adherence to the PCA-derived MD was significantly associated with an increase of 5% in serum 5-methyltetrahydrofolate (5-MTHF) concentration ($exp(\beta)=1.050$ ng/ml, $P=0.03$). There were significant interactions between GSTM1 deletion polymorphism and PCA-derived MD adherence on serum flavin mononucleotide (FMN) (P interaction=0.004; $exp(\beta)=1.069$, $P=0.004$ for GSTM1 null women) and 5-MTHF concentrations (P interaction=0.03; $exp(\beta)=1.087$, $P=0.01$ for GSTM1 null women). MTHFR rs1801133 interacted significantly with MD adherence in relation to serum 5-MTHF concentration (P interaction=0.04; $exp(\beta)=1.139$, $P=0.001$ for major 677C allele homozygotes).

Novel Aspect:
The simultaneous measurement of the serum levels of 14 metabolic markers by LC-MS/MS elucidates the beneficial health effects of SNPs-MD interactions.

Kalenius, Elina,
University of Jyväskylä

Contributing Authors:
Ondrej Jurcek, University of Jyväskylä; Pia Bonakdarzadeh, University of Jyväskylä; Michael Groessl Tofwerk AG.

Title:
Structural Characterization of Pt, Pd and Fe Coordination Cages by ESI-Q-TOF and Ion-mobility Mass Spectrometry

Introduction:
Over the last years, coordination-driven self-assembly has been demonstrated to be a highly efficient method to produce three-dimensional supramolecular architectures. Metal cages have recently arouse interest originating from their intrinsic features regarding 3-D structures, optical properties, potential in reactivity and storage of highly active chemicals. ESI-MS and IM-MS are potential methods.

Methods:
ESI-Q-TOF mass spectrometry (ABSciex QSTAR Elite) and Tofwerk drift tube IM-MS was used for characterisation M4L4, M6L4 and M4L6 coordination cages with Pt, Pd and Fe. The compounds are generally soluable in organic solvents and easily ionized. The mass spectrometric studies were supported by 1H and DOSY NMR, X-ray structure determinations and DFT calculations when applicable.

Results:
The formation of coordination compounds and structural monodispersity was followed by ESI-MS. In the case of superchiral M4L6 of ursodeoxycholic acid its reversible structural conversion into Pd3L3Cl6 metallocycles was also observed when chemical stimulants were added. The ESI-MS analysis was supplemented with CID experiments and collision cross section (CCS) determinations by drift tube IM-MS. The structural monodispersity was found largely to depend on flexibility of the organic ligand and the most flexible coordination complex M4L6 of ursodeoxycholic acid derivative displyed interesting temperature dependent behavior according to ion-mobilograms. For the spherical coordination cages, CCS values obtained by IM-MS were found to correlate well with hydrodynamic radii determined by DOSY NMR and dimensions measured from solid state structures.

Novel Aspect:
IM-MS has been rarely applied to studies of metal–organic cages. The comparison of CCS values with the DOSY NMR and crystal structures is current.

Kappelmann, Jannick,
Research Center Juelich

Contributing Authors:
Stephan Noack, Forschungszentrum Jülich (FZJ), Institute of Bio- and Geosciences.

Title:
Interpreting LC-MS/MS-spectra from primary metabolites for tandem MS-based 13C-Fluxomics

Introduction:
CID-fragments offer the opportunity to obtain positional information on 13C-label enrichment in primary metabolites. Improved positional 13C-enrichment information goes hand in hand with increased precision of 13C-based Fluxomics. However, for this purpose the positional identity of product ion carbon atoms needs to be known. This challenge is comprehensively tackled in this contribution.

Methods:
Exact mass spectrometry, enzymatically generated selectively-labeled standards and published fragmentation pathways are used to structurally annotate all intense product ion peaks of 68 primary metabolites. After structural annotation, we computationally deduce the maximal positional labeling information contained in corresponding LC-tandem MS spectra.

Results:
We show that several LC-MS/MS collisional fragments do not have a unique carbon ancestry, and some integral product ion peaks do not even have a unique elemental composition. By computing all arising labeling states of fragments neighboring intense fragments, we systematically identify all isotopic interferences which are shown to be rather pervasive. These obstacles notwithstanding, we were able to device a selection of 168 structurally annotated and intense product ions from 70 precursor ions of 68 metabolites that are intense and unique in carbon origin. We show that the positionally resolved information contained in the product ions that we structurally annotated allows to infer the entire isotopomer distribution of several central metabolism intermediates, which is experimentally demonstrated for malic acid using quadrupole-time-of-flight MS technology.

Novel Aspect:
This work provides the most accurate and detailed account of carbon atom fate of tandem MS collisional fragments of primary metabolites yet.

Kapron, Jim,
Bruker

Contributing Authors:
Karin Wendt, Bruker Daltonics, Bremen, Germany; Sebastian Goetz, Bruker Daltonics, Bremen, Germany; Carsten Baessmann Bruker Daltonics, Bremen, Germany.

Title:
Fast and Reliable Quantitation of Z-drug Hypnotics and Benzodiazepines in Serum using Accurate Mass LC-QTOF Mass Spectrometry

Introduction:
Psychoactive drugs like benzodiazepines and Z-drug hypnotics are used in a variety of medical therapies. This study evaluates the use of high resolution LC-QTOF mass spectrometry as a new approach. Suitability of LC-QTOF technology for the simultaneous detection, confirmation and quantitation of these highly relevant drugs in human serum will be evaluated.

Methods:
Fortified human serum samples & calibrators were extracted using liquid-liquid extraction and analyzed by reversed-phase, positive ESI using data independent broad-band-CID fragmentation. TASQ (Target Screen & Quan) software was used to analyze chromatograms using a database with known retention times & exact masses in MS and MS/MS modes for identification & quantitation at the same time.

Results:
A dilution series yielded calibration curves for 13 benzodiazepines and Z-drug hypnotics. All analytes & the 9 deuterated internal standards eluted within 8 minutes. High resolution extracted ion chromatograms (hrEIC) with <2mDa tolerance showed good separation & adequate chromatographic peak shape. The added selectivity of hrEIC resulted in linear calibration ranges from low-therapeutic to toxic levels (2-3 orders of magnitude). Observed retention times and MSMS fragment masses of the analytes were used to generate a screening and quantitation method. Serum samples with forensic-relevant concentrations were obtained from an externally-organized proficiency group. Analysis of these single-blind samples showed concentrations within ±20% of theoretical with confidence (z-values well below 1).

Novel Aspect:
Simultaneous detection, confirmation & quantification of benzodiazepine & Z-drug hypnotics in human serum by means of UHPLC-QTOF MS using bbCID.

Kast, Juergen,
University of British Columbia

Contributing Authors:

Savita Srinivasa, The Biomedical Research Centre, University of British Columbia; Juergen Kast, The Biomedical Research Centre, University of British Columbia.

Title:

Formaldehyde versus Established Cross-linkers in Examining a Transient Protein Complex using Mass Spectrometry

Introduction:

Chemical cross-linking along with mass spectrometry (MS) can elucidate protein geometry by using covalent linkages as distance constraints. Formaldehyde's (PFA) small size allows it to quickly permeate cellular membranes and stabilize transient protein interactions under physiological conditions. Despite its uses in biology and MS-compatibility, PFA has yet to be used to map protein structure.

Methods:

PFA along with four other established cross-linkers (EDC, sulfoDST, BS3, and sulfoEGS), varying in size and reactivity, were applied to the Ca2+ free calmodulin-melittin model system, a transient protein complex with an unknown binding conformation. Cross-linked mixtures were separated via SDS-PAGE, digested in-gel with trypsin and analyzed via tandem MS (MS/MS).

Results:

Cross-links generated by PFA and the four common cross-linkers were identified via MS/MS. A comparison between the MS and MS/MS detection of PFA versus conventional cross-linked species provided insight into what evidence is sufficient to confirm each cross-linked species. The relatively straightforward discovery of NHS ester cross-linking was offset by its large size and ambiguous distance constraints that may not be suitable for small proteins. Although EDC creates close proximity linkages, the high abundance of reactive sites in this model system produced diversified products, complicating MS detection. Similarly, the increased complexity in detecting PFA reaction products via MS was due to its semi-specific reactivity. Nevertheless, PFA cross-linked species were identified in a transient protein complex, supporting PFA's ability to stabilize weak interactions.

Novel Aspect:

This is the first report of PFA cross-links identified between non-covalent protein components, revealing its potential for mapping protein structure.

Kast, Juergen,
University of British Columbia

Contributing Authors:
Ru Li, The Biomedical Research Centre, University of British Columbia; Juergen Kast, The Biomedical Research Centre, University of British Columbia.

Title:
Quantitative analysis of induced reversible cysteine oxidation in an atherosclerotic model via modified biotin switch assays

Introduction:
Reversible cysteine oxidation, induced e.g. by reactive oxygen species (ROS) during oxidative stress, is emerging as a regulated signal transduction event. ROS production, triggered e.g. by platelet-monocyte interaction, is a key element in the development of atherosclerosis. Improving our understanding of these processes is crucial in designing novel intervention strategies for atherosclerosis.

Methods:
We used THP-1 monocytic cells exposed to LPA- or thrombin-induced platelet releasate as an atherosclerotic model. To quantify ROS-induced total reversible cysteine oxidation and sulfenic acid modification in this model, MS-based modified biotin-switch assays were developed. Identified redox-regulated proteins were validated on human primary monocytes by enzyme activity assays and flow cytometry.

Results:
NADPH oxidase (NOX)-derived ROS production was confirmed in our model and primary monocytes. Analysis of the total reversible cysteine oxidation proteome identified ~100 regulated cysteines involving processes such as glycolysis, cytoskeleton arrangement, and redox regulation. As this yielded no proteins linked to monocyte function, the assay was modified to isolate only sulfenic acid-modified cysteines, a specific type of reversible cysteine oxidation. This resulted in the quantitation of ~100 regulated sites at the protein level. Bioinformatic analysis highlighted additional proteins involved in monocyte transendothelial migration, including integrin β2. Flow cytometry of the two main integrin β2 complexes on monocytes revealed complementary regulation: LFA-1 (αLβ2) was activated by NOX-derived ROS, while Mac-1 (αMβ2) activation did not involve NOX and ROS, but P-selectin binding.

Novel Aspect:
A versatile biotin switch assay was developed, which identified differential redox-regulated receptor activation via a signaling pathway-like setup

Kauppila, Tiina,
University of Helsinki

Contributing Authors:
Heikki Räikkönen, University of Helsinki, Finland; Markus Haapala, University of Helsinki, Finland; Risto Kostiainen University of Helsinki, Finland; Anu Vaikkinen, University of Helsinki, Finland; Juha-Pekka Hieta, University of Helsinki, Finland.

Title:
Laser ablation atmospheric pressure photoionization mass spectrometry imaging. Instrumentation and applications.

Introduction:
Laser ablation atmospheric pressure photoionization (LAAPPI) [1] is an ambient ionization method, which can efficiently ionize both polar and nonpolar compounds. LAAPPI uses IR laser for ablation of analytes from the surface of a water-rich samples (e.g. tissue) and photoionization for ionization.

Methods:
LAAPPI uses an IR laser at 2940 nm for ablation of analytes from the sample. The ablated analytes are desolvated by a heated nebulized microchip and ionized by photons provided by a krypton discharge photoionization lamp. Dopant (e.g. toluene or acetone) vapor is supplied from the heated nebulizer microchip to enhance the ionization.

Results:
LAAPPI instrumentation, ionization mechanism, crucial parameters and their optimization, analytical parameters, and recent accomplishments in MS imaging are discussed. In LAAPPI, the ionization is initiated by photoionization, after which gas-phase reactions with dopant ions lead to ionization of the analytes. The main ionization reactions are proton transfer and charge exchange. The route of ionization depends on the dopant and the analyte properties. The imaging applications include plant and animal tissues. The results achieved with LAAPPI are compared to those with laser ablation electrospray ionization (LAESI). In the analysis of lipids, LAAPPI showed better sensitivity for e.g. cholesterol, glycerolipids and glycerophospholipids. References: [1] A. Vaikkinen et al., Anal. Chem., 2012, 84 (3), 1630–1636.

Novel Aspect:
The recent MS imaging results using LAAPPI.

Kellmann, Markus,
Thermo Fisher Scientific (Bremen) GmbH

Contributing Authors:
Andreas Kuehn, Thermo Fisher Scientific, Bremen, Germany.

Title:
Optimization of HRMS screening methods by Potential Interference Counting (PIC)

Introduction:
Currently there is a multitude of HRAM methods available, ranging from single stage HRAM-MS to data independent acquisition to targeted methods. These methods claim improved selectivity, reduced false positive rates and false negative rates. But it seems there are uncertainties which method or instrument setting is appropriate. Here we present a simple method to assess different parameters/methods

Methods:
Unspiked matrix samples were measured by orbitrap LC-MS/(MS) under varying conditions. A software tool was developed to determine potential interferences between a set of analytes and the matrix under different (simulated) resolution settings. Afterwards a mathematical model was applied to assess which conditions minimize the risk of false positive and false negative results.

Results:
Potential interferences of a representative set of pesticides and natural toxins were assessed under different conditions (targeted mode, wide isolation, All-Ion-Fragmentation) from non-spiked matrices. Starting from high resolving power measurements at R=240k different (lower) resolution was simulated for MS and MS/MS to assess the effects on the false negative rates. By applying simple statistical rules the selectivity of different methods could be quantified and compared to each other and to today's gold standard: MRM. Additionally, individual inspection of analytes with respect to potential false positive and false negative rates identified quickly unselective and unfeasible fragments/transitions and combinations of these. Relative comparison of the results can guide the method development with regard to selection of methods and parameters in HRAM screening applications.

Novel Aspect:
novel method to assess false positive and false negative rates of HRAM residue screening methods without the need of spiked standards

21ST INTERNATIONAL MASS SPECTROMETRY CONFERENCE
Toronto, Ontario, Canada
August 20 – 26, 2016

Kelly, Christina,
LECO Corporation

Contributing Authors:
Joseph E. Binkley, LECO Corporation; Lorne E. Fell, LECO Corporation; David E. Alonso LECO Corporation.

Title:
A Solution for Determination of High-Concentration Aromatic Compounds in Finished Gasolines Satisfying ASTM D5769 using a New Benchtop GC-TOFMS

Introduction:
ASTM D5769 is a widely accepted standard method used in the petroleum industry for determination of benzene, toluene, and total aromatics in finished gasolines by GC-MS. A common difficulty encountered with this method is the concern with saturation of ion source leading to nonlinearity, especially for the quantification of the high-concentration aromatic toluene.

Methods:
High split and low volume injections into the GC were utilized in addition to full mass range acquisition with a new benchtop time of flight mass spectrometer. Two methods of data handling were applied, with automatic peak finding and library search against commercially available NIST spectral libraries.

Results:
Method requirements for sensitivity, ion ratio, and calibration linearity were satisfied using a new benchtop GC-TOFMS system, providing a solution for analysis of the aromatic compounds listed in ASTM D5769. Calibration curves were built for the standard method analytes and then applied to samples with the addition of semi-quantification for similar analytes on a sample of 93-octane gasoline, as stipulated in the method.

Novel Aspect:
A new benchtop GC-TOFMS system is used to satisfy the method requirements for ASTM D5769.

Kelly, Christina,
LECO Corporation

Contributing Authors:
David E. Alonso, LECO Corporation; Jonathan D. Byer, LECO Corporation; Lorne E. Fell LECO Corporation; Joseph E. Binkley, LECO Corporation; , .

Title:
Battling Fuel-Washing: Identification of Accutrace S10 in Diesel Samples Using a New Benchtop GC-TOF MS System

Introduction:
Government-subsidized farm fuels are dyed to distinguish them from consumer diesels. "Fuel-washing," where dye is removed from fuel which is resold at higher price to fund terrorist activity, has been a major concern. Accutrace S10, a new fuel marker developed for use in the UK, is more difficult to remove and can be detected at low concentrations to prevent dilution and illegal reselling of fuel.

Methods:
Commercially available diesel samples were spiked with Accutrace S10 and analyzed with full mass range acquisition on a new benchtop GC-TOF MS system. ChromaTOF software was used to identify the fuel marker in matrix at low ppm concentrations, utilizing the Target Analyte Find feature.

Results:
Several spiked diesel samples were analyzed with the fuel marker, Accutrace S10, easily identified at multiple concentrations. Use of Target Analyte Find provided fast data processing with definitive results that allowed detection of the analyte of interest with little interference from the matrix.

Novel Aspect:
A new benchtop GC-TOFMS system was used to identify Accutrace S10 in spiked diesel samples.

Kelman, Megan,
Western University

Contributing Authors:
Justin B. Renaud, Agriculture and Agri-Food Canada; Ken K.-C Yeung, Western University; Amy McMillan Western University; Mark W. Sumarah, Agriculture and Agri-Food Canada.

Title:
Data independent acquisition-digital archiving mass spectrometry (DIA-DA MS) with next-generation sequencing for profiling fungal contaminated silage

Introduction:
Farmers are dependent on silage as a daily feed source for their livestock. However, silages are susceptible to spoilage from fungi such as Penicillium sp. and mycotoxin contamination. Combining next-generation sequencing (NGS) with liquid chromatography-MS data independent acquisition-digital archiving (DIA-DA), we wanted to compare metabolomic and microbial profiles from contaminated silage.

Methods:
Contaminated silage consisting of blends of corn, alfalfa, haylage and barley, was obtained from 26 Canadian farms. Predominant species of filamentous fungi were isolated, and silage was profiled for fungi and bacteria using NGS. Silage samples were extracted and analyzed using the LC-MS DIA-DA method previously developed for mycotoxins in maize, and compared to the archived digital library

Results:
15 individual species of filamentous fungi were isolated from the contaminated silage. The most commonly isolated mycotoxin-producing genera consisted of Penicillium, Aspergillus and Monascus, which have all been reported as common silage contaminants. P. nordicum, which is a known producer of ochratoxin A (OTA) and a common contaminant of cereal products, was isolated once. Using NGS profiling, 103 fungal species and 343 bacterial species were detected between the 26 sites. 46 mycotoxins were detected in the extracted silage from the archived digital library, and corresponded to both pre-harvest and post-ensiling fungal species. Principal Component Analysis (PCA) of the LC-MS DIA-DA data has demonstrated that not only are the associated mycotoxins farm-specific, but they are also specific to the silage substrate.

Novel Aspect:
DIA-DA provides digital archiving for retrospective analysis of samples as new mycotoxins are discovered, or as emerging mycotoxins become of concern

Kennedy, Joseph,
Prosolia

Contributing Authors:
Mariam ElNaggar, Prosolia; Justin Wiseman, Prosolia.

Title:
Imaging Phospholipid Fatty Acid Composition in Different Rat Brain Disease States using DESI and High Resolution Mass Spectrometry

Introduction:
Desorption electrospray ionization (DESI) is an ambient ionization technique not requiring high vacuum or matrix deposition and thus allows detection of low molecular weight species. Both lipids and metabolites exist as nominal mass isobars which can differ by less than 50 mDa. In this report, accurate mass and high mass resolution (> 40,000) experiments demonstrate the value of these attributes

Methods:
Coronal sections of three different type rat brains were flash frozen and 10 micron slices from the cerebellum, hippocampus, and frontal lobe were mounted on slides. The tissue sections were interrogated using DESI as well as liquid micro junction surface sampling. Both techniques were interfaced to a Thermo QExactive Focus mass spectrometer. Mass spectral images were visualized using MSiReader.

Results:
Arachidonic (m/z 303.23) and docosahexaenoic acid (m/z 327.23) were the most prevalent fatty acids in the rat brain samples. Comparison of the ratios of N-Acetyl-L-aspartic acid (m/z 174.04) and N-Acetylglutamine (m/z 187.04) indicated differences in distribution in the three species. A comparison of the distribution of m/z 888.62 (sHexCer (d42:2)) in SHRSP and Control rat brains indicated differences in the Hippocampus and Cortex regions. Images obtained for the distribution of isobaric lipids at m/z 834.5266 (PS (20:3)), 834.5734 (3-O-Sulfoglactosylceramide), and 834.6213 (PC (22:5)) illustrate the need for high resolution and accuracy mass (HRAM) in identification. The distribution of common amino acids and related free fatty acids in brain sections has not been extensively studied since matrix applied in MALDI experiments has interfering ions in the low molecular weight mass region.

Novel Aspect:
DESI-MS provides the capability to image amino acids or free fatty acids as well as traditional phospholipids with minimal sample preparation.

Kibbey, Richard,
Yale University School of Medicine

Title:
The perils, principles, and practice of high-throughput mitochondrial metabolic flux measurements

Introduction:
The high positional specificity of NMR 13C spin label analysis has been used for decades for measurement of metabolic fluxes in vivo. 13C can also be applied as mass labels that are amenable to the high sensitivity and throughput afforded by MS. A platform that could utilize both the sensitivity and positional specificity would enhance metabolic flux measurements.

Methods:
A recent advance has combined the advantages of NMR and MS is the Mass Isotopomeric Multi Ordinate Spectral Analysis (MIMOSA) platform. MIMOSA analyzes the dynamically generated labeling patterns in product ions of sequential intermediates in the TCA cycle to generate a more comprehensive high resolution picture of mitochondrial metabolism.

Results:
Positional labeling was determined even in symmetric or prochiral metabolites via biochemistry-guided, NMR validated analysis of product ions. Side by side analyses of position-specific isotopomeric vs. non-specific isotopologue approaches demonstrated the importance of knowing which carbons carry the tracer label. By capturing the position-specific transfer of label between sequential metabolites along a metabolic pathway, MIMOSA helps distinguish between net fluxes and exchange reactions. Importantly modeling dilutional exchanges corrects large errors in the calculation of important metabolic fluxes (e.g., pyruvate carboxylase and pyruvate kinase). Differential mobility ion prefiltration was compatible with high throughput analyses of key metabolic nodes. Dynamic tracking of positional isotopomers offers additional advantages for detailed mechanistic studies.

Novel Aspect:
MIMOSA identified challenges to core assumptions of many flux calculations and offers high throughput/content analyses of mitochondrial metabolism.

Kiehne, Andrea,
Bruker Daltonik GmbH

Contributing Authors:
Volker Auwärter, Institute of Forensic Medicine, Medical Center – University of Freiburg, Faculty of Medicine, University of Freiburg, Albertstraße 9, 79104 Freiburg, Germany; Rafaela Martin, Bruker Daltonik GmbH, Fahrenheitstraße 4, 28359 Bremen, Germany; Laura M. Huppertz Institute of Forensic Medicine, Medical Center – University of Freiburg, Faculty of Medicine, University of Freiburg, Albertstraße 9, 79104 Freiburg, Germany; Jürgen Kempf, Institute of Forensic Medicine, Medical Center – University of Freiburg, Faculty of Medicine, University of Freiburg, Albertstraße 9, 79104 Freiburg, Germany.

Title:
Quantitation of Synthetic Cannabinoids in Serum: A Comprehensive and Sensitive Multiplex Assay for 99 compounds by LC-MS/MS

Introduction:
Synthetic cannabinoids (SC) challenge toxicologists for nearly a decade now. With a constantly changing product portfolio, marketed e.g. as `incense´ and alleged legal alternative to cannabis products, SC keep on spreading onto the drug market with new compounds replacing those being scheduled by legal authorities. The presented multiplex assay allows the simultaneous analysis of 99 SC in serum.

Methods:
A mixture of 18 deuterated SC as internal standards was added to 1 mL of serum prior alkaline liquid-liquid extraction. The LC-MS/MS (Bruker Advance UHPLC coupled to an EVOQ TQ MS equipped with an HESI source) was operated in positive ESI. Chromatographic separation was performed on a Kinetex 2.6 µm column using 12-minute gradient elution with formic acid/acetonitrile.

Results:
Method validation was performed according to forensic guidelines. Matrix effects were assessed according to Matuszewski et al.. Limits of detection ranged from 5 to 25 pg/mL. For the majority of analytes, the assay is linear within the working range of 0.05 to 1.25 ng/ml. Accuracy, sensitivity and imprecision were within the required limits. For new designer drugs, the number of isotope labelled analogues available is usually limited. Therefore, validation criteria were not completely met for all of the analytes. For compounds not fulfilling all validation criteria, semi-quantitative results can be obtained. The calibration range of the assay reflects the concentrations usually found in blood samples after consumption of SC. The limits of detection are sufficient to prove a recent uptake of SC in forensic cases like roadside testing, abstinence control or post mortem analysis.

Novel Aspect:
The method allows for the rapid and sensitive quantitation of 99 SC including recently emerged substances like Cumyl-PINACA-5F, MDMB-CHMICA or EG018.

Kiehne, Andrea,
Bruker Daltonik GmbH

Contributing Authors:
Joop Hoogvliet, Synergy Health Pharmaceutical Laboratories (SHPL) Utrecht, The Netherlands; Zoltan Czentnar, Bruker Daltonik GmbH, Bremen, Germany; Rob van der Heijden Bruker Nederland BV, Leiderdorp, The Netherlands; Annemiek Kosse, Utrecht University, Department of Pharmaceutical Sciences, Utrecht, The Netherlands.

Title:
Use of LC-UV-ion trap MSn for comprehensive quality control of active pharmaceutical ingredients (API): a case study of gamma-irradiated oxybutynin

Introduction:
This case study was initiated after detection of an unknown degradation product in a gamma-irradiated sample of the anticholinergic drug oxybutynin HCl during routine drug substance QC based on HPLC with UV detection at 210 nm. The sample was qualified as being non-compliant with the specifications of the Ph. Eur. monograph 1354 which defines a limit of 0.1% for unspecified impurities.

Methods:
Oxybutynin control and irradiated sample were analyzed by LC-UV-ion trap MSn (Bruker Daltonics) with APCI in positive and negative mode. After differential analysis of the samples, the impurity was identified based on its MSn spectra and comparison with a reference standard. For comprehensive drug QC a fully automated LC-UV-MSn workflow was applied based on Compass OpenAccess QC software.

Results:
Differential analysis of the LC-APCI-MS data of oxybutynin control and irradiated sample enabled the detection of the impurity. MS/MS spectrum interpretation and comparison to a reference standard resulted in the identification of cyclohexyl phenyl ketone (CPK). For comprehensive QC of oxybutynin samples an automated LC-UV-MSn workflow was established based on OpenAccess QC software. Verification of the drug substance identity is based on its molecular formula and MSn spectral library matching. The purity of the oxybutynin samples is determined by UV detection at 210 nm; the identity of the impurities is confirmed by UV and MS spectra. The workflow was tested with spiked oxybutynin samples at levels of 0.1 and 1% CPK. Experimental results being at 0.1 and 1.3% respectively, showed good agreement. The relative abundance of CPK in the irradiated oxybutynin sample was determined at 0.13%.

Novel Aspect:
Fully automated LC-UV-ion trap MSn workflow including reporting for QC of active pharmaceutical ingredients using combined UV and MSn information.

Kiesilä, Anniina,
University of Jyväskylä

Contributing Authors:
Michael Groessl, Tofwerk AG; Lauri Kivijärvi, University of Jyväskylä; Filip Topic University of Jyväskylä; Elina Kalenius, University of Jyväskylä.

Title:
Simultaneous endo- and exo- complexation to pyridine[4]arenes in gas phase studied by ESI-QTOF and ion-mobility mass spectrometry

Introduction:
Pyridinearenes are analogous to resorcinarenes, which are widely studied in supramolecular chemistry. Both molecules self-assemble forming dimeric and hexameric capsules, but due to electronical differences pyridinearenes are expected to encapsulate anions instead of cations. In our recent study encapsulation properties and structures of pyridinearenes were studied by ESI-QTOF and IM-MS.

Methods:
Complexation was studied by ABSciex QSTAR ESI-Q-TOF MS using profile spectra, competition experiments, titrations and CID experiments. Samples were prepared in ACN, acetone, DCM or their mixtures. CID and drift tube IM-MS were used to study structural features of the complexes. Complexation was also studied by NMR spectroscopy (1H, 19F, DOSY), DFT calculations and X-ray crystallography.

Results:
Complexation properties of pyridinearenes were tested with different anions, and most abundant complexes were seen with non-coordinating anions BF4- and PF6-. Complexes with a solvent molecule and anion were also observed. Affinity of pyridinearene towards anions was studied by competition experiments, showing the preference for BF4-. By solvent titrations affinity was determined to favor DCM. Surprisingly, CID and IM-MS revealed endo-complexation of neutral molecules and exo-complexation of anions. Observation of exo-complexation is relatively common in solid state but not previously observed in gas phase. Crystal structures showed that similar structures can exist also in solid state. All gained data were further supported by NMR and theoretical calculations. Particularly IM-MS was found as invaluable tool to detect and separate supramolecules with endo/exo-complexation ability.

Novel Aspect:
Simultaneous endo/exo-complexation to pyridinearene was detected by MS. IM-MS was found as valuable tool to study structures of supramolecules.

KIM, Yoonhwan,
KIST(Korea Institute of Science and Technology)

Contributing Authors:
Jisun Oh, Molecular Recognition Research Center, Korea Institute of Science and Technology; Hong Seog Seo, Cardiovascular Center, Korea University Guro Hospital; Kyoungheon Kim Department of Biotechnology, College of Life Sciences and Biotechnology, Korea University; Bong Chul Chung, Molecular Recognition Research Center, Korea Institute of Science and Technology; Jeongae Lee, Molecular Recognition Research Center, Korea Institute of Science and Technology.

Title:
Urinary global metabolic profiling of Korean population with metabolic syndrome via UPLC–TOFMS

Introduction:
Over the past few decades, the prevalence of metabolic syndrome has been expanding significantly. Furthermore, it is related with diabetes, hypertension, hyperlipidemia and obesity which are becoming common dieases in Korea. However, its metabolic etiology was not fully understood. This study seek to know endogenous marker's fluctuation by metabolic syndrome and its related metabolism.

Methods:
To complementarily investigate polar and non-polar metabolites in metabolic syndrome, we used both reverse-phase chromatography (RPC) and hydrophilic interaction chromatography (HILIC) separation mode on UPLC/Q-TOF MS. The study population was recruited from Korea University Guro Hospital [metabolic syndrome patients (n=183), normal controls (n=163), 181 women and 183 men].

Results:
Multivariate data analysis were performed by PCA, PLS-DA and OPLS-DA. The result showed clear discrimination between metabolic syndrome patients and normal control groups. Compare to single RPC separation mode, we had got 1.4 times increased features from mutual supplementation analysis of RPC and HILIC separation mode [RPC: a total of 487 (VIP>4) out of 21827 variable ions, HILIC: a total of 210 (VIP>1) out of 13133 variable ions]. The most distinctive endogenous markers were phenylalanine, tyrosine, phenylacetylglutamine, 2-phenylacetamide, 4-hydroxyphenylacetate, hippurate, 5-oxo-proline, glutamine, creatinine from phenylalanine metabolism, glutamine and glutamate metabolism, arginine and proline metabolism. Especially, glutamine and glutamate metabolism was highly involved, such as glutamate is well known for insulin resistance, arteriosclerosis, obesity, and so on.

Novel Aspect:
In our study, metabolic different with metabolic syndrome patients and normal control group was found by UPLC-QTOF/MS and it will need further study.

Kim, Seung Yong,
Korea Basic Science Institute

Contributing Authors:
Hyun Sik Kim, Korea Basic Science Institute; Mo Yang, Korea Basic Science Institute; Wanseop Jeong Korea Basic Science Institute.

Title:
Low Power Consuming Handheld Portable Ion Trap Mass Spectrometer

Introduction:
Several innovative low power consuming technologies were adopted to develop a handheld portable mass spectrometer. Cold electron source, miniature vacuum system and nano-pulse injection valve were combines with a module RF amp and a low power electronics. The integrated portable mass spectrometer showed mass spectra with 0.7 u resolution, 0.5 ppm detection limit and 4.4 W average consuming power.

Methods:
The electron ionization source was generated with 2 microchannel plate(MCP) and one of MCPs were illuminated by UV photons to generate the secondary electrons. The miniature vacuum system used only ion getter pump without any rotary and turbo pump, but the initial vacuum state was generated with an initial vacuum system. To maintain the vacuum state, an accurate 50 nl pulse value has been adopted.

Results:
The cold electron source generates no heat, so that an ionization device can be small, low power consuming, and light polymeric materials can be used for accurate machining. The cold electron ionization consumes only 200 mA in pulsed mode, while a hot cathode electron gun generally draws more than 2 A. For a battery operated portable mass spectrometer(MS), the cold electron ionization provided a critical method to save power. The miniature high vacuum system uses only ion getter pump, but the initial vacuum state was generated with a rotary and a turbo pump system. After initial vacuum state was generated, the pumping system was separated from the MS. The pocket MS has been operated in the stable vacuum system around 10-6 torr. A prototype portable MS has been built with a volume ~1.83 L (211 X 117.5 X 73.8) mm3 and weight 1.6Kg. It showed 0.7 u mass resolution and 0.5 ppm detection limit.

Novel Aspect:
The smallest handheld mass spectrometer was fabricated with a low power consuming cold electron ionization source.

Kinumi, Tomoya,
National Metrology Institute of Japan (NMIJ)/ AIST

Contributing Authors:
Tomoya Kinumi, NMIJ/AIAT; Akiko Takatsu, NMIJ/AIST.

Title:
Quantification of serum C-reactive protein by LC-MS/MS using non-labeled internal standard protein

Introduction:
Absolute quantification of a specific protein in serum by mass spectrometry has been a crucial issue. We have developed the quantification method using an internal standard protein with amino acid substitution of the target protein to create the mass difference. We show here the application of serum C-reactive protein (CRP) quantification using this non-labeled internal standard protein by LC-MS.

Methods:
The internal standard protein having two point mutations in human CRP (mCRP, V20L and S53T) was prepared. The wild type CRP (target protein) and mCRP were collected by immobilized antibody, and digested by trypsin. The two-sets of signature peptides, CRP 14-23 (CRP1), CRP 48-57 (CRP2) and the mutated forms, CRP1 V20L and CRP2 S53T were quantified by LC-MS/MS.

Results:
Four signature peptides, CRP1/CRP1 V20L and CRP2/CRP2 S53T were successfully observed. The amino acid substitution did not affect the retention time on the separation. We prepared four CRP concentrations of serum sample by spiking certified reference material of CRP solution into CRP depleted serum to be 90, 465, 940, 1880 ng/mL. The quantification of these serums was performed by double ID method using mCRP as the internal standard and CRM of CRP. The quantification results were 119 (90), 352 (465), 731 (941), 1393 (1883) ng/mL, where the prepared concentration were indicated in parentheses. These results show that the quantification results were the range of 20-30 % of the prepared concentrations. This approach allows ID MS-like quantification using non-labeled internal standard protein, and is applicable to quantification of low concentration range of serum CRP.

Novel Aspect:
This is the first example of a quantification of serum protein using an non-labeled internal standard protein with amino acid substitution.

Kirk, Jayne,
Waters

Contributing Authors:
Mark Wrona, Waters Corporation, Milford, MA, US; Russell Mortishire-Smith, Waters Corporation, Wilmslow, UK; David Eatough Waters Corporation, Wilmslow, UK.

Title:
Towards an Automated Approach to Characterizing the Metabolism of Biotherapeutic Drugs

Introduction:
As the complexity and development of biotherapeutic drugs evolves, there is an increasing interest in, and need, for efficient characterisation of their metabolic fate. Here, we describe the incorporation of intelligent software tools to predict metabolic routes, and mine high resolution mass spectrometric data for the analysis of biotherapeutic drug metabolism.

Methods:
A variety of therapeutic peptides were incubated with human hepatocytes for 0, 60 and 120 min. Data were collected on an ACQUITY UPLC coupled to a benchtop IMS QToF using HDMSE. Scientific information system (UNIFI) was employed for acquisition, processing and reporting. A tailored cleavage strategy was used to allow non-specific amide cleavages of the peptides to take place.

Results:
In this study, several peptides were used to demonstrate the feasibility of employing intelligent software tools to characterize their metabolic fate. Through the use of a single integrated platform, we show that data can be collected routinely, processed and reviewed efficiently across time courses and treatments. Cleavage and fragment ion assignment tools, employed historically for small molecule identification, were modified to enable rule sets appropriate for various classes of biotherapeutics. Screening tools will be demonstrated which accurately screen, detect and group multiple charge states for catabolites. A tightly integrated, UPLC HRMS informatics platform with high chromatographic performance, high mass accuracy and fast scanning capabilities afforded by the IMS QTof enabled rapid and confident detection of peptides and any corresponding metabolites.

Novel Aspect:
Investigating informatics and HRMS for monitoring fate of biotherapeutics

Kirk, Jayne,
Waters

Contributing Authors:
Paul Doorbar, Waters Corporation, Wilmslow, UK; Martin Green, Waters Corporation, Wilmslow, UK; Kirsten Craven Waters Corporation, Wilmslow, UK; Witold Niklewski, Waters Corporation, Wilmslow, UK; Steven Lai, Waters Corporation, Beverly, MA, US; Keith Richardson, Waters Corporation, Wilmslow, UK; Jason L Wildgoose, Waters Corporation, Wilmslow, UK; Mark Wrona, Waters Corporation, Milford, MA, US.

Title:
Utilising a novel, data independent, mode of acquisition on a QTof instrument in combination with a scanning quadrupole mass filter for lipid analysis

Introduction:
Quadrupole time of flight (QTof) instruments have found great utility in data-dependent modes of operation. Here we describe a new mode of operation which a resolving quadrupole mass filter is scanned, concomitantly with MS/MS data acquisition, thereby enabling the specificity of MS/MS fragmentation whilst maintaining a powerful, unbiased, data-independent approach for analysis of phospholipids.

Methods:
The quadrupole mass filter of a QTof has been modified, to allow it to operate in a continuously and repetitively scanning mode. The instrument is switched between a post-quadrupole fragmentation and a non-fragmentation mode after each scan to provide fragment ion profiles, each of which can be assigned to the precursor ion profiles resulting from scanning the quadrupole transmission window.

Results:
Preliminary data show that fragment ions can be assigned to precursor ions with uncertainties in alignment that are significantly smaller than the resolving quadrupole transmission window. This improved specificity / selectivity leads to a demonstrable increase in confidence in the assignment of fragment ions to their corresponding precursor ion when compared with the standard MSE mode of acquisition or a stepped/fixed mode of MS/MS. The unbiased nature of this approach provides meaningful qualitative and quantitative information for all identified phospholipids. The validity of the approach was tested using a complex phospholipid rat brain sample which contained many closely eluting lipids, with precursor ions that fragment to generate the same (or similar) product ions. Here, we show examples of phosphatidylcholine (PC) lipids clearly resolved with clean and unambiguous MS/MS spectra.

Novel Aspect:
Novel, data independent, mode of acquisition strategy with high duty cycle and precursor ion specificity for the application of lipid analysis.

Klassen, John,
University of Alberta

Contributing Authors:

Ling Han, University of Alberta; Jun Li, University of Alberta; Elena Kitova University of Alberta.

Title:

New MS tools for the discovery and characterization of protein-glycan interactions

Introduction:

Protein interactions with glycans present in the plasma membrane represent an important class of recognition events that are implicated in a wide variety of cellular processes, including cell-cell adhesion, signalling, immune response and pathogen infection. Here, we describe new ESI-MS methods for the discovery and quantification of protein-glycan interactions.

Methods:

All of the ESI-MS and catch-and-release (CaR)-ESI-MS measurements were carried out using a Waters Synapt G2S quadrupole/ion mobility separation-ToF mass spectrometer equipped with a nanoflow ESI source.

Results:

The application of the CaR-ESI-MS assay, implemented using nanodiscs and picodiscs, to detect low and high affinity glycolipid interactions with the B subunit homopentamers of cholera toxin and shiga toxin, and a sub-fragment of toxin A from Clostridium difficile will be described. The reliability of the CaR-ESI-MS assay, implemented using micellar solutions of glycolipids, for detecting protein-glycolipid interactions will also be discussed. Finally, a newly ESI-MS assay, referred to as the proxy ligand ESI-MS method, for quantifying association constants for protein-GL interactions and investigating the influence of lipid environment thereon, will be described.

Novel Aspect:

Development of new ESI-MS methods for the discovery and quantification of protein-glycan interactions

Knox, Ruth,
NSERC CREATE York University and University of Toronto

Title:
Investigating Allosteric Hotpots and Drug Binding Affinity of TEM-1 β-Lactamase using TRESI-HDX-MS

Introduction:
We are interested in studying conformational changes in the non-binding regions of TEM-1 (Class A β-lactamase) to develop allosteric inhibitors for drug therapy and reduce antibiotic overuse. Our aim is to discover non-active site regions of high deuterium exchange upon substrate binding that can be modeled for drug therapies without using epitopes or fusion proteins.

Methods:
The binding effect of TEM-1 on β-lactam antibiotics was studied using Time-Resolved ElectroSpray Ionization Hydrogen Deuterium eXchange (TRESI-HDX) mass spectrometry. Experiments were performed on a Q-Star Elite quadropole time-of-flight (Q-TOF) instrument. The experimental deuterium uptake of each peptide obtained was calculated using in-house developed FORTRAN software.

Results:
Native TEM-1 protein sequence coverage was 54%. Deuterium uptake of the native protein has given evidence that regions associated with common mutant variants of TEM-1 exhibit high levels of deuterium uptake. HDX experiments of TEM-1 incubated with Ampicillin have provided evidence of peptide fragments not associated with active site binding that exhibit a dramatic increase in deuterium uptake upon substrate binding. Future studies will focus on a weak binding substrate (Cephalexin) and an inhibitor (Potassium Clavulanate) to compare conformational changes upon binding.

Novel Aspect:
We hope give new insight on conformational changes associated with substrate binding and provide new evidence to other drug binding pockets.

Koch, Annika,
University of Münster, Biomedical Mass Spectrometry

Contributing Authors:
Jens Soltwisch, Institute of Hygiene, University of Münster, Robert-Koch-Str. 41, 48149 Münster, Germany; Andreas Schnapp, Institute of Hygiene, University of Münster, Robert-Koch-Str. 41, 48149 Münster, Germany; Klaus Dreisewerd Institute of Hygiene, University of Münster, Robert-Koch-Str. 41, 48149 Münster, Germany.

Title:
Multiply Charged Peptides and Proteins from Liquid Matrices Using Laser Spallation at UV/VIS and Near-IR Wavelengths and Rapid Evaporative Ionization

Introduction:
Conventional MALDI-MS is characterized by the predominant generation of singly charged ions. However, some important mass analyzer types cannot be used for the detection of larger singly charged biomolecules. Here we used an AP ion source equipped with a wavelength-tunable laser and a heated ion transfer capillary to generate multiple charges of peptides and proteins from liquid glycerol matrices.

Methods:
The AP ion source [1] was mounted to a QSTAR mass spectrometer. Either an OPO laser (wavelength range: 260-1080 nm) or a Nd:YAG laser (l=1.064 µm) served to induce material ejection. Glycerol mixed with TFA or a composite system of glycerol and classical UV-absorbing dihydroxybenzoic acid MALDI matrices were investigated. Sample substrates were held at 1 kV and the inlet tube heated to ~225 °C.

Results:
Independent of wavelength and matrix system, laser ablation rapid evaporative ionization mass spectrometry (LA-REIMS) produced charge states that widely resembled ESI data. For melittin (2845 Da) the three and four times protonated species formed the base peaks, whereas cytochrome C (12384 Da) produced charge states between 7 and 16. Surprisingly, best results in terms of the analytical sensitivity were obtained if the glycerol/TFA system was used at near-IR wavelengths. For the chromophore-containing matrices best results were obtained at non-absorbing wavelengths, in notable contrast to conventional MALDI. We hypothesize that material ejection is caused by low-energy spallation of charged droplets and ionization of analytes is caused by rapid disintegration of the droplets in the heated transfer tube.

Novel Aspect:
High yields of multiply charged peptides are obtained from glycerol-based matrices using an AP-ion source and UV, visible or near-IR light

Krause, Eberhard,
Leibniz-Institut für Molekulare Pharmakologie Berlin

Contributing Authors:

Eberhard Krause, Leibniz-Institut für Molekulare Pharmakologie, Berlin, Germany; Martin Penkert, Leibniz-Institut für Molekulare Pharmakologie, Berlin, Germany; Dorothea Fiedler Leibniz-Institut für Molekulare Pharmakologie, Berlin, Germany; Michael Schümann, Leibniz-Institut für Molekulare Pharmakologie, Berlin, Germany; Lisa Yates, Department of Chemistry, Princeton University, Princeton, NJ, USA.

Title:

Analysis of protein pyrophosphorylation by CID and EThcD mass

Introduction:

Protein pyrophosphorylation is a non-enzymatic modification on serine amino acid residues (ppSer), which can be formed by the reaction of a pre-formed phosphoserine with the β-phosphoryl group of an inositol pyrophosphate. Although previous studies have identified pyrophosphorylated proteins in vitro, a direct method for the detection of pyrophosphorylated proteins in vivo is still missing.

Methods:

Pyrophosphorylated peptides were synthesized as recently described. Peptides were analyzed by reversed phase nanoLC (Dionex Ultimate 3000) ESI-MS/MS (Orbitrap Fusion). Doubly and triply charged peptide ions were fragmented using CID, ETD, and EThcD. CID MS/MS-spectra were acquired in the linear iontrap (collision energies 25 - 35%). EThcD spectra were acquired using an AGC target value of 1e5.

Results:

CID fragmentation of pyrophosphorylated peptides leads to a complete cleavage of the modified side-chain. Measurements with reduced collision energies provide pyrophospho-specific neutral loss patterns, which can be used to distinguish pyrophosphorylated from isobaric diphosphorylated peptides. In contrast, pyrophosphorylated side-chains remain intact during ETD measurements. In particular, EThcD MS/MS spectra of doubly and triply charged precursor ions are very suitable for sequence assignment of pyrophosphorylated peptides. The combination of the characteristic CID neutral loss pattern with a neutral loss-triggered EThcD acquisition method enables an unambiguous localization of the site of modification. We envision that this strategy will be the method of choice to identify protein pyrophosphorylation in vivo from complex phosphoproteome samples.

Novel Aspect:

Unambiguous identification of serine and threonine pyrophosphorylation using CID/EThcD mass spectrometry

Krogh, Erik,
Applied Environmental Research Labs, Department of Chemistry

Contributing Authors:

Ryan Bell, Beaver Creek Analytical; Larissa Richards, Applied Environmental Research Labs, Chemistry, Vancouver Island University and University of Victoria; Nicholas Davey Applied Environmental Research Labs, Chemistry, Vancouver Island University and University of Victoria; Chris Gill, Applied Environmental Research Labs, Chemistry, Vancouver Island University and University of Victoria.

Title:

Mobile membrane introduction mass spectrometry for real-time geo-spatial resolution of atmospheric trace organics

Introduction:

Volatile organic compounds (VOCs) in the atmosphere may arise from both natural and anthropogenic sources. While some VOCs are directly detrimental to human health (e.g., BTEX), they also indirectly contribute to poor air quality through their role in the formation of secondary organic aerosols and ground level ozone.

Methods:

A ruggedized cylindrical ion trap mass spectrometer modified with a polydimethylsiloxane membrane interface was continuously operated in a moving vehicle collecting tandem mass spectra for targeted quantitative analysis as well as full scan mass spectra for non-targeted qualitative analysis. Data streams are time and location stamped enabling both temporal and spatial resolution.

Results:

Recent advances in instrument miniaturization and portability have enabled real-time mobile environmental monitoring with applications to fence line monitoring, adaptive sampling strategies and source apportionment. These include the use of a continuously infused internal standard, magnetic shielding of the ionization source, and chemical ionization strategies to increase sensitivity and/or selectivity. We report the use of multivariate statistics on the full scan mass spectra of VOC mixtures in constructed and real-world contaminated air samples. Preliminary data suggests that we can discriminate between samples and identify potential sources. Atmospheric data from several field campaigns, viewable in Google Earth, will be presented.

Novel Aspect:

Geo-spatially resolved chemical fingerprinting of atmospheric VOCs using a ruggedized mobile membrane introduction mass spectrometer.

Krokhin, Oleg,
University of Manitoba

Contributing Authors:
Peyman Ezzati, Manitoba Centre for Proteomics and Systems Biology; Haley Neustaeter, Department of Chemistry, University of Manitoba; Victor Spicer Manitoba Centre for Proteomics and Systems Biology.

Title:
Optimal selection of separation chemistry for 3D-LC-MS/MS in bottom-up proteomics

Introduction:
The growing complexity of proteomics samples drives the development of advanced multi-dimensional separation schemes. RP HPLC offers superior peptide separation efficiency, but limits the choice of orthogonal separation chemistry. To address this issue we employed advanced peptide retention modeling for optimal selection of eluent conditions for 3D (RP-RP-RP) LC-MS bottom-up proteomics.

Methods:
1D, 2D and 3D LC-MS methods were compared based on identification output and utilization of MS instrument time. The 2D procedure was based on the well-established RP-RP high pH – low pH approach, with fraction concatenation. The 3D protocol used C18 separation with heptafluorobutyric acid (HFBA, 6 concatenated fractions) as the ion pairing modifier, followed by the 2D procedure for each fraction.

Results:
This approach permitted a detailed evaluation of the analysis output consuming 21x and 126x the analysis time and sample load compared to 1D (a tryptic digest of Jurkat cell lysate). Acquisition over 189 hours of instrument time in the 3D mode resulted in the acquisition of 2.4 million MS/MS spectra, identification of ~14,000 proteins and ~250,000 unique peptides. The uniform distributions of the number of acquired MS/MS, protein and peptide identifications across all 21 fractions in 2D and 126 fractions in 3D indicates a good separation orthogonality. The deep fractionation also permitted identification of a large number of post-translationally modified peptides (e.g. phosphorylated species). The protocol is scalable and amenable to the use of peptide retention prediction in all dimensions for improved confidence of MS/MS-based identifications.

Novel Aspect:
We developed first 3D LC-MS protocol based on C18 chemistry, with separation orthogonality exceeding previously described 3D methods in proteomics.

Kruve, Anneli,
University of Tartu

Contributing Authors:
Piia Liigand, University of Tartu; Karl Kaupmees, University of Tartu; Jaanus Liigand University of Tartu; Mari Sild, University of Tartu.

Title:
Predicting concentrations of small molecules without standard substances in LC/ESI/MS via ionization efficiency scales

Introduction:
The recently presented ionization efficiency (IE) scales for ESI that can be used for predicting concentrations for compounds for which the standard substances are not available. It is proven that using these scales concentrations can be predicted by two simple parameters – ionization degree of the molecule in the solvent and degree of charge delocalization in the formed ion.

Methods:
The ESI ionization efficiency (logIE) scale has been measured in different solvents in both positive and negative mode. Molecular properties (pKa, WAPS, etc) were calculated with COSMO-RS computational method. Multilinear regression models were compiled for predicting logIE from the molecular parameters. Overfitting was prevented by including only statistically significant parameters.

Results:
The model has been developed for more than 200 compounds in positive and 100 compounds in negative mode. The compounds range from small polar pesticides to oligopeptides (up to 1300 Da). The accuracy for logIE prediction for validation set was 0.3 logarithmical units. As an example, the best logIE prediction model obtained for negative ESI in 0.1% of formic acid in water / acetonitrile 20/80 was: logIE=+1.04+2.23alpha-0.51WAPS (alpha– dissociation degree, WAPS – charge delocalization parameter in anion). Similar models in both negative and positive mode have been obtained also for other solvent systems. The concentrations predicted (based on the estimated logIE values) were on the average 2 times different from the actual concentrations, making this approach significantly more accurate then could be achieved purely from peak area ratios.

Novel Aspect:
Universal model for predicting concentrations from ESI/MS responses was developed for both positive and negative mode.

Kryuchkov, Fedor,
University of Bergen

Contributing Authors:
Mathias Ziegler, University of Bergen; Frank Kjeldsen, University of Southern Denmark.

Title:
Application of dimetallic complex for mass spectrometric analysis of phosphate-containing metabolites

Introduction:
An UHPLC-MS/MS analysis of metabolites is a well-established method. Since highly polar water soluble metabolite cannot be analyzed on reverse phase column, HILIC columns become the most popular choice. This fact restricts using a nano-scale LC. Here we propose tagging of phosphate-containing metabolites which convert them to less hydrophilic ions, which can be retained on standard C18 columns.

Methods:
Targeted and untargeted metabolites were analyzed on Dionex UltiMate 3000 RSLC Systems coupled with QExactive mass spectrometer (Thermo Fisher Scientific). UHPLC system was equipped with SeQuant ZIC-pHILIC (Merck Millipore) and Acqiuity UPLC HSS T3 (Waters) columns. Nanoflow LC was performed on Agilent 1260 Infinity System equipped with ReproSil-Pur 120 ODS-3 (Dr. Maisch) column.

Results:
Wide range of phosphate-containing metabolite standards was investigated in accordance with Dimetal Phosphate Ester Stabilization (DIMPES) approach. As expected, Ga-containing dimetallic complex shows high selectivity to metabolite phosphate group. This modification improves ionization efficiency, which allows us to fragment not only singly charged, but also multiple charged ion species. The most important finding is the changing of chromatographic properties of polar metabolites, which allows their analysis on reverse phase columns without need of ion pairing reagents and accompanying undesired effect of ion suppression. We compared sensitivity of popular HILIC-based setup with conventional and nanoscale reversed-phase HPLC setups. We found that tagging of metabolites with Ga-complex can improve sensitivity to one-two orders of magnitude dependent on instrumental setup.

Novel Aspect:
Tagging of phosphate-containing metabolites allows their analysis on reverse phase columns and provides alternative structural information.

Kubo, Akiko,
Keio University

Contributing Authors:
Kentaro Fujii, Department of Internal Medicine, School of Medicine, Keio University; Hiroshi Itoh, Department of Internal Medicine, School of Medicine, Keio University; Kazutoshi Miyashita Department of Internal Medicine, School of Medicine, Keio University; Makoto Suematsu, Department of Biochemistry, School of Medicine, Keio University.

Title:
Dense accumulation of adenosine in the outer stripes of outer medulla in murine kidney identified by microscopic MALDI-imaging mass spectrometry

Introduction:
Adenosine is a well-known regulator of renal function and plays a protective role against ischemic kidney injury; however, the exact distribution of adenosine in the kidney has not been elucidated. MALDI - imaging mass spectrometry is a technology used for visualizing the in situ distributions of molecules and we visualized physiological distribution of adenosine in murine kidney.

Methods:
Kidneys from eight-week-old mice anesthetized were harvested and frozen by liquid nitrogen within a second (snap-frozen) or a minute (acute ischemia) after cutting blood vessels. 10-μm cryosections were mounted on ITO-slide followed by being coated with 2,5-DHB for positive mode, or coated with 9-aminoacridine for negative mode to conduct microscopic imaging mass spectrometry.

Results:
Physiologically important low-molecular weight metabolites including adenylates (ATP, ADP, AMP, adenosine, inosine, hypoxanthine and xanthine) were successfully visualized. High energy phosphorylated adenylates were detected from both renal cortex and outer stripes of outer medulla (OSOM) and remarkable accumulation of adenosine in the OSOM was identified in the snap-frozen kidneys. Although adenosine usually accumulates in the ischemic area where phosphorylated adenylates have degraded, the OSOM were paradoxically rich in both ATP and adenosine. In addition, the accumulation of adenosine disappeared in the ischemic sections, although ischemia usually increases adenosine. These facts suggested that the accumulation of adenosine was not caused by degradation of phosphorylated adenylates alone and implied the existence of unique mechanism to maintain adenosine accumulation in the OSOM.

Novel Aspect:
MALDI-IMS provides essential information for the exact understanding of renal physiology through defining the distribution of small molecules.

Kukacka, Zdenek,
Steinbeis Centre for Biopolymer Analysis and Biomedical Mass Spectrometry

Contributing Authors:
Anna Maria Papini, Interdepartmental French-Italian Laboratory of Peptide and Protein Chemistry and Biology, Fiorentina, Italy; Michael Przybylski, Steinbeis Centre for Biopolymer Analysis and Biomedical Mass Spectrometry, Rüsselsheim, Germany and University of Konstanz, Konstanz, Germany; Lorenzo Altamato Interdepartmental French-Italian Laboratory of Peptide and Protein Chemistry and Biology, Fiorentina, Italy; Marius Iurascu, Steinbeis Centre for Biopolymer Analysis and Biomedical Mass Spectrometry, Rüsselsheim, Germany and University of Konstanz, Konstanz, Germany; Jeff Bornheim, Reichert-Ametek Technologies, Buffalo, USA; Mary Murphy, Reichert-Ametek Technologies, Buffalo, USA.

Title:
Epitope Identifilcation and Affinity Characterization of Human Alpha-Galactosidase A to a Monoclonal Antibody by SPR-Affinity Mass Spectrometry

Introduction:
α-Galactosidase (aGal) hydrolyses the terminal galactose from glycolipids. Disorder caused by deficiency of human aGal A is called Fabry Disease and its treatment is carried out by enzyme replacement therapy. Since immune reaction is a therapy-limiting complication, the present study was focused on the identification and affinity characterization of the aGal epitope against a monoclonal antibody.

Methods:
Epitope determination was performed by SPR-MS, using a newly developed online combination of an SPR-biosensor coupled via a microfluidic interface to ESI-MS. The human aGal antibody was immobilized on a gold chip of the affinity interface and a tryptic peptide mixture was loaded onto the online SPR-MS system. After washing out nonbinding peptides, epitopes were eluted, desalted and analyzed by MS.

Results:
ESI-MS revealed the presence of a single epitope fraction, aGal(309-322). Synthesis of the epitope was performed and following SPR (Ametek-Reichert 7500-2Ch-SPR). analysis revealed the dissociation constant of interaction to be in the micromolar range These results show that the new online SPR-MS is a powerful tool to elucidate and structurally characterize the antibody interaction of human alpha-galactosidase, for the development of a potential immune hyposensitizing epitope in enzyme replacement therapy.

Novel Aspect:
New online SPR-MS system is capable to determine the dissociation constant of interaction between affinity partners and revealed their structure.

Kulyk, Kostiantyn,
Stockholm University

Contributing Authors:

Borys Palianytsia, Chuiko Institute of Surface Chemistry; Mats Larsson, Stockholm University; Michael Gatchell Stockholm University; Tetiana Kulik, Chuiko Institute of Surface Chemistry; Mykola Borysenko, Chuiko Institute of Surface Chemistry; Henning Zettergren, Stockholm University; John D. Alexander, Stockholm University.

Title:

Dimethylsilanone generation from pyrolysis of pure and filled polysiloxanes: a study by desorption mass spectrometry

Introduction:

Polydimethylsiloxane (PDMS) is a widely used polymer often employed in formulations with fine oxide particles for various high temperature applications. Although PDMS is considered to be thermally stable and chemically inert, it is not always clear how the filler influences its thermoresistance, decomposition chemistry and what reactive products are formed in the underlying thermal reactions.

Methods:

In this work we used a combination of temperature programmed desorption mass spectrometry (TPD MS) and thermogravimetric analysis (TGA) to study the pyrolysis of PDMS and its composites with nanosized silica and ceria/silica.

Results:

Our results suggest that the elusive organosilicon compound – dimethylsilanone (DMS) is generated from PDMS over a broad temperature range (in some cases starting at 70 °C). The presence of nano-oxides was found to catalyze this process. Ions characteristic of the fragmentation of DMS under electron ionization were assigned with the aid of density functional theory (DFT) structure calculations. Possible reaction mechanisms for generating DMS were discussed in the context of the calculated kinetic parameters. Observed accompanying products of PDMS pyrolysis, such as tetramethylcyclodisiloxane (TMCDS) and hexamethylcyclotrisiloxane (HMCTS), indicate that multiple channels are involved in the dimethylsilanone release. In the food processing PDMS/silica is used as an antifoaming agent for edible oils and fats – products that are generally designated for cooking and frying.

Novel Aspect:

The generation of such a reactive compound as DMS over a broad temperature range questions the general belief in the safety of using PDMS additives.

Kumar, Mukesh,
MPI-CBG, Dresden, Germany

Contributing Authors:
Martina Augsburg, Medical Systems Biology Medical Faculty, TU Dresden, Germany; Marc Gentzel, MPI-CBG, Dresden, Germany; Nadine Vastenhouw MPI-CBG, Dresden, Germany; Andrej Shevchenko, MPI-CBG, Dresden, Germany; Frank Buchholz, Medical Systems Biology Medical Faculty, TU Dresden, Germany; David Drechsel, MPI-CBG, Dresden, Germany; Shai Joseph, MPI-CBG, Dresden, Germany.

Title:
Mass spec Western: Target Absolute Quantification of Proteins and Post-translational modifications

Introduction:
Western blot owing to its simplicity is still preferred over mass spectrometry for quantification of a limited selection of biologically interesting proteins,although it suffers from specificity,dynamic range and multiplexing capabilities.We present a technically simple (like WB) GeLC-MS based method which can rapidly quantify a user-defined constellation of moles of proteins from cell-organism

Methods:
Method relies on the recombinant expression of 13C15N-Arg and 13C-Lys labelled chimeric protein, composed of multiple reporter peptides from target proteins. The chimeric and the target proteins gel band were co-digested and the recovered tryptic peptides were analysed by LC-MS/MS. The abundance of reference peptides determines the absolute content of the chimera and thereby of target proteins

Results:
Absolute quantification of protein constellations in moles opens up the attractive possibility to determine the stoichiometry of subunits of protein complexes, protein abundances within pathways, as well as molar ratios of proteins to non-proteinous ligands, such as DNA and lipids. We successfully implemented the method in quantifying the absolute amount and stoichiometry of histone proteins from developing zebrafish embryo at 10 developmental stages. We found that histones were initially present in large excess than actually required to pack the genomic DNA into nucleosomes. Histone excess was depleted at the onset of zygotic transcription suggesting that histones may act as repressors before zygotic genome activation. We also employed this method in flies to decipher, how the diet dependent change in lipidome influences the membrane proteome.The method could be extended to quantify PTM

Novel Aspect:
Accurate absolute quantification of protein constellations from cell to organism

Kung, Jocky,
University of Toronto

Contributing Authors:
Benjamin Schuler, University of Zurich; Rebecca Jockusch, University of Toronto; Martin Czar University of Toronto.

Title:
Multiple Pair Förster Resonance Energy Transfer (FRET) as a Distance Probe for Gas-phase Ubiquitin Ions

Introduction:
To further our understanding of the effects of desolvation on protein conformation, we develop combined optical spectroscopy with mass spectrometry techniques. Here, we employ Förster Resonance Energy Transfer (FRET) as a tool to estimate distances between multiple pairs of residues in gas-phase ubiquitin ions and to study the effects of different mass spectrometric factors on its conformation.

Methods:
Four variants of ubiquitin are labelled with pairs of FRET active dyes. Each variant is electrospray ionized, stored inside a quadrupole ion trap and irradiated by 446 nm light. Fluorescence emission spectra and lifetimes of the variants at different charge state and mass spectrometric conditions are measured. FRET efficiencies are used to estimate distances between the dye pair in each variant.

Results:
Initial steady-state fluorescence emission measurements of the isolated ubiquitin ions in the 5+, 6+ and 7+ charge states of two variants, which have dye pairs on residues (6,48) and (20,74), indicate significant energy transfer between the two dyes in each variant. The FRET efficiencies of the two variants are similar, where for both the 5+ and 6+ ions show around 90% FRET, compared with around 65% FRET for the 7+ ions. This suggests substantial unfolding of the 7+ ions in comparison to the lower charge states, which is consistent with earlier ion mobility data. Donor fluorescence lifetime measurements of the 7+ ions in both variants indicate the presence of at least two distinct conformations. Moreover, trap temperature is shown to affect measured FRET efficiencies, but variations in other MS parameters, such as ion accumulation time, result in no significant efficiency changes.

Novel Aspect:
Multiple FRET pairs used to determine multiple distance constraints for a gaseous protein, enabling separate tracking of distinct structural regions

Kunold, Elena,
TU Munich

Contributing Authors:
Philipp Kleiner, TU Munich; Stephan A. Sieber, TU Munich.

Title:
Identification of the molecular target of a growth inhibiting compound in Staphylococcus aureus using a chemical proteomics strategy.

Introduction:
Despite the huge arsenal of antibiotic drugs available these days, infections caused by S. aureus still represent a major health threat. This problem is mainly caused by resistances to established as well as newly introduced agents. As resistances are spreading rapidly there is an urgent need for the development of new classes of agents that address novel yet unexploited cellular targets.

Methods:
Based on a phenotypic screen of a library of small molecules, we discovered a compound which is highly active against S. aureus growth. We applied a chemical proteomic strategy to identify the molecular target of the compound using a photoreactive derivative thereof. The workflow involved affinity-based enrichment and gel-free mass spectrometry-based quantification using dimethyl labeling.

Results:
We identified a promising putative protein target and are currently investigating the detailed mechanism of action by further biochemical methods. Furthermore, we were able to optimize the hit scaffold to gain a 10-fold higher activity against S. aureus growth. The compounds are stable in mouse plasma und show little cytotoxicity in human cells, making them promising drug candidates.

Novel Aspect:
We present agents that potently inhibit staphylococcal growth and show activity against a novel protein target, for which no drugs are approved yet.

Kupervaser-Cohen, Meital,
Weizmann Institute of Science

Contributing Authors:
Alon Savidor, Weizmann Institute of Science; Yishai Levin, Weizmann Institute of Science; Alexandra Gabashvili Weizmann Institute of Science; Dalia Elinger, Weizmann Institute of Science.

Title:
On-Beads Digestion: Improved Proteins Identification Rate by Addition of a Chaotropic Agent

Introduction:
On-beads digestion is widely used for proteomic analysis of co-immunoprecipitation experiments and a key process in discovering new protein-protein or DNA/RNA-protein interacting partners. Since the goal of these experiments requires performing them in as close as possible to native conditions, many parts of the proteins are inaccessible to the digesting enzyme, resulting in poor identification.

Methods:
Here we introduced urea, a chaotropic agent widely used to denature proteins in other mass-spectrometry compatible processes, such as filter-aided sample preparation (FASP), into the on-beads digestion mixture prior to adding the digesting protease. We used 1M of urea to balance between the denaturation of the proteins in the mixture and retaining the activity of trypsin.

Results:
Different co-immunoprecipitation experiments were subjected to on-beads digestion in the presence of 1M urea or with no urea added. The identification rates of the 1M urea on-beads digested proteins were higher than the traditional protocol, even though there was a slight increase in the mis-cleavage rates.

Novel Aspect:
Achieving higher identification rates of proteins in co-immunoprecipitation experiments, allowing a wider and more comprehensive screen

Kurulugama, Ruwan,
Agilent Technologies

Contributing Authors:
George Stafford, Agilent Technologies, Santa Clara, CA; Brandon Ruotolo, University of Michigan, Ann Arbor, MI; Joseph Eschweiler University of Michigan, Ann Arbor, MI; John Fjeldsted, Agilent Technologies, Santa Clara, CA.

Title:
Protein Structural Studies using Ion Mobility MS and In-Source Ion Activation Techniques

Introduction:
Ion mobility coupled with mass spectrometry is becoming a widely used technique in protein structural analyses. Ion fragmentation and tandem mass spectrometry are universally applicable techniques in compound identification. In a similar fashion, collision induced unfolding technique coupled with IM-MS provides information for intact proteins in structural studies and pharmaceutical applications.

Methods:
An IM-QTOF (6560 IM-QTOF, Agilent Technologies) instrument was used for this study. For initial experiments Carbonic anhydrase and Bovine serum albumin proteins were prepared under native and denatured conditions. Samples were directly infused at 200 nL/min flow rate using a syringe pump and Agilent NanoESI source. Data were analyzed using IM-MS Browser software and CIUSuite software program.

Results:
For native carbonic anhydrase, charge state +8 - +11 were observed with very narrow single peak CCS profiles. For denatured protein, charge states +12 - +38 were observed with much broader peaks. For some charge states (e.g. +12 - +16) multiple conformers are observed. For native BSA charge states, +30 - +43 were observed, while for denatured state, +35 - +58 charge states were observed. For denatured BSA, protein structural changes were observed for charge states +27, +31-32, +35-38. For denatured carbonic anhydrase, charge states +15-23, and +27 showed structural changes with collision activation. Collision induced activation data for monoclonal antibody proteins as well as kinase/Inhibitor complexes will be presented. Further improvements to the source design for improved ion activation and fragmentation are being implemented and will be discussed.

Novel Aspect:
In-source ion activation coupled with uniform field ion mobility MS for protein structural analyses.

Kurulugama, Ruwan,
Agilent Technologies

Contributing Authors:
Bruce Wang, Agilent Technologies; Gregor Overney, Agilent Technologies; Frank Kuhlmann Agilent Technologies; George Stafford, Agilent Technologies; Joseph Roark, Agilent Technologies; John Fjeldsted, Agilent Technologies.

Title:
Complex Protein Digest Analyses Using Ion Mobility Drift Time Directed Quadrupole Wide-band Isolation for Improved Peptide Coverage

Introduction:
The benefit of applying quadrupole wide-band isolation (WBI) following the ion mobility traces during an ion mobility mass spectrometry experiment (IM-MS) is to significantly improve the specificity and coverage of analytes of interest. Directing the wide-band isolation based on IM drift times within the IM experiment significantly improves the duty cycle compared to conventional WBI method.

Methods:
Experiments are performed with IM trace aware quadrupole wide-band isolation and contrasted with corresponding IM All-Ions experiments. Quadrupole wide-band isolation (50 to 150 m/z) is accomplished with a new quadrupole driver. The timing map and isolation window position are calculated from the IM traces and entered into a table for a subsequent quadrupole wide-band isolation experiment.

Results:
With quadrupole wide-band isolation following ion mobility traces, we see significantly less ion signal interferences compared to IM All-Ions experiments while maintaining a high duty cycle. These improvements lead to enhanced protein sample coverage for complex protein digests. A bovine serum albumin tryptic digest was spiked in to Ecoli protein trypitic digest to evaluate this new approach. A standard flow LC separation method with a 150 mm long C18 reversed phase column was used prior to IM-MS analyses. Ion Mobility Feature Extraction (IMFE) software tool was used to identify the precursor ions from the MS1 spectra and combine those with the fragment ions data from the MS2 spectra. Spectrum Mill software was used for database search and peptide coverage results.

Novel Aspect:
IM separation directed quadrupole wide-band isolation within ion mobility experiment.

Kusinski, Matthew,
University of Toronto

Contributing Authors:
Rebecca A. Jockusch, University of Toronto.

Title:
Conformational Changes of Ubiquitin Upon Heating Assessed with FRET in a QIT Mass Spectrometer

Introduction:
Understanding molecular structure and the thermodynamics of its change is vital in the chemical and biological sciences. Intermolecular interactions influence structure. Here, changes in the efficiency of Förster resonance energy transfer (FRET) monitor thermally-induced conformational changes of molecular ions stored in a heated quadrupole ion trap mass spectrometer split from those influences.

Methods:
Modifications to a commercial QIT mass spectrometer (Bruker Esquire 3000+) enable heating of the helium bath gas, which in turn heats the ions stored within via collisions. This instrument has been previously modified to allow optical excitation of mass-selected trapped gaseous ions, and collection of their emitted fluorescence. Effective ion temperature is calibrated using thermometer molecules.

Results:
Proof of principle experiments employed a variant of ubiquitin, labelled at residues 20 and 74 with a fluorescent dye pair (Alexa Fluor 488 and 594) suitable for gas-phase FRET experiments. Efficiency of energy transfer between the donor and acceptor dye is strongly distance dependent. Therefore, changes in FRET efficiency are sensitive measurements of the change in distance between pairs. FRET efficiency of the 6+ charge state of ubiquitin, formed from aqueous solution by electrospray, was monitored at room temperature and at an increased trap temperature. At room temperature, ion mobility suggests that ubiquitin 6+ primarily adopts a compact structure under gentle conditions, but that more extended conformations are thermally accessible. The measured FRET efficiency decreased ~17% upon heating, indicating thermally induced unfolding of the conformation of ubiquitin +6 in the gas phase.

Novel Aspect:
Conformation change of gaseous ubiquitin induced using a novel heating design and monitored via changes in FRET efficiency.

Kuzdzal, Scott,
Shimadzu Scientific Instruments

Contributing Authors:
Jeff Dahl, Shimadzu Scientific Instruments, Columbia, MD.; Paul Winkler, Shimadzu Scientific Instruments, Columbia, MD.; Gordon Fagras Trace Anmalytics, Spokane, WA.; Simon Ashton, Shimadzu Corporation, Manchester, UK.; Julie Kowalski, Restek Corporation, Bellefonte, PA.; David Baker, Shimadzu Corporation, Manchester, UK.; Neil Loftus, Shimadzu Corporation, Manchester, UK.

Title:
Mass Spectrometry Applications in Medical Cannabis Contamination Testing.

Introduction:
Medical cannabis refers to the use of cannabis and its corresponding cannabinoids, as a therapy to treat diseases and alleviate symptoms. Mass spectrometry has emerged as a powerful tool for cannabis QC testing and research, and is particularly useful for the analysis of cannabis contaminants, including pesticides, mycotoxins/aflatoxins, heavy metals and residual solvents.

Methods:
Ground and homogenized cannabis was extracted with a modified QuEChERS and dispersive SPE cleanup. Detection was carried out by UHPLC-MS-MS using a Shimadzu Prominence HPLC with LCMS-8050 triple quadrupole mass spectrometer. Electrospray ionization was used with continuous polarity switching to allow analysis of pesticides in both positive and negative mode during the same run.

Results:
Approximately 200 pesticide analytes were measured with over 500 MRM transitions per run in 12 cannabis samples in Washington State. Five samples obtained at retail establishments contained no detectable pesticide residues, whereas pesticides were identified in seven samples obtained from individual growers known to have used pesticides. Matrix matched calibration curves were linear within the quantitation limits established for each pesticide, which was compound dependent, but ranged from as low as 1.5 ng/g at the lower limit to 2000 ng/g at the upper limit. A LC method was optimized for the top ten most commonly monitored cannabinoids. Additionally, a headspace GC-MS technique was developed for the analysis of cannabis residual solvents (including N-butane, i-butane, methanol, n-pentane, i-pentane, isopropanol, chloroform, benzene, toluene, and acetone).

Novel Aspect:
Ultra-fast mass spectrometry enables simultaneous quantitation of hundreds of pesticides by LC-MS/MS with exceptionally fast cycle times.

Kuzdzal, Scott,
Shimadzu Scientific Instruments

Contributing Authors:
Hikaru Shibata, Shimadzu Corporation, Kyoto, Japan.; Brian Field, Shimadzu Scientific Instruments, Columbia, MD.; Daisuke Kawakami Shimadzu Corporation, Kyoto, Japan.; Toshikazu Minohata, Shimadzu Corporation, Kyoto, Japan.

Title:
A Fully Integrated Sample Preparation Automation Module for LC-MS/MS with Demonstrated Application to Steroid Hormone Research.

Introduction:
Sample preparation for mass spectrometry workflows can often be tedious, time-consuming and user dependent. Automation allows laboratories to work more efficiently by removing the human factor from sample preparation, freeing up time for other tasks and eliminating mistakes. CLAM-2000 automates common sample preparation steps for LCMS analysis including pipetting, shaking, filtration and heating.

Methods:
Ten steroid hormones in sera were verified using CHS™ MSMS Steroids Kit (PerkinElmer). Serum samples were loaded directly into the automated sample preparation system, which was programmed to perform protein precipitation using acetonitrile followed by filtration and sample collection. Treated samples were trapped using a MAYI-ODS column and then separated by core-shell biphenyl HPLC column.

Results:
We evaluated this system using calibrator and control serum spiked with 10 steroids and carried out concurrent analyses over a range of concentrations for each steroid: cortisol (1.51-320 ng/mL), aldosterone (0.03-7.05 ng/mL), 11-deoxycortisol (0.08-18 ng/mL), corticosterone (0.29-62 ng/mL), 17-OHP (0.12-26 ng/mL), androstenedione (0.08-18 ng/mL), DHEA (0.31-65 ng/mL), DHEAS (12.9-2750 ng/mL), progesterone (0.12-26.5 ng/mL) and testosterone (0.03-7.2 ng/mL). The calibration curves that were generated had linear regression values of $r2 > 0.997$ for each curve. The reproducibility (N=3) at seven concentrations, including LLOQ of each compounds was excellent (CV<10%). We found that the sample preparation time was reduced from 60 minutes to 10 minutes by the automated system. The CLAM-2000 is for Research Use Only. Not for Use in Clinical Diagnostics.

Novel Aspect:
Automated clinical research sample prep now interfaces directly to a fast scanning LC-MS/MS for improved reproducibility and sample prep efficiency.

Kuzmich, Oleksandra,
Humboldt Universität zu Berlin

Contributing Authors:
Uschi Gruber, Caprotec bioanalytics GmbH Berlin; Matthias Hakelberg, Caprotec bioanalytics GmbH Berlin; Thomas Lenz Caprotec bioanalytics GmbH Berlin; Mathias Dreger, Caprotec bioanalytics GmbH Berlin; Michael W. Linscheid, Humboldt Universität zu Berlin; Hubert Köster, Caprotec bioanalytics GmbH Berlin; Simon Michaelis, Caprotec bioanalytics GmbH Berlin.

Title:
Metal labelling as a detection functionality for tri-functional Capture Compounds (CCs) to detect protein targets of small molecules

Introduction:
Accurate quantification and identification of proteins and peptides is challenging in modern proteomics. Metal tags have been reported as ones that allow to test this aspect. The usage of functionalized small molecule probes to target proteins that specifically bind to drug molecules, followed by detection of lanthanide tags, represents a novel approach in chemical proteomics.

Methods:
Tri-functional Capture Compounds composed of a selectivity function, a photo-reactive function, and a lanthanide containing detection function were synthesized to detect specifically binding proteins. As a readout, inductively coupled plasma mass spectrometry quantification (ICP-MS) was used. To prove the concept, labelling of an isolated protein and ICP-MS –based quantification was performed.

Results:
A first proof of concept experiment was performed on the methyltransferase M. TaqI from Thermus aquaticus, showed reproducible labeling (RSD 17%). M. TaqI was targeted by a tri-funtional CC containing S-adenosyl-L-homocysteine, a product inhibitor of many MTases, as the selectivity function, a photo-reactivity function, and a Tb-DOTA complex as detection function. Gel-based analysis allowed the detection of specific photocross-link of the CC to M. TaqI based on a mass shift of the labeled species, and ICP-MS based quantification allowed an estimation of the cross-link yield of the probe to the target protein. Another CC with the dopamine D2 receptor (DRD2) antagonist sertindole as selectivity function was used to target DRD2 expressed on the surface of living HEK293 cells. Results are compared to fluorescence readout using ananalogous CC with a fluorescent detection function.

Novel Aspect:
First synthesis and application of a tri-functional CC to target specific binders of small molecules, using a metal tag for detection by ICP-MS.

Kwok, Wai Him,
Racing Laboratory, The Hong Kong Jockey Club

Contributing Authors:
Zhong-Ping YAO, State Key Laboratory of Chirosciences, Food Safety and Technology Research Centre and Department of Applied Biology and Chemical Technology, The Hong Kong Polytechnic University; Timmy L. S. CHOI, Racing Laboratory, The Hong Kong Jockey Club; Terence S. M. WAN Racing Laboratory, The Hong Kong Jockey Club; Pui-Kin SO, State Key Laboratory of Chirosciences, Food Safety and Technology Research Centre and Department of Applied Biology and Chemical Technology, The Hong Kong Polytechnic University.

Title:
Simultaneous Detection of Xenon and Krypton in Equine Plasma by Gas Chromatography / Tandem Mass Spectrometry for Doping Control

Introduction:
Xenon is alleged to have been used in human and horse sports to enhance performance while Krypton, a potential and less-expensive Hypoxia-Inducible Factor activator, is recently advertised on the internet for enhancing performance. Our paper describes the first method to separate and detect both noble gases in horse plasma for doping control.

Methods:
Xenon and krypton in equine plasma were simultaneously detected by gas chromatography / triple quadrupole mass spectrometry using headspace injection. Three xenon isotopes (m/z 129, 131 and 132) and four krypton isotopes (m/z 82, 83, 84 and 86) were targeted in selected reaction monitoring mode with precursor-product ion pairs set at identical masses.

Results:
Xenon and krypton could be detected in equine plasma with limits of detection at 19 pmol/mL and 98 pmol/mL respectively. Precisions for both analytes were less than 15%. The method has good specificity as background signals were not observed in negative equine plasma samples (n = 73). Loss of analytes under different storage temperatures (-80 °C, 4 °C and ambient temperature ~25 °C) was also evaluated, and the results showed that 4 °C is the most suitable temperature to conserve xenon and krypton in equine plasma.

Novel Aspect:
This is the first method to separate and detect both xenon and krypton in horse plasma for doping control.

Lacoursiere, Jean,
Phytronix Technologies Inc.

Contributing Authors:
Alex Birsan, Phytronix Technologies Inc.; Serge Auger, Phytronix Technologies Inc.; Pierre Picard Phytronix Technologies Inc.

Title:
High Throughput Quantification of Tenofovir and Emtricitabine in human plasma using LDTD-MS/MS

Introduction:
Tenofovir (TFV) and emtricitabine (FTC) are prescribed anti-retroviral drugs used for the treatment of HIV infection. Various LC-MS/MS methods are reported in literature for the quantification of individual drugs or in a single method. We developed a novel extraction method using LDTD-MS/MS for a fast quantification of Tenofovir and Emtricitabine in plasma at 9 seconds sample to sample.

Methods:
150 µL of plasma sample, 25 µL IS and 250 µL FA were mixed and loaded on an SCX. After washing step, drugs are eluted using 0.6 mL of NH4OH (10%). Evaporation of elution and a quick derivatization at room temperature using TMPAH is performed to obtain methyled drug form. Samples are diluted and 4 µL is spotted onto LazWell and dried. The mass spectrometer is operated in positive MRM mode.

Results:
The following validation parameters were evaluated: linearity, intra-run and inter-run accuracy/precision, extracted sample stability and matrix effect. Blank plasma was spiked with Tenofovir and Emtricitabine at concentrations ranging from 25 to 5000 ng/mL in order to build calibration curves. Calibration curves had excellent linearity with R2 values between 0.9958 and 0.9988. The intra-run accuracy and precision across the calibration curves were between 95.4 to 104.1% and 2.8 to 15.6% for both compounds, respectively. Following the extraction procedure, all samples were stored at 4°C to evaluate the drugs stability in a wet state over time. After a time lapse, all samples were re-spotted and analyzed. A wet stability greater than 12h was obtained with accuracy between 87.4% and 98.8%. Precision between 6.0% and 14.1% for concentrations equivalent to the LLOQ is observed.

Novel Aspect:
The LDTD technology combined with a mass spectrometer system allows ultra-fast drug quantification in plasma samples with a fast turnaround time.

Ladak, Adam,
Waters

Contributing Authors:
Lauren Mullin, Waters; Rhys Jones, Waters.

Title:
Harnessing the Efficiency of N2 as GC Carrier Gas with an Atmospheric Pressure MS Source

Introduction:
Helium (He) is the most commonly used carrier gas in gas chromatography applications; however recent global shortages have resulted in increasing costs. Nitrogen (N2) is a more affordable and readily available option that has historically been less utilized as a GC carrier gas. Reasons for this include lower diffusivity than He or hydrogen and often longer run times to achieve similar separation

Methods:
Here we show GC coupled with an atmospheric pressure ionization mass spectrometry, which utilizes N2 for both ionization and make-up flow. This approach allows a single gas source to be used for chromatographic separation as well as ionization.

Results:
Following automated method transfer calculations available within the software, efficient and comparable chromatography using N2 carrier gas were achieved for pesticides, polybrominated diphenyl ethers (PBDEs), polychlorinated biphenyls (PCBs), and chlorinated dioxin/furans (PCDD/Fs). Unlike traditional electron ionization, atmospheric pressure ionization is resilient to the introduction of N2. Also, higher gas load allowance into the atmospheric pressure source afforded flexibility with regards to method translation such that appropriate velocities could be achieved for N2 carrier gas. System maintenance, robustness of analyte signal and separation as compared to He carrier gas are also described.

Novel Aspect:
N2 carrier gas for GC

Ladak, Adam,
Waters

Contributing Authors:
Anna Karmann, MTM Research Centre, Orebro University, Orebro, Sweden; Lauren Mullin, Waters.

Title:
DETERMINATION OF AQUEOUS FILM FORMING FOAM (AFFF) COMPONENTS USING A MULTIVARIATE ANALYSIS APPROACH

Introduction:
Aqueous film-forming foams (AFFFs) have been implemented in both military and commercial fire-fighting activities to extinguish flammable liquid fuels. However, uses of these formulations have resulted inadvertently in the release of contaminants into the environment due to migration from the site of application. The various formulations of AFFFs consist of numerous fluorocarbon and hydrocarbon

Methods:
Data was acquired using alternating high and low collision energy states across the full analytical mass range, such that product ions were also generated. Samples were diluted in methanol and chromatographic separation performed using an existing liquid chromatography method for the analysis of perfluorinated compounds. Instrumental performance with regards to mass accuracy (<5 ppm mass error),

Results:
Distinctive grouping from PCA plots could be observed for three AFFFs in both modes, with cluster of five of the seven in ESI+ and four of the seven in ESI-. Pool samples, which contain all ions used in the multivariate analysis, were clustered appropriately towards the middle. In order to arrive at specific marker associates, group-to-group comparisons were carried out with two foams at a time, resulting in the generation of S-Plots. Investigation of the labeled markers strongly associated with specific groupings using structural elucidation tools within MS software resulted in the identification of multiple sulfate, hydrocarbon and fluorinated compounds. Further interrogation of the markers using trend plots to indicate presence and abundance across all the injections yielded additional identifications that were either unique to specific formulations or in some cases common compoun

Novel Aspect:
Data independent acquisition HRMS analysis of complex AFFF mixtures with differentiation based on chemical fingerprints

Lai, Cheuk Kuen,
MS-ESE York University - NSERC CREATE

Contributing Authors:

Alan C. Hopkinson, Department of Chemistry and Centre for Research in Mass Spectrometry, York University, 4700 Keele Street, Toronto ON, Canada M3J 1P3; Justin Kai-Chi Lau, Department of Chemistry and Centre for Research in Mass Spectrometry, York University, 4700 Keele Street, Toronto ON, Canada M3J 1P3; Department of Chemistry and Biochemistry, University of Windsor, 401 Sunset Avenue, Windsor, ON, Canada N9B 3P4; Xiaoyan Mu Department of Chemistry, The University of Hong Kong, Hong Kong, China; K.W. Michael Siu, Department of Chemistry and Centre for Research in Mass Spectrometry, York University, 4700 Keele Street, Toronto ON, Canada M3J 1P3; Department of Chemistry and Biochemistry, University of Windsor, 401 Sunset Avenue, Windsor, ON, Canada N9B 3P4; Ivan K. Chu, Department of Chemistry, The University of Hong Kong, Hong Kong, China.

Title:

Nucleophilic substitution in the aromatic rings of [zn - H]•+ ions via amide nitrogen

Introduction:

Gas-phase fragmentation of peptides using mass spectrometry is important for peptide sequencing and protein identification. [zn - H]•+ ions are one of the common dissociation pathways of peptide radical cations originated from N–Cα bond cleavage of aromatic residues. Here, novel residue-specific product ions from dissociation of tyrosine- and tryptophan- containing [zn – H]•+ ions are reported.

Methods:

Molecular peptide radical cations (M•+) were generated through dissociative electron transfer in ternary metal–ligand–peptide complexes, [Cu(ligand)(M)]•2+. Two density functionals, B3LYP and M06-2X, with a 6-31++G(d,p) basis set have been used to explore possible nucleophilic substitution pathways.

Results:

The CID spectrum of the canonical peptide radical cation tyrosylglycylglycylglycine [Yπ•GGG]+ produced two unusual product ions, [c1 – 17]+ at m/z 162 and [b2 – H – 17]•+ at m/z 203. These two ions are found to be the major dissociation products of the [z4 – H]•+ ion generated from [Yπ•GGG]+. In addition, [c1 – 17]+ (m/z 185 and 146) and [b2 – H – 17]•+ (m/z 226 and 187) ions are observed in the CID spectra of [z3 – H]•+ ions generated from [Wπ•GG]+ and [FGGα•]+ respectively. Further investigation reveals the novel residue-specific fragment ions [c1 – 17]+ (m/z 162 and 185) and [b2 - H - 17]•+ (m/z 203 and 226) are present in the low-energy CID of a series of tyrosine- and tryptophan- containing [zn - H]•+ ions respectively. Isotopically labelled amino acids and density functional theory have been used to systematically study this novel residue-specific fragmentation.

Novel Aspect:

Novel nucleophilic attack by the backbone on the aromatic ring

Lam, Brian,
NSERC CREATE York University and University of Toronto

Contributing Authors:
K.W. Michael Siu, Department of Chemistry and Centre for Research in Mass Spectrometry, York University, 4700 Keele Street, Toronto, ON, M3J 1P3, Canada; Department of Chemistry and Biochemistry, University of Windsor, 401 Sunset Avenue, Windsor, ON, N9B 3P4, Canada; Alan C. Hopkinson, Department of Chemistry and Centre for Research in Mass Spectrometry, York University, 4700 Keele Street, Toronto, ON, M3J 1P3, Canada; Cheuk-Kuen Lai Department of Chemistry and Centre for Research in Mass Spectrometry, York University, 4700 Keele Street, Toronto, ON, M3J 1P3, Canada.

Title:
The Water Loss from Protonated Polyglycines

Introduction:
The Mechanism for the loss of water from protonated polyglycines is dependent on the length of the peptide sequence. There have been two mechanisms that have been studied in detail for triglycines and tetraglycines. The dominant mechanism in these systems are different. Here we study the mechanism for the dissociation of water from longer protonated polyglycines.

Methods:
Protonated polyglycines were studied using collision-induced dissociation. Isotopic labelling of various oxygens on the peptide backbone was used to determine from which amide there is loss of water. Esterfication of the carboxylic acid end of polyglycines was employed to determine whether there was loss of water from the C-terminus.

Results:
The trend for the CID of polyglycines suggest that water loss becomes more prominent with longer peptide chains. Also, longer polyglycines are more likely to lose multiple waters, all from amide groups. Early results suggest that the traditional mechanism for the loss of water via oxazolone formation at the C-terminus is never the dominant pathway as polyglycine length increases. Furthermore, the dominant loss of water from polyglycines comes from the residues located near the N-terminus.

Novel Aspect:
The long held belief that water loss comes from formation of oxazolones at the C-terminus is not the dominant pathway as polyglycine length increases.

Lamoliatte, Frederic,
Institute for Research in Immunology and Cancer, University of Montréal

Contributing Authors:
Clémence Robert Rinfret, Institute for Research in Immunology and Cancer, Québec, Canada; Francis P. McManus, Institute for Research in Immunology and Cancer, Québec, Canada; Pierre Thibault Institute for Research in Immunology and Cancer, Québec, Canada.

Title:
Crosstalk between SUMO and phosphorylation during the biogenesis of PML Nuclear bodies

Introduction:
PML Nuclear Bodies (NBs) are vesicles that are formed during various cellular stresses. NBs are a center for protein modification and control protein stability via proteasomal degradation. However, the protein content of NBs remains mostly unknown. Here, we describe a method to profile changes in protein SUMOylation and phosphorylation and asses their significance on the transformation of PML NBs.

Methods:
HEK293 6xHis-SUMO3-Q87RQ88N cells were grown in triple SILAC media. Cells that incorporate amino acids were either mock treated, treated with ATO or with a combination of ATO and MG132 for up to 8h. For phospho-enrichment, proteins were digested and enriched on titanium dioxide column. For SUMO-enrichment, proteins were enriched on NiNTA beads, digested and purified with a custom-made antibody.

Results:
Our preliminary results reveal over 200 kinetic profiles corresponding to SUMOylated or Phosphorylated peptides identified as significantly regulated by ATO treatment. The protein PML was increased in SUMOylation in the first 2 h and decreased in abundance beyond that time in ATO treated cells, while the SUMOylated form of PML was stabilized in presence of the proteasomal inhibitor MG132. Most of the other known NBs proteins showed a similar behavior, including SUMO itself, Heat shock factor protein 2, Histone deacetylase 1 or Poly [ADP-ribose] polymerase 1. However, this strategy also revealed new possible PML NBs partners, such as Zinc finger protein 106 and Exosome complex component RRP45, thus providing new insights in the dynamic regulation of PML NBs and its interacting partners.

Novel Aspect:
Parallel identification of protein SUMOylation and phosphorylation.

Langley, John,
University of Southampton

Contributing Authors:
Jim Barker, Innospec Ltd, UK; Energy Institute, UK; Maria Ashe, University of Southampton, UK; Julie M. Herniman University of Southampton, UK; Anastarsia Carter, University of Southampton, UK; Edward Wilmot, University of Southampton, UK.

Title:
Utilising unusual ionisation and SFC to solve a taxing problem

Introduction:
Fuel fraud costs worldwide governments ~$100 billion p.a. A colourless marker has been produced1 for 2.5 ppm SIR-GC-MS detection and sub ppm LOD achieved with new GC-MS instruments. Older instruments give ~1 ppm LOD and day-to-day operation compromised by sample matrix. A 3 min. UHPSFC-MS method separates the marker from fuel utilising unexpected ionisation to give 1 ppm SIR detection, S/N >400.

Methods:
A Trace™ GC 2000 quadrupole GC-MS was used, with a RTX-5MS capillary column, He carrier gas 1.0 mL/min, a two-step gradient, 40 to 240°C to 320°C and a 1 µL splitless injection. An Acquity UPC2™ TQD was used with scCO2 mobile phase, a 3 min. 2-40% modifier gradient at a flow rate of 1.5 mL/min. Positive ion API techniques were used and full scan data m/z 150-1500, SIR and MRM methods developed.

Results:
1 ppm LOD SIR-GC-MS was achieved using 40 minute method (molecular ion and base peak ion) with a linear dynamic range (R2 > 0.99) observed over 2 orders of magnitude. The carry-over of fuel matrix is high requiring a rigorous wash regime and daily cleaning of the ion source volume. The 3 minute UHPSFC-MS method delivers a solution that is free from the fuel matrix and carryover problems inherent with GC-MS. Further, the scCO2 is compatible with direct injection of the fuel into the chromatographic eluent to afford an effective method for separation and detection though advantageous and selective of sub ppm fuel marker in different fuel samples (including diesel) with linear dynamic range over 3 orders of magnitude (R2 > 0.99) even though the analyte contains only one heteroatom, an ether. Molecular modelling has been used to probe and understand the unusual ionisation process.

Novel Aspect:
Unusual API ionisation and UHPSFC-MS.

21ST INTERNATIONAL MASS SPECTROMETRY CONFERENCE
Toronto, Ontario, Canada
August 20 – 26, 2016

Langley, John,
University of Southampton

Contributing Authors:
Efstathios Elia, University of Southampton, UK; Tim J. Jenkins, Waters Corporation, UK; Julie M. Herniman University of Southampton, UK; Ed Sprake, Waters Corporation, UK; William Durnie, BP Exploration Operating Company Ltd., UK.

Title:
Ionisation enhancement through use of a supercritical fluid and a novel ionisation technique

Introduction:
Enhanced API signal response for a series of amines is observed through (1) change of sample introduction (RP-UHPLC to UHPSFC) and (2) use of a novel ion source design. Unispray™ is similar to ESI except it uses a charged impactor pin to ionise the spray. A range of 1°, 2° and 4° amines were used to map the change in sample introduction and ionisation, each showing increased MS sample signal.

Methods:
Optimised UHPLC-MS(/MS) and UHPSFC-MS(/MS) sub 2 µm particle size methods were developed for the separation of a range of amines. The novel UnisprayTM ion source was compared to the standard Waters ESI ion source on both instruments. Positive ion ESI and UnisprayTM mass spectra were recorded for all experiments and the relative performance of the two TQD instruments was verified.

Results:
A range of amines were used to probe the ionisation phenomena described above. Initial expectations were that differences between the amines and quaternary ammonium compounds would be observed. This proved not to be so. Approximately a 10-fold increase in sensitivity is was obtained for all samples introduced using a supercritical fluid compared to the reverse phase solvent from a UHPLC. This increase in signal response does not appear to be related to aqueous content. The sensitivity of a novel ion source (UnisprayTM) was compared to that of a standard Waters ESI ion source and this introduces approximately a 6-fold increase in sensitivity. A 60-100 fold increase in sensitivity can be achieved when the UnisprayTM ion source is coupled with UHPSFC compared to UHPLC with ESI. The formation of smaller spray droplets is hypothesised for this increase in ionisation response.

Novel Aspect:
Enhanced API ionisation, novel ionisation source and UHPSFC.

Langridge, David,
Waters Corporation

Title:
Calculation of stability diagrams and quadrupole mass filter performance for general digital waveforms

Introduction:
Digitally driven quadrupole mass filters offer a number of unique capabilities in comparison to conventional analogue devices. The flexibility inherent in the digital drive allows instantaneous changes in the duty cycle, frequency and drive waveform itself. The ability to calculate and compare stability diagrams for differing digital pulse waveforms is critical to understanding these devices.

Methods:
We have implemented the matrix method approach to solving the Mathieu/Hill equation for any given harmonic or digital waveform. This approach allows calculation of stability diagrams, ion trajectories and simulated mass filter spectral peaks. In addition to the matrix method SIMION calculations were used to examine the effects of voltage noise / pulse jitter on mass filter performance.

Results:
Stability diagrams were calculated for a wide variety of digital waveforms. Stable region tip locations were compared to those in the literature. One complication of digital waveforms is that the choice of parameters used to plot the stability diagram is not as well defined as in the harmonic case. The ability to plot stability diagrams based on any definition of the q/a parameters allows a comparison of the differing definitions. We have performed 2D calculations of quadrupole mass filter performance for a range of digital waveforms and stability regions, and compared the results to those for sinusoidally driven mass filters. SIMION simulations have also been performed to examine the effects of realistic pulse shapes, pulse voltage stability and pulse period accuracy on mass filter performance.

Novel Aspect:
Stability diagrams and simulated mass spectra were calculated for a range of different digital waveforms applied to quadrupole mass filters.

21ST INTERNATIONAL MASS SPECTROMETRY CONFERENCE
Toronto, Ontario, Canada
August 20 – 26, 2016

Langridge, David,
Waters Corporation

Contributing Authors:

Chris Hughes, Waters Corporation, Wilmslow, UK; Keith Richardson, Waters Corporation, Wilmslow, UK; Jason Wildgoose Waters Corporation, Wilmslow, UK; Johannes PC Vissers, Waters Corporation, Wilmslow, UK; Jim Langridge, Waters Corporation, Wilmslow, UK.

Title:

Method Optimization of a novel scanning quadrupole DIA method for system-level omics

Introduction:

A novel DIA mode of operation, SONAR, which provides both qualitative and quantitative information, has been developed and implemented on a hybrid quadrupole/oa-time-of-flight (Q-TOF) MS. Here we describe the optimization of analytical parameters to obtain the best qualitative and quantitative coverage of complex proteomic samples.

Methods:

Complex digest samples from E.coli and human cell lines were evaluated. MS data were collected using SONAR acquisition, where a continuously scanning quadrupole isolates specific m/z regions prior to the TOF. Data were processed and database searched using Symphony, incorporating PLGS, Skyline and Scaffold informatics toolsets, and Progenesis QI providing both qualitative and quantitative results.

Results:

LC separation was performed and optimised for 75uM and 300 μm C18 columns using gradients of 30, 45 and 90 minutes. Sample loading of the digest samples was optimised for both columns to proved optimal maximum protein identifications. Experiments with E. coli resulted in 962 proteins and ~12,700 peptides identified from an optimized 90 minute gradient. Human cell lines (K562, Promega) optimized with >4000 proteins and >39000 peptides identified. Loadings were 1ug on a 75uM column and 8ug on a 300uM column. Proteins were identified using a traditional DB search strategy, not searching against a spectral library. Reducing the gradient to 45 minutes for K562 resulted in a 25% reduction in information at the protein level, but still allowed > 3200 proteins to be identified. Further optimization of the method, including the quadrupole transmission window and MS scan rate will be presented.

Novel Aspect:

Optimization of a novel DIA (SONAR) workflow for high throughput Omics profiling

Lariviere, Dominic,
Laval University

Contributing Authors:

Charles Labrecque, Chemistry Department, Laval University; Pablo J. Lebed, Chemistry Department, Laval University.

Title:

ISOTOPIC SIGNATURE OF SELECTED RARE EARTH ELEMENTS FOR NUCLEAR ACTIVITIES PROFILING USING CLOUD POINT EXTRACTION AND ICP-MS/MS

Introduction:

We developed a procedure for the determination of isotopic ratios of three rare earth elements (Nd, Sm, Eu). These light REE are of interest in nuclear science since they could be produced during nuclear fission, and thus can impact the natural isotopic abundance.

Methods:

A new cloud point extraction (CPE) method was used as a pre-concentration/separation strategy prior to measurement by inductively coupled plasma tandem mass spectrometry (ICP-MS/MS). To minimise polyatomic interferences in MS, the combination of decontamination achieved by CPE and the use of the reaction/collision cell in He and NH3 mode was attempted.

Results:

Isotopic ratios were initially measured in samples from San Joaquin soil (NIST-2709a), an area with little contamination of nuclear origin. Finally, samples collected from three sites with known nuclear activities (Fangataufa Lagoon in French Polynesia, Chernobyl and the Ottawa River near Chalk River Laboratory) and they all exhibited altered isotopic ratios for 143/145Nd, 147/149Sm, and 151/153Eu. Through the combination of CPE/ICP-MS/MS precision down to the per mil level was achieved and was sufficient to see alteration in the SRM selected. Using Principal Component Analysis (PCA), it was possible to link the type of nuclear event (reactor release, atomic bombardment) to the isotopic ratios observed.

Novel Aspect:

These results demonstrate the potential of the combined use of CPE and ICP-MS/MS for the detection of altered isotopic ratio in environmental samples.

Larsen, Brett,
LTRI

Contributing Authors:

Joshua Sandor, LTRI Toronto; Meghan McFadden, LTRI Toronto; Anne-Claude Gingras LTRI Toronto.

Title:

Slightly Bigger is Better: Protein Interaction Analysis with 100um ID columns

Introduction:

Affinity-purified protein interaction samples remain challenging as their complexity and dynamic range results in significant risk of carryover. In this study, we explored flow rates between typical nano-flow and micro-flow to compare the sensitivity, throughput and overall performance on 150um, 100um and 75um column platforms.

Methods:

Fused silica emitters were packed with 3um C18 (Reprosil, Dr. Maish) at 1500psi to lengths of 10cm (75um) or 15cm (150um and 100um). Performance was assessed using an Eksigent nanoLC 425 and Sciex TripleTOF 6600 instrument, with flow rates ranging from 200nl/min to 1500nl/min. Protein interaction samples from genes involved in intracellular protein trafficking were analyzed in DDA and SWATH modes.

Results:

Ionization remained stable across the tested flow rates and the observed operating pressure dropped by as much as 70% with wider bore columns. The average peak area for targeted peptides decreased by 30–45% at 100um ID and decreased by 60-75% with 150um ID columns. However, using the identification of peptides and proteins as a metric for analysis of 500ng of human lysate, the performance between 100um ID columns and 75um ID columns was difficult to distinguish. In all cases, approximately 2000 proteins and 20000 peptides were detected. The reproducibility of 100um column chromatography was excellent, even when flow rates and pressure were changed across a set of samples. This supported the analysis of unknown samples at 400nl/min, and higher flow rates during loading, recovery, regeneration and system suitability testing methods. The final project workflow doubled sample throughput.

Novel Aspect:

100um ID nano-flow column; low micro-flow LC; APMS; BioID; system suitability; variable flow rate LCMS workflow

Laserna, Anna Karen,
National University of Singapore

Contributing Authors:
Feng Huatao, Department of Chemistry, National University of Singapore; Sam Li Fong Yau, Department of Chemistry, National University of Singapore.

Title:
LC-MS glycan fingerprinting for the species and serotype-specific determination of bacterial endotoxins

Introduction:
Endotoxins or lipopolysaccharides (LPS) are cell wall components of Gram-negative bacteria. They are the recognition targets of immune receptors and are responsible for the bacteria's toxicity. Current methods of endotoxin analysis are mostly generic, providing only total endotoxin levels. There is a need for a method that can allow the simultaneous and highly specific analysis of endotoxins.

Methods:
LPS samples were subjected to mild hydrolysis to obtain the glycan components and derivatized with a tagging reagent to improve MS ionizability. Different derivatizing agents will also be explored to identify one that can provide good sensitivity and allow high-throughput sample preparation prior to LC-MS analysis. Chemometric techniques will be used to identify characteristic glycan features.

Results:
Preliminary analysis demonstrated species-specific glycan profiles for some bacterial species using one of the derivatizing agents being evaluated. Subsequent analysis will then be performed on different serotypes of a few species for serotype fingerprinting. Results of the evaluation of derivatization methods, as well as the chemometric analysis for species and serotype glycan fingerprinting will be presented.

Novel Aspect:
Distinct LC-MS glycan features will allow the simultaneous species and serotype-specific analysis of endotoxin levels in different samples.

Laskin, Julia,
Pacific Northwest National Laboratory

Contributing Authors:
Ingela Lanekoff, Department of Chemistry-BMC, Uppsala University, Uppsala, Sweden; Son Ngyuen, Physical Sciences Division, Pacific Northwest National Laboratory, Richland, WA 99352.

Title:
Nanospray desorption electrospray ionization (nano-DESI) imaging of biological systems

Introduction:
We have developed nanospray desorption electrospray ionization (nano-DESI), an ambient ionization technique that relies on localized liquid extraction of analyte molecules from the sample followed by soft ionization of extracted species at a mass spectrometer inlet. Nano-DESI enables quantitative imaging of biomolecules in fully hydrated samples with minimal or no sample pre-treatment

Methods:
The nano-DESI probe is assembled using two fused silica capillaries (50 × 150 μm, ID × OD) each mounted on an XYZ stage. The sample is mounted on a custom-designed holder positioned on a high-resolution XYZ stage. Mass spectra are acquired using LTQ-Orbitrap while the sample is moved in lines under the nano-DESI probe. Quantification is achieved by adding appropriate standards to the solvent.

Results:
Nano-DESI has been used for or studying metabolite exchange between living microbial communities and quantitative imaging of lipids and metabolites in tissues. Several examples of this work will be presented. Matrix effects in mass spectrometry imaging (MSI) will be discussed and approaches for mitigating matrix effects and obtaining both absolute and relative quantification in MSI will be presented. Strategies for combining liquid extraction MSI with other scanning probe microscopy techniques will be also discussed.

Novel Aspect:
New approaches for ambient ionization imaging of biological samples.

Lau, Kai Chi,
MS-ESE York University - NSERC CREATE

Contributing Authors:
K.W. Michael Siu, York University and University of Windsor; Alan C. Hopkinson, York University; Ivan K. Chu University of Hong Kong; Xiaoyan Mu, University of Hong Kong.

Title:
Isomerizations and dissociations of FGW radical cations

Introduction:
The dissociation patterns of peptide radical cations are dictated by the position of the charge and the radical. Here, four types of radical cation isomers of FGW (pi-, indolyl-, alpha-, and zeta-radicals) have been generated in the gas phase and their dissociations examined. In particular isomerization among the isomers and how the position of the spin affects the dissociation has been studied.

Methods:
Molecular (hydrogen-deficient) peptide radical cations were generated by collision- induced dissociations of the metal/peptide complex and the nitrosylated peptide, as well as laser-induced dissociation of the iodo-phenylalanine-containing peptide. DFT calculations have been employed to study the energetics and dissociation mechanisms.

Results:
Dissociation spectra of the pi- and indolyl-radical ions show virtually identical product ions suggesting these two peptide radicals undergo isomerization prior to dissociations and their major fragmentation channel are the same loss of CO_2. In contrast, all N–C bonds in the alpha-radical peptide are fragile, leading to the formations of $[z_n - H]^{\bullet+}$ ions (n=1-3). The higher energy zeta-radical ion undergoes hydrogen-atom transfer prior to dissociation. Loss of NH_3 is the major fragmentation pathway. Interestingly, the $[z_3 - H]^{\bullet+}$ ions isolated in the dissociation of alpha- and zeta-radicals respectively, show distinctly different fragmentation patterns, indicating that the structures of these ions are different.

Novel Aspect:
Isomerization between the pi- and indolyl peptide radicals is facile. Isomeric non-interconvertible z-ions are produced in the experiments.

21ST INTERNATIONAL MASS SPECTROMETRY CONFERENCE
Toronto, Ontario, Canada
August 20 – 26, 2016

Laufer, Philipp,
Technische Universität Dresden - Institute for Aerospace Engineering

Contributing Authors:
Daniel Bock, Technische Universität Dresden, Institute for Aerospace Engineering; Martin Tajmar, Technische Universität Dresden, Institute for Aerospace Engineering; Christian Drobny Technische Universität Dresden, Institute for Aerospace Engineering.

Title:
Development of a MEMS Gas Analyzer and Mass Spectrometer using a Highly Efficient CNT Field Ionizer/Electron Source

Introduction:
Low-cost and highly-miniaturized mass spectrometers could open up new market for these kind of instruments e.g. in the chemical industry or to watch for gas leaks along pipelines. Recent studies on MEMS ion thruster technology for space applications leads to a very compact design of an ionization source as small as 1 cm^2 as the heart of a highly miniaturized mass spectrometer and gas analyzer.

Methods:
A combined MEMS-fabricated chip-size field ionizer and electron source has been manufactured. Its core consists of a silicon chip with laser drilled holes where carbon-nanotubes have been grown on the top surface. The gas flows through the chip and is field-ionized by a characteristic voltage which is applied between the CNT-plate and an extractor electrode.

Results:
Field ionizer chips down to a size of only 1 cm^2 have been manufactured that can efficiently ionize gases as well as provide electrons for electron bombardment ionization. Our chip design uses a novel manufacturing technique to grow CNTs in cluster arrays together with focusing electrodes for the electric field. This leads to ionization efficiencies up to 30% and currents up to 1 mA in such small dimensions. Our results are orders of magnitude better compared to other CNT based field ionizers so far. Our presentation will review the manufacturing process, results on both electron and ion emission capabilities as well as an outline of a mass spectrometer/gas analyzer based on our new chip-scaled source.

Novel Aspect:
A highly compact ionization/electron source was developed that can lead to highly compact and low cost mass spectrometers and gas analyzers.

Lauter, Simon,
Bruker Daltonics Germany

Contributing Authors:

Ole Hørning, Bruker Daltonics Scandinavia, Odense, Denmark; Alexandre Podtelejnikov, Bruker Daltonics Scandinavia, Odense, Denmark; Stephanie Kaspar-Schoenefeld Bruker Daltonik GmbH, Bremen, Germany; Peter A Nielsen, Bruker Daltonics Scandinavia, Odense, Denmark; Christoph Gebhardt, Bruker Daltonik GmbH, Bremen, Germany; Nicolai Bache, Bruker Daltonics Scandinavia, Odense, Denmark.

Title:

Development of a novel nano separation device for robust and accurate gradient delivery with intelligent diagnostics.

Introduction:

Matching modern HR/AM nano-ESI MS analysis robustness and reliability demands from the front-end LC separation the absolute highest quality and reproducibility of sample separation. To ensure such reproducible system performance, we developed a device with deeper system integration and novel approaches to quality control as well as intelligent system diagnostics.

Methods:

A predigested human cell line (Promega), dissolved in 0.1% TFA was used as a QC. 200ng were injected to avoid blank injections and to reduce carry-over. The tryptic peptides were separated on a novel nano-UHPLC system using a 50 min reverse phase gradient, on standardized 15 cm 75 µm bore columns with C18 resin. Data was acquired on an impact II MS (Bruker Daltonik) equipped with a nanoESI source.

Results:

The novel LC system configuration minimizes the risk of introducing leaks in the system and additionally automates trouble-shooting by pin-pointing the exact leak position. The design incorporates a valve configuration allowing software controlled switching between "trap and separation column mode" and "separation column mode". The elimination of manual intervention removes the risk of damaging a well-functioning system by mounting or unmounting a trap column. 3 different LCs instruments were installed in 3 different labs; all applying the same described setup with their own instrumentation. High reproducibility of identification results was achieved, showing only ±6.5% variation of protein identifications and ±16% variation of peptide identification between the labs when comparing average value of 2 to 4 analyses per lab.

Novel Aspect:

We have developed a novel nano-UHPLC system with outstanding robustness, deep system integration and ease-of-use without compromising top performance.

Lauter, Simon,
Bruker Daltonics Germany

Contributing Authors:
Annette Michalski, Bruker Daltonik GmbH, Bremen, Germany; Markus Lubeck, Bruker Daltonik GmbH, Bremen, Germany; Stephanie Kaspar-Schoenefeld Bruker Daltonik GmbH, Bremen, Germany; Pierre-Olivier Schmit, Bruker Daltonique S.A, France.

Title:
Investigation of Q-TOF instruments advantages for label-free proteome analysis

Introduction:
The capacity of detecting very accurately and reproducibly differences between peptide and protein ratios in different samples is key to understand the proteome dynamic. This study illustrates how the UHR-QTOF's combination of intra-spectral dynamic range, reduced space charge effects, speed and spray stability translate into label-free proteomics enhanced performances.

Methods:
Label-free quantitative performance of an impact II -instrument (Bruker Daltonik GmbH) was investigated using a 1:2 mixture of human cell line and yeast proteomes. A nanoLC was coupled to the Q-TOF via a CaptiveSpray ionization source. Peptides were separated using a 90 min gradient and analyzed with a Self-Adapting LCMS/MS method. Data have been analyzed applying the MaxLFQ algorithm.

Results:
The quantity of human cell line proteome was kept constant for both samples. The yeast proteome was spiked in two different amounts resulting in a 2-fold increase for all yeast proteins. Consequently, we created a highly complex sample set with the possibility to detect fold changes over the complete dynamic range. Results showed a high reproducibility between technical replicates ($R2 \sim 0.99$). Nearly all (~95%) of the non-changing human proteins had CV values below 20% and roughly 70% even below 10%. The overall difference between changing and non-changing population was 1.9, i.e. very close to the theoretical one of 2 and thereby representing excellent accuracy of the evaluated Q-TOF instrument for label-free proteomics. The data showed that the instrument can easily cover 4 orders of magnitude of dynamic range in the sample for identification as well as for quantitation.

Novel Aspect:
In-depth investigation of advantages of Q-TOF technology for label-free proteomics by mixing complex proteomes.

Lavold, Thorleif,
Biomotif AB

Title:
Towards a High Throughput Binding-Site Specific Drug Screening by HDX MS

Introduction:
Most high-throughput screening technologies are based on immobilization. However, immobilization procedure does not always produce a functional protein target, and fluorescent labeling can interfere with binding. Thus, there is an increased interest in label-free binding assays, such as HDX MS. However, HDX MS has so far suffered from low throughput. Here we solve this problem. i

Methods:
Standard ligand solutions are arranged in a n*m array. The ligands in each of the n rows and m columns are pooled together, giving m+n pools. Each pool's aliquot is added to the target protein, and m+n protein+ligands mixtures are subjected to HDX MS analysis with shortened LC time.

Results:
96 ligands were arranged into a 12*8 array (8 rows, coded A-H, and 12 columns, coded 1-12), with each well containing a single ligand. Then, 8+12=20 pooled samples were prepared by mixing aliquots in the same row as well as the same column. An aliquot of each pooled sample was mixed with that of the target protein. These 20 test samples, together with a mixture of the known binder and the target protein (positive control), and target protein only (negative control) were subjected to HDX MS analysis. With a 10 min LC gradient time, the total analysis time of one sample was 45 min. The masses of four peptides involved in binding and four peptides not involved in binding were monitored; the extracted signal was proportional to the mass difference between these two groups.

Novel Aspect:
For the first time, HDX MS demonstrated the capability to be used in high-throughput site-specific protein ligand screening.

Lavold, Thorleif,
Biomotif AB

Title:
Blood Proteins with isoelectric point close to 7.4 as Alzheimer Disease Biomarker

Introduction:
The pH of human blood is ≈7.4, normal forms of human proteins tend to eschew this specific pI region, thus avoiding charge neutrality, and as a consequence enhanced precipitation. Protein isoforms, which are potential biomarkers of diseases, are likely to be found everywhere in the pI distribution.Therefore, we hypothesize that neurodegenerative disease biomarkers are well detected around pI=7.4.

Methods:
To isolate this pI region, we applied the recently introduced pI Trap capillary isoelectric focusing (MJ-CIEF) device. Blood serum samples from 14 female patients, clinically diagnosed with probable Alzheimer's disease (AD), were divided into two groups according to the rate of memory decline (8 fast and 6 slow decliners). Each sample was pI-fractionated by MJ-CIEF device, followed by LC-MS/MS.

Results:
Analysis of all LC-MS/MS datasets resulted in quantification of 650 protein groups that passed 1 % FDR threshold at the both peptide and protein levels. The experimental isoelectric point of quantified proteins also calculated by means of spiking of the set of 7 synthetic peptides as pI markers in the sample. Results suggest that patients with fast memory decline had a higher concentration of proteins in the pI=7.4±0.1 region than those with a slow decline ($p<0.003$). Thus the protein concentration in this narrow pH region is a potential biomarker for predicting the memory decline speed in AD patients. Detailed investigation of proteins with elevated abundances in this region identified several previously reported AD biomarkers as well as a few novel biomarker candidates.

Novel Aspect:
The applied method enabled the study of biomarkers for early stage diagnosis and prognosis of Alzheimer disease in blood.

Lawson, Graham,
De Montfort University

Contributing Authors:
C Aldcroft, Advion Ltd; J Ogwu, De Montfort University; R Armitage De Montfort University; S Tanna, De Montfort University.

Title:
Fast identification of counterfeit medicines – a comparison of two MS systems

Introduction:
Comparison of a heated atmospheric pressure ionisation (APCI) MS with a vacuum solids probe MS method for the identification of the active pharmaceutical ingredients (API) in powdered fully formulated tablet samples was carried out. Thermal volatilisation allowed direct identification of the APIs at therapeutic levels, in a matter of minutes, in many tablet formulations.

Methods:
Reference spectra were recorded for pure APIs using the Advion ASAP (APCI) probe and the Bruker vacuum insertion probe. Powder samples in the probe cup were transferred to the MS systems and heated in flowing nitrogen at 300C (ASAP) and 40-250C (Bruker). Crushed globally sourced tablet samples were analysed as above. Comparison of the results with tablet labels identified some suspect tablets.

Results:
Thermal vaporization is needed for the analysis of these samples. In APCI the vaporized API is ionised by corona discharge to give M+H+ ions with little fragmentation. Observed signals were: acetaminophen m/z 152,110; caffeine m/z 195, 138; chloroquine phosphate m/z 320, 247 and atenolol m/z 267, 225. Aspirin was detected as a -ve ion at m/z 179, 137. The simplicity of these spectra meant that tablets containing several APIs could be readily identified: acetaminophen/caffeine/aspirin or artemether and lumifantrine in antimalarial tablets. Electron impact is more energetic giving some fragmentation. EI observed signals were: acetaminophen m/z 151,110; caffeine m/z 194, 109; chloroquine phosphate m/z 320, 247 and atenolol m/z 251, 223. Aspirin was detected as a +ve ion at m/z 138, 120. Tablet and reference data was directly comparable. Analysis time was only minutes per sample.

Novel Aspect:
Direct MS analysis of APIs in fully formulated tablet medicines to identify counterfeit or substandard products is novel.

Le, X. Chris,
University of Alberta

Contributing Authors:
Xiufen Lu, University of Alberta; X. Chris Le, University of Alberta; Xing-Fang Li, University of Alberta; Bin Hu, Wuhan University; Qingqing Liu, University of Alberta; Hanyong Peng, University of Alberta; Martin J. Zuidhof, University of Alberta.

Title:
Complementary mass spectrometry and chromatography techniques enabling the characterization of arsenic metabolites in chicken meat and liver

Introduction:
Human exposure to high concentrations of arsenic from water and food is a main health concern. Chicken meat has the highest per capita consumption among all meat types in North America. The practice of feeding 3-nitro-4-hydroxyphenylarsonic acid (Roxarsone) to chickens lasted for over 60 years. However, little is known about the metabolism of this arsenic species.

Methods:
Simultaneous detection with both inductively coupled plasma mass spectrometry (ICPMS) and electrospray ionization tandem mass spectrometry (ESI-MS/MS) was coupled to high performance liquid chromatography (HPLC) separation. The method was used to identify and quantify arsenic metabolites in a feeding study that involved 1600 chickens, half of which were fed a Roxarsone-supplemented diet.

Results:
HPLC separation with both ICPMS and ESI-MS/MS detection allowed for identification and quantification of eight arsenic species in chicken livers and breast meat, including arsenobetaine, inorganic arsenite, dimethylarsinic acid, monomethylarsonic acid, inorganic arsenate, 3-amino-4-hydroxyphenylarsonic acid, N-acetyl-4-hydroxyphenylarsonic acid (N-AHPAA), and Roxarsone. The concentrations of these arsenic species, with the exception of arsenobetaine, were significantly higher in the Roxarsone-fed chickens than those in the control chickens. Although arsenic species were excreted rapidly from the chickens, with the half-lives of elimination being 0.4-1 day, the residual arsenic concentrations in chicken breast meat 7 days after terminating Roxarsone feeding remained significantly higher in the Roxarsone -fed chickens than in the control chickens.

Novel Aspect:
The simultaneous detection by both ICPMS and ESIMS from the single HPLC separation allowed for identification of new arsenic metabolites.

Lebedev, Albert,
Lomonosov Moscow State University

Contributing Authors:

S.Pugh, LECO Corporation,3000 Lakeview Avenue, St. Joseph, Michigan, 49085, USA; M. Vaïtilingom, Université Clermont Auvergne, Université Blaise Pascal, Institut de Chimie de Clermont-Ferrand, BP 10448, F-63000 CLERMONT-FERRAND, FRANCE; D.M.Mazur Lomonosov Moscow State University, Chemistry Department, LeninskieGory 1/3, Moscow, 119991, Russia; J. Byer, LECO Corporation,3000 Lakeview Avenue, St. Joseph, Michigan, 49085, USA; I.Canet, Université Clermont Auvergne, Université Blaise Pascal, Institut de Chimie de Clermont-Ferrand, BP 10448, F-63000 CLERMONT-FERRAND, FRANCE; V.Artaev, LECO Corporation,3000 Lakeview Avenue, St. Joseph, Michigan, 49085, USA; J. Binkley, LECO Corporation,3000 Lakeview Avenue, St. Joseph, Michigan, 49085, USA; A-M. Delort, CNRS, UMR 6296, ICCF, F-63171 AUBIERE, FRANCE; O.V.Polyakova, Lomonosov Moscow State University, Chemistry Department, LeninskieGory 1/3, Moscow, 119991, Russia.

Title:

What's in a Cloud? A semi-volatile GC/MS study.

Introduction:

Understanding the chemical makeup of clouds is crucial, as clouds participate in the formation and transformation of chemical species, production and consumption of oxidants, droplet activation and growth, transport and deposition of pollutants. A large fraction of the organic compounds in clouds has not been characterized to date;this paper addresses that gap. SVOCs are the aim of this study.

Methods:

Cloud and rain samples were collected at the puy de Dôme station (France, 1465 m) in 2013-2016. The samples were filtered and kept frozen at -20°C before MS analysis. Dichloromethane extracts were analyzed with LECO prototype GC-HRT instrument in GC and GCxGC modes. The obtained EI mass spectra were used for identification by utilizing the tools present in the instrument's software package.

Results:

About 300 hundred individual organic compounds were reliably identified in each of the analyzed samples. GCxGC allowed separation of coeluting constituents, making the identification more reliable. The content of the samples differs from the usual picture for water or soil pollution. For example, the levels of dibutyl and bis(2-ethylhexyl)phthalates are rather low. Another peculiarity involves almost complete absence of the chlorinated compounds. Maybe chlorine is lost from the pollutants due to radical atmospheric reactions. Interestingly a wide assortment of oil hydrocarbons including alkanes, alkylated polycyclic aromatic and saturated hydrocarbons, alkylated pyridines and dibenzothiophenes have been identified. Phenols, aldehydes, dodecylbenzenes worth mentioning as well. The levels of SVOC are estimated using internal standards (deuterated PAH). They are in the range 0.1-10 ng/ml.

Novel Aspect:

Several hundred semi-volatile organic compounds were identified and roughly quantified in the samples of cloud water.

Lebedev, Albert,
Lomonosov Moscow State University

Contributing Authors:
S.S.Zhokhov, Department of Organic Chemistry, Moscow State M. V. Lomonosov University, 119991, Moscow, Russia; M.I.Tolpina, Department of Organic Chemistry, Moscow State M. V. Lomonosov University, 119991, Moscow, Russia; R.A.Zubarev Division of Physiological Chemistry I, Department of Medical Biochemistry and Biophysics, Karolinska Institutet, Scheelesväg 2, SE-17 177 Stockholm, Sweden; S.V.Kovalev, Department of Organic Chemistry, Moscow State M. V. Lomonosov University, 119991, Moscow, Russia; T.Y.Samgina, Department of Organic Chemistry, Moscow State M. V. Lomonosov University, 119991, Moscow, Russia.

Title:
Reliable discrimination of Leucine and Isolucine in high throughput peptide sequencing

Introduction:
Mass spectrometry is the best tool for peptide sequencing. However discrimination of isomeric Leu and Ile residues remains a problem deals. Recently we proposed an efficient ETD-HCD method to discriminate L/I in the isolated peptides. It involves isolation of primary z• ions and triggering their fragmentation to form w-ions. In these conditions 100% of the isomeric residues are correctly assigned.

Methods:
Orbitrap Fusion and Elite instruments (Thermo) were used in EThcD mode. Studying isolated peptides standard HCD mode was applied to the whole array of the ions formed in the optimized ETD conditions. Nano-ESI HPLC/MS/MS was applied to a cocktail of the natural peptides (crude secretion of R.ridibunda frog). Neither formation of the targeted z-ions, nor their fragmentation was optimized.

Results:
The first set of experiments involved ESI ionization of isolated standard and natural non-tryptic peptides. Optimised ETD conditions were used to produce primary c- and z-ions. HCD with the same collision energy was applied to the whole array of these ions, i.e. without preliminary isolation of the targeted z-ions. In these conditions I/L identification was successful for 19 out of 20 registered z-ions. In the second set of experiments a cocktail of natural peptides was studied in HPLC-ESI/EThcD mode in the standard conditions used for the tryptic peptides analysis. In the applied conditions the sequence coverage of the long frog peptides was rather poor. However for 26 detected z-ions the corresponding w-ions were detected in 19 cases (73%). The number of L/I residues in the peptide sequence and their position including neighboring sites does not interfere with the results.

Novel Aspect:
The proposed EThcD approach allows reliably discriminate Leu and Ile in high-throughput regime of peptide sequencing.

LeBlanc, André,
McGill University

Contributing Authors:
André LeBlanc, University of Victoria - Genome BC Proteomics Centre, Victoria, BC, CANADA; Sarah Michaud, MRM Proteomics, Inc., Victoria, BC, CANADA; Andrew Percy University of Victoria - Genome BC Proteomics Centre, Victoria, BC, CANADA; Darryl Hardie, University of Victoria - Genome BC Proteomics Centre, Victoria, BC, CANADA; Juncong Yang, University of Victoria - Genome BC Proteomics Centre, Victoria, BC, CANADA; Nicholas Sinclair, University of Victoria - Genome BC Proteomics Centre, Victoria, BC, CANADA; Jillaine Proudfoot, University of Victoria - Genome BC Proteomics Centre, Victoria, BC, CANADA; Adam Pistawka, University of Victoria - Genome BC Proteomics Centre, Victoria, BC, CANADA; Derek Smith, University of Victoria - Genome BC Proteomics Centre, Victoria, BC, CANADA; Christoph Borchers, University of Victoria - Genome BC Proteomics Centre, Victoria, BC, CANADA; Department of Biochemistry and Microbiology, University of Victoria, Victoria, BC, CANADA.

Title:
Plasma Protein Quantitation by MRM using Two Stable Isotope Labeled Peptides for Calibration: Assessing Accuracy for Endogenous Protein Assays

Introduction:
A new approach using two peptides with different labels, enabling an external calibration curve and quality control samples to be prepared in pooled human plasma is presented for the quantitation of endogenous protein by MRM. Samples of known concentration, with a second labeled peptide, in human plasma can evaluate assay accuracy, compare calibration strategies and simplify validation.

Methods:
Assay concentration ranges for 31 synthesized peptides were established based on the LOD in human plasma. Calibration strategies (7 calibration strategies, 5 surrogate matrices) were evaluated and compared based on precision and accuracy via quality control samples (n=6, 3 QC levels in plasma, 2 batches), using three isotopes (light, heavy, and double-labeled).

Results:
When comparing the accuracy of the different calibration strategies, the two labeled peptide method clearly provides consistently more accurate results when compared to the reverse curve methods and single point measurement strategies. Calibration curves in 5 different surrogate matrices were compared and all had comparably excellent accuracy at all QC levels as well as similar slopes. Curves were prepared in chicken plasma, dimethylated human plasma digest, phosphate buffered saline (PBS) and BSA solution (10 mg/ml in PBS), in addition to human plasma. Using two different labeled peptides also allowed for spiking standards and IS into human plasma prior to digestion. Such an approach could increase sample throughput and facilitate automation since unknown samples, calibration standards and QC samples would undergo identical sample processing steps.

Novel Aspect:
The use of two different stable-isotope-containing peptides for multiplexed MRM-based quantitation of endogenous protein in human plasma.

21ST INTERNATIONAL MASS SPECTROMETRY CONFERENCE
Toronto, Ontario, Canada
August 20 – 26, 2016

LEE, JUHYUN,
KOREA INSTITUTE OF NUCLEAR SAFETY

Contributing Authors:

Yong-Jae KIM, Byung-Uck Chang, Korea Institute of Nuclear Safety.

Title:

Evaluation of thorium and uranium concentrations and isotope ratios by quadrupole-ICP-MS in naturally occurred radioactive materials.

Introduction:

This paper describes a rapid pretreatment methods and determination of uranium and thorium elements concentrations by quadrupole inductively coupled plasma mass spectrometry (ICP-MS) in naturally occurred radioactivity materials (NORM).

Methods:

In this paper, prior to the pretreatment, it was using a potable TXRF in order to analyze the components of the substance suspected as NORM materials. The samples were dissolved by fusion method or acid digestion depending on the type of components. It was found that a mixture amount of fluxer and remover to separate the sample melted with the platinum crucible.

Results:

Pretreatment of the radioactivity materials is a lot of economical consumption, time and difficult. It was tested to get good performances and simplicity of the pretreatment method and determined the uranium and thorium elements concentrations and isotope ratios using by quadrupole ICP-MS. Depending on the ratio of isotope, it is determined whether the artificial materials or environment.

Novel Aspect:

The established method was applied to samples suspected as contaminant materials in the field survey based on NORM regulation in South Korea.

Lee, Chuping,
Institute of Atomic and Molecular Sciences, Academia Sinica

Contributing Authors:
Yuan Tseh Lee, Institute of Atomic and Molecular Sciences, Academia Sinica; Chi-Kung Ni, Institute of Atomic and Molecular Sciences, Academia Sinica; Li-Hao Chen Institute of Atomic and Molecular Sciences, Academia Sinica.

Title:
Comparison of different soft ionization matrices in matrix-assisted laser desorption ionization (MALDI)

Introduction:
MALDI is one of the soft ionization methods in mass analysis for biomolecules, such as glycan. However, ionization by MALDI is not soft enough for some fragile molecules, such as the glycan with sulfates or sialic acid residues, which tend to lose the sulfate and sialic acid during the ionization process. The intensity of intact ion is usually small,and sometime no intact ions are observed at all.

Methods:
Recently, many studies purposed new matrices to solve this problem. Compared matrices include (1) frozen sample (CW Liang et al., 2012) (2) trilayer sample (CL Wu et al., 2013) (3) HgTe nanostructures (MF Huang et al., 2012) (4) 5-methoxysalicylic acid (MSA) (D Lee et al., 2015) (5) super-DHB (sDHB) (D Lee et al., 2015) (6) 2,5-dihydroxybenzoic acid butylamine (DHBB) (M Mank et al., 2004).

Results:
Because different analytes were used in these studies, it is difficult to compare the softness of ionization for these matrices. In this study, we used these matrices and several delicate analytes to investigate softness of these matrices in the ionization process. These analytes include heparin disaccharide I-S, ganglioside GD1a, pullulan and dextran. We demonstrated that HgTe and MSA are not as soft as the other matrices in the ionization process.

Novel Aspect:
The frozen sample reduces the in-source and post-source decays as well as to increase the sensitivity of intact molecules.

Lee, Hui-Ling,
Department of Chemistry, Fu Jen Catholic University

Title:
Determination of association oxidative stress biomarker in metal oxides nanomaterial handling workers using on-line solid phase LC-MS/MS

Introduction:
The nanotechnology industry continues to grow rapidly. The most commonly used NMs are, titanium dioxide (TiO2), silica (SiO2) and zinc oxide. Nanomaterial-induced reactive oxygen species (ROS) play a key role in cellular and tissue toxicity. 8-oxodG as a biomarker to detect systemic oxidatively damaged DNA to estimate oxidative stress induced by occupational and environmental exposures.

Methods:
In the present study, we examined oxidative stress markers of nucleic acids (8-oxodG) in various body fluids (urine, plasma and white blood cell (WBC)) were associated with nanomaterial exposure using on-line solid phase extraction LC-MS/MS.

Results:
This study assessed oxidative DNA damage in workers occupational exposure to metal oxides nanomaterials (NM). 8-hydroxydeoxyguanosine (8-OHdG) was measured in white blood cells (WBC), plasma, and urine. Exposure to TiO2, SiO2 and indium tin oxide (ITO) resulted in significant higher oxidative biomarkers 8-OHdG than comparison workers. Statistically significant correlations were noted between plasma and urine 8-OHdG (Spearman correlation $r = 0.396$, $p < 0.001$), between WBC and urine 8-OHdG ($r = 0.230$, $p < 0.001$), and between WBC and plasma 8-OHdG ($r = 0.156$, $p = 0.019$).

Novel Aspect:
The results showed that urinary 8-oxodG may be considered to be better biomarker.

LEE, DaeYoung,
Department of Herbal Crop Research, National Institute of Horticultural and Herbal Science, Rural Development Administration

Contributing Authors:
Young-Sup Lee, Department of Herbal Crop Research, National Institute of Horticultural and Herbal Science, RDA, Eumseong 27709, Republic of Korea; Geum-Soog Kim, Department of Herbal Crop Research, National Institute of Horticultural and Herbal Science, RDA, Eumseong 27709, Republic of Korea; Yurry Um Department of Herbal Crop Research, National Institute of Horticultural and Herbal Science, RDA, Eumseong 27709, Republic of Korea; Jae Won Lee, Department of Herbal Crop Research, National Institute of Horticultural and Herbal Science, RDA, Eumseong 27709, Republic of Korea; Dae Young Lee, Department of Herbal Crop Research, National Institute of Horticultural and Herbal Science, RDA, Eumseong 27709, Republic of Korea.

Title:
Comprehensive Profiling of Various Platycosides in Platycodi Radix by UPLC-QTOF/MS

Introduction:
The benefits of Platycodi Radix (PR) for health and biological activities have been previously reviewed. Platycosides, are the primary constituents of PR, having pharmacological activities including anti-oxidant, anti-inflammatory, anti-obesity, and anti-cancer effects. Thus, the determination of platycosides is critical for the quality control and clinical use of PR.

Methods:
UPLC-QTOF/MS and an in-house library were used to comprehensively profile the platycosides in PR. The optimal UPLC provided the fastest and most effective separation of various platycosides, including structural isomers, in 15 min. Exact mass measurements based on QTOF/MS were also effective for the reliable identification of individual platycosides.

Results:
The qualitative and quantitative analysis of 44 platycosides in Jangbaek PR was successfully carried out. A high-throughput and high-sensitive analytical method was applied to profile high and low abundant platycosides. Furthermore, the 7 pairs of isomers ($C42H68O17$, $C54H86O25$, $C59H94O28$, $C59H94O29$, $C63H102O32$, $C63H102O33$, $C65H104O34$) were determined based on two separation rules using the ODS column. The results demonstrate that the proposed method based on UPLC-QTOF/MS and the in-house library of platycosides, which includes the compound's name, molecular formula, observed m/z, RT, and adducts can be used as an effective tool for determining the detail characterization of various platycosides and the quality control of the PR.

Novel Aspect:
This study indicated that the proposed method based on the UPLC-QTOF/MS can be an effective tool for identifying the platycosides

Lee, Jason,
University of Oxford

Contributing Authors:
Claire Vallance, University of Oxford; Mark Brouard, University of Oxford; Jaya John John University of Oxford; Renato Turchetta, Rutherford Appleton Laboratory; Iain Sedgwick, Rutherford Appleton Laboratory; Daniel Pooley, Rutherford Appleton Laboratory.

Title:
A Next-Generation Ultrafast Detector for Imaging Mass Spectrometry: the Pixel Imaging Mass Spectrometry (PImMS) Sensor

Introduction:
This poster outlines a new ultrafast event-triggered CMOS image sensor, which records the position and arrival time of each detected ion. The sensor opens a variety of new experimental techniques, including 2D surface imaging, investigating the chemical dynamics of multiple ions and probing specific fragmentation pathways through covariance analysis.

Methods:
The PImMS sensor operates by timestamping detected events when signal intensity crosses a specified threshold. The sensor has a spatial resolution of 324x324 pixels and a temporal resolution of 12.5 ns over a 50 μs TOF. Four memory registers in each pixel allow multiple ions to be recorded each TOF cycle. Post-processing allows reconstruction of the full TOF mass spectrum and all ion images.

Results:
This poster focusses on describing the hardware and firmware capabilities of the PImMS sensor. This includes a schematic representation of the pixel circuitry architecture, a description of how ion events are detected, as well as the advantages of having flexible reprogrammable firmware. This poster couples with another poster showing examples of the experimental use of the sensor, including surface imaging and dynamics experiments at laboratory and synchrotron level. Further, this poster also demonstrates the flexibility of the sensor in the field of time-resolved particle imaging by presenting results obtained in the field of neutron imaging. These experiments involve the creation of packet of neutrons, separated in energy via time-of-flight, leading to material identification, tomography reconstruction and stress-strain analysis.

Novel Aspect:
The technology of a new ultrafast multi-mass spatial imaging sensor is outlined and demonstrated with imaging mass spectrometry applications.

Lee, Richard,
Advanced Chemistry Development, Inc. (ACD/Labs)

Contributing Authors:

Vitaly Lashin, Advanced Chemistry Development, Moscow, Russia; David Adams, Advanced Chemistry Development, Toronto, ON, Canada; Alexandr Sakarov Advanced Chemistry Development, Moscow, Russia.

Title:

Data and Prediction Driven Convergent Analysis for Metabolite Identification in Drug Discovery

Introduction:

Analysis of metabolic "hotspots" through interpretation of LC/MS data plays a critical role in drug discovery and development. There have been significant advancements but the main challenge is still structure elucidation of metabolites from the parent. Here, we describe a new automated software protocol for prediction and data driven detection and structure verification via LC/MS analysis.

Methods:

A software algorithm was investigated for batch processing mass spectrometry data acquired from Orbitrap and Q-Tof analyzers. A regio-selectivity model was employed to predict metabolites which were used as a target list. To complement, a data driven untargeted analysis was also performed. In cases where the structures were ambiguous, the software was able to provide Markush notation.

Results:

Data files representing a study across incubation time points were automatically processed post-acquisition. Structures of possible phase 1/2 metabolites were generated from an assembly metabolism model. An unexpected metabolite detection algorithm was integrated into the overall processing routine, using a fractional mass filter. Metabolites were initially detected based on accurate mass and isotopic pattern, subsequent MSn spectra were extracted. The software was able to assign fragment ions of the parent and metabolites to their respective MS2 spectra. Structures were verified and scores were provided by comparing assigned and common neutral loss fragments between it and the parent. The protocol then combined both predicted and unexpected metabolites into a single biotransformation map. Components were tracked across the study and graphically represented as a stability/kinetic plot.

Novel Aspect:

Automated software routine combining prediction and data driven metabolite analysis for drug discovery and development.

Lee, Chuping,
Institute of Atomic and Molecular Sciences, Academia Sinica

Contributing Authors:
Yuan Tseh Lee, Institute of Atomic and Molecular Sciences, Academia Sinica; Chi-Kung Ni, Institute of Atomic and Molecular Sciences, Academia Sinica; Li-Hao Chen Institute of Atomic and Molecular Sciences, Academia Sinica.

Title:
Comparison of different soft ionization matrices in matrix-assisted laser desorption ionization (MALDI)

Introduction:
MALDI is one of the soft ionization methods in mass analysis for biomolecules, such as glycan. However, ionization by MALDI is not soft enough for some fragile molecules, such as the glycan with sulfates or sialic acid residues, which tend to lose the sulfate and sialic acid during the ionization process. The intensity of intact ion is usually small,and sometime no intact ions are observed at all.

Methods:
Recently, many studies purposed new matrices to solve this problem. Compared matrices include (1) frozen sample (CW Liang et al., 2012) (2) trilayer sample (CL Wu et al., 2013) (3) HgTe nanostructures (MF Huang et al., 2012) (4) 5-methoxysalicylic acid (MSA) (D Lee et al., 2015) (5) super-DHB (sDHB) (D Lee et al., 2015) (6) 2,5-dihydroxybenzoic acid butylamine (DHBB) (M Mank et al., 2004).

Results:
Because different analytes were used in these studies, it is difficult to compare the softness of ionization for these matrices. In this study, we used these matrices and several delicate analytes to investigate softness of these matrices in the ionization process. These analytes include heparin disaccharide I-S, ganglioside GD1a, pullulan and dextran. We demonstrated that HgTe and MSA are not as soft as the other matrices in the ionization process.

Novel Aspect:
The frozen sample reduces the in-source and post-source decays as well as to increase the sensitivity of intact molecules.

Lee, Ronald Fook Seng,
Ecole Polytechnique Federale de Lausanne

Contributing Authors:
Stéphane Escrig, Swiss Federal Institute of Technology Lausanne (EPFL); Graham Knott, Swiss Federal Institute of Technology Lausanne (EPFL); Kai Johnsson Swiss Federal Institute of Technology Lausanne (EPFL); Paul Dyson, Swiss Federal Institute of Technology Lausanne (EPFL); Anders Meibom, Swiss Federal Institute of Technology Lausanne (EPFL.

Title:
NanoSIMS for probing intracellular distribution of a isotopically labelled ruthenium (II) anti-cancer drug.

Introduction:
Nano scale secondary ion mass spectrometry (NanoSIMS) is useful for metallodrug imaging as it can separately image the metal and its ligands with high lateral resolutions of up to 50nm. However, imaging ruthenium metallodrugs that ionize weakly via NanoSIMS is challenging. Here, we describe the distribution and ligand state of an isotope labelled ruthenium (II) drug, RAPTA-T via Nano-SIMS.

Methods:
RAPTA-T was 13C labelled on its arene ligand and 15N labelled on its phosphine ligand. A2780 cisplatin resistant ovarian cancer cells were dosed with 500µM of RAPTA-T for 24 hours. Cells were then chemically fixed, resin embedded and cut into semi-thin sections of 500nm and mounted onto glass slides. They were then gold coated and mounted onto sample holders for Nano-SIMS analysis.

Results:
The cellular distribution of RAPTA-T seemed to be concentrated on the cell membrane and intercellular spaces between cells. This correspond well with the observed anti-metastatic effects of RAPTA-T where the drug is known to bind to extracellular cell adhesion proteins such as fibronectin and collagen IV. There was a co-accumulation of Ru and 15N with little accumulation of 13C observed. This seemed to indicate the phosphine ligand stays intact and at least some of the arene detaches, which could serve in addition to aquation, an additional activation mechanism of RAPTA-T. Overall, these results represents a step forward in the use of NanoSIMS for imaging metallodrugs with metal centers that ionize poorly, and also showcases the applicability of the technique for understanding the ligand state of metallodrugs in biological systems with multiple isotope labelling.

Novel Aspect:
NanoSIMS imaging of ruthenium (II) drugs and use of multi isotope labels for understanding ligand state of metallodrugs in vitro

Lee, Mi Kyeong,
Department of Food Technology and Services, Eulji University

Contributing Authors:
Min-Ho Lee, Department of Food Technology and Services, Eulji University, Seongnam, Gyeonggi 13135, Republic of Korea; Dae Young Lee, Department of Herbal Crop Research, National Institute of Horticultural and Herbal Science, RDA, Eumseong 369-873, Republic of Korea; Mi Kyoung Lee Department of Food Technology and Services, Eulji University, Seongnam, Gyeonggi 13135, Republic of Korea.

Title:
Metabolomic analysis of the enhanced effect of ginsenoside F1 and doxorubicin in human cancer cells using UPLC-QTOF/MS and chemometric analysis

Introduction:
Ginseng, the root of Panax ginseng C.A. Meyer, is a medicinal plant used worldwide and has been reported to have various biological effects. Ginsenosides are the major effective ingredients in ginseng. Among them, ginsenoside F1 with a dammarane skeleton has been proven to have a remarkable potentiality of anti-proliferation and pro-apoptosis.

Methods:
In this study, metabolomic technique was employed to evaluate the mechanism of F1 and DOX of anti-cancer activities using human breast cancer cells (MCF-7) by LC-QTOF/MS analysis. LC-QTOF/MS spectra of cell extracts compared with those of F1 and DOX treated cell extracts gave bio marker candidates using chemometric analysis.

Results:
Ten bio marker candidates found and assumed to be involved metabolism of MCF-7 cells. And ginsenoside F1 and DOX treated cell extracts gave a marker which was appeared as different intensity of TIC. An metabolite with masses m/z 611.1533 [M-H]- was identified as oxidized glutathione. Therefore, we investigated the enhanced effect of ginsenoside F1 and DOX in metabolic level and may bring benefit to clinical chemotherapy. F1+DOX induced high accumulation of sub-G1 phase. MCF-7 cells were treated with no 3 (A), 10 uM (B), 15 uM (C), 20 uM (D) for 24h. Cells were stained with PI, and DNA contents were analyzed with flow cytometry

Novel Aspect:
We investigated the enhanced effect of F1 and DOX in metabolic level and may bring benefit to clinical chemotherapy.

LEE, JAE-JIN,
EWHA woman's univ

Contributing Authors:
Kong-Joo Lee, EWHA womans univ; JIN-HWAN Cho, EWHA womans univ.

Title:
Proteomic analyses of Nm23-H1/NDPKA activator induced cellular responses in triple negative breast cancer cells

Introduction:
Nucleoside diphosphate kinases (NDPK) are house keeping enzymes that catalyze the reversible phosphotransfer of γ-phosphates between di- and triphosphonucleosides. NDPKA/Nm23-H1 is the first identified tumor metastasis suppressor that reverse correlates with its expression and metastatic stage in breast, prostate, and lung cancer.

Methods:
We screened and identified NDPK catalytic activator that activates Nm23-H1 and H2. To understand Nm23-H1/NDPKA activator induced cellular responses in triple negative breast cancer cells, NDPK activator treated MDA-MB-231 cells were analyzed by 2DE-LC-MS/MS.

Results:
We identified allosteric activator of NDPKs that activated catalytic activity of Nm23-H1 and H2. NDPK activator significantly reduced invasion and migration of MDA-MB-231 cell through Nm23-H1. We identified 18 spots which altered by NDPK activator treatment in MDA-MB-231 cell. Cytokeratin 8, 9, 18, HSP27 and LASP-1 was shifted in 2DE PAGE indicating truncation or post translational modification (PTM) of these proteins. We further identified its PTM by UPLC-ESI-Q-Tof-MSMS.

Novel Aspect:
Chemical proteomics of NDPK activator revealed that novel mechanisms of Nm23-H1 in metastasis suppression of breast cancer cells.

Lehmann, Rainer,
University Hospital Tuebingen / Central Laboratory

Title:
Unexpected characteristics and artificial changes of body fluids: Crucial factors in clinical metabolomics

Introduction:
Sample collection and/or selection is one of the most critical steps in clinical metabolomics. Collection and handling of body fluids, although regulated by sophisticated standard operating procedures (SOP), are not infallibly. Other factors, which may affect the metabolome and consequently the study outcome, are patient instructions as well as the choice of sample material (e.g. serum or plasma).

Methods:
Non-targeted and targeted UPLC-MS analyses were applied to a) investigate changes in whole blood under daily clinical routine conditions aiming to identify a sample quality biomarker; b) check the robustness of this marker and analyze biobank sample quality; c) perform a comparison of standard plasma with platelet-free plasma and serum; d) study differences between first and second morning urine.

Results:
Non-targeted UPLC-QTOF-MS revealed a biomarker suitable for the quality assessment of serum and plasma without having knowledge of the preceding processing of blood. Using 650 samples this marker was rigorously validated including evaluation of the robustness. Confidence intervals enabling the classification of clinical and biobank samples in excellent, good, fair and questionable quality for clinical metabolomics studies were defined. Furthermore, detected differences in the metabolomes of standard plasma, serum and platelet-free plasma and consequences for sample material selection are presented. Finally similarities and differences between the most common spot urine samples used in clinical metabolomics (first and second morning urine) and important factors affecting the metabolome therein are demonstrated.

Novel Aspect:
Novel insights in causes of misleading results and pitfalls in clinical metabolomics are given and strategies to avoid these confounding effects.

Lento, Cristina,
MS-ESE York University - NSERC CREATE

Contributing Authors:
Derek Wilson, York University; Gerald F. Audette, York University.

Title:
Functional and HDX-MS Analysis of the Type 4 Secretion System Protein TraF from the Escherichia coli F-plasmid

Introduction:
Escherichia coli are able to adapt to changing environmental conditions and develop antibiotic resistance through a process called F-plasmid conjugation. Conjugative DNA transfer by the F-plasmid is achieved through a type IV secretion system (T4SS) encoded within the plasmid's transfer region; TraF is one of several F-T4SS proteins essential for F-pilus assembly.

Methods:
The TraF protein was purified as a recombinant GST-TraF fusion by affinity chromatography. For HDX experiments, 70 µM of the fusion was reacted with deuterium in a microfluidic-chip based rapid mixing device with labelling times ranging from 200 to 2000 milliseconds. The labelled protein is then quenched upon entering a proteolytic chamber at pH 2.4 followed by pepsin digestion and ESI detection.

Results:
The isotopic distribution of each peptide was analyzed for each reaction time with combined runs yielding an overall sequence coverage of 75%. The C-terminal region contains the majority of peptides observed to have low deuterium exchange levels (0-55%). This is expected since it is the region of the protein predicted to contain a defined thioredoxin-like fold. The N-terminal region of unknown structure contains most of the peptides observed to take up moderate-high (66-75%) and high deuterium levels (76-100%). This indicates that the region is much more dynamic relative to the C-terminus, either due to lack of structure or by being more exposed to the solvent. In addition, peptides falling within the region hypothesized to interact with TraH (a.a. 47-70 and 68-73) show moderate-high and high deuterium uptakes respectively in the absence of its binding partner.

Novel Aspect:
Using time-resolved hydrogen-deuterium exchange mass spectrometry to study the solution-phase dynamics of proteins with domains of unknown structure.

Leonards, Pim,
VU University

Contributing Authors:
Marta Llansola, Centro de Investigación Principe Felipe; Vicente Felipo, Centro de Investigación Principe Felipe; Belen Gomez-Gimenez Centro de Investigación Principe Felipe; Sara Tufi, Waters; Stephan Jung, Proteome Sciences.

Title:
In vivo metabolomic approach to study developmental neurotoxicity of pesticides: explaining cognitive and motor function effects with metabolomics

Introduction:
Worldwide, serious concern has arisen about the increased incidence of learning and developmental disorders in children shown by recent epidemiological studies which indicated exposure to low doses of environmental contaminants can have adverse effects on cognitivity in childhood. The EU DENAMIC investigated neurotoxic effects of low-concentrations of pesticides using animal and humans studies.

Methods:
Effects of pesticide exposure on cognitive and motor function in rats was investigated using targeted and untargeted metabolomics to understand the underlying molecular mechanisms. Cross platform metabolomics were carried out using LC-MS/MS, LC-HRTOF-MS, GC-HRTOF-MS, and shotgun HRTOF-MS. Biochemical networks were made for males, females and brain tissues. Also proteomics studies were carried out.

Results:
The cognitive and behaviour tests showed that pesticide had effects on anxiety, spatial learning, motor coordination, motor activity, active avoidance, and working memory. These effects were pesticide specific and some effects were gender specific. Metabolomics showed brain tissue and gender specific effects on different metabolic and neuronal pathways. Some of the down and upregulated metabolites were correlated to the observed behaviour and cognitive effects. For example, a decrease in the brain neurotransmitters in males was found, but not in females, which correlates to the impairment of spatial learning observed in males but not in females. Pesticide exposure also caused increased neuroinflammation and biomarkers for this effect were found. We showed that omics approaches can explain gender specific differences and molecular pathways are affected by pesticide exposure.

Novel Aspect:
Metabolomics/molecular mechanisms. showed weight of evidence of observed gender specific effects on cognitive and motor function of pesticide exposure

Letarte, Sylvain,
Phytronix Instruments inc.

Contributing Authors:

Pierre Picard, Phytronix Technologies inc.; Serge Auger, Phytronix Technologies inc.

Title:

Fully automated on-line extraction procedure using DBS In-Tips with RTC-LC-MS/MS

Introduction:

Dry blood spot (DBS) sampling has received considerable attention over the last few years. In this study, we have evaluated a fully automated method for the blood volume on DBS In-Tip with a direct injection by tips in the LC injection valve. The extracted samples were then analyzed by standard LC-MS/MS analysis.

Methods:

Human blood was spiked with Dextrorphan from 10 to 1000 ng/mL. Calibration curves and QCs were added on DBS In-Tips. The filling tip material was first wet with 2 µL of internal standard solution followed by 1 µL of blood calibration curve. The RTC loaded the spiked DBS In-Tip, aspirated 24 µL of MeOH:Water with IS, aspirated 16 µL air volume, waited for 20 seconds and injected in ESI-LC-MS/MS.

Results:

Analyzing the extracts from the DBS In-Tips gave linearity over the calibration range with r2 higher than 0.998. Measured QC showed accuracy and precision within the acceptance criteria for the intra and inter run assay. With this method, the entire sample was extracted and blood distribution onto paper was no longer a variable. Without any human intervention, this method is able to extract and inject samples to evaluate the Dextrorphan concentration in blood samples added in DBS In-Tips. The same tip is used to collect the blood, prepare the sample and the extraction and directly inject it into the LC-MS/MS. Tip injection yielded an accuracy of 5 to 10% RSD and analysis results matched the values of traditional DBS card methods.

Novel Aspect:

Fully automated evaluation of Dextrorphan with an In-Tips extraction method and a direct injection with same Tips for LC-MS/MS analysis.

Li, Haiyang,
Dalian Institute of Chemical Physics, Chinese Academy of Sciences

Contributing Authors:

Keyong Hou, Dalian Institute of Chemical Physics, Chinese Academy of Sciences, China; Yan Wang, Dalian Institute of Chemical Physics, Chinese Academy of Sciences, China; Haiyang Li Dalian Institute of Chemical Physics, Chinese Academy of Sciences, China; Qingyun Li, Dalian Institute of Chemical Physics, Chinese Academy of Sciences, China; Yuanyuan Xie, Dalian Institute of Chemical Physics, Chinese Academy of Sciences, China; Wei Huang, Dalian Institute of Chemical Physics, Chinese Academy of Sciences, China; Lei Hua, Dalian Institute of Chemical Physics, Chinese Academy of Sciences, China.

Title:

Sensitive and rapid analysis of biogenic isomers and perfluoro-compounds by high pressure ion mobility spectrometry/time-of-flight mass spectrometry

Introduction:

The combination of mass spectrometry (MS) with ion mobility spectrometry (IMS) can provide information on ion structure and formation that is not possible with either method alone. In this talk, we will present some new IMS-MS methods and instruments developed in our laboratory for rapid analysis of biogenic isomers (BVOCs) and perfluoro-compounds (PFCs).

Methods:

Two kinds of IMS-TOFMS coupled with a dopant-assisted positive photoionization source and a negative 63Ni ionization source were designed and applied for rapid analysis of complex monoterpene isomer mixtures and SF6, respectively. In-source collision-induced dissociation (CID) in the region between the sample cone and the skimmer was carried out to obtain the characterized spectra of each analyte.

Results:

For monoterpene isomers analysis, an algorithm based on Matlab utilizing least-square fit method was developed for quantification of each isomeric monoterpene in the mixtures by analyzing the 3D ion mobility-fragmentation mass spectra. As a result, monoterpene mixtures containing six kinds of isomers could be analyzed qualitatively and quantitatively in 3 min with the quantitative error of no more than 5%. For SF6 analysis, the ionization mechanism of SF6 under different ionization conditions was investigated. SF6- was the main ion formed with low humidity, while SOF4- ion was produced massively in the presence of water vapor in the 63Ni ionization source.

Novel Aspect:

A high pressure and high resolution drift tube IMS/MS coupled with in-source CID is developed for sensitive and rapid analysis of isomeric mixtures.

li, jian jun,
university of calgary

Contributing Authors:
Michelle Forgeron, Chemistry Instrumentation Facility, Department of Chemistry University of Calgary, 2500 University Drive NW Calgary, AB T2N 1N4, Canada; Jian Jun Li, Chemistry Instrumentation Facility, Department of Chemistry University of Calgary, 2500 University Drive NW Calgary, AB T2N 1N4, Canada.

Title:
Accurate Analysis of Small Molecules with Conventional HCCA (α-cyano-4-hydoxycinnamic acid) MALDI-TOF-MS

Introduction:
A variety of small molecules (molecule weight < 500 Da) were analyzed accurately using MALDI-TOF-MS with conventional HCCA used as a matrix in ground steel target. It was found that samples containing specific functional groups tend to form specific ions and samples containing less functional groups do not form any recognizable species (see results).

Methods:
Experiments were conducted on a Bruker Autoflex III Smartbeam MALDI-TOF (Na:YAG laser, 355nm), set in positive reflective mode. Standard operation settings were used: laser frequency 200 Hz, and number of shots 300, and MTP 384 ground steel plate target. Chemical acetanilide and HCCA matrix were used as calibration standards. The concentration ratio of sample to matrix was approximately 1:250.

Results:
Samples containing organic onium (e.g. tetraphenylphosphonium chloride) form [M-X] + ion; samples containing multiple conjugated double bonds (e.g. anthracene, pyrene) form M+ ion; samples containing amine or phosphino functional groups (e.g. diphenylamine, triphenylphosphine) form [M+H]+ ion; samples containing multiple hydroxyl, ester groups (e.g. sucrose, dimethyl phthalate) form both [M+ Na]+ and [M+K]+ ions; samples containing nitrogen and oxygen functional groups (e.g. caffeine, glutamic acid) form [M+H]+ , [M+ Na]+ and [M+K]+ ions; samples containing less functional groups (e.g. octacosane, cholesterol) did not form any recognizable species. With a proper mass calibration and laser power adjustment, all involved ion peaks were readily identified from matrix-related background and measured with an accuracy of ±0.002 Da (Dalton).

Novel Aspect:
Good sensitivity and accuracy Easy sample preparation Routine analysis Low cost

Li, Jianjun,
National Research Council of Canada

Contributing Authors:
Kenneth Chan, National Research Council of Canada; Jacek Stupak, National Research Council of Canada.

Title:
Combination of reductive isotope labeling, exoglycosidase digestions and chromatographic separation for glycomics

Introduction:
The development of glycomics requires large-scale analysis of the structure and function of diverse glycans, and ultimately of whole glycomes. Therefore, the development of glycomics requires highly sensitive techniques for the fast and consistent detection and quantification of large numbers of glycans across multiple samples, which is only partially achieved by current glycomics approaches.

Methods:
N-glycans were prepared from RNase B and rabbit sera using similar procedures as described previously. PNGase F-released glycans were purified by using a non-PGC cartridge. The purified glycans were derivatized with aniline or aniline-2,3,4,5,6-d5. The mixture of aniline and aniline-d5 derivatized glycans were subjected to exoglycosidase digestions and analysed with LC-MRM.

Results:
To demonstrate the proof of principle, we first applied the strategy to the analysis of N-glycans released from RNase B. Aniline was chosen as the derivative reagent because stable isotopic aniline is inexpensive and easily available. Most importantly however, aniline-derivatization offers good detection sensitivity and superior separation resolution. In order to demonstrate that targeted glycomics is capable of identifying a wide range of glycans in a specific sample; we applied the method to pancreatic cancer cells. We further examined the targeted glycomics strategy with the linkage analysis of N-glycans with the combination of with exoglycosidase digestions and chromatographic separation. The mixture of aniline and aniline-d5 derivatized glycans were subjected to exoglycosidase digestions that can minimize the variability of exoglycosidase digestion procedures.

Novel Aspect:
LC-MRM for N-glycan quantitation and linkage analysis

Li, Xin,
Sphere Fluidics Limited

Contributing Authors:
Xin Liu, Sphere Fluidics Ltd, The Jonas Webb Building, Babraham Research Campus, Cambridge, CB22 3AT, UK; Marian Rehak, Sphere Fluidics Ltd, The Jonas Webb Building, Babraham Research Campus, Cambridge, CB22 3AT, UK; Vinayaka Pawate Sphere Fluidics Ltd, The Jonas Webb Building, Babraham Research Campus, Cambridge, CB22 3AT, UK; Anthony Dossang, GSK Medicines Research Centre, Gunnels Wood Road, Stevenage, Hertfordshire, SG1 2NY, UK; Clive A. Smith, Sphere Fluidics Ltd, The Jonas Webb Building, Babraham Research Campus, Cambridge, CB22 3AT, UK; Angela Bridges, GSK Medicines Research Centre, Gunnels Wood Road, Stevenage, Hertfordshire, SG1 2NY, UK; Murray J. Brown, GSK Medicines Research Centre, Gunnels Wood Road, Stevenage, Hertfordshire, SG1 2NY, UK; Bill Leavens, GSK Medicines Research Centre, Gunnels Wood Road, Stevenage, Hertfordshire, SG1 2NY, UK; Xin Li, Sphere Fluidics Ltd, The Jonas Webb Building, Babraham Research Campus, Cambridge, CB22 3AT, UK; Gareth Cooper, GSK Medicines Research Centre, Gunnels Wood Road, Stevenage, Hertfordshire, SG1 2NY, UK.

Title:
Picodroplet Mass Spectrometry for Miniaturized High Throughput Analysis of Synthetic Biology Microbial Clones

Introduction:
Mass spectrometry (MS) measures native molecules but is low throughput. Synthetic biologists construct multiple gene pathway libraries, which require efficient analytical tools. We have developed a Microfluidic Picodroplet (i.e. droplets of pL volume) MS method as a high-throughput tool (i.e. measures up to 200,000 samples/day) for characterizing synthetic biology-derived cell libraries.

Methods:
Sphere Fluidics has developed a high-throughput MS system using proprietary picodroplet technology for single cell analysis, which can test up to 200,000 samples (e.g. small molecular metabolites) per day. This uses miniaturized volumes of 600 pL (i.e. 1,400 fold lower than conventional MS) from single cell phenotypes, whilst retaining replicate samples of the genetic information/cells.

Results:
Individual engineered ketoreductase (KRED) secreting B. subtilis cells were encapsulated in picodroplets (~300pL) using a Pico-Gen™ biochip, and grew to 5,000 cells per picodroplet. Substrate molecules 2-(4-oxopiperidin-1-yl) acetic acid, introduced by picodroplet fusion, were converted into 2-(4-hydroxypiperidin-1-yl) acetic acid (HPAA) after incubation. The latter picodroplets were reinjected in a microfluidic MS emitter device, and split into two daughter picodroplets. One daughter picodroplet stream flowed via a plastic capillary emitter, which interfaced with a Perkin Elmer AxION 2 MS, and was analyzed by MS giving a saw-tooth trace of selected ion, HPAA (M+=160.0881). The other daughter picodroplet stream was sorted at 2 – 3Hz for inline retrieval of 'hits' based on MS analysis of the first daughter picodroplet and use of a novel picodroplet coding and decoding algorithm.

Novel Aspect:
High throughput and label free microfluidic picodroplet mass spectrometry for profiling valuable molecules produced by synthetic biology libraries.

Li, Liang,
University of Alberta

Contributing Authors:
Jaspaul Tatlay, University of Alberta; Xian Luo, University of Alberta; Wei Han University of Alberta; Kevin Hooton, University of Alberta; Shuang Zhao, University of Alberta; Yunong Li, University of Alberta; Liang Li, University of Alberta; Dorothea Mung, University of Alberta; Nan Wang, University of Alberta.

Title:
New Advances in High-Performance Chemical Isotope Labeling LC-MS for Metabolomics

Introduction:
Detailed analysis of the metabolome can provide holistic insight into a biological process, as well as potentially facilitate the discovery of disease biomarkers. However, there are still several technical challenges in metabolome profiling. In this presentation, new advances in the development of the high-performance chemical isotope labeling (CIL) LC-MS metabolomics platform will be described.

Methods:
In differential CIL, a 12C-labeling reagent is used to label an individual sample, while a pooled sample produced from mixing equal amounts of aliquots of all individual samples is labeled with the 13C-labeling reagent. The 13C-labeled pool is spiked into an individual 12C-labeled sample and the mixture is then subjected to LC-MS analysis. The labeled metabolites are detected as peak pairs.

Results:
We will highlight several new advances in the development of CIL LC-MS for metabolomics. These include the discussion of two labeling reagents targeting the analysis of hydroxyl metabolites and aldehyde/ketone metabolites, respectively. With the previously reported dansyl chloride labeling for analyzing the amine/phenol submetabolome and the DmPA labeling for analyzing the carboxylic acid submetabolome, the two new reactions significantly extend the metabolomic coverage by CIL LC-MS. We will also describe a cellular metabolomics workflow based on glass-bead for efficient cell lysis and CIL LC-MS for in-depth metabolomic coverage. Finally, research on identification of labeled metabolites will be presented to illustrate that the combination of labeled standards library search and metabolome database search allows us to perform efficient metabolite identification.

Novel Aspect:
New labeling chemistries and workflow combined with efficient compound library search improve the performance of CIL LC-MS for metabolomics.

Li, Xing-Fang,
University of Alberta

Contributing Authors:
Lindsay Jmaiff, University of Alberta; Guang Huang, University of Alberta; Yanan Tang University of Alberta; Wei Wang, University of Alberta.

Title:
New approaches for characterization of unknown precursors of disinfection byproducts in water

Introduction:
Drinking water disinfection effectively prevents waterborne diseases. However, disinfectants (e.g. chlorine) can react with natural organic matter in water to produce disinfection byproducts (DBPs) of health concerns. The majority (>70%) of DBPs remain unknown. Identifying and detecting toxicologically relevant DBPs is challenging.

Methods:
We developed new appraoches using multiple SPEs and HPLC separations with tandem MS techniques. Combined targeted and nontargeted analysis strategies using HPLC-MS/MS techniques are applied to characterize precursors and new DBPs. Water samples including plant influents and effluent of each treatment step were collected from nine water plants. The samples were chlorinated to produce HBQs.

Results:
The levels of HBQ precursors are indirectly determined based on measurements of four HBQs produced. The results show that HBQ precursors are widely present in all plant influents and that the current treatments do not substantially reduce HBQ precursors in water. We then demonstrate nontargeted analysis of raw water and treated water for identification of unknown precursors and DBPs using HPLC-QTOF. This study demonstrates the integration of multiple SPE and HPLC-MS/MS analysis with precursor ion exclusion to enhance the detection of chemicals in raw water >17000 and in fininshed water >13000.

Novel Aspect:
New approaches for characterization of unknown precursors of DBPs

Li, Frederick,
IonSense, Inc.

Contributing Authors:
Joseph Tice, IonSense, Inc.; Stephen Shrader, Shrader Software Solutions, Inc.; Brian Musselman IonSense, Inc.; Rachel Beck, IonSense, Inc.

Title:
Utilizing New Reverse Search Library Program and DART-MS for Targeted Screening of Food and Supplements for Adulterants

Introduction:
Mass spectral libraries are now critical to increasing the confidence in compound identification and confirmation. Along with DART-MS, reverse search libraries provide in-source CID spectra for identification and confirmation of various compounds. In an effort to increase the confidence in the screening of food for adulterants, libraries of toxicological compounds were created and evaluated.

Methods:
A DART source was interfaced to a single quadrupole mass spectrometer with the capability of in-source CID. Standards of compounds were analyzed by varying the voltage applied to the sample cone to induce fragmentation. A reverse search algorithm along with libraries consisting of four mass spectral entries with the most characteristic precursor and product ions for every compound was created.

Results:
The library consists of four entries for every compound, one for each of four cone voltages, which is modeled after the DART Forensic library for drugs of abuse developed by NIST. We have determined that voltages of +15V, +30V, +50V and +70V in the positive polarity provided sufficient molecular ion and fragmentation data required for identifying and differentiating 170 drug-related compounds. A total of 120 samples consisting of adjudicated case samples and training samples were analyzed and searched against the library consisting of 170 toxicological compounds. This process, requiring approximately 1-2 minutes per sample, resulted in a 90% positive identification with a probability match of greater than 0.70 when searched against the library. Although the 10% error rate was attributed to multi-component samples, only one of the components was not detected.

Novel Aspect:
A new reverse library search program and economical platform for targeted screening of food and supplements for adulterants utilizing DART-MS.

Liddle, Jen,
University of Colorado Boulder

Contributing Authors:
Jeremy Balsbaugh, University of Colorado Boulder; Johannes Rudolph, University of Colorado Boulder; Natalie Ahn University of Colorado Boulder.

Title:
Analysis of ligand binding by hydrogen exchange mass spectrometry reveals novel coupling between protein kinase subdomains

Introduction:
The MAP kinase ERK2 has become a valuable drug target in melanoma treatment, but current drug development strategies focus primarily on catalytic inhibition and affinity, overlooking global dynamic and conformational effects that can also alter cellular signaling. Hydrogen exchange mass spectrometry allows whole protein dynamic characterization as a function of inhibitor identity.

Methods:
Both active and inactive ERK2 in complex with nucleotide or inhibitory ligands were analyzed by injection into a temperature-controlled Waters nanoAcquity HDX Manager UPLC system for controlled proteolysis and high-resolution separation. Peptides were ionized by electrospray into a Waters Synapt G2 HDMS Q-TOF in positive ion MSe mode. Data were processed using PLGS 3.0 and DynamX 3.0.

Results:
Timecourse data comparing a nucleotide, moderate-affinity, and high-affinity ligand to apo protein show ligand-induced exchange reduction in selected regions of ERK2, but with strongly distinctive effects, immediate and distal to any contacts visible by crystallography. Differential effects do not correlate with number of contacts made or with solvent accessibility. An expanded analysis comparing 10 different ligands to apo protein reveals regions of ERK2 that co-vary in response to any ligand. These co-varying regions offer an experimental basis for protein community definitions. In addition, ligand analysis of global ERK2 effects show unexpected clustering only partially consistent with affinity. These clusters reveal a structural motif in small molecules important for coordinating the two kinase lobes.

Novel Aspect:
This work provides both an experimental basis for defining allosteric connectivities within ERK2, and a novel tool for interrogating new drugs.

Lieblein, Tobias,
Goethe-University Frankfurt

Contributing Authors:
Jun.-Prof. Nina Morgner, Institute of physical and theoretical chemistry, Goethe-University Frankfurt/Main; Tina Stark, Institute of organic chemistry and chemical biology, Goethe-University Frankfurt/Main; Prof. Michael Göbel Institute of organic chemistry and chemical biology, Goethe-University Frankfurt/Main.

Title:
Monitoring the oligomerization of beta-amyloid (1-42) and its inhibition with LILBID mass-spectrometry

Introduction:
Alzheimer (AD) is a widespread illness that diminishes the mental skills of its patients. Developing a drug for curing AD is of great relevance. The formation of oligomers of the beta-amyloid peptide (Aß) is known to play a key role in AD which can be analyzed with laser-induced-liquid-bead-ion-desorption (LILBID). LILBID allows to quantify the effect of molecules on the Aß oligomerization.

Methods:
In LILBID droplets are transferred into vacuum and irradiated by IR laser pulses leading to a droplet explosion and thus release of analyte ions into gas phase. The ions are analyzed with a reflectron time-of-flight mass spectrometer. Samples of Aß with and without small molecule inhibitors were incubated for several days; the oligomerization was studied at different time points.

Results:
LILBID is a useful tool to investigate the oligomerization of Aß. With this technique we can monitor oligomers as large as 20mers, which allows the scanning of different molecules regarding their effectiveness in inhibiting and reversing the Aß aggregation. We confirm that the known inhibitor molecule OR1 can inhibit Aß aggregation. Additionally, we can show that it is possible for OR1 to disaggregate oligomers of aggregated Aß. We also found a new inhibitor molecule (TSP25); TSP25 is non-toxic, non-degradable by peptidases because of its retro-inverso form, and more easily transported through the blood-brain barrier due to its smaller size and hydrophobicity. LILBID demonstrates the ability of TSP25 to block the process of Aß aggregation. Like OR1, TSP25 also has the ability to disaggregate oligomers of aggregated Aß. This proves LILBID is suitable to analyze small molecule inhibitors.

Novel Aspect:
LILBID has the ability to detect Aß oligomers as large as 20mer. This makes it possible to scan for inhibitors for the Aß oligomerization.

Lienert, Ian,
Biognosys AG

Contributing Authors:
Tobias Treiber, BiognoSYS AG; Oliver M. Bernhardt, BiognoSYS AG; Tejas Gandhi BiognoSYS AG; Roland Bruderer, BiognoSYS AG; Lukas Reiter, BiognoSYS AG; Magdalena Bober, BiognoSYS AG.

Title:
Delivering FDR control for MS1 alignment through the targeted analysis of MS1 as applied to shotgun

Introduction:
One of the major challenges of analysis of shotgun data is to overcome the limitations of reproducibility due to its semi-stochastic nature. Currently, this limitation can be partially overcome by MS1 alignment. Most current algorithms, however, do not deliver any FDRs for aligned peaks. Here, we present targeted analysis of MS1 as a possible solution and show the performance on various data sets.

Methods:
Our method performs per-run mass and iRT/RT (Escher et al. 2012) calibration using shotgun search identifications. Drawing on ideas from the mProphet algorithm (Reiter et al. 2011), we use the calibrations to inform our dynamic peak selection, ranking and scoring strategies at MS1-level to recover the set of missing precursors and their respective FDRs. Homogenous quantitation is also performed.

Results:
We first tested our algorithm on technical triplicates of samples with varying complexity: E. coli, yeast, human plasma and HEK-293. Median CVs for all overlapping precursors changed from 24.2% to 24.0%, 10% to 7%, 15.1% to 14.7% and 10% to 8%, respectively, while data completeness improved from ~80% to ~100% for these samples. To consider signals over a wide intensity range, we used a well-controlled sample set consisting of a HEK background with 12 non-human proteins spiked in at known absolute amounts and varying levels (Bruderer et al. 2015). Using this data set, we could benchmark how quantification of peptides/proteins that are differentially abundant improved. We found a significant improvement in quantification accuracy. Finally, we investigated performance on different forms of variance including technical, sample prep, and biological (rat brain, liver, and kidney) variance.

Novel Aspect:
Targeted analysis of MS1 to overcome the semi-stochastic nature of DDA data and additionally providing FDR control

Liigand, Jaanus,
University of Tartu

Contributing Authors:
Asko Laaniste, University of Tartu; Anneli Kruve, University of Tartu.

Title:
pH effects on electrospray ionization efficiency

Introduction:
Electrospray ionization mass-spectrometry is an increasingly used analytical technique. Although ESI is the most used ionization source, the mechanism of ESI is still not fully understood. The ionization efficiency (IE) dependence on different variables has been studied, but there is no comparative study of pH effects. This work aims to systematically study the pH effect on ionization efficiency.

Methods:
ESI ionization efficiencies of 28 compounds with pKa(aq) from -4 to 8.5 were measured on ion trap mass spectrometer. Two solvent compositions acetonitrile/buffer 80/20 and 20/80 (v/v) were studied. The compounds that showed statistically significant ionization efficiency differences at pH=2.1 and 7 were studied in detail. 11 solvents with pH= 2.1 to 7 with increment of 0.5 pH units were studied.

Results:
The studied 28 compounds can be divided according to behavior in solvent phase and in ESI source into four groups. The obtained linear discrimination function was able to predict the pH dependence with prediction precision of 80%. Quadratic discriminant analysis was performed as well and the logP of neutral form can predict with 71% the discrimination. Additionally interesting phenomena arises analyzing 2-nitroaniline, 3-nitroaniline and 4-nitroaniline. The 3-nitroaniline and 2-nitroaniline are in the pH-dependent group while 4-nitroaniline belongs to pH independent group. It is very interesting that 2-nitroaniline and 4-nitroaniline have similar resonance behavior but they behave differently in ESI ionization. Also was interesting that the ionization degree change in solution occurs at 0.5 pH units lower pH than in ESI source.

Novel Aspect:
Detailed comparative study of water pH effects on electrospray ionization efficiency were carried out.

Liigand, Piia,
University of Tartu

Contributing Authors:
Karl Kaupmees, University of Tartu; Anneli Kruve, University of Tartu.

Title:
Ionization efficiency of oligopeptides and small hydrophilic molecules in ESI/MS

Introduction:
ESI/MS is widely used for qualitative and quantitative analysis of various types of compounds. We have previously developed a model for predicting ionization efficiencies of singly charged analytes and studied multiple charging for small model compounds in negative ESI. Here we present a study of oligopeptides and some drug-like compounds, rationalizing ionization efficiency and multiple charging.

Methods:
Ionization efficiencies of 21 amino acids, 10 drug-like compounds and 24 oligopeptides were measured in positive ESI/MS using 80/20 MeCN/0.1% HCOOH (v/v) solution. Ionization efficiencies of amino acids and oligopeptides were measured in negative mode ESI using 80/20 MeCN/0.1% NH4OH (v/v) solution. pKa, logPow, WANS, Klamt parameters were calculated and correlated with ionization efficiencies.

Results:
For 14 peptides doubly charged ions and for 1 peptide also triple charged ions were observed in ESI/MS. The scope in ionization efficiencies is ca 5 logIE units. Higher ionization efficiencies were observed for larger more hydrophobic compounds. Additionally basicity or acidity (for ESI positive or negative, respectively) in solvent were found to be relevant descriptors. In a series of peptides with the same amino-acid "backbone" the position of easily ionizable amino acids becomes important, e.g. being in the ends of peptide chain increases the ionization efficiency relative to the placement in the middle of the chain. Also, ionization efficiency prediction model, employing degree of dissociation α and charge delocalization parameter WANS, was obtained. In validation set the root mean square error of this model was 0.67.

Novel Aspect:
Ionization efficiencies of hydrophilic compounds are studied, that give multiply charged ions (up to 3 charges) in ESI/MS.

Lin, Hou-Yu,
Institute of Atomic and Molecular Sciences Academia Sinica

Contributing Authors:

Sheng-Ping Liang, Institute of Atomic and Molecular Sciences, Academia Sinica; Jui-Yang Wang, Institute of Atomic and Molecular Sciences, Academia Sinica; Yuan-Tseh Lee Institute of Atomic and Molecular Sciences, Academia Sinica; Chi-Kung Ni, Institute of Atomic and Molecular Sciences, Academia Sinica.

Title:

Time dependent desorption process in MALDI

Introduction:

The ion generation by MALDI includes the success of ionization and desorption processes. Because neutrals are the dominant desorbed species in MALDI and ions have many collisions with neutrals during the desorption process, ions must follow some desorption properties of neutrals, such as desorption velocity distribution, angular distribution, and time dependent desorption.

Methods:

This work includes two experiments.First, time-delay post ionization laser and TOF mass spectrometer were used to measure the time-dependent desorbed neutrals from single crystal of 2,5-DHB in MALDI. Second, we used a modified crossed-molecular beam apparatus to measure the velocity distribution of the desorbed 2,5-DHB produced by the excitation of laser pulse with various pulse durations.

Results:

The post-ionization experiment reveals that the desorption takes less than 50 ns. The measurements of neutral velocity distributions show that the velocity distributions are the same for the pulse duration of 150 ps, 400 ps, 600ps, and 1.5 ns, indicating that the desorption process occurs after 1.5 ns.

Novel Aspect:

The result show the neutral desorption does not depends on the laser pulse duration from 200 ps to 1.5 ns.

Linford, Bryan,
Memorial University

Contributing Authors:
Travis Fridgen, supervisor.

Title:
Fragmentation of BMAA clusters with Alkali and Alkali earth metals using SORI-CID

Introduction:
BMAA is speculated to be a contributing factor to ALS and ALS like diseases. Recently BMAA has received more interest as it has been demonstrated to be produced by cyanobacteria, which are ubiquitous in most environments. No fundamental data has yet to be reported on BMAA structures or energetics, which can make it difficult to determine how much of a factor it has in these diseases.

Methods:
Dilute aqueous BMAA solutions were prepared with cations (H+ to Cs+, and Mg2+ to Ba2+) in a 50/50 methanol/water solvent, and ionized by electrospray ionization. Ions were then analysed using a Bruker Apex Qe 7T FTICR. For H to Cs, [(BMAA)M]+ clusters are probed, while for Mg to Ba [M(BMAA)2-H)]+ clusters are isolated and studied. Fragmentation patterns are observed using low energy SORI-CID.

Results:
The primary fragmentation pathways from SORI-CID for H-Na are the loss of 17 (NH3), and 31 (CH3NH2). For (BMAA)Rb+ and (BMAA)Cs+ BMAA is lost, and for the K+ complex there is no observable product ion, however loss of BMAA is speculated which leaves the unobservable K+ ion. For Mg-Ba the primary fragments are loss of 43 (CH3NCH2) and 86 (two CH3NCH2), as well as 60 (C2H8N2), 74 (C3H10N2), and 86 (C4H10N2). MS/MS of the first fragment (loss of CH3NCH2) gives loss of 17 (NH3), 31 (CH3NH2), and 43 (CH3NCH2). Further CID on the fragment from loss of NH3 gives loss of 44.

Novel Aspect:
Fragmentation patterns of previously unexplored BMAA complexes are presented

Liu, Mine-Yine,
National Changhua University of Education

Contributing Authors:
You-Sian Lin, Department of Chemistry, National Changhua University of Education.; Wen-Jie Li, Department of Chemistry, National Changhua University of Education.; Yu-Qing Chen Department of Chemistry, National Changhua University of Education.; Yun-Hsun Hsu, Department of Chemistry, National Changhua University of Education.; Yen-Yi Chen, Department of Chemistry, National Changhua University of Education.; Mine-Yine Liu*, Department of Chemistry, National Changhua University of Education.

Title:
Profiling of oxidized 1-palmitoyl-2-arachidonoyl-sn-glycero-3-phosphocholine products on lipoproteins by ESI-MS analysis

Introduction:
1-palmitoyl-2-arachidonoyl-sn-glycero-3-phosphocholine (PAPC) is a very important structural component of lipoproteins. PAPC contains a polyunsaturated fatty acyl chain, and is highly susceptible to oxidation. Oxidation of PAPC produces full-length fatty acyl chains inserted by several oxygen atoms, or shortened fatty acyl chains. Many of the oxidized PAPC products are pro-inflammatory.

Methods:
A simple and fast ESI-MS method in combination with liquid-phase and solid-phase extractions was developed to profile ox-PAPC products on lipoproteins. Various oxidation methods including Fenton reaction, Cu2+ oxidation and air auto-oxidation were examined to find the best oxidation conditions. Bindings between ox-PAPC products and apolipoproteins AI, AII were also investigated by ESI-MS analysis.

Results:
A total of 9 long-chain and 10 short-chain ox-PAPC products were identified. The ion intensities of ox-PAPC products were higher for Fenton reaction (Fe2+: H2O2 = 2:1) than for Cu2+ oxidation and air auto-oxidation. The nineteen ox-PAPC products have been analyzed for in vitro oxidized very low-density lipoprotein (VLDL), low-density lipoprotein (LDL) and high-density lipoprotein (HDL). For VLDL and LDL, the ion intensities of ox-PAPC products were higher for oxidized particles than for native particles. For HDL, the ion intensities of ox-PAPC products were similar for oxidized and native particles. Moreover, all of the nineteen ox-PAPC products have been found to bind to apolipoproteins A-I and A-II by ESI-MS analysis, but they appeared to bind stronger to apo A-II than to apo A-I.

Novel Aspect:
Liquid-phase and solid-phase extractions followed by ESI-MS analysis is a simple and fast method for profiling ox-PAPC products on lipoproteins.

Liu, Xiaoyun,
Peking University

Title:
Discovery of novel PTMs mediated by bacterial effectors

Introduction:
Bacterial pathogens can deliver a set of virulence factors (effectors) directly into host cells. These effectors often modulate key host signaling pathways to promote bacterial invasion as well as survival and proliferation. Two Legionella effectors, AnkX and SdeA, were recently discovered to mediate unique post translational modifications of host small GTPases such as Rab1.

Methods:
Host cells were either transfected with effector proteins or infected with different bacterial strains, and then candidate substrate(s) were affinity-purified and subjected to LC-MS analysis. Peptide signals were compared between unmodified and modified samples. MS2 and/or MS3 were used to characterize the molecular nature of the modifications and determine the cite(s) as well.

Results:
Legionella effector AnkX possesses the FIC domain, which is known to mediate AMPylation. Yet MS/MS analysis showed that AnkX modified Rab1 in an unknown manner instead of AMPylation. High-resolution measurements and MS/MS/MS analysis revealed that the unknown moiety that was attached to Rab1 is phosphocholine. Recently, we further discovered that Rab1 was also targeted by another Legionella effector SdeA. Though SdeA possesses a putative mono-ADP-ribose transferase motif, it was found to mediate ubiquitination of Rab1. Strikingly, in-vitro assays revealed that SdeA carried out ubiquitination independent of E1 or E2, using NAD as a cofactor instead of ATP. Further MS/MS analysis indicated the ligation between Rab1 and ubiquitin is not via the C-terminal glycine but arginine 42 of ubiquitin, suggesting an ADP-ribosylated ubiquitin is activated before attachment to its substrates.

Novel Aspect:
Two novel PTMs were discovered. Such PTMs were utilized by bacterial pathogens to modulate important host cellular functions during infection.

Liu, Chang,
SCIEX

Contributing Authors:

Chang Liu, Sciex; German Augusto Gomez Rioz, University of Waterloo; Don W. Arnold Sciex; Janusz Pawliszyn, University of Waterloo; Thomas R. Covey, Sciex.

Title:

Coupling solid phase micro-extraction (SPME) to mass spectrometry via an open-port probe (OPP) sampling interface

Introduction:

SPME is a green technology that combines sampling, sample preparation and extraction in one step thus simplifying the analytical process. Recently, several approaches have been developed for direct SPME-MS coupling. However, substantial modifications to the instrument frontend were required. Here, we introduce an electrospray based open-port probe as a direct sampling interface for SPME devices.

Methods:

An OPP that couples to a commercial ESI source was constructed to accept SPME fibers. Analytes in biological samples were extracted with a SPME fiber by a 2 min immersion. After a 10s wash with water, the fiber was inserted into the continuously flowing methanolic stream of the OPP entrance for 5s to desorb and transfer to the pneumatically assisted ESI electrode. LOQ and LDR were determined.

Results:

OPP sampling interface used a vertically aligned, co-axial tube arrangement enabling solvent delivery to the sampling end through the tubing annulus and aspiration down the center tube into the ion source driven by the nebulizer gas. Here, the SPME fiber was inserted into the open-port for 5s to touch the continuous flowing stream that transports the desorbed material into the ion source directly. LC separation was avoided by utilizing two different techniques to gain selectivity, MRM3 and differential mobility spectrometry (DMS). For the clenbuterol analysis in urine, MRM3 was used to eliminate the interference from matrix. DMS was used to differentiate codeine and hydrocodone for their simultaneous quantification from plasma. Good accuracy (>90%) and linearity (R2>0.998) was observed with the linear range of 0.1-100 ng/mL for clenbuterol, and 1-500 ng/mL for codeine and hydrocodone.

Novel Aspect:

Coupling of a new open port ESI probe with SPME devices for ambient sampling mass spectrometry.

Liuni, Peter,
Sanofi Pasteur/York University

Contributing Authors:
Bruce Carpick, Analytical Research & Development NA, Sanofi Pasteur Limited, Toronto, Ontario M2R 3T4, Canada; Andrew James, Analytical Research & Development NA, Sanofi Pasteur Limited, Toronto, Ontario M2R 3T4, Canada; Difei Sun Analytical Research & Development NA, Sanofi Pasteur Limited, Toronto, Ontario M2R 3T4, Canada; Derek Wilson, Department of Chemistry, York University, 4700 Keele Street, Toronto, Ontario M3J 1P3, Canada.

Title:
Structural differences between Diphtheria toxin and a detoxified single-point mutant CRM197 revealed by HDX-MS

Introduction:
Hydrogen/Deuterium exchange mass spectrometry (HDX-MS) is evolving into a powerful tool in the biopharmaceutical industry for evaluating the structure and consistency of manufactured proteins. In this study, we apply an HDX-MS workflow to compare the structures of two related proteins: Diphtheria toxin (DTx), and the genetically detoxified single-point mutant of DTx, CRM197 (mutation at G52E).

Methods:
A Waters Synapt G2-S quadrupole time-of-flight (QTOF) mass spectrometer equipped with a nanoAQUITY UPLC system and an HDX module was used to perform all deuterium exchange experiments. A 4.7 µL aliquot of DTx (8 µM) was incubated with 70.3 µL of deuterium oxide buffer (10 mM phosphate buffer in D2O, pD 7.00) at 25 ºC, for times ranging from 20 s to 4 h.

Results:
Preliminary comparison of DTx to CRM197 indicates large differences in dynamics across the entire protein sequence, with strong structural destabilization in CRM197 localized to the catalytic domain. Recent x-ray crystallographic and molecular dynamics work by Malito and coworkers identified a highly flexible active site loop (amino acids 38-49), but could only comment on increased overall protein instability observed from differential scanning fluorimetry data (1). Our HDX results provide a higher-resolution map of dynamics observed in CRM197, and may further explain its lack of toxicity. (1) Malito, E., et al., PNAS 2012, 109 (14) 5229-5234.

Novel Aspect:
HDX-MS as a method for structural characterization of toxin and nontoxic protein variants.

Lobodin, Vlad,
National High Magnetic Field Laboratory

Contributing Authors:
Alan G. Marshall, National High Magnetic Field Laboratory, 1800 East Paul Dirac Drive, Tallahassee, FL 32310; Tessa E. Bartges, Department of Chemistry and Biochemistry,95 Chieftain Way, Florida State University, Tallahassee, FL 32306.; Jie Lu Future Fuels Institute, 1800 East Paul Dirac Drive, Tallahassee, FL 32310; Markus Huettel, Department of Earth, Ocean, and Atmospheric Science, Florida State University, Tallahassee, FL 32306; Ian R. MacDonald, Department of Earth, Ocean, and Atmospheric Science, Florida State University, Tallahassee, FL 32306; Phoebe Z. Ray, National High Magnetic Field Laboratory, 1800 East Paul Dirac Drive, Tallahassee, FL 32310; Ekaterina V. Maksimova, College of Marine Science, University of South Florida, St. Petersburg, FL 33701; Caroline Johansen, Department of Earth, Ocean, and Atmospheric Science, Florida State University, Tallahassee, FL 32306; David C. Podgorski, Future Fuels Institute, 1800 East Paul Dirac Drive, Tallahassee, FL 32310; Yuri E. Corilo, Future Fuels Institute, 1800 East Paul Dirac Drive, Tallahassee, FL 32310; Logan C. Krajewski, Department of Chemistry and Biochemistry,95 Chieftain Way, Florida State University, Tallahassee, FL 32306.; Ryan P. Rodgers, Future Fuels Institute, 1800 East Paul Dirac Drive, Tallahassee, FL 32310.

Title:
Characterization of the Gulf of Mexico Deepwater Horizon Oil Spill and Natural Oil Seeps by FT-ICR MS

Introduction:
Crude oil is the main fossil fuel produced on the planet and is the source of widespread environmental pollution. It is estimated that about half of crude oil in the marine environment is due to oil spills and leaks during petroleum exploration, production, transportation, refining, and storage. The other half comes from natural seeps, making them important factor in the Gulf of Mexico ecosystem.

Methods:
Environmental samples collected from the Gulf of Mexico were extracted prior to analysis. High resolution (+)ESI, (-)ESI, and (+)APP I mass spectra were collected with a custom-built 9.4 T FT-ICR mass spectrometer. APGC/MS-MS analysis was conducted with a Xevo TQ-S instrument (Waters) equipped with an APGC ion source, and GCxGC/MS analysis was performed with a LECO Pegasus IV.

Results:
Crude oil, on entry to the environment, undergoes biotic and abiotic transformations (weathering) that, in turn, alter the original petroleum composition. FT-ICR MS reveals increased compositional complexity of environmental samples collected after the Macondo oil spill and enable detection of thousands of compounds formed upon oil weathering, containing one to eight oxygen atoms per molecule that cannot be detected by conventional GC/MS analysis. Among the products, we identified ketones, phenols, and carboxylic acids. FT-ICR MS analysis of natural oils seeps shows improved (over conventional GC/MS) compositional coverage of aromatic and heteroatomic (nitrogen, sulfur, oxygen) compounds, which comprise a significant mass fraction of petroleum and need to be taken into account in oil spill characterization. Work supported by NSF DMR-11-57490, FSU-FFI, GOMRI, and the State of Florida.

Novel Aspect:
First combination of FT-ICR MS and APGC-MS/MS for a comprehensive analysis of the Macondo oil spill and natural oils seeps from the Gulf of Mexico.

Longnecker, Krista,
Woods Hole Oceanographic Institution

Title:
Environmental metabolomics in marine ecosystems: impact of viral lysis and predation on marine organic matter

Introduction:
The marine environment holds one of the largest pools of reduced organic carbon on Earth. Within this organic carbon are metabolites produced as a result of microbial activity. This project considers the effects of predation and viral lysis on the metabolites found in the surface ocean.

Methods:
Using seawater from the surface ocean, we experimentally manipulated levels of predation and viral lysis. At the beginning and end of the experiment, we sampled the intracellular and extracellular metabolites. The extracts were analyzed using liquid chromatography-based targeted and untargeted metabolomics methods that we have modified for use with marine samples.

Results:
The untargeted metabolomics data revealed a complex mixture of organic compounds in all of the experimental treatments. However, changes in the majority of the features could not be linked to viral lysis or predation. Within the targeted metabolomics data, increased intracellular levels of osmolytes (glycine betaine, dimethylsulfoniopropionate, proline, and ectoine) were observed under conditions with limited grazing or viral lysis. The molecular insights derived here will explicitly inform our understanding of microbial processes in the surface ocean and the subsequent impacts on the global carbon cycle.

Novel Aspect:
This is the first analysis of the impact of grazing and viral lysis on the metabolites observed in marine ecosystems.

Loo, Joseph,
University of California, Los Angeles

Contributing Authors:
Huilin Li, University of California, Los Angeles; Piriya Wongkongkathep, University of California, Los Angeles; Rachel R. Ogorzalek Loo University of California, Los Angeles; Iain Campuzano, AMGEN.

Title:
Native Mass Spectrometry of Protein-Protein and Protein-Drug Complexes

Introduction:
Native mass spectrometry of proteins and protein assemblies reveals size and stoichiometry, but elucidating their structures is more challenging. Membrane proteins are involved in many biological functions, but their characterization by a variety of techniques is challenging. We show that MS can be effective for deriving structural information for soluble and membrane protein complexes.

Methods:
Proteins were analyzed by nanoESI-MS under native buffer conditions using a Waters Synapt ion mobility QTOF and a 15-Tesla Bruker solariX Fourier transform ion cyclotron resonance (FT-ICR) system. The FT-ICR-MS instrument is equipped with electron capture dissociation (ECD; and 30 eV EID) and a CO2-laser operating at 10.6 µm for activated ion-ECD and infrared multiphoton dissociation (IRMPD).

Results:
ESI's gift for transforming solution-phase macromolecules into gas-phase ionized counterparts without disrupting weak noncovalent interactions is key for applying MS to study protein complexes. We are using native MS to investigate the molecular action of compounds that prevent amyloid fibril formation in neurodegenerative diseases such as Alzheimer's and Parkinson's disease. Solubilizing membrane proteins in MS-compatible detergents or lipid structures are effective for delivery into the mass spectrometer, but require efficient means to unwrap the native membrane protein from their cocoon for detection. We show that FT mass spectrometers can be especially effective for "shaking off" detergents and lipids from native membrane proteins for subsequent MS and top-down MS measurements. The importance of salt bridges in the dissociation behavior of gas phase proteins will be presented.

Novel Aspect:
Native MS and top-down MS of protein-drug complexes and membrane proteins yield information of utility to structural biology.

Loo, Joseph,
University of California, Los Angeles

Contributing Authors:

Huilin Li, University of California, Los Angeles.

Title:

Top-Down Mass Spectrometry of Proteins and Protein Complexes as a Tool for Structural Biology

Introduction:

There is still a growing need for the development of new activation/dissociation techniques in the biological MS field for obtaining structural information for protein complexes. Top-down mass spectrometry can be an effective tool for protein sequencing. We use top-down high resolution MS to probe ligand-binding sites and to generate topological information for large proteins and complexes.

Methods:

A Bruker Solarix 15-Tesla Fourier transform ion cyclotron resonance mass spectrometer (FT-ICR) with an infinity cell and with nanoelectrospray ionization (nanoESI) was used in this study. Electron capture dissociation (ECD), infrared multiphoton dissociation (IRMPD), and electron ionization dissociation (EID; bias voltage of 30 V) were used with the FT-ICR system.

Results:

Native top-down MS generates information on the surface topology, ligand binding sites, and post-translational modifications of protein complexes. The various activation methods available with FT-ICR MS allow membrane proteins to be measured with high accuracy. Top-down native MS with CAD, IRMPD, and EID of 27 kDa bacteriorhodopsin yields mostly products from the N-terminal A- and B-helices and reveals the site of retinal binding. EID employs irradiation of cations by fast electrons with energy at least 10 eV higher than the ionization threshold of the cations; such irradiation causes simultaneous backbone bond cleavages, which makes it suitable for top-down analysis of folded gas-phase proteins. We present the unique capabilities of EID for the characterization of native protein-ligand and protein-protein complexes, and its similarity to 193 nm ultraviolet photodissociation (UVPD).

Novel Aspect:

Native top-down MS should be one of several MS-based approaches to provide important data for an integrated approach to structural biology.

LOPEZ MUÑOZ, GERSON DIRCEU,
University of los Andes

Contributing Authors:
EDGAR RODRIGUEZ, Universidad Nacional de Colombia; MARTHA ISABEL MURCIA, Universidad Nacional de Colombia; CHIARA CARAZZONE University of Los Andes.

Title:
Analysis of Phthiocerol Dimycocerosates (PDIM) as Potential Biomarkers of Rifampicin Resistant Tuberculosis Using HPLC-MS/MS and Shotgun Lipidomics

Introduction:
Tuberculosis (TB) is a pulmonar infectious disease produced by Mycobacterium tuberculosis (Mtb) , representing the second cause of mortality in the world after HIV, which accounts for 2% of the total. Among the new cases of symptomatic TB, 5% is resistant to the treatment with rifampicin and isoniazid.

Methods:
The aim of this study is to evaluate the lipid phthiocerol dimycocerosate (PDIM) as a potential biomarker of rifampicin resistant Mtb through a mass spectrometric approach. This study represents a novel approach to finding a biomarker related to the pathogen instead of the host. Three different strains of Mtb were cultivated: Mtb: H37Rv (control strain), Mtb Beijing Rif-resistant and Mtb Erdman.

Results:
Liquid-liquid extraction with CHCl3 was performed, CHCl3 was evaporated, and the sample was dissolved in hexane and finally cleaned by solid phase extraction (SPE). Lipid extracts were subjected to flow injection analysis (FIA) in Ion trap mass spectrometer and to HPLC-DAD-ESI-MS/MS equipped with electrospray ionization (ESI) operated in positive and negative mode. Similar profiles and characteristic patterns of fragmentation of PDIM could be observed in both flow injection analysis and HPLC-MS/MS. The preliminary results obtained, show a difference in the appearance and the intensities of the peaks of PDIM in the different strains of Mtb under consideration, thereby suggesting that PDIM could represent a potential biomarker for rif-resistant tuberculosis. The analytical aspects of simple preparation methods and determination should be further standardized to provide a reproducible.

Novel Aspect:
The search for a potential biomarker of the pathogen Mtb represents a novel approach towards the diagnosis and eventually treatment of TBRif-resistant

Louarn, Essyllt,
Paris Sud University

Contributing Authors:
Hubert Latappy, LCP - CNRS/Université Paris Sud - Orsay, France; Michel Héninger, LCP - CNRS/Université Paris Sud - Orsay, France; Hélène Mestdagh LCP - CNRS/Université Paris Sud - Orsay, France; Stéphane Pasquiers, LPGP - CNRS/Université Paris Sud - Orsay, France; Julien Leprovost, AlyXan, Juvisy-sur-Orge, France; Joel Lemaine, LCP - CNRS/Université Paris Sud - Orsay, France; Elsa Bauchard, AlyXan, Juvisy-sur-Orge, France.

Title:
Real-time analysis with a transportable high-resolution PTR-FTICR-MS

Introduction:
Analysis of complex environmental samples are a challenging operation. The development of direct and in-field analytical systems is crucial. For that purpose, a team including the start-up AlyXan and the LCP Laboratory developed an innovative transportable FTICR/MS. The apparatus is combined to a direct introduction system for air analysis and to a membrane introduction system for water analysis.

Methods:
The FTICR is based on a permanent magnet of 1 to 1.5 Tesla allowing high resolution in the range 15-300 amu. Volatile organic compounds (VOC) are analyzed in a sample thanks to chemical ionization, using H3O+ as precursor ion. MH+ is formed through proton transfer reaction. All the ions are detected simultaneously using a dedicate software. When needed other precursors may be used.

Results:
The instruments are able to achieve exact mass measurement up to 0.005amu after proper calibration thanks to two known compounds. The use of chemical ionization (CI) and continuous measurment of pressure inside the cell makes it possible to quantify directly the amount of reactive species introduced. All ions are monitored simulatneously in 2-4s. Detection limit range from 500ppb to 20ppb according to species, magnetic field and detection length time. Those instruments were successfully used for a number of different applications such as wastewater analysis, polymer degradation studies, depollution system efficiency studies and car's exhaust fumes temporal variation studies.

Novel Aspect:
We demonstrate the use of an enhanced transportable FTICR using PTR for direct and in-situ detection of VOC for temporal variation studies.

Lu, I-Chung,
Department of Chemistry, Wayne State University

Contributing Authors:
Charles N. McEwen, Universirty of the Sciences, Philadelphia, PA; Sarah Trimpin, Wayne State University, Detroit, MI; Milan Pophristic MSTM, Newark, DE.

Title:
Development of a High Sensitivity Source for Matrix-Assisted Ionization in Vacuum (MAIV) and Approaches to Fundamental Studies

Introduction:
MAIV spontaneously produces abundant analyte ions when matrix:analyte crystals are introduced to sub-atmospheric pressure. The minimum ion source requirements for MAI were determined by eliminating elements from an IP-MALDI source. The outcome is a highly sensitive minimalistic ion source with fast matrix:analyte sample introduction. Fundamental studies and applications are presented.

Methods:
A SYNAPT G2 (Waters Corporation) IP-MALDI source housing was modified, removing the laser, extraction lens and hexapole ion guides and installing a solids probe vacuum lock to allow rapid sample introduction. The source chamber and vacuum lock is pumped only through an aperture separating the housing from the mass spectrometer. 3-Nitrobenzonitrile was used as the matrix with a variety of analytes.

Results:
The ionization process in MAI occurs when sample is introduced into vacuum, which we have reduced from ca. 2 minutes using MALDI sample plate introduction to ca. 10 seconds using the solid probe. Additionally, the distance between sample probe and ion entrance have been reduced while improving MAI sensitivity relative to using the IP-MALDI source. A sample containing a few pmol of analyte (drug, peptide, and small protein) produced continuous abundant analyte signals for several minutes and reloading only the matrix solution multiple times further extended the time abundant ions were observed. Extended ionization aids MS/MS analysis of complex tryptic digests. In addition to applications, we have studied the correlation between temperature/pressure and the time-dependent sublimation providing evidence of desolvation from solid state particles producing bare gas phase ions.

Novel Aspect:
A minimalistic ion source designed specifically for matrix-assisted ionization is promising for fundamental and application studies.

Luethy, Lucas,
CTC Analytics AG

Contributing Authors:
Tom Flug, CTC Analytics, Zwingen, Switzerland; Timm Hettich, University of Applied Sciences and Arts Northwestern Switzerland FHNW, Muttenz, Switzerland; Guenter Boehm CTC Analytics, Zwingen, Switzerland; Christian Berchtold, University of Applied Sciences and Arts Northwestern Switzerland FHNW, Muttenz, Switzerland; Goetz Schlotterbeck, University of Applied Sciences and Arts Northwestern Switzerland FHNW, Muttenz, Switzerland; Reto Bolliger, CTC Analytics, Zwingen, Switzerland; Irene Wegner, University of Applied Sciences and Arts Northwestern Switzerland FHNW, Muttenz, Switzerland.

Title:
Fully Automated Sample Preparation of Whole Blood for TDM and Diagnostics Procedures by LC/MS/MS using a Commercial AutoSampler with Tool Change.

Introduction:
Although automation is not completely new to therapeutic drug monitoring (TDM) programs and diagnostic procedures it has typically involved expensive specialized pipetting robots and off-line steps to handle the sample preparation of whole blood. Traditionally, only select steps have been automated, resulting in a segmented and often error prone path from sample preparation to detection.

Methods:
A completely integrated system using a PAL RTC autosampler, vortexer, centrifuge and LC/MS/MS was used to automate protein precipitation, vortexing, centrifugation and injection into a LC/MS/MS. Amino acids, acyl-carnitins and steroids as analytes were chosen as models for diagnostics. Diclofenac, carbamazepine and immunosuppressives were used as examples for TDM applications.

Results:
Steps: On-line sample preparation of whole blood (1) Protein Precipitation: 75% organic solvent for precipitation such as methanol, acetonitrile and 0.1M $ZnSO_4$/methanol solution proved effective; while mixtures using isopropanol and water proved to be insufficient. (2) Vortexing: For most metabolites, vortexing of 90 seconds (at 1400 rpm) was sufficient. The exceptions was seen with highly bound protein drugs such as immunosuppressants which needed prolonged vortexing. (3) Centrifugation: A centrifugation time of at least 2 minutes was necessary at 3000 g to create an acceptable pellet. (4) Removing aliquots of supernatant: The needle penetration of 22 mm into a 1.5 mL vial allowed us to collect up to a 200 µL aliquot of supernatant. This allowed enough volume to inject and extra for further analysis.

Novel Aspect:
Complete automation workflow of whole blood sample preparation for LC/MS/MS for TDM applications and diagnostic procedures.

Luo, Xi,
Japan advanced institute of science and technology

Contributing Authors:
Yuzuru Takamura, Japan Advanced Institute of Science and Technology; Phan Trong Tue, Japan Advanced Institute of Science and Technology.

Title:
Negative ion behavior in pulse-heating ion source mass spectrometry

Introduction:
A pulse-heating ion source has been established to peptides and proteins detection in positive mode by only applying thermal energy from a single pulse without laser or high voltage. In this study, the negative ions are first generated and detected by negative mode. Differences and similarities between both positive and negative modes are discussed aiming to understand the mechanism of ionization.

Methods:
The mass spectrometry consists of a fabricated pulse-heating ion source, a home-built time-of-flight (TOF) mass filter and a commercially available detector. The proposed system can work in both positive and negative detection modes simply by switching voltage polarity of the TOF. Detection of Cytochrome c using the negative mode with only single pulse-heating for ionization was demonstrated.

Results:
The NaCl was chosen for testing and calibrating the negative mode, which aims to avoid the fragment signal caused by organic reactions and decomposition. For the positive mode, mass spectrum of only matrix (2,5-dihydroxybenzoic acid, DHB) sample showed no signal, whereas that of the Cytochrome c sample also exhibited no peak in a low molecular weight region ($m/z < 2000$). On the contrary, for the negative mode, mass spectrum of the matrix sample showed signals associated with DHB contents. For protein sample, fragment signals were observed at both high and low molecular weight regions. These are originated from the fact that DHB contains phenolic hydroxyl and carboxylic groups which can donate protons. As a result, the residue of DHB molecule might be negatively charged and detected in the negative mode, which appeared as the fragment signals at the low molecular weight region.

Novel Aspect:
The pulse-heating on-chip ion source was successfully operated using negative mode for the first time for miniaturized mass spectrometry.

Ma, Bin,
University of Waterloo

Contributing Authors:

Bin Ma, University of Waterloo; Lian Yang, Rapid Novor Inc.; Mingjie Xie Rapid Novor Inc.; Johnathan Krieger, Hospital for Sick Children; Paul Taylor, Hospital for Sick Children; Qixin Liu, Rapid Novor Inc.

Title:

Software for Monoclonal Antibody Protein Sequencing with LC-MS/MS

Introduction:

Antibodies are important biopharmaceuticals. Being able to routinely and directly sequence the antibody protein molecules enables a number of important applications. We developed an LC-MS/MS workflow and automated software to sequence the antibody protein molecule directly, routinely, and in-depth. In addition, the modifications (including glycosylation) are characterized and quantified.

Methods:

The antibody is reduced, alkylated, and digested with multiple enzymes. LC-MS/MS is performed on each digest. Novor software is used for de novo peptide sequencing. An internal database search software is used to find reference sequences from an antibody database. The de novo peptides are mapped to the reference, and the consensus of the de novo sequences are taken as the real protein sequence.

Results:

The workflow is tested with the Waters' IgG-I antibody standard (product number 186006552). The full sequences for both the heavy and light chains are fully recovered de novo. Compared to the sequence provided by Waters, two variations in the variable regions of the heavy chain were discovered. Both changes are on the variable region of the heavy chain, and are supported by strong MS/MS signal peaks. Since the variations do not change the intact mass of any tryptic peptide, they can only be picked up with de novo peptide sequencing. Additionally, 6 glycosylation forms were identified. The largest glycan form conforms to the published structure. The other forms lack one or more Hex and HexNac units compared to the published structure. 5 of the 6 forms have a slightly different retention time, suggesting that they are in the original sample and not the result of in-source fragmentation.

Novel Aspect:

Automated software for de novo sequencing antibody proteins with LC-MS/MS. PTMs (including glycosylation) are characterized in the same workflow.

Mackenzie, Stuart,
University of Oxford

Title:
Infrared-driven chemistry on isolated transition metal clusters: Catalytic reaction barriers from IR-MPD and BIRD

Introduction:
We explore the use of various techniques including infrared multiple photon dissociation (IR-MPD) spectroscopy, BIRD and collisional excitation to investigate catalytically relevant chemical reactions performed on the surface of isolated, size-selected transition metal clusters.

Methods:
Infrared multiple-photon dissociation spectroscopy, FT-ICR Mass spectrometry, BIRD, velocity map imaging.

Results:
We demonstrate how IR-MPD can be used to gain information not only on the structures of entrance-channel complexes but also on relative barrier heights for desorption and/or reaction of molecules adsorbed on size-selected metal clusters. Cluster size- and cooperative binding effects are observed in a range of systems. Where reactive barrier heights are sufficiently small, black-body infrared radiative dissociation is demonstrated for a small decorated transition metal cluster system. Results are compared with those from collisional activation experiments. In other experiments we illustrate the use of velocity map imaging to study small metal-containing molecules produced by laser ablation.

Novel Aspect:
New approaches to reactivity of size-selected clusters as model systems for heterogeneous catalysis including rare example of BIRD for a small metal c

Madugundu, Guru Swamy,
PERFORM Centre, Concordia University

Contributing Authors:
J. Richard Wagner, Nuclear Medicine and Radiobiology, Université de Sherbrooke, Sherbrooke, CANADA.

Title:
Ozone-induced DNA damage: A pandora box of oxidatively modified DNA bases.

Introduction:
Oxidatively-induced DNA damage can result from environmental factors, by various oxidizing agents. Air pollution is associated with an increased risk of lung cancer indeed; the ozone has been associated with a wide spectrum of adverse health effects. Here we have developed an isotopically labeled mass spectrometry based assay to measure the oxidation products of DNA bases by exposed to ozone.

Methods:
The oxidation was carried out by the addition of CT-DNA to an O3 saturated aqueous solution of PBS. A solution of 0.2 mL containing CT-DNA and ozone was allowed to react for 60 min at room temperature with different concentrations of ozone. Analysis of products, DNA was completely digested to nucleosides by P1 nuclease and SVP, and the mixture was separated by LC using reverse phase chromatography

Results:
In this work, we focused on the analysis of products arising from oxidation of cytosine, thymine and 5-methyl cytosine in CT-DNA. We found that the oxidation of these DNA bases by ozone gave a highly selective distribution of damage with overwhelming formation of hydantion products from the reaction with pyrimidines. The first step in the formation of hydantoin products, involves the initial attack of ozone at the 5,6-double bond leading to the formation of an Criegee ozonide intermediate, followed by ring opening to an open chain hydroperoxide. The stable products are then formed by decomposition of hydroperoxides, which in the case of thymine depending on the conditions, led mainly to the formation of the corresponding hydanotin, formamide, formylurea and urea fragmented products of pyrimidine bases. In addition a small yield formation of methyl oxidation (aprox 4-10% of total damage).

Novel Aspect:
A mechanism of oxidation of pyrimidines by ozone exposer in DNA is elucidated by comprehensive analysis of the product by LC-MS/MS

Madugundu, Guru Swamy,
PERFORM Centre, Concordia University

Contributing Authors:
Marie-Eve Rivard, Clinical analysis suite, PERFORM Centre, Montreal, CANADA.; Dajana Vuckovic, Clinical analysis suite, PERFORM Centre, Montreal, CANADA.

Title:
A liquid chromatography-tandem mass spectrometry method for quantitative analysis of curcumin and curcumin metabolites in human plasma.

Introduction:
Alzheimer's disease starts from over-accumulation of amyloid β (Aβ) monomers caused by imbalanced clearance. Curcumin is known as a fluorescent probe by linking specifically Aβ plaques which may make it useful for the detection of Alzeheimer's disease. However, curcumin usefulness is compromised by its poor bioavailability and the lack of analytical methods to monitor its absorption.

Methods:
In this work, we have developed and validated a selective, fast and highly sensitive method employing ultra-high performance liquid chromatography coupled with tandem mass spectrometry (UPHLC-MS/MS) for simultaneous quantification of curcumin and its major metabolites; tetrahydrocurcumin, demethoxycurcumin and bisdemethoxycurcumin in the linear range of 0.5-50 ng/mL in human plasma.

Results:
Here we used triple quadrupole mass detection with positive mode multiple reaction monitoring (MRM) to monitor the ion transitions, m/z of 369 > 285 for curcumin, 373 > 137 for tetrahydrocurcumin, 339 > 255 for demethoxycurcumin, 309 > 225 for bisdemethoxycurcumin and m/z of 240 > 148 for salbutamol (internal standard) respectively. The chromatography separation achieved by reverse phase chromatography using C18 column and analytes were extracted by an LEE technique using ethyl acetate solvent. The method was validated for recovery, accuracy, precision, linearity, and applicability. The inter-day and intra-day precisions and accuracy values were within the Food and Drug Administration (FDA) acceptance criteria, for plasma matrix. This method was capable of detecting curcumin and its metabolites in plasma samples from one healthy volunteer after an oral ingestion of curcumin.

Novel Aspect:
This method provides a comprehensive analytic method to quantify the curcumin and its metabolites in human plasma after oral ingestion.

Maedler, Stefanie,
Ontario Ministry of the Environment and Climate Change

Contributing Authors:
H.M. "Skip" Kingston, Duquesne University; James P. Sherry, Environment Canada; Robert J. Tooley Ontario Ministry of the Environment and Climate Change; Vasile I. Furdui, Ontario Ministry of the Environment and Climate Change; Matt Pamuku, Applied Isotope Technologies.

Title:
Can glutathione species be used as biomarkers for environmental metal exposure in fish?

Introduction:
The tripeptide glutathione can act as a free radical scavenger to maintain the redox status within cells. Under oxidative stress, reduced glutathione (GSH) is oxidized to glutathione disulfide (GSSG), and the GSH/GSSG ratio can be used as an indicator of pathophysiological conditions. The analysis of glutathione species is hampered by oxidation reactions induced by sample preparation and analysis.

Methods:
After exposure to environmental toxins, aliquots of fish blood were spiked with isotopically enriched analogues of the analytes (310GSH and 616GSSG) and the protein-precipitated supernatant analyzed by liquid chromatography tandem mass spectrometry (SCIEX 4000QTrap). The analytes were quantified by simultaneously tracking and correcting auto-oxidation and the results compared to control samples.

Results:
Rainbow or lake trout and white suckers were exposed to Cr, Cu and/or Li. When analyzing juvenile rainbow trout brain samples after 4-day exposure to Li, SIDMS analysis comparing control and exposed fish brain samples showed a significant decrease in GSSG and GSH levels after 96h ($p<0.01$), resulting in about half of the original levels. The GSH/GSSG ratios increased in the exposed fish compared to the control samples ($p<0.05$). Both trends have been observed with Li-fed rats as well. Exposure to Cr(VI) for 21 days resulted in a significant change of GSH levels and GSH/GSSG ratios in lake trout blood compared to control samples ($p<0.05$ and $p<0.01$, respectively). Exposure to Cu(II) for the same time frame and fish species did not yield any significant differences in GSH and GSSG levels. SIDMS can quantify GSH and GSSG and thus provides a great tool to investigate oxidative stress.

Novel Aspect:
Absolute quantification of reduced and oxidized glutathione in fish with simultaneous correction for auto-oxidation using molecular-speciated isotope

Magni, Fulvio,
University Milano Bicocca

Contributing Authors:
Manuel Galli, Dipartimento di Medicina e Chirurgia, Univ Milano Bicocca, Monza (Italy); Andrew Smith, Dipartimento di Medicina e Chirurgia, Univ Milano Bicocca, Monza (Italy); Clizia Chinello Dipartimento di Medicina e Chirurgia, Univ Milano Bicocca, Monza (Italy); Martina Stella, Dipartimento di Medicina e Chirurgia, Univ Milano Bicocca, Monza (Italy); Niccolò Mosele, Dipartimento di Medicina e Chirurgia, Univ Milano Bicocca, Monza (Italy); Fabio Pagni, Dipartimento di Medicina e Chirurgia, Univ Milano Bicocca, Monza (Italy); Italo Zoppis, Dipartimento di Informatica, Sistemistica e Comunicazione, Univ Milano Bicocca, Monza (Italy); Giancarlo Mauri, Dipartimento di Informatica, Sistemistica e Comunicazione, Univ Milano Bicocca, Monza (Italy) .

Title:
Combining MALDI-Imaging and LC-ESI-MS/MS in Thyroid TMAs for the discrimination of different thyroid lesions

Introduction:
MALDI-imaging has been successfully used to differentiate benign and malignant thyroid lesions. Protein markers to support the diagnosis of thyroid specimens are needed due to the high prevalence of undetermined reports, which are treated as malignant with total thyroidectomy and lifelong hormone therapy. To identify new markers, we investigated a wide histological TMA-based thyroid lesion series.

Methods:
After deparaffinization, tissue rehydration and antigen retrieval, TMAs were tryptic digested and analyzed by MALDI-TOF in reflector positive mode using HCCA as matrix. Protein identification was performed analyzing the extracted tryptic peptides, after matrix removal, by nanoLC-ESI-MS/MS and by Mascot. Discriminant features were selected using the K-Nearest Neighbor and the VizRank algorithms.

Results:
Results showed specific protein signatures for malignant and benign specimens and enabled us to build clusters comprising several proteins with discriminant capabilities, by matching MALDI-imaging data with protein identifications through the mass values. Among them, Fibronectin (FINC), Cytoplasmic Actin 1 (ACTB1), Prelamin-A/C (LMNA), Heat Shock cognate 71KDa Protein (HSP7C) and Adenylate Cyclase Isoenzyme 1 (KAD1) were identified by LC-ESI-MS/MS and found to be up-expressed in malignant lesions.

Novel Aspect:
These results could be translated into clinical practice, e.g. for new immunohistochemical tools, and used for better understanding thyroid lesions.

Magni, Fulvio,
University Milano Bicocca

Contributing Authors:
Andrew Smith, Department of Medicine and Surgery, Proteomics and Metabolomics Unit, University of Milano-Bicocca, Monza, Italy; Vincenzo L'Imperio, Department of Medicine and Surgery, Pathology, University of Milano-Bicocca, San Gerardo Hospital, Italy; Niccolò Mosele Department of Medicine and Surgery, Proteomics and Metabolomics Unit, University of Milano-Bicocca, Monza, Italy; Elena Ajello, Department of Medicine and Surgery, Pathology, University of Milano-Bicocca, San Gerardo Hospital, Italy; Franco Ferrario, Department of Medicine and Surgery, Pathology, University of Milano-Bicocca, San Gerardo Hospital, Italy; Martina Stella, Department of Medicine and Surgery, Proteomics and Metabolomics Unit, University of Milano-Bicocca, Monza, Italy; Manuel Galli, Department of Medicine and Surgery, Proteomics and Metabolomics Unit, University of Milano-Bicocca, Monza, Italy; Clizia Chinello, Department of Medicine and Surgery, Proteomics and Metabolomics Unit, University of Milano-Bicocca, Monza, Italy; Federico Pieruzzi, Department of Medicine and Surgery, Nephrology Unit, University of Milano-Bicocca, Monza, Italy; Goce Spasovski, Department of Nephrology, Medical Faculty, University of Skopje, Skopje, Macedonia; Fabio Pagni, Department of Medicine and Surgery, Pathology, University of Milano-Bicocca, San Gerardo Hospital, Italy.

Title:
The putative role of MALDI-MSI in the study of Membranous Nephropathy

Introduction:
Membranous Nephropathy (MN) is an immunocomplex mediated renal disease that represents the leading cause of nephrotic syndrome in adults and is one of the most frequent glomerulopathies. This disease can manifest as primary (iMN) or secondary and this distinction is crucial when choosing the most appropriate management of patients, with markers to further stratify this disease seeming mandatory.

Methods:
MALDI-MSI was applied to formalin-fixed paraffin-embedded (FFPE) renal biopsies from histologically diagnosed primary and secondary MN patients (n=20). Briefly, deparaffinisation, tissue rehydration, antigen retrieval and in situ trypsin digestion were performed prior to the deposition of CHCA. All the mass spectra were acquired in reflectron positive mode in the mass range of m/z 750 to 3500

Results:
Results showed a good level of analytical reproducibility, both in terms of the number of signals as well as their intensity, when analytical reproducibility was evaluated. MALDI-MSI was able to generate molecular signatures of primary and secondary MN, with one particular signal (m/z 1459), identified as Serine/threonine-protein kinase MRCK gamma, being over-expressed in the glomeruli of primary MN patients with respect to secondary MN. Furthermore, this proteomic approach detected a number of signals that could differentiate the different forms of iMN that were positive to PLA2R or IgG4 as well as a further set of signals (m/z 1094, 1116, 1381 and 1459) that distinguish these patients from those who were negative to both. These signals could potentially represent future targets to be investigated as proteomic markers for the further stratification of iMN patients.

Novel Aspect:
This preliminary study represents the first example of MALDI-MSI being applied to FFPE renal biopsies in order to study the tissue proteome of MN.

Makarov, Alexander,
Thermo Fisher Scientific

Contributing Authors:

Cristian Cojocariu, Thermo Fisher Scientific, Tudor Road, Manor Park, Runcorn, Cheshire, WA7 1TA, UK; Alexander Makarov, Thermo Fisher Scientific, Hanna-Kunath Str. 11, Bremen 28199 Germany; Paul Silcock Thermo Fisher Scientific, Tudor Road, Manor Park, Runcorn, Cheshire, WA7 1TA, UK; Dominic Roberts, Thermo Fisher Scientific, Tudor Road, Manor Park, Runcorn, Cheshire, WA7 1TA, UK.

Title:

ORBITRAP GC-MS: ONE YEAR ON

Introduction:

Since the middle of the 20th century GC-MS has made a long journey towards its current status as one of the major techniques in a diverse range of applications. One year on from the first commercial introduction of Orbitrap GC-MS in 2015, its unique capabilities in both targeted and untargeted analysis were utilized in a diverse range of GC-MS applications, as reviewed in this presentation

Methods:

Thermo Scientific Q Exactive GC was used to analyze samples in the areas of environmental, food safety, authenticity and pharmaceuticals analysis. The main modes of operation were electron ionisation and chemical ionisation in full scan and MS/MS with resolving powers between 15,000 and 120,000 (@ m/z 200) with sufficient scan speed to accurately represent GC peaks

Results:

For environmental, broad scope screening of POPs using targeted and untargeted screening workflows yielded low-level detections in biota samples(tuna). Analysis of iodinated disinfection by-products in water treatment resulted in some new compounds reported for the first time. In food analysis, screening, identification and quantitation of pesticides to regulatory levels was possible in full scan, providing new opportunity to optimise workflows and broaden the scope of analysis. Other experiments focused on chemical differences in whiskies based upon their production and origin, with potential to help detect fraud. In pharmaceutical experiments, HR/AM analysis allowed impurities to be observed and identified. Additionally, untargeted full scan analysis in combination with unique software algorithms help identify compounds that could leach into final products from container O-rings

Novel Aspect:

First overview of Orbitrap GC-MS experiments for a broad range of targeted and untargeted applications

Malkawi, Abeer,
Jordan University of Science and Technology

Contributing Authors:
Minnie Jacob, Department of Genetics, King Faisal Specialist Hospital and Research Center (KFSHRC), Riyadh, Saudi Arabia; Goran Matic, Department of Comparative Medicine, King Faisal Specialist Hospital and Research Center (KFSHRC), Riyadh, Saudi Arabia; Asmaa Ali Department of Comparative Medicine, King Faisal Specialist Hospital and Research Center (KFSHRC), Riyadh, Saudi Arabia; Falah Al-Mohanna, Department of Comparative Medicine, King Faisal Specialist Hospital and Research Center (KFSHRC), Riyadh, Saudi Arabia; Karem H. Alzoubi, Department of Clinical Pharmacy, Faculty of Pharmacy, Jordan University of Science and Technology (JUST), Irbid, Jordan; Anas M. Abdel Rahman, Department of Genetics, King Faisal Specialist Hospital and Research Center (KFSHRC), Riyadh, Saudi Arabia.

Title:
Mass Spectrometry metabolomics approach reveals the metabolic side effects of dexamethasone

Introduction:
Dexamethasone (Dex) is a synthetic glucocorticoid and classified as steroid hormones drug. Dex has anti-inflammatory and immunosuppressant effects, and it is used mainly in asthma and severe allergy. Patients under Dex, either at a high dose or for a long time, develop several side effects due to its non-selectivity inside the body such as hyperglycemia, weight change, and osteoporosis.

Methods:
A rat model developing Dex side effect was adopted in this study, where these side effects were monitored over the experiment such as weight gain, lipid profile, and glucose changes. The bone changes was proven in several ways. A targeted metabolomics study was performed on the serum and tissues collected from these animals after 14 weeks using liquid chromatography tandem mass spectrometry.

Results:
Several metabolic changes were observed, explained and connected to Dex's side effects. In particular, most of the amino acids were reduced significantly in Dex group ($P<0.0001$), which suggests gluconeogenesis pathway is the main source of hyperglycemia and high level of protein catabolism that lead to severe weight loss. The level of phenylalanine was reduced while tyrosine and tryptophan were increased and found to be associated with osteoporosis. The significant increase in some acylcarnitines indicates an elevation in serum fatty acids, and degradation in adipose tissue, which supports the reduction weight gain. The serum hydroxyproline, produced due to collagen breakdown, was increased in Dex and introduced a sensitive marker of bone resorption in the earlier stage before even shown in routine testing such as Beta-crosslap marker.

Novel Aspect:
Mapping the Dex side effects with metabolism and finding potential biomarkers and therapeutic targets for better diagnosis, prevention, and treatment

21ST INTERNATIONAL MASS SPECTROMETRY CONFERENCE
Toronto, Ontario, Canada
August 20 – 26, 2016

malosse, christian,
institut pasteur

Contributing Authors:
julia chamot-rooke, institut pasteur cnrs3528 paris france; dominique clermont, institut pasteur cip paris france; sylviane hamon institut pasteur cip paris france; mathieu dupre, institut pasteur cnrs3528 paris france; chantal bizet, institut pasteur cip paris france; thomas ottone, institut pasteur cnrs3528 paris france.

Title:
Characterization of virulent Bacillus cereus species using Top-Down Proteomics

Introduction:
Bacillus cereus is a pathogen known to cause a diarrheal-type of foodborne illness. However more serious and even lethal cases have been described. The current challenge for B. cereus diagnostics is thus the development of a rapid method to identify virulent strains with specific characteristics, such as toxin production. We show here that top-down proteomics can be used for this purpose.

Methods:
Bacillus cereus, B. thuringiensis, B. mycoides, B. pseudomycoides, B. weihenstephanensis, B. cytotoxicus. were cultured on Petri dishes and proteins were extracted after sonication. Intact proteins were analysed in LC-MS/MS (monolithic column, Data Dependent Analysis in HCD, ETD or EThcD, ProsightPD for data analysis) using an Orbitrap Tribrid Lumos (Thermofisher).

Results:
Although ranging from innocuous to toxic, all the strains under study (Bacillus cereus, B. thuringiensis, B. mycoides, B. pseudomycoides, B. weihenstephanensis, B. cytotoxicus) led to nearly identical MALDI-TOF MS profiles. We therefore decided to test whether top-down proteomics could differentiate them. For this purpose, we used an integrated LC-MS/MS top-down proteomics workflow, previously optimized for the analysis of E. coli intact proteins. The top-down proteomics analysis of all B. cereus strains led to the identification of about 200 proteins and 500 proteoforms. Most of them were found to be common to all strains under study. However, a few proteoforms were found to be specific of the most toxic strains, indicating that top-down could be of high interest to achieve an accurate identification of virulent strains involved in foodborne illness.

Novel Aspect:
First use of top-down proteomics in food safety.

Mangal, Vaughn,
Trent University

Contributing Authors:
Dr. Celine Gueguen, Trent University.

Title:
Identification of Mercury Dissolved Organic Matter Complexes by High Resolution Mass Spectrometry

Introduction:
Dissolved organic matter (DOM) is a ubiquitous mixture of complex organic molecules that is operationally defined as material <0.45μm. DOM has the ability to interact and complex with Hg (II), influencing the overall bioavailability of the complex. Understanding the interactions between Hg and DOM in freshwater environments is crucial for understanding mobility and toxicity to aquatic biota.

Methods:
Our study utilizes high resolution mass spectrometryto examine DOM produced by phytoplankton in freshwater settings and how it interacts with environmentally relevant concentrations of Hg (II). By identifying common Hg isotopic shifts, Hg peaks can be identified and corresponding organic ligands and heteroatoms bound to Hg can be simultaneously identified.

Results:
Laboratory grown phytoplankton displayed a high degree of lipid, carbohydrate and protein structures based on O/C and H/C ratios. The abundance of low molecular weight material (<600Da) present in phytoplankton-derived DOM suggests highly labile compounds being produced. At environmentally relevant ratios of DOM to Hg, a large proportion of Hg was found bound to nitrogen and sulfur containing heteroatoms in phytoplankton DOM. Examining Hg-DOM interactions and both equilibrium and pseudo-equilibrium conditions displayed an overall change in Hg binding. Understanding the interactions with phytoplankton-derived DOM upon immediate atmospheric deposition of Hg can aid in predicting Hg mobility and bioavailability in freshwater systems.

Novel Aspect:
While a few studies have utilized HRMS to examine Hg interactions this is the first study that examines Hg-dissolved organic matter interactions.

21ST INTERNATIONAL MASS SPECTROMETRY CONFERENCE
Toronto, Ontario, Canada
August 20 – 26, 2016

March, Raymond,
Trent University

Contributing Authors:

Raymond E. March, Department of Chemistry, Trent University, 1600 West Bank Drive, Peterborough, Ontario, Canada K9J 7B8; Cole J. French, Department of Chemistry, Trent University, 1600 West Bank Drive, Peterborough, Ontario, Canada K9J 7B8; Michael A. Hayes Department of Chemistry, Trent University, 1600 West Bank Drive, Peterborough, Ontario, Canada K9J 7B8; Naomi L. Stock, Water Quality Centre, Trent University, 1600 West Bank Drive, Peterborough, Ontario, Canada K9J 7B8; David A. Ellis, Water Quality Centre, Trent University, 1600 West Bank Drive, Peterborough, Ontario, Canada K9J 7B8.

Title:

An LC-MS/MS study of glyphosate and aminomethylphosphonic acid in soil and the reactivities of mono- to triply-deprotonated glyphosate

Introduction:

Introduced in 1974, glyphosate (Round-Up) is the most widely-used herbicide in the world; it was said to be safe. Glyphosate use is under scrutiny due to reports of glyphosate-resistant weeds, increased application rates, and of carcinogenicity (by the European Community). Examination of soil samples, requested by a farm owner near Whitby, ON, revealed evidence of prior glyphosate application.

Methods:

Tandem mass spectrometry (MS/MS) has been combined with liquid chromatography for the analysis of glyphosate extracted from soil samples. The MS/MS method is of high specificity due to the isolation of deprotonated glyphosate and dissociation to m/z 150 and 63 that are detected in a 10:1 ratio. Soil extracts plus internal standard pass through IC-Chelate and IC-RP SPE cartridges prior to analysis.

Results:

Using a 5500 Qtrap mass spectrometer (AB Sciex, Concord, ON, Canada), LC-MS/MS has been employed for the detection of glyphosate in soil samples at the several parts per billion (ppb) level. While the soil came from a farm with an unknown glyphosate history, clearly the glyphosate has persisted in the soil for some years. The optimum eluting mobile phase was found as 20:80 acetonitrile: 0.1 M ammonium acetate. The precise concentrations await completion of a calibration curve. Calibration standards were prepared using 18.2 MW water dilutions from a 500 mg L-1 glyphosate stock standard together with additions of 10% phosphoric acid solution. Doubly isotopically-labeled glyphosate (2×13C), the internal standard, has been detected by MS/MS from a 10.6 ppb solution of the labeled compound. A number of soil samples have been extracted ready for analysis.

Novel Aspect:

The doubly- and triply-charged glyphosate ions of m/z 83.5 and 55.3, respectively, have been observed for the first time in the pH 6-11 range.

Marchand, Adrien,
Inserm, ARNA Lab

Contributing Authors:

Valerie Gabelica, Inserm, ARNA Lab, Univ. Bordeaux.

Title:

Deciphering G-quadruplex DNA folding landscape with mass spectrometry, ion mobility and circular dichroism

Introduction:

A G-quadruplex (G4) is a nucleic acid structure made of stacked guanine quartets. To fold, G4s require monovalent cations (physiologically: K+). The structural polymorphism of the human telomeric G4 sequences is what makes their biophysical study difficult. Even less is known about G4 folding pathways, which are nevertheless crucial to understand for regulating biological events.

Methods:

Using mass spectrometry (ESI-MS), we determine the equilibrium constants (KDs) and kinetics of K+ binding to several variants of the human telomeric sequence. Ion mobility spectrometry (IMS) confirmed the number of distinct structures and their rates of appearance. Decomposition of the circular dichroism (CD) signal was done using the KDs or kinetic parameters obtained from the MS data.

Results:

From the KDs, we classify K+ binding according to cooperativity. For the shortest variants, the 1:1 K+ stoichiometry prevails at low K+ concentration and the binding of the second K+ is non-cooperative. Longer sequences bind 2 K+ cooperatively. The folding of the cooperative 2-K+ complex is however much slower (several minutes at 500 µM K+) than for the non-cooperative sequences. This is due to competing parallel reactions. We found that six ensembles account for all the kinetic data: one unfolded species, two complexes with 1-K+ and three structures with 2-K+. The best kinetic models involve the presence of a misfolding pathway involving antiparallel 1-K+ (2-quartet) topologies. The folding pathway of telomeric sequences is transposable to other systems, including a c-myc gene promoter or one artificially designed G4, suggested this is a general tenet of G-quadruplex folding.

Novel Aspect:

We used IMS-MS to identify and quantify individual K+ binding events to G-quadruplexes and reveal their folding mechanism.

Marchand, Adrien,
Inserm, ARNA Lab

Contributing Authors:
Dominika Strzelecka, Division of Biophysics, Institute of Experimental Physics, Faculty of Physics, University of Warsaw; Valerie Gabelica, Inserm, ARNA Lab, Univ. Bordeaux.

Title:
Selective and Cooperative Ligand Binding to Telomeric DNA G-quadruplex targets

Introduction:
G-quadruplex nucleic acid structures (G4) have become a very attractive research field as they can have effects on gene expression or telomere maintenance. Their structural polymorphism brings challenges for characterizing potential anti-cancer drugs, which would bind with high affinity and specificity to one structure. It is particularly important to conceive molecules with specific effects.

Methods:
Native electrospray mass spectrometry (ESI-MS) screening of over 21 DNA sequences reveals a high affinity binding of Cu-tolylterpyridine (Cu-ttpy) complexes to G-quadruplexes (G4). Circular dichroism (CD) indicated that the binding of the ligands was accompanied with a conformational change of the G4. Collision induced dissociation (CID) MS/MS was used to assess the ligand binding site.

Results:
We estimated the equilibrium constants for the binding of the first two Cu-ttpy ligands to 8 human telomeric G4, 11 other G4 topologies (parallel and antiparallel), and 2 non-G4 structures. Cu-ttpy's affinity is higher for the human telomeric G4 for which two ligands bind with unprecedented cooperativity: the 2:1 L:DNA stoichiometry prevails and the 1:1 complex is almost nonexistent upon titration. The cooperativity is attributed to conformational changes in the polymorphic telomeric G4, which, as indicated by CD, convert to an antiparallel topology upon binding two ligands. The G4-specific potassium cations bound is also affected: the ligands bind to a 2-K+ structure. CID experiments on bimolecular G-quadruplexes indicate that the two ligands bind on two distal positions, presumably to each of the two external G-quartets, to which they π-π stack thanks to the planar aromatic ttpy moiety

Novel Aspect:
We elucidated the binding mode of the high affinity Cu-ttpy ligand to G-quadruplex structures using MS, CID and CD.

Marshall, David,
Queensland University of Technology

Contributing Authors:
Michael J. Kelso, School of Chemistry, University of Wollongong, Wollongong, Australia; Ramesh Ailuri, School of Chemistry, University of Wollongong, Wollongong, Australia; Todd W. Mitchell School of Medicine, Illawarra Health and Medical Research Institute, University of Wollongong, Australia; Stephen J. Blanksby, Central Analytical Research Facility, Queensland University of Technology, Brisbane, Australia.

Title:
Discriminating isomeric (O-acyl)-hydroxy fatty acids (OAHFAs) by tandem mass spectrometry

Introduction:
OAHFAs are a recently discovered class of endogenous lipids, of substantial interest for their correlation with enhanced glucose tolerance, thus exhibiting anti-diabetic effects. Interestingly, the ω-OAHFA sub-class play a role in stabilizing the human tear film. Developing tools for rapid and unambiguous structure elucidation of OAHFAs is essential to understanding their physiological functions.

Methods:
OAHFA standards (1 μM) were prepared in MeOH:CHCl3 with 1.5 mM ammonium acetate. Negative ion collision-induced dissociation (CID) MSn spectra were acquired using chip-based nano-ESI coupled to an Orbitrap Elite mass spectrometer (Thermo Fisher Scientific). Crucially when analyzing isomeric lipids, a new pipette tip and chip nozzle is used for each sample, thus minimizing carryover.

Results:
ESI of OAHFAs produces abundant [M – H]– ions, which upon CID yield 3 major product ions. The fatty acid (FA), hydroxy-fatty acid (HFA), and dehydrated HFA ions observed readily identify the OAHFA components, however further detail (e.g., ester position) is based solely on subtle differences in ion abundances. Further interrogation of the HFA in an MS3 experiment produces diagnostic fragment ions that readily pinpoint hydroxylation position and thus, the OAHFA ester location. By contrast, these ions are characteristically absent in the MS3 spectra of HFAs derived from ω-OAHFAs. Unimolecular dissociation mechanisms are proposed that draw parallels to fragmentation pathways for anionic phospholipids, consistent with isotopic labeling experiments. The general applicability of the suggested mechanisms is such that the structure of unknown OAHFAs could be assigned based on their MSn spectra.

Novel Aspect:
Mechanistic understanding of gas phase ion dissociation enables differentiation of isomeric OAHFAs from MSn spectra in shotgun lipidomics approaches.

Martin, Esther,
University of Leeds

Contributing Authors:
Martin Gamerdinger, University of Konstanz; Elke Deuerling, University of Konstanz; Alison. E Ashcroft University of Leeds; Sheena E. Radford, University of Leeds.

Title:
A mass spectrometry approach to investigate the structure and function of nascent chain associated complex (NAC)

Introduction:
Newly synthesised polypeptides emerge from the ribosome and interact with chaperones at the exit tunnel to prevent misfolding. Nascent chain associated complex (NAC) is a ribosome-associated heterodimer that is highly conserved among eukaryotes. Recently, it has been shown that NAC depletion reduces the lifespan of C. elegans nematodes through impaired protein homeostasis.

Methods:
Native MS and ion-mobility mass spectrometry (IM-MS) were used to compare the conformations of WT NAC and the heterodimer with the UBA region deleted (ΔUBA-NAC). To explore the structures in further detail, fast photochemical oxidation and limited proteolysis were employed. Crosslinking experiments revealed information about the heterodimer formation and also how NAC interacts with other proteins.

Results:
Native mass spectrometry of C. elegans NAC revealed that compact and extended forms of the heterodimer exist in the gas phase. Deletion of the UBA domain did not influence the conformers observed by IM-MS but did lower the stability of the dimer as assessed by collision induced unfolding (CIU). ΔUBA NAC produced more tryptic fragments than the wild type NAC when subjected to limited proteolysis. Crosslinking experiments demonstrated that NAC interacts with disordered proteins and in-gel tryptic digests were performed to locate the substrate binding site. Hydroxyl radical labelling of the complex revealed that there are more surface accessible residues on alpha-NAC than beta-NAC and therefore this could be used as a tool to map interaction sites. Work is currently ongoing to investigate how NAC affects the aggregation of amyloidogenic proteins.

Novel Aspect:
Using crosslinking coupled with mass spectrometry to locate substrate binding sites on NAC.

Martin, Roy,
Waters

Contributing Authors:
Yvonne Woolerton, Centre for Proteome Research, Institute of Integrative Biology, University of Liverpool, Liverpool, UK; Lee Gethings, Waters Corporation, Wilmslow, United Kingdom; Jose Castro-Perez Waters Corporation, Milford, MA; James Langridge, Waters Corporation, Wilmslow, United Kingdom; Lynn McLean, Centre for Proteome Research, Institute of Integrative Biology, University of Liverpool, Liverpool, UK; Johannes PC Vissers, Waters Corporation, Wilmslow, United Kingdom; Robert Beynon, Centre for Proteome Research, Institute of Integrative Biology, University of Liverpool, Liverpool, UK.

Title:
IMS-DIA-MS Characterisation and IMS-MRM QconCAT Quantitation of the Lipidome and Apolipoprotein Complements of Obesity and Diabetes Cohorts

Introduction:
Obesity is a metabolic syndrome risk factor causing accumulation of excess body, which can lead to type 2 diabetes. The exact mechanisms are not well understood. A lipidomic approach is described to reveal factors involved in biomolecular processes. The results were complemented with quantitation of the Apolipoproteins using concatenated signature peptides to identify disease associated pathways

Methods:
Six control, diabetic and obese plasma samples each were used. Extracts were LC separated, lipidomics data acquired using TWIMS assisted DIA, and CCS values determined. The data were searched with compound databases providing normalized quantitation results. The samples were also digested and targeted proteomics data obtained using combined IMS/mass selected oa-ToF MRM

Results:
Label-free lipid LC-MS data were acquired in ESI(-) and ESI (+) modes with an IMS oa-QTof platform using an IMS assisted DIA workflow. Unsupervised MVA of the data showed clear distinction between cohorts. OPLS-DA was used to filter for features prior to identification. IMS-derived CCS measurements allowed for improved specificity with the inclusion of drift time, providing additional confidence in the identifications. The results highlighted differential expression of fatty acids, phosphatidylcholines, triglycerides and phosphatidylserines. The identified networks were extended and confirmed by targeted IMS-MRM experiments of the most common alleles of the apolipoprotein plasma complement. Inclusion of IMS in the targeted oaToF MRM workflow increased quantitative precision and accuracy with the results in agreement with expected changes in relation with disease and/or phenotype.

Novel Aspect:
Combined discovery IMS-DIA-MS lipidomics and targeted IMS-MRM proteomics workflow

Martin, Roy,
Waters

Contributing Authors:
Billy Molloy, Waters Corporation, Wilmslow, United Kingdom; Donald JL Jones, Department of Cancer Studies, University of Leicester, and Cardiovascular Research Centre, Glenfield Hospital, Leicester, United Kingdom; James I Langridge Waters Corporation, Wilmslow, United Kingdom; Johannes PC Vissers, Waters Corporation, Wilmslow, United Kingdom; Donald P Cooper, Waters Corporation, Wilmslow, United Kingdom.

Title:
Evaluation of a Novel Tandem Quadrupole Mass Spectrometer for Quantitative Peptides Analysis Using a Multi-Point Internal Standard Calibration Method

Introduction:
Large cohort translational studies are challenged in that they entail high throughput, sensitivity, and resolution, as well as good selectivity over a large dynamic range. Targeted LC-MS/MS affords reproducible quantification to rapidly assess biomarker performance. We present a multi-point internal calibration method for peptide quantitation using microfluidics coupled to a novel Tandem quad MS.

Methods:
Separation was performed with 0.15 × 100 mm C18 1.7 μm micro-fluidics using a 2-30% B gradient over 45 min at 1 μL/min. Samples were injected on-column or using a 0.3 × 50 mm 5 μm C18 precolumn. MRM experiments were performed using a prototype tandem quadrupole MS. Dwell and interscan delay times were automatically calculated based on a minimum number of data points across a chromatographic peak.

Results:
Benchmarking of the MS system which was fitted with a novel ion guide for improved and more selective transmission with additional detector advances was performed by comparing sensitivity and precision against four other MS systems. S/N was found to be two-fold better compared to the next most sensitive instrument and %CV similar to other tandem quadrupole instruments and about two fold better than time-of-flight instruments. A multi-point internal calibration method was used for within-sample calibration and assessed with regard to quantitative precision and throughput gains determined. Performance metrics of the method will be demonstrated and the results compared with conventional MRM quantitation approaches for the direct analysis of three putative human cardiovascular disease plasma peptides.

Novel Aspect:
High sensitivity detection of peptides in biological matrices using a novel tandem quadrupole MS system and quantitation method.

Masui, Olena,
MS-ESE York University - NSERC CREATE

Contributing Authors:

Fernando Ausar, Sanofi Pasteur; Vladimir Baranov, Fluidigm; Roger Brookes Sanofi Pasteur; Andrew James, Sanofi Pasteur; Olena Masui, York University; Nausheen Rahman, Sanofi Pasteur; Michael Siu, York University; Derek Wilson, York University.

Title:

Identification of biomarkers related to potential critical quality attributes (pCQA) for a new vaccine formulation.

Introduction:

Sanofi Pasteur is developing an in vitro human whole blood (hWB) approach to complement animal models and formulate new vaccines. In this study two approaches will be used to identify biomarkers associated with an adjuvant-modulated antigen-specific response. The first will use mass spectrometry (MS) while the second will use mass cytometry. Biomarkers identified will be used to optimize pCQAs.

Methods:

The fresh hWB will be restimulated in vitro using vaccine formulations with predefined modification to pCQAs. Supernatants recovered after 10 days will be used to identify proteomic biomarkers by a label-free quantitative method using online nanoLC-MS (LTQ Orbitrap, Thermo Fisher). In parallel, the phenotypic biomarkers of responding cells will be evaluated by mass cytometry (CyToF, Fluidigm).

Results:

Stimulation of hWB with specific vaccine formulations results in an IFN gamma response that is both adjuvant-modulated and antigen-specific. Investigation of the proteomic and phenotypic biomarkers associated with vaccine driven innate and acquired immunity will be assessed through an industrial-academic partnership. Results of the study are not yet available, here we report the strategy.

Novel Aspect:

New immuno-modulatory biomarkers identified in hWB cultures, driven by vaccine formulations, will define attributes critical to vaccine manufacture.

Matondo, Mariette,
Institut Pasteur

Contributing Authors:
Pascale Pescher, Molecular Parasitology and Signaling unit; Thibault Chaze, Proteomics Platform, Structural Mass Spectrometry and Proteomics; Thibaut Douché Proteomics Platform, Structural Mass Spectrometry and Proteomics; Veronique Hourdel, Proteomics Platform, Structural Mass Spectrometry and Proteomics; Gerald F. Späth, Molecular Parasitology and Signaling unit.

Title:
Identification of proteins and phosphorylation events associated with Leishmania stage differentiation

Introduction:
The life cycle of the human pathogen Leishmania comprises a promastigote stage that develops inside the sand fly insect vector, and a disease-causing amastigote stage that infects host macrophages. Only little is know on signaling events underlying differentiation. To overcome this limitation we established and compared the global and phosphoproteome of both Leishmania life cycle stages.

Methods:
We used label-free quantitative analysis to assess changes in proteome and phosphoproteome between avirulent culture-derived axenic amastigotes (ax-ama), virulent hamster-derived amastigotes (h-ama), and promastigotes (pro). Phosphopeptide enrichment was performed using the TiO_2 approach. Tryptic peptides or phosphopeptides were analyzed by nanoLC-MS/MS on a Q Exactive HF instrument.

Results:
Whole proteome analysis identified over 5,000 proteins, which represents approximately 70% of the predicted proteome of Leishmania. Statistical analysis revealed proteins that change significantly in abundance between (i) ax-ama versus h-ama (891), (ii) ax-ama versus pro (1087), and h-ama versus pro (2300). Phosphoproteomic analysis identified 20,000 phosphorylation sites of which 1781 exhibited stage-dependent phosphorylation changes. Both global and phosphoproteome data are currently being compared to establish stage-specific signaling pathways. Genomic, transcriptomic, and metabolomic analyses have been performed, and are currently being integrated with the proteomics data sets establishing a system-wide analysis of Leishmania stage differentiation.

Novel Aspect:
Our systems-level analysis allows new insight into Leishmania-specific signaling events that can inform on novel intervention strategies

21ST INTERNATIONAL MASS SPECTROMETRY CONFERENCE
Toronto, Ontario, Canada
August 20 – 26, 2016

Mayer, Paul M,
University of Ottawa

Contributing Authors:
Christine Joblin, Universite de Toulouse and CNRS; Eduardo Solano, University of Ottawa; Brandi West University of Ottawa.

Title:
Reaction Dynamics of Ionized PAHs

Introduction:
We have explored PAHs by examining their dissociation, especially the loss of hydrocarbon fragments, typically C2H2, one of the commonly found building blocks of PAHs. This not only provides information on reaction barriers, but also allows us to make predictions of the lifetime of these smaller species under interstellar radiation conditions.

Methods:
The imaging photo-electron photo-ion coincidence (iPEPICO) spectrometry experiments were conducted on the VUV beamline of the Swiss Light Source. In the experiment, PAHs are photoionized by monochromatic synchrotron radiation. The ions are then extracted towards a time-of-flight (TOF) mass spectrometer while the ejected electrons are velocity map imaged on an imaging multichannel plate.

Results:
iPEPICO data for an extensive series of unsaturated and hydrogenated PAHs have been modeled with RRKM theory to extract reaction energies, notably for H and C2H2 loss. The wide range of environments explored permit us to make general conclusions about the energetic cost of these reactions for large PAHs of astrochemical significance. Results will be summarized for the unsaturated PAHs: naphthalene, anthracene, pyrene, perylene, acenaphthylene, fluoranthrene, coronene, and two isomeric benzopyrenes. Representative hydrogenated PAHs are: dihydronaphthalene and phenanthrene, fluorene and cyclopentaphenanthrene.

Novel Aspect:
Reaction energies extrapolated to PAHs of astrochemical significance

Mayrand-Provencher, Laurence,
Algorithme Pharma, an Altasciences Company

Contributing Authors:
Mingluan Chen, Algorithme Pharma; Milton Furtado, Algorithme Pharma; Anahita Keyhani Algorithme Pharma.

Title:
A Novel Determination for Menthol in Rat Plasma Using Salting-Out LLE and Derivatization Coupled with LC-MS/MS Detection

Introduction:
Menthol is widely used as a flavoring aid in consumer products and inflammatory diseases. There is current demand for menthol assays in biological matrices. However, poor electrospray ionization can limit the achievable detection limit. Therefore, an LC-MS/MS method coupled with a Salting-out Liquid-Liquid Extract and derivatization allowed the quantitation of menthol in rat plasma samples.

Methods:
A 25 µL aliquot of rat plasma, internal standard (menthol-d4), buffer, and acetonitrile were combined, vortexed, and centrifuged to generate two distinct phases. A 100 µL aliquot of the upper phase was derivatized with dansyl chloride containing pyridine. The resultant extract was further diluted and separated on a Waters XBridge C18 column and detected on a SCIEX API 5000.

Results:
Due to low ionization efficiency by electrospray, derivatization was essential to improve menthol detection limit. Menthol was successfully derivatized with dansyl chloride utilizing pyridine as a catalyst. Derivatization conditions were optimized for reaction temperature, reaction time, derivatizing agent and catalyst concentration furnishing a 90% yield in 30 min (at 70°C). SALLE was optimized by the addition of a potassium phosphate buffer. Post-column infusion results from the SALLE extracts indicate the absence of matrix effect. Menthol was found to be stable in rat plasma for three freeze-thaw cycles, 17hr at 4°C short-term and 4 days at -80°C long-term stability. The bioanalytical method for menthol extracted from rat plasma was successfully validated for a concentration range of 40.0 to 1,000.0 ng/mL.

Novel Aspect:
Quantitation of menthol in rat plasma samples coupling Salting-out Liquid-Liquid Extraction with derivatization and LC-MS/MS detection.

Mazigh, Bihac,
University of Ottawa

Contributing Authors:
Paul M Mayer, Corresponding Author.

Title:
Dual-spray Hydrogen Deuterium Exchange (HDX) Reactions: A New Method of Probing Protein Structure

Introduction:
Traditionally, hydrogen deuterium exchange (HDX) reactions are done in solution phase. A more efficient method may be to conduct the reaction within the ion source of a mass spectrometer and subsequently analyze the products.

Methods:
Using the two electrospray emitters equipped on the Waters Synapt G1 mass spectrometer, HDX reactions were conducted within the ion source region in a controlled fashion ("dual-spray"). Peptide and protein solutions were electrosprayed through one emitter and the deuterating agent D2O through the secondary electrospray emitter.

Results:
Clear shifts in isotope distributions indicated hydrogen-deuterium exchange occurring within the ion source. By ion mobility, simultaneous deuterium exchange for two isobaric species, the oxytocin monomer and dimer, was observed. For denatured ubiquitin, the 12+ and 13+ charge states have a lower average number of exchanges relative to the lower charge states which indicates that these charge states have segments which restrict the access of D2O. Lysozyme has a linear relation between the charge state and the average number of exchanges, indicating that lysozyme becomes increasingly unfolded as the charge state increases. The dual-spray HDX method was paired to HPLC to demonstrate the applicability of the technique for probing gas phase structures in protein mixtures.

Novel Aspect:
Deuterium labeling can occur on the sub-millisecond time scale and the acid quenching step involved with solution phase HDX can be surpassed.

McCullough, Bryan,
LGC Ltd.

Contributing Authors:
Chris Hopley, LGC.

Title:
A Validated High Resolution Mass Spectrometry Method for the Quantification of T-2 and HT-2 Toxins in Cereals

Introduction:
Recent advances in QToF technology mean it is now a viable alternative to QqQ for quantitative mass spectrometry. Here we describe the development and validation of a LC-QToF-MS method for the measurement of the fusarium mycotoxins T-2 toxin and HT-2 toxin in cereals.

Methods:
The method was developed using two matrices: oat flour and corn flour. T-2 and HT-2 toxins were spiked into the flour at 10, 25 and 100 µg/kg (N=6) before extraction using a modified QuEChERS method. The resulting extracts were analysed by LC-QToF-MS operating in full scan mode using confirmation criteria based on SANCO guidelines (mass accuracy <5 ppm, resolution >10,000, etc).

Results:
The validation criteria for this method were taken from the EFSA guidance on EC Recommendation 2013/165/EU which requires repeatability of ≤30% and recoveries between 60 and 130% with maximum limits of quantitation of 10 to 20 µg/kg. The LODs for the developed method were 5 µg/kg for both analytes in both matrices with LOQs of 10 µg/kg. Obtained recoveries and repeatability were all within the required limits in each matrix and the method was therefore validated for the analysis of T-2 and HT-2 toxin in cereals. A current concern in food safety is the issue of masked mycotoxins due to the potential for underestimating the total amount of mycotoxin present in a sample. A major advantage of acquiring data in a non-targeted manner is that it can retrospectively interrogated for other analytes. The method was therefore investigated as a potential screening tool for masked mycotoxin species

Novel Aspect:
QToF quantitation of mycotoxin species. Accurate mass screening for masked mycotoxins

McGregor, Nicole,
NSERC CREATE York University and University of Toronto

Contributing Authors:
Derek Wilson, York University; Andrew Fairman, York University.

Title:
The Role of Conformational Dynamics in Differing Activity and Substrate Specificity amongst Isozymes of Glutathione S-Transferase.

Introduction:
GSTs are a superfamily of enzymes involved in a wide range of detoxification and metabolism reactions. Despite isozymes having very similar sequences and structural motifs, the functional parameters and substrate affinities are substantially different. This conserved nature suggest conformational dynamics plays a significant role in known variations of activity between isozymes.

Methods:
Isozymes of the Pi, Alpha and Mu class human cytosolic GSTs were purified from E. coli strains and prepared for time resolved(TR) electrospray ionization mass spectrometry (ESI-MS) through buffer exchange into NH4Ac. The dynamics were probed using global and site specific hydrogen deuterium exchange (HDX), coupled to on-line acid quenching and protein digestion with a microfluidic chip.

Results:
Site specific TRESI HDX analysis showed the binding of GSH among isozymes had some commonality, with a general stabilization of the dimer interface and parts of the G-domain where conserved residues across all isozymes are involved in GSH binding interactions. Variability in the G-domain conformations were mostly seen in the alpha-2 helix region, and larger translatative differences between the apo and holo forms were seen in regions involved in the H-domain of the enzyme, where the electrophilic substrates bind. These differences appeared to correlate with the respective isozyme functionality, with more solvent exposure and increased dynamics seen in holo form of promiscuous isozymes compared with increased stability and rigidity found in holo forms of specific GST binders compared with the respective apo isozyme conformations.

Novel Aspect:
The TRESI set-up allows HDX rates and amounts to be seen on the never before done millisecond to second timescale.

McIndoe, Scott,
University of Victoria

Title:
Catalytic Reaction Mechanism Elucidation via Real Time Mass Spectrometric Analysis

Introduction:
Continuous monitoring of catalyzed reactions using infrared spectroscopy (IR) measures the transformation of reactant into product; while mass spectrometry delineates the dynamics of the catalytically-relevant species present at much lower concentrations, a holistic approach that provides mechanistic insight into reaction components whose abundance spans five orders of magnitude.

Methods:
Electrospray ionization mass spectrometry Pressurized sample infusion Infrared spectroscopy Real-time analysis

Results:
The combination of a traditional flow method (FTIR) in conjunction with an ESI-MS variant that enables continuous solution monitoring gives a composite picture of the changes in speciation during a catalytic reaction across many orders of dynamic range. At highest concentration, we see (with IR) just the reactant and product. Delving deeper (with ESI-MS), we see the catalyst precursor and the resting state. However, by leaving ESI-MS to deal only what it's good at – detecting low abundance charged species – we have tons of dynamic range left to burn, and can zoom in on impurities, intermediates, byproducts and decomposition products.

Novel Aspect:
Simultaneous orthogonal analysis across orders of magnitude of concentration in an evolving catalytic reaction

McIndoe, Scott,
University of Victoria

Title:
Kinetic analysis of chemical reactions using electrospray ionization mass spectrometry

Introduction:
Pressurized sample infusion (PSI) ESI-MS provides a straightforward means of continuously analyzing chemical reactions. Data density is high, variability low, and reactions can be conducted under a wide range of conditions, and provided certain criteria are met, high quality kinetic data can be obtained.

Methods:
Electrospray ionization mass spectrometry Pressurized sample infusion Kinetic analysis Parameter estimation Reaction simulation

Results:
Examples of a range of different reactions in which we have obtained high quality kinetic data for reaction analysis will be presented, including simple, complex and competitive substitution reactions, catalytic hydrogenation, palladium-catalyzed cross-coupling, and problems in mass transfer. Methodologies and pitfalls will be discussed.

Novel Aspect:
Mass spectrometric kinetic analysis of catalytic reactions has been underused despite obvious strengths.

McLean, John,
Vanderbilt University

Contributing Authors:

Katrina L. Leaptrot, Vanderbilt University; Jody C. May, Vanderbilt University; James N. Dodds Vanderbilt University.

Title:

Conformational atlas of lipid structure for directed characterization in lipidomics and molecular phenomics

Introduction:

Ion mobility-mass spectrometry (IM-MS) is one such technique which has previously demonstrated utility in analysis of biomolecules, complex mixtures, and lipid structural analysis. Structural differences in lipids determine biological function, and techniques such as IM-MS allow the generation of structural atlases for integrated omics measurements as described in this report.

Methods:

Dry lipid standards, purchased from Avanti Lipids (Birmingham, AL, USA), were reconstituted and structurally analyzed using an Agilent 6560 IM-MS. Lipid identification of singly charged species was based on exact mass measurements, Lipid Metabolites and Pathways Strategy Structural Database, and CCS.

Results:

For lipid features of common class, modification, and charge carrier differing either in the length of the fatty acid chains or the degree of unsaturation, unique trends were found that were consistent in slope across the listed variables, suggesting that not only can the CCS values found in this work be used for future identification, but that size-to-mass trend information is widely applicable and generalizable towards the identification of lipids not observed in this study. This research is currently being expanded to investigate separation capabilities of ion mobility at more descriptive levels of identification, specifically with isobaric lipid standards varying in fatty acid chain structure and position of attachment. New strategies will include a weighting function for lipid identification in broad scale analysis initiatives.

Novel Aspect:

High confidence lipid characterization of lipids for broadscale untargeted lipidomics from complex biological samples.

McLean, John,
Vanderbilt University

Title:
Structural mass spectrometry for molecular phenomics in systems, synthetic, and chemical biology

Introduction:
One of the predominant challenges in systems-wide analyses to assess drug impact and mechanism of action is the broad-scale characterization of the molecular inventory in cells, tissues, and biological fluids. Advances in structural mass spectrometry have mitigated this challenge through the ability to make untargeted omics measurements on the basis of molecular structure and mass.

Methods:
Structural mass spectrometry is performed using ion mobility-mass spectrometry and a combination of fast condensed-phase separations such as GC, SFC, and UPLC (min) for enhanced separation selectivity as the sample complexity increases. These measurements are combined with new advances in bioinformatics and biostatistics that will be described to approach biological queries from an unbiased view.

Results:
This report will describe recent advances in IM-MS integrated omics measurement strategies in the analyses of complex biological samples of interest in systems, synthetic, and chemical biology. Specific examples will demonstrate the ability to monitor the impact of drug perturbation on human organotypic constructs, or so-called "human-on-a-chip" for the rapid assessment of drug-adduct conjugates and the braodscale molecular impact from drug exposure. A new methodology for rapid quantitation of the abundances of these species in vitro will be shown, which allows description on a copies per cell basis. This is shown for several cellular intoxicants on human cell lines as models for elucidating drug mode of action. Informatics strategies for understanding and validating these species in a networking construct will be shown to recapitulate canonical mechanism of action and discovery.

Novel Aspect:
New structural mass spectrometry strategies for rapid elucidation of molecular perturbation by drugs and intoxicants on human constructs.

Mcmahon, Terry,
University of Waterloo

Contributing Authors:

Michael Lecours, University of Waterloo; Neil Keddie, University of St. Andrews; Vincent Steinmetz Laboratoire de Chimie Physique, Orsay; David O'Hagan, University of St. Andrews; Rick Marta, University of Waterloo; Eric Fillion, University of Waterloo; Joshua Featherstone, University of Waterloo; Scott Hopkins, University of Waterloo.

Title:

Remarkable Cation and Anion Interactions of all-cis 1,2,3,4,5,6-hexafluorocyclohexane in the Gas Phase

Introduction:

A recently synthesized , very polar, substituted cyclohexane molecule has been found to have a facial polarization such that remarkable cation and anion binding ability is demonstrated to opposite faces of the ring.

Methods:

IRMPD and Ion Trap Mass Spectrometry

Results:

Experiments have been carried out in which electrospray ionization has been used to generate ionic complexes of all-cis 1,2,3,4,5,6 hexafluoro-cyclohexane. These complexes were subsequently mass isolated in a quadrupole ion trap mass spec-trometer and then irradiated by the tunable infrared output of a free electron laser in the 800-1600 cm−1 range. From the frequency dependence of the frag-mentation of the complexes, vibrational signatures of the complexes were obtained. Computational work carried out in parallel reveals that the complexes formed are very strongly bound and are among the most strongly bound complexes of Na+ and Cl− ever observed with molecular species. The dipole moment calculated for the heaxafluorocyclohexane is very large (~7 D) and it appears that the bonding in each of the complexes has a significant electrostatic con-tribution

Novel Aspect:

Very strong ion-molecule interactions of a small molecular species.

Mehl, John,
Bristol-Myers Squibb

Contributing Authors:
Lorell N. Discenza, Bristol-Myers Squibb; Bogdan G. Sleczka, Bristol-Myers Squibb; Timothy V. Olah Bristol-Myers Squibb; Eugene F. Ciccimaro, Bristol-Myers Squibb.

Title:
Analytical Approaches for Quantification of Surrogate mAbs in Rodent Biological Matrices using LC-MS

Introduction:
Surrogate mAbs are "tool molecules" used during drug discovery in animal models. The bioanalysis of surrogate mAbs can present challenges for traditional ligand binding assay development due to species similarity of the mAb and animal matrix and the limited availability of reagents. LC-MS can provide an alternative bioanalytical path forward due to the advantage of mass selectivity.

Methods:
Affinity enrichment (AE) techniques were used for preparation of serum and tissue samples. Extracted mAbs are denatured and digested either in solution or in situ. UPLC combined with either high resolution MS or triple quadrupole MS was used for detection of peptides unique to the surrogate mAb of interest.

Results:
Generic approaches for AE of surrogate mAbs from serum and tissue homogenates will be discussed. Use of generic reagents requires consideration of sample volume, otherwise competition with matrix proteins will limit the quantitative range. In cases where the surrogate mAb sequence is unknown, LC-MS enables quantitation based on marker peptides that are unique to the surrogate mAb. These approaches have been applied to cases where unknown rat and mouse surrogate mAbs are dosed in mouse models. High-resolution mass spectrometry (HRMS) is applied during method development to enable selection of unique marker peptides. Optionally, triple quadrupole MS in MRM mode can be applied for quantification. This presentation will provide examples of how generic reagents have been applied for the quantification of surrogate mAbs dosed in mouse.

Novel Aspect:
Quantification of surrogate mouse and rat mAbs in mouse biological matrices using AE and marker peptides in the absence of sequence information.

MELENDEZ, JOSE JAVIER,
UNIVERSITY OF CAMPINAS

Contributing Authors:
Marcos Nogueira Eberlin, University of Campinas.

Title:
CHOS compounds in dissolved organic matter (DOM) are formed from inorganic S-containing ions and lignin derivates?

Introduction:
A recent study performed by Sleighter et al. (2014) suggest that DOM plays a role in the sulfur cycle since they found a correlation between bisulfide concentration and the abundance of sulfur polyoxygenated (CHOS) compounds. This study aims to show that CHOS compounds can be formed in aquatic environments by the addition of sulfide/bisulfide ions to polyoxygenated lignin-derived compounds.

Methods:
A sample of water from a lacustrine system was selected to perform experiments of sulfurization of DOM. The sample was filtered and separated into three portions, the first one was used as control, the second one was enriched with sodium sulfide and the third was enriched with sodium sulfate. After two days, DOM of each portion was extracted with SPE cartridges and analyzed by ESI(-)-FT ICR MS.

Results:
Bar chart of classes of compounds (BCC) and 3D van krevelen (vK) diagrams were constructed from FT ICR MS spectra. The control show high content of lignin-like CHO compounds (H/C, 0.5–1.5; O/C, 0.3–0.7) and low content of CHOS compounds. BCC was not different between control and the sulfate enriched sample. Moreover, the BCC of the sulfide enriched sample yields a mass spectrum with higher abundance of CHOS compounds and these compounds appear in both diagrams with the same distribution than in other water samples with natural high content of CHOS compounds. Both, CHOS compounds (natural and synthesized) and CHO compounds have very similar distribution and maximum location in BCC and were placed in the same region in vK diagrams indicating than CHOS compounds can be formed by the addition of sulfide ions to non-S-containing DOM such as lignin like CHO compounds.

Novel Aspect:
This study shows new insights regading the source, structure and formation of CHOS compounds in DOM, proving its inlfuence in the global sulfur cycle

Menezes, Riya Christina,
Max Planck Institute for Chemical Ecology

Contributing Authors:
Purva Kulkarni, Research Group Mass Spectrometry, Max Planck Institute for Chemical Ecology, Hans-Knöll-Straße 8, 07745 Jena, Germany; Aleš Svatoš, Research Group Mass Spectrometry, Max Planck Institute for Chemical Ecology, Hans-Knöll-Straße 8, 07745 Jena, Germany; Riya C. Menezes Research Group Mass Spectrometry, Max Planck Institute for Chemical Ecology, Hans-Knöll-Straße 8, 07745 Jena, Germany; Paolina Garbeva, Department of Microbial Ecology, Netherlands Institute of Ecology, Droevendaalsesteeg 10 6708 PB Wageningen, The Netherlands.

Title:
LESA – a novel method to disentangle the bacterial metabolome

Introduction:
Bacteria produce a wealth of structurally diverse specialized metabolites with a remarkable range of activity. Recently augmented knowledge about bacterial genomes has revealed that many bacteria have far greater potential to produce secondary metabolites. This potential, however, has been hampered by the fact that many of these metabolites are difficult to extract or remain unexpressed.

Methods:
In this study, we applied Liquid Extraction Surface Analysis (LESA) combined with high-resolution mass spectrometry to analyse the secondary metabolites produced by 12 phylogenetically different bacteria isolated from soil and rhizosphere. The bacteria were grown under 3 different nutrient conditions namely, nutrient-rich TSBA;and nutrient-poor Water Agar with and without artificial root exudates.

Results:
Our results revealed that LESA with high-resolution mass spectrometry is a powerful tool for high-throughput extraction and detection of secondary metabolites produced by different bacteria under different culture conditions. Of all the solvents tested, ethyl acetate/acetone (65:35, v/v) containing 0.1% formic acid revealed the most effective extraction. Pyrrolnitrin, a chlorinated phenylpyrrol antibiotic was detected in Serratia plymuthica on nutrient-rich 1/10 TSBA and Water Agar supplemented with Artificial Root Exudates but not on nutrient-poor Water Agar. Violacein, a violet pigmented indole derivative antibiotic was detected in Janthinobacterium sp. AD80 only under nutrient-rich growth condition. Besides known secondary metabolites, several unknown metabolites were detected in bacterial cultures but not in any of the controls consisting of only media.

Novel Aspect:
LESA-HRMS is a fast, sensitive and cost-effective high-throughput screening device for direct detection of bacterial secondary metabolites.

Metwally, Haidy,
University of Western Ontario

Contributing Authors:
Vlad Popa, University of Western Ontario; Lars Konermann, University of Western Ontario; Robert G. McAllister University of Western Ontario.

Title:
Mechanism of Protein Supercharging by Sulfolane and m-NBA: Molecular Dynamics Simulations of the Electrospray Process

Introduction:
Electrospray ionization allows the production of desolvated ions from proteins and protein complexes. Natively folded conformers generate low charge states. However, supercharging agents such as sulfolane or m-nitrobenzyl alcohol (m-NBA) are used to produce ions with high charge states. We use molecular dynamics simulations and IMS to gain insights into the supercharging process.

Methods:
MD simulations were carried out using Gromacs 5 with the CHARMM36 force field. Simulations were conducted on holo-myoglobin (hMb). The initial droplet radius was 3.5 nm, corresponding to about 5000 TIP4P/2005 waters with the protein at its center. Sulfolane or m-NBA molecules were inserted at random positions. Experimental ESI-MS and IMS data were acquired on a Synapt HDMS instrument (Waters).

Results:
The MD simulations reveal major differences in charge carrier (Na+) behavior for droplets with and without SCAs. In pure water droplets Na+ moves freely. Droplet shrinkage is accompanied by Na+ ejection via IEM. In the presence of SCAs, differential evaporation causes SCA enrichment. Solvent segregation leads to an outermost SCA layer that encapsulates an aqueous core. This core traps the protein and Na+. The ejection of charge carriers is disfavored by unfavorable Na+/SCA interactions, thereby forcing the charge carriers to associate with protein. hMb remains compact within the droplet throughout the simulation. SCA evaporation releases the protein into the gas phase. As a result of the high charge state values, Coulombically driven unfolding takes place after release into the gas phase. These MD-generated charge states agree remarkably well with experiments.

Novel Aspect:
We provide the first atomistic view of the ESI/supercharging mechanism. Supercharging is attributed to a charge trapping mechanism

21ST INTERNATIONAL MASS SPECTROMETRY CONFERENCE
Toronto, Ontario, Canada
August 20 – 26, 2016

Meyer, Sven,
Bruker Daltonik GmbH

Title:
High resolution trapped ion mobility mass spectrometry analysis of isomeric compounds

Introduction:
Ion mobility spectrometry (IMS) mass spectrometry allows extending the identification and characterization of compounds by separation of isobaric compounds by collision cross sections (CCS). Here trapped IMS separation of isomeric compounds was used to enable their clear identification by MS/MS. The examples used were isobaric trisaccharides or lipids, which are otherwise difficult to analyze.

Methods:
Isomer mixtures were prepared from standard compounds. These were trisaccharides (raffinose, melezitose, maltotriose) or lipids (18:1 Δ9-cis PC and 18:1 Δ6-cis PC, as well as 16:0-18:0 PC, 16:0-18:1 PC and 16:0-18:2 PC). In direct infusion, TIMS was used to resolve isomer mixtures and retrieve clean MS and MS/MS spectra with good mass accuracys and isotopic pattern qualities.

Results:
Often, isomeric compounds are co-eluting in standard screening LC-gradients hampering their identification by MS and MS/MS in standard analyses. In case of lipids, the interpretation of fragment spectra is further impaired by partially overlapping isotopic patterns due to additional or fewer double bonds. Here, TIMS allowed obtaining clean MS and MS/MS spectra for compound identification. All isobaric compounds were baseline separated in the mobility domain; CCS values were determined and compared to published values. The sub ppm QTOF data delivered exact masses in the ppm range as well as true isotopic patterns of the pure MS and MS/MS spectra. For lipid analysis the mobility separation simplified the identification due to clean isotopic patterns of fragment peaks.

Novel Aspect:
Investigation of the use of HR trapped ion mobility MS to identify isomeric lipids and trisaccharides by pure MS/MS analysis.

Meyer, Sven,
Bruker Daltonik GmbH

Title:
Analysis of propolis extracts and isomeric flavonoid mixtures using trapped ion mobility QTOF-MS

Introduction:
Flavonoids are a structural class of plant secondary metabolites exceeding 8000 compounds and LC- MS/MS analysis is frequently complicated by co-eluting isomers. Ion Mobility coupled with QTOF mass spectrometry gives an additional dimension of separation for identification of the co-eluting isomers. Here, high resolution trapped ion mobility spectrometry was used for the analysis .

Methods:
A high resolution trapped ion mobility (TIMS) QTOF was used for analysis of bee propolis. Ethanol extracts of propolis were spiked with a mixture of isomeric flavonoids at a concentration of 1 µg/mL. The pure standard mixture as well as the complex samples were investigated. The flavonoid ions were identified via HPLC- TIMS-QTOF-MS and -MS/MS, isotopic information supported the identification.

Results:
The 10 compound mixture consisted, e.g., of quercetin and morin (C15H10O7), 3 C21H20O11, 4 C15H10O6 and 2 C15H10O5 isomers; it was investigated in direct infusion experiments, first. Ion mobility with high resolving power (150-250) often provided for baseline separation of the analytes and thus the acquisition of separate spectra of the respective isomers. "TIMS-purified" MS/MS spectra yielded a high confidence for flavonoid isomer identification. The collision cross section values (CCS) were obtained in addition and were within 0.5 % in agreement with published values. HPLC could not separate several of the isomeric analytes well enough. However, isobaric flavonoids were baseline separated in the mobility domain at 10 s LC peak widths. LC-TIMS-MS was shown to be capable to support substance identification of highly isomeric compound classes and in very complex matrices.

Novel Aspect:
Separation and identification of isobaric flavonoids in complex samples by high resolution trapped ion mobility QTOF-MS.

Michalik-Onichomowska, Aleksandra,
Bundesanstalt für Materialforschung und -prüfung (BAM)

Contributing Authors:
Hans-Gerd Löhmannsröben, Physical Chemistry, University of Potsdam; Jens Riedel, Federal Institute for Materials Research and Testing; Ulrich Panne Federal Institute for Materials Research and Testing.

Title:
Real time monitoring of photoreactions performed within levitated droplets by LA-DBD-MS

Introduction:
The upscaling from small scale academic reactors to large industrial processes typically suffers from a large change in surface-to-volume ratio. A promising approach is the general avoidance of surfaces as in levitated droplet techniques. However, up to now, no mass spectrometric interface for online reaction monitoring in levitated droplets has been provided.

Methods:
As model reaction the photoinitiated thiol-ene coupling between N-boc cysteine methyl ester and allyl alcohol was studied. A droplet of 5 µL reactand solution was provisioned into an acoustic trap aligned lateral to the MS inlet. Contactless sampling by laser ablation (LA) is followed by dielectric barrier discharge (DBD) postionization. The latter is needed to address non polar reaction partners.

Results:
The formed C-S bond can be unambiguously identified by LA-DBD-MS. The obtained spectra show the formation of singly protonated monomers and dimers. Addition of an internal standard allows the quantification of the involved reaction partners. The results are quantitatively consistent with data achieved by NMR measurements proving the applicability of the introduced technique for online monitoring of complex chemical reactions. Levitated droplets in combination with LA-DBD-MS provide powerful capabilities to predict optimized reaction conditions for large scale production while ruling out surface effects. Application of LA for sampling followed by DBD as post ionization results in a robust and nonselective technique, suitable for a large number of reactions, providing flexibility in operation not achievable for ionization method solely based on LA.

Novel Aspect:
Application of levitated droplets for contactless, non-selective monitoring of photoreactions by MS.

Michel, Deborah,
University of Saskatchewan

Contributing Authors:
Shaokun Pang, Sciex 1201 Radio Road Redwood City CA 94064; Jane Alcorn, College of Pharmacy and Nutrition, University of Saskatchewan, SK, Canada; Richard Tang-Wai Pediatrics, Edmonton Clinic Health Academy, University of Alberta, AB, Canada; Richard Huntsman, Pediatrics, College of Medicine, University of Saskatchewan, SK, Canada; Ahmed Almousa, College of Pharmacy and Nutrition, University of Saskatchewan, SK, Canada.

Title:
MRM3 versus differential mass spectrometry (DMS) in background noise reduction for complex matrices like human plasma

Introduction:
We wish to develop and validate a rapid and sensitive mass spectrometry (MS) method for quantitating cannabidiol (CBD) and tetrahydrocannabinol (THC) in small volume human plasma samples, but at present high background noise confounds the robustness of the method. We compared MRM3 with DMS for their ability to improve the signal to noise ratio in normal and hyperlipidemic human plasma.

Methods:
Spiked plasma (50 μL) first undergoes protein precipitation (acetonitrile, 150 μL) and 2 μL of supernatant is introduced onto an Agilent 1290 UPLC with a zorbax eclipse column to separate analytes for MS detection. CBD and THC are quantitated by MRM, MRM3, and DMS using a Sciex QTrap 6500 equipped with SelexION. Analysis is performed on Multiquant by Sciex.

Results:
We monitored CBD and THC to compare the MS methods. Addition of an MS stage to the LC-MSMS analysis (i.e. MRM3) did not improve the signal to noise ratio. Differential ion mobility caused a significant improvement in background noise with a small loss in analyte detection sensitivity. Compared with MRM3 DMS resolved the high background noise while demonstrating acceptable sensitivity, linearity, precision, and accuracy.

Novel Aspect:
LC-DMS-MSMS is robust versatile method for the quantification of CBD and THC in special populations where complexity and small volumes are expected.

Mikaia, Anzor,
National Institute of Standards and Technology

Contributing Authors:
Stephen E. Stein, National Institute of Standards & Technology; Nino G. Todua, National Institute of Standards & Technology.

Title:
Identification of amino acids and dipeptides in biological samples by GC-MS

Introduction:
Metabolic actions and therapeutic potential of small peptides can be justified by considering them as metabolic intermediates in biological interconversions between their structural analogs – amino acids and proteins. Moreover, certain classes of small peptides have received considerable attention of doping control authorities because of their up-regulating effect on growth hormone.

Methods:
Essential amino acids and fifty di-, tri- and tetra-peptides and amides are examined. Commercially available alkyl chloroformates, alkanol, pyridine and chloroform are employed for N-alkoxycarbonylation and esterification. Selective and quantitative derivatization is achieved by slow addition of a 20× molar excess of reactants to the peptides and SRMs 2389 and 1950 at 15oC.

Results:
Ions due to the loss of methoxycarbonyl radicals from M+. are successfully used for molecular mass determination of methyl esters of N-methoxycarbonyl-peptides. Peaks of a ions containing the derivatized N-terminal amino groups are outstanding in spectra of all peptide derivatives. The nature of the amino acid residue located at the N-terminus end determines these specific decomposition patterns of a ions in the cases of 4-hydroxyproline, cysteine, serine, tryptophan, tyrosine and phenylalanine derivatives; alkene molecules are eliminated from the side chains of a ions when valine, leucine, isoleucine and norleucine residues are located at the N-terminus end. The other significant peaks in the spectra correspond to b and y ions while peaks of x ions do not show significant intensities. Identified fragmentation directions are used for qualitative analysis of these compounds in NIST SRMs.

Novel Aspect:
A GC-MS method incorporating preliminary sample derivatization is developed for the study of amino acids and peptides.

Mikaia, Anzor,
National Institute of Standards and Technology

Contributing Authors:
Nino G. Todua, National Institute of Standards & Technology; Levan A. Megutnishvili, National Institute of Standards & Technology; Stephen E. Stein National Institute of Standards & Technology.

Title:
Fragmentation of thebaol chemical modification products: a violation of the 'even-electron rule'

Introduction:
Chemical modification is a valuable technique effectively used in GC-MS to extend the limits of its application. Selection of suitable reactions and obtaining appropriate derivatives leads to achieving desired volatility for a separation, sufficient sensitivity, selectivity and specificity of an analyte under electron ionization. Derivatization is applied to Thebaol that is found in opium poppy.

Methods:
Methyl, ethyl, trimethylsilyl, trifluoroacetyl, pentafluoropropionyl, heptafluorobutyryl, and alkoxycarbonyl (alkyl = methyl, ethyl, isopropyl, n-propyl) derivatives of thebaol are prepared using well established chemical modification procedures. EI mass spectra were acquired on GC/MS systems with quadrupole mass analyzers.

Results:
The mass spectrum of trimethylsilyl ester of thebaol exhibits the base peak of [M-30]+. radical cations originating from consecutive loss of two methyl radicals. Perfluoroacyl theabols show outstanding peaks of [M - CnF2n+1 – CH3]+. radical cations. The observed ortho effect is valuable for reliable determination of theabol within mixtures. The other significant peaks in the spectra of perfluoroacyl derivatives correspond to nitrilium cations due to oxygen rearrangement. Molecular ions of alkoxycarbonyl derivatives of thebaol undergo hydrogen rearrangements under electron ionization and they do not violate the 'even-electron rule'. The driving forces for major fragmentation pathways for each type of derivative will be presented.

Novel Aspect:
A GC-MS method incorporating preliminary sample derivatization methods is developed for the study of drugs of abuse.

Miller, Ken,
Thermo Fisher Scientific

Contributing Authors:
Romain Huguet, Thermo Fisher Scientific; Michael Wenz, Thermo Fisher Scientific; Steve Williams Thermo Fisher Scientific; Andreas Krupke, Thermo Fisher Scientific; Andreas FR Huhmer, Thermo Fisher Scientific; Achim Karger, Thermo Fisher Scientific; Shiaw-Min Chen, Thermo Fisher Scientific; Daniel Lopez-Ferrer, Thermo Fisher Scientific; Monica Carrera, Marine Research Institute, Vigo, Spain; Chad R. Weisbrod, Thermo Fisher Scientific; Chien-Hsun Chen, Thermo Fisher Scientific; Aran Paulus, Thermo Fisher Scientific.

Title:
Ultra-Fast Analysis of Allergens using capillary electrophoresis coupled to mass spectrometry and Ultraviolet Photodissociation

Introduction:
The identification of proteins in food matrices that cause sensitization in individuals or allergic reactions in those individuals already sensitized, represent a major concern for the food industry. We explore the application of CE/MS/MS as an accurate and sensitive way to rapidly test for protein allergens in food.

Methods:
The workflow is based in a simple protein extraction step and the use of capillary electrophoresis (CE) hyphenated to an Orbitrap Fusion Tribrid mass spectrometer with an UVPD source to do the analysis of the extracted proteins.

Results:
We first purified the thermo-stable proteins from the sarcoplasmic protein fraction of the reference salmon species. The second step, consist in a top-down proteomic approach of the purified proteins. Samples were cleaned up and loaded onto CE system. A 30 min separation was performed and the proteins were then analyzed doing a top-down msx in the HCD cell of a benchtop quadrupole orbitrap mass spectrometer. EThcD and UVPD fragmentation modes were also performed. For all the samples, unique mass to migration time maps were built representing a unique signature for each of the fish allergens. The most abundant protein in the electropherograms was a 11 kDa protein, identified as parvalbumin (PRVB). This thermo-stable protein is considered as the major fish allergen. We were able to efficiently separate several PRVB protein isoforms.

Novel Aspect:
Characterization of food allergens in minutes using a high throughput CE-MS/MS platform and top-down proteomics

Miller, Ken,
Thermo Fisher Scientific

Contributing Authors:
Daniel Lopez-Ferrer, Thermo Fisher Scientific; Michael Blank, Thermo Fisher Scientific; Romain Huguet Thermo Fisher Scientific; Dave Horn, Thermo Fisher Scientific; Ignacio Ortea, IMIBIC, Cordoba, Spain; Andreas FR Huhmer, Thermo Fisher Scientific.

Title:
Improvements in LFQ for reproducible quantification of proteomic experiments: how DDA outperforms DIA

Introduction:
While DIA has been receiving a lot of attention lately, DDA remains gold standard for label-free quantitation (LFQ) proteomics. Here we put QE DDA, QE DIA, and QTOF DIA head-to-head to evaluate the sensitivity and number of peptides identified and quantified and demonstrate that QE DDA outperforms DIA analyses significantly in proteome coverage and quantitative reproducibility.

Methods:
Hela cell lysates spiked with different amounts of spiked BSA were analyzed using either DIA or DDA method on QE HF mass spectrometer or a Sciex TripleToF 5600 QTOF keeping the same chromatography conditions. Data were analyzed using pre-release version of PD 2.2 and MaxQuant software. DIA experiments were processed using OpenSwath and Skyline.

Results:
Data presented shows that DDA outperforms DIA in number of peptide identifications. DDA used with the LFQ node in PD 2.1 allows for quantification of ~90% of identified peptides in the study using the MS1 scan to integrate the peptide signals. In addition, over 80% of quantified peptides have CVs below 15%. Also threshold to potentially identify and quantify a peptide is higher in DDA analyses compared to the DIA as is demonstrated by proteins identified. The main reason is that by using orbitrap technology, the inherent accumulation of ions before scanning provides bias towards the most intense ions in MS2 scan, making it difficult to use specific transitions. However, this bias seems to have an invaluable advantage compared to pulsed based scanning methods such the ones provided by QTOFs, because the accumulation of ions increases the signal of the less abundant peptides.

Novel Aspect:
DDA analyses in Orbitrap instruments coupled to sophisticated label free software outperform DIA analyses in any type of instrument

Mirabelli, Mario Francesco,
ETH Zurich

Contributing Authors:
Renato Zenobi, ETH Zurich.

Title:
Observing proton transfer reactions inside the MALDI plume: experimental and theoretical insight into MALDI gas-phase processes

Introduction:
Although matrix-assisted laser desorption/ionization at ultraviolet wavelengths (UV-MALDI) is widely used for routine and research applications in the field of analytical chemistry and biochemistry, the underlying ionization mechanisms are still not fully understood. We evaluated the relevance of gas-phase in-plume proton transfer reactions to the ion formation in the MALDI process.

Methods:
The experiments were performed on a Bruker AutoFlex and an internal MALDI FTICR. A split sample holder was used to simultaneously desorb/ionize two different samples by the same laser pulse, and to avoid their mixing during sample preparation. Using a combination of deuterated and non-deuterated matrices and analytes, it was possible to selectively observe the plume proton transfer processes.

Results:
Because of the complexity of the MALDI process, relevant information about the plume reactions can be obtained only when gas-phase reactions are isolated from the ones happening in the condensed phase. We collected an extensive set of data that clearly shows the relevance of proton transfer processes in the gas-phase to the final ion distribution. The observed HDX reactions were found to be energetically feasible by DFT calculations, and although several reaction pathways are possible, we found that a simultaneous double H/D transfer is favored over a 2–steps process. We also show the relevance of neutral-to-neutral and ion-to-neutral reactions. The experimental results, supported by theoretical simulations, confirm that the plume is a very reactive environment, with HDX reactions observed up to 400 ns after the laser pulse.

Novel Aspect:
Separation of condensed phase from gas-phase reactions. Evaluation of plume proton transfer energetics, kinetics and relevance to the final ion yield.

Mirabelli, Mario Francesco,
ETH Zurich

Contributing Authors:
Renato Zenobi, ETH Zurich; Emanuela Gionfriddo, University of Waterloo; Janusz Pawliszyn University of Waterloo.

Title:
Direct SPME-ambient MS couplings: towards the non-chromatographic era

Introduction:
Direct mass spectrometry (MS) coupling methods, such as those employing solid-phase microextraction (SPME), have shown promise for high throughput chemical analyses, and represent a viable alternative to standard chromatographic methods. One of the main challenges to overcome with non-chromatographic approaches is the occurrence of extensive matrix effects for complex samples.

Methods:
We employed an ambient dielectric barrier discharge ionization (DBDI) source to directly couple SPME to MS. Different from DART or DESI, the SPME desorption step in our setup is completely separated from the ionization event. Several SPME devices, such as commercial SPME fibers, the novel matrix compatible PDMS/DVB/PDMS coatings, and PDMS/DVB thin films were employed.

Results:
Pesticides and illicit drugs were quantified in several matrices, including water samples, beverages, urine and small insects. The matrix compatible SPME coatings showed enhanced antifouling capability and reduced tendency to degradation when directly exposed to complex matrices, compared to conventional coatings. The combination of these devices with our recently developed active capillary DBDI source allowed sub pg/g sensitivity to be reached. Complete (>99%) thermal desorption of the compounds from the SPME devices usually occurred within 1 minute, avoiding the need for any cryofocusing system. In most cases chromatography could be avoided, enabling quantitative results to be available within minutes rather than hours. Strategies to reduce suppression effects and a comprehensive outlook of these approaches for bioclinical, environmental, forensic and food analyses will be discussed.

Novel Aspect:
Non-chromatographic quantitation of analytes in complex samples, using ambient mass spectrometry and SPME extraction.

Miura, Daisuke,
Kyushu University

Contributing Authors:
Yoshinori Fujimura, Innovation Center for Medical Redox Navigation, Kyushu University; Hiroyuki Wariihi, Innovation Center for Medical Redox Navigation, Kyushu University; Mitsuru Shindo Departments of Molecular and Material Sciences, Kyushu University; Daichi Yukihira, Innovation Center for Medical Redox Navigation, Kyushu University; Takanori Ishii, Innovation Center for Medical Redox Navigation, Kyushu University.

Title:
Structure-Property Relationship of MALDI Matrix for Low-Molecular-Weight Metabolites Analysis

Introduction:
In MALDI-MS analysis, the molecular coverage of detectable compounds is strongly associated with the molecular species of the matrix. Herein, we report structure-property relationship between novel synthesized 9-aminoacridine (9-AA) analogues and ionization efficiency of representative low-molecular-weight metabolites in negative ionization mode MALDI-MS analysis.

Methods:
All 9-AA analogues were chemically synthesized by ourselves, and solved in 100% methanol (10 mg/mL or saturated). All metabolite standards were dissolved in water, and mixed with matrix solution at 1:1 just before MALDI-MS analysis, then these were spotted onto stainless MALDI plate. All analysis data were obtained in negative ionization mode by using AXIMA Performance (Shimadzu, Japan).

Results:
Herein, we newly synthesized 80 molecular species of 9-AA analogues, and attempted to acquire the information about structure-property relationship between 9-AA analogues and ionization efficiency of representative low-molecular-weight metabolites (328 species) via brute force MALDI-MS analysis in negative ionization mode. As the result, 233 species (71%) of metabolites were ionized by using at least one of 9-AA analogues at the concentration of 5 pmol/well. We also found the potential relationship between the main skeleton of 9-AA analogues and ionizable metabolite. Interestingly, some matrix candidates having naphthalene/anthracene skeleton were able to ionize several amino acids or flavonoids, which was not ionized by 9-AA. On the other hands, such analogues were not able to ionize phosphorylated metabolites such as sugar phosphate or nucleotides, which was easily ionized by 9-AA.

Novel Aspect:
We found the structure-property relationships of 9-AA-based matrix for the analysis of low-molecular-weight metabolites.

Mobbs, Helene,
Department of Biochemistry, University Cambridge

Contributing Authors:
Liz Bentley, MRC Harwell; Roger D. Cox, MRC Harwell; Tertius A. Hough MRC Harwell; Lee Moir, MRC Harwell; John P. Shockcor, Waters Ltd; Julian L. Griffin, University of Cambridge; Steve A. Murfitt, University of Cambridge.

Title:
The development of a high throughput metabolomic screen for mouse models of type 2 diabetes using high resolution mass spectrometry

Introduction:
As part of the International Mouse Phenotyping Consortium (IMPC), sites across the world are generating and phenotyping 20,000 knockout mouse strains. One disease area that this phenotyping is focussed on is type 2 diabetes. This abstract outlines how we have developed an LC-MS screen for diabetic mouse models, largely focussed on lipids, guided in part by literature on obese and diabetic mice.

Methods:
Sample analysis methods were streamlined to allow for a balance between ease of sample preparation and analysis time. A methanol precipitation was used to prepare 15 µL plasma samples. Separation was achieved by reserve phase chromatography with a C18 column. The mobile phase pH was modified to optimise signal in either positive or negative MS modes and ions detected using a Waters Xevo Q-ToF MS.

Results:
Currently, 28 KO mouse models have been tested across 443 blood plasma samples taken routinely during the IMPC phenotyping screen. To date the KO mouse line that was most aberrant in lipid profile was Nedd4l, which has been associated with human hypertension. The blood plasma of the mouse was associated with alterations in a variety of free fatty acids including 14:0, 17:0 and 18:0, with similar detected changes associated with human insulin resistance and type 2 diabetes in the literature. In addition a number of specific triglycerides (TG) appear important in the classification of diabetic mouse models, with alterations in the concentration of TG(56:9) being important in discriminating Bbs5 and Lyplal1 KO mice, and increases in TG(52:6) in Nbr1, Glra2 and Plac9 KO mice.

Novel Aspect:
The MRC Harwell knockout mice are a world class resource, allowing us to investigate lipid metabolism changes associated with disease.

Moghaddam, Ghazaleh,
Université du Québec à Montréal, Chemistry Department

Contributing Authors:

Lekha Sleno*, Université du Québec à Montréal, Chemistry Department.

Title:

Global metabolomics approach for identifying acetaminophen dose-dependent metabolites using HRMS and MRM

Introduction:

Acetaminophen is the main cause for acute liver failure in North America due to its reactive metabolite-related hepatotoxicity. Also the use of metabolomics to identify small molecule biomarkers of acetaminophen toxicity is novel. In order to better characterize the metabolic perturbations occurring with acetaminophen dosing, a global metabolomics approach was employed in this study.

Methods:

Four dosing levels of acetaminophen were administered to rats. Metabolites were extracted and injected onto a Nexera UHPLC-AB Sciex 5600 TripleTOFTM system. Samples were separated on PFP and T3HSS columns (in +/-). Samples were compared to controls by statistical analysis. MRM was then optimized using a 5500 QTRAP® platform to analyze time-course samples for metabolites of interest.

Results:

The goal for this study was to develop a high-throughput LC/MS-MS based metabolomic assay to rapidly investigate acetaminophen metabolites profile in rat plasma samples. Many features have been found to have interesting profiles, including increasing as well as decreasing with dosing level, as well as discriminating between endogenous metabolites and APAP-related metabolites.

Novel Aspect:

Acetaminophen safety could be evaluated by quantifying metabolic profiles.

Mohammed, Yassene,
Center for Proteomics and Metabolomics, Leiden University Medical Center

Contributing Authors:
Dana Ohana, LUMC; Oleg A. Mayboroda, LUMC; Hans Dalebout LUMC; Magnus Palmblad, LUMC; Anthonius A. M. Heemskerk, LUMC.

Title:
Method and Software Workflow for Integrating Paired CE-MS and LC-MS Bottom-up Proteomics Data from SDS-PAGE Pre-fractionated Samples

Introduction:
In mass spectrometry based proteomics, pre-fractionation that is orthogonal to the second dimension separation based on hydrophobicity (RPLC) or charge and size (CE) increases the separation power. The complementarity use of CE and RPLC for in-depth analyses is challenging due to, foremost the absence of a data processing strategy. We present a workflow for "paired" RPLC-MS and CE-MS proteomics.

Methods:
We used a whole-cell protein extract from E.coli, fractionated by SDS-PAGE, followed by tryptic in-gel digestion and analysis by RPLC- and CE-MS/MS. Paired datasets from CE-MS and RPLC-MS were acquired, i.e. with the same effective time for tandem mass spectrometry. We built an automatic software workflow for peptide and protein identification followed by integration of the CE-MS and RPLC-MS data.

Results:
The identification in CE-MS data was better for peptides in the higher molecular weight range. We saw an increase in identified peptides in the 1,000 to 1,300 Da range for RPLC-MS and for CE-MS in the 1,500 to 2,500 Da range. There was a clear difference in average GRAVY score between CE-MS and RPLC-MS for peptide identifications. The difference in hydrophobicity and the discrepancies between the masses of the identified peptides shows strong complementarity of CE-MS and RPLC-MS. The numbers of confidently identified peptides between the two techniques is relatively similar with 2,846 peptides by CE-MS and 2,469 peptides by RPLC-MS of which 1,363 could be found in both. Visualization of retention/migration time model parameters gave further insight into the nature of the different separations and their dependencies on physico-chemical parameters such as pH.

Novel Aspect:
Automated data processing of paired RPLC- and CE-MS acquisitions. The two methods show complementarity with around 50% non-shared peptides.

Mohammed Saeid, Waleed,
Drug Design & Discovery Research Group, College of Pharmacy & Nutrition, University of Saskatchewan

Contributing Authors:
Deborah Michel, Drug Design & Discovery Research Group, College of Pharmacy & Nutrition, University of Saskatchewan; Ildiko Badea, Drug Design & Discovery Research Group, College of Pharmacy & Nutrition, University of Saskatchewan; Anas El-Aneed Drug Design & Discovery Research Group, College of Pharmacy & Nutrition, University of Saskatchewan.

Title:
Rapid and simple flow injection analysis-tandem (FIA–MS/MS) mass spectrometric method for the quantification of melphalan in drug delivery systems

Introduction:
The use of the anticancer drug melphalan (Mel) is limited due to its poor water solubility. To address such limitation, it is incorporated within a novel delivery system using β-cyclodextrin-gemini surfactant (β-CD-gemini). Two fast and simple FIA-MS/MS methods are developed for the quantification of Mel within the drug delivery system so that solubilization efficiency and stability are assessed

Methods:
Method 1: Mel was formulated with [β-CD-gemini] as drug delivery system. Method 2: Mel was prepared in aqueous solution using 10% HCl ethanol. . FIA-MS/MS Methods are developed using a triple quad linear ion trap MS equipped with electrospray ionization (ESI) in positive mode. Deuterated form of Mel (Mel-d8) was used as an internal standard (IS). The methods were validated according to the FDA

Results:
Developed methods were fully validated according to FDA bioanalytical validation criteria. Both methods were validated with linearity (r=0.9986, 0.9971 respectively) in the range of 1–200 ng/mL with accuracy and precision below 15% for all standard points and quality control samples. The intra- and inter-day variations, freeze-thaw stability and long term stability (1 month) were within the acceptable criteria according to the FDA guidelines. The developed methods were applied successfully in determining the encapsulation/solubilisation activity of [Mel:β-CD-gemini] delivery system and to evaluate the stability of the developed formulations. In addition, these new analytical methods will be applied to evaluated the influence of [Mel:β-CD-gemini] on the drugs' anti-neoplastic activity and pharmacokinetic profile in rat animal model

Novel Aspect:
Simple FIA-MS/MS methods were developed without the need for tedious chromatographic separation. Solubilization of the target drug was enhanced

Moller, Ingvar,
Department of Health and Medical Sciences, University of Copenhagen

Contributing Authors:
Patrick S. Merkle, Department of Pharmacy, University of Copenhagen, Copenhagen, Denmark; Alan Kádek, Institute of Microbiology, Academy of Sciences of Czech Republic, Prague, Czech Republic; Petr Man Institute of Microbiology, Academy of Sciences of Czech Republic, Prague, Czech Republic; Claus J. Løland, Department of Neuroscience and Pharmacology, University of Copenhagen, Copenhagen, Denmark; Rasmus R. Svejdal, Department of Pharmacy, University of Copenhagen, Copenhagen, Denmark; Eliska Pospisilova, Institute of Microbiology, Academy of Sciences of Czech Republic, Prague, Czech Republic; Marika Slivacka, Department of Pharmacy, University of Copenhagen, Copenhagen, Denmark; Kasper D. Rand, Department of Pharmacy, University of Copenhagen, Copenhagen, Denmark; Anne K. Nielsen, Department of Neuroscience and Pharmacology, University of Copenhagen, Copenhagen, Denmark; Jiři Hausner, Institute of Microbiology, Academy of Sciences of Czech Republic, Prague, Czech Republic.

Title:
Improving digestion efficiency of problematic membrane proteins in hydrogen/deuterium exchange compatible workflows

Introduction:
Analysis of protein dynamics by HDX-MS requires effective digestion of target proteins at low pH and temperature, suboptimal conditions for most proteases. Here we investigate the use of alternative proteases (Rhizopuspepsin and Nepenthesin I and II), immobilized both on column and microchip formats to specifically improve the HDX-MS analysis of a challenging class of integral membrane proteins.

Methods:
Four integral membrane proteins were subjected to conditions often tested in bottom-up HDX-MS workflows. The initial optimization of protease activity was performed in parallel on Phosphorylase B. Proteins were diluted into buffers which simulated quench conditions for HDX-MS. Digestions were either performed in solution or using immobilized proteases packed into columns or microfluidics chips.

Results:
Digestion using alternative proteases resulted in over 80% sequence coverage for Phosphorylase B which included certain regions not covered by digestion with pepsin and an increase in amino acid redundancy. Digestion optimizations of integral membrane proteins yielded mixed results when using immobilized pepsin in a column format. The sequence coverage of a prokaryotic transporter (LeuT) was almost complete using HDX-MS compatible workflows developed with immobilized pepsin on a NanoAcquity UPLC system coupled to Synapt G2-HDMS(Waters). Eukaryotic transporter homologues of LeuT have, however, resulted in below 50% coverage using similar setups but the complimentary digestion patterns of the alternative proteases have shown regions not before covered with pepsin. The proteases were also immobilized on a microfluidics chip by photo click chemistry to compare to a standard column format.

Novel Aspect:
Alternative proteases immobilized on both column and microchip formats to improve the digestion of complex membrane proteins.

Moon, Myeong Hee,
Yonsei University

Title:
On-line Coupling of Field-Flow Fractionation with Mass Spectrometry for Lipoproteins and Metalloproteins

Introduction:
Introdued are the on-line coupling of flow field-flow fractionation (FlFFF) with MS for the top-down lipidomic analysis of lipoproteins via electrospray ionization (ESI) and for the direct metal analysis of metalloproteins in blood plasma via inductively coupled plasma (ICP). FlFFF is an elution-based separation method that is capable of size-sorting particles and biological macromolecules.

Methods:
FlFFF is an elution-based separation method that is capable of size-sorting particles, proteins, cells, and etc, in an unobstructed channel without shear induced deformation or degradation. For the direct coupling of FlFFF with MS, a miniaturized asymmetrical FlFFF (mAF4) was developed to reduce operating flow rate regime at few uL/min, so that effluent from FlFFF can be directly fed into MS.

Results:
Plasma lipoproteins are separated by sizes first using mAF4 and continuously fed into MS (mAF4-ESI-MS/MS) so that lipids in different lipoproteins can be directly characterized and quantified using selected reaction monitoring (SRM). Results are compared with those from a comprehensive analysis of lipids in lipoproteins by off-line combination of FlFFF with nLC-MS/MS. mAF4 can be interfaced with ICP-MS (mAF4-ICP-MS) in such a way that metals in metalloproteins can be selectively analyzed. It will show that mAF4-ICP-MS provides size separation of proteins in their intact states in FlFFF and simultaneous analysis of metals contained in metalloproteins in blood plasma. Optimizations in the analysis of blood plasma will be discussed together with applications to lung cancer plasma samples.

Novel Aspect:
Covered topics include the demonstration of mAF4-ESI-MS/MS for on-line lipidomic analysis of lipoproteins and mAF4-ICP-MS for metaloprotein analysis.

Morais, Damila,
Unicamp

Contributing Authors:

Damila R de Morais, ThoMSon Mass Spectrometry Laboratory, University of Campinas (UNICAMP), Campinas-SP, Brazil; Rafael Lanaro, Poison Control Center, University of Campinas (UNICAMP), Campinas-SP, Brazil; Vinicius V. Hernandes ThoMSon Mass Spectrometry Laboratory, University of Campinas (UNICAMP), Campinas-SP, Brazil; Elias. P. Tessaro, ThoMSon Mass Spectrometry Laboratory, University of Campinas (UNICAMP), Campinas-SP, Brazil; Deleon N. Correa, ThoMSon Mass Spectrometry Laboratory, University of Campinas (UNICAMP), Campinas-SP, Brazil; Cezar S. Gomes, Brazilian Federal Police, Natal-RN, Brazil; Rodrigo Borges, National Institute of Metrology, Quality and Technology—INMETRO, 25250-020, Rio de Janeiro, RJ, Brazil; Marcos N. Eberlin, ThoMSon Mass Spectrometry Laboratory, University of Campinas (UNICAMP), Campinas-SP, Brazil; Jose Luiz Costa, Faculty of Pharmaceutical Sciences and Poison Control Center, University of Campinas (UNICAMP), Campinas-SP, Brazil.

Title:

Mass spectrometry application in a real case of acute intoxication after alternative natural therapy treatment

Introduction:

Alternative health treatments using Ayahuasca beverage and Kambo (the secretion collected from the skin of the frog Phyllomedusa bicolor applied into the skin of the patient) has been used by patients suffering from serious diseases. In the present work, we used mass spectrometry to evaluate biological fluids of a patient submitted by simultaneous treatments using Ayahuasca and Kambo.

Methods:

The patient was admitted to emergency service (ES) with hallucination, agitation, skin lesions and seizures. Diazepam was administered by ES and was effective to control the hallucination, but failed to control agitation and seizures. Patient samples (urine and serum) were submitted to rapid analysis using liquid/liquid extraction, GC-MS, LC-MS/MS and high resolution ESI-MS/MS.

Results:

In patient's sample, it was possible to detect N,N-dimethyltryptamine (3.2 ng/mL serum, 487.7 ng/mL urine), harmine (12.3 ng/mL serum, 60.6 ng/mL urine), harmaline(0.3 ng/mL serum, 49.4 ng/mL urine) and tetrahydroharmine (182.2 ng/mL serum, 567.4 ng/mL urine). No other drug of abuse or pharmaceutical (except diazepam) was detected viageneral unknown screening. In Ayahuasca, it was possible to detect N,N-dimethyltryptamine, harmine, harmaline and tetrahydroharmine at concentrations of 678, 597, 132 and 1235 mg/L, respectively. High-resolution and accuracy MS analysis performed in Kambo sample detected the presence of two abundant ions, identified as the protonated molecules of the active peptides phyllomedusin and phyllocaerulein.

Novel Aspect:

Mass spectrometry strategies in order to diagnostic a real case of intoxication with active substances from Ayahuasca and Kambo.

Morais, Damila,
Unicamp

Contributing Authors:
Damila R de Morais, ThoMSon Mass Spectrometry Laboratory, University of Campinas (UNICAMP), Campinas-SP, Brazil; Ana Carolina de Aguiar, ThoMSon Mass Spectrometry Laboratory, University of Campinas (UNICAMP), Campinas-SP, Brazil; Mariana Baptistão ThoMSon Mass Spectrometry Laboratory, University of Campinas (UNICAMP), Campinas-SP, Brazil; José L. Paz, ThoMSon Mass Spectrometry Laboratory, University of Campinas (UNICAMP), Campinas-SP, Brazil; Lygia A. Marques, ThoMSon Mass Spectrometry Laboratory, University of Campinas (UNICAMP), Campinas-SP, Brazil; Mirela B. Coelho, ThoMSon Mass Spectrometry Laboratory, University of Campinas (UNICAMP), Campinas-SP, Brazil; Claudio L. Aguiar, Hugot Sugar Technology Laboratory, University of São Paulo (ESALQ-USP), Piracicaba-SP, Brazil; Marcos N. Eberlin, ThoMSon Mass Spectrometry Laboratory, University of Campinas (UNICAMP), Campinas-SP, Brazil.

Title:
Lipid characterization of annatto seeds by EASI-MS and GC-MS – An evaluation of health-promoting compounds

Introduction:
Annatto (Bixa orellana L.) is originated from tropical America and Brazil is one of the main producers. The larger annatto cultivation is found in the state of São Paulo. Apart from the colored bixins, the annatto seeds also contain high concentrations of other bioactivity compounds. Mass spectrometry (MS) is a powerful tool for characterize and quantify bioactive and other valuable components.

Methods:
23 lots of annatto seeds from different farmers in SP state (Brazil) were obtained at august of 2015. Seeds were washed in tap water and dried for 24h at ambient temperature (25 °C). Without sample preparation, annatto seeds were direct evaluated by EASI-MS. For lipids analysis (total lipids, TL), a Bligh and Dyer extraction was performed, followed by a BSTFA derivatization for GC-MS analysis.

Results:
All annatto seed oils presented bioactive constituents such as triacylglycerol, fatty acid, tocopherols and phytosterols. It was noted that TAG were detected by EASI-MS along the ca. m/z 850-1000 range and mainly as [M+Na]+, but a minor [M+K]+ series separated by 16 m/z units was also observed, with major TAG ions constituted of m/z 853 (PPL), 877 (PLL), 877 (POO), 893 (PLL), 895 (PLO), 901 (LLL), 917 (LLL). Palmitic acid (16:0), linoleic acid (18:2n-6) and stearic acid (18:0) were present in all samples evaluated by GC-MS. Squalene, tocopherols and phytosterols were quantified by GC-MS and the amount presented were respectively: squalene (787.23 ± 351.65 µg100g-1 of TL), gamma-Tocopherol (126.52 ± 34.52 µg100g-1 of TL), colestane (484.04 ± 313.96 µg100g-1 of TL), stigmasterol (646.95 ± 231.35 µg100g-1 of TL) and beta-sitosterol (938.78 ± 378.99 µg100g-1 of TL).

Novel Aspect:
Characterization of TAGs in annatto seeds by EASI-MS, and quantification by GC-MS. Fast and simple method for product quality control.

Morgan, Chris,
Sanofi Genzyme

Contributing Authors:
Yves Sabbagh, Sanofi Genzyme; Kate Zhang, Sanofi Genzyme; Bill Brondyk Sanofi Genzyme; Yangde Chen, Sanofi Genzyme.

Title:
Epitope binning by hydrogen deuterium exchange mass spectrometry: Understanding epitope-driven effects

Introduction:
Hydrogen deuterium exchange (HDX) mass spectrometry (MS) has become an established technique for detecting protein:protein interactions. In the biotech industry, arguably the most important of these interactions is that of a monoclonal antibody and its antigen. Epitope mapping by HDX MS can be used for several purposes including classifying antibodies into epitope bins.

Methods:
Hydrogen deuterium exchange mass spectrometry was used to characterize >15 monoclonal antibodies targeting the same antigen into epitope bins.

Results:
While epitope binning by HDX MS has lower throughput than other techniques such as surface plasmon resonance (SPR), it does have higher (peptide level, ~5-15 residues) resolution. The higher resolution enables the differentiation of antibodies that appear to be in the same bin by SPR because of a small level of overlap. This added knowledge allows the intelligent direction of crystallography efforts to focus on unique epitopes. The resulting data is useful to rationalize species cross-reactivity issues and epitope-driven biological effects.

Novel Aspect:
Demonstrate that higher resolution epitope binning can rationalize biological effects like species cross-reactivity and function (agonist/antagonist)

Morris, Michael,
Waters Corporation

Contributing Authors:
Chris Hughes, Waters Corporation; Keith Richardson, Waters Corporation; Jason Wildgoose Waters Corporation; Lee Gethings, Waters Corporation; James Langridge, Waters Corporation; Johannes P C Vissers, Waters Corporation.

Title:
Lipidomics Quantitation Using a Novel Scanning Quadrupole DIA Method

Introduction:
Targeted LC-MS assays are common in lipidomics with an emphasis on the quantitative measurement of specific target analytes. A novel DIA mode of operation, which provides both qualitative and quantitative information, has been developed for a hybrid quadrupole/oa-time-of-flight (ToF) MS and applied to targeted lipidomics using transition extraction lists and compound library based approaches.

Methods:
LC separation was performed using a Waters Acquity UPLC equipped with a 2.1mm x 100mm column. The m/z isolation range of the quadrupole was continuously scanned with MS data acquired up to a frequency of 2000 spectra/s. Alternate MS scans comprised precursor and CID product ions. The resulting data were processed using Progenesis and Skyline informatics.

Results:
Quantitative proof-of-principle data were acquired by serially diluting lipid standards in protein precipitated plasma. Analysis indicates that scanning quad DIA enables over an order of magnitude more specificity than a static quad operated with the same resolution and it was found that a quadrupole transmission window of approx 10 Da provided optimum identifications. Qualitative information was obtained by extracting lipid class information based on neutral loss or product ion extraction. Control, diabetic and obese human plasma samples were treated with isopropanol and centrifuged to precipitate proteins and the lipid-containing layer collected. 2DMS data were collected and differentially expressed lipid classes across the three conditions were quantified. The obtained results agreed with previous discovery studies and the expected changes in relation to disease and/or phenotype.

Novel Aspect:
Targeted quantitation in lipidomics using a novel precursor quadrupole scanning based DIA method.

Morris, Michael,
Waters Corporation

Contributing Authors:
Lee Gethings, Waters Corporation; Chris Hughes, Waters Corporation; Keith Richardson Waters Corporation; Jason Wildgoose, Waters Corporation; Johannes P C Vissers, Waters Corporation; Jim Langridge, Waters Corporation.

Title:
Metabolomic workflow utilizing a rapid microbore metabolic processing (RAMMP) in conjunction with a novel scanning quadrupole DIA method

Introduction:
Previous studies utilizing rapid microbore metabolic profiling (RAMMP) have shown comparable group discrimination and improved selectivity over conventional UPLC chromatography. Here we demonstrate further improvements of the workflow by coupling RAMMP with a novel DIA method (SONAR), providing qualitative and quantitative information for metabolomic based applications.

Methods:
Urine from pregnant women over 3 trimesters was prepared and LC separated using a 1mm or 300 µm column. MS data were collected using SONAR acquisition, based on a m/z isolation range of the quadrupole being continuously scanned. Alternate MS scans comprised precursor and CID product ions. Data were processed and database searched using Progenesis QI, providing normalized quantitation results.

Results:
Urine samples were prepared and analyzed in a randomized order, with pooled QC's constructed from individual samples. LC separation was performed with 1mm and 300 µm columns over 12, 5 and 2 min gradients. Utilizing the scanning quad operation of SONAR provided an order of magnitude specificity when compared with other MS acquisition methods. A quadrupole transmission window of 10 Da provided optimum data. Processing with Progenesis QI provided qualitative and quantitative results, showing higher feature detection with 300 µm based data. Comparison with earlier RAMMP experiments, show that implementation of SONAR increases specificity and feature detection. This is particularly prevalent in cases were short gradients were investigated for high throughput analyses. Data analysis also indicates a high level of robustness with good technical reproducibility and consistent quantitation.

Novel Aspect:
Implementation of RAMMP with a novel DIA (SONAR) workflow results in higher numbers of metabolite identifications and increased sample throughput.

Mosely, Jackie,
DURHAM UNIVERSITY

Contributing Authors:
Hannah M. Britt, Department of Chemistry, Durham University, Durham, UK; John M. Sanderson, Department of Chemistry, Durham University, Durham, UK; Vian S. Ismail Department of Chemistry, Durham University, Durham, UK.

Title:
LC-MSMS offers new insights on the lipidation of peptides and proteins in lipid membranes

Introduction:
Naturally occurring peptide and protein modifications by acylation is potentially of great importance in biological systems. It is demonstrates the intrinsic reactivity between lipid membrane and protein, increasing protein hydrophobicity and as a result increasing its affinity toward the lipid membrane. Analytically, detecting such low abundant hydrophobic natural products is highly challenging.

Methods:
The acylated and multi-acylated peptides were analysed by reverse phase LC ES+ MS and MSMS (using a Thermo LTQFT and a Waters Synapt G2s). The multi-acylated protein was enzymatically digested prior to analysis. The LC gradient was modified to enable optimal detection of the hydrophobic precursor ions and high resolution MS/MS required to provide full characterisation.

Results:
LC-MS/MS has been used to show that peptides undergo intrinsic lipidation reactions in lipid membranes (doi: 10.1016/j.jmb.2013.07.013), involving acyl transfer from lipid constituents of the membrane to acceptor sites on the peptide. LC-MS/MS has has now been used to show lipidation by-products, lyso-lipids, serve as acyl group donors for further lipidation and at a faster reaction rate than diacyl-lipids. Intrinsic lipidation is not restricted to peptides: proteins and small molecules also undergo the process. We have examined the lipidation profiles of aquaporin-0 and shown that it can be accounted for with knowledge of the fatty acyl composition of lens membranes, thereby supporting the hypothesis that their lipidation arises from membrane lipid precursors. Current work to understand the factors that dictate this reactivity, and the consequences of lipidation, will be presented.

Novel Aspect:
LC-MS/MS to characterise low abundant co-eluting peptides made strongly hydrophobic through biological modifications.

Moyano, Encarna,
University of Barcelona

Contributing Authors:
E. Moyano, Department of Analytical Chemistry (University of Barcelona); J. Saurina, Department of Analytical Chemistry (University of Barcelona); D. Filatova Department of Analytical Chemistry (University of Barcelona); Ó. Núñez, Department of Analytical Chemistry (University of Barcelona) .

Title:
LC-HRMS METABOLOMICS AND POLYPHENOLICS IN THE CLASSIFICATION AND CHARACTERIZATION OF OLIVE OILS

Introduction:
Olive oils are rich in phenolic compounds, which act as natural antioxidants and may contribute to the prevention of several human diseases. Nowadays, there are some concerns that some oil products labelled as derived only from olives could contain other vegetable oils. The aim of this work was to explore suitable methods to characterize and classify olive and vegetable oils to prevent frauds.

Methods:
Full scan LC-MS raw data was considered as metabolomic fingerprint (untargeted approach) to be treated by principal component analysis (PCA). On the targeted approach (polyphenolic profiles), polyphenolic signals were employed for oil characterization. In order to achieve this, MS data was processed by Exact Finder v2.0 software by applying a customized target database of polyphenols.

Results:
The plot of scores obtained after PCA of both metabolomic fingerprints and polyphenolic profiles revealed patterns that were correlated to sample characteristics such as the type of fruit or seed employed, allowing a clear differentiation between olive oils in comparison with other vegetable oils (sunflower, soy, corn and mixtures of fruits and/or seeds). Based on polyphenolic profile, the most remarkable polyphenols and phenolic acids on each type of oil were identified, making possible to differentiate them according to their different content of polyphenols. Olive oils resulted being more enriched in polyphenols than other vegetable oils. Focusing on olive oils, metabolomics and polyphenolics also allowed the classification of analysed samples regarding the type of olives (Arbequina, Picual). These results will be useful to guarantee sample authentication and the prevention of frauds.

Novel Aspect:
New analytical approach for the classification and characterization of olive and vegetable oils based on metabolomics and polyphenolics.

Moyano, Encarna,
University of Barcelona

Contributing Authors:
E. Berdalet, Department of Marine Biology and Oceanography (ICM-CSIC) (Barcelona, Spain); O. Palacios, MS Laboratory/Organic Pollutants (IDAEA-CSIC) (Barcelona, Spain); M. Vila Department of Marine Biology and Oceanography (ICM-CSIC) (Barcelona, Spain); L. Viure, Department of Marine Biology and Oceanography (ICM-CSIC) (Barcelona, Spain); J. Farràs, Department of Organic Chemistry (University of Barcelona) (Barcelona, Spain); E. Moyano, Department of Analytical Chemistry (University of Barcelona) (Barcelona, Spain); J. Caixach, MS Laboratory/Organic Pollutants (IDAEA-CSIC) (Barcelona, Spain); É. Alechaga, Department of Analytical Chemistry (University of Barcelona) (Barcelona, Spain); S. Lacorte, Department of Environmental Chemistry (IDAEA-CSIC) (Barcelona, Spain) .

Title:
UHPLC-HRMS determination of palytoxin and ovatoxins from Ostreopsis cf. ovata, a marine microalgae blooming in a Northwestern Mediterranean beach.

Introduction:
Blooms of Ostreopsis cf. ovata (palytoxin/ovatoxins-producing microalgae) are spreading to coastal areas of temperate latitudes and constitute a health and environmental emergent problem. The risks posed by these events will only be determined if toxin patterns are identified and quantified. Nevertheless, the mass spectrometry characterization of palytoxin/ovatoxins is already a challenge.

Methods:
Ultra-high performance reversed phase liquid chromatography coupled to electrospray-high resolution mass spectrometry (UHPLC-HRMS) was applied for the reliable identification and determination of palytoxin/ovatoxins. Extraction, UHPLC and HRMS protocols were optimized to quantify the different toxins along an Ostreopsis cf. ovata bloom in Sant Andreu de Llavaneres beach (Barcelona, Spain).

Results:
Ovatoxins are complex macromolecules closely related to Palytoxin, whose mass spectrometry characterization is already a challenge. The complex electrospray mass spectra show multiply charged ions, mixed cationized species and adducts, thus requiring efficient chromatographic separations of potential interfering species and the careful selection of diagnostic ions for their analysis. The relative intensity variation of cationized species from sample-to-sample and from standards hinders quantitative analysis, thus entailing the use of integrated signals from selected non-interfering ions. Microalgae samples from a Mediterranean beach were analyzed during an Ostreopsis bloom in 2015. Putative palytoxin and ovatoxins were detected, being ovatoxin-a the dominant component. The highest concentration per cell almost coincided with maximum Ostreopsis abundances.

Novel Aspect:
The improved UHPLC-HRMS method evidences the presence of putative palytoxin and ovatoxins in the Ostreopsis bloom in a NW Mediterranean beach.

Muck, Alexander,
Waters Corp.

Contributing Authors:
Robert Birdsall, Waters Corp., Milford, MA, USA; Sean M. McCarthy, Waters Corp., Milford, MA, USA; Alain Beck Pierre-Fabre, Feigéres, France; John Gebler, Waters Corp, Milford, MA, USA; Weibin Chen, Waters Corp, Milford, MA, USA.

Title:
A sensitive two dimensional LC-MS approach for the characterization of free drug impurities in antibody-drug conjugates

Introduction:
Conjugation methods and stability studies associated with the production and shelf life of antibody-drug conjugates (ADCs) can result in free (non-conjugated) drug species. These can increase the risk to patients and reduce the efficacy of the ADCs. The recently reported multidimensional LC-UV methods had limited sensitivity. We report here a novel mass spectrometry coupled SEC-RPLC procedure.

Methods:
Solid phase extraction (SPE) was coupled to reversed-phase ultra performance liquid chromatography (UPRPLC) via programmed valve events for a specific and sensitive single quadrupole mass spectrometric (MS) analysis. Selected ion recordings were acquired in Masslynx in positive polarity and scan range of 30-1250 m/z, ESI voltage of 0.8kV and source temperature 400 °C.

Results:
The proposed method was evaluated using an antibody-fluorophore conjugate (AFC) as an ADC surrogate to brentuximab vedotin. The dynamic range for the single quadrupole was determined to span 2.5 orders of magnitude for the mal-linker-DSEA standard and 2.5 orders of magnitude for the NAc-linker-DSEA standard with LOQs of 1.35 and 1.65 pg on-column at SNR = 9.97, respectively. The current MS-based method was 150-fold more sensitive compared to previous SEC-RPLC/UV based methods. AFC samples were analyzed as received which was not possible with UV detection. Triplicates were ran with a blank prior to each sample run to assess carry-over. The concentration of the free-drug components were determined to be 7.19 ng/mL and 3.82 ng/mL for the NAc-linker-DSEA and mal-linker-DSEA components, respectively. Triplicate SIR spectrum overlays of the blank injections showed no observable carry-over.

Novel Aspect:
The SPE-RPLC/MS method provides specificity and control over each dimension with sensitivity at concentration levels not detectable by UV alone

Muddiman, David,
North Carolina State University

Title:
IR MALDESI: A Novel Molecular Microscopy Method

Introduction:
The intersection of mass spectrometry and imaging technology of biological specimens is an exciting and important avenue for unraveling physiological pathways involved in human health and disease and determining drug distributions. Evolving instrumentation, researching new ionization methods, and developing informatics will enable cellular spatial resolution of analyte classes.

Methods:
This presentation will outline our efforts in developing a novel molecular microscopy approach called infrared matrix-assisted laser desorption electrospray ionization (IR-MALDESI) mass spectrometry imaging (MSI). IR-MALDESI is a disruptive technology which we continue to drive and evolve its key advantages, including: minimal sample preparation and quantitative ablation of material.

Results:
This presentation will report on our recent innovations of the technology and how it can be used to solve a wide variety of biological problems. This includes coupling the imaging data with laser microdissection (LMD) proteomics data to investigate biological pathways, quantification of drugs in tissues, as well as a host of interesting attributes that IR-MALDESI when coupled to high resolving power high mass accuracy detection can provide. Finally, MSiREADER will be described as a vendor neutral, open-source software for the entire mass spectrometry imaging community.

Novel Aspect:
An innovative mass spectrometry imaging approach coupled with other OMICS information to understand biology

Mueller, Miriam,
Institute of Bioanalytical Chemistry

Contributing Authors:
Dietrich A. Volmer, Institute of Bioanalytical Chemistry, Saarland University, 66123 Saarbrücken, Germany.

Title:
Metabolic distributions of vitamin D metabolites in dried blood spots after chemical derivatization and LC-MS/MS

Introduction:
Most current assays for vitamin D are limited to determination of main metabolite 25(OH)D. No assay has been specifically developed to examine multiple vitamin D metabolites from dried blood spots (DBS), which are common for sampling blood from infants, but have also found widespread use as storage medium. We report the simultaneous profiling of wide range of vitamin D compounds from neonatal DBS.

Methods:
DBS punches 5 mm were extracted using H2O/ACN and isotope-labelled internal standards added to extracts without cleaning procedure.Compounds were derivatized by Diels-Alder reaction using a commercial Amplifex diene derivatization reagent to enhance ionization properties.Metabolites were separated by UHPLC-MS/MS using MRM. Method validation was carried out in accordance with FDA guidelines.

Results:
Presentation describes a proof-of-concept study for a LC-MS/MS method for quantification of multiple vitamin D metabolites from neonatal and adult dried blood spots. Metabolites were vitamin D3, 3-epi-vitamin D3, 25(OH)D3/2, 3-epi-25(OH)D3, 24,25(OH)2D3 and 1,25(OH)2D3/2.The implemented derivatization reaction provided significant improvements of ionization efficiency (~500) for the investigated metabolites under ESI conditions.For the assay, only a small volume of whole blood was required (4.8 µL) was sufficient from a 5mm DBS punch. Derivatized metabolites were successfully separated by UHPLC on a reversed-phase column.Limits of quantification were in the lower picomolar range, allowing quantification of metabolites at their physiological levels. DBS offered several advantages over serum including the minimally invasive nature of blood collection, stability of compounds and transport.

Novel Aspect:
Novel dried blood spot (DBS) assay readily demonstrated quantitative profiling of multiple vitamin D metabolites from neonatal samples.

Muntel, Jan,
Biognosys

Contributing Authors:

Oliver Bernhardt, Biognosys; Lukas Reiter, Biognosys; Tejas Gandhi Biognosys; Roland Bruderer, Biognosys.

Title:

High precision iRT retention time prediction in targeted analysis of data-independent acquisition: implications on identification and quantitation

Introduction:

DIA is a powerful approach for comprehensive and precise proteome quantification. The targeted analysis requires a spectral library comprising empirically determined fragment masses and predicted retention times (RTs). For RT prediction indexed RTs (iRTs) based on 11 peptides are crucial. We extended this set of peptides, analyzed the iRT precision and the influence of 9 experimental factors.

Methods:

To extend the iRT set, HeLa digests were analyzed by a DDA and DIA method on a Q Exactive MS. Database search was performed by MaxQuant and DIA data analysis in Spectronaut 8 (Biognosys). Based on the DIA data, precise iRT values were calculated. Afterwards mobile phase acid, gradient length, sample, loading, flow rate and column stationary phase, length, replicates and temperature were varied.

Results:

The usage of >20,000 HeLa peptides as high precision iRTs greatly improved the RT prediction. Compared to the classical iRT approach, the median delta iRT was 10 fold better resulting in ~20% more identification with higher detection reproducibility. To analyze the experimental influence on the iRT precision we varied 9 experimental factors and analyzed the iRT standard deviations (SD). The mobile phase acid had the largest influence (SD>2), followed by gradient length and the stationary phase. Only a small influence was noted for different samples (human cell lines), column replicates or sample loading (SD<0.5). In summary the usage of high precision iRTs significantly improved the data quality and we show that iRTs are generally transferable across experimental conditions. However, best results are achieved if library generation and DIA experiments are performed on the same system.

Novel Aspect:

Novel method for high precision prediction of RTs to significantly improve the quality of targeted DIA data analysis

21ST INTERNATIONAL MASS SPECTROMETRY CONFERENCE
Toronto, Ontario, Canada
August 20 – 26, 2016

Murray, Kermit,
Louisiana State University

Contributing Authors:
Kelin Wang, Louisiana State University; Fabrizio Donnarumma, Louisiana State University.

Title:
Infrared Laser Ablation Microdissection for Imaging Mass Spectrometry

Introduction:
Mass spectrometry imaging creates ions directly from tissue sections, which limits the ability to detect low abundant molecules and identify and quantify biomolecules in the sample. This project uses ambient infrared laser ablation microdissection for liquid chromatography tandem mass spectrometry in conjunction with matrix-assisted laser desorption ionization (MALDI) imaging mass spectrometry.

Methods:
Regions of interest are identified with tissue staining and MALDI imaging mass spectrometry using a commercial mass spectrometer. A 3 µm mid-infrared laser with 1 mJ pulse energy is used to ablate material from selected areas of tissue sections that are collected in a multiwell plate for separation by nano liquid chromatography tandem mass spectrometry.

Results:
Infrared laser ablation microdissection was demonstrated with nano electrospray tandem mass spectrometry analysis with an ion trap mass spectrometer. Rat brain tissue sections 10 to 50 µm thick were sectioned and mounted on a conductive microscope slide. Matrix was added by pneumatic spray and images were obtained with a commercial MALDI time-of-flight mass spectrometer. Regions of interest are identified using the heat-map mass image and laser ablation sample transfer was used to sample material from selected regions on the tissue. Either MALDI analyzed or consecutive tissue sections were used. Multi-compartment proteomic analysis was performed from a 0.25 square mm sampling area. Ongoing work is aimed at automation of the region selection and laser ablation sample transfer and developing workflows for quantitative imaging mass spectrometry.

Novel Aspect:
A combined proteomics method using MALDI imaging for region of interest selection and LC-MS/MS for identification and quantification.

Murray, Kermit,
Louisiana State University

Contributing Authors:
Fabrizio Donnarumma, Louisiana State University; Roshan Shetty, Anasys Instruments; Jonathan I Brauer Anasys Instruments; Fan Cao, Louisiana State University.

Title:
Atomic Force Microscope Tip Enhanced Laser Ablation Mass Spectrometry

Introduction:
The goal of this research is to bring large molecule mass spectrometry into the regime of single cell analysis. Atomic force microscope (AFM) tip-enhanced laser ablation is used to remove material from a sub-micrometer region of a cell or tissue sample for transportation to the ionization region and then to the mass spectrometer for direct ionization in an electrospray source.

Methods:
A pulsed visible wavelength laser is focused onto a metal-coated AFM tip for near-field ablation. Cell and tissue samples are deposited a glass cover slip mounted on a circular metallic disc and ablated from 500 nm and 1 μm diameter spots. The ablated material is sampled into a high-efficiency nanoelectrospray ion source where the ablated material is post-ionized for mass spectrometry

Results:
The AFM system has been used to ablate biomolecules from cell and tissue samples for direct analysis by off-line mass spectrometry. The system has been demonstrated for soft ablation of material for mass spectrometry analysis up to the size of small proteins and ablation of lipids from small regions of tissue samples. We are developing this method for on-line electrospray post-ionization mass spectrometry with an ultra-low flow electrospray merged ionization source to create a mass spectrometry system with sub-micrometer sampling resolution and mass range up to the size of small proteins.

Novel Aspect:
Ultra-low flow laser ablation electrospray post-ionization combined with sub-micrometer laser ablation sampling for ambient cell and tissue sampling.

Nagurnaja, Zoja,
University College London

Contributing Authors:
Flemming Hansen, Institute of Structural and Molecular Biology, University College London, UK; Konstantinos Thalassinos, Institute of Structural and Molecular Biology, University College London, UK.

Title:
Exploring the dynamics and function of the flexible domain of Human Histone Deacetylase 2 using mass spectrometry

Introduction:
We are using multiple mass spectrometry techniques to delineate the ensemble of conformations present in the sample and to probe the position of the unstructured region relative to the globular domains of HDAC2. This will help us understand how the function of the protein is regulated by the flexible C-terminal region and will guide future discovery of HDAC2-specific inhibitors.

Methods:
Native, ion mobility, and collision induced unfolding experiments were performed on a Synapt G1. Top down ETD experiments were performed on isolated peaks on a Synapt G2Si. Full-length protein was chemically-crosslinked using BS3 or DSS cross-linker and peptides were fractionated on an ÄKTAmicro system. Cross-links were analysed using xQuest and used for further modelling analysis.

Results:
Our native mass spectrometry data shows that HDAC2 exists in two different conformational charge state distributions: one is more native-like with a small number of peaks over a narrow range, and the other with more charges and wider charge state range. Upon proteolytic digestion of the full length protein, we see mass loss equal to the mass of the flexible domain, and the highly charged state distribution disappears from the native spectra. Interestingly, protein activity assay measurements of the HDAC2 showed that upon proteolytic digestion, the enzyme becomes approximately 3 times more active compared to the full-length. Our further cross-linking LC-MS experiments showed that the unstructured region of the protein has multiple contact sites with the globular area of the activity domain of HDAC2 leading to the zinc-binding pocket of the active site.

Novel Aspect:
We are using multiple mass spectrometry techniques to study the dynamics of the unstructured region in relation to the globular domain.

Nakamura, Takemichi,
RIKEN CSRS, Molecular Structure Characterization

Contributing Authors:
Yasuaki Suda, RIKEN, CSRS, Molecular Structure Characterization; Futoshi Sato, Nihon Waters K.K.

Title:
Utility of Energy-Resolved Tandem Mass Spectrometry in Identification of Isomeric Small Organic Molecules Including Diastereomers

Introduction:
Both MS data and retention indices are commonly used for identification of small molecules including isomeric metabolites. Some isomers can also be identified based on 70V EI spectra. However, differentiation of isomers solely based on collision-induced dissociation (CID) tandem-MS data is a challenge. We explore a strategy for isomer differentiation in future "no-chromatography" situations.

Methods:
Difficulty of isomer differentiation is attributable to the condition/platform-dependent variability of CID spectral patterns. To get around the problem, we established a standardized procedure to acquire energy-resolved (ER) MS2 data set for each compound by using a Synapt G2 Q-TOF instrument. The data for standards including common metabolites were evaluated for isomer-discriminating power.

Results:
Isomers often give similar CID-MS2 patterns as they share common substructures/structural features that would give same fragment ions. Although some differences in fragment-ion abundances are commonly observed, reliability of isomer differentiation would be compromised due to the condition-dependent spectral variability. In contrast, each ER-MS2 data set involves a series of variable spectral patterns. Hence if a "representative pattern" associated to each data set was defined appropriately, such pattern would be immune to changes in experimental conditions. As each fragmentation channel has structure-dependent characteristics including critical energy in each reaction pathway, breakdown diagrams associated to those channels would become characteristic for each isomer. A few examples including diastereomers have shown the utility of such ER-MS2 patterns in isomer differentiation.

Novel Aspect:
Energy-resolved MS2 data sets were shown to be useful for differentiation of isomeric small organic molecules including diastereomeric metabolites.

Nam, Myung Hee,
Korea Basic Science Instiyute

Contributing Authors:
Herym Lee, Korea Basic Science Institute; Myung Hee Nam, Korea Basic Science Institute; Jae-Gu Seo Cell Biotech Co; Joong-Su Lee, Cell Biotech Co; Joo-Hyun Shin, Cell Biotech Co.

Title:
Effect of probiotics on the lipid metabolism in a high fat diet-induced obese rat model

Introduction:
Obesity is a major public health concern. Although the etiology is multi-factorial, alterations in intestinal microbiota and immune system are commonly observed. The aim of the present study was to investigate the effect of probiotics feed with high fat diet on morphometric and metabolic parameters in a high-fat diet (HFD)-induced obese rat model

Methods:
Rats received either a normal diet (ND) or HFD with or without water containing each 6 kind of probiotics over 8 weeks. HFD-induced adiposity, inflammatory cytokines and chemokine in serum were analyzed. Non-targeted metabolite profiling was performed using UPLC-Q/TOF MS in serum of rats.

Results:
Among 6 kinds of probiotics administrated with HFD, three kinds were effectively decreased HFD-induced body weight and modulated morphometric and metabolic parameters. Metabolic profiling in serum of 5 groups: ND, HFD, HFD + each 3kinds of probiotics revealed that HFD diet induced the change of serum metabolites including bile acids, saturated and unsaturated fatty acid containing triacylglycerol (TAG). The HFD induced change of these metabolites in serum was attenuated by co-administration of probiotics. In conclusion, administration of probiotics appeared to have potential to alleviate HDF-induced obesity through modulation of immune response and metabolism of the host.

Novel Aspect:
Using LC-MS based metabolomics approach; effect of probiotics on the HFD induced lipid metabolism was elucidated.

Nam, Myung Hee,
Korea Basic Science Instiyute

Contributing Authors:
Hyerim Lee, Korea Basic Science Institute; Hwang-Bae Sohn, Highland Agriculture Research Institute; Kyoungwon Cho Korea Basic Science Institute; Myung Hee Nam, Korea Basic Science Institute; Kwang-Soo Cho, Highland Agriculture Research Institute.

Title:
Integrated approach of metabolomics and transcriptomics to Reveals Regulatory network of Potato Pigmentation

Introduction:
Study on regulatory networks of anthocyanin pathways is important for the elucidation of regulatory genes involved in plant pigmentation and plant metabolism involved in tolerance to biotic and abiotic stresses. An integration of transcript and metabolite datasets can be a useful approach for the for the elucidation of regulatory networks of anthocyanin pathways related to potato pigmentation.

Methods:
An integrated analysis of the transcriptome and metabolome performed in white, light-red and dark purple colored potato cultivars: Anthocyanin derivatives were selected using specific features of their MS2 spectra. Pearson correlation coefficients were calculated for metabolome and transcriptome data integration. Metabolome and transcriptome relationships were visualized using Cytoscape.

Results:
The compositions of flavonoids on the downstream of phenylpropanoid-pathway were highly different between light-red colored Hongyoung and dark purple Jayoung, indicating that these metabolites might play a crucial role in determining the pigmentation in potato. Apigenin-, kaempferol-, dihydrokaempferol-, and pelargonidin-derivatives were shown to be most abundant in the Hongyoung, whereas. Peonidin-, petunidin-, delphinidin-, and a malvidin-derivative were shown to be the more abundant in the Jayoung. Correlation tests showed 823 strong correlations between 22 anthocyanin compounds and 119 transcripts categorized into flavonoid metabolism, hormones, transcription regulation and signaling. The relative changes of genes and metabolites between Hongyoung and Jayoung in the correlation network showed a regulatory connection involved in the differential pigmentation of potato cultivars.

Novel Aspect:
Using an integrated approach, new insights into the regulatory mechanism underlying the differential accumulation of anthocyanin were obtained.

21ST INTERNATIONAL MASS SPECTROMETRY CONFERENCE
Toronto, Ontario, Canada
August 20 – 26, 2016

Nasif, Ammar,
University of Southampton

Contributing Authors:
Ammar Nasif, Chemistry, University of Southampton Southampton UK; Christianne Wicking, BP Pangbourne Technology Centre Pangbourne UK; Mike Hodges BP Sunbury ICBT Sunbury UK; John Couves, BP Sunbury ICBT Sunbury UK; Huang Zeng, BP Houston Houston TX; John G Langle, Chemistry, University of Southampton Southampton UK.

Title:
Structural elucidation of N-containing compounds in crude oil using positive ion ESI FT-ICR MS/MS

Introduction:
Crude oil is a complex mixture of hydrocarbon and heteroatom containing compounds. Elucidation of the structure of compounds in crude oil can have applications in the development and optimisation of crude oil processing, such as designing of more effective catalysts. A tandem MS approach has been developed where fragment ions related to different precursor ions of the CxHyN1 were investigated.

Methods:
Data were acquired using a Bruker 4.7 T FT-ICR MS. Isolation of the precursor ions was undertaken in the quadrupole. Different collision energies were used at 40 and 50 V. Data acquisitions were performed using solariXcontrol version 1.5.0.and data processing using Bruker DataAnalysis. Sample: Crude oil solution was prepared at 0.5 mg/mL with 0.1% formic acid in toluene:methanol ratio of 6:4.

Results:
Mixed precursor ions were isolated at m/z 340 ± 0.5 that are related to different classes of compounds such as CxHyN1 and CxHyN1S1. Fragments of precursor ions for the CxHyN1 class at DBE 7 and 14 are discussed. Acquiring the data at average spectra (AS) 800 detect fragments that are not detected when AS 300 is used but with increase in the analysis time from 15 min to 40 min respectively. A reduction of the analysis time to 4.5 min can be achieved through increasing the ion accumulation time with AS 40 without sacrificing ion intensity and with sufficient resolution for elemental assignment of fragment ions. The developed method shows that fragments related to N expulsion from the aromatic ring by losing HCN and acetonitrile are observed at collision energy 50 V. This supports the postulation that the CxHyN1 DBE 7 and 14 precursor ions identified using positive ion ESI are pyridinic.

Novel Aspect:
Positive ion ESI FT-ICR MS/MS method has been developed for structural elucidation of different classes of compounds in crude oils.

Navakauskiene, Ruta,
Institute of Biochemistry, Vilnius University

Contributing Authors:
Artūras Serackis, Electronic Systems Department, Faculty of Electronics, Vilnius Gediminas Technical University, LT-03227 Vilnius, Lithuania; Dovilė Kurpytė, Electronic Systems Department, Faculty of Electronics, Vilnius Gediminas Technical University, LT-03227 Vilnius, Lithuania; Dalius Matuzevičius Electronic Systems Department, Faculty of Electronics, Vilnius Gediminas Technical University, LT-03227 Vilnius, Lithuania; Dalius Navakauskas, Electronic Systems Department, Faculty of Electronics, Vilnius Gediminas Technical University, LT-03227 Vilnius, Lithuania.

Title:
The key point extraction technique for amniotic fluid proteome characterization

Introduction:
The protein composition and their expression levels in amniotic fluid (AF) vary throughout gestation. Thus, the proteome characterization in AF of normal and abnormal fetuses is important for prenatal diagnosis. In this study, proteins isolated from AF were subjected to 2DE fractionation, coupled with mass spectrometry (MS) analysis.

Methods:
The preliminary matching of warping key points in AF protein 2DE maps was not satisfactory. Thus for the analysis of protein expression level changes the key point extraction technique is developed. Accordingly, candidate protein spot detection, spot matching by the asymmetrical model, and clustering by the neighboring distances and model parameters, was done.

Results:
Protein spot detection and segmentation algorithm is based on Watershed transformation combined with symmetrical feature detection. After pairing and segmentation of the protein spots in 2DE gel images, a ratio of spots normalized quantities among the image groups were calculated to estimate changes of spot abundance between proteins identified in normal and pathological AF groups. The majority of identified proteins of AF belong to cellular signaling – 24%, pregnancy and embryo development – 23%, immune response and inflammation – 16%, transcription – 14%, structural – 12% and other proteins. Computer assisted analysis empowers to evaluate and describe the changes in protein levels typical for AF of normal fetuses and fetuses with chromosomal and/or structural abnormalities. This research was funded by a grant (No. MIP-083/2015) from the Research Council of Lithuania.

Novel Aspect:
The key point extraction technique reduces the protein spots matching error. The spot modeling stipulates prediction of the protein quantity changes.

Navakauskiene, Ruta,
Institute of Biochemistry, Vilnius University

Contributing Authors:
Jurate Savickiene, Department of Molecular Cell Biology, Institute of Biochemistry, Vilnius University, Sauletekio av. 7, LT-10224 Vilnius, Lithuania; Grazina Treigyte, Department of Molecular Cell Biology, Institute of Biochemistry, Vilnius University, Sauletekio av. 7, LT-10224 Vilnius, Lithuania; Sandra Baronaite Department of Molecular Cell Biology, Institute of Biochemistry, Vilnius University, Sauletekio av. 7, LT-10224 Vilnius, Lithuania; Dalius Matuzevicius, Electronic Systems Department, Faculty of Electronics, Vilnius Gediminas Technical University, LT-03227 Vilnius, Lithuania; Dalius Navakauskas, Electronic Systems Department, Faculty of Electronics, Vilnius Gediminas Technical University, LT-03227 Vilnius, Lithuania.

Title:
Mass spectrometry and computational approach for proteomic characterization of amniotic fluid in polyhydramnios pregnancies

Introduction:
Mass spectrometry-based proteomics has become a valued tool for conducting comprehensive analyses in amniotic fluid samples with pathologies. Our research interest is in finding and characterizing proteins related to polyhydramnios vs normal pregnancy.

Methods:
Proteomic analysis was performed by using total proteins isolated from fresh amniotic fluid samples of normal and pathological pregnancies. Proteins isolated from amniotic fluid were analyzed by 2-DE (pI 3-11 and pI 4-7), coupled with MALDI-TOF-MS and ESI-TOF-MS analysis. Changes in identified protein expression levels defined in different status of pregnancy were analyzed by bioinformatics tools.

Results:
We identified and characterized highly expressed proteins that are mostly associated with heart function and circulatory system and other proteins associated with abnormalities in prenatal medicine. Proteins identified in polyhydramnios are related to cardiovascular, anemia, growth restriction, metabolic disorders (i.e. serotransferrin, prothrombin, transthyretin, zinc-alpha-2-glycprotein, haptoglobin, kininogen-1, hemopexin, clusterin, lumican, afamin, gelsolin etc.). Computational approach empowers to compare protein maps (fractionated in pI 3-11 and pI 4-7 range) describing proteins characteristic for normal and pathological pregnancy.

Novel Aspect:
Proteome analysis of amniotic fluid creates the possibility to identify protein biomarkers important for prenatal tests diagnosing fetal development.

Neuweger, Heiko,
Bruker Daltonik GmbH

Contributing Authors:
Emmanuel Gaquerel, Centre for Organismal Studies Heidelberg, University of Heidelberg, Germany; Ian T. Baldwin, Max Planck Institute for Chemical Ecology, Department for Molecular Ecology, Jena, Germany; Aiko Barsch Bruker Daltonics, Bremen, Germany; Sven Heiling, Max Planck Institute for Chemical Ecology, Department for Molecular Ecology, Jena, Germany.

Title:
Metabolomics to study the "war" between plants and insects - Systemic defense induction and post-ingestive rearrangement of plant toxins in insects.

Introduction:
17-hydroxygeranyllinallool diterpene glycosides (HGL-DTGs) are secondary metabolites in N. attenuata defending tissue against the herbivore M. sexta. Gene manipulation followed by software based metabolite identification enabled characterization of glycosylation in the HGL-DTG pathway in tobacco and clarified the impact on chemical rearrangements with other pathways when ingested by M. sexta.

Methods:
Leaf tissue from genetically silenced lines of N. attenuata and frass of 2nd instar M. sexta larvae were extracted with 80% methanol. Extracts were separated with an RSLC (Thermo Fisher). MS analysis was performed by Q-TOF MS (impact II, Bruker). MetaboScape was applied for identification of novel secondary metabolites, automatic identification of target compounds, statistics, and pathway mapping

Results:
We characterized silenced rhamnosyl and glucosyltransferase lines IRrt1, IRgt1 and IRgt2 impaired in the glycosylation of HGL-DTGs. Metabolite profiling rapidly determined a complete loss of rhamnosylated HGL-DTGs in IRrt1 and the appearance of 12 novel intermediate products in IRgt1 and IRgt2. Feeding assays and screening excreta of M. sexta elucidated the defensive capabilities of HGL-DTGs. 55 novel HGL-DTG-like compounds with different decorations (like caffeoyl moieties) and further deglycosylation products were found. HGL-DTGs with caffeoyl moieties could only be detected in frass of caterpillars directly feeding on N. attenuata plants and not in frass of caterpillars feeding on detached leaves. This indicates that other systemically-induced defensive pathways play a crucial role in the process of detoxification or that HGL-DTGs are used to detoxify potential poisonous metabolites.

Novel Aspect:
MS based metabolite profiling strategies elucidate post-ingestive rearrangement detoxification mechanisms of plant secondary metabolites in insects.

Neuweger, Heiko,
Bruker Daltonik GmbH

Contributing Authors:

Matthias Szesny, Bruker Daltonik GmbH, Bremen, Germany; Aiko Barsch, Bruker Daltonik GmbH, Bremen, Germany; Stephan Mävers Bruker Daltonik GmbH, Bremen, Germany; Marcus Persicke, Bielefeld University, Bielefeld, Germany; Nikolas Kessler, Bruker Daltonik GmbH, Bremen, Germany; Jörn Kalinowski, Bielefeld University, Bielefeld, Germany; Frederik Walter, Centre for Biomolecular Sciences, The University of Nottingham, Nottingham, UK.

Title:

Boosting compound identification confidence in Metabolomics by exploiting all available Q-TOF high resolution accurate mass spectral information.

Introduction:

Confident compound identification is still one of the major bottlenecks in metabolomics. Ongoing efforts to refine the definitions and levels of metabolite identifications that were first proposed by the MSI initiative stress that higher levels of identification confidence can be reached by joining accurate measurement technology, orthogonal molecular features, and sophisticated software tools.

Methods:

Methanolic extracts of three mutant and one wildtype C. glutamicum strain were diluted and separated using an UltiMate RSLC (Thermo Scientific). The LC was coupled to an impact II QToF-MS (Bruker). HRAM MS & MS/MS spectra were acquired by data dependent MS/MS. MetaboScape was used for automatic identification of known target compounds, identification of unknowns, statistics, and pathway mapping.

Results:

We could highly improve the quality of compound identifications in a study investigating the arginine biosynthesis in C. glutamicum conducted by LC-QTOF non-targeted metabolomics. The developed tools were used to create annotations throughout increasing confidence levels: First, for all compound spectra molecular formulas were generated based on accurate masses, true isotopic patterns, and in-source fragmentation patterns, applying metabolomics-tailored rules and filters. Afterwards, public chemical databases were queried to find structural candidates for the generated molecular formulas of interesting features. Then, a customized analyte target list was applied to additionally exploit retention times of expected features. Lastly, MS/MS spectral library comparisons and in-silico fragmentations enabled to create and verify identifications based on MS/MS fragmentation patterns.

Novel Aspect:

Strategy to achieve highest confidence in compound identifications based on LC-HRAM-MS/MS spectral information using an integrated "turnkey" solution.

newman, carla,
The University of Nottingham

Contributing Authors:
Rasmus Havelund, National Physical laboratory (UK); Melissa Passarelli, National Physical laboratory (UK); Ian Gilmore National Physical laboratory (UK); Peter Marshall, GlaxoSmithKline (UK); Andy West, GlaxoSmithKline (UK); Morgan Alexander, The University of Nottingham.

Title:
Probing compound intracellular distribution using ToF-SIMS

Introduction:
Time-of-flight Secondary ion mass spectrometry (ToF-SIMS) allows imaging of unlabelled pharmaceutical compounds in a single cell using physiological doses. As many targets are intracellular the ability to visualize compounds inside the cell can offer valuable insight on the compound behavior both on and off-target, this information is critical for successful drug discovery.

Methods:
This work comprises a number of experiments in which we have identified drug molecules inside biological matrices such as mammalian cells and bacteria. We show evidence of drugs inside 5 different types of mammalian cells and a bacterial tuberculosis model using TOF-SIMS (TOF-SIMS IV, ION-TOF).

Results:
The cell lines studied were chosen due to their relevance either to the pharmaceutical industry or the disease model. The mycobacterium model was selected as tuberculosis is still a huge healthcare problem, as 1/3 or the work is thought to be infected and antibiotic resistant strains are emerging in every WHO (World Health Organization) surveyed country. Depending on the analyte the analyses were performed in either high mass or high lateral resolution mode. Data will be shown from 3D experiments, which were acquired in non-interlaced mode to generate 2D image representations of 3D datasets.

Novel Aspect:
The work presented shows the potential applications of ToF-SIMS as complementary technique to other imaging techniques in drug discovery and biologica

Nicol, Gordon,
MilliporeSigma

Contributing Authors:
Kevin Ray, MilliporeSigma; Amber Henry, MilliporeSigma; Brian Gau MilliporeSigma.

Title:
A comparison of three orthogonal methods of glycosylation analysis

Introduction:
The characterization of glycoproteins is critical to therapeutic biomolecule control and to glycomics-driven biomarker discovery. For monoclonal antibodies, assay by released glycan or intact mass may be sufficient, as there is only one site for glycosylation. Many other glycoproteins have multiple glycosylation sites, each with micro-heterogeneity, and must be assayed at the glycopeptide level.

Methods:
Released glycans by HILIC-FLR-MS analysis, using a procainamide chromophore, is used for quantitative analysis. Intact mass of the protein or antibody fragment, using SEC-MS, is used for high throughput screening of the major glycans. Glycopeptide analysis using HILIC-MS is used to assess micro-heterogeneity of multiple glycan sites. These three methods of glycan analysis will be compared.

Results:
The ability to separate glycopeptides from non-glycopeptides with HILIC chromatography simplifies their analysis. Glycopeptide analysis enabled the profiling of individual glycosylation sites, which was essential for the analysis of complex glycoproteins and fusion proteins used in biotherapeutics. We show that the glycoform profile from the glycopeptide and the intact mass workflows correlate well to the released glycan profiles of the same molecules. The orthogonality of the methods provides increased confidence in the validity of glycosylation analyses. Plots comparing glycan profiles from the released glycan workflow to both the intact and glycopeptide workflows show correlation greater than 0.99 for several glycoproteins, such as ribonuclease and a monoclonal antibody. More complex biotherapeutics, such as Enbrel and Erbitux, are currently being compared.

Novel Aspect:
The quantitative comparison of three orthogonal methods of analysis of the glycan profile for glycoproteins and biotherapeutics.

Nikolaev, Evgeny,
Institute of Energy Problems of Chemical Physics Russian Acad. Sciences

Contributing Authors:

Alexander Potapov, Burdenko Neurosurgical Institute, Moscow, Russian Federation; Evgeny Kukaev, Moscow Institute of Physics and Technology, Moscow, Russia; Nikita Levin Moscow Institute of Physics and Technology, Moscow, Russia; Igor Popov, Moscow Institute of Physics and Technology, Moscow, Russia; Yury Kostyukevic, Skolkovo Institute of Technology, Skolkovo, Moscow region, Russia; Alexey Kononikhin, Emanuel Institute of Biochemical Physics, Moscow, Russia; Evgeny Nikolaev, Skolkovo Institute of Technology, Skolkovo, Moscow region, Russia; Evgeny Zhvansky, Moscow Institute of Physics and Technology, Moscow, Russia; Vsevolod Shurkhay, Burdenko Neurosurgical Institute, Moscow, Russian Federation; Maria Indeykina, Emanuel Institute of Biochemical Physics, Moscow, Russia.

Title:

Ambient MS based analytical platform for rapid identification of brain cancer tumor tissues

Introduction:

Rapid and reliable tumor tissue identification during surgical operations is a challenging problem. Especially in neurosurgery, where the tissue is formlessness and precision of tumor/normal tissue border determination is extremely important. Ambient ionization mass spectrometry based on situ fingerprinting of tissue is an attractive solution.

Methods:

A special spray-from-tissue ambient ionization method was developed, where liquid extraction is immediately followed by ionization. 200 tumor samples were analized. For each sample, mass spectra were measured in both ICR and in ion trap mode. MS/MS was used for deeper analysis and identification. Final validation of database profiles was performed using ultrahigh resolution FTICR.

Results:

It was demonstrated that MS profile analysis of brain tissue samples gives sufficient information for the identification of tissue and the type of pathology. One of the most important methodological results of the database accumulation stage of the project is the conclusion that in the boundary regions of each tumor the MS profiles can be represented as a linear combination of two and more profiles from different types of cells (necrotic tissue with necrotized vessels, necrotic tissue with tumor stain, tumor with necrosis (with a predominance of tumor tissue), necrotized tumor (with a predominance of necrotic masses), parts of tumor cells). In our work we demonstrate the possibility to use high and ultrahigh resolution data (accurate mass and fine isotopic structure of mass spectra) to identify lipids, metabolites and other compounds giving molecular signatures of the whole tissue.

Novel Aspect:

Novel analytical MS-based platform for brain tissue classification validated by ultrahigh resolution MS data.

Nikolaev, Evgeny,
Institute of Energy Problems of Chemical Physics Russian Acad. Sciences

Contributing Authors:
Gökhan Baykut, Bruker Daltonik GmbH, Bremen, Germany; Igor Popov, Moscow Institute of Physics and Technology, Moscow, Russia; Oleg Kharybin Skolkovo Institute of Technology, Moscow, Russia; Roland Jertz, Bruker Daltonik GmbH, Bremen, Germany; Gleb Vladimirov, Skolkovo Institute of Technology, Moscow, Russia.

Title:
Performance limits in ultrahigh resolution FT ICR mass spectrometry and the main limiting factors.

Introduction:
The implementation of the new technologies in FT ICR MS such as dynamically harmonized cell caused a jump in performances of this type of MS. Resolving power of 10 million is now routinely available even on low tesla instruments which is not reachable on any other type of MS. Deaper undestanding of the physical nature of performance limiting factors is needed for futher progress.

Methods:
Dynamically harmonized cell was used for the measurement of close doublet mass spectra (6 mDa m/z difference) on 7 tesla FT ICR mass spectrometer with 107 resolving power at different number of ions in the cell (until peak coalescence). Supercomputer simulations based on Particle In Cell algorithm has been performed at close to the real experiment conditions to explain experimental observations.

Results:
Mass spectra of the isotopically labeled peptides having the same nominal masses with equal concentration in the mixture were measured as a function of the number of ions in the cell N in relatively large dynamic range. It was found that doublet peak ratio (intensity of the smaller m/z to larger) is significantly increasing with N. More than 1ppm mass measurement error was observed at large number of ions. Supercomputer simulations are showing that the reason for such deviations from ideality is the Coulomb interaction of ion clouds in the close doublets. At large N this interaction causes the formation of comets in small m/z ion cloud similar to what was observed in not harmonized cells. This phenomenon has a general character and could limit further increase of FT ICR MS performances of low tesla instruments even with harmonized cell in the absence of strict control of ion population.

Novel Aspect:
Explanation of the physical effects, which limit ultrahigh resolution FT ICR performance, was given.

Noestheden, Matthew,
SupraTHC Services Inc

Contributing Authors:
Simon Roberts, SCIEX; Chunyan Hao, Ontario Ministry of the Environment and Climate Change.

Title:
Nitenpyram Degradation in Finished Drinking Water

Introduction:
Neonicotinoid pesticides and their metabolites have been indicated as contributing factors in the decline of honey bee colonies. A thorough understanding of neonicotinoid toxicity requires knowledge of their metabolites and environmental breakdown products. This work investigated the rapid degradation of the neonicotinoid nitenpyram in finished drinking water.

Methods:
Nitenpyram reaction products were characterized using liquid chromatography/quadrupole time-of-flight mass spectrometry (LC/QTOFMS). A software algorithm that compared degraded and control samples was utilized to facilitate efficient data reduction. Fragmentation pathways for six reaction products and nitenpyram were proposed using predictive software and manual product ion analysis.

Results:
Similarities between the identified nitenpyram reaction products and imidacloprid metabolites highlight the importance of this study, as the toxicity of neonicotinoids to pollinators has been linked to their metabolites. Based on the proposed reaction mechanisms, the identified nitenpyram reaction products in finished drinking water could also be present in the environment and water treatment facilities. As such, identifying these degradation products will aid in evaluating the environmental impact of neonicotinoid pesticides.

Novel Aspect:
First report and in-depth characterization of nitenpyram reaction products in drinking water.

Noestheden, Matthew,
SupraTHC Services Inc

Title:
Analysis of Cannabinoids by uHPLC-MS/MS: High Flow-Rates for Increased Throughput

Introduction:
Deficiencies in existing cannabis potency methods result in low throughput and/or an incomplete complement of phytocannabinoids. This study addressed these analytical deficiencies by leveraging uHPLC separations for the development of a fast quantitative method for the analysis of 11 phytocannabinoids cannabinoids.

Methods:
The merits of superficially porous versus < 2 µm fully porous particles, as well as the selectivity and retentivity differences that exist between seemingly similar phases for the selected phytocannabinoids were explored. Tandem mass spectrometry (MS/MS) was utilized for analyte detection. Method performance was demonstrated in a variety of cannabis extracts.

Results:
The experimental results showed that a 2.6 µm superficially porous particle yielded the lowest plate heights, suggesting that theoretically this would be the best column for this application. However, an evaluation of the column behavior at higher flow-rates suggested that with similar throughput potential a 1.3 µm superficially porous particle was more desirable due to the reduced solvent consumption required during routine analyses. Ion suppression using electrospray ionization at the empirically chosen optimal uHPLC conditions demonstrated that, even though suppression exists due to the high flow-rates, the analytical method was still acceptable due to the high endogenous levels of the studied phytocannabinoids in cannabis extracts.

Novel Aspect:
1) uHPLC-MS/MS analysis of phytocannabinoids. 2) Ion suppression at very high flow-rates using ESI. 3) Optimal vs acceptable column performance.

Novak, Petr,
Institute of Microbiology

Contributing Authors:
Lukas Slavata, Institute of Microbiology, Prague, Czech Republic; Alan Kadek, Institute of Microbiology, Prague, Czech Republic; Hynek Mrazek Institute of Microbiology, Prague, Czech Republic; Michal Rosulek, Institute of Microbiology, Prague, Czech Republic; Daniel Kavan, Institute of Microbiology, Prague, Czech Republic; Petr Man, Institute of Microbiology, Prague, Czech Republic.

Title:
Structural mass spectrometry: Novel tool to study transcription factor/DNA interaction.

Introduction:
Transcription factors (TF) regulate gene expression through interactions with DNA. Our understanding of this regulation depends on the knowledge of the TF-DNA complex structure. Forkhead-box transcription factor (FOXO4) and DAF-16 DNA native response element served as a model complex. A palette of MS based approaches were applied to investigate the FOXO4/DNA complex structure in solution.

Methods:
Construct covering residues 82-207 of FOXO4 was expressed in E. coli and oligonucleotide was synthesized double stranded nature. Hydrogen/deuterium exchange was followed for FOXO4 alone or in 1:1 molar mixture with the DAF-16. Chemical cross-linking included both quantitative protein-protein (amino reactive reagents) and protein-DNA cross-linking (trans-Platinum(II)diammine dichloride).

Results:
H/D exchange results were partially in agreement with the known high resolution structures obtained on free (NMR) and DNA bound (X-ray) FOXO4. However the most significant changes (stabilization of the helix H1) upon interaction with DNA lacked any explanation from the known structures. On the other hand these results were in excellent agreement with the molecular dynamic simulations. Protein-protein chemical cross-linking provided 17 cross-links showing marked quantitative differences in several regions, especially in the unstructured part where the data span further beyond the known high-resolution structures. Initial experiments with protein-DNA cross-linking by trans-platinum were performed. These data clarified the interaction interface region and promises a good start point for developing new in vivo protein-DNA interaction identification approach.

Novel Aspect:
Structural mass spectrometry sheds a light on dynamics of transcription factor/DNA interaction.

Novak, Ondrej,
Laboratory of Growth Regulators, IEB & UP Olomouc

Contributing Authors:
Miroslav Strnad, Laboratory of Growth Regulators, Centre of Region Haná for Biotechnological and Agricultural Research, Palacký University & Institute of Experimental Botany AS CR, Olomouc, Czech Republic; Jan Buček, Department of Chemical Biology and Genetics, Centre of Region Haná for Biotechnological and Agricultural Research, Palacký University & Institute of Experimental Botany AS CR, Olomouc, Czech Republic; Jana Oklestkova Laboratory of Growth Regulators, Centre of Region Haná for Biotechnological and Agricultural Research, Palacký University & Institute of Experimental Botany AS CR, Olomouc, Czech Republic; Kristýna Pospíšková, Regional Centre of Advanced Technologies and Materials, Palacký University, Olomouc, Czech Republic; Radek Zbořil, Regional Centre of Advanced Technologies and Materials, Palacký University, Olomouc, Czech Republic; Lenka Plačková, Laboratory of Growth Regulators, Centre of Region Haná for Biotechnological and Agricultural Research, Palacký University & Institute of Experimental Botany AS CR, Olomouc, Czech Republic; Karel Doležal, Department of Chemical Biology and Genetics, Centre of Region Haná for Biotechnological and Agricultural Research, Palacký University & Institute of Experimental Botany AS CR, Olomouc, Czech Republic; Marek Zatloukal, Department of Chemical Biology and Genetics, Centre of Region Haná for Biotechnological and Agricultural Research, Palacký University & Institute of Experimental Botany AS CR, Olomouc, Czech Republic; Ivo Šafařík, Regional Centre of Advanced Technologies and Materials, Palacký University, Olomouc, Czech Republic; Kateřina Poláková, Regional Centre of Advanced Technologies and Materials, Palacký University, Olomouc, Czech Republic.

Title:
Microscale isolation of cytokinins by magnetic microparticle-based immunopurification

Introduction:
Cytokinins (CK), pivotal plant hormones, play crucial roles in plant growth and development. Their isolation and quantification are challenging tasks because of extremely low CK levels in plant tissues (pmol/g fresh weight). Herein, we have developed a new microscale magnetic immunoaffinity-based method for selective one-step isolation of CKs from minute amounts of plant tissue (less than 0.1 mg).

Methods:
The immunoaffinity extraction based on group-specific monoclonal anticytokinin antibodies immobilized onto four different microparticles (magnetic cellulose beads and magnetic chitosan) was combined with a highly-sensitive ultra-fast mass spectrometry using multiple reaction monitoring (MRM) modes to profile a wide range of naturally occurring cytokinins (bases, ribosides and N9-glucosides).

Results:
The capacity of immunosorbent and the effect of a complex plant matrix on the yield of the rapid one-step purification were tested using a wide dynamic range of CK concentrations. The total recovery range of a new microscale isolation procedure was found to be 30–80%, depending on individual CKs. These results nicely demonstrate the microscale immunopurification step as a powerful one-step high-throughput method, which increases the selectivity and sensitivity of subsequent ultra-high performance liquid chromatography-tandem mass spectrometry analysis. Over 4.0 min, an ultrafast separation of isoprenoid and aromatic cytokinins with the method detection limits close to 1 attomole was obtained. Finally, we show the versatility of our method in CK quantification from 10-days-old Arabidopsis thaliana seedlings to just 0.5 mm of the apical part of the Arabidopsis roots.

Novel Aspect:

The magnetic immunoaffinity separation is a fast and simple procedure markedly improving miniaturized tissue/cell-specific phytohormone profiling.

Novak, Petr,
Institute of Microbiology

Contributing Authors:
Jiri Hausner, Institute of Microbiology, Prague, Czech Republic; Oldrich Benada, Institute of Microbiology, Prague, Czech Republic; Petr Pompach Institute of Microbiology, Prague, Czech Republic; Viktor Ruzicka, Biovendor, a.s., Brno, Czech Republic; Michael Volny, AffiPro, s.r.o., Mratin, Czech Republic; Petra Darebna, Institute of Microbiology, Prague, Czech Republic; Michal Rosulek, Institute of Microbiology, Prague, Czech Republic.

Title:
Performance and Bioanalytical Applicability of MALDI-compatible Protein Chips Prepared by Ambient Ion Soft Landing

Introduction:
The presented technology allows preparation of MALDI compatible protein chips by ambient ion landing. The electrosprayed proteins are immobilized on conductive surfaces. The protein chips prepared by protein ambient ion landing are used for different types of bioanalytical assays. The reaction takes place directly on the protein chip and is followed by in-situ analysis by MALDI mass spectrometry.

Methods:
Conductive and semiconductive surfaces were modified with proteases, lectins and antibodies by ambient ion landing utilizing a nanospray to produce protein ions and heated tube to dry them. The beam of charged species was deposited on the vertically mounted ITO slides that were kept at high voltage of the opposite polarity with respect to the spray.

Results:
The ion landing of electrosprayed protein molecules was performed under atmospheric pressure by automated ion landing apparatus on dry metal and metal oxide surfaces. Compare to other materials, non-reactive surfaces suffer minimal nonspecific interactions with chemical species in the investigated sample and are thus an ideal substrate for selective protein chips. Our results show possibility to in-situ digest proteins very effectively and fast with 100% sequence coverage and with many overlapping peptides. Identification of several glycopeptides shows their efficient enrichment on lectin modified MALDI protein chips. We further demonstrate that utilization of antibody modified MALDI chips opens the door to immunoMALDI assays that allow sensitive determination of antigens in serum and leads to fast clinical assays that can be easily automated for high-throughput screening.

Novel Aspect:
Ambient ion landing represents an efficient technology for preparation of mass spectrometry compatible protein chip arrays.

Ogata, Kosuke,
Graduate School of Pharmaceutical Sciences, Kyoto University

Contributing Authors:

Kosuke Ogata, Graduate School of Pharmaceutical Sciences, Kyoto University, Kyoto, Japan; Chia-Feng Tsai, Graduate School of Pharmaceutical Sciences, Kyoto University, Kyoto, Japan; Masaki Wakabayashi Graduate School of Pharmaceutical Sciences, Kyoto University, Kyoto, Japan; Naoyuki Sugiyama, Graduate School of Pharmaceutical Sciences, Kyoto University, Kyoto, Japan; Yasushi Ishihama, Graduate School of Pharmaceutical Sciences, Kyoto University, Kyoto, Japan.

Title:

High sensitivity and high throughput phosphoproteomics with micro-scale TMT labeling.

Introduction:

Current phosphoproteomics allows identifying thousands of phosphopeptides in single analysis. However, it is still hard to achieve deep and high throughput quantitation in phosphoproteomics for a small sample amount. In this study, we developed a micro-scale isotope tagging coupled with high resolution LCMS using a meter-long monolithic silica column for deep and high throughput quantitation.

Methods:

Tryptic digests from HeLa lysates were used for phosphopeptide enrichment by titanium dioxide chromatography with lactic acid. Tandem Mass Tags (10-plex TMT) were employed for the enriched phosphopeptides according to a new micro-scale labeling protocol and were measured by one-shot LC-MS/MS using a 2-meter long C18 monolithic silica capillary column to identify and quantify phosphopeptides.

Results:

We found that the recovery in TiO2-based phosphopeptide enrichment for TMT-labeled phosphopeptides was 4.6-fold lower than that for unlabeled phosphopeptides on average, showing that TMT labeling should be conducted after TiO2 chromatography. We screened chromatographic particles for solid phase TMT reaction and found that hydrophobic polymer beads together with ion pair reagents at neutral pH gave the best results among the published protocols in terms of the number of identified phosphopeptides, labeling efficiency and the recovery. Compared with the in-solution TMT protocol given by the supplier, the new protocol gave 5-times more identified phosphopeptides with 8.7-fold higher reporter ion intensity on average. This workflow is useful for deep and high-throughput quantitative phosphoproteomics with limited sample amounts.

Novel Aspect:

Multiplexed phosphoproteomics with a novel micro-scale labeling protocol for deep and high throughput analysis.

21ST INTERNATIONAL MASS SPECTROMETRY CONFERENCE
Toronto, Ontario, Canada
August 20 – 26, 2016

Ogrinc Potocnik, Nina,
M4I Institute, Maastricht University

Contributing Authors:
Julie Hamaide, Bio-Imaging Lab, University of Antwerp, Wilrijk, Belgium; Annemie Van Der Linden, Bio-Imaging Lab, University of Antwerp, Wilrijk, Belgium; Wesley Suntjens M4I Institute, Maastricht University, Maastricht, The Netherlands; Ron M.A. Heeren, M4I Institute, Maastricht University, Maastricht, The Netherlands.

Title:
A molecular insight of the song learning behaviour of Zebra Finch (Taeniopygia guttata) songbirds during ontogeny with mass spectrometry imaging

Introduction:
The neuro-processes associated with ontogeny offer an opportunity to study their influence on behaviour and learning. The avian forebrain is interconnected with discreet song control nuclei which undergo a series of anatomical and neurochemical changes. We employed a multimodal MSI approach to link molecular signatures within the song control system to the plasticity changes of the finch songbird.

Methods:
Fresh frozen brain tissues were collected at 20, 40, 65 and 120 dph. 12µm sections were coated with 7mg/mL (2:1=CHCl3:MeOH) Norharmane matrix and analysed using RapifleX or Synapt G2Si at 50µm pixel size. Neurotransmitters were detected by performing on-tissue derivatisation with pyrylium salts and analysed with the 9.4T SolariX FT-ICR. The SIMS experiments were performed on a nano-TOF II MS/MS.

Results:
Intact lipids were studied using the dual polarity mode with emphasis on the composition and distribution within the discreet nuclei and the neuronal connections of the song-learning pathway. In positive ion mode, specific ions such as m/z 806.6 [PC (16:0/22:6)+H]+ and m/z 834.6 [PC (18:0/22:6)+H]+ revealed specific enrichment in the discreet nuclei while in negative ion mode m/z 904.6 [ST(d18:1/24:1(OH))-H]- and m/z 888.6 [ST(d18:1/24:1)-H]- are present in the laminae partitioning the brain. The high resolution SIMS images differentiate subareas of the nuclei where the sulfatides and cholesterol show an increase in the interconnecting circuits due to the presence of myelin. Further data analysis revealed simultaneous changes in lipid signatures linked to ontogeny. In addition several visual changes of the nuclei were observed complementing data of prior Diffusion Tensor Imaging (DTI).

Novel Aspect:
We show for the first time the distribution of specific intact molecular signatures associated to the song learning pathways in Zebra Finch ontogeny.

Oh, Jooyeon,
ASTA Inc.

Contributing Authors:
Joohee Jung, College of Pharmacy, Duksung Women's University; Jooyeon Oh, ASTA Inc; Hyung-Kyung Lim College of Pharmacy, Duksung Women's University; Yangsun Kim, ASTA Inc.

Title:
Near infrared fluorescence (NIRF) guided MALDI mass imaging of hypoxia region in tumor tissue

Introduction:
Matrix assisted laser desorption ionization (MALDI) - mass spectrometric imaging (MSI) is a powerful technique showing molecular distributions in tissue sections. Also, near infrared fluorescence (NIRF) imaging provides high dimensional information and sensitive imaging decreased self-fluorescence. Here, two imaging techniques are combined for tumor tissue analysis.

Methods:
Gastric cancer cell and lung cancer cell-derived tumors were isolated and determined. Hypoxia-inducible factor-1 and pimonidazole were used as endogenous and exogenous hypoxia markers, respectively. Anti-HIF-1 antibody and anti-pimonidazole antibody and mouse Cy7 antibodies were incubated on tissue sections. After matrix sprayed on the tissue, NIRF and MALDI images were measured sequentially.

Results:
Immunostaining of HIF-1 and pimonidazole was observed at the core of tumor tissue. Hypoxia-drived lipids were detected in pimonidazole-stained region and HIF-1 expressed region by MALDI-MS. Distribution of phosphatidylcholine (especially, PC (16:0/18:1) and PC (18:0/18:1)) species are different between hypoxia region and normoxia region.

Novel Aspect:
NIRF imaging for hypoxic areas were obtained quickly and then analyzed specific molecular distributions of NIRF-guided hypoxic areas by MALDI-MS.

Oh, Han-bin,
Sogang University

Title:
TEMPO-assisted free radical initiated peptide sequencing (FRIPS) mass spectrometry

Introduction:
Last decade, we have witnessed the development of a variety of radical-based peptide tandem mass spectrometry methods. Among many, our group have focused on an approach in which a radical precursor is covalently attached to the peptides of interest, i.e., the so-called FRIPS (free radical initiated peptide sequencing) approach. We've upgraded this approach into a form of practical omics tools.

Methods:
A new linker group was introduced into the original o-TEMPO-Bz-C(O)-NHS. This new reagent enhanced the solubility in DMSO and thus peptide conjugation efficiency. A newly conjugated peptide underwent collisional activation in a linear ion trap mass spectromer with its normalized collisional energy being optimize to yield a large number of peptide radical fragments.

Results:
With the original o-TEMPO-Bz-C(O)-NHS reagent, two-step collisional activations (CA) were required to see peptide backbone fragments. The first CA step generates a radical site and the second CA step induces extensive radical-driven peptide backbone fragmentations. This two-step procedure is not welcomed in practical applications. However, with the upgraded new reagent, the two-step CA processes can be reduced into only one step collisional activation. In the case of the original reagent, the generated radical is stabilized through the resonance effect of the carbonyl group on the benzene ring, thus requiring the additional collisional activation. However, in the upgraded design, this resonance stabilization is eliminated, i.e., thus destabilized. The second peptide backbone fragmentations proceeded without the application of the second collisional activations.

Novel Aspect:
Two-step FRIPS processes can be reduced into a single step FRIPS process with the use of the upgraded design concept in the free radical reagent.

21ST INTERNATIONAL MASS SPECTROMETRY CONFERENCE
Toronto, Ontario, Canada
August 20 – 26, 2016

Okuda, Koji,
JEOL USA, Inc.

Contributing Authors:
Makoto Okano, JEOL Ltd. Tokyo Japan; Tetsuo Higuchi, JEOL Ltd. Tokyo Japan; Masahiro Hashimoto JEOL Ltd. Tokyo Japan; Akihiko Kusai, JEOL Ltd. Tokyo Japan.

Title:
Analysis of evolved hydrogen gas from industrial materials by using a quadrupole mass spectrometer

Introduction:
Stainless steel is used in high-vacuum products for airtightness and corrosion resistance. However, adsorption of hydrogen causes embrittlement. The optimum treatment temperature to remove hydrogen is determined empirically instead of measuring desorbed H2. Here, we measured low-molecular-weight gas compounds including hydrogen from the stainless steel by a GC–quadrupole mass spectrometer (QMS).

Methods:
GC-QMS (JEOL Ltd.) was used with low energy (17eV) electron ionization (EI). GC-QMS coupled with pyrolysis equipment (Py) was used to detect the evolved gas from a piece of stainless steel. Evolved gas from the sample piece was directly introduced into QMS through a short blank capillary. The temperature dependence of gas evolution from the sample piece was measured over the range 100 to 800°C.

Results:
Two broad peaks corresponding to the hydrogen gas were observed in the extracted ion chromatogram EIC (m/z 2). The first broad peak appeared at the temperature range 150–400°C. The second broad peak was observed at the temperature range 400–800°C. The peak area of the second broad peak was 17 times larger than the peak area of the first broad peak. We interpret this to mean that the amount of evolved hydrogen gas has a high correlation with thermal expansion of stainless steel. The stainless steel sample used for these measurements has a high thermal expansion coefficient. These results suggest that hydrogen gas was evolved from the surface at 150–400°C and from inside the steel in the heating temperature range 400–800°C. The analysis results of evolved hydrogen gas from other metals will be presented.

Novel Aspect:
Evolved hydrogen gas from stainless steel was measured by GC-MS using a quadrupole mass spectrometer.

21ST INTERNATIONAL MASS SPECTROMETRY CONFERENCE
Toronto, Ontario, Canada
August 20 – 26, 2016

Okuda, Koji,
JEOL USA, Inc.

Contributing Authors:
Robert B. Cody, JEOL USA Inc., Peabody, MA, USA; Jean-François Focant, University of Liège, Liège, Belgium; John Dane JEOL USA Inc., Peabody, MA, USA; Anupam Giri, University of Liège, Liège, Belgium.

Title:
GCxGC-HRTOFMS Analysis of Base Oils with Chemical Ionization, Field Ionization and Photoionization

Introduction:
GCxGC-MS is a powerful method for characterizing complex mixtures such as base oils. We examined soft ionization methods (CI, FI, and PI) that can provide molecular weight information that is weak or absent in EI mass spectra. Here, we compare mass spectra of base oils analyzed by GCxGC-TOFMS with soft-ionization methods.

Methods:
Two base oils with low and high boiling points were measured by using GCxGC-TOFMS. Samples of base oil were prepared at 2 different dilution levels in hexane (1:25, 1:100). Data were acquired with a JEOL AccuTOF-GCV mass spectrometer and an Agilent 78890B GC fitted with a Zoex thermal modulator. The 1st and 2nd columns were Phenomenex ZB-XLB-HT and ZB-50, respectively.

Results:
Saturated and unsaturated hydrocarbons were detected in both samples. 2D maps for the photoionization (PI) data were similar to the 2D maps for the electron ionization (EI) data for both samples. Higher PCI source pressures influenced the retention times for high-boiling compounds. PI gave stronger signals than field ionization (FI). In the positive chemical ionization (PCI) data, saturated hydrocarbons exhibited hydride abstraction to produce [M - H]+. Unsaturated hydrocarbons exhibited proton attachment to produce [M + H]+. Therefore, molecular weights were determined from the retention times. Molecular ions M+• were detected in the PI and FI data, but fragment ions were observed in the mass spectra for saturated hydrocarbons. We will present analysis results for these and other components, and compare the characteristics of each ionization method.

Novel Aspect:
Analysis of base oils by GCxGC–HRTOFMS with several ionization methods (EI, PCI, PI and FI).

Oliva, Petra,
Sanofi Genzyme

Contributing Authors:
Rena Beak, Biologics Pharmacology and Pharmacokinetics, Sanofi Genzyme, Framingham, MA; Michelle Busch, Translational Sciences/ Biological Mass Spectrometry, Sanofi Genzyme, Framingham, MA; Hanlan Liu Drug Metabolism and Pharmacokinetics, Sanofi, Waltham, MA USA; Dinesh Bangari, Pathology, Sanofi Genzyme, Framingham, MA USA; Katherine Klinger, Translational Sciences/ Biological Mass Spectrometry, Sanofi Genzyme, Framingham, MA; Kate Zhang, Translational Sciences/ Biological Mass Spectrometry, Sanofi Genzyme, Framingham, MA; Niklas Boess, Translational Sciences/ Biological Mass Spectrometry, Sanofi Genzyme, Framingham, MA; Leslie Jost, Biologics Pharmacology and Pharmacokinetics, Sanofi Genzyme, Framingham, MA; Sue Ryan, Pathology, Sanofi Genzyme, Framingham, MA USA.

Title:
Tissue mass spectrometry imaging (tMSI) of biomarkers for Niemann-Pick disease

Introduction:
Niemann-Pick disease Types A and B is a lysosomal storage disorder caused by a deficiency of an acid sphingomyelinase (ASM) resulting in the abnormal accumulation of sphingomyelin. In this study, we utilized tissue MS imaging to determine special distribution of potential protein and lipid biomarkers in the ASM knockout mouse model that recapitulates hallmark biochemical features of ASMD.

Methods:
Tissues (liver and spleen) from ASMKO mice were sectioned and deposited on ITO slides. For the lipid workflow, the tissue was dried following matrix application and MALDI-TOF analysis (200-2000 Da range). Later, the tissue was washed in series of organic solvents with subsequent matrix re-application for the protein workflow and re-analyzed (2-20 kDa range). Finally, H&E staining was performed.

Results:
Each workflow was optimized for the best sensitivity/signal. Various MALDI matrices including 2,5-dihydroxybenzoic acid (DHB), α-cyano-4-hydroxycinnamic acid (CHCA), sinapinic acid (SA), and 5-chloro-2-mercaptobenzothiazole (CMBT). CMBT and CHCA provided the best sensitivity in positive ion mode for lipid and protein analysis, respectively. Application of sequential washing steps (isopropanol/buffer) significantly helped to improve signal intensity and the overall number of peaks for protein analysis. tMSI showed differential expression of glycosphingolipids, including sphingomyelin, specific to ASMD compared to wild type mice. Protein analysis is currently pending.

Novel Aspect:
Utilizing tMSI to determine special distribution of biomarkers for Niemann-Pick disease

Ollikainen, Elisa,
Faculty of Pharmacy, University of Helsinki

Contributing Authors:

Dongfei Liu, Division of Pharmaceutical Chemistry and Technology, Faculty of Pharmacy, University of Helsinki, Finland; Arttu Kallio, Division of Pharmaceutical Chemistry and Technology, Faculty of Pharmacy, University of Helsinki, Finland; Ermei Mäkilä Laboratory of Industrial Physics, Department of Physics and Astronomy, University of Turku, Finland; Jarno Salonen, Laboratory of Industrial Physics, Department of Physics and Astronomy, University of Turku, Finland; Hélder A. Santos, Division of Pharmaceutical Chemistry and Technology, Faculty of Pharmacy, University of Helsinki, Finland; Tiina Sikanen, Division of Pharmaceutical Chemistry and Technology, Faculty of Pharmacy, University of Helsinki, Finland.

Title:

Cytochrome P450 Inhibition Screening of Pharmaceutical Polymers and Nanoparticles by Liquid Chromatography-Mass Spectrometry

Introduction:

The growing use of polymers and nanoparticles in pharmaceutics has increased the concerns about their effects on human. Recent studies have shown that both silver and gold nanoparticles, and block co-polymers can interfere with cytochrome P450 (CYP) enzymes, responsible for metabolizing the majority of drugs. Here, we investigated the CYP interactions of selected nanoparticles and polymers.

Methods:

The effects of three porous silicon nanoparticles (APTES-TCPSi, Alkyne-THCPSi, and TCPSi) and three polymers (F68, F127 and polyvinyl alcohol, PVA) on the activities of four CYP isoenzymes (1A2, 2A6, 2D6 and 3A4) was evaluated on human liver microsomes with selective probe reactions. The metabolites formed were analyzed by Waters Acquity UPLC and Xevo QTOF mass spectrometer in positive ion mode.

Results:

The maximum reaction rates (Vmax) and Michaelis constants (Km) were determined separately for all of the probe reactions in the presence of each polymer (1 mg/mL) and nanoparticle type (1 mg/mL). Compared with controls (without polymers or nanoparticles), increased activity of CYP1A2 and CYP2A6 was observed in the presence of F127 and of F127 or F68, respectively. CYP3A4 was not affected by any of the polymers, whereas CYP2D6 was inhibited by F68 and PVA. In addition, CYP2D6 was inhibited by all nanoparticle types. Furthermore, minor decrease of CYP2A6 activity was observed with Alkyne-THCPSi and of CYP3A4 with APTES-TCPSi. CYP1A2 was not affected by any of the nanoparticle types. Based on these tentative findings, the future research on nanoparticles and polymers as pharmaceutical additives should also address their effects on drug metabolism via CYP enzymes.

Novel Aspect:

This study reveals that nanoparticles and polymers, as pharmaceutical additives, may interfere with drug metabolism via CYP enzymes in vitro.

Olsen, Jesper,
University of Copenhagen

Contributing Authors:
Anna Secher, Novo Nordisk; Christian Kelstrup, University of Copenhagen, Denmark; Jan Refsgaard University of Copenhagen, Denmark; Alicia Lundby, University of Copenhagen, Denmark; Lars Juhl Jensen, University of Copenhagen, Denmark; Dorte Bekker-Jensen, University of Copenhagen, Denmark; Chiara Francavilla, University of Copenhagen, Denmark.

Title:
Phosphotyrosine interaction proteomics decodes signaling network architecture by functional annotation of in-vivo phosphorylation sites

Introduction:
Tyrosine phosphorylation represents only a few percent of the total phosphoproteome, but regulates complex and multi-layered signaling networks with broad implications in human (patho)physiology. Here we describe a large-scale mass spectrometry-based phosphotyrosine interaction proteomics technology to assign function to regulated phosphorylation sites and decipher cellular signaling networks.

Methods:
We apply our developed quantitative technology to examine the epidermal growth factor (EGF)-dependent signaling network in-vivo in lung tissue and delineate the dynamic protein complexes wired by tyrosine phosphorylation.

Results:
We quantify more than one thousand phosphotyrosine sites of which less than ten percent are significantly regulated by 5 minutes of EGF stimulation. Using large-scale phosphotyrosine interaction proteomics we identify the recruited protein complexes to all regulated phosphotyrosine sites and reveal that EGF induces assembly of 11 major dynamic signaling modules in lung tissue. Finally, we demonstrate how a somatic lung cancer mutation in EGFR switches interaction at phosphorylated tyrosine. Our phosphotyrosine interaction proteomics provide a general technological platform to assign function to individual phosphotyrosine sites from large-scale phosphoproteomics and Our phosphotyrosine interaction proteomics provide a general technological platform to assign function to individual phosphotyrosine sites from large-scale phosphoproteomics.

Novel Aspect:
Large-scale assignment of function to in-vivo phosphotyrosine sites by mass spectrometry-based interaction proteomics

Orejas, Jaime,
University at Buffalo

Contributing Authors:
Alfredo Sanz-Medel, University of Oviedo; Jorge Pisonero, University of Oviedo; Steven J. Ray University at Buffalo; Nerea Bordel, University of Oviedo.

Title:
Negative ion generation in the FAPA source: formation processes and auxiliary gas addition effect.

Introduction:
Ambient Desorption/Ionization (ADI) sources in combination with mass spectrometric detection are a powerful pair. This work presents fundamental studies on negative ionization mode for the Flowing Atmospheric Pressure Afterglow (FAPA) source that can be useful for certain compounds detection such as organohalogens, fatty acids, and other compounds difficult to ionize with conventional approaches.

Methods:
The FAPA source used in this work use a pin-to-plate electrode geometry enclosed in a quartz and brass cell. The exit hole in the plate electrode is surrounded by a concentric auxiliary gas outlet. He was used as plasma gas, while He, N2, Ar, O2 and N2O where used as auxiliary gases. A quadrupole mass spectrometer was used as a detector.

Results:
Negatively charged reagent ions generated by the FAPA source in the open atmosphere were detected and identified, and optimized for use as reagent ions for ambient mass spectrometry. Several ions were detected (OH-, O2-, CO3-, NO2-, NO3- and hydrated species of them). Their production pathways and ionization mechanisms are discussed and contrasted with conventional Atmospheric Pressure Chemical Ionization (APCI). The effect of inert and reactive gases and gas flow geometries on the ion signals of the reagent ions was also studied. Inert gases tend to reduce reagent ion signals. On the contrary, reactive gases (O2 and N2O) increase noticeably reagent ion signals, while new ions are registered in the mass spectra, such as O3-. Finally, ionization pathways for potential analytes were investigated through the detection of several organohalogens, fatty acids and other compounds.

Novel Aspect:
Although negative mode is used to detect certain compounds, few specific fundamental studies on the ion chemistry for ADI sources have been presented.

Ortiz, Xavier,
Ontario Ministry of the Environment and Climate Change

Contributing Authors:
Marie Meyer, Ontario Ministry of the Environment and Climate Change; Eva Korenkova, Ontario Ministry of the Environment and Climate Change; Karen MacPherson Ontario Ministry of the Environment and Climate Change; Karl Jobst, Ontario Ministry of the Environment and Climate Change; Eric Reiner, Ontario Ministry of the Environment and Climate Change.

Title:
An automated method for microcystins analysis using two-dimensional liquid chromatography-quadrupole time-of-flight mass spectrometry (2DLC-QTOFMS).

Introduction:
Microcystins are cyclic heptatpeptide hepatotoxins produced by cyanobacteria, including over 130 different variants. In order to develop non-targeted methods for the characterization of these compounds, the present study describes an automated method for the analysis of microcystins by two-dimensional liquid chromatography-quadrupole time-of-flight mass spectrometry (2DLC-QTOFMS).

Methods:
In order to reduce the sample size without compromising the method detection limits, a 2DLC in the trap and elute configuration was used: a large volume of sample was injected and trapped in the first column, using at column dilution to focus the analytes. Microcystins were subsequently desorbed with reversed flow and injected to the analytical column, prior to the QTOFMS analysis.

Results:
The online automated 2DLC system allows water samples to be prepared for analysis while the instrument analyzes the previous sample, improving method productivity. When the QTOF was operated in high resolution full scan mode (RP > 25.000), the instrument proved to be very sensitive for microcystin-LR (100 fg on column with S/N > 10). Combined with the 2DLC, the system could detect 100 ppq of microcistyn-LR in water with S/N>10. Mass accuracy (< 1 ppm) allowed assigning elemental compositions for unknown compounds with confidence. MS/MS mode monitoring the characteristic microcystin ion at m/z=135.0804 was useful to provide quantitative results of targeted compounds in complex samples. Advanced acquisition methods such as Data Dependant Acquisition (DDA) and Data Independent Acquisition (DIA) proved to be useful tools when looking for non-targeted microcystin variants.

Novel Aspect:
Development of an automated method for the non-targeted analysis of microcystins by (2DLC-QTOFMS with low detection limits.

Ouyang, Zheng,
Tsinghua University

Contributing Authors:

Wengpeng Zhang, Purdue University; Yu Xia, Purdue University; R. Graham Cooks Purdue University; Ran Zou, Purdue University.

Title:

Point-of-care analysis of lipids in biological samples using miniature mass spectrometry systems

Introduction:

We have been developing miniature mass spectrometry systems for point-of-care analysis of biological samples. Currently, we are pushing the frontier of this field toward the lipid analysis, a class of biological compounds that are intrinsically related to biological processes and also can be readily sampled directly from biological samples using ambient ionization.

Methods:

A suite of technologies has been developed to enable the direct analysis and quantitation of the lipids in blood and tissue samples using miniature MS systems. The mass range of the miniature ion trap mass spectrometer has been extended to m/z 1000 with the combination of wider range of RF voltage scan and lower AC frequency for resonance ejection.

Results:

Probes with polyporous organic polymer coatings have been developed to collect the lipids from blood and tissue samples, followed by an extraction ionization for analysis. Miniature MS system with an endoscopic sampling probe of 1.5 m length has also been developed for in-vivo sampling and analysis of tissue surfaces. In addition to the intensity profile that can be obtained through MS analysis, quantitation of the unsaturated lipid isomers has also been enabled by implementing an online photochemical ((Paternò–Büchi) reaction. The C=C bonds were modified into an oxetane ring that can be easily fragmented during collisional induced dissociation. Quantitation of the unsaturated fatty acids in blood has also been developed using the combination of PB reaction and neutral loss scan. Comparative study involving normal and diseased samples have been successfully done and will be reported.

Novel Aspect:

Direct sampling and quantitative analysis of lipids using miniature mass spectrometry systems, enabling POC applications.

Ouyang, Jin,
College of Chemistry, Beijing Normal University

Title:
Application of Ambient Mass Spectrometry to High-throughput Detection of Drugs Binding to Proteins

Introduction:
High-throughput screening (HTS) is a powerful strategy for drug discovery. With this strategy, scientists can quickly identify the active compounds and study the interaction of drugs binding to proteins. Mass spectrometry has been applied now as a routine tool for HTS, due to the high-speed and molecule-identification abilities. However, most of MS method required complicated sampling procedures.

Methods:
In this report, ambient MS methods have been developed by our groups for HTS analysis. The details of the experimental parameters were not shown in this abstract but it could be found mainly in our publications[1-4]. The ambient MS we used including DESI-MS, Venturi easy ionization MS and MALDI-TOF-MS. The high-throughput was achieved in analyzing 96 samples at the level of seconds per sample.

Results:
In DESI-MS method [1], the drug–protein complex was deposited on desired substrate in an array type. Drugs were dissociated and liberated from the complex when the DESI screening across the samples. Rapid screening was achieved by screening 21 types of drugs against HSA. Monitoring binding affinity between drug and α1- acid glycoprotein in real time by Venturi easy ambient sonic-spray ionization MS has also been achieved [2]. We also demonstrated this ambient MS method for enzyme inhibitors[3]. The high-throughput was achieved in analyzing 96 samples at 1.6s per sample. The calculated IC50 agreed well with reported values. To detect the small molecules like drugs by MALDI-TOF-MS, we developed the new aggregation-induced emission (AIE) matrices[4]. The results demonstrated that the AIE matrices produce lower background while retaining a high sensitivity.

Novel Aspect:
The results in this abstract are powerful for monitoring drug–protein binding and suggested the potential of HTS in drug discovery by the ambient MS.

Pagel, Kevin,
Freie Universitaet Berlin

Contributing Authors:
Pauline M. Rudd, National Institute for Bioprocessing Research and Training, Dublin; Heung Sik Hahm, Max Planck Institute of Colloids and Interfaces, Potsdam; Hannes Hinneburg Max Planck Institute of Colloids and Interfaces, Potsdam; Johanna Hofmann, Fritz-Haber Institute of the Max Planck Society, Berlin; Peter H. Seeberger, Max Planck Institute of Colloids and Interfaces, Potsdam; Daniel Kolarich, Max Planck Institute of Colloids and Interfaces, Potsdam; Weston B. Struwe, Oxford Glycobiology Institute, Oxford.

Title:
Glycomics in a Wind Tunnel - Ion Mobility-Mass Spectrometry of Isomeric Glycans and Glycopeptides

Introduction:
Currently, the majority of glycans are characterized using MS-based techniques. Measuring the molecular weight of a sugar, however, immediately poses a fundamental problem: entire classes of monosaccharide building blocks exhibit an identical atomic composition and mass. Therefore, glycan MS data can be highly ambiguous and often it is not possible to clearly assign a particular structure.

Methods:
A promising approach to overcome the above-mentioned limitation is to implement an additional gas-phase separation step using ion mobility-mass spectrometry (IM-MS). Here, ions travel through a gas-filled cell aided by a weak electric field and are separated according to their collision cross section (CCS).

Results:
Here, we demonstrate the potential of IM-MS to be used as a tool for the separation and identification of isomeric glycans and glycopeptides. First, six synthetic oligosaccharide isomers that differ with respect to their composition, connectivity, or configuration were analyzed. Our data reveal that linkage- and stereoisomers, which are difficult to distinguish using established techniques, can be separated and unambiguously identified on basis of their CCS. When mixed, even minor isomeric components with concentrations as low as 0.1% are still clearly detectable. Second, we extended our investigations to glycopeptides. Our data show that glycopeptides, which merely differ in the regiochemistry of the attached glycan can be distinguished using fragmentation and subsequent IM-MS analysis.

Novel Aspect:
Separation and identification of oligosaccharide and glycopeptide isomers using ion mobility-mass spectrometry

Palmblad, Magnus,
Leiden University Medical Center

Contributing Authors:
Arzu Tugce Guler, Leiden University Medical Center, Center for Proteomics and Metabolomics, Leiden, The Netherlands; Dmitrii Travin, Faculty of Bioengineering and Bioinformatics, Lomonosov Moscow State University, Moscow, Russia.

Title:
Scientific workflows for data analysis and visualization in quantitative proteomics

Introduction:
Scientific workflows are important tools for automation and excellently suited for integrative mass spectrometry (MS) and omics data analyses. Domain-specific ontologies and controlled vocabularies, make condition-decision making feasible within such workflows. We here demonstrate how to build workflows for integrating and visualizing proteomics and gene expression data in anatomical contexts.

Methods:
The workflow for integrative omics analysis was created in Taverna, a scientific workflow manager The data was converted to mzML. The Proteomics Standards Initiative controlled vocabulary was used to embed additional metadata. This was used to guide the processing. Datasets were anatomically encoded and the spatial distribution of expression integrated and visualized using an anatomical ontology.

Results:
The scientific workflow consists of 8 modules, each performing a discrete analysis within a well-defined scope. By default, the workflow reads the metadata and selects the most optimal parameters for the analysis. However, the expert user can also apply specific parameters, overriding the defaults. The workflow performs format conversions, peptide and protein identification, chromatographic alignment and label-free quantitation across all datasets. The results were visualized spatially to reveal biologically interesting differences in abundance levels, or unexpected correlations between gene and protein expression. We used the Zebrafish Anatomical Ontology to align data with a novel Web-based tool for visualizing quantitative gene and/or protein expression, simplifying multi-omics studies in zebrafish and carp model systems.

Novel Aspect:
First combination of ontologies from different domains for annotating samples and datasets, facilitating integration and visualization of multi-omics

Palmblad, Magnus,
Leiden University Medical Center

Contributing Authors:
Kristian Davidsen, Center for Biological Sequence Analysis, Technical University of Denmark, Copenhagen, Denmark; Anna-Lena Lamprecht, Lero - The Irish Software Research Centre, University of Limerick, Ireland; Arzu Tugce Guler Center for Proteomics and Metabolomics Leiden University Medical Center, Leiden, the Netherlands; Veit Schwämmle, Department of Biochemistry and Molecular Biology, University of Southern Denmark, Odense, Denmark; Jon Ison, Center for Biological Sequence Analysis, Technical University of Denmark, Copenhagen, Denmark.

Title:
Functional software annotation and automatic workflow generation for mass spectrometry data processing

Introduction:
Many software utilities operating on mass spectrometry (MS) data have been described in the literature. Finding that which one needs is often hard, however. To help, the EDAM ontology of concepts, including formats and operations, was created in 2010. Here we show how new EDAM terms and annotation of MS software support automatic generation and validation of data processing workflows.

Methods:
We have added MS-related terms to EDAM and annotated 238 software tools in the public domain, including those on http://ms-utils.org, in the ELIXIR Tools and Data Services Registry. We used these annotations to automatically generate workflows in two use cases: the calculation of amino acid retention coefficients and pathway enrichment analysis, both starting from tandem mass spectrometry data.

Results:
Using the EDAM ontology, we annotated 238 software packages, libraries and utilities in the public domain already described in the ms-utils.org wiki. We then returned to the wiki to ensure the categorization of the tools, corresponding to EDAM operations, are consistent with the controlled vocabulary. Annotating the ms-utils.org content with EDAM terms elevates it to the biotoolsXSD standard, supporting the exposure of these resources in the bio.tools registry, bringing the utilities to a broader audience. The ms-utils.org tools emphasizes modular rather than monolithic design. Such small utilities performing one operation with well-defined inputs and outputs are ideally suited for assembly into scientific workflows. We will give examples of such workflows, generated using the EDAM ontology annotations and functionally validated in two use cases.

Novel Aspect:
First demonstration of using the EDAM ontology to annotate mass spectrometry software utilities and in turn generate workflows for MS data processing.

PAPANASTASIOU, DIMITRIS,
Fasmatech Science & Technology

Contributing Authors:
Athanasios Zacharos, Fasmatech Science & Technology; Michael Burt, The Chemistry Research Laboratory, Department of Chemistry, University of Oxford; Claire Vallance The Chemistry Research Laboratory, Department of Chemistry, University of Oxford; Emmanuel Raptakis, Fasmatech Science & Technology; Mark Brouard, The Chemistry Research Laboratory, Department of Chemistry, University of Oxford.

Title:
A Computational Study of a Multi Capillary Inlet System using Gas Dynamics and Ion Optical Simulations

Introduction:
Computational methods in gas dynamics combined with ion tracing algorithms are a powerful tool for vacuum interface design affording a high level of exploration and optimization of complex geometries. A multi-capillary inlet system is characterized here using different solvers and ion transfer is monitored from atmospheric pressure to medium vacuum.

Methods:
The Reynolds-Averaged Navier-Stokes (RANS) and the Direct-Simulation Monte Carlo (DSMC) methods are employed to solve the flow through a capillary inlet system and in the near-field region of the free jet expansion. The DSMC method is further applied to compute the flow emerging from the multi capillary inlet and throughout the entire geometry. Flow fields are imported in SIMION.

Results:
Computations are initially limited to the capillary inlet designed with 100 µm id capillary including the near field region of the free jet expansion at 1 and 10 mbar background pressures. The RANS method is applied from high pressure throughout the capillary into the low pressure region of the domain. The properties of the flow determined by the RANS solver at the capillary exit are imposed as inlet boundary conditions for the DSMC method initiated at lower pressure. The free jet flows are in perfect agreement, demonstrating the applicability of RANS at pressures as low as 1 mbar. The DSMC method applied in 3D includes a multi capillary inlet, aerolens optics arranged coaxially with each of the free jet flows and terminating skimmer-shaped apertures to enter into the low pressure region of the system. Ion tracing is performed over a pressure range extending four orders of magnitude.

Novel Aspect:
Gas dynamics and ion tracing solvers combined to investigate complex vacuum interface systems.

Parakra, Rinky,
University of Otago

Contributing Authors:
Dr. Torsten Kleffmann, Center for Protein Research, Univeristy of Otago.; Ass. Prof. Guy Jameson, Department of Chemistry, University of Otago.; Dr. Elizabeth Ledgerwood Department of Biochemistry, University of Otago.

Title:
Investigating the peroxidase reaction mechanism of human cytochrome c (cyt c).

Introduction:
Cyt c has peroxidase activity that is enhanced upon binding to the mitochondrial cardiolipin (CL). Native cyt c has a 6-coordinated Fe atom and loss of Met80-Fe coordination increases this activity. We showed that naturally occurring (G41S) and recombinant (G41T) cyt c variants have increased activity albeit intact coordination. Our aim is to elucidate the reaction mechanism of cyt c and CL-cyt c.

Methods:
Oxidation of cyt c was determined by HPLC and LC-coupled LTQ Orbitrap tandem MS. Quantification of modifications was performed using the Sequest HT search engine with the SIEVE program for the alignment of ion chromatograms. CL oxidation was analysed in negative ion mode using direct injection Orbitrap-MS and MALDI TOF-MS. Peroxidase activity was assayed using tetramethylbenzidine (TMB).

Results:
The increased peroxidase activity of free G41S and G41T cyts c correlated with increased susceptibility to H2O2–induced oxidation of multiple amino acids and heme Fe loss. Combined LC-MS/MS analysis of tryptic digests and HPLC analysis of CNBr digests showed that Met80 was irreversibly oxidized to Met80-SO2 and was more susceptible than the surface located Met12. Catalysis of H2O2-dependent oxidation of TMB formed Met80-SO and there was no oxidation of Met12 or loss of heme. This shows that Met80 Fe coordination is lost during the peroxidase reaction cycle due to specific oxidation of the Met80 S atom, and is a critical step in determining the peroxidase activity of cyt c. Surprisingly using CL:DOPC liposomes,oxidation of CL when incubated with WT or variant cyt c in the absence of H2O2. This implies a distinct catalysed mechanism for oxidizing CL compared to the small molecule TMB.

Novel Aspect:
The heme Fe of free cyt c is converted to a penta-coordinate state during the peroxidase reaction, residue 41 contributing to Met80 monooxidation.

Parakra, Rinky,
University of Otago

Contributing Authors:
Torsten Kleffmann, Center for Protein Research, Univeristy of Otago.; Guy Jameson, Department of Chemistry, University of Otago.; Elizabeth Ledgerwood Department of Biochemistry, University of Otago.

Title:
Investigating the peroxidase reaction mechanism of human cytochrome c (cyt c).

Introduction:
Cyt c has peroxidase activity that is enhanced upon binding to the mitochondrial cardiolipin (CL). Native cyt c has a 6-coordinated Fe atom and loss of Met80-Fe coordination increases this activity. We showed that naturally occurring (G41S) and recombinant (G41T) cyt c variants have increased activity albeit intact coordination. Our aim is to elucidate the reaction mechanism of cyt c and CL-cyt c.

Methods:
Oxidation of cyt c was determined by HPLC and LC-coupled LTQ Orbitrap tandem MS. Quantification of modifications was performed using the Sequest HT search engine with the SIEVE program for the alignment of ion chromatograms. CL oxidation was analysed in negative ion mode using direct injection Orbitrap-MS and MALDI TOF-MS. Peroxidase activity was assayed using tetramethylbenzidine (TMB).

Results:
The increased peroxidase activity of free G41S and G41T cyts c correlated with increased susceptibility to H2O2–induced oxidation of multiple amino acids and heme Fe loss. Combined LC-MS/MS analysis of tryptic digests and HPLC analysis of CNBr digests showed that Met80 was irreversibly oxidized to Met80-SO2 and was more susceptible than the surface located Met12. Catalysis of H2O2-dependent oxidation of TMB formed Met80-SO and there was no oxidation of Met12 or loss of heme. This shows that Met80 Fe coordination is lost during the peroxidase reaction cycle due to specific oxidation of the Met80 S atom, and is a critical step in determining the peroxidase activity of cyt c. Surprisingly using CL:DOPC liposomes,oxidation of CL when incubated with WT or variant cyt c in the absence of H2O2. This implies a distinct catalysed mechanism for oxidizing CL compared to the small molecule TMB.

Novel Aspect:
The heme Fe of free cyt c is converted to a penta-coordinate state during the peroxidase reaction, residue 41 contributing to Met80 monooxidation.

Park, Melvin,
Bruker Daltonics, Inc.

Contributing Authors:

Mark E. Ridgeway, Bruker Daltonics, Inc; Joshua Silveira, Bruker Daltonics, Inc.

Title:

Ion Mobility Mass Spectrometry Coming of Age

Introduction:

With commercialization, IMMS has become more widespread. The increased interest in IMMS in recent years has thus spurred the development of new technology, bringing IMS to a level of maturity worthy of consideration in many analytical fields. Trapped ion mobility spectrometry (TIMS) is one such new technology. Here we describe TIMS-MS and some of its potential applications.

Methods:

TIMS analyzers were incorporated into the first vacuum stage of prototype ESI-QqTOF, and FTICR mass spectrometers (Bruker Daltonics, Billerica, MA). Ubiquitin (Sigma-Aldrich, St. Louis, MO) was diluted to a concentration of 1 uM in 70:30 H2O:acetonitrile. Samples of isobaric peptides were prepared at concentrations of 1 to 5 uM in 50:50 H2O:acetonitrile. Petroleum samples were ionized by APPI.

Results:

Results from the analysis of ubiquitin on both the TIMS-QqTOF and TIMS-FTICR spectrometers demonstrate that TIMS can be operated in a "soft" manner. Mobility resolving powers exceeding 200 allow for not only the separation of compact, partially unfolded, and extended states of ubiquitin ions, but also various conformations within these broad states. Results from TIMS on QqTOF and on FTICR are consistent in conformations observed and their collision cross sections. Specifically, it was found that the ubiquitin cross sections measured on the two types of instruments agreed to within 0.3%. Similarly, the analysis of peptides, shows previously unresolved conformations. The analysis of PTM peptides (specifically PFGK modified with propionyl v propanal carbonyl acyl groups) shows that such isobaric regioisomers can be nearly baseline resolved at mobility resolving powers of about 160.

Novel Aspect:

Trapped ion mobility spectrometry hyphenated with OTOF and FTICR and its use in the analysis of PTM peptides, proteins, and petroleum

21ST INTERNATIONAL MASS SPECTROMETRY CONFERENCE
Toronto, Ontario, Canada
August 20 – 26, 2016

PARK, HYE YOUNG,
Ministry of food and drug safety

Contributing Authors:
Mi-Hyune Oh, Ministry of Food and Drug Safety, Gyeongin Regional Food and Drug Administration; Bomi Kim, Ministry of Food and Drug Safety, Gyeongin Regional Food and Drug Administration; Ga-Young Nam Ministry of Food and Drug Safety, Gyeongin Regional Food and Drug Administration; Chul-Joo Lim, Ministry of Food and Drug Safety, Gyeongin Regional Food and Drug Administration; Sun-Ok Choi, Ministry of Food and Drug Safety, Gyeongin Regional Food and Drug Administration.

Title:
Determination of the antibiotic residues in meat by optimized liquid chromatography tandem mass spectrometry

Introduction:
Antibiotics are widely used in livestock management for the treatment of disease and as dietary supplements. The livestock animals may be administered orally as feed additives or directly by injection. The widespread use of antibiotics in livestock animals may result in the presence of antibiotic residues in meat.

Methods:
While microbial screening tests are commonly used to detect the presence of antibiotic residues in meat, more accurate chromatographic methods are required by government regulatory agencies to identify and confirm of antibiotic residues present. The multi-residues methods have been developed for the chromatographic determination of antibiotic residues at Ministry of Food and Drug Safety in Korea.

Results:
This method will be used to play an important role in the confirmation of penicillins, sulfonamides, quinolones, cephalosporins, peptides, amphenicols and polyethers antibiotic, as well as various other groups including tetracyclines, aminoglycosides, macrolides antibiotic.

Novel Aspect:
The applicability of this method will be used in the chromatographic determination test of 90 antibiotic residues to supply safe livestock products in

Parkinson, Erika,
University of Southampton

Contributing Authors:
Maja Aleksic, SEAC, Unilever plc, Sharnbrook, UK; Paul Skipp, Centre for Proteomic Research, Centre for Biological Sciences, University of Southampton, Southampton UK; Richard Cubberley SEAC, Unilever plc, Sharnbrook, UK.

Title:
Quantifying the effect of exposure to the model sensitiser DNCB on the human skin proteome

Introduction:
Allergic Contact Dermatitis is a delayed-type hypersensitivity response to small chemicals occurring in two stages; initiation, where sensitization is acquired; and elicitation, characterized by an inflammatory skin reaction. Both phases are mediated by the covalent modification of skin proteins by a chemical sensitizer; however the identities of these modified proteins are currently unknown.

Methods:
Human skin cells in culture were exposed to a non-toxic, clinically relevant dose of DNCB (50:50 mix of unlabeled and stable isotopically labelled DNCB) for a maximum of 4 hours, before removing the chemical and allowing the cells to grow for up to 48 hours. Samples were taken across a range of time points, spiked with an equal concentration of SuperSILAC reference standard and analysed by HDMSE.

Results:
Confident assignment of sensitizer modified peptides and the site of modification is often hampered by high background ions from unmodified peptides, which can mask the low abundance sensitizer modified peptides. To address this, we developed a novel methodology using mass spectrometry and stable isotope labelling that significantly increases the confidence in identifying protein modification by a chemical sensitizer. To further enhance our understanding of the mechanisms of the allergic response, we created a SuperSILAC reference standard containing 4 SILAC labelled skin cell lines and quantified this using targeted SRM. By spiking in the SuperSILAC reference into sensitizer exposed cells, we were able to obtain a quantitative measure of protein modification over time, in addition to determining the cellular response of human skin cells when exposed to the extreme sensitizer DNCB.

Novel Aspect:
By combining HDMSE, stable isotope labelling and SuperSILAC we are able to characterise the effect of protein modification by sensitizers in vitro.

Peake, David,
Thermo Fisher Scientific

Contributing Authors:
Reiko Kiyoanami, Thermo Fisher Scientific; Julian Saba, Thermo Fisher Scientific; Sergei Snovida Thermo Fisher Scientific; Devin Drew, Thermo Fisher Scientific; David Peake, Thermo Fisher Scientific; Andreas Huhmer, Thermo Fisher Scientific; Ken Miller, Thermo Fisher Scientific.

Title:
Facilitating Biomarker Discovery Using an Integrated "Omics" Differential Analysis with High Resolution Accurate Mass LC/MS Approach

Introduction:
LC-MS platforms are increasingly used for proteomics, lipidomics, metabolomics and glycomics experiments to discover potential biomarkers for early disease diagnosis. Since each "omics" data set provides valuable insights into biomarker candidates, integrating multiple "omics" data will help to understand the system behavior as a whole for unravelling biological regulatory mechanisms.

Methods:
Plasma from 3 lean and 3 Zucker fatty rats were obtained from Bioreclamation. Total lipid extracts were profiled by LC-MS using a C30 column and quadrupole Orbitrap MS. Protein digests and permethylated glycans from the rat plasma was analyzed by nanoLC-MS and an Orbitrap Fusion mass spectrometer. Data analysis was performed using LipidSearch 4.1, Proteome Discoverer 2.2 and Simglycan software.

Results:
Proteomics, lipidomics and glycomics studies were conducted with Zucker diabetic fatty (ZDF) vs. normal rat plasma using high resolution Orbitrap based LC/MS systems. The lipid identification and simultaneous relative quantitation were performed using LipidSearch software. Results were then filtered by adduct ion, match score, ID quality, signal-to-noise, peak areas and coefficient of variation. After filtering and manual validation, more than 700 lipid species from 15 different lipid classes were identified and quantified from the rat plasma with high confidence and analytical precision. Many lipid species including PC, LPC, SM, DG and TG lipid classes showed significant relative concentration differences between the ZDF and lean rat plasma. Further analysis of proteomics and glycomics differential profiling experiments are being carried out to complete this study.

Novel Aspect:
Rapid parallel and integrated biomarker discovery by systematic proteomic, lipidomic and glycomic plasma profiling of a rat model of Type II Diabetes.

Peake, David,
Thermo Fisher Scientific

Contributing Authors:
Mandy Bowman, Thermo Fisher Scientific; Reiko Kiyonami, Thermo Fisher Scientific; Andreas Huhmer Thermo Fisher Scientific; Danni Li, Univerisity of Minnesota.

Title:
Identification of Phospholipid Species Implicated in Dementia by Untargeted High Resolution LC/MS and Data Dependent MS/MS

Introduction:
Alzheimer's disease (AD) is the leading cause of dementia in the elderly. Recently, Mapstone et al. reported plasma phospholipids that predicted cognitively normal adults who later progressed to either mild cognitive impairment or dementia due to AD (Mapstone 2014). This study utilized a Biocrates targeted metabolomics p180 kit measured by flow injection and selected reaction monitoring (SRM).

Methods:
Plasma lipid extracts (2µL) were separated by ultra-high pressure liquid chromatography using a C30 Acclaim column (2.1 X 150 mm, 3.0 um) and analyzed with a Orbitrap Fusion Lumos MS system. LC/MS and MS/MS data were obtained at 120,000 and 30,000 mass resolving power, respectively in duplicate positive and negative ion runs. LipidSearch 4.1 software was used for molecular lipid identification.

Results:
The objective of this study is to identify seven phosphatidylcholine (PC) lipids: PC 36:6, PC 38:0, PC 38:6, PC 40:1, PC 40:2, PC 40:6, and PC 40:6e (ether) in plasma by using liquid chromatography and high-resolution accurate mass and tandem mass spectrometry (LC/MS/MS). LipidSearch software identified that the plasma PC 36:6 is primarily PC 14:0_22:6; PC 40:2 is PC 22:0_18:2; and PC 40:6 is PC 18:0_22:6. The PC ae 40:6 is a mixture of 18:1e_22:5 and 20:2e_20:4; and PC aa C38:6 is a mixture of 16:0_22:6, 18:1_20:5 and 18:2_20:4 PC species. However, we could not confirm the fatty acid molecular identity for PC 38:0 or PC 40:1. The PC 38:0 species was in fact identified as PC 18.1e_22:6 (m/z 818.6058, M+H) instead of PC 38:0 (m/z 818.6633, M+H). This was confirmed by MS/MS of the negatively-charged formate adduct of 40:7e (m/z 862.5967, M+HCOO).

Novel Aspect:
Identification of these particular phospholipids will enable us to develop quantitative methods in order to understand their potential roles in AD.

Pedro, Liliana,
Eskitis Institute for Drug Discovery, Griffith University

Contributing Authors:
Ronald J Quinn, Eskitis Institute for Drug Discovery, Griffith University, Brisbane, Queensland, Australia.

Title:
Native state mass spectrometry in natural product and fragment-based drug discovery against malaria protein targets

Introduction:
Genomics and proteomics are identifying new therapeutic target proteins. While many of these proteins are successfully expressed and purified, not all get their structures determined by X-ray or NMR. Mass spectrometry allows the identification/characterization of protein-small molecule interactions providing ways of understanding protein function/dysfunction in the settings of different diseases.

Methods:
Native state mass spectrometry was used to screen an in-house library of natural product-based fragments against 33 malaria protein targets. The resulting hits that bound to Plasmodium vivax guanylate kinase were studied using native state mass spectrometry in combination with other biophysical methods to obtain inhibitory activities, binding affinities and binding sites.

Results:
Plasmodium vivax guanylate kinase has 41% sequence identity to Homo sapiens guanylate kinase and constitutes a good malaria drug target. 26 natural product-based fragments were identified as hits against this protein. Half are specifically binding to this protein, while the other half represent connecting points between this protein and other screened malaria protein targets. Over half of the hits have inhibitory activities above 30% at 5mM concentration. Information pertaining to the active hits binding affinities and binding sites is now being used to guide future directions in rational fragment elaboration. Natural product-based fragments offer higher 3D content than their synthetic or commercially-obtained counterparts. As a result, they are excellent starting points for fragment elaboration in all directions of space into potent drug-sized molecules.

Novel Aspect:
Native state mass spectrometry is often overlooked as a tool for fragment-based drug discovery. Here we show that it can be central to the process.

Peetz, Oliver,
Goethe-University of Frankfurt

Contributing Authors:
Nina Morgner, Goethe-University Frankfurt; Jan Hoffmann, Goethe-University Frankfurt.

Title:
Mass Spectrometry of non-covalent bound membrane proteins in nanodiscs with LILBID-MS

Introduction:
Particularly challenging to investigate with MS are membrane protein complexes as they require e.g. detergents for solubilization. However some complexes are not stable in detergents. For those, we focused on nanodiscs (ND) consisting of lipids which more closely mimic the lipid bilayer of the membrane. We used our homebuilt LILBID-MS (Laser-Induced-Liquid-Bead-Ion-Desorption) for investigations.

Methods:
A piezo-driven droplet generator is used to produce droplets of 30 µm diameter. The droplets are transferred to high vacuum and irradiated by an IR laser operating at 2.94 µm, a vibrational absorption wavelength of water. This leads to an explosive expansion of the sample droplet. The solvated analyte ions are released and analyzed by a homebuilt reflectron-time-of-flight (TOF) setup.

Results:
As a proof of principle of the use of ND as carriers for membrane proteins we investigated the well known dimer EmrE, the monomer LspA and the tetramer KcsA. Results show that LILBID is able to reveal the expected stoichiometry for all of this membrane proteins. Interestingly, for all investigated membrane proteins, peaks in all mass spectra can be assigned to the complexes with or without the scaffold proteins. MS from ND allowed us to investigate the factors involved in a protein dimerization process for e.g. the membrane embedded domain of hHv1-VSD. We used LILBID to proof that MraYs lipid dependent activity is due to the lipids influence on the oligomeric state. Further on we can follow the self-assembling process of proteorhodopsin into NDs in a cell free expression system. In a native environment, PR preferably assembles into pentameric and hexameric oligomers.

Novel Aspect:
We show that LILBID-MS is able to answer different questions on membrane proteins in NDs which even keeps sensitive complexes biologically active.

Pelivan, Karla,
Institute of Inorganic Chemistry, University of Vienna

Contributing Authors:
Bernhard K. Keppler, Institute of Inorganic Chemistry, Faculty of Chemistry, University of Vienna, Waehringer Strasse 42, 1090 Vienna, Austria; Uwe Karst, Institute of Inorganic and Analytical Chemistry, University of Münster, Corrensstrasse 28/30, 48149 Münster, Germany; Gunda Koellensperger Institute of Analytical Chemistry, Faculty of Chemistry, University of Vienna, Waehringer Strasse 38, 1090 Vienna, Austria; Christian Kowol, Institute of Inorganic Chemistry, Faculty of Chemistry, University of Vienna, Waehringer Strasse 42, 1090 Vienna, Austria; Lisa Frensemeier, Institute of Inorganic and Analytical Chemistry, University of Münster, Corrensstrasse 28/30, 48149 Münster, Germany.

Title:
INVESTIGATION OF TRIAPINE METABOLISM BY APPLICATION OF ELECTROCHEMISTRY AND LIQUID CHROMATOGRAPHY COUPLED TO MASS SPECTROMETRY (EC/LC/ESI-MS)

Introduction:
Thiosemicarbazone-based anticancer drugs belong to the most effective ribonucleotide reductase inhibitors with Triapine as the most promising compound. With regard to clinical trials, Triapine showed promising activity against leukemia, but ineffectiveness against solid tumors. Therefore, it is of high interest to gain more insight into pharmacological characteristics and metabolism of Triapine.

Methods:
Electrochemistry (EC) coupled to liquid chromatography (LC) and mass spectrometry (MS) is a promising tool in metabolism research and the application enables the simulation of many oxidative liver reactions. Complementary to this purely instrumental technique, drug metabolism was investigated by in vitro incubations with human liver microsomes (HLM) followed by identification via LC/MS.

Results:
The metabolic transformation of Triapine was investigated by means of EC/ESI-MS. First, a mass voltammogram was recorded by plotting mass spectra against the applied potential and the oxidation products of Triapine were identified. Dehydrogenation was the main oxidation reaction obtained in addition to hydroxylation. In order to characterize the products regarding their polarity and isomer formation, a reversed-phase liquid chromatographic separation coupled to ESI-MS was carried out. For comparison, in vitro incubations with human liver microsomes were performed and the obtained metabolites were separated via LC/ESI-MS, resulting in the same hydroxylated and dehydrogenated species. Thus, electrochemistry, as the instrumental approach, and in vitro incubations with HLM, can both be applied for simulation of metabolic pathways of Triapine.

Novel Aspect:
The metabolism of the clinically relevant anticancer drug Triapine was investigated using EC/LC/MS and compared with oxidation using HLM.

Pellegrini, Davide,
NEST, Scuola Normale Superiore

Contributing Authors:

Marialaura Dilillo, Fondazione Pisana per la Scienza - ONLUS, Pisa (Italy); Erik de Graaf, Fondazione Pisana per la Scienza - ONLUS, Pisa (Italy); Rima Ait-Belkacem Fondazione Pisana per la Scienza - ONLUS, Pisa (Italy); Liam McDonnell, Center for Proteomics and Metabolomics and Department of Pathology, Leiden (The Netherlands) ; Fondazione Pisana per la Scienza - ONLUS, Pisa (Italy).

Title:

Exploring microfluidics for ultrasensitive proteomic analysis of laser-capture microdissected fresh frozen tissues

Introduction:

The number of proteins that can be identified or quantified by MS decreases with smaller sample amounts, partly due to the lack of suitable microsample preparation protocols. To minimize sample losses one-pot solutions have been proposed. In this work we further explored the adaptation of these methods to a microfluidic scale.

Methods:

Laser-capture microdissection was used to cut 1mm2 regions of mouse brain tissues. The total protein content of each sample was determined with a modified microBCA assay. The SP3 protocol was applied for protein digestion and clean-up. nLC-MS (Orbitrap Fusion) was used for peptide separation and identification.

Results:

We optimized the microBCA assay for the determination of protein concentration in low concentration samples in the presence of detergents and chelating agents. The NanoQuant plate provided a high sensitivity and reproducibility for only 1µL of sample. The protein content was found to be extremely variable among different tissue samples, indicating a high heterogeneity of brain tissue and thus the need to quantify protein amount for each sample before digestion. The SP3 protocol was particularly suited for the manipulation of low sample amounts identifying more than 5000 proteins from 1mm2 tissue samples. In order to further minimize the handling of small sample amounts and increase throughput, the entire protocol will be adapted into a microfluidic setup and combined with immobilized trypsin.

Novel Aspect:

An ultrasensitive workflow provides the sensitivity and reproducibility required for the in-depth proteomic analysis of small tissue biopsies.

Penning, Renske,
Utrecht University

Contributing Authors:
Albert JR Heck, Utrecht University; Harold D MacGillavry, Utrecht University; AF Maarten Altelaar Utrecht University; Charlotte AGH van Gelder, Utrecht University; Casper C Hoogenraad, Utrecht University.

Title:
In-depth study of the molecular events underlying mGluR-LTD by combining pulsedSILAC/AHA labeling and phosphoproteomics: implications for ASD.

Introduction:
At neuronal synapses, activation of metabotropic glutamate receptors (mGluR1/5) triggers a form of long-term depression (mGluR-LTD) that relies on new protein synthesis and the internalization of AMPA-type glutamate receptors (AMPARs). Dysregulation of these processes has been implicated in the development of mental disorders such as autism spectrum disorders (ASD).

Methods:
Rat hippocampal neurons (both wildtype and with reduced expression of Shank proteins) were stimulated with the mGluR1/5 agonist DHPG and harvested at different time points. A highly sensitive phosphoproteomics workflow with Fe(III) IMAC was applied to study the phosphoproteome. Newly synthesized proteins upon stimulation were enriched and analyzed by applying a pulsed SILAC and AHA-LC-MS strategy.

Results:
Stimulation of mGluR1/5 triggered a long-lasting decrease in AMPAR surface expression in control, but not Shank knockdown neurons. We observed a very high correlation between phosphorylation dynamics and protein synthesis upon mGluR stimulation, allowing the study of mGluR-LTD in a detailed and comprehensive manner. Knockdown of Shank3 revealed in a previous study that dysregulation of PP2a and the Akt-mTORC pathway are involved in the display of autistic features. We identified several phosphorylation changes in pathways leading to mTORC regulation. Also changes in the AMPA receptor endocytosis pathway in both protein synthesis and phosphorylation upon LTD were identified. Combining these results enhanced the understanding of cellular pathways leading to mGluR-LTD in general, but also specifically in an ASD-linked genotype.

Novel Aspect:
The first application of pSILAC/AHA in primary neurons combined with phosphoproteomics to study the processes underlying mGluR-LTD related to ASD.

Pfammatter, Sibylle,
IRIC - Université de Montréal

Contributing Authors:
Pierre Thibault, Institute of Research in Immunology and Cancer, Université de Montréal, Montréal, Canada and Department of Chemistry, Université de Montréal, Montréal, Canada.; Chantal Durette, Institute of Research in Immunology and Cancer, Université de Montréal, Montréal, Canada.; Eric Bonneil Institute of Research in Immunology and Cancer, Université de Montréal, Montréal, Canada.

Title:
Targeted proteomic analyses of MHC peptides using differential ion mobility mass spectrometry

Introduction:
The analysis of Major Histocompatibility Complex (MHC) class I peptides present a challenge for proteomic analysis due to their low abundance at the cell surface. We used differential ion mobility (FAIMS) to facilitate the detection of low abundance peptides by transmitting MHC I peptide ions specific mobilities while reducing the contribution of interfering ions typically observed in LC-MS.

Methods:
Synthetic peptides and cell extracts isolated from human B lymphocyte cells were analyzed by LC-MS/MS with and without FAIMS on a LTQ-Orbitrap Elite instrument. The FAIMS interface was operated at a dispersion voltage (DV) of -5000 V using 100% N2 as carrier gas (2.3 L/min) with compensation voltage (CV) values optimised for maximal transmission of target peptide ions.

Results:
Targeted LC-MS/MS experiments using FAIMS enabled the identification of 14 synthetic MHC peptides present at low fmoles levels in HEK293 cell extracts whereas only 9 peptides were detected without FAIMS. We analyzed MHC peptide extracts isolated from B-cell lymphocytes that contained 15 putative minor histocompatibility antigens (MiHA) harboring single nucleotide polymorphisms. Heavy synthetic peptides of these MiHAs were spiked in the extract to confirm identity. LC-MS/MS analyses only identified 3 MiHA peptides vs. 10 peptides using LC-FAIMS-MS/MS. MiHAs were expressed at the cell surface at levels ranging from 10E2 to 10E4 peptides/cell. The enhanced sensitivity of LC-FAIMS-MS/MS enables the detection of MiHAs, and opens up interesting avenues for better personalized medicine where prior knowledge of MiHAs in leukemic patient samples is required for adoptive T-cell immunotherapy.

Novel Aspect:
FAIMS improves the identification and quantification of Major Histocompatibility Complex peptides

Picard, Pierre,
Phytronix Technologies Inc

Contributing Authors:
Jean Lacoursiere, Phytronix Technologies; Serge Auger, Phytronix Technologies; Alex Birsan Phytronix Technologies.

Title:
Now a reality: Sub second quantitative mass spectrometry analysis with Acoustic sample deposition and fiber coupled LDTD-MS/MS

Introduction:
During the past several years, LDTD-MS/MS methods were developed in different fields. Fastest published analysis reports 2.14 seconds sample to sample in ADMET. Other publications show 6-9 seconds for biological matrix. New gas dynamic design combined to high performance fiber coupled laser diode permits a reduction of desorption time cycle under one second sample to sample analysis.

Methods:
Carrier gas flow stabilisation time and dynamics are compared with previous model. Real sample analysis uses 40nL of test molecule mixed with 10µL of buffer, 10µL CYP/NADPH/Buffer and 10µL of substrate solution. Reactions were quenched with 20µL Water:MeCN:Formic acid containing IS. 100 nL of Samples are transferred via acoustic sample deposition onto 384-LazWell plates using ECHO550.

Results:
Stabilisation of the gas flow occurs within 100msec after the completion of the piston movement. Heating uniformity with the use of fiber coupled laser desorbs simultaneously a zone from the center of the well to 90% of the outer edge. The perimeter of the well desorbs 30msec later as the thermal conductivity reduces the heating ramp slope. To evaluate application to real assays, we optimized the analysis method resulting in a sample analysis speed of less than 0.9sec (sample to sample) per well. The analysis of twelve 384-well plates was completed in about 1.15 hours. Another important advantage is that it is amendable to post-reaction sample multiplexing; where 3-6 reaction samples can be transferred into the same well (same LazWell plate). LDTD coupled with acoustic sample deposition and multiplexing may significantly decrease analysis time to 0.15-0.30 seconds/compound.

Novel Aspect:
New High throughput fiber coupled LDTD ion source tested with IC50 evaluation of CYP inhibition assay using low volume (100 nL)

Piotrowski, Mary,
Pfizer

Contributing Authors:
Richard Kibbey, Yale School of Medicine; Julie Keefer, Pfizer; Darren Dumlao Pfizer; John Janiszewski, Pfizer.

Title:
Differential Mobility Spectrometry/Mass Spectrometry Methodology Used for the Quantitation of Glycolysis/Sugar Phosphates in Metabolic Flux Studies.

Introduction:
In metabolic flux studies, cells are fed stable label precursors (13C-Glucose) resulting in formation of isotopically labeled metabolic intermediates wherein the isotopic fragmentation pattern can reveal changes occurring in biochemical pathways (e.g. Glycolysis). The resulting isotopomers and positional isomers with concurring masses were quantified using DMS technology.

Methods:
All analyses performed on a Sciex 6500+ triple quadrupole mass spectrometer equipped with the with SelexION+ (differential mobility spectrometry interface) in negative electrospray ionization mode. Liquid chromatography performed on an Imtakt Scherzo SM (75x2.1mm, 3u) chromatography column. Data was acquired with Analyst (version 1.6) using multiple reaction monitoring (MRM).

Results:
DMS parameters were optimized to separate the molecules that were not chromatographically resolved and have identical MRM transitions. Currently, the following isomers, 2-PG and 3-PG were resolved in the DMS cell by using different compensation voltages. 2-propranol was found to be an adequate chemical modifier in the DMS cell to aid in this separation. In addition, isomers Glucose 6-Phosphate and Glucose 1-Phosphate were also completely resolved, as were ribose 5-phosphate and xylose 5-phosphate. Further work will include testing different chemical modifiers (1-propanol, acetone, acetonitrile and methanol) to improve separation. Additional analytes of interest will be evaluated. Methods developed will be evaluated for necessary sensitivity, linearity, precision and accuracy for sample sets to be analyzed.

Novel Aspect:
Differential mobility spectrometry/LC/MS/MS to separate co-eluting isomer components for the quantitation of key glycolysis and sugar phosphates.

Plath, Logan,
Carnegie Mellon University

Contributing Authors:
Zongyu Wang, Carnegie Mellon University; Mark Bier, Carnegie Mellon University; Krysztof Matyjaszewski Carnegie Mellon University.

Title:
Characterization of ZnO Nanoparticles using Superconducting Tunnel Junction Cryodetection MS

Introduction:
MALDI-TOF-STJ MS was used to characterize synthetic ZnO nanoparticles with varying ligand coatings. Not only can a single MS analysis provide high m/z determination, dispersity, and size, but also ligand loading via STJ energy resolved metastable fragmentation. This methodology demonstrates applicability to many areas of nanoparticle characterization through a single experiment.

Methods:
ZnO nanoparticles (octylamine coated) were synthesized by nonhydrolytic decomposition methods. Polystyrene acrylonitrile coated ZnO was synthesized by ligand exchange methods. MALDI-TOF-STJ MS experiments were performed on a modified Macromizer instrument with a 16-channel STJ detector array held at 0.34 K and biased at 350 µV. Data was processed on a custom Mathematica program developed in-house.

Results:
Ligand coated ZnO nanoparticles were analyzed intact by both LDI and MALDI-TOF-STJ MS. Intact molecular weights for the nanoparticles spanned from ~100-600 kDa in agreement with average TEM diameters of ~6 nm. Use of STJs for ion detection affords the advantage of measuring the relative ion energy deposited into the detector. The 3D-spectra (m/z, energy and intensity) generated from the analysis of these ZnO nanoparticles revealed a broad peak corresponding to the singly-charged ZnO species and a band lower than the singly charged energy. Through a combination of LDI and MALDI experiments, the band of ions registered at lower energy was assigned to metastable fragments resulting from ligand removal from the nanoparticles. Using this STJ energy measurement, ligand loading was determined to account for ~25% of the mass of the particle. These results support previous TGA studies.

Novel Aspect:
A single STJ cryodetector MS experiment can be used to determine nanoparticle molecular weight, dispersity, size, and ligand loading.

Pöhö, Päivi,
Division of Pharmaceutical Chemistry and Technology, Faculty of Pharmacy, University of Helsinki

Contributing Authors:
Niina Kärkkäinen, Division of Pharmaceutical Chemistry and Technology, Faculty of Pharmacy, University of Helsinki; Risto Kostiainen, Division of Pharmaceutical Chemistry and Technology, Faculty of Pharmacy, University of Helsinki; Markus Haapala Division of Pharmaceutical Chemistry and Technology, Faculty of Pharmacy, University of Helsinki.

Title:
Analysis of steroids in human urine by gas chromatography-capillary photoionization-tandem mass spectrometry

Introduction:
Steroids are hydrophobic compounds, which have many important roles in the biological processes. The concentrations of steroids in biological samples as well as the sample volume available for analysis can often be very limited, requiring highly sensitive methods for analysis. We have applied gas chromatography-capillary photoionization-tandem mass spectrometry to analyze steroids from urine.

Methods:
Steroids were extracted with diethyl ether and derivatized with silylation prior analysis. Separation was done applying gas chromatography and two analytical columns concatenated. Analytes were transferred via heated transfer line and coaxial dopant gas flow to capillary photoionization interface developed in-house. For detection we applied triple quadrupole and selected reaction monitoring mode.

Results:
Gas chromatographic separation and mass spectrometric detection was optimized for in total 17 steroids. Precursor ions for the most of the derivatized steroids were molecular ions with few exceptions. The most abundant precursor as well as two most abundant fragments were selected for further method development. Separation optimization focused to overlapping isobaric species and all were at least partially separated in the final method. Method was validated with artificial urine for linearity, limit of detection, limit of quantification, intraday repeatability, liquid-liquid extraction recovery and intra- and interday repeatability of the retention times. Promising results were obtained with good sensitivity and analytical performance. The methods applicability for urine analysis was demonstrated with urine samples from healthy volunteers.

Novel Aspect:
Sensitive atmospheric ionization technique with efficient ion transmission and suitability for non-polar analytes.

Polasky, Daniel,
University of Michigan - Ann Arbor

Contributing Authors:

Philip Andrews, University of Michigan, Department of Biological Chemistry; Brandon Ruotolo, University of Michigan, Department of Chemistry.

Title:

Chemical Modification as a Path to Improved Sequencing and Structural Analysis of Proteins and Protein Complexes

Introduction:

Direct analysis of protein complexes by collision-induced dissociation (CID) is limited by ejection of highly charged monomers and poor sequence coverage. We are developing chemical modification of protein complexes as a means to reveal unique peptide and protein fragment ions that cover a greater portion of the protein sequence and more fully reflect the structure of the assembly.

Methods:

Chemical modifications were performed on model proteins including Avidin using homobifunctional primary amine-targeting agents BS3 and DC4, a monofunctional, methylated analog of DC4 dubbed M-tag, and 2,4,6-trimethyl pyrylium (TMP). Proteins were buffer exchanged into 0.2M ammonium acetate prior to MS analysis. All data were acquired on a Synapt HDMS G2 Q-IM-TOF instrument (Waters, Milford, MA).

Results:

Tetrameric Avidin complexes modified with fixed positive charges are shown to dissociate unique peptide products not seen in CID of unmodified Avidin, resulting in improved sequence coverage. In addition, both covalent crosslinking and charge fixing modifications led to significant yields of dimeric subcomplexes of the Avidin tetramer that are not seen in CID of the unmodified tetramer, despite the fact that Avidin possesses a dimer of dimers topology. Here, we discuss the use of these improvements in both sequence coverage and subunit dissociation pathways to build an optimized workflow for the top-down analysis of intact protein complexes. We discuss the mechanism and capabilities of these alternate dissociation pathways and next-generation informatics tools developed to interpret the output of top-down chemically-modified protein complex CID experiments in an integrated fashion.

Novel Aspect:

New workflows to improve the top-down sequencing and structural analysis of protein complexes using covalent chemical modifications.

Politis, Argyris,
King's College London

Contributing Authors:
Andy Lau, King's College London.

Title:
Ion mobility mass spectrometry uncovers the conformational states of a supercomplex

Introduction:
The integration of MS with modelling has emerged as a pivotal tool for studying the structure and dynamics of protein complexes. Combining native MS with ion mobility (IM) enables structural information for interrogating computer-based models. Information about their dynamics can be inferred from the distribution of mobilities measured by IM-MS, adding a dimension in model-building methods.

Methods:
We use IM-MS in conjunction with modelling capabilities to study the conformational dynamics of heterogeneous and multi-protein complexes. CCSs and mobility distributions derived from IM-MS are encoded into restraints and subsequently used by an integrative workflow to interrogate protein assembly models. This method highlights the ability of IM-MS to build ensemble models of macromolecules.

Results:
Here, we investigate the structure and dynamics of Cop9 signalosome (CSN) - Cullin Ring Ligase 2 (CRL2) supercomplex (~500kDa, 13 proteins). IM-MS experiments enabled shape and flexibility information of the CSN-CRL2 and after in-solution disruption of many other (sub)complexes. IM-MS measurements revealed flexible regions within the supercomplex, while the addition of the ubiquitin like protein Nedd8 prompted an increase in its CCS consistent with significant structural rearrangement. Computational integration of the structural information derived from native MS and IM-MS allowed us to build an ensemble of model structures of CSN-CRL2 supercomplex corresponding to distinct conformational states. Such models demonstrate the power of integrative MS-based methods to unravel the molecular architectures and dynamics of large and heterogeneous assemblies.

Novel Aspect:
The novelty lies in the integration of ion mobility MS with modelling to probe the conformational states of native protein complexes

Politis, Argyris,
King's College London

Contributing Authors:

Ken Wei Tan, King's College London; Carla Schmidt, University of Halle-Wittenberg.

Title:

Surface accessibility and dynamics of macromolecular assemblies captured by covalent labelling mass spectrometry and molecular simulations

Introduction:

The combination of MS and computational methods is increasingly used to study protein complex structures. A recent breakthrough was the integration of complementary MS methods enabling atomic-level characterization of large and heterogeneous complexes. Here we use a new combination of labeling MS methods allowing mapping of the solvent accessible residues on the surface of protein complexes.

Methods:

To study protein complex structure and dynamics, we introduce a method that computationally integrates structural information from covalent labeling and cross-linking MS with MD simulations. This method highlights the ability of labeling of accessible residues to map the protein surface enabling us to define the location and orientation of individual subunits within the assemblies.

Results:

We assessed our method on a tangible target, tryptophan synthase, with known structure. We then used it to characterize a challenging multi-subunit complex, the ATP-synthase form spinach chloroplasts (cATPase). We performed covalent labeling using diethylpyrocarbonate and cross-linking with BS3 allowing us to label serines, threonines, tyrosines and histidines on the surface of the complexes and map cross-linked lysines, respectively. We labelled 75 residues in all protein subunits. Together with cross-linked lysines we unambiguously defined the orientation and proximities of the subunits, while simulations enabled us to assign flexible regions and interconverting states within the enzyme consistent with its nature as a molecular motor. We identified enhanced flexibility in the central rotor subunit of cATPase in line with its conformational heterogeneity during catalytic rotation.

Novel Aspect:

The novelty lies in the use of covalent labelling and cross-linking MS to interrogate models of multi-protein assemblies built by computational tools.

Pompach, Petr,
Institute of Biotechnology

Contributing Authors:
Viktor Ruzicka, BioVendor, a.s., Brno, Czech Republic; Oldrich Benada, Institute of Microbiology, v.v.i., Czech Academy of Sciences, Prague, Czech Republic; Jana Novakova Institute of Microbiology, v.v.i., Czech Academy of Sciences, Prague, Czech Republic; Daniel Kavan, Institute of Microbiology, v.v.i., Czech Academy of Sciences, Prague, Czech Republic; Michael Volny, AffiPro, s.r.o., Mratin, Czech Republic; Petr Novak, Institute of Microbiology, v.v.i., Czech Academy of Sciences, Prague, Czech Republic.

Title:
Functionalized Surfaces for Direct Immuno-Affinity Mass Spectrometry - Detection of Haptoglobin Phenotypes

Introduction:
Antibody-functionalized MALDI surfaces prepared by ambient ion landing allow efficient immuno-affinity enrichment of antigen from complex samples directly on the plate. The effectivity of antibody-functionalized MALDI surfaces to enrich protein antigen was demonstrated on determination of haptoglobin phenotype - an important biomarker in patient survival.

Methods:
MALDI plates were functionalized by ambient ion landing of polyclonal human anti-haptoglobin antibody using the lab-made apparatus. One microliter of serum was applied on the anti-haptoglobin spot and incubated for one hour. After several washing steps, samples were in-situ reduced and mixed with matrix. Automated sample analysis was used to acquire spectra of intact proteins.

Results:
The MALDI slides were functionalized by anti-haptoglobin antibody using the ambient ion landing procedure. To detect the alpha subunits, it was necessary to reduce the protein after the enrichment, which was achieved by applying the TCEP reagent directly on the spots. The DHAP matrix was used for identification of alpha 1 or alpha 2 haptoglobin subunits. For the haptoglobin phenotype Hp 1-1, only an ion at m/z 9192 corresponding to molecular mass of alpha 1 subunit, was observed in the spectra. The spectrum of haptoglobin phenotype Hp 2-1 contained ions at m/z 9192 and at m/z 15945 (alpha 2 subunit). The spectrum of haptoglobin phenotype Hp 2-2 contained the dominant ion at m/z 15945 corresponding to the alpha 2 subunit. The method was tested using 116 human serum samples. The validation was performed by using SDS electrophoresis followed by western blot with chemiluminescent detection.

Novel Aspect:
Combination of planar functionalized surfaces with mass spectrometric detection represents a robust and fast method for antigen screening.

Popp, Robert,
University of Victoria - Genome BC Proteomics Centre

Contributing Authors:
Mark Basik, Jewish General Hospital, Lady Davis Institute, Department of Oncology, McGill University, Montreal, QC, Canada; Adriana Aguilar-Mahecha, Jewish General Hospital, Lady Davis Institute, Department of Oncology, McGill University, Montreal, QC, Canada; Gerald Batist Jewish General Hospital, Lady Davis Institute, Department of Oncology, McGill University, Montreal, QC, Canada; Oliver Pötz, Natural and Medical Sciences Institute (NMI) at the University of Tübingen, Reutlingen, Germany; Andrew G. Chambers, University of Victoria - Genome BC Proteomics Centre, Victoria, BC, Canada; Yassene Mohammed, University of Victoria - Genome BC Proteomics Centre, Victoria, BC, Canada; Christoph H. Borchers, University of Victoria - Genome BC Proteomics Centre, Victoria, BC, Canada.

Title:
Quantifying signalling pathway activity from tissue lysates using immuno-MALDI

Introduction:
Novel anti-cancer drugs target signaling pathways which, however, only work in a minority of patients. We therefore set out to develop immuno-matrix assisted laser ionization/desorption (iMALDI) assays based on our phosphatase-based phosphopeptide quantitation (PPQ) method to measure PI3K/AKT/mTOR pathway activity by quantitating AKT1 and AKT2 in cancer tissues as predictive biomarkers.

Methods:
Following trypsin digestion of cell lysate, stable isotope-labeled standard (SIS) peptides are added. After splitting the solution, one aliquot is treated with phosphatase. Bead-coupled antibodies enrich the target peptides, are washed and spotted onto a MALDI plate, and acidic MALDI matrix elutes the peptides. Light/heavy ratios of both aliquots allow calculation of phosphorylation stoichiometry.

Results:
We quantified the C-terminal target peptides of AKT1 and AKT2 from SW480 and HCT116 colon cancer cell lines, MDA-231 breast cancer cell lines, as well as flash-frozen and formalin-fixed paraffin-embedded (FFPE) HCT116 mouse xenograft colon cancer tissues and breast cancer tissues. Per sample, ~50 µg protein are required (10 µg protein per affinity-capture). Target peptide levels ranged from ~0.1-1 fmol/µg protein lysate with phosphorylation levels of 0-~50%. The lower limit of detection was determined to be 100 amol of peptide, with a linear range of 0.5-10 fmol of peptide on the MALDI plate. CVs for iMALDI assays were consistently <10%. In conclusion, we have developed iMALDI assays for quantitation of AKT1 and AKT2 expression levels and phosphorylation from tumor samples. Next steps include a full assay validation study.

Novel Aspect:
First iMALDI assays to quantify PI3K/AKT/mTOR signaling pathway members from tumor samples.

Porcari, Andréia,
ThoMSon Mass Spectrometry Laboratory, Institute of Chemistry, University of Campinas, UNICAMP, Campinas, SP, Brazil

Contributing Authors:

Alexandre Nowill, Integrated Research Center for Oncohematology of Infancy – CIPOI, UNICAMP, Campinas, SP, Brazil; Gustavo Duarte, ThoMSon Mass Spectrometry Laboratory, Institute of Chemistry, University of Campinas, UNICAMP, Campinas, SP, Brazil.; Eduardo C. Meurer Federal University of Parana, Jandaia do Sul, PR - Brazil; Marcos N. Eberlin, ThoMSon Mass Spectrometry Laboratory, Institute of Chemistry, University of Campinas, UNICAMP, Campinas, SP, Brazil.; Danilo Pereira, Waters Corporation, SP – Brazil; Andreia M. Porcari, ThoMSon Mass Spectrometry Laboratory, Institute of Chemistry, University of Campinas, UNICAMP, Campinas, SP, Brazil.; Rosana M. Alberici, ThoMSon Mass Spectrometry Laboratory, Institute of Chemistry, University of Campinas, UNICAMP, Campinas, SP, Brazil.

Title:

MAKING ESI-MS/MS NEWBORN SCREENING SIMPLIER AND MUCH LESS EXPENSIVE

Introduction:

Newborn screening (NBS) detects inborn errors of metabolism and it has been expanded to many diseases by the use of ESI-MS/MS. However, a large set of standards is required. Here we show that, using the response factor of the analytes, a set of only four standards is required to achieve proper quantitation and accuracy.

Methods:

Standards were acquired from Cambridge Isotope Laboratories. Samples from the Center for Disease Control (CDC, USA) quality assurance program and proficiency test from the years of 2015 and 2016 were used. Derivatization with acid butanol was employed. A Xevo TQD (Waters), in the MRM positive ion mode, connected to an I-Class UPLC (Waters) was used in the flow injection mode.

Results:

The proposed method uses 4 internal standards (IS) instead of the original 20-24 recommended IS. Linearity, precision and accuracy were evaluated for alanine, valine, leucine, methionine, tyrosine, arginine, phenylalanine and citruline by using a single labeled compound as IS. Linearity reached R=0.99 for all compounds. Accuracy ranged from 80-120% for these aminoacids (AA) and with-in and inter-run precisions averaged 15%. Analyses for acylcarnitines are in progress. By making use of 4 IS and considering their present individual costs, the cost per sample should not exceed 0.10 USD even after importing fees. Nowadays, NBS cost is around 13.00 USD when using the commercial IS kits and 1.57 USD when using in house methods. This great cost reduction should encourage governments thought the whole world particularly in developing and third world countries to implement NBS using MS/MS to all

Novel Aspect:

Simplification and price reduction of MS/MS NBS.

Postigo, Cristina,
Institute for Environmental Assessment and Water Research (IDAEA-CSIC)

Contributing Authors:
Paul Silcock, Thermo Fisher Scientific, Runcorn, UK; Cristian I. Cojocariu, Thermo Fisher Scientific, Runcorn, UK; Susan D. Richardson Department of Chemistry and Biochemistry, University of South Carolina, Columbia, SC, USA; Damia Barcelo, Catalan Institute for Water Research (ICRA), Girona, Spain.

Title:
Orbitrap™ based gas chromatography-mass spectrometry to characterize semi-volatile and volatile disinfection by-products in water

Introduction:
The recent development of the Q Exactive™ GC Orbitrap MS system provides a unique opportunity to qualitatively and quantitatively characterize small volatile and semi-volatile molecules in environmental samples. In the present work, this novel technology combination was applied to characterize low molecular weight iodinated disinfection by-products (iodo-DBPs) in disinfected waters.

Methods:
High resolution and high mass accuracy electron ionization - full scan data provided by the Q Exactive™ GC were used to characterize the main low molecular weight iodo-DBPs present in lab-scale generated chlorination and chloramination DBP mixtures. Two different waters with different natural organic matter, bromide and iodide content were tested.

Results:
Automatic peak detection, spectral deconvolution and compound identification was performed in the full scan chromatograms with the Trace Finder software. Iodine containing DBPs were picked after applying a mass filter (iodine mass: m/z 126.90392). Elucidation of iodo-DBP molecular structures was performed according to the elemental compositions obtained for molecular and fragment ions (mass error < 1 ppm) and fragment rationalization. The workflow applied allowed the identification of up to 11 different iodo-DBPs; however, the investigation of softer ionization techniques may help emerge additional iodo-DBPs in these extracts. In the light of the results, and as expected, iodo-DBPs were more abundant in chloraminated extracts, and the type of iodo-DBP formed was found to be related to the physical-chemical characteristics of the waters tested (the aromaticity content of NOM, Br-, and I-)

Novel Aspect:
First application of the novel Orbitrap based GC-MS technology to characterize semi-volatile and volatile iodinated molecules in waters

Pourhaghighi, Reza,
University of Toronto

Contributing Authors:
Florian Goebels, Donnelly Centre for Cellular and Biomolecular Research, University of Toronto, Toronto, Ontario M5S 3E1, Canada; Gary Bader, Donnelly Centre for Cellular and Biomolecular Research, University of Toronto, Toronto, Ontario M5S 3E1, Canada; Andrew Emili Donnelly Centre for Cellular and Biomolecular Research, University of Toronto, Toronto, Ontario M5S 3E1, Canada.

Title:
A novel two-dimensional (2-D) proteomic platform for proteome-wide protein-protein interaction studies

Introduction:
We developed a powerful 2-D proteomic platform, based on hyphenation of IEF and IEX chromatography followed by tandem Mass Spectrometry analysis of biochemical fractions for large-scale identification of protein complexes and protein-protein interactions (PPIs). The method has been positively employed and validated by identification of annotated PPIs in Yeast, C. elegans and C. briggsae.

Methods:
Isoelectric focusing (IEF) was carried out on a MicroRotofor cell and 5 fractions were collected. Subsequently, IEX was performed using an Agilent 1260 HPLC equipped with a CATWAX column (PolyLC) using Tris pH 8 and MES pH 6 (with increasing NaCl gradient) as buffers. Following trypsin digestion, peptides were analyzed by nESI-LTQ Orbitrap MS.

Results:
MS analysis of 240 soluble protein 2D fractions resulted in identification of 2708, 4345, 4035 unique proteins in S. cerevisiae, C. elegans and C. briggsae, respectively. Co-fractionation of complex subunits was evaluated by comparing co-elution profiles against a reference gold standard constructed by mapping conserved CORUM protein complexes to our target organisms. Compared to our previously reported 1D approach [Havugimana et al, Cell 2012], the 2-D method predicts more PPIs at a higher precision mainly due to higher dynamic range and decreased chance co-elution of non-related proteins. Finally, by combining the 2D co-fractionation data with complementary functional association information, a high confidence protein interaction network was created for each species.

Novel Aspect:
Our method is the first 2-D protein complex separation platform that markedly improves the quality of large-scale PPI studies.

Poutsma, John,
College of William and Mary

Contributing Authors:
Vicki Wysocki, Ohio State University; Hannah Smith, College of William and Mary; Arpad Somogyi Ohio State University.

Title:
IRMPD spectroscopy of fragments from proline-analog-containing peptides

Introduction:
Proline is the only secondary amine among the 20 amino acids. The structural rigidity of the five-membered ring of proline is thought to contribute to the "proline effect" in tandem mass spectrometry, which causes selective fragmentation N-terminal to the Pro residue. IRMPD was performed on fragments from peptides containing Pro or one of its analogs to better understand the proline effect

Methods:
IRMPD of peptide fragments were obtained at the CLIO free-electron laser (FEL) facility. Peptide ions are formed by ESI and allowed to undergo CID in the ion trap. Fragment ions are then isolated and allowed to interact with tunable IR radiation from the FEL. IR action spectra of photofragmentation yield versus IR wavelength are obtained and compared with vibrational spectra from DFT calculations.

Results:
The b2+ ion from APA peptide was previously shown to be an oxazolone ion with a strong stretch near 1900 cm-1. We show here that the b2+ ions from A(Aze)A, A(Pip)A, and A(D-Pro)A, where Aze is a 4-membered ring analog of proline, Pip is a 6-membered ring analog of proline, and D-Pro is the D-enantiomer of L-Proline, are also oxazolones. We will also present data from aza-proline containing peptides. Azaproline has been shown to prefer a cis-configuration in small peptides and should therefore be more likely to form diketopiperazine b ions.

Novel Aspect:
Direct vibrational spectroscopy of peptide fragments containing novel amino acids.

Poyer, Salomé,
University of Evry - LAMBE

Contributing Authors:
Juan XIE, PPSM, ENS Cachan, CNRS, Université Paris-Saclay, 94235, Cachan, France; Jean-Yves Salpin, LAMBE, Université d'Evry Val d'Essonne, CNRS, UMR 8587, Evry, France; Nicolas Bogliotti PPSM, ENS Cachan, CNRS, Université Paris-Saclay, 94235, Cachan, France; Salomé Poyer, LAMBE, Université d'Evry Val d'Essonne, CNRS, UMR 8587, Evry, France; Claire Deo, PPSM, ENS Cachan, CNRS, Université Paris-Saclay, 94235, Cachan, France.

Title:
Kinetic study of azobenzene E/Z isomerization using ion mobility-mass spectrometry

Introduction:
Azobenzene molecular switches can interconvert both photochemically and thermally. Due to their important development, their fast identification has become a challenge. Usually identified using UV-VIS spectrophotometry, we propose different coupling to MS such as ion mobility spectrometry (IMS) and liquid chromatography (LC) to study the thermal Z to E back relaxation kinetic of isomers.

Methods:
Different synthetized isomerically pure azobenzenes were analysed in MS/MS using different cation adducts. LC experiments were performed using a 1260 series Infinity system packed with a Kinetex XB C-18 column coupled to an ion trap mass analyser (AmaZon speed ETD). Ion mobility-mass spectrometry experiments were carried out on a hybrid instrument (Synapt G2-Si HDMS) equipped with an ESI source.

Results:
Analyses of isomers by tandem mass spectrometry were carried out on different azobenzene pairs and an azobenzene sample consisting of three isomers. Differentiation of isomers was achieved for isomer couples using MS/MS. An upstream separation dimension was necessary to characterise the sample with three isomers. For these compounds, IMS-MS and LC-(UV)-MS were evaluated in parallel to study the kinetic of thermal Z to E isomerisation. Baseline separation of two isomers was achieved in IMS-MS from [M+Ag]+, while the third isomer was differentiated from [2M+Ag]+. In addition, LC separation of the three isomers was achieved in less than 10 minutes. These methodologies can be applied to different systems in order to determine their thermal back relaxation kinetic. IMS-MS and MS/MS methods will be preferred for fast systems as they allowed differentiation in the millisecond scale.

Novel Aspect:
For the first time, the thermal back relaxation kinetic of azobenzene isomers is studied using hyphenated techniques (IMS and LC) to MS and MS/MS.

Preisler, Jan,
Masaryk University

Contributing Authors:
František Zelenák, Masaryk University, Brno, Czech Republic; Tomáš Vaculovič, Masaryk University, Brno, Czech Republic; Kristýna Dlabková Masaryk University, Brno, Czech Republic; Viktor Kanický, Masaryk University, Brno, Czech Republic; Iva Benešová, Masaryk University, Brno, Czech Republic.

Title:
Laser-Assisted Sample Introduction for Nanoparticle Characterization using ICP MS

Introduction:
Applications of nanoparticles (NPs) raise an elevated concern in environmental scientific community because of their rapid uptake, unknown distribution pathways and potential interactions in biological systems. We report a new sample introduction technique into single particle inductively coupled plasma mass spectrometry (SP-ICP MS) for characterization of NPs.

Methods:
Suspension of hundreds Au NPs in 200 nL droplet is deposited and dried on a strongly absorbing plastic plate (polyethylene terephthalate glycol) which is ablated with 213 nm laser (UP 213, New Wave Research) at a relatively low fluence (0.15 J·cm-2). The generated aerosol is delivered to an ICP quadrupole mass spectrometer (7500 CE, Agilent Technologies).

Results:
We describe a new approach of NP introduction to ICP MS from solid surfaces: substrate-assisted laser desorption (SALD), a technique originally developed for analysis of sub-microliter volumes of liquid samples. After drying, a plastic plate with samples in the form of dried droplets is ablated with a UV laser. We demonstrate that individual intact NPs can be efficiently desorbed from the surface and carried to plasma with transport efficiency of 61 %. Experimental conditions, such as the laser fluence, sample scan rate, carrier gas type and flow rate were carefully optimized to gain the highest NP signal and transport efficiency and to reduce NP disintegration within the desorption process. SALD ICP MS was validated using well characterized, commercially available Au NPs. The results are compared with routine nebulizer SP-ICP MS and possible applications of the technique are mentioned.

Novel Aspect:
New laser-assisted technique for nanoparticle introduction to ICP MS. Qantitative nanoparticle characterization with high transport efficiency (>50%).

Pruski, Pamela,
Imperial College London

Contributing Authors:
Trevor Hansel, Imperial College London; Phillip Bennett, Imperial College London; David MacIntryre Imperial College London; Holly Lewis, Imperial College London; Paolo Inglese, Imperial College London; Elaine Holmes, Imperial College London; Goncalo DS Correia, Imperial College London; Zoltan Takats, Imperial College London.

Title:
Medical swab analysis using desorption electrospray ionization mass spectrometry (DESI-MS) – a novel, non-invasive approach for mucosal diagnostics.

Introduction:
Medical swabs are used clinically for microbiological screening of mucosal surfaces. Rapid metabolic and microbial profiling of medical swabs can improve the diagnostic potential of this routine procedure. We show that DESI-MS enables rapid and direct metabolic profiling of mucosal tissue sampled with rayon medical swabs, offering a promising analytical platform for novel POC applications.

Methods:
Mucosal samples were collected from the urogenital tract (n=65), oral (n=25) and nasal (n=20) cavities using medical rayon swabs, and their surface directly analysed by DESI-MS. Liquid-liquid extraction was then used on the swabs to allow metabolite analysis by HILIC and RP LC-MS of the same swabs. DESI-MS was performed on a LTQ Orbitrap Discovery and Xevo G2-S Q-TOF MS and LC MS on Acquity UPLC.

Results:
Multivariate statistical analysis of the DESI-MS shows that the technique can readily distinguish between different mucosal sites (oral, nasal and vaginal) and even detect biochemical differences in the vaginal mucosa induced by pregnancy. To establish a link between mass spectrometry features and microbial presence, DESI-MS was also performed on cultivated microbe species. Spectral features were matched against vaginal swab samples, revealing the presence of C. albicans and Lactobacillus features. Since DESI-MS analysis minimally disrupts the sample, swabs were reused for LC-MS analysis, to benchmark the performance of DESI surface analysis with liquid-liquid extraction. Since DESI-MS spectral profiles can be obtained within minutes, the approach provides a rapid, culture independent method for mucosal diagnostics readily applicable to clinical settings.

Novel Aspect:
Direct swab analysis by DESI-MS is relevant for a wide range of clinical POC applications, such as faster microbiology diagnostics of mucosal samples.

Przybylski, Michael,
Steinbeis Centre for Biopolymer Analysis and Biomedical Mass Spectrometry

Contributing Authors:

Zdenek Kukacka, Steinbeis Centre for Biopolymer Analysis and Biomedical Mass Spectrometry, 65428 Rüsselsheim, Germany; Yannick Baschung, Steinbeis Centre for Biopolymer Analysis and Biomedical Mass Spectrometry, 65428 Rüsselsheim, Germany; Loredana Lupu Steinbeis Centre for Biopolymer Analysis and Biomedical Mass Spectrometry, 65428 Rüsselsheim, Germany; Jeff Bornheim, Reichert-Ametek Technologies, Buffalo, N.Y., USA; Mary Murphy, Reichert-Ametek Technologies, Buffalo, N.Y., USA; Hendrik Rusche, Steinbeis Centre for Biopolymer Analysis and Biomedical Mass Spectrometry, 65428 Rüsselsheim, Germany.

Title:

Online Biosensor- Mass Spectrometry for Epitope identification and affinity quantification of protein- ligand interactions

Introduction:

Biosensor analysis is established technique for detection and quantification of biomolecular interactions, but a major limitation is their lack of providing chemical structure information. Here we report a continuous online biosensor-MS combination with ESI that enables the simultaneous affinity isolation, structure identification and affinity quantification of biopolymer ligand complexes.

Methods:

Key tool of the online biosensor-MS epitope analyzer is a new, automated microfluidic interface that provides sample concentration and in-situ desalting for the direct MS analysis of the ligand eluate. ESI-MS systems from all major manufacturers can be coupled. Antibodies are immobilized on a gold chip and a tryptic peptide mixture loaded onto the SPR-MS, and the epitope eluate analyzed by ESI-MS.

Results:

The broad application potential of the online-SPR-MS epitope analyzer has been shown by recent studies of an unusual mixed-disulfide antibody epitope of the rheumatoic target protein, HLA-B27; and the interaction site identification of chaperone complexes of lysosomal enzymes. Interaction epitopes as diverse as antigen-antibody and lectin- carbohydrate complexes, and affinity binding constants (KD) from milli- to nanomolar ranges can be directly analysed. First applications to the direct analysis of biological samples, such as cell lysate and tissue homogenate, were successfully performed, as illustrated by the identification of a structure modification of the Parkinson protein alpha-synuclein from rat brain homogenate.

Novel Aspect:

New applications of the online-SPR-MS epitope analyzer include biomarker identification, protein epitope analysis and antigen quantification.

Pulliam, Christopher,
Purdue University

Contributing Authors:
Stephen T. Ayrton, Purdue University; Shannon A. Raab, Purdue University; Dalton T. Snyder Purdue University; Ryan M. Bain, Purdue University; R. Graham Cooks, Purdue University; Heather L. Osswald, Purdue University.

Title:
Multi-sample monitoring on the Mini 12 mass spectrometer

Introduction:
On-line reaction monitoring is frequently demonstrated using a single full-size mass spectrometer to monitor one reaction. Herein we will describe advances in miniature mass spectrometry and how this technology can be used as a miniature process analytical tool by demonstrating single sample reaction monitoring and multi-sample reaction monitoring while only using one instrument.

Methods:
On-line reaction monitoring was carried out by pressurizing a reaction vessel causing solution to flow through a capillary which then filled a nanoESI emitter for ionization and analysis by a Mini 12. Multi-sample monitoring was performed with multiple pressurized flasks flowing to multiple emitters. The nanoESI probes were automatically placed in front of MS inlet using a rotary actuator.

Results:
Multi-sample reaction monitoring was explored using variants of a hydrazone formation reaction involving an aldehyde and a substituted hydrazine. Up to four reactions were simultaneously monitored using a Mini mass spectrometer. Because nanoESI emitters were used, cross contamination was not evident even after hours of continuous monitoring at synthetically relevant concentrations. Also because the samples are time-shared, the time resolution scales as a function of the number of samples (n) being analyzed. Time resolution = 20 seconds x n Although the time resolution decreases, there is still adequate sampling frequency to extract dynamic profiles for species of interest. All the data was processed using Matlab to extract the ion chronograms. This allows visualization of the starting material, intermediates, and product as they change over the course of the reaction(s).

Novel Aspect:
Multi-sample reaction monitoring performed on a miniature mass spectrometer for on-line chemical analysis.

Purves, Randy,
University of Saskatchewan

Contributing Authors:
Albert Vandenberg, University of Saskatchewan.

Title:
Semi Quantitative LC-SRM workflow for the analysis of over 100 polyphenols in pulse crops

Introduction:
Pulse crops (e.g. lentil) are being promoted in crop rotations. They provide environmental benefits through nitrogen fixation, and health benefits as food through quality protein and low glycemic index. Polyphenols have positive effects on human health and plant health. Understanding how polyphenol profiles affect these properties is important in genetic improvement and crop economics.

Methods:
Various tissues from a variety of pulse crops were freeze dried. Polyphenols were extracted from ~100 mg of plant tissues by pulverizing using an acetone/water mixture. An Agilent 1290 UPLC equipped with PDA detection was coupled with a Thermo Vantage triple quadrupole mass spectrometer to detect the polyphenols present. A 30 minute gradient was used with a variety of poroshell columns.

Results:
The number of compounds in the workflow is continually being expanded and currently uses 79 standards and 6 internal standards to examine more than 100 polyphenols covering a variety of polyphenol classes. Several different columns were tested and due to the variety of structures and pKa values, typically the anthocyanidins gave very wide peaks unless high acid conditions were used. The Agilent Poroshell PFP was chosen because it was the only column tested that gave narrow peaks for these compounds using 1% formic acid. Ionization conditions were also found to be critically important as in-source CID for several glycosylated compounds can greatly complicate the spectrum in positive mode and therefore negative ionization was preferred for the majority of analyes. The workflow will be described in detail and results from several pulse crops will be presented.

Novel Aspect:
This workflow enables the rapid characterization of a large number of polyphenols in various tissues for a wide range of pulse crops.

Raaijmakers, Linsey,
Utrecht University

Contributing Authors:
Gianluca Maddalo, Biomolecular Mass Spectrometry and Proteomics, Bijvoet Center for Biomolecular Research and Utrecht Institute for Pharmaceutical Science, Utrecht University, Padualaan 8, 3584 CH Utrecht, The Netherlands; Albert J.R. Heck, Biomolecular Mass Spectrometry and Proteomics, Bijvoet Center for Biomolecular Research and Utrecht Institute for Pharmaceutical Science, Utrecht University, Padualaan 8, 3584 CH Utrecht, The Netherlands; Liesbeth Hoekman The Netherlands Cancer Institute, Plesmanlaan 121, 1066 CX Amsterdam, The Netherlands; A.F. Maarten Altelaar, Biomolecular Mass Spectrometry and Proteomics, Bijvoet Center for Biomolecular Research and Utrecht Institute for Pharmaceutical Science, Utrecht University, Padualaan 8, 3584 CH Utrecht, The Netherlands; Kristel Kemper, The Netherlands Cancer Institute, Plesmanlaan 121, 1066 CX Amsterdam, The Netherlands; Onno Bleijerveld, The Netherlands Cancer Institute, Plesmanlaan 121, 1066 CX Amsterdam, The Netherlands; Daniel S. Peeper, The Netherlands Cancer Institute, Plesmanlaan 121, 1066 CX Amsterdam, The Netherlands; Violette Gautier, Biomolecular Mass Spectrometry and Proteomics, Bijvoet Center for Biomolecular Research and Utrecht Institute for Pharmaceutical Science, Utrecht University, Padualaan 8, 3584 CH Utrecht, The Netherlands.

Title:
Visualizing mechanisms of BRAF-mutant melanoma drug resistance using PhosphoPath

Introduction:
Melanoma is the most aggressive form of skin cancer. The discovery of the BRAFV600E mutation has resulted in the development of targeted therapies. However, all melanomas eventually become resistant. To study regulated pathways, we should examine the proteome as well as regulation on PTM level. To analyze these data and integrate it with knowledge stored in databases, we developed PhosphoPath

Methods:
Biopsies from BRAF mutant melanoma patients were taken before treatment with a BRAF inhibitor, treatment sensitive, and upon drug resistance. Samples were analyzed by shotgun (phospho)proteomics and changing proteins and phosphosites were taken for further analysis in PhosphoPath, where most up-to-date information was retrieved from PhoshositePlus, Wikipathways and Biogrid.

Results:
Using PhosphoPath we were able to identify pathways significantly regulated upon resistance for BRAFi treatment. We observe activation of RAS and mTOR signaling on both proteome and phosphorylation level, driven by the NRAS mutation identified in the RNA-seq data, leading to cell survival, proliferation and inhibition of cell-extracellular matrix adhesion. Increased cell migration and invasiveness is observed through the decreased activity of the focal adhesion pathway. PhosphoPath is a valuable tool for the analysis of phosphoproteomics data. It integrates information from resources and visualizes available data on dynamic regulation in one network. By visualizing pathways of interests and adding quantitative information, a dynamic picture is constructed of regulated proteins and phosphorylation sites.Overall, we identified several critical signaling switches maintaining drug resistance

Novel Aspect:
Phosphopath is specifically developed to visualize dynamic quantitative regulation at PTM level over multiple conditions, interactions, and pathways

RABIN, Clémence,
INSERM U1212 - ARNA Lab

Contributing Authors:
Adrien Marchand, INSERM U1212 - ARNA Lab; Joséphine Abi-Ghanem, INSERM U1212 - ARNA Lab; Valérie Gabelica INSERM U1212 - ARNA Lab.

Title:
A native MS study of RNA kissing complexes and their interaction with magnesium cations

Introduction:
Native MS uses gentle conditions to maintain complex structures from solution to gas phase. Folding preservation has however not been fully validated for RNA. Here we studied RNA kissing complexes (small RNA hairpins dimerizing through their loops) and their interaction with Mg2+ cations. The TAR RNA element of HIV1 viral genome and its aptamer R06, and RNAI-RNAII from E.Coli were used as models.

Methods:
An Agilent 6560 ESI-IMS-QTOF was used. We deduce the number of cation adducts specifically pertaining to the loop-loop kissing motif, using reference sequences forming a duplex RNA of the same length and base pairs than the kissing complexes. By mathematical subtraction of the non-specific adducts, we can deduce the number of specific cations. Collision cross sections were also determined.

Results:
ESI-MS allowed us to detect specifically RNA-RNA kissing complexes present in solutions doped with Mg(OAc)2 up to 1 mM. We show that Mg2+ cations help to detect larger amount of dimers and hence stabilize them. We also determined that there are 2 Mg2+ specific ions into the kissing loop complex. We used the same method with Mn(OAc)2 in order to study the cation effect. We also investigated the link between the number of bases in the loop, the length of the sequence and the number of specific cations. We show here that sequences with different number of bases in the loop have different cation binding patterns. The results about specific cations binding will help us to determine possible mechanisms and binding constants for the formation of kissing complexes. Moreover, IMS allows us to determine the potential impact of cations on the preservation of the gas-phase RNA structures.

Novel Aspect:
We observe Mg2+-bound RNA complexes using native mass spectrometry and present novel insight in the structure of kissing loop motifs.

Radicova, Monika,
Comenius University in Bratislava, Faculty of Natural Sciences, Department of Analytical Chemistry

Contributing Authors:
Marian Marton, Slovak University of Technology in Bratislava, Faculty of electrical engineering and information technology, Institute of Electronics and Photonics, Ilkovicova 3, SK-81219 Bratislava, Slovakia; Robert Bodor, Comenius University in Bratislava, Faculty of Natural Sciences, Department of Analytical Chemistry, Mlynska Dolina, Ilkovicova 6, SK-84215, Bratislava, Slovakia; Marian Vojs Slovak University of Technology in Bratislava, Faculty of electrical engineering and information technology, Institute of Electronics and Photonics, Ilkovicova 3, SK-81219 Bratislava, Slovakia; Robert Redhammer, Slovak University of Technology in Bratislava, Faculty of electrical engineering and information technology, Institute of Electronics and Photonics, Ilkovicova 3, SK-81219 Bratislava, Slovakia; Miroslav Behul, Slovak University of Technology in Bratislava, Faculty of electrical engineering and information technology, Institute of Electronics and Photonics, Ilkovicova 3, SK-81219 Bratislava, Slovakia; Kvetoslava Mladoniczka, Comenius University in Bratislava, Faculty of Natural Sciences, Department of Analytical Chemistry, Mlynska Dolina, Ilkovicova 6, SK-84215, Bratislava, Slovakia; Andrea Vojs Stanova, Comenius University in Bratislava, Faculty of Natural Sciences, Department of Analytical Chemistry, Mlynska Dolina, Ilkovicova 6, SK-84215, Bratislava, Slovakia.

Title:
Using of a high resolution mass spectrometry for the identification of electrochemical degradation products of selected drugs

Introduction:
Pharmaceuticals are widely used in human and veterinary medicine. Consequently, it is common to find diverse pharmaceuticals in effluents from municipal wastewater treatment plants, which are generally unable to effectively remove pharmaceuticals from wastewaters. The aim of our work was to identify products of electrochemical degradation of selected drugs using high resolution mass spectrometry.

Methods:
Electrochemical degradation of analytes was performed in a stirred solution using two parallel BDD electrodes (1 cm2 active surface) connected to a power supply. Identification of degradation products was performed by LCMS-IT-TOF analyzer (Shimadzu) in a direct infusion mode, with ESI ionization in a positive and negative ionization mode and with a mobile phase flow rate 0.2 ml/min.

Results:
Electrochemical degradation of selected drugs was performed in a 10 ml electrochemical cell with two parallel BDD electrodes. Because of the compatibility of EC conditions with the MS detection, ammonium acetate buffer (0.1 mol/L) has been selected as a working electrolyte. For the identification of degradation products the direct infusion - HRMS was used. The important part of this work was an optimization of the mobile phase composition in order to obtain a sufficient ionization of analytes. In most cases, the combination of 10 mmol/L formic acid with acetonitrile (50/50, v/v) seems to be the optimal. After a proper degradation time it was possible to identify individual degradation products using full scan data with a high mass accuracy and MSn experiments. Acknowledgement: This work was financially supported by the grant of the Slovak Research and Development Agency (APVV-0365-12).

Novel Aspect:
Electrochemical degradation of selected drugs on BDD electrodes and identification of degradation products using a high resolution mass spectrometry.

Ramsey, J. Michael,
University of North Carolina at Chapel Hill

Title:
Microfabricated Technologies for Addressing Molecular Measurements: Microchip CE-nESI and Highly Pressure Mass Spectrometry

Introduction:
The first microfabricated fluidic research tools were commercially launched over 15 years ago for characterizing DNA, RNA, and proteins. An overwhelming number of these tools utilize fluorescence probes/tags for quantification of the target species. Mass spectrometry is a label free measurement technique that provides primary structure in addition to quantification.

Methods:
We utilize microfabricated fluidic circuits and surface covalent coatings to achieve state of the art electrophoretic separation performance with monolithically integrated ESI emitters. This capability has been coupled to miniaturized ion trap mass analyzers that have the ability to generate mass spectra at pressures exceeding 1 Torr.

Results:
We have demonstrated ESI-HPMS analysis of biological compounds ranging amino acids to proteins. Focusing on the twenty common amino acids, we have shown previously that we can perform CE-ESI-HPMS to monitor amino acid content in LB cell growth medium and monitor the consumption of amino acids by E. coli over time. Using this method, components of growth medium can be analyzed in under 3 minutes. With such a short time for analysis and the small size of the analysis platform, this strategy could be used for on-line monitoring of cell cultures. We have also previously shown detection of intact proteins and are working to improve protein analysis for several applications, including monitoring glycated hemoglobin in blood as an on-site diagnostic tool for diabetes and at line monitoring of biopharmaceutical materials.

Novel Aspect:
Microchip CE and high pressure mass spectrometry are coupled via a custom interface for the analysis of biomolecules.

Rand, Kasper,
University of Copenhagen

Contributing Authors:
Patrick S. Merkle, Department of Pharmacy, University of Copenhagen; Ingvar R. Møller, Department of Pharmacy, University of Copenhagen; Kamil Gotfryd Department of Neuroscience and Pharmacology, University of Copenhagen; Ulrik Gether, Department of Neuroscience and Pharmacology, University of Copenhagen; Claus J. Løland, Department of Neuroscience and Pharmacology, University of Copenhagen; Kasper D. Rand, Department of Pharmacy, University of Copenhagen.

Title:
Probing the conformation and ligand binding of neurotransmitter symporter (NSS) proteins by HDX-MS

Introduction:
NSS proteins, such as the Serotonin Transporter (SERT) are pivotal in terminating synaptic signaling. SERT is a major drug target, yet insufficient knowledge is available on its conformation and mechanism of action. We have used HDX-MS to study the conformational dynamics of SERT and its homologue LeuT and provide insights into how NSS proteins translocate ligands and bind drugs.

Methods:
SERT and LeuT were expressed recombinantly in HEK and E.Coli cells respectively, and purified. To induce distinct functional states LeuT or SERT was allowed to incubate with the respective substrates. Protein samples were labeled in deuterated buffer for various time intervals. Quenched samples were analyzed by an optimized HDX-MS workflow employing a cooled UPLC system coupled to a Synapt G2 MS.

Results:
We have measured the HDX of the inward-facing (apo state), outward-facing open (Na+-bound state) or substrate-occluded conformation (Na+/leucine-bound state) of the prokaryotic NSS protein LeuT, an established structural model of the NSS family. Upon binding substrate, several regions on the extracellular side of LeuT became more dynamic whereas substrate binding sites and regions on the intracellular side were stabilized. Interestingly, regions thought to be critical in the outward-to-inward transition showed exchange indicative of slow and correlated unfolding motions (EX1), modulated by the addition of substrates. Our findings thus provide the first detailed view of the solution-phase dynamics associated with the outward-to-inward transition. We are currently performing HDX-MS analyses on SERT to compare how LeuT and the actual human drug target SERT bind and translocate ligands.

Novel Aspect:
HDX-MS analysis of NSS proteins provide a dynamic view into the mechanism of action and drug binding of this challenging family of membrane proteins

Rashid, Shaan,
University of Ottawa

Contributing Authors:
Paul M. Mayer, Corresponding author.

Title:
Dual-spray reactions: A new method of synthesizing inorganic complexes within the electrospray source

Introduction:
Electrospray ionization is known to produce non-covalent complexes that are not present in solution, something that is often seen as an unwanted complication in sample analysis. We have recently also started to probe it as a method to synthesize non-covalent organometallic species by using two simultaneous electrospray emitters containing different solutions ("dual-spray").

Methods:
Dual-spray formation of transition metal complexes was achieved using a Synapt G1 mass spectrometer which is equipped with a dual electrospray ionization (ESI) source. Tris(2,2'-bipyridine)metal(II) and similar complexes containing the 1,10-phenanthroline ligand were formed by spraying a ligand solution and the solution containing transition metals through two independent ESI emitters.

Results:
Formation of inorganic complexes were confirmed by comparing ion mobility drift time, mass spectra, and CID fragmentation with the reference standard compounds. Tris(2,2'-bipyridine)ruthenium(II) and tris(1,10-phenantroline)ruthenium(II) were formed by the dual-spray reaction. Tris(2,2'-bipyridine)iron(II), tris(1,10-phenantroline)iron(II) and mixed ligand complexes of iron(II) formed by dual-spray showed two additional hydrogens bonded to the complex. These unique complexes showed similar initial ligand loss to the reference standards however the secondary ligand loss is showed dissimilar dissociation channels and energetics.

Novel Aspect:
This technique has also been used to bind iron(II) ions to the heme moiety and will be discussed.

Raupers, Björn,
Institute of pysical chemistry, University of Kiel

Contributing Authors:
Jürgen Grotemeyer, Institute of physical chemistry, University of Kiel.

Title:
Cluster analysis by means of low temperature plasma fourier transform ion cyclotron resonance mass spectrometry (LTP-FT-ICR-MS) and Cassinian ion trap

Introduction:
Our previous LTP ion mobility spectrometry studies have shown that volatile organic compounds have an affinity to form large clusters inside of a LTP ion source. To further investigate the formation processes of these clusters under atmospheric conditions we coupled a LTP-source to a FT-ICR mass spectrometer. Its high resolving power enabled us to systematically analyze the observed clusters.

Methods:
Mass analysis were performed on a 7.05 T FT-ICR mass spectrometer and a Cassinian ion trap by Bruker Daltonik. A LTP torch analog to the work of Nørgaard et al. with slight variations was used as an ion source. Plasma gas and sample gas were Argon. As a model system (S)-(-)-Limonene was chosen.

Results:
The observed spectra show a great variety of formed species with a total of around 550 signals. The resolving power of an ICR experiment provides the possibility to terminate the exact composition of each signal. Beside the clusters consisting of Hydrogen and Carbon there are also clusters containing Oxygen and Nitrogen atoms present. This is a direct effect of working under atmospheric conditions. The oxidation and amination reactions allow the system to undergo further reactions like dehydration inside the LTP. This can lead to stacked and multiple charged clusters. Multiple charged clusters with the theoretical composition [2M-4H+H]+, [2M-4H+2H]2+ and [2M-4H+3H]3+ have been identified. Due to their composition it is likely that these are bond over a π-system formed during the ionization process. Our next goal is coupling the LTP source to a Cassinian ion trap mass spectrometer.

Novel Aspect:
Investigation of clustering processes by LTP-FT-ICR and Cassinian ion trap mass spectrometry of volatile organic compounds at atmospheric conditions.

Ravikumar, Vaishnavi,
Chalmers University of Technology

Contributing Authors:
Vaishnavi Ravikumar, Systems and Synthetic Biology, Chalmers University of Technology; Alyaa Mohamed, Computational Bioscience Research Center, King Abdullah University of Science and Technology; Takashi Gojobori Computational Bioscience Research Center, King Abdullah University of Science and Technology; Sun Nyunt Wai, Department of Molecular Biology, Umeå University; Åsa Sjöling, Department of Microbiology, Tumor and Cell Biology, Karolinska Institutet; Ivan Mijakovic, Systems and Synthetic Biology, Chalmers University of Technology.

Title:
Phosphoproteomic analysis of the mechanisms behind Vibrio cholerae and Enterotoxigenic Escherichia coli pathogenesis

Introduction:
Diarrhoea causes millions of illnesses and deaths in children worldwide and is a major threat to global health. Vibrio cholerae and Enterotoxigenic Escherichia coli (ETEC) collectively cause more than 50% of all cases of diarrheal disease globally. Our objective is to understand the molecular mechanisms of the complex interplay between the two pathogens, ETEC and V. cholerae, and the host.

Methods:
We propose to study ETEC and V. cholerae interaction by performing a comprehensive proteome analysis focused on protein abundance and specific post-translational modifications (PTMs), namely phosphorylation and acetylation. The quantitative proteome with PTM analysis will be performed for single bacterial cultures and co-cultures, in the presence or absence of human cell cultures.

Results:
Preliminary results, based on killing assays, indicate complete dominance of V. cholerae over specific ETEC strains, under laboratory conditions. It was observed that the growth of ETEC E36 was being efficiently suppressed by V. cholerae ("sensitive") while ETEC E616 was able to co-survive even in the presence of V. cholera ("resistant"). A ΔlysA strain has been generated for employing the quantitative Stable Isotope Labelling by Amino acids in Cell culture (SILAC) strategy in order to ensure that the cells incorporated the isotopically labelled lysine amino acid provided in the growth medium. Preliminary genome-scale metabolic model predictions indicated the possible metabolite conversions that occur by either of the pathogens into a more favourable compound that can be utilized by the other, thus supporting the mutually synergistic hypothesis.

Novel Aspect:
This work is expected to improve our understanding of the mechanisms governing co-infection by bacterial pathogens.

Ray, Steven,
State University of New York at Buffalo

Contributing Authors:
Gary Hieftje, Indiana University; David Koppenaal, Pacific Northwest National Laboratory; Christie Enke University of New Mexico.

Title:
Matrix-Assisted Laser Desorption Ionization Distance-of-Flight Mass Spectrometry

Introduction:
Distance of Flight Mass Spectrometry is a new strategy for mass spectrometry wherein ions are separated in space and their m/z determined according to the distance each ion travels over an identical flight period. The potential advantages the DOFMS holds for analytical methodologies based on MALDI (and other ionization sources) is considered.

Methods:
A home-built DOFMS constructed around the MALDI experiment is employed. Ionization is produced with a conventional UV-laser source. The MALDI-DOFMS uses a very brief high voltage pulse to accelerate all ions to the same momentum. Ions are focused at the same instant at a distance dependent on the m/z. A position-sensitive ion detector measures flight distance.

Results:
The new DOFMS instrument has been designed and constructed based upon the MALDI experimental architecture. The MALDI-DOFMS closely resembles the MALDI-TOFMS architecture, which allows both experiments to be housed in the same instrument and operated in close synchrony. Here, several standard samples were measured by both techniques and the analytical performance compared. MALDI-DOFMS has thusfar shown comparable analytical performance to conventional TOFMS systems, however, the value of the DOFMS approach lies in the unique attributes of the ion focusing system. While DOFMS physically separates ions in space according to m/z, it also permits the spatial distribution of the ions upon extraction to be imaged. The instrumental composition and parameters will be discussed.

Novel Aspect:
The DOFMS is a new approach to velocity-based mass spectrometry, and it is used here for MALDI-MS.

Reid, Gavin,
University of Melbourne

Title:
Functional Group Selective Derivatization for Comprehensive Shotgun Lipidome Analysis and Multiplexed Lipid Quantitation

Introduction:
A major goal of lipidomics is the identification, characterization and quantitation of the diverse range of individual lipid species that may be present within a sample of interest, under different physiological or pathological conditions. However, the presence of isobaric and isomeric mass lipids imposes significant limitations on the utility of 'shotgun' MS-based lipidome analysis strategies.

Methods:
To address these limitations, a workflow will be described that incorporates functional group selective chemical derivatization of aminophospholipids and/or O-alk-1'-enyl ether-containing lipids following 'global' monophasic lipid extraction, together with direct infusion nano-ESI, ultra-high resolution accurate mass spectrometry and 'targeted' MS/MS analysis, and automated database searching [1].

Results:
This workflow is demonstrated to enable the unambiguous 'sum-composition' level identification and quantification of 500-600 individual lipid species across 4 lipid categories and from 36 lipid classes and sub classes, including isobaric mass lipids and certain categories of isomeric mass lipids, from within crude lipid extracts with minimal sample consumption, only 1-2 minutes data acquisition time and no requirement for chromatographic separation prior to analysis. Additional benefits include enhanced ionization (i.e., greater coverage and detection sensitivity) and improved MS/MS based 'molecular level' structural characterization of the derivatized lipid species, as well as the capability for 'multiplexed' quantification when using isomeric mass stable isotope labelled variants of the derivatization reagents. [1] Ryan and Reid. Acc. Chem. Res. 2016. In Press.

Novel Aspect:
Covalent labeling facilitates comprehensive 'sum composition level' and 'molecular level' lipidome analysis using shotgun UHRAMS and MS/MS

Reiter, Lukas,
Biognosys AG

Contributing Authors:
David Gomez-Valera, Max Planck Institute of Experimental Medicine, Goettingen, Germany; Oliver M. Bernhardt, Biognosys AG, Schlieren, Switzerland; Roland M. Bruderer Biognosys AG, Schlieren, Switzerland; Tejas Gandhi, Biognosys AG, Schlieren, Switzerland.

Title:
Extending the limits of deep DIA in human cell line and mouse brain tissue samples

Introduction:
Targeted analysis of DIA combined with label free quantification (LFQ) is a reproducible and precise quantification method. There are no limits on the number of samples per experiment in contrast to label based methods such as TMT. However, fractionation, to increase proteome depth, in combination with LFQ is not trivial. Therefore, it is of interest where the limits of single shot deep DIA are.

Methods:
Human HeLa/HEK-293 cell line and mouse cerebellum digests were spiked with the Biognosys iRT kit and analyzed by DDA/DIA on a Q Exactive HF with 2h/4h gradients and 50cm/1m columns. DDA data was searched with MaxQuant, spectral libraries were generated and DIA was analyzed with Spectronaut 9 (Biognosys). Two large DDA data sets (M. Mann lab) were downloaded and used to make a spectral library.

Results:
The human cell line spectral library comprised 210'000 peptide precursors or 8'500 protein groups (1% PSM and protein FDR). The mouse spectral library (Azimifar et al. 2014, Sharma et al. 2015) comprised 548'000 peptide precursors or 13'500 protein groups (1% PSM and protein FDR). When analyzing the human cell line data acquired with optimized 2h DIA methods we could identify more than 100'000 peptide precursors corresponding to close to 7'000 protein groups with median CVs below 10% (1% FDR). When analyzing the 4h DIA triplicate runs we could identify on average 157'407 peptide precursors and 7'394 protein groups in HeLa and 157'847 and 7'555 protein groups in HEK-293 with median CVs of 13% (1% FDR). The analysis of the complex 4h DIA cerebellum tissue sample using the large spectral library gave rise to 120'000 peptide precursors and 9'100 protein groups in a single shot DIA analysis.

Novel Aspect:
To our knowledge highest reported single shot quantification in human cell line and mouse tissue using DIA.

Remes, Philip,
Thermo Fisher Scientific

Title:
Multiplexed Targeted Assays Using Ion Trap Waveform Isolation

Introduction:
Multiplexing based on ion trap waveform based isolation has the potential to increase sampling efficiency by as much as 2-10x. In this study, waveform multiplexing is applied to LC/MS targeted assays, with the goal of characterizing the potential advantages of increased sampling efficiency with the potential disadvantages of decreased selectivity.

Methods:
Single precursor isolation was performed with a quadrupole filter or with ion trap waveforms. Multiplexed precursors were isolated with waveforms in the ion trap. Precursors were fragmented either with beam CID or resonance CID. Various performance characteristics were evaluated as a function of multiplexing parameters, such as number of precursors, and other precursor grouping parameters.

Results:
For a set of 40 pesticides and 10 ms dwell, the instrument scans at 53Hz for quad isolation. Due to latency of data transfer, waveform isolation incurs a 14ms delay, and was capable of 18Hz. Without this transfer the rate was 25 Hz. As the number of precursors was increased to 5, the scan rate stayed constant, for an effective 81Hz (108Hz no latency). The Promega peptide isotopologues were used to quantify 4 orders of magnitude concentration. Precursors were paired in concentrations of either 1:10 or 1:100. For the 1:10 ratio, 9 of 12 pairs exhibited good linearity, while for the 1:100 ratio, only 2 of 12 pairs gave good linearity. This result demonstrated a key to effective waveform multiplexing, that precursors be grouped with similar intensities. Future work will improve the data transfer latency, and increase the number of multiplexed precursors for quantitative pesticide analysis.

Novel Aspect:
Multiplexing with ion trap based waveform isolation, applied to real world LC/MS assays.

Reubsaet, Leon,
Department of Pharmaceutical Chemistry, School of Pharmacy

Contributing Authors:
Trine Grønhaug Halvorsen, Department of Pharmaceutical Chemistry, School of Pharmacy, University of Oslo, Oslo, Norway; SIlje Torsetnes, Department of Pharmaceutical Chemistry, School of Pharmacy, University of Oslo, Oslo, Norway; Elisabeth Paus Department of Medical Biochemistry, Oslo University Hospital, Radiumhospitalet, Oslo, Norway; Marianne Nordlund Broughton, Department of Medical Biochemistry, Oslo University Hospital, Radiumhospitalet, Oslo, Norway; Hanne Lund, Department of Pharmaceutical Chemistry, School of Pharmacy, University of Oslo, Oslo, Norway; Bjørn Winther, Department of Pharmaceutical Chemistry, School of Pharmacy, University of Oslo, Oslo, Norway; Maren Levernæs, Department of Pharmaceutical Chemistry, School of Pharmacy, University of Oslo, Oslo, Norway; Siri Valen Egeland, Department of Pharmaceutical Chemistry, School of Pharmacy, University of Oslo, Oslo, Norway.

Title:
Fundamental aspects and diagnostic potential of immunocapture in bottom-up LC-MS/MS

Introduction:
Immunocapture using magnetic beads has become an important sample preparation technique in targeted protein analysis using LC-MS/MS. It allows to determine diagnostic proteins robustly at the picomolar level as well as it through a multiplexing approach decreases the number of assays necessary for a diagnosis to be made.

Methods:
Antibodies covalently coupled to magnetic beads are incubated with human serum to capture the following diagnostic proteins: ProGRP, NSE (both small cell lung cancer) and hCG (ovarium and testicular cancer). After proteolysis the generated peptides are determined using a triple quadrupole (MRM mode) or an Orbitrap mass spectrometer (data dependent MS/MS mode).

Results:
This presentation will give an overview of our efforts on strategy development of immunocapture based targeted analysis of selected diagnostic proteins. It will include optimization of immunocapture with LC-MS/MS in routine targeted protein determination, the potential in multiplex analysis and the possibility of isoform differentiation. Also fundamental experiments to improve the methodology will be discussed.

Novel Aspect:
Insight in fundamental aspects of immunocapture LC-MS/MS based targeted protein analysis.

Reyes-Garces, Nathaly,
University of Waterloo

Contributing Authors:
German A. Gomez-Rios, University of Waterloo; Janusz Pawliszyn, University of Waterloo; Ezel Boyacı University of Waterloo; Barbara Bojko, University of Waterloo.

Title:
Untargeted analysis of brain tissue using solid phase microextraction

Introduction:
Bio-SPME is a sample preparation tool that has demonstrated its great applicability for analysis of biological matrices. In this work we present the suitability of this technology for untargeted analysis of brain tissue. Results showed that SPME is able to offer a broad range of metabolites coverage, exceptional sample clean-up, and significant advantages for in vivo analysis.

Methods:
7 and 4 mm (mixed mode coating) SPME fibres were used. Fibres were preconditioned, rinsed, and subsequently introduced into the tissue for 30 min. A quick washing step was done afterwards. All the SPME probes were desorbed in 1:1 acetonitrile:water. This procedure was followed to sample brain homogenate, human brain tumors (6) and rat brains. Extracts were run using LC-HRMS (Orbitrap).

Results:
In the first case of study, six brain tumors were sampled using 7 mm fibres. Depending on the tumor size, two or more fibres were used to sample each case. Results indicated that satisfactory coverage of a broad range of polar and non-polar compounds was achieved after 30 min extraction. Furthermore, multivariate statistical analysis showed good clustering among fibres according to the tumor sampled. For the in vivo sampling, 4 mm mixed mode fibres were employed. Animals were monitored before and after 3 hours of deep brain stimulation (DBS). Several metabolic features were found to be up-regulated or down-regulated after the DBS treatment. Baseline versus after-DBS measurements were compared using both paired and unpaired analysis. Advantages of SPME as an approach that allows for monitoring the same animal at different time points are certainly demonstrated.

Novel Aspect:
The performance of SPME for untargeted brain tissue analysis is investigated. Important advantages of the technology are demonstrated.

Richardson, Susan,
University of South Carolina

Contributing Authors:
Joe Binkley, LECO Corporation, 3000 Lakeview Ave; St Joseph, MI, 49085, USA; Judith E. Schmid, National Health and Environmental Effects Research Laboratory, U.S. Environmental Protection Agency, Research Triangle Park, NC 27711, USA; Hannah K. Liberatore Department of Chemistry and Biochemistry, University of South Carolina, Columbia, SC 29208, USA; Sridevi Anduri Ravuri, Student Services Authority, U.S. Environmental Protection Agency, National Exposure Research Laboratory, Athens, GA 30605, USA; Amy A. Cuthbertson, Department of Chemistry and Biochemistry, University of South Carolina, Columbia, SC 29208, USA; Jonathan D. Byer, LECO Corporation, 3000 Lakeview Ave; St Joseph, MI, 49085, USA; Alexis Thompson-Klemish, Department of Chemistry and Biochemistry, University of South Carolina, Columbia, SC 29208, USA; Ernest R. Blatchley III, Lyles School of Civil Engineering, Purdue University, 550 Stadium Mall Drive, West Lafayette, IN 47907, USA; Mehrnaz Z. Afifi, Lyles School of Civil Engineering, Purdue University, 550 Stadium Mall Drive, West Lafayette, IN 47907, USA; David M. DeMarini, National Health and Environmental Effects Research Laboratory, U.S. Environmental Protection Agency, Research Triangle Park, NC 27711, USA; Eric J. Daiber, Student Services Authority, U.S. Environmental Protection Agency, National Exposure Research Laboratory, Athens, GA 30605, USA.

Title:
Comprehensive Chemistry and Mutagenicity of Disinfection By-Products in Swimming Pools and Spas

Introduction:
Pools and spas are enjoyed throughout the world for exercise and relaxation. However, there are no previous studies on mutagenicity of spa (hot tub) waters or comprehensive identification of disinfection byproducts in spas. We report the first integrated mutagenicity and comprehensive analytical chemistry of spas and pools treated with chlorine, bromine, or ozone.

Methods:
We collected 28 water samples from 7 sites and used gas chromatography with high-resolution mass spectrometry, two-dimensional gas chromatography high resolution time-of-flight mass spectrometry (GC×GC-HR-TOF-MS), and membrane-introduction mass spectrometry to comprehensively identify and quantify DBPs and other contaminants. Mutagenicity was assessed by the Salmonella mutagenicity assay.

Results:
Most of the 27 DBPs quantified were found at levels ranging from 0.9 µg/L for cyanogen chloride to 3658 µg/L for dichloromethylamine. More than 100 DBPs were identified, including several new classes of DBPs. Brominated pool/spa waters were 1.8X more mutagenic than chlorinated ones; spa waters were 1.7X more mutagenic than pools. Mutagenic potencies of samples from a chlorinated site correlated best with brominated haloacetic acid concentrations (Br-HAAs) ($r = 0.98$) and nitrogen-containing DBPs (N-DBPs) ($r = 0.97$) and the least with Br-trihalomethanes ($r = 0.29$) and Br-N-DBPs ($r = 0.04$). The mutagenic potencies of samples from a brominated site correlated best ($r = 0.82$) with the concentrations of the 9 HAAs, Br-HAAs, and Br-DBPs. Human use increased significantly the DBP concentrations and mutagenic potencies for most pools and spas.

Novel Aspect:
Evidence that human precursors can increase mutagenicity of pools and spas and that this increase is associated with increased DBP concentrations

Richter, Florian,
Goethe University, Medical School

Contributing Authors:
Karl-Friedrich Beck, Pharmazentrum Frankfurt/ZAFES, Universitätsklinikum Frankfurt, Frankfurt am Main, Germany; Ilka Wittig, Functional Proteomics, SFB 815 Core Unit, Goethe University Frankfurt, Frankfurt, Germany; Sebastian Longen Pharmazentrum Frankfurt/ZAFES, Universitätsklinikum Frankfurt, Frankfurt am Main, Germany; Josef Pfeilschifter, Pharmazentrum Frankfurt/ZAFES, Universitätsklinikum Frankfurt, Frankfurt am Main, Germany; Florian Richter, Functional Proteomics, SFB 815 Core Unit, Goethe University Frankfurt, Frankfurt, Germany; Yvette Köhler, Pharmazentrum Frankfurt/ZAFES, Universitätsklinikum Frankfurt, Frankfurt am Main, Germany.

Title:
Quantitative Persulfide Side Identification (qPerS-SID) Reveals Protein Targets of H2S Releasing Donors in Mammalian Cells

Introduction:
H2S is an important signalling molecule which provokes the formation of cysteine persulfides (R-S-SH). Like thiols, they show reactivity towards electrophiles and can hence be labelled with iodoacetyl-PEG2-biotin (PEO-Biotin). Here, we report a proteome-wide isolation of those peptides for indirect and direct detection. We also show the efficiency of diverse H2S donors (GYY4137 and Polysulfides).

Methods:
SILAC-labelled cells were treated 4 h with 1 mM GYY4137 or 30 min with 200 µM Na2S4, NaSH or Na2S. Treatment with 12 % TCA denatured and precipitated proteins and avoided other thiol reactions but labelling with PEO-Biotin. Resulting Peptides were enriched on streptavidin agarose beads. Persulfides were specifically eluted with TCEP. Finally, all remaining Cysteine peptides were eluted by boiling.

Results:
Unless the in-vitro-Assay with pyruvate kinase M2 as a model protein, that showed the formation of persulfides for Na2S4 (poysulfides), but not for GYY4137, the proteomics approach revealed that persulfide formation can be a response to another type of cysteine modification (for GYY) or can be induced non-enzymatically (Na2S4). Extensive data analysis shows that persulfide formation affects all subcellular compartments and cellular processes. The neighborhood of negatively charged amino acids appears beneficial or their formation. Currently, we are investigating the detailed mechanism of persulfide formation and the role in different cell types. Taken together, we present here a method that is highly suitable to freeze unstable cysteine modifications and enrich them specifically. It can be seen as a starting point for a more complex scheme to identify different PTM and proteome changes.

Novel Aspect:
Direct detection of labelled persulfide peptides. Freezing instable Cys-modifications. SILAC based study reveals efficiency of different H2S donors.

Richter, Florian,
Goethe University, Medical School

Contributing Authors:
Anfisa Solovyeva, Juniorprofessor "RNA Regulation in Higher Eukaryotes", Cluster of Excellence Macromolecular Complexes (CEF), Goethe University Frankfurt, Institute for Cell Biology and Neuroscience, Max-von-Laue-Str. 13, 60438 Frankfurt am Main; Ilka Wittig, Functional Proteomics, Goethe University Frankfurt, Medical School, Theodor-Stern-Kai 7, 60590 Frankfurt am Main; Michalela Müller-McNicoll Juniorprofessor "RNA Regulation in Higher Eukaryotes", Cluster of Excellence Macromolecular Complexes (CEF), Goethe University Frankfurt, Institute for Cell Biology and Neuroscience, Max-von-Laue-Str. 13, 60438 Frankfurt am Main.

Title:
A single approach to quantify protein-protein, protein-RNA interactions and posttranslational modifications

Introduction:
SR-Proteins are known splicing factors and transcriptional regulators. Some function as mRNA export adapters for the main export receptor NXF1. Binding is presumably regulated by phosphorylation and length of the RS domains. Different GFP-tagged baits were produced for stringent pulldown experiments in toder to devise the direct protein-protein and indirect RNA associated interaction partners.

Methods:
SRSF5, SRSF2, SRSF4 and NXF1 were produced as a GFP tagged mouse protein in P19 or HeLa cells, which were either UV254nm crosslinked or not. A brief RNaseI digestion removed the distant RNA binders. We established then a two-step digestion protocol with (i) RNAse T1A to elute RNA associated proteins and (ii) digestion on the beads of residual proteins. LC-MSMS analysis is followed by data mining.

Results:
Ratios +UV/-UV revealed proteins that were stabilized in complex on the bait by binding the same piece of RNA. We furthermore eluted a small subset of RNA binding proteins with RNAse digestion. SRSF5 interacts with NXF1 in P19 cells, while NXF1 is absent in the HeLa interactomes. SRSF2 is the most highly enriched SR protein in UV-crosslinked SRSF5 IPs. However, very strong RNA binders like HNRPs are located in the middle of the ratio-vs-significance plot since UV crosslinking did not affect their binding. Though we expected a major contribution of the phosphorylated RS domain, only few peptides from adjacent regions were detected. More interestingly we found different degrees of methylation to play a major role in the binding of SRSF5 with more robust di- and tri-methyl residues in P19, and more mono-methyl in HeLa. Finally, a large amount of UV-induced RNA-peptides were identified.

Novel Aspect:
A two-step digestion protocol is established or RNA mediated and direct binders.UV254nm stabilized a weak protein-RNA complex for pulldown conditions.

Rijs, Nicole,
Karlsruhe Institute of Technology

Title:
Ion/molecule chemistry of isomeric (multi-)metal complexes using ESI-TWIMS-MS

Introduction:
Ion/molecule reactions are an ideal platform to study the intrinsic reactivity of catalytic species. Bond-breaking or making events can be targeted and studied in isolation. One limitation of standard mass spectral analysis is the inability to separate isomers. However, by coupling ion/molecule reactions with TWIMS-MS, gas phase isomers can have their individual reactivity probed.

Methods:
A Synapt G2 traveling wave ion-mobility mass spectrometer was modified to allow for ion/molecule reactions at various locations in the ions' path, prior to TOF analysis. The ion/molecule reactions can be carried out before or after- TWIMS separation. The latter allows individual ion/molecule mass spectra of individual isomers to be generated, revealing direct structure vs reactivity relationships.

Results:
Preliminary data has revealed that the modified instrument is able to be utilized to separate model reactive species, and even those originating from reaction solutions involving metal containing reagents or catalysts. The versatility of the set-up in measuring structure (via collisional cross sections), probing unimolecular and bimolecular reactivity and direct comparison to computational results has allowed the measurement of (i) nobel metal containing reagents for C-C formation, (ii) metal-oxo and hydroxo- complexes capable of bond activations. Several recent highlights will be presented, including some of the ongoing challenges of measuring the collisional cross sections of metal containing species.

Novel Aspect:
An newly modified ion/molecule and TWIMS instrument is presented; separations conducted on species relevant to catalysis and metal mediated reactions.

Rijs, Nicole,
Karlsruhe Institute of Technology

Contributing Authors:
Manfred M. Kappes, Physical Chemistry of Microscopic Systems, Karlsruhe Institute of Technology, Karlsruhe, Germany; Nicole J. Rijs, Institute of Nanotechnology, Karlsruhe Institute of Technology, Eggenstein-Leopoldshafen, Germany; Jessica Holmes School of Chemistry, University of Melbourne, Parkville, Australia; Richard Robson, School of Chemistry, University of Melbourne, Parkville, Australia; Brendan F. Abrahams, School of Chemistry, University of Melbourne, Parkville, Australia.

Title:
The gas phase structures of self-assembled cyclotricatechylene cation-pi and anion-pi complexes

Introduction:
Molecular building blocks which engage in multiple non-covalent interactions are promising candidates for the construction of functional assemblies. Of these, cyclotricatechylene (ctc) is a bowl shaped molecule consisting of three catechol units, which can form "clam-like" hydrogen bonded structures. These structures are known to encapsulate molecules, e.g. metal cations and solvent molecules.

Methods:
In order to systematically investigate their structures, ctc-X complexes were formed by dissolving synthesised CTC ligand and appropriate salt and introduction to the gas phase via nanoESI. Travelling wave ion mobility (Synapt G2 Si), along with other mass spectrometry approaches, were used to examine the structure of the resulting cationic and anionic CTC complexes, [(ctc)X]+/- and [(ctc)2X]+/-.

Results:
Cation-pi clam complexes were previously observed in the condensed phase via x-ray crystallography. Here, the intriguing formation of both cation-pi and anion-pi complexes in solution was observed. A systematic examination of the alkali metal cation reveals the periodic trends in bonding, i.e. preference of bonding with the aromatic pi bowl versus the catechol oxygens, resulting in different measurable structures. Due to competitive nature of the pi versus O- bonding, the degree of hydrogen bonding and protonation state of ctc, and the association of both cations and anions, perhaps unsurprisingly, the structures formed are highly dependent on solution conditions. The measured collisional cross sections of the ctc complexes [(ctc)X]+/- and [(ctc)2X]+/- are thus compared with computed cross sections and reveal the role of the various bonding effects for the first time.

Novel Aspect:
Discrete anion-pi ctc structures were observed for the first time. TWIMS has revealed the competitive formation of anion-pi and cation-pi complexes.

Rocha, Daniele F. O.,
post doc student

Contributing Authors:
Robert H. Hinz, Empresa de Pesquisa Agropecuária e Extensão Rural de Santa Catarina- EPAGRI, Itajaí –SC, Brazil; Ester Wicket, Empresa de Pesquisa Agropecuária e Extensão Rural de Santa Catarina- EPAGRI, Itajaí –SC, Brazil; Katia R. A. Belaz ThoMSon Mass Spectrometry Laboratory, University of Campinas -UNICAMP, Campinas-SP, Brazil; Cristiane M. S. Cunha, Flora Biotecnologia Ltda, Itajaí-SC, Brazil; Marcos N. Eberlin, ThoMSon Mass Spectrometry Laboratory, University of Campinas -UNICAMP, Campinas-SP, Brazil; Fabio N. dos Santos, ThoMSon Mass Spectrometry Laboratory, University of Campinas -UNICAMP, Campinas-SP, Brazil; Claudio A. O. Nascimento, Chemical Engineering Department of Polytechnic School of University of São Paulo - Brazil; Adriana Pereira, Empresa de Pesquisa Agropecuária e Extensão Rural de Santa Catarina- EPAGRI, Itajaí –SC, Brazil; Alexandre Visconti, Empresa de Pesquisa Agropecuária e Extensão Rural de Santa Catarina- EPAGRI, Itajaí –SC, Brazil; Lidiane M. Andrade, Chemical Engineering Department of Polytechnic School of University of São Paulo - Brazil.

Title:
MALDI-TOF MS protein fingerprinting as a tool to explore Fusarium oxysporum f. sp. cubense adaptation and aggressiveness

Introduction:
Fusarium oxysporum f. sp. cubense (Foc) causes Panama disease, which destroys banana plantations around the world. Resistant cultivars are the best control strategy because pesticides does not end the infestation. Since low molecular weight (LMW) proteins can be related to cell division and stress responses, we investigated how LMW protein fingerprinting is affected by the aggressiveness behavior.

Methods:
Foc isolates were cultured in BDE medium for 14 days at 25ºC in triplicate. Mycelium was extracted with 70% formic acid/acetonitrile. CHCA as MALDI matrix. MALDI-TOF MS Bruker UltrafleXtreme SmartbeamTM in positive ion mode and linear detector, from m/z 2 to 20 kDa. Multivariate analysis: MetaboAnalyst 3.0.

Results:
The host cultivars Pome and Cavendish divides the samples in two distinct groups. Considering that Cavendish cultivar have only recently became Foc susceptible, this feature indicates that LMW proteins participate in this fungus adaptation to a more resistant host. The ions at m/z 2017, 2036, 2174, 2514, 2642, 2603, 2862, and 3323 are typical of isolates from Cavendish and may play a role in this process. Aggressiveness level showed a tendency to discriminate LMW protein profile and the statistically significant ions at m/z 3294, 6068, 6113, 6151, 6566, and 6753. A survey in UniProt database showed that most of Foc LMW proteins were not characterized yet. Among the identified proteins, some enzymatic activities are listed and could be related to host-pathogen relationship. Even without identification of these proteins, LMW proteins seems to be important to defeat host immune response.

Novel Aspect:
MALDI-TOF MS fingerprinting is presented as a convenient tool to select LMW proteins related to Foc adaptation and virulence.

Rodrigues Peres, Raquel Mary,
CAISM/UNICAMP

Contributing Authors:
Marcos N. Eberlin, ThoMSon Mass Spectrometry Laboratory, Institute of Chemistry, State University of Campinas - UNICAMP, Campinas, SP, Brazil.; Luis Otávio Z. Sarian, CAISM – Women´s Hospital, State University of Campinas - UNICAMP, Campinas, SP, Brazil.; Juliana K. Heinrich-Muçouçah CAISM – Women´s Hospital, State University of Campinas - UNICAMP, Campinas, SP, Brazil.; Andréia M. Porcari, ThoMSon Mass Spectrometry Laboratory, Institute of Chemistry, State University of Campinas - UNICAMP, Campinas, SP, Brazil.

Title:
Effects of AlCl3 in the lipid profile of a breast cancer cell line detected by Desorption Electrospray Ionization Mass Spectrometry

Introduction:
Aluminum compounds may have implications in human cells and its linkage to breast cancer has been hypothesized. In vitro and in vivo studies show that aluminum might bind estrogen receptors, interfering with gene expression or other phenotype characteristics. We used DESI-MS to verify the differential expression of lipids in a breast cancer cell line cultivated with and without aluminum chloride.

Methods:
Breast cancer cell line T47D was grown in HAM F-10 medium containing 10% FBS (control), 10, 15 and 25% of AlCl3. Pellets were extracted (Bligh & Dyer) and analyzed using a DESI source connected to a Q-Exactive (Thermo) at a resolution of 70,000. The sprayed solvent was MeOH (2 uLmin-1) and the negative ion mode was used in 100 – 1500 m/z range. Metaboanalyst 3.0 was used for chemometrics.

Results:
Mass spectra were mainly composed of ions up to m/z 700. Fat acids molecules and dimmers, as well as other small lipids formed the predominant ions. When partial least square discriminant analysis (PLS-DA) was applied for the m/z 290 - 470 range, full differentiation of the control and treated groups was achieved. Such discrimination was mainly attributed to the ions of m/z 350.246, 364.226, 346.215, 367.215. Based on their exact mass values, these ions were assigned as [M + Cl]- adducts of acylcarnitines, sphingosines, prostaglandins and steroidal compounds. The examination of fragments from these ions are in progress. These findings corroborate the assumption that aluminum might bind estrogen receptors, affecting lipids' expression in breast cancer cells. Studies should continue to be done to elucidate the possible effects of aluminum in cell pathways or its microenvironment.

Novel Aspect:
To verify the influence of aluminum compounds in a breast cancer cell line using DESI-MS.

Rodriguez, Sarah,
Institut de Recherche en Astrophysique et Planétologie (IRAP)

Contributing Authors:
Hassan Sabbah, Institut de Recherche en Astrophysique et Planétologie (IRAP-CNRS-UPS); Christine Joblin, Institut de Recherche en Astrophysique et Planétologie (IRAP-CNRS-UPS); Paul M Mayer Department of Chemistry and Biomolecular Sciences, University of Ottawa; Aude Simon, Laboratoire de Chimie et Physique Quantiques (LCPQ-UPS); Junfeng Zhen, Institut de Recherche en Astrophysique et Planétologie (IRAP-CNRS-UPS); Giacomo Mulas, Institut de Recherche en Astrophysique et Planétologie (IRAP-CNRS-UPS.

Title:
H/H2 loss of Dibenzopyrene isomer cations: Combining linear ion trap measurements and Density Functional Theory calculations

Introduction:
Polycyclic Aromatic Hydrocarbons (PAHs) are key component in astrochemistry. Their interaction with vacuum ultraviolet (VUV) irradiation impacts on the energy balance and the chemistry of the gas in the interstellar medium, including the formation of the most abundant molecule, H2. To evaluate these processes, laboratory experiments and theoretical calculations are essential.

Methods:
We studied the photo-processing of two isomers of C24H14+ under VUV irradiation (photon energy range: 7- 20 eV) using action spectroscopy with the LTQ linear ion trap setup available in the DESIRS beamline of the synchrotron SOLEIL. To analyze the results, density functional theory (DFT) calculations were carried out with Gaussian 09 using the UB3LYP functional along with the 6-31(d,p) basis set.

Results:
The two studied isomer cations, Dibenzo(a,e)pyrene (AE) and Dibenzo(a,l)pyrene (AL) , were produced in gas phase by an atmospheric pressure photo-ionization (APPI) source and thermalized though He atoms in the trap before being exposed to synchrotron radiation. Branching ratio of the main fragments, H, 2H/H2 and C2H2, were obtained as a function of the photon energy. In this contribution, we focus on the H and 2H/H2 loss, where noticeable differences were found at low energy: the “closed”structure (AL) dehydrogenates at substantially lower energy than the “open” one (AE). In addition, geometry optimizations and vibrational calculations were performed to quantify the dehydrogenation mechanisms. They reveal a lower energy path for both direct H loss and H2 formation in the case of AL, which is induced by the non-planarity of this molecule.

Novel Aspect:
The results reveal an efficient H2 formation channel for AL, demonstrating the importance of the structure in the fragmentation path of PAHs.

Roithova, Jana,
Charles University in Prague

Title:
Listening in on reaction intermediates in solution

Introduction:
Electrospray ionization represents a great inter-phase between a solution and the gas-phase environment of a mass spectrometer. Inter alia, ESI-MS is used for investigation of ionic intermediates generated from a reaction mixture. We are developing a method that would relate the original intermediates in solution with their gas-phase prints.

Methods:
The methods is referred to as Delayed Reactant Labeling. A given reaction is set up with isotopically labeled as well as unlabeled reactant(s), one of them being added with a time delay. ESI-MS monitoring of this reaction mixture shows pairs of peaks corresponding to the labeled and unlabeled reactants, intermediates and products.

Results:
Time evolution of the pairs of peaks corresponding to the labeled and unlabeled reactants, intermediates and products is related to the reaction kinetics in solution. We have developed several models and applied it to the problems in gold, palladium, and copper catalysis. I will show the application for the determination of half life of reaction intermediates in solution. I will also demonstrate the application of the method for qualitative evaluation of consecutive and parallel reactions.

Novel Aspect:
Delayed Reactant Labeling represents a direct feedback between ESI-MS spectra and the solution.

Roithova, Jana,
Charles University in Prague

Title:
Helium tagging infrared photodissociation spectroscopy

Introduction:
Infrared photodissociation spectroscopy became one of the standard tools in determination of the structure of the ions in the gas phase. Next to the often applied infrared multiphoton dissociation spectroscopy, many approaches based on the messenger technique have been developed. We are using helium tagging which permits investigation of highly reactive ions.

Methods:
Helium tagging infrared photodissociation spectroscopy is performed with a tandem mass spectrometer containing a wire quadrupole trap cooled to 3 K. Mass-selected ions are trapped with helium buffer gas and form helium complexes. Spectroscopy can be performed with tunable light sources in IR or VIS spectral range.

Results:
I will show application of our approach for the investigation of hyper-valent metal-oxo complexes. We can routinely detect metal-oxo stretching vibrations (among others, I will show the first detection of copper(III)-oxo stretching mode). Recently, we have also demonstrated that we can spectroscopically distinguish triplet and quintet states of the iron(IV)-oxo complexes.

Novel Aspect:
Helium tagging is the method of choice for IRPD of highly reactive ions.

ROLANDO, Christian,
Université Lille 1, Sciences et Technologies

Contributing Authors:
Lionel CHIRON, CASC4DE, Strasbourg, France; Julien BOUCLON, Ecole Normale Supérieure, Paris, France; Fabrice BRAY Université Lille 1, Villeneuve d'Ascq, France; Marc-André DELSUC, IGBMC, Illkirch-Graffenstaden, France; Geoffrey BODENHAUSEN, Ecole Normale Supérieure, Paris, France.

Title:
Non Uniform Sampling (NUS) for bidimensional (2D) FT-ICR MS: how many points we can skip and what resolution we can achieve in a given time?

Introduction:
Until now 2D FT-ICR was limited to unit resolution for parent ion selection. Using non-uniform sampling, i.e. skipping points randomly, on the first dimension gives access, after data processing, to all structural information contained in a complex mixture at high resolution. The achievable NUS compression will be discussed in function of the classical dimension size and sample complexity.

Methods:
Experiments were performed on a Solarix 12 T or an Apex 9.4 T FT-ICR MS from Bruker-Daltonics with a nanoESI ion source using IRMPD/ECD for fragmentation. Random NUS sequence has uniformly sampled gaps. Deconvolution were performed using an adapted version of our uRQrd algorithm (PNAS, 2014, 1385). Simulated 2D spectrum were obtained with a software developed in Python language.

Results:
To test the limits of the technique we firstly recorded the 2D FT-ICR MS of substance P ionized by ESI and fragmented by ECD at 2 k (precursor) × 128 k (fragments) points using NUS factors of 2, 4, 8, 16, 32 i.e. skipping 1 point over 2 to 31 points over 32 on the first dimension. The resolution of the doubly charged parent ion follows the expected increase of the NUS factor in the precursor dimension and surprisingly up to NUS 32 the overall 2D spectrum remains very similar. The robustness of FT-ICR compared to NMR which is limited to a NUS 4 seems due to the much higher size of FT-ICR spectra. For testing this hypothesis we recorded 2D spectra of different F1 size (up to 16k points) and F2 size (up to 4M points) of substance P and complex mixtures we were able to deconvolute thanks to innovative algorithms. A limit on achievable NUS factor and hence resolution will be given.

Novel Aspect:
Non Uniform Sampling (NUS) was adapted for 2D FT-ICR MS allowing high resolution for parent selection. The achievable resolution has been explored.

ROLANDO, Christian,
Université Lille 1, Sciences et Technologies

Contributing Authors:
Fabrice BRAY, Université Lille 1, Villeneuve d'Ascq, France; Matthias WITT, Bruker Daltonics, Bremen, Germany; Julien BOUCLON Ecole Normale Supérieure, Paris, France; Geoffrey BODENHAUSEN, Ecole Normale Supérieure, Paris, France.

Title:
Bidimensional FT-ICR MS: unravelling structural information from a complex sample without separation

Introduction:
Although mass information of all compounds is acquired at the same time leading to high throughput parallel acquisition, the structural information (MS/MS spectrum) is obtained one compound by one leading to a slow serial acquisition. Two-dimensional FT-ICR solve this problem as it records all MS/MS the correlations, in only one experiment, regardless of the sample complexity.

Methods:
Experiments were performed on a Solarix 12 T or an Apex 9.4 T FT-ICR MS from Bruker-Daltonics with a nanoESI ion source using IRMPD/ECD for fragmentation. Reproducible 2D_FT-ICR mass spectra were recorded by applying our improved pulse sequence and a precise tuning. 2D_FT ICR experiments were performed using non uniform sampling (NUS) which allows to obtain high resolution for precursors.

Results:
The LC-MS/MS approach is not possible for complex sample, the more well-known being petroleum, that cannot be separated by liquid chromatography. 2D FT-ICR provides a new and complementary way for such complex mixtures as it allows to obtain all the relevant structural information, for all precursor molecules, in a single experiment. 2D FT-ICR also provides high resolution selection of the precursor ions (up-to 10,000 at m/z 400) which allows separation of molecules with isobaric masses that have different chemical formulae. Here, we will compare results obtained by 2D FT-ICR of digested peptide samples of increasing complexity ranging between Cytochrome C calibration mixture to full yeast digest with classical LC-MS proteomics. First examples of quantitative 2D FT-ICR using Aqua strategy will be presented. We will present examples from other fields like alcoholic drink analysis.

Novel Aspect:
2D FT-ICR MS is a fully Data Independent Acquisition regardless of sample complexity with high resolution on both precursor and fragment ions.

ROSSATO, Maxime,
University of Montpellier

Contributing Authors:

Gilles SUBRA, Université de Montpellier; Sonia CANTEL, Université de Montpellier; Christine ENJALBAL Université de Montpellier; Jean MARTINEZ, Université de Montpellier; Muriel AMBLARD, Université de Montpellier.

Title:

MALDI mass spectrometry and HCCA labeling as a powerful tool for peptide quantitation in biology.

Introduction:

Since peptides represent a class of biomolecules involved in many biological processes, they are extensively studied for pharmaceutical developments. MALDI is well-known to provide very sensitive detection but not commonly used for quantitative measurements. Using MALDI for such purposes is challenging and has been successfully applied to track peptides at low concentrations in various media.

Methods:

We developed a methodology associating HCCA-targeted peptide to be analyzed by MALDI-MS in a matrix such as HCCE. This original approach allowed to selectively enhance and discriminate the MALDI-MS signals of targeted peptides. This concept was successfully applied to protein structure issues illustrated by the cross-linking of a model protein.

Results:

The cornerstone of the presented analytical methodology relies on the production of very abundant molecular ions upon DI conditions related exclusively to the compound of interest. We enlarged the scope of the designed analytical tool to applications which require low concentration measurement. We investigated peptide quantitation commonly performed in biological samples through fluorescent or radioactive labeling for pharmacological binding assays. The highly sensitive and specific measurement of HCCA-labeled peptide ligands affinity for their receptor through isotope dilution strategy allowed the determination of Kd, Ki and IC50 through saturation and competitive assays.

Novel Aspect:

HCCA-Peptide tagging combined with MALDI-MS technology represents an alternative to fluorescence and radioactivity to perform quantitative measurement

Rosu, Frederic,
Univ. Bordeaux. CNRS ums3033

Contributing Authors:
Massimiliano Porrini, ARNA Laboratory (Inserm U1212 / CNRS UMR 5320 / Univ. Bordeaux), France; Valerie Gabelica, ARNA Laboratory (Inserm U1212 / CNRS UMR 5320 / Univ. Bordeaux), France.

Title:
Structure of Single Stranded DNA in Vacuo. Ion Mobility Spectrometry and Spectroscopic Signature

Introduction:
The conformation of polyanions of DNA single strands is probed by ion mobility spectrometry (IMS), infrared IR and UV action spectroscopy. H-bond interactions between nucleic bases and stacking interactions are responsible of the formation of nucleic acid secondary structures. Here we will show how combining IMS and action spectroscopy provides information on the structure of nucleic acid ions.

Methods:
Samples were analyzed on an Agilent 6560 ESI-IMS-QTOF (drift tube in He) and the CCS was determined. The DNA anions were irradiated in a Bruker ion trap through the ring electrode (CLIO FEL facility (France). IRMPD action spectra are recorded in the range 1560-1760 cm-1. UV spectra (200-300 nm) were recorded on a Bruker Amazon coupled to an OPO laser. Molecular modeling: IMS: T-REMD, IR: DFT-D3.

Results:
Collision cross section (CCS) obtained by IMS-MS were compared with CCS of candidate structures resulting from T-REMD. Clustering and filtering allowed us to define candidate structures for DFT calculations. IRMPD spectra of dG6, dA6, dC6, dT6 were recorded to evaluate the contribution from each base (A, T, G, C) in a DNA strand. The wavelength range was chosen according to our previous IRMPD works on G-quadruplex and i-motif structures, which showed that C=O stretching and NH2 scissoring were sensitive to the secondary structure. Multiple conformations are detected in IMS for all the single strands and good matches of the candidate structures are obtained with the IRMPD spectra. We found that gas-phase ssDNA readily form POH...OP hydrogen bonds that did not exist in solution, that protons are mobile, and that each base has a different propensity to form H-bonds with the backbone.

Novel Aspect:
Combination of IMS, action spectroscopy (IR and UV) and molecular modeling to determine the conformation of single strand DNA ions.

Rosu, Frederic,
Univ. Bordeaux. CNRS ums3033

Contributing Authors:
Marie-Paule Teulade-Fichou, Institut Curie (UMR9187/U1196), Orsay, France; Sophie Bombard, Institut Curie (UMR9187/U1196), Orsay, France; Valérie Gabelica ARNA Laboratory (Inserm U1212 / CNRS UMR 5320/ Univ. Bordeaux), France; Daniela Verga, Institut Curie (UMR9187/U1196), Orsay, France; Dababrata Paul, ARNA Laboratory (Inserm U1212 / CNRS UMR 5320/ Univ. Bordeaux), France; Adrien Marchand, ARNA Laboratory (Inserm U1212 / CNRS UMR 5320/ Univ. Bordeaux), France.

Title:
Top-Down Sequencing by Electron Photodetachment Dissociation (EPD) to Locate Ligand and Cation Binding Sites on G-Quadruplexes.

Introduction:
Mass spectrometry provides exquisite detail on ligand and cation binding stoichiometries with a nucleic acid target. The next important step is to develop reliable methods to determine the cation and ligand binding sites in each complex separated by mass. Here we used EPD to locate cations and noncovalently bound ligands, and compare its merits with ligand-cross linking approaches.

Methods:
EPD was carried out on a Bruker Amazon SL quadrupole ion trap mass spectrometer coupled to an OPO laser set at 245 nm. A single pulse detaches an electron from the multiply charged negative ions, and CID on the radical provides top-down sequencing on the 22/23-mer DNA sequences. Ion mobility experiments were carried out on an Agilent 6560 IMS-Q-TOF. Sample preparation was as in JACS, 2015, p750.

Results:
We used sulfolane to produce slightly more negative intact complexes, with enough Coulomb repulsion to enhance the sequence coverage thanks to separation of the fragments, but not too much so as to avoid gas-phase unfolding (as verified with ion mobility MS). EPD sequencing showed a clear suppression of fragmentation when cations are bound at a specific location of the DNA backbone. This allowed to locate the cations and ligands on the central loop of the G-quadruplex structure. We compared these results on unlabeled non-covalently bound ligands with those obtained by traditional sequencing of cross-linking ligand derivatives. The binding sites are consistent, but the derivatized ligands have altered binding affinities compared to the original molecules. EPD is therefore advantageous to localize non-covalent binding sites of ligands, without requiring derivatization.

Novel Aspect:
EPD is applied for the first time to characterize nucleic acid complexes. The method is transposable to other multiply charged anions like proteins.

Roux-Dalvai, Florence,
CHU Quebec Research Center

Contributing Authors:
Benjamin Nehmé, CHU Quebec - Laval University - Proteomics lab; Isabelle Kelly, CHU Quebec - Laval University - Proteomics lab; Maurice Boissinot CHU Quebec - Laval University - Infectiology Research Center; Frédéric Fournier, CHU Quebec - Laval University - Proteomics lab; Michel G. Bergeron, CHU Quebec - Laval University - Infectiology Research Center; Arnaud Droit, CHU Quebec - Laval University - Proteomics lab.

Title:
Towards a new diagnostic tool for human urinary tract infection using LCMS Data Independent Acquisition (DIA) analysis and specific spectral libraries

Introduction:
Bacterial infections cause millions of deaths each year, hence the need for rapid and accurate diagnosis. MALDI-TOF analysis is largely used in microbiology laboratories but requires pure colonies obtained after a long step of bacterial culture. We assumed that new proteomic strategies using DIA should allow the detection of a large number of bacteria species without bacterial culture.

Methods:
We initially generated specific spectral libraries on 3 bacteria species representing 60-80% of all urinary tract infections in human and we used them to extract bacterial peptide signals acquired in DIA mode on the Orbitrap Fusion in urine samples. As a proof of concept we used urine samples spiked with bacteria at various concentrations and prepared them using an optimized short protocol.

Results:
Spectral libraries from pure colonies of E.coli, E.feacalis and K.pneumoniae were generated in Data Dependent Acquisition mode. Each library (≈ 20000 peptides) was filtered out for specificity over 18 bacterial genera. Among the remaining peptides (≈ 100), the best intensity ones were selected. The MS sample preparation protocol for urine was optimized. Standard protocol (17h) and optimized protocol (1h) displayed the same results in terms of protein identification and peptide signal intensities. Urine samples were spiked with various concentrations of the 3 different bacteria, treated using with the optimized protocol and analyzed in DIA mode. We were able to detect bacterial species specific signals at concentrations < 1e5 CFU/mL in urine (limit for significant infection). The method was also proved to be linear ($R2>0.99$) and reproducible (<20% CV on 3 biological replicates).

Novel Aspect:
By drastically reducing the time for bacterial ID, we showed that DIA analysis might be an alternative to conventional infection detection strategies

21ST INTERNATIONAL MASS SPECTROMETRY CONFERENCE
Toronto, Ontario, Canada
August 20 – 26, 2016

Rule, Geoffrey,
ARUP Laboratories

Contributing Authors:
Alan L. Rockwood, Department of Pathology, University of Utah School of Medicine; Silviana Ruci, IsoSciences; Stephen Jones IsoSciences.

Title:
Isotope effects in stability of MS analyte/internal standard ratio measurements

Introduction:
We examine sources of assay variability when using isotopically labeled internal standards (IS). The effect of chromatographic separation of deuterated IS from analyte is known. We focus on measurement of analyte/IS (A/IS) ratio when varying MS parameters, without chromatography. We conclude, for utmost assay stability, there may be additional reason to choose alternatives to 2H in selecting ISs.

Methods:
We categorize sources of LC-MS assay variability and consider their contributions to total variability. We focus on the MS (QQQ) alone and stability of A/IS ratio measurements when using either 13C- or 2H-labeled analogs of three compounds. The influence of three MS parameters (collision energy, collision gas thickness, declustering potential) on ratio measurement is evaluated by infusion.

Results:
Ideally, A/IS remains constant during the analytical process but recent studies show that chromatographic separation of 2H internal standards can result in differential suppression effects and nonlinear curves. To evaluate sources of variability in assay response we examined how isotope labeling affects A/IS ratio measurements when varying several MS parameters. Data shows the MS may not yield perfectly constant ratios, particularly when 2H ISs are used. This is believed due to isotopic effects involved in fragmentation. For example, the A/IS ratio was found to vary with declustering potential, collision gas thickness, and collision energy when a [9-2H] label was used for tolbutamide whereas the 13C label was better behaved. Choice of appropriate IS labeling will reduce assay variability and help enable reduction of calibration curve preparation and analysis.

Novel Aspect:
The influence of a QQQ mass spectrometer on analyte/internal standard ratio measurement is studied with respect to choice of 2H versus 13C labels.

Russo, Mariana,
Concordia University

Contributing Authors:

Alexandra Paquet, Concordia University; Dajana Vuckovic, Concordia University; Mariana Russo Concordia University.

Title:

Evaluation of the derivatization of organic thiols with N-substituted maleimides

Introduction:

The ratio between reduced and oxidized thiols in biological samples is an important indicator of oxidative stress. However, sample preparation techniques expose thiols to an oxidative environment, and the resulting oxidation of reduced thiols can skew the measured ratio. To avoid this problem, the reduced thiols can be stabilized by derivatization using reagents such as maleimides.

Methods:

Three common reagents, iodoacetamide, N-ethylmaleimide and (R)-(+)-N-(1-phenylethyl) maleimide (NPEM) were compared for the derivatization of common organic thiols prior to analysis by liquid chromatography-mass spectrometry (LC-MS). MS/MS was used to elucidate the structure of the derivatized compounds and to investigate the selectivity of the derivatization reaction at optimum derivatization pH.

Results:

NPEM was found to be the best derivatizing agent with high derivatization efficiency and good derivative stability, but its derivatives eluted in two chromatographic peaks, which reduces precision in quantification. pH 7 is considered in literature as ideal for selective thiol derivatization. Surprisingly, we found clear evidence of double derivatization for several multi-functional thiols we evaluated, as well as hydrolytic opening of the maleimide ring. We also confirmed lysine which contains no thiol moiety, also underwent derivatization. These results indicate that maleimide derivatization is not as specific to thiols as described in literature and that other functional groups such as amines can undergo partial derivatization despite pH control of the reaction.

Novel Aspect:

First in-depth investigation of reaction selectivity with thiols at optimal pH and elucidation of double peaks due to maleimide derivatization in MS.

Ryzhov, Victor,
Northern Illinois University

Contributing Authors:
Michael Lesslie, Northern Illinois University; John Lawler, Northern Illinois University; Gardenia Pacheco Northern Illinois University; Justin K.-C. Lau, York University; Michael K.-W. Siu, York University; Alan Hopkinson, York University.

Title:
Structure and reactivity of metal ion adducts of cysteine-containing peptide radicals

Introduction:
An important property of Cys radicals in proteins is their ability to abstract an H atom from a Cα position. This process is pH-dependent and is not modeled well by protonation. Using metal ions as charge sources offers a different perspective. Here we characterize Cys-containing radicals cationized by various metal ions in amino acids and peptides by a combination of gas-phase techniques.

Methods:
Radical Formation: XxxCys(SH) + R'ONO ⇌ XxxCys(SNO) + R'OH XxxCys(SNO) + M+ ⇌ M+ XxxCys(SNO) +ESI -->CID M+ XxxCys(SNO) --> M+XxxCys(S•) + •NO Ion-Molecule Reactions (IMRs): - Modified Bruker Esquire 3000 ESI QIT mass spectrometer - M+XxxCys(S•) + •NO, CH2=CH-CH2-I, CH3-S-S-CH3 Density Function Theory (DFT) Calculations B3LYP functional and 6-311++G(d,p) level of theory

Results:
We have previously characterized cysteine radical complex with Li+, Na+, and K+ as the charge bearing species by IMR and IRMPD. They remain sulfur-based radicals with tridentate (N,O,S) coordination of the metal ion. Moving to homocysteine radical/alkali metal ion complexes, all three (Li, Na, K) displayed fast HAT to form bidentate (S, O) metal ion-coordinating Cα radicals, in contrast to the protonated homocysteine radical. We have also studied metal complexes of glutathione thiyl radical with numerous metal ions (group IA, IIA, transition metals) and charge states (+1 /+2). Some of the systems showed rapid HAT similar to the protonated system (glutathione radical cation). Others remain sulfur-based radicals or displayed much slower HAT, most likely due to strong coordination of the doubly charged ion to oxygen atoms. Explanations based on DFT calculations will be provided.

Novel Aspect:
Metal ions modulate hydrogen atom transfer from a Cα position to S in Cys thiyl radicals in amino acid and peptide systems in the gas phase

Saba, Julian,
Thermo Fisher Scientific

Contributing Authors:
Sergei Snovida, Thermo Fisher Scientific; Christa Feasley, Thermo Fisher Scientific.

Title:
Comparison and Optimization of First and Second Generation Quadrupole Dual Cell Linear Ion Trap Orbitrap Hybrid MS for Glycopeptide Analysis

Introduction:
Large scale intact glycopeptide analysis remains challenging. Not only must one sequence the peptide, but site localization and glycan composition are required for analysis. Here we present performance evaluation of first and second generation tribrid mass spectrometer for glycopeptide analysis. Workflows will be presented that highlight large scale glycoproteomics.

Methods:
Glycopeptides were enriched from human serum and Hela lysates digests using SAX columns. The enriched glycopeptides were analyzed using an EASY nLC 1000 with a C18 PepMap column (2um, 100A, 75umx50 cm) on an Orbitrap Fusion and Orbitrap Fusion Lumos mass spectrometer. Data analysis were performed using Byonic software (Protein Metrics).

Results:
In total 11 parameters were tested with 21 individual runs to maximize performance. After optimization of parameters, experiments were conducted on both platforms to examine performance relative to each other. Overall, Lumos identified 9% more unique glycopeptides relative Orbitrap Fusion by ETD, 43% more by EThcD and 49% by EThcD over ETD. The increase in identification came from large glycopeptides which are challenging in mass spectrometry. The primary focus of our experiments was on ETD and EThcD, however, we observed that the quality of HCD spectra were superior to spectra acquired on other platforms for intact glycopeptides. In these instruments we could easily use HCD for sequencing. Glycosylation, specifically changes in glycosylation, is a hallmark of cancer. Using Hela as model we explored the possibility of sequencing these modifications in glycoproteomics experiment.

Novel Aspect:
Instrument performance improvement for glycoproteomics data on a novel quadrupole dual cell linear ion trap Orbitrap hybrid mass spectrometer

Saba, Julian,
Thermo Fisher Scientific

Contributing Authors:

Chris Etienne, Thermo Fisher Scientific; Ryan Bomgarden, Thermo Fisher Scientific; Albert Heck Utrecht University, the Netherlands; Fan Liu, Utrecht University, the Netherlands; Erum Raja, Thermo Fisher Scientific; Mathias Mueller, Thermo Fisher Scientific; Rosa Viner, Thermo Fisher Scientific.

Title:

Optimization of crosslinked peptide analysis on an Orbitrap Fusion Lumos mass spectrometer

Introduction:

Chemical crosslinking with MS is powerful method to determine protein-protein interactions. Here we evaluated traditional and MS-cleavable crosslinkers for crosslinked peptide analysis using an Orbitrap Fusion Lumos. Different types of fragmentation and levels of MSn were compared. The data generated provides insights to improving identification of protein-protein interaction sites.

Methods:

Different amine-reactive, homobifuctional crosslinkers were used to crosslink BSA and E. coli whole cell lysates. Samples were reduced, alkylated and digested. Concentration determined using protein and peptide assays. All analysis were done on Orbitrap Fusion Lumos. E. coli samples were fractionated to enrich for crosslinked peptides before analysis. Data were analyzed with Xlink node in PD 2.2.

Results:

We compared two MS-cleavable crosslinkers, DSSO and BuUrBu to DSS for protein crosslinking labeling efficiency and identification using MSn. For both DSSO and BuurBu, we identified over 40 BSA crosslinked inter peptides using MS2-MS3 approach compared to less than 20 using MS2 CID. We also compared these crosslinkers using an E. coli whole cell lysate that was fractionated after crosslinking using SCX chromatography. Our preliminary results show an increase of confidently identified peptides after crosslinking using the MSn method compared to MS2 method. Overall, the combination of MS-cleavable crosslinking with MS3 acquisition methods enables the analysis of protein-protein interactions in complex samples. - DSSO: 26 interpeptidal crosslinks, 12 intrapeptidal, 19 hydrolized - BuUrBu: 39 interpeptidal, 26 intrapeptidal, 21 hydrolized

Novel Aspect:

Comparison of MS-cleavable crosslinkers and optimization of crosslinked peptide analysis workflows using the Orbitrap Fusion Lumos mass spectrometer

Sabbah, Hassan,
University Paul Sabatier - IRAP/CNRS

Contributing Authors:
Anthony Bonnamy, Institut de Recherche en Astrophysique et Planétologie - Université de Toulouse (UPS) & CNRS, Toulouse, France; Christine Joblin, Institut de Recherche en Astrophysique et Planétologie - Université de Toulouse (UPS) & CNRS, Toulouse, France; Jose-Angel Martin-Gago ICMM-CSIC, Madrid, Spain; Dimitris Papanastasiou, Fasmatech Science + Technology, Athens, GR; Jose Cernicharo, ICMM-CSIC, Madrid, Spain.

Title:
Characterization of PAH Content and Distribution in Cosmic Dust Analogues

Introduction:
AROMA (Aromatic Research of Organics with Molecular analyzer) is a new experimental set-up developed in the frame of the Nanocosmos ERC synergy project. The aim of Nanocosmos is to understand the physical and chemical processes leading to the formation of cosmic dust. The AROMA main purpose is to analyze, with micro-scale resolution, the molecular content of cosmic dust analogues.

Methods:
The experimental set-up consists of a microprobe laser desorption ionization chamber and a segmented linear quadrupole ion trap connected to an orthogonal time of flight mass spectrometer (LQIT-oTOF). The LQIT-oTOF platform was developed and constructed by Fasmatech Science and technology, Greece.

Results:
The ion source offers the possibility of probing polycyclic aromatic hydrocarbons (PAHs) in solid samples by performing Laser Desorption Ionization (LDI) in a single step as well as using the L2MS technique, where desorption and ionization are separated in time and space and each step is performed with a different laser. A comparison between the two techniques will be demonstrated on a variety of PAHs and derivatives (methyl sidegroups and heteroatoms). The LQIT also allows trapping and isolating desired species and further performing collision induced dissociation experiments as well as photodissociation studies. Exploiting the versatility of the LQIT the molecular structure of PAHs desorbed from the selected samples can be identified. The limits of detection of the apparatus and the ability to detect the PAH content in dust analogues using the L2MS technique is demonstrated.

Novel Aspect:
The L2MS and LDI techniques are performed in segmented LQIT-oTOF platform for the analysis of PAHs.

Sabo-Attwood, Tara,
University of Florida

Contributing Authors:

L. Cody Smith, University of Florida; Candice Lavelle, University of Florida; Cecilia Silva-Sanchez University of Florida; Fahong Yu, University of Florida; Natalia Garcia-Reyero, Mississippi State University; Nancy D. Denslow, University of Florida.

Title:

Phosphoproteomic Pipeline for the Identification of Non-Genomic Signaling Pathways Activated by Contaminant Exposure in Aquatic Species

Introduction:

Protein phosphorylation by environmental contaminants alter signaling pathways related to adverse health outcomes. Examining phosphorylation events on a global scale in ecotoxicology is challenging due to insufficient databases. Here we optimized a phosphoproteomic pipeline using a fathead minnow protein sequence database to identify rapid signaling in brains of chemically exposed fish.

Methods:

Brain tissue was collected from fathead minnows (FHM) after 30 minutes of exposure to 5 ng/L ethinylestradiol or ¬100 ng/L levonorgestrel. Phosphopeptides were enriched using TiO2 columns and samples analyzed by LC-MS/MS. Peptides were searched against our FHM database, analyzed using Scaffold, quantified based on average total ion current, and subnetworks enriched using Pathway Studio.

Results:

Our enrichment method was efficient as evidenced by a high percentage of phosphorylated peptides relative to non-phosphorylated peptides identified in the enriched fraction (60-72%) compared to the fraction not bound by TiO2 (1.6-2.17%). Interestingly, EE2 and LNG exhibited overlapping but distinct profiles with chief pathways related to neuronal function including neurite outgrowth, neurogenesis, transmission of nerve impulse, neuronal migration and learning and memory. Using pathway enrichment analysis we identified several key proteins, including TP53 and CBP, that are differentially phosphorylated by EE2 and LNG that play important roles in human brain diseases that involve nerve degeneration. These optimized phosphoproteomic methods support investigations of rapid signaling cascades mediated by phosphorylation events in response to environmental contaminants in fish.

Novel Aspect:

This study highlights the utility of fish to serve as model systems to study the role of PTMs in rapid signaling initiated by environmental stress.

Sahoo, Sarata,
National Institute of Radiological Sciences

Contributing Authors:
Zenon Palacz, IsotopX Ltd, Middlewich, Cheshire CW 10 0GE, United Kingdom; Tatsuo Aono, Fukushima Research Headquarter, National Institute for Quantum and Radiological Science and Technology, 4-9-1 Anagawa, Inage-ku, Chiba 263-8555, Japan; Norbert Kavasi Fukushima Research Headquarter, National Institute for Quantum and Radiological Science and Technology, 4-9-1 Anagawa, Inage-ku, Chiba 263-8555, Japan.

Title:
Novel method for determination of 90Sr using Thermal Ionization Mass Spectrometry

Introduction:
Measurement of 90Sr (t1/2 =28.8 y) is carried out since it is one of the most common and hazardous fission product after a nuclear reactor accident. 90Sr has a long biological half-life (~18y) in human body. Due to its chemical similarity to calcium, accumulates in bones and irradiates bone marrow, causing high radio-toxicity. Therefore, to assess 90Sr is important in case of a nuclear disaster.

Methods:
Thermal ionization mass spectrometry is the technique of choice because of its inherent high precision and accurate measurement of isotopic ratio. Typical abundance sensitivity for strontium isotope ratio measurement with conventional TIMS is about 10-7. By the use of special lense like WARP filter for TIMS, an upper limit of 1.0 x 10-10 was achieved.

Results:
The fission yield of stable 88Sr (~3.5×10-2) isotope in thermal neutron fission is comparable with the radioactive 90Sr (~5.8×10-2) isotope. Therefore, it is important to check if there is any variation of the stable strontium isotope ratio (88Sr/86Sr) in environmental samples after a nuclear accident. Recent developments in analytical techniques have resulted in use of TIMS as a substitute for β counting method of strontium. In our laboratory, a Phoenix X62 TIMS has been set up with nine Farday cup collectors as well as Daly ion counting system to measure minor isotopes e.g. 90Sr. The range of the tested 90Sr activity varied from 0.12 mBq to 660 mBq. Isotope ratios of 90Sr as well as stable Sr isotope ratios in a few contaminated soil samples with radiocaesium and some plants from Fukushima prefecture will be presented.

Novel Aspect:
Mass spectrometric method is rapid over radiometric methods since two week equilibration is not required.

Salzet, Michel,
University of Lille - France

Contributing Authors:
Jusal Quanico, PRISM INSERM U1192, University of Lille, France; Isabelle Fournier, PRISM INSERM U1192, University of Lille, France; Stephanie Devaux PRISM INSERM U1192, University of Lille, France; Lena Hauberg-Lotte, Center for industrial mathematics, University of Bremen, Bremen, Germany; Dasa Cizkova, Institute of Neurobiology, Slovak Academy of Sciences, Center of Excellence for Brain Research, Kosice, Slovakia; Peter Maas, Center for industrial mathematics, University of Bremen, Bremen, Germany; Jan H. Hobart, Center for industrial mathematics, University of Bremen, Bremen, Germany.

Title:
Novel Therapeutic Targets of Spinal Cord Injury Based on Spatio-Temporal Proteomics Analysis

Introduction:
Spinal cord injury (SCI) represents a major debilitating health issue with a direct socioeconomic burden on the public and private sectors worldwide. Although several studies have been conducted to identify the molecular progression of injury sequel due from the lesion site, still the exact underlying mechanisms and pathways of injury development have not been fully elucidated.

Methods:
A balloon compressive technique was used to produce SCI at thoracic Th8-9 spinal level in adult rats. Proteins identification was performed by Shot-Gun analysis in each segment-derived conditioned medium (CM) along the rostral-caudal axis after SCI with time course. 3D MALDI-MS Imaging were undertaken to specify the molecular environment combined with in-vitro and in-vivo tests.

Results:
Based on proteomics and 3D MS imaging we establish for the first time that molecular and cellular processes occurring after SCI are altered differently according to the lesion proximity, i.e. rostral and caudal segments nearby the lesion (R1-C1) express inflammatory factors whereas distant segments (R2-C2 and R3-C3) co-express factors implicated in neurogenesis. We also observe a discrepancy between R1 and C1 related to the delay in T-regulators recruitment and the presence in C1 of neurites outgrowth inhibitors (MEMO1, lectins, GFAP fragments and IgGs) that are absents in R1. The presence of IgG in neurons at very early stage of SCI may represent an additional factor contributing to the limited regeneration. Treatment in-vivo with anti-CD20 after SCI does not improve locomotor functions whereas using alginate filled with RhoA inhibitor we can observe a stimulation of regeneration.

Novel Aspect:
These results open the door of a novel view of the SCI treatment by considering the C1 as the therapeutic target.

Sansoucy, Maxime,
Université du Québec à Montréal

Contributing Authors:

Lekha Sleno, Université du Québec à Montréal; Makan Golizeh, Université du Québec à Montréal; Pat Forgione Concordia University.

Title:

A Quantitative Approach using LC-MS in Hepatitis C Virus NS2-3 Protease Autolysis Inhibition

Introduction:

As a public health problem, estimation states that around 120-200 million of people worldwide were affected by the hepatitis C virus. Although a large diversity of antiviral mixtures used as actual treatments, no specific drug was design against this virus. A high-resolution tandem mass spectrometry approach was developed to evaluate potential peptide-based inhibitors over the NS2-3 protease.

Methods:

The autolysis assay was done and reaction kinetics was followed over three-time points. Once quenched and digested by trypsin, protein digests were analyzed by LC-MS to then quantify the relative concentration of the cleaved and uncleaved peptides. Further experiments were conducted with increasing zinc concentration as well as assays in the presence of inhibitors, iodoacetamide and phenanthroline

Results:

A method was developed to monitor the HCV NS2-3 protease autolysis reaction employing Qq-TOF-MS. Kinetics assay has demonstrated an increasing trend over the incubation time presenting a maximum cleavage around 7%. As reported by literature, incubations in a presence of 50 µM of zinc had, however, decreased the autolysis activity whilst the protein has been categorized as a zinc-dependant protease. Examinations with iodoacetamide and 1,10-phenanthroline, known to affect cysteine and metalloprotease activity, respectively, were employed as model inhibitors to validate the LC-MS based inhibition assay. Percentages around 1.4 and 1.2 testified to decrease the protease activity in a presence of cysteine and metalloprotease. Subsequently, peptide-based inhibition efficiency will be quantified according to the percentage of cleavage.

Novel Aspect:

High-resolution mass spectrometry to study the NS2-3 protease inhibition and aid to design specific antiviral drugs against the hepatitis C virus.

Sant'Ana, Débora,
McGill University

Contributing Authors:
Mark Lefsrud, Bioresource Engineering Dept, McGill University, Canada.

Title:
Analysis of light stress in tomato plants by proteomics.

Introduction:
The use of LED (Light Emitting Diodes) for growing plants in greenhouses presents several advantages compared to traditional light, from the economic and environmental perspectives. Plant proteomics can lead to crop improvement as it can aid to increase yield and nutritional aspects and can also improve the knowledge on plants abiotic stress responses in plants.

Methods:
Tomato leaf tissue damaged and undamaged by intense LED lights were analysed using the MudPit methodology in an LTQ XL mass spectrometer. Spectra were searched by the Comet algorithm against Solanum lycopersicum (H1706 tomato) protein sequence database (Uniprot) and the PatternLab bioinformatics pipeline was used to associate spectra to the peptide sequence and to validate the results.

Results:
Around 100 proteins were validated in the tissue corresponding to the undamaged area and ten proteins were identified in samples of intense light damaged tissues. Primary results from undamaged plant tissues suggest the expression of stress proteins as chlorophyll a-b binding protein 1A, 1B, 3A, 3B, 1C, 1D, 8 chloroplastic; heat shock cognate protein 80; molecular chaperone Hsp90-1; pathogenesis-related leaf protein 6 and 4; and 11 uncharacterized proteins. While only phytochrome, short-chain dehydrogenase/reductase family protein and <10 uncharacterized proteins were able to be identified from the damaged tissue area. Although an initial result, the identified proteins from the undamaged tissue could confirm the cross-linking of D1-protein with surrounding polypeptides (D2, CP43 and PsbE) under photoinhibitory conditions in vivo.

Novel Aspect:
The study of the effect of LED lights in crops utilizing proteomics and the use of LED lights setup with lenses to achieve spots on plant tissue.

Santos, Fabio,
University of Campinas

Contributing Authors:
Edna Dora Martins Newman Luz, Cacao Research Center; Dilze Maria Argôlo Magalhães, Cacao Research Center; Marcos Nogueira Eberlin University of Campinas.

Title:
Comparative metabolite profiling of cocoa genotypes with contrasting resistance to ceratocystis cacaofunesta phytopathogen

Introduction:
Ceratocystis wilt is a lethal disease of cacao and plants of any age are susceptible to the pathogen. To avoid plant losses resistant genotypes have been searched to manage the disease. A study of metabolites expression from infected cacao genotypes is required in order to better understand the different levels of resistances from CCN51, CEPEC 2002 and TSH1188 genotypes toward C. cacaofunesta

Methods:
The metabolites of stems extracted with methanol were analyzed by UHPLC-MS using an Agilent 1290 Infinity Binary LC system coupled to an Agilent 6550 Accurate-Mass Q-TOF system with Agilent Jet Stream electrospray ionization. Total ion spectra were collected over a mass range of m/z 100–1,000, in both negative and positive modes. Peaks of metabolites were further investigated using MS/MS analysis

Results:
The results showed clear differences in the metabolite profiles of the different cacao genotypes, CCN51 (susceptible), CEPEC2002 (moderately susceptible) and TSH1188 (resistant). Based in metabolites, the cacao genotypes infected were clearly distinguishable into three clusters by principal component analysis (PCA)-derived loading plots. The metabolites identified by (PCA)-derived loading plots as responsible for the differentiation of the three clusters were phenylpropanoids, flavonoids, fatty acids, lipids, alkaloids, and amino acids. The fatty acids, linolenic, linoleic, oleic, stearic, arachidonic and asiatic acid were identified as major metabolites of both infected susceptible and resistant genotypes. Interestingly, methyl jasmonate, jasmonic and hydroxylated jasmonic acid as well as caffeine and theobromine were identified as major in the resistant genotypes.

Novel Aspect:
Metabolites related to response of cacao genotypes susceptible, moderately susceptible and resistant to Ceratocystis cacaofunesta were identified.

Santos Júnior, Júlio César,
Department of Clinical Pathology, School of Medical Sciences, State University of Campinas — UNICAMP

Contributing Authors:
Marco Antônio Tartarella, Team of Forensic Medicine West, Medico-Legal Institute, Police Technical Scientific Superintendence — SPTC, 307 Dr. Gastão Vidigal Avenue, 05314-000, São Paulo, SP, Brazil; Pedro Henrique Vendramini, ThoMSon Mass Spectrometry Laboratory, Institute of Chemistry, State University of Campinas — UNICAMP, 13085-970 Campinas, SP, Brazil; Pedro Carlos Mollo Filho Team of Forensic Medicine West, Medico-Legal Institute, Police Technical Scientific Superintendence — SPTC, 307 Dr. Gastão Vidigal Avenue, 05314-000, São Paulo, SP, Brazil; Marcos Nogueira Eberlin, ThoMSon Mass Spectrometry Laboratory, Institute of Chemistry, State University of Campinas — UNICAMP, 13085-970 Campinas, SP, Brazil; Simone Mesquita, Department of Clinical Pathology, School of Medical Sciences. State University of Campinas — UNICAMP, 181 Alexander Fleming St, 13083-881 Campinas, SP, Brazil; Nelci Fenalti Höehr, Department of Clinical Pathology, School of Medical Sciences. State University of Campinas — UNICAMP, 181 Alexander Fleming St, 13083-881 Campinas, SP, Brazil.

Title:
Vitreous humour: real-time analysis of opiates using Exactive FTMS

Introduction:
Opiates refers to substances derived from the opium (Papaver somniferum). Alternative samples are being applied in forensic toxicology as vitreous humor. Fourier transform ion cyclotron resonance mass spectrometry (FT-ICR-MS) provides ultra high resolution in mass analysis. In this work we evaluated opiates and their metabolites by real-time direct FT-ICR-MS analysis of vitreous humors.

Methods:
Vitreous humor were obtained by forensic medicine from the West zone from the Medico-Legal Institute of Police Technical Scientific Superintendence of São Paulo, Brazil. For the analysis in the positive ion mode, 10 µL was injected through a Advion TriVersa NanoMate® Robotic system. The m/z were then monitored by FT-ICR-MS in a LTQ-FT Ultra (Thermo Scientific, Bremen, Germany) mass spectrometer.

Results:
In the present work, opiates and their metabolites were detected in post-mortem vitreous humour analyzed. Thus, for its analytical stability and sterility for a large number of days after death, vitreous humour is a good sample for toxicological analyzes, therefore is an important and sometimes exclusive target. Finally, the proposed method based on real-time analysis of opiates by Exactive FTMS proved to be satisfactory, constitutes an important tool for "fingerprinting" in the post-mortem toxicological analysis, which is essential to the toxicology and forensic pathology routine.

Novel Aspect:
Metabolomics allied to forensic toxicology is extremely promising and innovative, even using Exactive FTMS.

Sathyanarayanan, Gowtham,
University of Helsinki

Contributing Authors:

Gowtham Sathyanarayanan, University of Helsinki; Tiina Sikanen, University of Helsinki; Markus Haapala University of Helsinki.

Title:

Integrating digital microfluidic sample preparation with desorption atmospheric pressure photoionization-mass spectrometry

Introduction:

Digital microfluidics (DMF) based on electrowetting-on-dielectric is an established tool for diverse sample manipulation on surfaces, in small volumes, typically in combination with optical or electrochemical detection. This work describes the coupling of DMF to desorption atmospheric pressure photoionization mass spectrometry (DAPPI-MS), which allows direct analysis of solid samples on surfaces.

Methods:

The DMF chips featured a ~8-µm-thick SU-8 layer (as the dielectric) and ~100-nm-thick fluoropolymer coating on top of Cr electrodes used for sample actuation by electrowetting. DAPPI-MS was performed on an Agilent 6410 QQQ-MS by using a heated nebulizer microchip for vaporizing the solvent, chlorobenzene (jet T ~250 °C), similarly as in Haapala et al., Anal Chem, 79, 2007, 7867.

Results:

The possibility to combine DMF sample manipulation with direct desorption/ionization-MS analysis would greatly increase the flexibility of e.g., reaction monitoring. In this study, the feasibility of DAPPI-MS for direct analysis on DMF chips was examined using paracetamol, naproxen, and testosterone as model compounds; each of which gave good S/N ratios at 10 mg/L concentration (at best, of 230, 81, and 21, respectively). Chip-to-chip and within chip repeatability was, however, tens of percent due to the fact that proper alignment of the heated jet with the dried sample spot was complicated because of the very dense precipitate formed during solvent evaporation on a very hydrophobic fluoropolymer surface. Thus, various solvent compositions were tested to ensure more homogenous and disperse sample precipitation prior to DAPPI-MS. Future work focuses on improving the repeatability.

Novel Aspect:

Coupling of digital microfluidics based sample manipulation to DAPPI/MS analysis is described for the first time.

Sato, Hiroaki,
National Institute of Advanced Science and Technilogy (AIST)

Contributing Authors:
Reiko Tanaka, Medical Mycology Research Center, Chiba University; Hiroki Takahashi, Medical Mycology Research Center, Chiba University; Takashi Yaguchi Medical Mycology Research Center, Chiba University; Sayaka Nakamura, National Institute of Advanced Industrial Science and Technology, AIST.

Title:
Rapid Discrimination of Aspergillus section Fumigati Based on Ribosomal Protein Profiling Observed by MALDI-TOFMS

Introduction:
Because most Aspergillus species are fungal pathogens, rapid and accurate identification is important to select appropriate antifungal agent. However, characterization of closely related species within genus Aspergillus has been difficult. We have applied ribosomal protein-based MALDI–TOF MS to discriminate five species belonging to Aspergillus section Fumigati.

Methods:
About 30 strains of clinical and environmental isolates were used as samples. They have been identified as A. fumigatus, A. lentulus, A. viridinutans, A. udagawae, and Neosartorya fischeri, where genus Neosartorya is teleomorph of Aspergillus section Fumigati. The mass spectra of the ribosomal protein fractions purified from fungal cells were observed by an AXIMA CFR MALDI-TOFMS in a linear mode.

Results:
Amino acid sequences of ribosomal proteins of genome-sequenced Aspergillus strains were obtained from the UniProtKB database. The theoretical mass of ribosomal proteins were calculated, and the actual mass of the expressed proteins were verified by MALDI-TOF MS measurements. During this verification, the correction of sequence errors and speculation of post translational modifications were also performed. A total of 29 types of ribosomal subunit proteins were selected as reliable biomarkers. A phylogenetic tree were constructed based on the profiles of ribosomal protein biomarkers. The sample strains were divided into clusters reflecting the species. These results demonstrated ribosomal protein profiling method using MALDI-MS would be a sufficiently high-throughput technique for rapid discrimination of closely related Aspergillus species.

Novel Aspect:
Ribosomal protein-based MALDI–TOF MS can successfully perform rapid identification of the species belonging to Aspergillus section Fumigati.

Savitski, Mikhail,
EMBL

Title:
Tracking drug action in living cells by thermal profiling of the proteome

Introduction:
The thermal stability of proteins can be used to assess ligand binding in living cells. We have generalized this concept by determining the thermal profiles of more than 7000 proteins in human cells by means of quantitative mass spectrometry.

Methods:
Combination TMT10 reagents and novel bioinformatics approaches enabled determination of melting curves of proteins across a broad temperature range.

Results:
Monitoring the effects of small-molecule ligands on the profiles delineated more than 50 targets for the kinase inhibitor staurosporine. We identified the heme biosynthesis enzyme ferrochelatase as a target of kinase inhibitors and suggest that its inhibition causes the phototoxicity observed with vemurafenib and alectinib. Thermal shifts were also observed for downstream effectors of drug treatment. In live cells, dasatinib induced shifts in BCR-ABL pathway proteins, including CRK/CRKL. The methodology is also applicable to other target classes such as histone deacetylases. This is illustrated by profiling the HDAC inhibitor panobinostat which lead to identification of hitherto unknown targets of this drug. Validation of novel targets will be presented.

Novel Aspect:
Thermal proteome profiling provides an unbiased measure of drug-target engagement and identification of markers for drug efficacy and toxicity

Schäfer, Mathias,
University Cologne, Department of Chemistry

Contributing Authors:
Albrecht Berkessel, University of Cologne, Department of Chemistry, Germany; Katrin Peckelsen, University of Cologne, Department of Chemistry, Germany; Mathias Paul University of Cologne, Department of Chemistry, Germany; Giel Berden, Institute for Molecules and Materials, Radboud University, Nijmegen, NL; Jos Omens, Institute for Molecules and Materials, Radboud University, Nijmegen, NL; Anthony J. H. M. Meijer, Department of Chemistry, University of Sheffield, Sheffield, UK; Jonathan Martens, Institute for Molecules and Materials, Radboud University, Nijmegen, NL.

Title:
Structure and Reactivity of Phenyl-Hydroxycarbene studied in the Gas Phase: Hydrogen Tunneling at Room Temperature

Introduction:
Carbenes form Breslow intermediates, which are of pivotal importance for the understanding of the organo-catalytic umpolung of reactivity. Many enzymes like alpha-keto decarboxylases, which generate Breslow intermediates of hydroxycarbenes, follow this concept and the general question arises whether quantum effects are relevant in enzymatic hydrogen transfer processes at room temperature (RT).

Methods:
An electrospray-quadrupole ion trap tandem-MS approach is employed to investigate the structure and reactivity of charge-tagged hydroxycarbenes and isomeric aldehydes. IRMPD-spectra, acquired with the wavenumber tuneable light of a free electron laser and of an OPO-laser, are compared to computed spectra of ion structures identified by high-level theory (B3LYP-GD3BJ / cc-pVTZ).

Results:
The nature and reactivity of a charge-tagged phenyl-hydroxycarbene, which is accessible in quadrupole ion trap product ion experiments by conventional collision induced dissociation CID tandem MS at room temperature, is investigated and discussed. The charge-tagged phenyl-hydroxycarbene and other related photo-fragments were characterized by IRMPD-spectroscopy and DFT computations with confirmation of the data sets of reference aldehyde compounds (deuterated and undeuterated). Most importantly, the monomolecular isomerization of the phenyl-hydroxycarbene to the respective aldehyde was investigated by monitoring individual fingerprint bands and the isomerization kinetics of this first order reaction could be determined. In doing so, we present ample evidence that hydrogen atom tunneling is responsible for the isomerization reaction of the phenyl-hydroxycarbene at room temperature.

Novel Aspect:
Identification and characterization of hydroxycarbenes at RT and the first unambiguous confirmation of hydrogen atom tunnelling at RT.

Schäpe, Kaija,
Justus-Liebig-University of Giessen

Contributing Authors:
Tessa Bartges, Justus-Liebig-University of Giessen, Institute of Physical Chemistry, Heinrich-Buff-Ring 17, D-35392 Gießen, Germany.; Jürgen Janek, Justus-Liebig-University of Giessen, Institute of Physical Chemistry, Heinrich-Buff-Ring 17, D-35392 Gießen, Germany.; Janina Werner Justus-Liebig-University of Giessen, Institute of Veterinary Anatomy, -Histology and -Embryology, Frankfurter Str. 94, D-35392 Gießen, Germany.; Dhaka R. Bhandari, Justus-Liebig-University of Giessen, Institute of Inorganic and Analytical Chemistry, Heinrich-Buff-Ring 17, D-35392 Gießen, Germany.; Thimo Schaaf, Justus-Liebig-University of Giessen, Institute of Veterinary Anatomy, -Histology and -Embryology, Frankfurter Str. 94, D-35392 Gießen, Germany.; Alena-Svenja Wagner, Justus-Liebig-University of Giessen, Institute of Veterinary Anatomy, -Histology and -Embryology, Frankfurter Str. 94, D-35392 Gießen, Germany.; Marcus Rohnke, Justus-Liebig-University of Giessen, Institute of Physical Chemistry, Heinrich-Buff-Ring 17, D-35392 Gießen, Germany.; Julia Kokesch-Himmelreich, Justus-Liebig-University of Giessen, Institute of Physical Chemistry, Heinrich-Buff-Ring 17, D-35392 Gießen, Germany.; Bernhard Spengler, Justus-Liebig-University of Giessen, Institute of Inorganic and Analytical Chemistry, Heinrich-Buff-Ring 17, D-35392 Gießen, Germany.; Sabine Wenisch, Justus-Liebig-University of Giessen, Institute of Veterinary Anatomy, -Histology and -Embryology, Frankfurter Str. 94, D-35392 Gießen, Germany.

Title:
Lipidomics in Osteoporosis: From sample preparation to mass spectrometry (imaging) of lipids in cells and bone sections by MALDI and ToF-SIMS

Introduction:
Lipids play a crucial role in human diseases such as cancer, diabetes, and osteoporosis. The goal of the study was to adapt sample preparation, measurement parameters and procedures for mass spectrometric (imaging) analysis of lipids in human bone cells and sections via ToF-SIMS or a combination of ToF-SIMS and AP-SMALDI, respectively.

Methods:
Different sample preparation and storage procedures for cells were compared via ToF-SIMS surface spectra and Principal Component Analysis to optimize lipid analysis. For complementary lipid imaging in bone sections via ToF-SIMS and AP-SMALDI, a cryosectioning protocol was established. The optimized protocols were applied to osteoporotic and control donor samples.

Results:
Among the different fixation/preparation methods for cell culture samples analyzed by ToF-SIMS, chemical fixation using aldehydes proved beneficial in terms of reproducibility and lipid preservation, while the freeze-fracturing/ frozen-hydrated method had disadvantages in terms of reproducibility. Wet storage in buffer or water preserved the membrane lipids more efficiently than dry storage procedures. Application of this chemical fixation method in combination with wet storage to differentiated bone marrow stem cells of osteoporotic and control donors revealed differences in their fatty acid profiles via ToF-SIMS. For bone sections, a method consisting of freeze-embedding and -sectioning with a cryotape was adapted from Kawamoto et al., enabling consecutive imaging via ToF-SIMS followed by AP-SMALDI, preliminary experiments provided localization of different lipids.

Novel Aspect:
The focus on sample preparation for MSI of lipids in cell cultures and human bone sections is a novel and crucial point addressed by this study.

Scharikow, Leanne,
Public Health Agency of Canada

Contributing Authors:
Stuart McCorrister, National Microbiology Laboratory, Public Health Agency of Canada; Patrick Chong, National Microbiology Laboratory, Public Health Agency of Canada; Morag Graham National Microbiology Laboratory, Public Health Agency of Canada and Department of Medical Microbiology, University of Manitoba; Derek Davlut, National Microbiology Laboratory, Public Health Agency of Canada; Garrett Westmacott, National Microbiology Laboratory, Public Health Agency of Canada.

Title:
Development of a targeted proteomic assay for rapid detection of Shiga toxins (Stx) 1 and 2 in Shiga toxin-producing E. coli (STEC) bacterial cultures

Introduction:
STEC can cause severe renal and central nervous system damage due to Stx, and antibiotic treatment of STEC can increase the damage caused by the toxin. Current Stx detection techniques are time consuming, expensive, and lack sensitivity. We developed and evaluated a novel mass spectrometry-based assay for detection of signature Stx peptides using parallel reaction monitoring (PRM).

Methods:
Purified Stx1a and Stx2a were analyzed by shotgun proteomics, and signature target tryptic peptides were chosen to develop a targeted proteomics assay. The parallel reaction monitoring assay was developed and tested on a Q-Exactive Plus mass spectrometer. Results were compared to PCR and Vero cell assay gold-standard methods.

Results:
Eleven target tryptic peptides were selected based on signal intensity and unique amino acid sequence. Of these 11 peptides, four can be mapped specifically to Stx1, six to Stx2, and one peptide is common to both. Moreover, each subtype of Stx2 has a unique peptide fingerprint, enabling Stx2-subtype identification in a single assay. Synthetic copies of these peptides were diluted into digested cell lysate. No interference from the cell lysate was observed. The lower limit of detection for individual target peptides using PRM ranged from 46 attomole to 3.7 femtomole. The PRM assay will be compared to PCR and the Vero cell assay using 90 E. coli isolates (60 STEC).

Novel Aspect:
Applying a targeted proteomics assay to detect Stx signature tryptic peptides for all subtypes of Stx in bacterial cultures.

Schnapp, Andreas,
University of Münster

Contributing Authors:
Ann-Christin Niehoff, Institute of Inorganic and Analytical Chemistry, University of Münster, Germany; Klaus Dreisewerd, Institute for Hygiene, University of Münster, Germany; Annika Koch Institute for Hygiene, University of Münster, Germany.

Title:
Laser Mass Spectrometry Imaging of Cuticular Lipids from Insects Using Etched Silver Substrates

Introduction:
UV-LDI mass spectrometry can be used to analyze lipids directly from insect cuticles. However, saturated hydrocarbons (HC) cannot be detected. Previous studies showed that these lipids can be visualized as [M+Ag]+ ions after addition of Ag nanoparticles. Here we present a modified approach using etched silver foils as substrates. These targets also enable imaging of insect surfaces after blotting.

Methods:
Etching of silver substrates was achieved with HNO3. Either 0.1 µL of extracts were spotted or solid samples (e.g., fly wings) pressed onto the substrate. MS experiments were performed with a modified MALDI Synapt G2-S mass spectrometer (Waters). The buffer gas pressure in the ion source was set to 0.7 mbar. An AP-MALDI instrument (iMScope TRIO; Shimadzu) was used in a few comparison experiments.

Results:
Using the Synapt LODs were determined to fall into the 100 fmol range, with the exact value depending on chemical composition of the lipids and degree of unsaturation. Next to saturated and unsaturated HCs, more polar lipids, such as alcohols, free fatty acids, wax esters, sterols and triacylglycerols were sensitively detected as [M+Ag]+ species. This feature provided a "comprehensive" profile of the rich cuticular lipid bouquet of insects such as fruit flies, worker bees, and the butterfly Bicyclus anynana within a single MS analysis. MS images recorded after pressing a hind wing of the male butterfly demonstrated the potential of the method for visualization the lateral distribution of sex pheromones. After pressing a fingertip onto the silver substrate MS images of cholesterol, squalene and numerous further squamous compounds was possible at about 20 µm resolution.

Novel Aspect:
Novel method for silver-assisted UV-LDI mass spectrometry and MS imaging of several lipid classes including long to very long chain alkanes

Schofield, James,
University of Southampton, UK

Contributing Authors:
James Schofield, Centre for Proteomic Research, Centre for Biological Sciences, University of Southampton, UK; Paul Skipp, Centre for Proteomic Research, Centre for Biological Sciences, University of Southampton, UK; Peter Sterk Dept. Respiratory Medicine, Academic Medical Centre, University of Amsterdam, The Netherlands; Ian Adcock, National Heart & Lung Institute, Imperial College & Biomedical Research Unit, Royal Brompton & Harefield NHS Trust, London, UK; Julie Corfield, Areteva Ltd, Nottingham, UK; Doroteya Staykova, Centre for Proteomic Research, Centre for Biological Sciences, University of Southampton, UK; Dominic Burg, Centre for Proteomic Research, Centre for Biological Sciences, University of Southampton, UK; Ratko Djukanović, NIHR Southampton Respiratory Biomedical Research Unit, University Hospital Southampton, Southampton, UK; Fan Chung, National Heart & Lung Institute, Imperial College & Biomedical Research Unit, Royal Brompton & Harefield NHS Trust, London, UK; Patient Input Platform and patient representatives from the Ethics Board and Safety Management Board, U-BIOPRED.

Title:
Stratifying asthma disease in the U-BIOPRED severe asthma project: large-scale MSE and TDA to define sub-phenotype biomarker fingerprints.

Introduction:
Severe asthma is not a single disease and likely consists of several subtypes (phenotypes), but the molecular basis for this diversity is poorly understood. The pan-European U-BIOPRED consortium utilized state of the art omics technologies to characterise the molecular biology of severe asthma sub-phenotypes as 'omic fingerprints and combined 'omic handprints by systems biology.

Methods:
270 induced sputum samples from asthma and healthy study participants were analysed by label-free quantitative data independent mass spectrometry. Similarity in abundance profiles of proteins between participants were analysed using topological data analysis (TDA) supported by standard statistics.

Results:
Sputum samples were precipitated, lipid extracted, processed to peptides and analysed using a label-free approach combining data independent MSE acquisition with Hi-3 quantitation. Data were pre-processed and cleaned using a custom in-house developed pipeline. High quality data were analysed using a range of statistical and machine learning techniques; TDA and biological context analysis. 5574 sputum proteins were identified and quantified. In-depth analysis of these data using a novel high-dimensional data analysis approach combining TDA with machine learning has revealed sub-phenotypes of severe asthma characterised by panels of protein biomarkers. In severe asthmatic participants, 10 new phenotypic groups were identified with distinct proteome fingerprints.

Novel Aspect:
Large-scale proteomic characterisation in combination with TDA reveals phenotype-specific biomarkers of asthma; a step towards stratified medicine.

Schoof, Erwin,
Princess Margaret Cancer Centre, University Health Network

Contributing Authors:
Eric R. Lechman, Princess Margaret Cancer Centre; Gabriela Krivdova, Princess Margaret Cancer Centre; Ruth Isserlin University of Toronto; Amanda Mitchell, Princess Margaret Cancer Centre; Gary D. Bader, University of Toronto; Bo T. Porse, Copenhagen University; John E. Dick, Princess Margaret Cancer Centre; Aaron Trotman-Grant, Princess Margaret Cancer Centre; Thomas Kislinger, Princess Margaret Cancer Centre; Mark Minden, Princess Margaret Cancer Centre; Kerstin Kaufmann, Princess Margaret Cancer Centre; Stephanie Xie, Princess Margaret Cancer Centre.

Title:
Identifying key stem cell-associated proteins in normal blood development and leukemia

Introduction:
Hematopoietic stem cells (HSC) are the origin of all cell lineages in the blood hierarchy. Acute myeloid leukemia (AML), a perturbed state of hematopoiesis, is also hierarchically organized, with leukemia stem cells (LSC) at the apex. Successful eradication of AML will likely depend on specifically targeting these tumor-initiating cells, which in turn requires their molecular characterization.

Methods:
Flow Cytometry is used to sort phenotypically defined sub-populations from the human blood hierarchy and primary AML samples. Using highly optimized single-reaction-chamber digestion protocols, we are able to obtain broad, quantitative proteome coverage out of 100,000 cells, essentially overcoming a major obstacle that precluded this work thus far on human primary cellular sub-populations.

Results:
By combining animal models and quantitative proteomics, we aspire to define and functionally validate key proteins that confer the unique HSC/LSC phenotypes. We have characterized the global protein landscapes defining the various hematopoietic sub-populations in humans, especially in terms of protein signaling networks distinguishing HSC and LSC from each other and their respective progenitor populations. Computational modeling of the proteomics data uncovered key proteins that were enriched in the HSC- and LSC-containing subpopulations, which are deemed potential regulators of stemness. Additionally, we identified LSC-specific proteins, thereby providing potential targets for therapeutic intervention that would spare HSC. Through functional validation using genetic gain- or loss-of-function assays, we are interrogating their role in maintaining stemness of HSC/LSC populations in vivo.

Novel Aspect:
This work comprises unprecedented protein-level interrogation of human HSC and LSC populations, and functionally investigates potential novel targets.

Schreiber, Andre,
SCIEX

Contributing Authors:
Jean-Daniel Berset, Water and Soil Protection Laboratory, Bern, Switzerland; Michael Scherer, SCIEX.

Title:
Identification and Quantitation of Legal and Illicit Drugs in Wastewater Samples Collected during a Dance Party Event

Introduction:
Drug abuse is a global problem with major negative impacts on human health and social welfare. Illicit drugs are substances for which nonmedical use is prohibited by national or international laws. Important groups of illicit drugs are opioids, cocaine, cannabis, amphetamines and ecstasy (MDMA). Among those, amphetamines and MDMA currently demand the most attention by law enforcement agencies.

Methods:
LC-MS/MS has become the method of choice for the quantitation of drugs in aqueous samples. Large volume injection and high sensitivity MS/MS allow limits of quantitation in the low ng/L range. The Street Parade Zürich is an interesting event to study trends because the wastewater is strongly affected due to the relationship between number of inhabitants (~300000) and visitors (~950000).

Results:
An LC-MS/MS method for the quantitation of Morphine, Codeine, Amphetamine, Monoacetylmorphine, Methamphetamine, MDMA, Mephedrone, Benzoylecgonine, Monoacetylcodeine, Cocaine, EDDP, Methadone was developed. The column dimension 100 x 4.6 mm allowed large volume direct injection without the breakthrough of polar compounds like morphine. LODs of 1 ng/mL or lower with good accuracy and precision were achieved. Matrix effects have been investigated by T-infusion experiments and were found to be negligible when wastewater samples are diluted 10x. The method was successfully applied to the determination of drugs of abuse wastewater during a dance party event (Street Parade Zürich 2014). Different profiles were observed for different drugs indicating different consumption during the time period of the event.

Novel Aspect:
Quantitation of drugs in wastewater during a dance party event

Schreiber, Andre,
SCIEX

Contributing Authors:
Wen Jin, SCIEX.

Title:
Identification and Quantitation of Underivatized Glyphosate and Other Polar Pesticides in Water and Beer Samples using LC-MS/MS

Introduction:
Glyphosate is a common broad-spectrum systemic herbicide used widely to kill weeds and grasses. Glyphosate is a topic with an extraordinary degree of public attention and concerns since the International Agency for Research on Cancer (IARC) classified glyphosate as a probable human carcinogen. Traces of glyphosate have been found in many foods and also in human urine.

Methods:
Here we present 3 different LC-MS/MS methods to detect underivatized Glyphosate and other polar pesticides in water and beer. MS/MS detection was performed using a SCIEX QTRAP 6500+ system operated in MRM mode in -ESI. Different published LC methods (using porous graphitic carbon, ion exchange, and mixed mode) were evaluated and compared with respect to sensitivity, selectivity and routine use.

Results:
Three different LC methods and MS/MS detection were evaluated for the detection of underivatized glyphosate, glufosinate, AMPA and MMPA. All 3 methods achieved low to sub ng/mL detection limits and good repeatability. High selectivity and confidence in identification was achieved by monitoring 4 MRM transitions per compound. The method using a porous graphitic carbon column required extensive priming and was considered less fit for routine use. The other 2 methods using ion exchange and mixed mode phases with acidic or basic mobile phase had better performance and also allowed a larger injection volume to reach an LOD for glyphosate of 100 ng/L matching challenging requirements for drinking water testing. The method was successfully applied to quantify Glyphosate in beer. Glyphosate was detected at concentration of up to 20 ng/mL confirming recent ELISA results of a German study.

Novel Aspect:
Direct injection of water and beverage samples to identify and quantify underivatized glyphosate and other polar pesticides

Schuhmann, Kai,
MPI CBG Dresden, Germany

Contributing Authors:
Henrik Thomas, MPI-CBG Dresden; HongKee Moon, MPI-CBG Dresden; Andrej Shevchenko MPI-CBG Dresden.

Title:
Accurate quantification of polyunsaturated glycerophospholipids by shotgun lipidomics

Introduction:
Carboxylate anions (CA) are employed for quantifying molecular species of glycerophospholipids (GPL) by MS2. While CA of saturated fatty acids are stable, CA of polyunsaturated fatty acids (PUFA) rapidly degrade which biases lipid quantification. We developed a method that circumvents collisional instability of CA and supports accurate quantification of PUFA containing GPL by shotgun lipidomics.

Methods:
Using a collection of shotgun HCD FT MS2 spectra of GPLs we correlated collision energy, fragment ion abundances, position and number of double bonds (DB) and the number of carbon atoms. We generalized these correlations in a multidimensional mathematical model that enabled a simple and accurate way of adjusting CA abundances of both saturated and highly unsaturated fatty acid moieties.

Results:
We observed that a higher number and a closer proximity of double bonds to the carboxyl group drastically reduced CA abundances during MS2, particularly at elevated collision energies (CE). While lower CE provided more equal response it drastically reduced sensitivity. Therefore, we developed a mathematical model that associated the abundance of CA with CE and structural features of FA moieties and implemented this into a software to correct for abundance losses. The accuracy of quantification was validated using model mixtures of synthetic lipids. It was further applied for profiling PUFA containing GPLs in total lipid extracts from mouse retina and brain. Overall, we found that the correction increased the abundance of physiologically important polyunsaturated lipids by more than 4x and the deviation from expected values was less than 20%.

Novel Aspect:
Accurate quantification of polyunsaturated glycerophospholipids by shotgun lipidomics

Schumacher, Katherine,
University of Nebraska

Contributing Authors:

Eric D. Dodds, University of Nebraska - Lincoln.

Title:

Site-Specific N-glycoform Analysis of Human Alpha1-Acid Glycoprotein: Towards an Integrated Approach for Complete Molecular Characterization

Introduction:

Alpha1-acid glycoprotein (AGP) is an acute phase protein with five N-glycosylation sites and is an attractive potential biomarker. However, little is known about AGP's glycosylation in molecular detail. This study aims to develop an integrated approach to fully characterize AGP and its glycoforms, with the goal of providing a better understanding of the disease-induced changes in glycosylation.

Methods:

AGP was subjected to either proteolytic digestion or glycan release, and the resulting glycopeptides and glycans were enriched and analyzed by mass spectrometry. Experiments were carried out using a Waters Synapt G2-S HDMS. Collision-induced dissociation (CID) was carried out in the trap cell at various collision energies (0 to 65 V) for structural characterization.

Results:

An integrated approach of AGP analysis is being performed, focusing on glycan centered and site-specific glycoproteomic analysis, with plans to explore other approaches later. For our site-specific studies, glycopeptides produced by trypsin and pronase digestion of AGP were subjected to CID at varying collision energies and energy-resolved CID plots were created. Currently, data has been gathered to characterize glycopeptides from four of AGP's five glycosylation sites, with the intention of identifying more glycopeptide structures from all five sites. Glycans released from AGP were also analyzed. These released glycan studies will provide a better understanding of the specific molecular structures of the glycan moieties, which in turn will provide insight on which glycopeptides may be most abundant or likely. At this point, several different glycan compositions have been identified.

Novel Aspect:

This project aims to provide a full molecular characterization of human AGP glycoforms through integration of multiple glycoanalytical approaches.

Schumacher, Katherine,
University of Nebraska

Contributing Authors:
Eric D. Dodds, University of Nebraska - Lincoln; Yuting Huang, University of Nebraska - Lincoln; Lauren Petrosh University of Nebraska - Lincoln.

Title:
Probing Carbohydrate Isomerism through Transition Metal Ion Adduction, Gas-Phase Ion Chemistry, and Collision-Induced Dissociation

Introduction:
The potential for numerous isomers within a given carbohydrate composition can lead to difficulty in structural analysis. However, the use of metal ion charge carriers can allow isomer differentiation by tandem mass spectrometry. Additionally, the application of gas-phase ion chemistry has been found to probe differences in the conformations or fragmentation behaviors of carbohydrate isomers.

Methods:
Several groups of carbohydrate isomers were selected for this analysis. Each carbohydrate was studied as the adduct of transition metal ions (divalent Fe, Co, Ni, Cu, and Zn) and their corresponding gas-phase electron transfer (ET) products. Experiments were performed using a Waters Synapt G2-S HDMS. MS/MS spectra were obtained using various vibrational activation energies.

Results:
Metal ions can be used to probe oligosaccharide isomerism, as metal ion binding can bring about conformational differences among isomers and can lead to unique fragmentation behaviors. Transition metals have been chosen to study how metal ion adduction can affect the MS/MS analysis of isomeric carbohydrates. It has been observed that carbohydrate/metal ion adducts show different fragmentation behavior than their corresponding gas-phase ET products. For this project, carbohydrate isomers as metal ion adducts and their analogous ET products were the focus. With the appropriate selection of charge carrier and gas-phase ion chemistry, isomer-distinguishing fragmentation pathways were accessed, allowing for the differentiation of multiple isomers. This approach will provide an additional tool for probing carbohydrate structure.

Novel Aspect:
This approach will provide an additional tool for probing carbohydrate structure through metal ion adduction combined with gas-phase ion chemistry.

Schwarz, Helmut,
TU Berlin

Contributing Authors:

Jilai Li, TU Berlin; Shaodong Zhou, TU Berlin; Maria Schlangen TU Berlin.

Title:

Bond Activation by Metal-Carbene Complexes in the Gas Phase

Introduction:

Novel ion/molecule reactions of metal carbenes will be presented, demonstrating how these species act as catalysts.

Methods:

Tandem mass spectrometry and FTICR are combined with state of the art computational methods.

Results:

(i) Coupling of gold carbene with methane to form C2Hx (x = 6, 4). (ii) Network of reactions in the Pt-mediated coupling of methane and ammonia. (iii) Thermal carbon-atom extrusion from halobenzenes and its coupling with methylene to form acetylene.

Novel Aspect:

The combined experimental/computational approach permits to identify the active site of catalysts.

Scupakova, Klara,
M4I Institute, University of Maastricht

Contributing Authors:
S.S. Rensen, Department of General Surgery, School of Nutrition and Translational Research in Metabolism (NUTRIM), University of Maastricht, The Netherlands; G. Ertaylan, The Maastricht Centre for Systems Biology (MaCSBio), University of Maastricht, The Netherlands; R.M.A. Heeren Maastricht Multimodal Molecular Imaging Institute (M4I), University of Maastricht, The Netherlands; S.W.M. Olde Damink, Department of General Surgery, School of Nutrition and Translational Research in Metabolism (NUTRIM), University of Maastricht, The Netherlands; Z. Soons, Department of General Surgery, School of Nutrition and Translational Research in Metabolism (NUTRIM), University of Maastricht, The Netherlands; K.A. Pierzchalski, Maastricht Multimodal Molecular Imaging Institute (M4I), University of Maastricht, The Netherlands.

Title:
Lipid profiling using integrative Mass Spectrometry Imaging (MSI) for Nonalcoholic Fatty Liver Disease (NAFLD) in human liver biopsies.

Introduction:
NAFLD, characterized by excessive lipid accumulation in hepatocytes, has been suggested to drive disease progression. MSI provides sensitive spatial information for lipid species to distinguish the potential spectrum of liver damage. Impaired lipid metabolism accompanies NAFLD, thus we aim to establish specific lipid profiles by MSI to reliably distinguish different NAFLD stages.

Methods:
Frozen liver biopsies of 15 obese patients were provided by MUMC+. Categorical stratification determined no steatosis, simple steatosis and steatohepatitis by pathology annotation. DESI and MALDI MSI techniques were applied to visualize lipid profiles. Tissues sections (10um) were sliced at -20oC. DHB and norharmane were sublimed, then analyzed in positive and negative MALDI modes respectively.

Results:
We use DESI (Xevo G2-XS, Waters) - an ambient imaging tool with soft ionization requiring no sample preparation - to detect complex intact lipid species. Because ionization efficiency can be greatly influenced by ionization techniques, we also analyzed tissues on the Rapiflex MALDI TissueTyper (Bruker) as a complementary technique. Species identification confirmation was analyzed by MS/MS on the Synapt G2Si (Waters) and MALDI-FTICR (9.4 T SolariX, Bruker) to provide high mass resolution information. Free fatty acids (m/z 175.1023, 253.2993, 279.3176, 281.3333, 303.3184), phospholipids (m/z 789.54, 824.5609, 853.7238), sphingomyelins (m/z 542.123, 617.5146, 789.603, 815.6349) and ceramides (m/z 832.5873, 876.741) differed between the disease stages determined by in-house statistical analysis software, such as PCA or hierarchical clustering, which indicates unique lipid metabolism.

Novel Aspect:
Complementary multimodal MSI techniques were applied to provide spatial visualization of the lipid metabolism specific to NAFLD stages.

Seale, Brendon,
NSERC CREATE York University and University of Toronto

Contributing Authors:

Charis Lam, University of Toronto; Aaron R. Wheeler, University of Toronto; Darius G. Rackus University of Toronto.

Title:

Digital microfluidic immunoprecipitation for the microscale targeted analysis of proteins from biological samples using mass spectrometry

Introduction:

Immunoprecipitation (IP) is often used to identify proteins in biological samples coupled with MS as it is an effective at reducing matrix interference. But the combination of IP & MS requires complex sample preparation making it demanding and time consuming. We present a microscale IP platform using digital microfluidics (DMF) which automates and miniaturizes IP for MS analysis of proteins.

Methods:

We use a technique in which magnetic particles coated in antibodies specific to human serum albumin (HSA) are delivered onto a DMF device and incubated with a biological sample. HSA is then eluted from the particles using citric acid. We optimized the technique using fluorescent HSA and then evaluate its applicability to direct injection MS analysis.

Results:

DMF-IP was possible with blood serum sample sizes as small as 4 µL. Each chip could process 4 such samples simultaneously. Optimization using fluorescent protein showed that 80% of the capture HSA was eluted in one minute and subsequent elution steps did not greatly increase the recovery of HSA. Using DMF-IP with mass spec, HSA signal intensity was improved by 2.4 times over no IP procedures. Pre-concentration of magnetic particles is demonstrated by capturing particles from 100 µL of serum into a 4 µL droplet. Pre-concentration resulted in an HSA signal increase of 11 times over non-pre-concentrated DMF-IP samples, allowing a far greater range of target protein concentrations to be analyzed.

Novel Aspect:

We perform immunoprecipitation using digital microfluidics, greatly reducing the manual effort and sample size necessary in performing IP workups.

Selevsek, Nathalie,
ETH Zurich

Contributing Authors:
Laura Kunz, Functional Genomics Center Zurich, Zurich , Switzerland; Olivia Clayton, Roche Innovation Center, Basel, Switzerland; Witold Wolski Functional Genomics Center Zurich, Zurich , Switzerland; Ralph Schlapbach, Functional Genomics Center Zurich, Zurich , Switzerland; Ramona Nudischer, Roche Innovation Center, Basel, Switzerland.

Title:
Comprehensive molecular profiling of 3D cardiac and liver cells in the context of toxicogenomics

Introduction:
Aiming at developing integrative in silico tools for predicting human liver and heart toxicity, computational chemistry and systems toxicology are combined to model toxic perturbations in liver and heart across multiple scales. Models are populated with data generated using advances Omics technologies applied to in vitro liver and heart spheroids that are challenged with prototypical toxicants.

Methods:
Cardiac and liver spheroids were incubated with multiple cardiotoxicants and harvested at different time points. After cell lysis, protein extracts were subjected to in solution digestion. Peptides were analyzed on an Orbitrap Fusion instrument in data dependent mode. MS/MS spectra were searched using Mascot and validated using Scaffold. Protein quantification was performed using Progenesis QI.

Results:
More than two hundreds cardiac samples were measured by mass spectrometry and were processed using bioinformatics pipeline. Using state-of-the-art mass spectrometric analysis, 3'000 proteins can be identified and quantified from as little as the protein equivalent of 1'000 cells per measurement. Combining the data from multiple treated spheroids allows to quantitatively map out more than thousands of proteins of a 3 D microtissue proteome covering the relevant molecular protein pathways. Together with data generated at the epigenome and the transcriptome level, the resulting proteomics data allows for a comprehensive mapping of molecular regulation and the precise description of cellular behaviour.

Novel Aspect:
Comprehensive analysis of drug mechanisms and toxicity is feasible at the molecular scale and can be applied to various types of 3D tissue models.

Seró Llor, Raquel,
University of Barcelona

Contributing Authors:
Amelia Camprubi, IRTA, Protecció Vegetal Sostenible, Crta. De Cabrils Km 2, E-08348 Cabrils, Barcelona; Josep Manuel Grasses, IRTA, Protecció Vegetal Sostenible, Crta. De Cabrils Km 2, E-08348 Cabrils, Barcelona; Cinta Calvet IRTA, Protecció Vegetal Sostenible, Crta. De Cabrils Km 2, E-08348 Cabrils, Barcelona; Javier Saurina, Department of Analytical Chemistry, Faculty of Chemistry, University of Barcelona, Martí I Franqués 1-11, 08028, Barcelona; Encarnación Moyano, Department of Analytical Chemistry, Faculty of Chemistry, University of Barcelona, Martí I Franqués 1-11, 08028, Barcelona; Oscar Núñez, Department of Analytical Chemistry, Faculty of Chemistry, University of Barcelona, Martí I Franqués 1-11, 08028, Barcelona; Raquel Seró, Department of Analytical Chemistry, Faculty of Chemistry, University of Barcelona, Martí I Franqués 1-11, 08028, Barcelona.

Title:
DESI-HRMS leaf analysis for classification and characterization of mycorrhizal rosemary plants

Introduction:
Rosemary is a shrub from Mediterranean area used as culinary spice and in phytotherapy for its antioxidant properties. Long-term experiments have shown that rosemary plants colonized with arbuscular mycorrhizal fungi (AMF) contribute to plant community development and soil stabilization and remediation. The effects of this symbiosis between plants and AMF on chemical composition are still unknown.

Methods:
DESI-HRMS metabolomics and polyphenolics were explored to assess characterization of rosemary AMF inoculated plants. DESI geometrical parameters and instrumental conditions were optimized for leaf analysis. Q-Orbitrap was employed in negative ion mode working in both HRMS and MS/HRMS modes. DESI-HRMS data was subjected to principal component analysis (PCA) to achieve sample classification.

Results:
DESI geometrical parameters showed to be critical for method performance. The use of an adequate internal standard (e.g. arbutin) was mandatory to guarantee robustness on DESI geometrical parameters and to improve DESI-HRMS data reproducibility. DESI-HRMS metabolomics and polyphenolics data were employed to discriminate between non-inoculated and AMF inoculated rosemary plants and to identify the most important chemical compositional changes in plant metabolome. Characterization and identification of discriminant compounds was achieved by DESI-MS/HRMS, UHPLC-HRMS and in-silico fragmentation software. Enhancement on several polyphenols content (e.g. rosmanol, hesperidin and ursolic acid) was observed on AMF inoculated rosemary plants, while other metabolites showed reduced levels. PCA using DESI-HRMS data provided acceptable sample classification according to plant metabolome changes.

Novel Aspect:
Novel DESI-HRMS leaf analysis for the characterization of non-inoculated and AMF inoculated rosemary plants.

Seró Llor, Raquel,
University of Barcelona

Contributing Authors:
Josep Manuel Grases, IRTA, Protecció Vegetal Sostenible, Crta. De Cabrils Km 2, E-08348 Cabrils, Barcelona; Maria Teresa Galceran, Department of Analytical Chemistry, Faculty of Chemistry, University of Barcelona, Martí I Franqués 1-11, 08028, Barcelon; Raquel Seró Llor Department of Analytical Chemistry, Faculty of Chemistry, University of Barcelona, Martí I Franqués 1-11, 08028, Barcelon; Encarnación Moyano, Department of Analytical Chemistry, Faculty of Chemistry, University of Barcelona, Martí I Franqués 1-11, 08028, Barcelon; Marta Vidal, IRTA, Protecció Vegetal Sostenible, Crta. De Cabrils Km 2, E-08348 Cabrils, Barcelona; Jaume Bosch, IRTA, Protecció Vegetal Sostenible, Crta. De Cabrils Km 2, E-08348 Cabrils, Barcelona; Pilar Rodríguez, IRTA, Protecció Vegetal Sostenible, Crta. De Cabrils Km 2, E-08348 Cabrils, Barcelona.

Title:
Identification of organotin compounds in a molluscicide product by DESI-HRMS

Introduction:
Triphenyltin compounds (TPhT), widely used in agriculture as biocides, are highly toxic due to their endocrine disrupting capabilities, resulting in adverse effects on the environment and on human health. Thus, the control of the presence of TPhT compounds in phytosanitary products is mandatory. In this context, TPhT maximum levels were established by EU under the Directive 2009/425/EC.

Methods:
The applicability of Desorption Electrospray Ionization for the analysis of TPhT compounds in a phytosanitary product is explored. A fast and easy sample manipulation consisting in a filter paper impregnated with the sample was used. DESI-HRMS (17.500-70.00 FWHM at m/z 400) data were acquired in a Q-Exactive MS in positive ion mode working in both full-scan and target MS2 acquisition modes.

Results:
The DESI-HRMS analysis of a filter paper impregnated with the molluscicide provided a full scan mass spectrum with isotope pattern distributions typical of organotin compounds with several Sn atoms. The accurate mass measurements and isotope pattern fits combined with the structural information obtained by DESI-MS/HRMS allowed identifying the presence, above legislated levels, of bis(triphenyltin) oxide (BTPhT), a biocide frequently used in agricultural and responsible of widespread pollution. In order to obtain information about the isotope pattern distribution of product ions, the full isotopic cluster of precursor ion was isolated for MS/HRMS experiments. The DESI-HRMS and DESI-MS/HRMS analysis of the BTPhT standard confirmed the identification of this compound in the sample and suggested that the presence of TPhT in mass spectrum might be due to in- source fragmentation of BTPhT.

Novel Aspect:
To best of our knowledge, this is the first time organoestannic compounds have been analysed by ambient mass spectrometry techniques.

Seulen, Sarah,
Christian-Albrechts-Universität zu Kiel

Contributing Authors:
Sarah Seulen, Christian-Albrechts-Universität zu Kiel; Jürgen Grotemeyer, Christian-Albrechts-Universität zu Kiel.

Title:
Kinetic Energy Release and Fragmentation Pattern of Substituted Anilines after Collision Activation

Introduction:
Molecules with diethylamino sidegroups show unusual fragmentation patterns. The undergoing mechanisms are not completely understood. Distinctions between consecutive or concerted machanisms are not easily feasible. A model system based on different diethylaniline derivatives is characterized. Small changes in the molecular structure lead to a significant change of the observed fragments.

Methods:
All spectra are recorded with a Micromass ZAB-2F from Vacuum Generators, a double-focussing sector field instrument. The samples including isotopic labelling are ionized via electron ionization. Mass-analysed-Ion-Kinetic-Energy-(MIKE)-spectrometry provides values for the kinetic energy release (KER). Collisional activation in the second field free region is performed using argon as collision gas.

Results:
All diethylaniline derivatives show the sequential loss of a methyl radical and ethylene. The substitution of methyl groups on the aromatic system lead to a second fragment in the MIKE-spectra of the [M-15]+, the loss of a methyl radical, the intensity depending on the number of methyl substituents. For both losses small KER values are determined, indicating simple bond cleavages. With collision activation in the second field free region additional losses of methane and ethane from the [M-15]+ occur, the intensity depending on the number of methyl substituents on the aromatic system. Increased KER values are found for both fragments indicating rearrangement reactions. The intensity of the fragments from the [M-15]+ is highly dependent on the number of collisions and the flight time of the ions. This behavior indicates different fragmentation reactions leading to the proposed mechanisms.

Novel Aspect:
KER values for alkyl fragments for fragmentation reactions of diethylaniline derivates are determined and mechanisms proposed.

Sharon, Michal,
Weizmann Institute of Science

Contributing Authors:
Maria G. Fuzesi-Levi, Weizmann Institute of Science; Alexander A. Makarov, Thermo Fisher Scientific, 28199 Bremen, Germany; Michal Sharon Weizmann Institute of Science; Mikhail E. Belov, Thermo Fisher Scientific, 28199 Bremen, Germany; Gili Ben-Nissan, Weizmann Institute of Science.

Title:
Uncovering the complexity of protein complexes: A modified Q Exactive Plus Orbitrap instrument for multi-level analysis of protein complexes

Introduction:
Today it is being realized that protein complexes are not single units but rather variable assemblies. This increased complexity is enabled through multiple cellular diversification mechanisms. Yet, despite the importance of this layer of complexity, little is known about the compositional plasticity of protein complexes, mostly due to technical barriers of studying such closely related species

Methods:
To analyze all levels of protein organization from the intact assembly, through its constituent subunits, to the primary sequence of each protein, a standard Thermo ScientificTM Q ExactiveTM Plus Orbitrap MS has been modified to allow extended mass range of analysis. Specifically, a dual ion-funnel and a unique inject flatapole device were introduced, enabling MS3 analysis of protein complexes.

Results:
To explore the applicability of the system for unraveling the compositional heterogeneity of endogenous protein complexes isolated directly from cells, we have analyzed two heterogeneous protein complexes, the yeast 20S proteasome (720 kDa) and the human COP9 signalosome (CSN, 320 kDa). The highly resolved MS1 spectra of the protein assemblies demonstrate the presence of several populations. MS2 data, acquired after trapping and activating the ions in the inject flatapole, enabled the dissociation of individual subunits. MS3 data enabled sequential analysis and post-translational modification mapping of individual selected subunits. By repeating the analysis on complexes isolated from cells exposed to various stimuli we aim to map their cross-regulation. Overall, our data is very promising as it reflects the feasibility of this approach to obtain multiple levels of structural information

Novel Aspect:
This experimental advancement exposes the multilevel diversity of cellular-protein assemblies, an aspect that has up-to-now been barely explored.

shen, xuejing,
NCS Testing technology CO., LTD

Contributing Authors:

Zhigang Yang, NCS Testing Technology Co.,LTD; Shaocheng Hu, NCS Testing Technology Co.,LTD; Peng Wang NCS Testing Technology Co.,LTD; Wei Ga, NCS Testing Technology Co.,LTD; Qianqian Yang, NCS Testing Technology Co.,LTD; Hongquan Ma, NCS Testing Technology Co.,LTD; Haizhou Wang, NCS Testing Technology Co.,LTD; Xuejing Shen, NCS Testing Technology Co.,LTD.

Title:

The method for correcting mass spectral interference of CO2+ from N+ signal and its application in simultaneous analysis of O and N in metals

Introduction:

It is not easy for a normal inorganic mass spectrometer to directly analyze gaseous carbon monoxide (CO) and nitrogen (N2) simultaneously for the overlapping between CO+ and N2+ (m/z 28). Fragment peaks, C+ (m/z12) and N+ (m/z14) are usually selected for analysis of CO and N2, respectively. However the interference of the double-charged ion of CO, (CO2+(m/z 14)) on N+ (m/z14) couldn't be ignored.

Methods:

Firstly, the linear equation between the ion intensity of CO2+ and C+ was established by measuring a series of known content standard gas of CO in Helium. For an unknown content gas sample,with the same experimental condition, the contribution of CO2+ could be calculated by the above linear equation and the ion intensity of C+ acquired on the analyzer.

Results:

On a self-patented pulse-heating time of flight mass spectrometric gas elements analyzer, the method has been applied for simultaneous and precise analysis of oxygen(O), nitrogen(N), hydrogen(H) and Argon(Ar) in metal samples. The result has shown that the correction is significant and effective, especially when the ratio of the content of O to N is higher than 2:1.

Novel Aspect:

It is the first time that the proposed method for correcting the interference of CO2+ from N+ is applied for analysis of oxygen and nitrogen in metals

Shevchenko, Ganna,
Uppsala University

Contributing Authors:
Sravani Musunuri, Uppsala University; Sami Abu Hamdeh, Uppsala University; Jia Mi Uppsala University; Niklas Marklund, Uppsala University; Jonas Bergquist, Uppsala University.

Title:
Differential proteomic analysis reveals characteristic protein expression changes in traumatic brain injury

Introduction:
Brain injury results in complex alterations of biological systems within the central nervous system. Proteomic profiling of CNS samples from brain-injured patients can provide a better understanding of the mechanisms and offer new insight into therapeutics. Mass spectrometry-based proteomics has become a powerful tool to analyze complex biological samples and detection of disease-specific proteins

Methods:
We used shotgun MS-based label free (LF) and stable isotope dimethyl labeling (DML) approaches on fresh cortical brain biopsies, surgically removed from the frontal lobe of patients with traumatic brain injury and controls at time for surgical implantation of a shunt or intracranial monitoring device. Samples were analyzed using a LTQ-FT MS. The proteins were quantified using MaxQuant software.

Results:
Fresh, surgically removed small biopsy brain sections require optimization of sample preparation which has been performed in this study. A total of 555 proteins were quantified, out of which 544 and 246 proteins were found in LF and DML approaches with 235 (58%) proteins common for both approaches. Using both quantitation approaches, a total of 58 proteins (49 were increased and 9 decreased) displayed significant quantitative changes ($p<0.05$) between TBI and normal pressure hydrocephalus in more than 80% patient samples. These proteins represent a wide variety of pathways suggested to be involved in the TCA cycle, calcium signaling, oxidative phosphorylation, apoptosis of neurons, etc. Apart from leading to new insights into the mechanisms that TBI triggers Alzheimer's disease, the findings also provide us with possible novel candidates for future therapy target for TBI complications.

Novel Aspect:
This is the first proteomic study to use seemingly uninjured brain tissues from living traumatic brain injury and normal pressure hydrocephalus patien

Shevchenko, Andrej,
MPI of Molecular Cell Biology and Genetics

Title:
Healthy Plasma Lipidome

Introduction:
Lipidomics of human blood plasma is an emerging biomarker discovery approach that compares lipid profiles under pathological and physiologically normal conditions. However the gender impact on healthy lipidome composition is poorly understood and, in contrast to common clinical indices, no reference values for levels of individual lipids were established.

Methods:
We assembled a cohort of 71 Caucasians under the age of 33 years. Each individual had clean medical record with no factors comorbid to common metabolic disorders and 35 clinical blood test and anthropometric indices were within the medical norm. Molar concentrations of 281 molecular species from 27 major lipid classes were quantified by shotgun profiling and LC-MS/MS.

Results:
Gender is a major and previously underestimated lipidomic factor that affects 112 out of the total of 281 plasma lipids and the total abundance of 21 out of 27 lipid classes in an ethnically homogeneous healthy young cohort. Mean values of anthropometric (BMI, WHR) and clinical (blood pressure, LDL, HDL) indices are not covariate to gender discriminative lipids. Gender impact is strongly enhanced by hormonal contraceptives, mostly affecting the concentrations of GPL, Gl and cholesterol and was mediated by sex hormone-binding globulin. Within gender-restricted cohorts lipidomics revealed a compositional signature corroborating clinical indices trends towards developing a metabolic syndrome. We propose that healthy plasma lipidome is a valuable resource for biomarker discovery that helps to recognize early and relatively minor signs of emerging metabolic abnormalities.

Novel Aspect:
Gender, hormonal medication and individual metabolic predisposition are major lipidomic factors in healthy individuals

21ST INTERNATIONAL MASS SPECTROMETRY CONFERENCE
Toronto, Ontario, Canada
August 20 – 26, 2016

Shiea, Jentaie,
National Sun Yat-Sen University , Taiwan

Title:
Rapid Determination of Pharmacokinetic Profiles of Caffeine and Its Metabolites Collected on Skin by Ambient Mass Spectrometry

Introduction:
Thermal desorption electrospray ionization/mass spectrometry (TD-ESI/MS), a type of ambient mass spectrometry (AMS), is capable of performing high throughput analysis of samples. In this study, caffeine and the metabolites released on the skin were characterized with TD-ESI/MS. Pharmacokinetic profiles of caffeine and one of its metabolites were rapidly obtained.

Methods:
The TD-ESI/MS device consists of a heated oven and an ESI source. A stainless steel probe was used to scrape skin surfaces. After sampling, the probe was inserted into the heated oven to thermally desorb analytes and reacted with charged species to form analyte ions. A mass analyzer was used to detect the ions. After analysis, the probe was cleaned by burning it with a flame.

Results:
The samples collected on the probe were analyzed with TD-ESI/MS without pretreatment. The signals of caffeine and one of its metabolites (paraxanthine) were first detected at 90 min, reached their highest intensities at 3 hours, and gradually decayed with time. Pharmacokinetic profiles of caffeine were obtained and matched well with those obtained by conventional HPLC/MS/MS. In addition to the forehead, samples collected from other skin areas such as the arms, legs, hands, fingers, chest, back, and nose were also examined. Forehead and nose skin displayed higher caffeine ion signals than those of the hands, arms and legs. The advantages of using the analytical strategy developed in this study for drug analysis is that pharmacokinetic profiles of drugs and their metabolites were rapidly determined without blood withdrawal or urine collection.

Novel Aspect:
Ambient mass spectrometry can rapidly determine the pharmacokinetic profiles of drugs and metabolites from surface skin samples.

Shields, Sam,
Carleton University

Contributing Authors:

Jeffrey C. Smith, Department of Chemistry, Carleton University, Ottawa, Ontario, Canada; Jeffrey M. Manthorpe, Department of Chemistry, Carleton University, Ottawa, Ontario, Canada; Stewart T. Bigelow Department of Chemistry, Carleton University, Ottawa, Ontario, Canada.

Title:

Solution phase charge inversion of methylated phosphatidylcholine for the identification of fatty acyl components via collision induced dissociation

Introduction:

Fatty acid identification of phosphatidylcholine (PC) is commonly performed via CID of PC-acetate (OAc) ion pairs; however, suffers from low sensitivity and incompatibility with TrEnDi-modified lipids. Recently, McLuckey et al. have reported gas phase charge inversion via the addition of a dicarboxylic acid (DCA). Here we optimize charge inversion of PC and TrEnDi-PC in solution with various DCAs.

Methods:

Numerous DCAs were investigated. Thiodipropionic acid (TDPA), suberic acid (SA), and tricarballylic acid (TCBA) were determined to have optimal compatibility. 100 uM PC was mixed with 10 mM DCA and 1 uM of sodium tetrafluoroborate in a solution of 49.5% MeCN, 49.5% MeOH and 1% NH4OH. The mixture analyzed in negative ion mode via ESI using a hybrid quadrupole TOF mass spectrometer.

Results:

Many of the DCA acids were not soluble in the solvent mixture used (MeCN:MeOH:NH4OH), and thus limited the formation of PC ion pair during ESI-MS. However, TDPA was not only soluble in the solvent mixture, but formed ion pairs with methylated PC as well as unmodified PC. Upon CID of methylated and unmodified PC-TDPA ion pair myristate and oleate fatty acyl components could be observed. To compare traditional [PC+OAc] induced charge inversion against [PC+TDPA] and [PC+Me+TDPA], sodium tetrafluoroborate was used as an external standard. The sensitivity of both [PC+TDPA] and [PC+Me+TDPA] was observed to be dramatically greater than the traditional acetate ion pair. In addition, fatty acyl fragments produced from CID experiments were qualitatively observed to increase in sensitivity. Current experiments are being explored to apply this methodology to biological samples.

Novel Aspect:

This is the first study to systematically explore charge inversion agents for gas phase fatty acid identification of PC in the solution phase.

Shimma, Shuichi,
Osaka University

Contributing Authors:
Murata Atsuki, Graduate school of engineering, Osaka University.

Title:
Quantitative imaging mass spectrometry (q-IMS) of tyrosine kinase inhibitor in mouse tissues

Introduction:
Quantitative imaging mass spectrometry (q-IMS) for drug analysis is one of the forefront subjects. Our approach is combination of quantitative information obtained from liquid chromatography-tandem mass spectrometry (LC-MS/MS) and distribution information obtained using IMS. Based on this idea, we compared measured drug amount and estimated drug amount in IMS.

Methods:
Tissue sections of erlotinib administrated mouse liver, kidney and brain were prepared. Serial sections were prepared. q-IMS of erlotinib was performed using iMScope. Using residual tissue sections, LC-MS/MS quantification was performed. After obtaining both data, we combined those data under the hypothesis of $Itotal \propto Wtotal$. We compared expected and experimental drug amount.

Results:
We performed imaging MS and LC-MS/MS quantification of erlotinib inside the mouse brain, liver and kidney. We compared quantification results of estimation values and experimental values. According to our animal model study results, the estimated values agreed in experimental values within 15% (CV). Since the combination of conventional LC-MS/MS methodology and highly reproducible imaging MS methodology, we considered our method was rigid q-IMS method.

Novel Aspect:
We can provide reproducible and rigid q-IMS method using LC-MS/MS and IMS combination.

Shion, Henry,
Waters Corperation

Contributing Authors:
Ying-Qing Yu, Waters Corporation, Milford, MA, USA; Kai Gao, 2National Institutes for Food and Drug Control, Beijing, China; Chunfei Yu 2National Institutes for Food and Drug Control, Beijing, China; Min Du, Waters Corporation, Milford, MA, USA; Lan Wang, 2National Institutes for Food and Drug Control, Beijing, China; Weibin Chen, Waters Corporation, Milford, MA, USA.

Title:
Monitoring the Critical Quality Attributes of Antibody Drug conjugates (ADCs) as Part of Biosimilar Development: Case Studies of T-DM1

Introduction:
Because of the level of ADC sample complexity, it is very challenge yet crucial to implement efficient, routine and robust analytical capabilities to characterize the heterogeneity of conjugation and other PTM modification sites as well as to demonstrate lot-to-lot consistency or product stability. We present here a LC/MS accurate mass screening method for monitoring ADC CQAs at the peptide level.

Methods:
Five batches of T-DM1 (reference and biosimilar candidate) were trypitcally digested (LeuEnk added as IS) and then analyzed by ESI UPLC–QTof MS. The peptide mapping data were acquired in data independent acquisition (MSE) mode under a linear 60-min gradient (at 0.2 mL/min) and processed in UNIFI with targeted component quantified across samples for key ADC attributes assessment.

Results:
A LC/MSE method was used to monitor the relative abundance of conjugated and PTM peptides across all ADC samples with improved sensitivity and selectivity comparing to LC/UV method. In this approach, the reference ADC sample was first assessed using precursor and MSMS fragment ions to identify drug conjugated, deamidation, oxidation and glycosylation peptides. The identified peptide list was then used toquantify all samples, in which the percent of modifications were automatically calculated and reported. Different levels of conjugation, deamindation and oxidation among reference and biosimilar candidate batches were found. A detailed comparison of the conjugation, deamidation and oxidation levels across samples will be presented. Collectively, we will demonstrate the utility of the accurate mass screening workflow in peptide multiple attributes monitoring and quantification.

Novel Aspect:
Data independent acquisition and unique data processing approach for T-DM1 ADCs peptides multiple attributes monitoring and quantification.

Shion, Henry,
Waters Corporation

Contributing Authors:
Catalin Doneanu, Waters Corporation, Milford, MA, USA; Asish Chakraborty, Waters Corporation, Milford, MA, USA; Alain Beck Centre d'Immunologie Pierre Fabre, France; Weibin Chen, Waters Corporation, Milford, MA, USA; William Alley, Waters Corporation, Milford, MA, USA; Ying Qing Yu, Waters Corporation, Milford, MA, USA; Jing Fang, Waters Corporation, Milford, MA, USA.

Title:
Innovator and Biosimilar Infliximab: Comparability Assessment of the Host Cell Proteins and Protein Higher Order Structure

Introduction:
Protein drugs are inherently more variable than small molecules. Small structural changes can have significant functional consequences and any impurities can affect the final product as well. In this study, we report a platform for a comparison of Remicade® with its biosimilar product "Inflectra®" in terms of higher-order-structure (HOS) and host cell protein investigations (HCP).

Methods:
HDX: samples were prepared by dilution of protein 15-fold with a labeling-buffer for certain time scales and quenched. Digestion was performed by online pepsin column. HCP: samples were denatured with RapiGest, and digested with a mixture of LysC/Trypsin. A 2DLC-RP/RP method with a pH-based separation was used. Data independent were performed on an ion-mobility enabled TOF-MS for both components.

Results:
In HDX MS, the deuterium incorporation as a function of time for each peptic peptide was determined for all samples. A total of 191 heavy-chain and 89 light-chain peptides were identified and measured, which covers ~100% of the sequence. Overall, the structures of the innovator and biosimilar products are comparable. Only minute differences were observed in Fc-CH2 domain. A HCP assay was applied to the same samples to identify and quantify the HCPs. The HCP assay relies on: proteolytic digestion; 2D chromatographic separation; high-resolution MS detection coupled with ion-mobility separation; drift-time specific fragmentation of peptide precursors; and database searching for HCP identification and quantification. Two HCPs (epidermal growth-factor-like-protein 8 and WD-repeat-containing protein 37) were identified in both samples with concentrations ranging from 10 to 100 ppm.

Novel Aspect:
Comparison of innovator vs biosimilar mAb products using HOS and HCP analysis reveals minute differences.

Shrestha, Bindesh,
Waters Corp.

Contributing Authors:
Miguel Sena-Esteves, University of Massachusetts Medical School; Khaja Muneeruddin, University of Massachusetts Medical School; Bindesh Shrestha Waters Corporation; Hernando Olivos, Waters Corporation; Scott A. Shaffer, University of Massachusetts Medical School.

Title:
Molecular Imaging of Gangliosides to Investigate Lysosomal Storage Diseases using Mass Spectrometry Imaging with Ion Mobility Separation

Introduction:
Gangliosides (e.g., GM1 and GM2) are glycosphingolipids with a ceramide base and sialated carbohydrate chain. GM1 and GM2 gangliosidosis are autosomal recessive disorders characterized by the accumulation of gangliosides in the central nervous system. Here, we use MALDI-MS with ion mobility separation for spatial distributions of gangliosides in murine brain model of lysosomal storage disease.

Methods:
Coronal murine brain sections were thaw-mounted onto a regular glass slide and coated with 5-chloro-2-mercaptobenzothiazole matrix. MALDI imaging with ion mobility separation was performed with a Q-TOF mass spectrometer with traveling wave ion mobility separation (SYNAPT G2Si MALDI, Waters Corp.). Molecular images were rendered in High Definition Imaging (HDI) software (Waters Corp.).

Results:
Spatial distribution of gangliosides in murine brain were mapped using MALDI-MS with ion mobility separation. The orthogonal gas phase separation by ion mobility helped to build tissue maps with better molecular specificity by filtering out matrix and chemical interferences. The highly reproducible drift time can be converted to collisional cross section to increase confidence of molecular assignment. Molecular images of wild-type, enzyme deficient knock out (KO) models and adeno-associated virus type 6 treated brains showed well resolved differential ganglioside expression. In β-galactosidase KO mice, an 8 fold increase of GM1 (measured by normalizing with average intensity of matrix peaks) was observed relative to wild-type. This correlates well with LC-MS/MS of tissue homogenates (12 fold increase). We anticipate using MS imaging for preclincal study on effectiveness of gene therapy.

Novel Aspect:
Investigation of spatial distribution and fold-changes in gangliosides in diseased and treated murine brains using MALDI with ion mobility separation.

Shrestha, Bindesh,
Waters Corp.

Contributing Authors:

Nicholas Morris, Protea Biosciences; Sylwia Stopka, George Washington University; Trust Razunguzwa Protea Biosciences; Bindesh Shrestha, Waters Corp.; Akos Vertes, George Washington University.

Title:

Molecular Imaging of Tissue Sections by Matrix-free Nanophotonic Laser Desorption Ionization Mass Spectrometry with Ion Mobility Separation

Introduction:

Mass spectrometry imaging (MSI) is dominated by MALDI that requires an organic matrix. Application of uniform and precise amount of matrix continues to be challenging. Nanophotonic structures, e.g., silicon nanopost arrays (NAPA), can produce ions with low internal energy without the addition of external molecules. Here, we present NAPA-MSI of tissues integrated with ion mobility separation.

Methods:

Silicon NAPAs were prepared by projection lithography followed by reactive ion etching. The MSI data were collected from the NAPA platform by a Q-TOF mass spectrometer with ion mobility separation (SYNAPT G2Si MALDI, Waters Corp.). Ion mobility separation was carried out by a traveling wave cell. MSI experiments were analyzed using High Definition Imaging (HDI 1.4) and MassLynx software.

Results:

Untargeted spatial mapping of multiple biomolecules can be done by many modalities of MSI, e.g., MALDI, DESI. Due to its ability to directly detect metabolites and lipids in tissues, NAPA offers complementary capabilities to MALDI. To accommodate tissue, larger surface NAPA platforms were nanofabricated with up to 100 mm diameter. Kidneys from six-month-old nude mice were sectioned to 10 µm thickness using a cryostat microtome. The sections were thaw-mounted directly onto the NAPA chips. In the lower m/z range where metabolites are prevalent, spectral interferences were reduced due to the absence of matrix and the incorporation of IMS. The drift-times of ions during IMS, converted to collision cross sections (CCS), also aided in their identification. The novel matrix-free NAPA-based MSI has shown promise of high-spatial resolution (< 20 µm) imaging without elaborate preparations.

Novel Aspect:

Mass spectrometry imaging of tissue was demonstrated using matrix-free nanophotonic laser desorption ionization with ion mobility separation.

Shvartsburg, Alexandre,
Dept. of Chemistry, Wichita State University

Contributing Authors:
Andrew Entwistle, Shimadzu Research Laboratory Europe; Julia Kaszycki, Wichita State University; Pavel V. Shliaha University of Southern Denmark; Roger Giles, Shimadzu Research Laboratory Europe; Ole N. Jensen, University of Southern Denmark; Matthew Baird, Wichita State University.

Title:
Differential Ion Mobility Separations at the Extremes of Resolution and Field Strength

Introduction:
FAIMS has grown from a niche security screening technique into a broad analytical tool in conjunction with mass spectrometry. High orthogonality to the m/z dimension has made FAIMS an attractive approach to separate biomolecular isomers in proteomics and metabolomics. A virtually unlimited complexity of biological samples is now driving the efforts to raise the FAIMS resolving power further.

Methods:
For maximum resolution, a planar-gap unit at ambient gas pressure was coupled to an LTQ XL ion trap with electron transfer dissociation (ETD) stage via an ion funnel. Separations at extreme fields were explored at low pressure in a multipole (within quadrupole and time-of-flight mass spectrometers) that switches between a focusing field for ion transmission and dipolar field for FAIMS analyses.

Results:
Operating near the electrical breakdown threshold with buffers rich in He or H_2 enabled FAIMS resolving power up to ~500 for multiply-charged peptides, permitting broad separation of PTM localization isomers. A downstream ETD stage now allows identifying them without standards, using a new procedure that considers both present and missing fragments for each FAIMS feature. With this capability, we have completely resolved the full acetylated and phosphorylated histone tails (5 - 6 kDa) and peptide isotopomers labeled on different residues. Expanding FAIMS to effective fields over 300 Td (reached at lower pressure) is of major fundamental and technological interest. Although reducing the pressure decreases resolution, multiple localization variants were distinguished. The lack of "abnormal" bands for large proteins found at atmospheric pressure confirms the dipole alignment hypothesis.

Novel Aspect:
Frontiers of FAIMS advanced with separations of full histone tail isomers, peptide isotopomers, and analyses using extreme fields at low pressure.

SIBILLE, Estelle,
ChemoSens - INRA (FRANCE)

Contributing Authors:
Elodie A.Y. MASSON, UMR CSGA - 1324 INRA - 6265 CNRS – University of Burgundy Franche-Comté, Dijon, France; Estelle SIBILLE, UMR CSGA - 1324 INRA - 6265 CNRS – University of Burgundy Franche-Comté, Dijon, France; Olivier BERDEAUX UMR CSGA - 1324 INRA - 6265 CNRS – University of Burgundy Franche-Comté, Dijon, France.

Title:
Sample preparation and analysis of gangliosides and phospholipids from human plasma by liquid chromatography coupled to mass spectrometry

Introduction:
Gangliosides (GG) and phospholipids (PL) are two classes of complex and polar lipids exhibiting a wide structural heterogeneity. LC/MS offers an access to this huge diversity but also faces some challenges. Efficient extraction and purification sample methods are thus required to perform accurate and reliable analyses.

Methods:
To this end, GG and PL were extracted from human plasma with three solvent mixtures ($CHCl_3$/MeOH/H_2O) of increasing polarity, in comparison with the reference Folch method. GG were further purified with a Phree Phospholipid Removal column, whose sorbent is designed to selectively remove PL. GG and PL classes were then separated under HILIC conditions before analysis using a QqQ mass spectrometer.

Results:
Our method enabled a fast, straightforward and efficient extraction of both complex lipids, PL and GG. The extraction efficiency was much higher than the Folch method for the especially polar lipids that are GG and Lyso-PC, and similar for the various PL classes (PG, PI, PE, PS, PC, Sphingomyelin). The relative quantitative analysis of GG, performed by single reaction monitoring thanks to their characteristic sialic acid fragment (m/z 290), revealed a massive ion-suppression effect (up to 90%) due to co-elution of PC with the less polar GG. Removal of PL is thus essential for an accurate GG analysis by LC/MS and the Phree Phospholipid Removal column appeared as a powerful tool to achieve this purification step.

Novel Aspect:
Our sample preparation method allows the accurate analysis of both GG and PL classes from a single biological extract.

Sickmann, Albert,
ISAS

Contributing Authors:
R. P. Zahedi, Leibniz-Institut für Analytische Wissenschaften – ISAS – e.V.; J.M. Burkhart, Leibniz-Institut für Analytische Wissenschaften – ISAS – e.V.; S. Malchow Leibniz-Institut für Analytische Wissenschaften – ISAS – e.V.; F. Beck, Leibniz-Institut für Analytische Wissenschaften – ISAS – e.V.

Title:
From systems biology towards a diagnostic tool for platelet function

Introduction:
Anti-platelet treatment is of fundamental importance in combatting functions/dysfunction of platelets in the pathogenesis of cardiovascular and inflammatory diseases. Owing to their anucleate nature platelets have only limited protein synthesis and dysfunction of platelets is likely to be completely attributable to alterations in protein expression patterns and post-translational modifications.

Methods:
Quantitative proteomics was used to study dynamics of different signaling pathways during activation and inhibition of platelets. Key proteins were selected for the design of stable isotope labeled reference peptides and targeted in parallel reaction monitoring assays (PRM) with the final goal to validate candidate proteins in > 600 clinical samples.

Results:
Combining elaborate protocols for platelet isolation from fresh blood donations in conjunction with quantitative mass spectrometry, our recent data indicate that the platelet proteome, representing the complete set of expressed proteins, comprises approximately 5,000 proteins and is remarkably stable between different healthy individuals. The protein composition and the phosphorylation patterns of platelets will be useful to understand certain activation/inhibition states and therapeutic interventions. In particular, quantitative phosphoproteomic studies will pave the way for a refined understanding of platelet properties. The detection of candidate proteins from different pathways in clinical samples as biomarker for platelet activity is further refined with the objective to achieve DIN ISO 15189 accreditation.

Novel Aspect:
Quantification of > 100 molecules involved in signaling pathways for the prediction of platelet activity in individual blood samples.

Silva, Manori,
Centers for Disease Control and Prevention

Contributing Authors:
Xiaoyun Ye, Centers for Disease Control and Prevention; Trevor W. Bontke, Centers for Disease Control and Prevention; Antonia M. Calafat Centers for Disease Control and Prevention.

Title:
Identification of potential biomarkers of exposure to diundecyl phthalate

Introduction:
Diundecyl phthalate (DUP) is a high production volume (HPV) chemical used as a plasticizer in polyvinyl chloride and other plastics. Specific biomarkers of DUP would be useful for human exposure assessment. To identify such biomarkers, we investigated the in vitro metabolism of DUP with human liver microsomes.

Methods:
Invitro metabolism of DUP was performed using human liver microsomes to generate DUP specific metabolites. We used online solid phase extraction coupled to HPLC-high resolution mass spectrometry to conclusively identify the metabolites.To evaluate the potential usefulness of the newly identified DUP metabolites as exposure biomarkers for DUP, we analyzed 36 human urine samples.

Results:
Using high resolution mass spectrometry (HRMS), we identified four DUP specific metabolites: monoundecyl phthalate (MUP), mono-hydroxyundecyl phthalate (MHUP), mono-oxoundecyl phthalate (MOUP), and mono-carboxydecyl phthalate (MCDP). HRMS also isolated MCDP and MHUP from co-eluting isobaric metabolites of diisodecyl phthalate and diisododecyl phthalate, that could not be separated with low resolution tandem mass spectrometry. In human urine we detected MHUP and MCDP in >83% of the samples tested; median concentrations were 0.21 ng/mL and 0.36 ng/mL, respectively. MOUP was detected in 14% of the samples, and MUP was not detected. All three metabolites eluted as peak clusters likely because of the presence of multiple oxidation sites and multiple isomers in DUP technical mixtures.

Novel Aspect:
Metabolites of DUP, an HPV chemical, were identified for the first time to be used as potential biomarkers of exposure to DUP.

Simmons, Denina,
University of Waterloo

Contributing Authors:
John R. Cosgrove, AXYS Analytical Services Ltd., 2045 Mills Road West, Sidney, British Columbia; Sigal Balshine, Department of Psychology, Neuroscience & Behaviour, McMaster University, 1280 Main St. West, Hamilton, Ontario; Bharat Chandramouli AXYS Analytical Services Ltd., 2045 Mills Road West, Sidney, British Columbia; Derek Muir, Aquatic Contaminants Research Division, Water and Science Technology Directorate, Environment Canada, Burlington, Ontario; Erin McCallum, Department of Psychology, Neuroscience & Behaviour, McMaster University, 1280 Main St. West, Hamilton, Ontario; James P. Sherry, Aquatic Contaminants Research Division, Water and Science Technology Directorate, Environment Canada, Burlington, Ontario.

Title:
The use of plasma and integrated OMICs approaches for Environmental Toxicology

Introduction:
OMICs methods can be used to assess the health of animals, to determine protein biomarkers that are specific to environmental exposures, and also to characterize unique mechanisms of action by contaminants. Plasma offers many unique advantages: it provides a snapshot of whole organism health, sample preparation is easy, and there is potential for non-lethal sampling methods.

Methods:
A method for analyzing the plasma proteome using formic acid protein digestion and liquid chromatography mass spectrometry was developed. Non-targeted proteomic and targeted metabolomics analyses were performed on plasma from fish in 2 studies: 1. rainbow trout exposed to a flame retardant in the laboratory and 2. caged goldfish exposed to pharmaceutical drugs in field deployments.

Results:
Formic Acid digestion of plasma was comparable to trypsin digestion. The plasma proteome is more variable compared to other tissues. There was a time-dependent and dose-response relationship for proteins and metabolites in rainbow trout exposed to the flame retardant. The expression of several proteins and metabolites were significantly altered in the plasma of goldfish which corresponded to site-specific alterations in fish behavior and levels of plasma pharmaceutical drug contaminants. These cases are good examples of how integrated OMICs approaches provide powerful insight into the molecular mechanistic effects of environmental contaminant exposures.

Novel Aspect:
Integrated OMICs of non-model organisms which are linked to environmentally relevant exposures and higher level biological effects including behavior.

Simons, Brigitte,
SCIEX

Contributing Authors:
Paul Baker, SCIEX, Redwood City, CA, USA; Larry Campbell, SCIEX, Concord, ON, Canada; Leo Wang SCIEX, Redwood City, CA, USA

Title:
Reducing Isobaric Species in the Lipidomics Analysis of Gycerol Lipids

Introduction:
Direct-infusion lipidomics is a powerful mass spectrometry technique for broad-scale lipidomic profiling of biological samples. Speed and simplicity make this workflow very appealing to researchers, but isobaric overlap from multiple lipid species remain a major weakness. Here, we demonstrate the use of the SelexION differential mobility spectrometry (DMS) combined with direct-infusion lipidomics

Methods:
The DMS-MS method was first optimized with LipidoMix ion mobility kit from Avanti Polar Lipids, which contained a mixture of synthetic lipid standards from different lipid classes (glycerolipids, phospholipids, sphingolipids and cholesterol esters). Lipid extracts were prepared from human pooled plasma (HPP) using a modified Bligh-Dyer extraction to evaluate the DMS method.

Results:
The DMS device is located at the outside of the instrument, so ion selection by the DMS occurs before mass analysis, which eliminates isobaric interference among target molecules.Individually infused pure TAG 50:1 standard (TAG 16:0/16:0/18:1 and TAG 16:0/18:1/16:0) display only one peak in CoV ion chromatograms, while the mixed two TAG 50:1 isomers clearly separate into two distinct peaks with characteristic CoV value in CoV ionogram. Similar results were observed for TAG 52:2 (TAG 18:1/18:1/16:0 and TAG 18:1/16:0/18:1), showing the reproducible differentiation capability of DMS. Parameters affecting DMS resolution, such as separation voltage, chemical modifier type and flow rate, and resolution enhancement gas were explored, and their effects are presented.

Novel Aspect:
Reducing isobaric interferences in neutral lipid classes for better lipid species quantitation of regioisomers.

Simons, Brigitte,
SCIEX

Contributing Authors:
Paul Baker, SCIEX, Framingham, USA; Baljit Uhbi, SCIEX, Redwood Shores, USA; Steve Watkins Metabolon, NC, USA; Alex Connor, Metabolon, NC, USA.

Title:
Resolving the Complexity of the Lipidome Using Differential Ion Mobility Separations

Introduction:
The analysis of lipids by mass spectrometry can provide in-depth characterization for many forms of biological samples. However, such workflows can also be hampered by challenges like low chromatographic resolution for lipid separations and the convolution of mass spectra from isomeric and isobaric species. To address these issues, we describe the use of differential mobility spectrometry (DMS) as

Methods:
Applying the kit for simplified sample extraction and preparation, a serum matrix was used following the protocols provided. A triple quadrupole with DMS was used for targeted profiling of over a thousand lipid species from 10 different lipid classes allowing for comprehensive coverage. Two methods covered ten lipid classes; one injection with DMS on and then off.

Results:
This system allows for (1).Quantitative results for each lipid class as a sum of individual species; (2).Mole percent composition was obtained computationally from lipid molecular species data; (3).Accurate lipid species compositions. The data will be compared with historical data generated by alternative methods. A total of 1153 lipid species are measured and quantified using a specific sample correction strategy against one of 68 lipid class specific lipid standards that carry a D9-palmitate fatty acid group. The triglycerides data shows that 99.6% of total TAG class mass contributors have robust RSDs of 10% or less, with 88.3% of the individual TAG species that have a RSD of 20% or less. Twenty-five human serum samples with known total CE and CE fatty acid compositions were profiled using the this lipid platform and quantified total CE with less than 10% bias.

Novel Aspect:
Routine lipid quantitation with a kit is presented, covering 13 lipid classes and use of ion mobility for high throughout lipidomics.

Šimura, Jan,
Laboratory of Growth Regulators & Department of Chemical Biology and Genetics, Centre of the Region Haná for Biotechnological and Agricultural Research, Faculty of Science, Palacký University & Institute of Experimental Botany ASCR

Contributing Authors:
Jitka Široká, Laboratory of Growth Regulators & Department of Chemical Biology and Genetics, Centre of the Region Haná for Biotechnological and Agricultural Research, Faculty of Science, Palacký University & Institute of Experimental Botany ASCR, Šlechtitelů 27, CZ-783 71, Olomouc, Czech Republic; Danuše Tarkowská, Laboratory of Growth Regulators & Department of Chemical Biology and Genetics, Centre of the Region Haná for Biotechnological and Agricultural Research, Faculty of Science, Palacký University & Institute of Experimental Botany ASCR, Šlechtitelů 27, CZ-783; Ioanna Antoniadi Umeå Plant Science Centre, Department of Forest Genetics and Plant Physiology, Swedish University of Agricultural Sciences, SE-90183 Umeå, Sweden; Karin Ljung, Umeå Plant Science Centre, Department of Forest Genetics and Plant Physiology, Swedish University of Agricultural Sciences, SE-90183 Umeå, Sweden; Ondřej Novák, Laboratory of Growth Regulators & Department of Chemical Biology and Genetics, Centre of the Region Haná for Biotechnological and Agricultural Research, Faculty of Science, Palacký University & Institute of Experimental Botany ASCR, Šlechtitelů 27, CZ-783.

Title:
A novel targeted metabolomic approach in plant hormone analysis

Introduction:
Plant hormones are highly bioactive signalling molecules, often present in very low concentrations, which are acting as chemical messengers involved in physiological processes during plant growth and development as well as in plant stress responses. The occurrence and levels of these compounds strongly depend on plant organ, plant age, developmental stage and environmental conditions.

Methods:
Due to complex composition of the plant extract this analysis poses a difficult task. Isotopically labeled internal standards were added during extraction for precise quantification. SPE was used for purification of crude plant extract. The analyte analysis was based on selected MRM transition in specific retention time and was carried out using UHPLC system combined with triple quadrupole MS.

Results:
In this study, we present a new method based on ultra-high liquid chromatography-tandem mass spectrometry (UHPLC-ESI-MS/MS) for targeted profiling of more than 100 analytes, members of the main classes of plant hormones namely cytokinins, auxins, brassinosteroids, gibberellins, jasmonates and abscisic acid (active compounds, biosynthetic precursors and primary metabolites) in one analytical run utilizing ESI (+) and ESI (-) modes simultaneously or in separate runs with total time, for both approaches, up to 35 minutes per analysis.

Novel Aspect:
We bring a new generalized analytical screening method useful for phytohormonal studies, dealing with inter- and intra-cellular crosstalk in plants.

Sinha, Ankit,
Princess Margaret Cancer Centre

Contributing Authors:
Alexandr Ignatchenko, Princess Margaret Cancer Center, University Health Network, Toronto, Canada; Kwan Ho Tang, Princess Margaret Cancer Center, University Health Network, Toronto, Canada; Vladimir Ignatchenko Princess Margaret Cancer Center, University Health Network, Toronto, Canada; Ali Hussain, University of Toronto, Department of Medical Biophysics, Toronto, Canada; Ankit Sinha, University of Toronto, Department of Medical Biophysics, Toronto, Canada; Thomas Kislinger, Princess Margaret Cancer Center, University Health Network, Toronto, Canada; Ben Neel, Perlmutter Cancer Center, Langone Medical Center, New York City, USA; Laurie Ailles, Princess Margaret Cancer Center, University Health Network, Toronto, Canada.

Title:
Novel biomarkers for detecting recurrence of ovarian cancer by combining patient derived xenografts with N-glycosylation enrichment strategies.

Introduction:
The majority of high-grade serous ovarian cancer (HGSOC) patients (>70%) will experience recurrence within 5 years of diagnosis and eventually die of their disease. Poor disease management is related with late detection of HGSOC recurrence; we hypothesize that a combined strategy of patient derived xenograft and N-glycoproteomics will identify novel markers for early detection of HGSOC recurrence.

Methods:
We used patient derived xenografts (using primary HGSOC human tumours) as a source for tumour tissue and serum, which were then processed for N-glycosylated protein enrichment and shotgun proteomics. The comparison of sera and tumours identified secreted proteins. Subsequently, we developed targeted PRM assays for quantifying candidates in longitudinally collected serum of HGSOC patients.

Results:
Tumour tissue and serum were obtained from PDXs (n=12), and the serum of un-engrafted mice (n=12) served as negative control. Comparison of serum and tumour lead to identification of 351 human N-glycosylated proteins detected exclusively in the xenograft tumour and serum, including the current biomarker CA125. PRM assays were developed using 392 isotope labelled peptides targeting the 351 tumour secreted proteins. The assays were optimized for interference using negative controls (mouse serum, normal human serum) and positive controls (ascites fluid from HGSOC patients). We evaluated chromatography, retention times, and propensity of variable modifications to ensure maximum sensitivity. The presented work will show performance of our targeted assays and the candidate markers in longitudinally collected serum from HGSOC patients through various stages of disease progression.

Novel Aspect:
The project utilizes a novel strategy for combining N-glycoproteomics, patient derived xenografts and validation in longitudinally collected serum.

Siu, C K Andy,
City University of Hong Kong

Title:
Reaction dynamics of small sulfur-containing molecules with a hydrated electron in the gas phase

Introduction:
Ionizing radiation can generate hydrated electrons leading to biological radiation damage. Thiol groups (—SH) contribute to tertiary protein structures through the formation of disulfide linkages (S—S) in cystine residues. Improved understanding of the electron-transfer processes with small model systems provides fundamental insights into their reduction chemistry at the molecular level.

Methods:
Dissociative electron transfers of CH3SH and CH3SSCH3 with a hydrated electron are examined using DFT molecular dynamics methods performed with CP2K package. Gaussian-planewave basis set is employed for the electronic wavefunctions. Revised PBE functionals with self-interaction correction for the local spin density is used. Dispersion correction for non-covalent interaction is also included.

Results:
With hydrated electrons, two reactions with relevance for biological radiation damage were observed, including reduction of the disulfide bond of CH3SSCH3 and the thiol group of CH3SH followed by a hydrogen atom loss. Our theoretically predicted reaction energies are in good agreement with experimental values estimated from nanocalorimetry measured with a FTICR mass spectrometer. The present DFT-MD simulations revealed the dynamics of reductive electron transfer from a water cluster to the σ* orbital of the dissociating bonds, monitored by the evolution of the electron spin density. Such electron transfer happens upon a solvent reorganization with a small activation barrier. For the reaction of CH3SH, several short-life intermediates involving different hydrogen bonding interactions are observed which are characterized by their reaction energies, bond distances and power spectra.

Novel Aspect:
Dissociative electron transfer of some sulfur-containing functional groups in aqueous media is characterized with gas-phase models using DFT-MD method

Skeene, Kirsty,
University of York, UK

Contributing Authors:
Daniel Ungar, Department of Biology, University of York, UK; Ed Bergström, Centre of Excellence in Mass Spectrometry, University of York, UK; Jane Thomas-Oates Centre of Excellence in Mass Spectrometry, University of York, UK; Kirsty Skeene, Department of Chemistry, University of York, UK; Graham Clarke, Bristol-Myers Squibb, Moreton, UK.

Title:
One sample, one pot and the whole glycome: towards a system to study disorders of protein glycosylation

Introduction:
Glycans are important in cellular processes and disease. We aim to understand the dependence of the glycan biosynthetic code on the presence/location of biosynthetic enzymes. Alterations in these can manifest in subtle changes to glycan profiles. To determine which of these changes are statistically significant, analysis of multiple biological samples in parallel is needed.

Methods:
We present a membrane filter-based, one-sample-one-pot method, to release and isolate N- and O-glycans from glycoproteins solubilised from whole cell lysates, for MS profiling, with the ability to subsequently also analyse residual protein. We have applied the method to mesenchymal stromal cell (MSC) glycans on differentiation of MSCs into adipocytes, and in genetically manipulated cultured cells.

Results:
The one-pot method enables N- and O-glycomes to be compared across sample replicates, from small numbers of cultured cells; in depth glycan profiling is irrelevant to such comparisons – instead, the ability to assess the often subtle variation and perform statistical analysis is critical. We have developed our approach for this purpose and used it to analyse glycomes of MSCs before and after differentiation into adipocytes, demonstrating that changes in the N-glycan profile are much less marked than those of the O-glycans on this differentiation. In genetically manipulated cells, both glycan types show changes compared with the wild-type cells. Our approach offers insight into the functional role of glycans in cellular processes such as differentiation and disease. In addition, our approach also illustrates the importance of analysing glycan profiles of both N- and O-glycans.

Novel Aspect:
A one-pot-one-sample method for analysis of the whole glycome, with the use of sonication to reduce length of incubation for O-glycan release.

Smith, Sara,
MilliporeSigma

Contributing Authors:
Emily Barrey, MilliporeSigma; Craig Aurand, MilliporeSigma; Dave Bell MilliporeSigma; Candace Price, MilliporeSigma.

Title:
Rapid Determination of Protein Binding Affinity Using Solid Phase Micro Extraction (SPME)

Introduction:
Protein binding affinity is commonly determined using micro dialysis based methods. Although accurate, processing times are often greater than 6 hours. In this study, a novel biocompatible micro extraction device is evaluated as a rapid means of extracting free drug from plasma. Initial studies demonstrate that protein binding can be determined in less than 30 minutes.

Methods:
Protein binding affinities for propranolol, warfarin and diclofenac were determined by comparative micro extraction from plasma and phosphate buffered saline. Plasma and buffer samples were plated into 96-well plates and extracted by immersion of the micro extraction fibers. The fibers were then directly desorbed into solvent followed by direct LC/MS analysis and quantitation.

Results:
Protein binding affinities determined using the novel micro extraction technique were found to be consistent with reported literature values. Correlation between the micro extraction sampling technique and published micro dialysis data was demonstrated to be within 1%. Therefore, the developed micro extraction procedure is capable of accurately determining drug protein binding affinities within 30 minutes. This offers a more rapid approach than the traditional micro dialysis methods. Additional studies to evaluate drugs with a broader range of protein binding affinities are being conducted.

Novel Aspect:
This biocompatible micro extraction technique provides faster determination of protein binding affinities compared to previously established methods.

Söderström, Martin,
University of Helsinki

Contributing Authors:

Martin Söderström, University of Helsinki; Olli Kostiainen, University of Helsinki; Jacek Beldowski The Institute of Oceanology of the Polish Academy of Sciences; Ullastiina Hakala, University of Helsinki; Paula Vanninen, University of Helsinki.

Title:

Monitoring of Baltic Sea sediments for Sea-Dumped Chemical Warfare Agents using GC–MS/MS and LC–MS/MS

Introduction:

After WWII, 40,000 tons of chemical weapons (CW) were dumped in Baltic Sea. Chemicals include mustard gas and arsenicals. Over the last decade several multinational projects have been mapping contamination in the dumping areas. A three-year NATO funded project – "Towards the monitoring of dumped munitions threat" (MODUM) – aims at collecting results and establishment of monitoring network.

Methods:

Baltic Sea sediment samples were taken from research vessels using, e.g. box corers and ROVs. The main sample preparation steps were non-polar extraction for GC–MS/MS analysis and polar extract for LC–MS/MS analysis. The instrumental methods were based on SRM using triple quadrupole instrumentation. A hydrid orbitrap instrument was also applied for screening and characterization of unknowns.

Results:

Since 2006, 20+ analytes have analyzed in ca. 1000 sediment samples. There have been several aims for the projects: search for relevant degradation products for the dumped CW agents, development of methods, estimation of bioavailability of the chemicals, verification and mapping of known dump sites, search for suspected dump sites as well as environmental monitoring. Strict identification criteria have been applied to the MS analysis. Based on results, a map of contamination has been established. MODUM project aims at establishment of continuous monitoring network using both repeated and passive sampling. Lately, a high-resolution instrument has been applied to search for new degradation products and metabolites. At least two novel chemicals have been identified in dump site sediments. Main findings of MODUM and previous projects will be presented with established analytical methods.

Novel Aspect:

Survey of 20+ targets over large area was carried out with reliable MS methods. Using HRMS, novel chemicals were found at low concentration level.

Sokratous, Kleitos,
The Cyprus Institute of Neurology & Genetics, Nicosia, Cyprus

Contributing Authors:

Orthodoxia Nicolaou, Cyprus School of Molecular Medicine, Cyprus Institute of Neurology & Neurology; Andreas Hadjisavvas, Cyprus School of Molecular Medicine, Cyprus Institute of Neurology & Neurology; Bernard R. Lauwerys Rheumatology Department, Cliniques Universitaires Saint-Luc, Pôle de Pathologies Rhumatismales Inflammatoires et Systémiques, Université catholique de Louvain, Bruxelles, Belgium; Kyriacos Kyriacou, Cyprus School of Molecular Medicine, Cyprus Institute of Neurology & Neurology.

Title:

Proteomic analyses of mouse kidney with systemic lupus erythematosus using mass spectrometry and ion mobility drift-time dependent collision energies

Introduction:

Systemic lupus erythematosus (SLE) is a chronic autoimmune disease affecting different organs. Despite extensive research efforts, SLE still displays a complex pathogenesis and has an unpredictable course. Using mass spectrometry (MS)-based proteomics, we compared proteomic profiles in the lupus-prone B6.Sle123 mice of various ages in order to gain insights into the mechanisms underlying SLE.

Methods:

Mice used were of 12-, 20- and 36-weeks old, representing three different stages of SLE. Proteins were extracted from kidneys, purified, reduced, alkylated and digested by trypsin. Purified peptides were separated by nanoflow ultra-performance liquid chromatography (UPLC) and analysed on a Synapt G2-Si MS instrument using the UDMSE approach. Data were processed by the Progenesis QI software tool.

Results:

The proteomic profiles of kidney tissues from the lupus-prone B6.Sle123 triple congenic strain were characterized by label-free MS approach using the UDMSE strategy, a data-independent acquisition (DIA) MS method that uses ion mobility drift time-specific collision energy profiles.More than 2000 non-redundant proteins were identified in all samples. Comparing the proteomic profiles obtained at the three different stages of disease progression, several hundred proteins including immunoglobulins, histones and glycoproteins such as complement component C1q, were found to be altered over disease progression. Intriguingly, further pathway analyses showed that the identified dysregulated proteins were involved in multiple pathways including, mitochondrial electron transport chain, cell redox homeostasis and actin-binding, which are known to be associated with autoimmune diseases.

Novel Aspect:

LC/UDMSE –based proteomic analyses offers new insights into further investigating and better understanding the molecular mechanisms underlying SLE.

Solano, Eduardo,
University of Ottawa

Contributing Authors:
Paul M Mayer, University of Ottawa; Sabria Mohammed, University of Ottawa.

Title:
Scattering resonance fingerprints in CID mass spectra of PAHs?

Introduction:
The internal energies transferred when projectile molecular ions of naphthalene, anthracene and pyrene collide with argon gas atoms are extracted from the APCI-CID (atmospheric-pressure chemical ionization collision-induced dissociation) mass spectra of the corresponding compound, acquired at different collision energies, using basic equations from physical chemistry.

Methods:
Ion abundances are calculated by microcanonical integration of the differential rate equations for the parent ion decomposition using RRKM rate constants and thermal energy distributions p_M (E,T_char). The mean vibrational energy excess is characterized by the parameter T_char ('characteristic temperature'), determined by fitting the theoretical ion abundances to the experimental ones.

Results:
The APCI ion source produces species with an internal energy excess of 3.26 – 5.84 eV above the ground state of the corresponding ion. Subsequent collisions heat the ions up further, giving rise to a sigmoid curve of T_char as a function of E_com (center-of-mass-frame kinetic energy). As a consequence, the kinetic-to-vibrational energy conversion obeys to a well-shaped function that brings on two fundamental properties, namely: the existence of some limit for the energy stored by the ions and the preferential absorption at some specific kinetic energy Kres. This result suggests that collisional activation in PAHs might be interpreted as a scattering resonance problem, a phenomenon in which the absorption exhibits a pronounced increase around some specific value of energy.

Novel Aspect:
No collision model assumptions are required to calculate energy transfers in CID mass spectrometry.

Soltwisch, Jens,
Institut for Hygiene, University of Muenster

Contributing Authors:
Klaus Dreisewerd, Institute for Hygiene, University of Münster; Andreas Bauwens, Institute for Hygiene, University of Münster; Simeon Vens-Cappell Institute for Hygiene, University of Münster; Hans Kettling, Institute for Hygiene, University of Münster.

Title:
Cool-MALDI: Matrix-free mass spectrometric imaging of metabolites in snap-frozen tissue sections using IR-laser ablation and UV-laser postionization

Introduction:
UV-MALDI-MS Imaging (MSI) of small molecules is often hindered by matrix-derived background ions. Another problem is caused by the ongoing rapid metabolic turn-over in dissected tissue, preventing recording the "true" metabolic situation. Here we introduce a novel MALDI-MSI modality, termed cool-MALDI, that enables matrix-free MSI analysis of metabolites directly from snap-frozen tissue sections.

Methods:
A prototype of a new cooling sample stage for a Synapt G2-S mass spectrometer (Waters) was developed, enabling MS analysis at −150 °C. An IR-OPO-laser (GWU Lasertechnik), operated at 2.94 µm, served for matrix-free ablation directly from hydrated samples. A wavelength-tunable UV-OPO-laser (GWU) boosted metabolic profiles by postionization at ~1 mbar in a fine-vacuum MALDI-2 ion source.

Results:
Next to introducing the concept enabling MSI at −150 °C, we will present selected application examples and studies investigating the fundamentals of the involved processes. Application examples will include MALDI-MS imaging of metabolites (e.g., vitamins) from mouse tissue and visualization of the interaction of bacteria with antibiotics and competitors. To shed light on the involved ablation/ionization mechanisms, we investigated, e.g., the influence of the postionization laser wavelength, delay between ablation and postionization laser pulses, and buffer gas pressure on the generated ion signals. Comparison with the UV-absorption spectra of identified compounds allowed distinguishing between direct resonance-enhanced multi-photon-ionization (REMPI) and secondary charge transfer processes, similar to the recently introduced UV-MALDI-2 method (Soltwisch et al., Science 348 (2015) 211).

Novel Aspect:
New laser-MS-imaging method enabling unaltered metabolic profiling directly from snap-frozen tissue sections and bacterial cultures.

SONG, In-kang,
Ewha womans university

Contributing Authors:
JAE-JIN LEE, Ewha womans university; Jin-hwan CHO, Ewha womans university; Kong-joo LEE Ewha womans university.

Title:
Redox-Sensitive Proteins are Regulated by Oxidative Modification Induced Conformational Changes

Introduction:
ROS (Redox oxidative species) are not only consequence of respiration, but also key molecules of cell survival including signaling, proliferation, differentiation and apoptosis. However, cellular targets of ROS are not well understood.

Methods:
In this study, we identifed the ROS target proteins in signaling pathway and their regulation mechanism employing 2D-gel electrophoresis and nanoUPLC-ESI-q-TOF tandem MS. The molecular mechanism was investigated by confirming the ROS-induced structural changes employing hydrogen/deuterium exchange (HDX) mass spectrometry.

Results:
We identifed the ROS target proteins in signaling pathway and their regulation mechanism. We also found that redox sensitive proteins are readily oxidized to diverse modifications at reactive Cys residues, which modulate their stability. Oxidized redox sensitive proteins are readily ubiquitinated and then degraded via proteasome and autophagy. The molecular mechanism of protein degradation was investigated by confirming the ROS-induced structural changes.

Novel Aspect:
Redox sensitive proteins are regulated by oxidation induced ubiquitination and degradation as another means of quality control.

song, lijiang,
University of Warwick

Title:
High resolution mass spectrometry for genomics-driven microbial drug discovery

Introduction:
Wide spread antimicrobial resistance (AMR) has been recognised as one of the major threats to human health in the last decade. Meanwhile, traditional bioactivity-based screening methodologies have failed to deliver new antibiotics. A new approach for the discovery of antimicrobials, genome mining, has been developed in recent years, which heavily rely on modern mass spectrometry instrumentation.

Methods:
The genome of Burkholderia ambifaria and gladioli are sequenced using Pacific Biosciences SMRT. We can identify and analyse gene clusters encoding novel polyketide synthase and nonribosomal peptide synthetase biosynthetic assembly lines and predict partial or complete structure for the products, which are then identified through comparative metabolite profiling using high resolution LC-MS/MS.

Results:
We have discovered that enacyloxin IIa is the product of an unusual cryptic trans-acetyltransferases polyketide biosynthetic gene cluster, identified via transposon mutagenesis and correlation to the genome of Burkholderia ambifaria AMMD. We have also shown it is active on multidrug resistant Gram-negative bacteria such as B. dolosa, B. multivorans, B. anthina and B. pyrrocinia and Acinetobacteria baumannii. A second potent polyketide macrolide antibiotic gladiolin is discovered from Burkholderia gladioli. Gladiolin possess strong activities towards multi-drug resistant gram-positive pathogens, while showing very low cytotoxicity towards normal mammalian cells. High resolution high accuracy mass spectrometry has been the major tool in the discovery and the following biosynthetic investigation of these novel antibiotics for the analysis of both small organic molecules and proteins.

Novel Aspect:
Rapid de-replication with high resolution mass spectrometry is critical to the success of the genomics-driven microbial drug discovery

Song, Xiaomin,
Australian Proteome Analysis Facility

Contributing Authors:
Jemma Xiaobing Wu, Australian Proteome Analysis Facility; Dana Pascovici, Australian Proteome Analysis Facility; Natasha Care Australian Proteome Analysis Facility; Christoph Krisp, Australian Proteome Analysis Facility; Mark P Molloy, Australian Proteome Analysis Facility; Thiri Zaw, Australian Proteome Analysis Facility.

Title:
SWATH data analysis using local and extended libraries

Introduction:
To extract quantitative MRM-like peak areas from MS/MSAll data such as SWATH, peptide spectral library can be generated locally through data dependent acquisition or obtained from shared data repositories for commonly studied organisms. This study evaluated how locally-generated or repository-based spectral libraries affect SWATH performance for proteomics studies.

Methods:
We developed a software workflow, SwathXtend, which generates extended peptide spectral libraries using a local seed library and delivers statistical analysis of SWATH-based sample comparisons. SWATH-MS data with 2, 5 and 10% of yeast peptides spiked into the peptides of human K562 cell lysate were assessed using local and repository-based spectral libraries.

Results:
our study shows that by using high quality extended spectral ion libraries, the number of peptides and proteins extracted as well as the number of correctly identified differentially abundant proteins can be increased while minimising quantitative false discovery rate. The set of proteins detected as differentially expressed by using the extended spectral library and the local comprehensive library are mostly consistent, and the new set of differentially expressed proteins detected by the extended libraries typically show the correct expression pattern. The quantification accuracy of the SWATH data extracted using the extended libraries is similar to that obtained when using the local comprehensive library.

Novel Aspect:
SWATH data were evaluated systematically using local and extended libraries

Soyk, Matthew,
LECO Corporation

Contributing Authors:

David Borton, LECO Corporation; Viacheslav Artaev, LECO Corporation.

Title:

Sub-Picogram Detection and Identification of Analytes in Complex Matrices Using a Novel GC-TOFMS

Introduction:

Improved detection limits in GC-MS analyses are beneficial in a number of ways. Detection of lower concentration analytes that were previously undetectable, a reduction in the amount of sample to be prepared for analysis, or the ability to use split injections to reduce stress on the system due to reduced matrix loading. A novel benchtop GC-TOFMS with improved detection limit has been developed.

Methods:

Replicate injections of multi-analyte standards, with and without being spiked into complex matrices, were acquired on a novel, nominal resolution benchtop GC-TOFMS to show detection limit and dynamic range capabilities on a variety of analytes. Instrument detection limits are calculated by IDL = t99% x (RSD/100%) x amount on column.

Results:

The GC-TOFMS uses a novel direct extraction, semi-open ion source design capable of generating higher resolution and higher ion intensity than open design sources. The robustness of the ion source will be demonstrated with a comparison of the signal intensity of standards before and after hundreds of injections of complex matrix. A novel signal processing scheme and a proprietary data compression algorithm are used in real-time during data acquisition before the data is written to disk. Combining the novel ion source optics and advanced data processing results in sub-picogram detection and quantitation of analytes present in complex matrix. The ability to acquire full mass spectra (m/z 10-1500) at maximum detection limit allows confident identification of the trace analytes. The novel benchtop GC-TOFMS is achieving sub 50fg detection limits and >4 orders of linear dynamic range for OFN.

Novel Aspect:

High efficiency ion optics and state-of-the-art signal processing achieve sub-picogram detection limit for analytes in complex matrices on a GC-TOFMS.

Spacil, Zdenek,
Masaryk University

Contributing Authors:
Veronika Vidova, Research Centre for Toxic Compounds in the Environment, Masaryk University in Brno, Czech Republic; Andrea Zalesakova, Research Centre for Toxic Compounds in the Environment, Masaryk University in Brno, Czech Republic; Jana Klanova Research Centre for Toxic Compounds in the Environment, Masaryk University in Brno, Czech Republic; Jiri Kohoutek, Research Centre for Toxic Compounds in the Environment, Masaryk University in Brno, Czech Republic.

Title:
Quantitative profiling by SFC and UHPLC-MS/MS reveals metabolic phenotype of internal exposure to phthalates

Introduction:
The exposure to widely used chemicals such as phthalates is associated with numerous adverse health effects in human. However, adaptive metabolic phenotypes consistent with early onset of certain disorders were marginally studied in exposed humans. We present the strategy combing quantitative profiling of endogenous metabolites and human biomonitoring to reveal the body burden of toxic chemicals.

Methods:
Convergence chromatography system (ACQUITY UPC2, Waters) with QDa mass detector and UHPLC system (1290 Infinity II, Agilent) with triple quadrupole MS (6495, Agilent) were used to screen chemical library of 172 metabolite standards and to develop SRM assays. Chromatography conditions were optimized and different sub 2 micron column chemistries were compared (BEH, Torus and CSH, all from Waters).

Results:
The sub-2-micron sorbents and innovative column chemistries transformed supercritical fluid chromatography (SFC) into highly versatile tool; particularly coupled with MS. SFC system which utilizes CO2 for a normal phase separation was used to screen a panel of structurally diverse primary metabolites. The same panel was screened by a reversed phase UHPLC-MS/MS as a reference method. A comparable percentage of metabolites was detectable with SFC-MS (41%) and UHPLC MS/MS (51%), while 12% and 21% respectively were mutually exclusive. The complementary nature of SFC-MS and UHPLC-MS/MS is demonstrated by median logP values of mutually exclusive metabolites, which were 6.6 (range 1.0-10.5) and 1.5 (range -1.0-6.3) respectively. The internal exposure to phthalates and corresponding metabolic phenotypes were characterized using SRM assays in urine collected within CELSPAC_TNG birth cohort study.

Novel Aspect:
Application of SFC-MS in metabolic profiling; assessment of internal exposure to phthalates and quantitative profiling of endogenous metabolites

Spacil, Zdenek,
Masaryk University

Contributing Authors:
Veronika Vidova, Research Centre for Toxic Compounds in the Environment, Masaryk University in Brno, Czech Republic; Jana Klanova, Research Centre for Toxic Compounds in the Environment, Masaryk University in Brno, Czech Republic.

Title:
SRM assays for endogenous biochemical markers of internal exposure

Introduction:
The human metabolome comprising all endogenous metabolites, xenobiotics and their metabolites expresses an individual's metabolic phenotype. The internal exposure to chemicals from diet and environment is accompanied by perturbations in metabolic phenotype represented by immediate biomarkers of effect. Observed biomarkers may be consistent with early indications of CVD, type II DM or cancer.

Methods:
A MS metabolite library of 634 standards was used to develop SRM assays for a wide range of biochemical molecules of primary metabolism using UHPLC system (1290 Infinity II, Agilent) with triple quadrupole MS (6495, Agilent). Dried blood spots (DBS) and urine samples were processed in 96 well plate format using extraction to MeOH and lyophilization/extraction respectively.

Results:
A panel of endogenous metabolic markers of energy metabolism pathways (i.e. glycolysis, TCA cycle, beta oxidation, ornithine cycle) and disease conditions were screened by UHPLC MS/MS. SRM assays were optimized and deployed to biological samples (urine and DBS) collected via CELSPAC_TNG birth cohort study from pregnant women and their newborns within the South Moravia Region. Metabotypes were evaluated in biological samples based on abundance and mutual ratios between key metabolic markers. The internal exposure to chemicals was assessed using specific markers of exposure and correlated with the metabolic profile. This study contributes data to the area of health related exposure factors, particularly to emerging field of the exposome (i.e. non genetic factors affecting human health), which complementary with molecular epidemiology.

Novel Aspect:
Simultaneous evaluation of biomarkers of exposure and effect to study health related exposure factors in the emerging field of the exposome.

Spahr, Chris,
Amgen, Inc.

Contributing Authors:
Opas Nuanmanee, Therapeutic Discovery, Amgen Inc.- Thousand Oaks, CA 91320; Ken Walker, Therapeutic Discovery, Amgen Inc.- Thousand Oaks, CA 91320; Grace Li Therapeutic Discovery, Amgen Inc.- Thousand Oaks, CA 91320; Stone D.-H. Shi, Therapeutic Discovery, Amgen Inc.- Thousand Oaks, CA 91320.

Title:
HR-MS characterization of Glc-Gal-Hyl-, O-xylosylation, phosphorylation, and O-linked glycosylation in GLP1-Fc expressed in CHO cells

Introduction:
Glucagon-like-peptide-1 (GLP-1) has been fused with human Fc and expressed in CHO cells for use as a reagent for various metabolic disorders assays. A number of post-translational modifications (PTMs) in the CEX fraction post peaks were discovered during analysis of in-process samples. The characterization of these PTMs can be used to optimize purification conditions to generate more pure reagent.

Methods:
By LC-MS, cation-exchange (CEX) fractions of the GLP1-Fc preparation showed PTMs of +80 Da, +340 Da, and + 948 Da. Peptide level analysis using high-resolution mass spectrometry on a Q-Exactive reveals Glc-Gal-Hyl- modification of Lys, various O-xylosylated species, Ser phosphorylation, and O-linked glycosylation within 21 amino acids spanning the C-terminus of GLP1 and the linker of the molecule.

Results:
Lys modified with Glc-Gal-Hyl- glycan doesn't cleave with Lys-C or trypsin. Higher energy collision dissociation (HCD) yields the labile loss of glycan components, but retention of the hydroxyl group. O-xylosylation has been observed in (G4S)x linker repeats. Initial placement of xylose onto Ser via a xylosyltransferase is followed by glycan extension of galactose and sialic acid to yield di- and tri-saccharides. Ser phosphorylation yields a neutral loss of ~98 Da, but can be confidently assigned using HCD fragmentation. Fragment ions ([y2]; m/z 234.14) critical for localizing the phosphate are present by HCD, but would be lost in a 3D ion trap due to the 1/3 rule. The site of the sialylated O-linked glycan can be determined using alkaline β-elimination of the peptide. The reaction removes the sugar and leaves an 18 Da mass loss that can be used to assign the glycosylation site.

Novel Aspect:
HR-MS, HCD fragmentation, and β-elimination were combined to identify closely spaced glycan and phosphate PTMs in fractions of GLP1-Fc.

Spellman, Daniel,
Merck Research Laboratories

Title:
LC/MS-based Approaches to Therapeutic Protein Quantitation to Support Discovery Stage Pharmacokinetic Studies

Introduction:
The prevalence of biotherapeutics (antibodies, fusion proteins, and peptides) has increased rapidly over the last decade within the pharmaceutical industry, along with the need to evaluate their pharmacokinetic and pharmacodynamic properties. Here, we present efforts standardize and streamline the approach to quantitation of biotherapeutics using automated sample preperation and LC-MS/MS.

Methods:
Automated approaches to sample preperation were implemented on multiple systems including the Agilent AssayMap. Analytes were quantitated using MRM approaches on a Waters Acquity UPLC coupled Xevo TQ-S mass spectrometer equipped with an IonKey source. Quantitation was based on curves generated by spiking various concentrations of analyte and a fixed concentration of a labeled internal standard.

Results:
We have developed approaches to LC/MS-based quantitation of biotherapeutics in plasma, tissues, and alternative matrices such as dried blood spots. Such approaches offer great potential value to the biopharmaceutical development process, particularly during early preclinical animal studies wherein multiple structural variants of a drug candidate may need to be evaluated in various animal species to enable prioritization for further development. The bulk of our methodology focuses on a "generic" approach in preclinical species using anti-human IgG immunocapture and a commercially available, stable labeled mAb internal standard. We also present here, interesting findings related to difference between ELISA and LC-MRM-MS results indicated that these two methods can provide complementary information (free drug vs total drug, bound vs. unbound, ..) regarding the drug's PK profile.

Novel Aspect:
Implementation of therapeutic protein quantitation using automated techniques and SILUMab internal standards on a triple-quadrupole mass spectrometer.

Spengler, Bernhard,
Justus Liebig University Giessen

Contributing Authors:
Dhaka Ram Bhandari, Justus Liebig University, Giessen, Germany; Mario Kompauer, TransMIT GmbH, Giessen, Germany; Bernhard Spengler Justus Liebig University, Giessen, Germany; Kerstin Strupat, Thermo Fisher Scientific (Bremen) GmbH; Karl Christian Schaefer, Justus Liebig University, Giessen, Germany; Sven Heiles, Justus Liebig University, Giessen, Germany.

Title:
Improvements in AP-SMALDI Orbitrap mass spectrometry imaging of biological tissue

Introduction:
Improvements in spatial resolution and performance will be reported for atmospheric-pressure scanning microprobe MALDI Orbitrap MS imaging (AP-SMALDI OT-MSI). The method has gained significant attention due to its capability to disclose morphologic distributions of substances in complex biological samples with high sensitivity and without vacuum-induced analyte losses or morphological artifacts.

Methods:
A commercial imaging system (AP-SMALDI10®, TransMIT GmbH) was modified in order to further increase spatial resolution. A new objective lens was developed and integrated into the ion source. Both, the commercial and the modified imaging source were coupled to a Q Exactive HFTM (Thermo Fisher Scientific) orbital trapping mass spectrometer, operated at a mass resolution of up to 240,000 at m/z 200.

Results:
Biological tissue samples were prepared from fresh frozen material and matrix-coated by various methods, to compare analytical sensitivities. The ablation spot size on tissue after AP-SMALDI MSI was determined as 1.4±0.2 µm in diameter. Results demonstrate routine capability of the instrumentation to obtain high-quality, highly sensitive images without oversampling down to a step size (pixel resolution) of 2 µm. Oversampling was performed down to 1 µm step size successfully with good signal-to-noise ratio. The spatial resolution was calculated based on King-Smith contrast function and error function evaluation. A very high imaging selectivity of ±5 ppm, mass accuracies of better than 2 ppm over the entire image, and a mass resolving power of up to 240,000 at m/z 200 were achieved routinely by direct coupling of the imaging system to the Q Exactive HF orbital trapping mass spectrometer.

Novel Aspect:
AP-SMALDI MSI of endogenous and exogenous compounds from tissue and single cells at ultrahigh lateral resolution of 1.4 µm and very high mass accuracy

Spivak-Lavrov, Igor,
Aktobe Regional State University

Contributing Authors:

Igor F. Spivak-Lavrov, Aktobe K. Zhubanov Regional State University, Aktobe, Kazakhstan; Orda A. Baisanov, T.Ya. Begeldinov Military Institute of Air Defense Forces, Aktobe, Kazakhstan; Guldana E. Gilimova Aktobe K. Zhubanov Regional State University, Aktobe, Kazakhstan.

Title:

Static mass spectrometers based on the cone-shaped achromatic prism (CAP)

Introduction:

In the work (H. Wollnik NIM 1971, 95, 453) it was shown that the Q-parameter defining the quality of static mass analyzers is proportional to a flux of a magnetic field through the section of an ion beam («magnetic flux theorem»). There appeared an idea of a beam expansion before entering the magnetic sector with the help of quadrupole lenses in the device was realized (H. Matsuda, NIMA, 1995).

Methods:

We have proved «the magnetic flux theorem» for any form of an axial trajectory and combined electric and magnetic field. In the cone-shaped achromatic prism (CAP) the idea of expansion of a beam before entering a magnet due to the refraction in the prismatic electric field was offered. Such field not only expands the beam before entering a magnet, but also provides a focus on energy.

Results:

In this context the unique design is CAP possessing record angular dispersion equal to about 50 radian per 100% of mass variation and small chromatic aberrations [1]. The parallel ion beam coming into CAP first expands due to the refraction in the electric field, and then its additional expansion in a non-homogeneous magnetic field takes place. Since the sector of a magnetic field in CAP is bigger than 180o, therefore the magnetic flux is very large. It leads to the fact that all trajectories of particles of the homogeneous flat parallel beam entering the CAP move in the mean plane by similar trajectories and keep parallelism at the exit from the CAP too. We will note that in CAP energy focusing is carried out, and parallelism of a volumetric beam is preserved owing to its vertical telescoping. [1] Spivak-Lavrov I.F. Nucl. Instr. and Meth. in Phys. Res. A, 1995, 363, 485-490.

Novel Aspect:

The CAP realized the electric and magnetic fields of the type 1/r, so it can be used as a prism in a static mass spectrometers.

Spivak-Lavrov, Igor,
Aktobe Regional State University

Contributing Authors:

Orda A. Baisanov, Military Institute of Air Defence Forces named after T.Begeldinov, Aktobe, Kazakhstan; Evgeniy M. Yakushev, RSE Institute of Nuclear Physics of Ministry of Industry and New Technologies of RK, Almaty; Raziay B. Bizhanova Aktobe Regional State University named after K.Zhubanov, Kazakhstan; Igor F. Spivak-Lavrov, Aktobe Regional State University named after K.Zhubanov, Kazakhstan; Leonid M. Nazarenko, RSE Institute of Nuclear Physics of Ministry of Industry and New Technologies of RK, Almaty.

Title:

Orthogonal ion accelerator on the basis of the wedge-shaped mirror with 2D electrostatic field as simple direct-flow TOF MS

Introduction:

TOF MS reflection type, which conduct a time-of-flight focusing of an ion packet by energy. The main drawback of these TOF MS, which limits their resolution and sensitivity, is the use of meshes or array of based apertures to create accelerating field (H. Haberland et. al., Rev. Sci. Instr. 62, 1991, 2621). Wedge-shaped mirror with 2D electric field [1] can also be used as an ion accelerator.

Methods:

In the present paper it is shown that this mirror can be used as a compact TOF mass analyzer with high resolution. Ions from the ion source by means of impulse activation of accelerator's electric field are extracted. The ion packet is pushed out into a drift space for the analysis. The detector is located in plane where time-of-flight and space focus is.

Results:

Due to the fact that ions are extracted from different points of the ion source on exit from the field into the drift space their speeds are different. The particles that passed the greater potentials difference obtain greater speed. As a result, the lagging particles of the packet catch up with the front ones in the drift space and the ion pack constricts along-track direction of movement ions. Having placed the plane of a ion detector where the greatest constriction of the packet takes place, we obtain the time-of-flight mass spectrometer with high resolution. Options of the mass spectrometer in which the beam along-track direction of movement ions is constricted by about more 100 times, which allows to obtain resolution more 10000 at the small device size have been computed. [1] Spivak-Lavrov I.F. Advances in Imaging and Electron Physics. Vol. 193, 2016, 45-128.

Novel Aspect:

The inhomogeneous accelerating electric 2D field in the wedge-shaped mirror without using of fine meshes and array of based apertures is created.

Steffen, Johnathan,
University of Waterloo

Contributing Authors:
Stephen Walker, University of Waterloo; Richard Marta, University of Waterloo; W. Scott Hopkins University of Waterloo; Denzel Huang, University of Waterloo; J. Larry Campbell, AB Sciex; Mike Lecours, University of Waterloo; Terry McMahon, University of Waterloo; Eric Fillion, University of Waterloo.

Title:
Measuring Ion-Solvent Binding Energies with DMS

Introduction:
The tricarbastannatrane cation is a cage structure stabilized by a transannular Sn-N bond. Of interest is how the cage structure responds to complexation with Lewis basic solvents. IRMPD was used to probe the structures of several Lewis acid/base complexes involving tricabastannatrane. Here, we detail how DMS was used to assess the strength of interaction between the cage and solvent molecules.

Methods:
Tricarbastannatrane was dissolved in acetonitrile (10 ng/ml) and infused into a DMS cell via ESI(+). The DMS N2 environment was seeded with 1.5% (v/v) solvent vapour and dispersion plots were recorded at 150 °C and 300 °C. Relative solvent binding energies were extracted from DMS results using trajectory simulations and curve fitting, and were calibrated against calculated dissociation energies.

Results:
Tricarbastannatrane has been studied using IRMPD and DMS in several different chemically modified N2 environments to probe its clustering propensity with a number of Lewis basic solvent molecules. Complexation can be monitored spectroscopically by observing the wavenumber shifts of tricarbastannatrane vibrations upon interaction with a solvent molecule. In DMS experiments at T = 150 °C, tricarbastannatrane is observed to cluster strongly with Lewis basic solvent molecules. At T = 300 °C, dispersion curves shift and several modifiers exhibit Type B (weak clustering) behavior. The observed DMS clustering behavior correlates with DG(dissociation) density functional theory calculations for the tricabastannatrane-solvent complexes. DMS simulations are currently underway to model tricarbastannatrane trajectories within the DMS cell from an atomistic, first principles standpoint.

Novel Aspect:
IRMPD is used to assess Lewis base donicity and DMS is used to quantify Lewis acid/base complexation energies.

Steimling, Justin,
Restek Corporation

Contributing Authors:
Frances Carroll, Restek Corporation; Sharon Lupo, Restek Corporation; Shun-Hsin Liang Restek Corporation; Ty Kahler, Restek Corporation.

Title:
Affecting Selectivity and HILIC Retention on a FluoroPhenyl Stationary Phase

Introduction:
In this study we have set out to explore the retention mechanisms of the FluoroPhenyl stationary phase. The FluoroPhenyl phase may be described as having Mixed-Mode and/or HILIC retention and selectivity. HILIC and Mixed-Mode offer retention mechanisms that vary, or are orthogonal to, typical reversed-phase columns like C18. Herein we explore strategies to utilize this unique selectivity.

Methods:
Our focus in this presentation is on method changes including temperature, mobile phase composition, and acid strength and concentration and how these changes affect the selectivity of targeted analytes. By demonstrating the influence of method changes on this phase we aim to gather a better understanding of the interactions provided by the FluoroPhenyl phase.

Results:
The FluoroPhenyl phase allows retention based on hydrophobicity and cation-exchange which results in a "U-shaped" retention profile for bases. This can be exploited when used in conjunction with electrospray ionization mass spectrometry (ESI-MS) as sensitivity will increase with the high percentage of organic solvent in the mobile phase due to increased desolvation efficiency. Acid strength and concentration can be used to alter selectivity based on the cation-exchange mechanism. As the acid strength and concentration increase, the retention of bases decreases.

Novel Aspect:
Understanding the retention characteristics of the FluoroPhenyl stationary phase.

21ST INTERNATIONAL MASS SPECTROMETRY CONFERENCE
Toronto, Ontario, Canada
August 20 – 26, 2016

Steimling, Justin,
Restek Corporation

Contributing Authors:
Ty Kahler, Restek Corporation; Shun-Hsin Liang, Restek Corporation; Frances Carroll Restek Corporation; Sharon Lupo, Restek Corporation.

Title:
A Multi-Class Drug and Metabolite Screen of 231 Analytes by LC-MS/MS

Introduction:
Many drug testing facilities are turning to liquid chromatography coupled with mass spectrometry (LC-MS/MS) for its increased speed, sensitivity, and specificity to overcome low cut-off levels, matrix interferences, and isobaric drug compounds. Herein, a method was developed for a multi-class drug and metabolite screen containing 231 compounds using a Restek Raptor™ Biphenyl column.

Methods:
Experiments performed included mobile phase considerations, sample diluent, isomer resolution, drug interferences, and instrumentation capabilities. Analytes were diluted in water and injected into a Shimadzu Nexera UHPLC equipped with an AB SCIEX API 4500™ MS/MS. Detection was performed using electrospray ionization in positive and negative ion modes using scheduled MRM.

Results:
During mobile phase investigations, it was found that methanol provided the best retention of early eluting analytes, such as morphine, oxymorphone, nicotine and norcotinine. Sample diluent was optimized to improve the peak shape of the early eluting compounds. Scan rates and retention time windows were optimized to ensure enough data points per peak were collected.

Novel Aspect:
231 compounds with a final optimized separation gradient run time of 10 minutes, with a total cycle time of 12 minutes.

21ST INTERNATIONAL MASS SPECTROMETRY CONFERENCE
Toronto, Ontario, Canada
August 20 – 26, 2016

Stein, Stephen,
nist

Contributing Authors:
Yuir Mirokhin, NIST; Yamil Simon, NIST; Dmitrii Tchekhovskoi NIST.

Title:
Annotating Unidentified LC-MS/MS Spectral Libraries

Introduction:
Identifying compounds by matching acquired spectra to spectra in a reference library is routine in gc/ms and finding increased use in lc/ms experiments. An ever present concern is whether the compound generating a measured spectrum is represented in the library. We an approach for building libraries from annotated unidentified spectra from urine and plasma reference materials.

Methods:
We describe a method for clustering and creating consensus spectra from unidentified spectra in LC-MS/MS experiments. This also involves annotation of spectra using special library search and analysis procedures, including in-source analysis and a new similarity search method using both neutral losses and directly detected ions, which we term the 'hybrid' search..

Results:
Our new method increases by several fold numbers of metabolite identifications in analyses by LC-tandem MS analysis of various NIST standard reference materials, including urine and plasma, against the NIST Tandem Mass Spectral Library. These new identifications are, in effect, members of the fragmentation families of molecules already present in the library. In many cases, these fragmentation classes closely correspond to classes and sub-classes of chemical groups of compounds common to biological fluids, including carnitines, lipids, amino acid derivatives, lipids, nucleosides and others. By using this method to annotate a library of recurring, unidentified spectra derived from urine we demonstrate a major extension of libraries that cover a large fraction of ions observed in metabolic fluids.

Novel Aspect:
A method is presented that expands the coverage of spectral libraries using annotated unidentified spectra derived biological fluids.

21ST INTERNATIONAL MASS SPECTROMETRY CONFERENCE
Toronto, Ontario, Canada
August 20 – 26, 2016

Stock, Naomi,
Water Quality Centre

Contributing Authors:
Michael Doran, Water Quality Centre, Trent University; John Guchardi, Faculty of Science - Aquatic Toxicology, University of Ontario Institute of Technology; Shane de Solla Ecotoxicology and Wildlife Health Division, Environment and Climate Change Canada.

Title:
Development of Extraction and LC-MS/MS Methods for the Quantitation of Finasteride, Levonorgestrel and Melengestrol Acetate in Water and Sediment

Introduction:
Synthetic steroids are emerging environmental contaminants of concern due to potential toxicity, including endocrine disruption. This study focuses on three steroids: finasteride, a synthetic antiandrogen; and melengestrol acetate and levonorgestrel, both synthetic progestins. Extraction and analysis methods were developed to quantify these steroids in spiked sediment and water samples.

Methods:
Finasteride, levonorgestrel and melengestrol acetate were concentrated in water samples using Oasis HLB solid phase extraction cartridges and extracted from sediment by sonication with organic solvent. The synthetic steroids were separated by liquid chromatography on a C18 column, and analyzed using multiple reaction monitoring mode on a hybrid triple quadrupole/linear ion trap mass spectrometer.

Results:
Robust extraction and liquid chromatography-tandem mass spectrometry (LC-MS/MS) methods for the quantitation of finasteride, levonorgestrel and melengestrol acetate have been developed and applied to the analysis of spiked water and sediment samples used for toxicity testing. Stable isotopically labelled standards for all three synthetic steroids were used for quantitation. The methods were validated, including evaluation of matrix effects and determination of spike and recoveries and limits of detection. Fragmentation of the three synthetic steroids was also investigated. Possible applications of this method for the extraction and analysis of environmental samples is also discussed.

Novel Aspect:
This is the first report of an extraction and analysis method for finasteride in sediment and water.

Straus, Rita,
NSERC CREATE York University and University of Toronto

Contributing Authors:
Rebecca A. Jockusch, University of Toronto.

Title:
Probing the Gaseous Structure of a β-hairpin Peptide with Hydrogen/Deuterium Exchange (HDX) and Electron Capture Dissociation (ECD)

Introduction:
Tools available for examining structure of large gaseous biomolecules fail to provide a high level of detail. Here, the gaseous structure of a β-hairpin peptide, derived from the protein GB1 (GB1p) and related peptides are probed using gas-phase hydrogen/deuterium exchange (HDX) with subsequent electron capture dissociation (ECD) to obtain site specific structural information.

Methods:
Experiments were performed in a Bruker Apex-Qe ion cyclotron resonance mass spectrometer. GB1p and variant ions were transferred into the gas-phase via nano-electrospray ionization, and sent into a collision cell for reaction with ND3(g). Ions were then transferred into the ICR cell for ECD. Deuterium levels were calculated by comparing masses of unlabeled and deuterium-labeled ions.

Results:
GB1p exhibited high levels of protection, with only 48% of labile hydrogens undergoing exchange. In contrast, a truncated variant which lacks the hydrogen bonding pairs necessary to form the hairpin in solution, exchanged 67% of labile hydrogens. These results suggest that the full length peptide contains a significant amount of secondary structure, and is consistent with the retention of a hairpin conformation in the vacuum. Site specific exchange analysis reveals that each residue of the truncated peptide has a higher deuteration level than the corresponding residue on full length GB1p. Four mutants, in which residues comprising a hydrophobic core were replaced with alanine, exhibited nearly identical H/D exchange profiles to wild type GB1p. Although these mutants have less stable structures in solution, the data suggests that they all collapse to similar structures upon desolvation.

Novel Aspect:
Development of a new method to probe gaseous peptide conformation by analysing residue-specific uptake levels following gas-phase HDX and ECD.

Sun, Wei,
Sanofi

Contributing Authors:
Thomas O'Shea, DMPK Department, Sanofi US, Waltham, MA; Stacy Ho, DMPK Department, Sanofi US, Waltham, MA; Walter Korfmacher DMPK Department, Sanofi US, Waltham, MA.

Title:
Determination of Rat Synovial Fluid Lavage Dilution Factor Using Urea as Dilution Biomarker

Introduction:
The collection of rat knee synovial fluid (SF) is typically conducted using lavage due to the small volumes involved. The challenge for accurate drug quantitation in these samples is the determination of the dilution factor due to the lavage solution. The objective of this study is to develop a method to correct for the dilution factor SF during lavage of a rat knee joint.

Methods:
Urea, an endogenous compound that maintains a fixed concentration relationship between synovial fluid and plasma, has been used as dilution biomarker. Dilution factor is determined by the ratio of measured urea concentration in plasma versus measured urea concentration in synovial fluid lavage (SFL)

Results:
Combining hydrophilic interaction chromatography and tandem mass spectrometry, a simple and sensitive method (calibration range of 0.1 to 100 µg/mL) has been developed for the determination of urea in rat synovial fluid lavage (SFL) using a 20µL sample, which allows accurate measurement up to approximately 5000-fold of dilution during joint lavage. The urea SFL method was modified and also applied to the analysis of urea in rat plasma (calibration range of 100 to 800 µg/mL) using a 5µL sample volume. The developed methodology was successfully applied to SFL samples from a rat pharmacokinetic study for the determination of drug concentration in synovial fluid. The high variability of the dilution factors observed demonstrated the necessity of a dilution marker to ensure accurate drug quantitation in SFL.

Novel Aspect:
Methodology developed that enables accurate measurement of drug concentration in rat synovial fluid after joint lavage.

21ST INTERNATIONAL MASS SPECTROMETRY CONFERENCE
Toronto, Ontario, Canada
August 20 – 26, 2016

Suter, Marc,
Eawag

Contributing Authors:
Adrian Ammann, Eawag.

Title:
Multimode Gradient HPLC-MS Method for Metabolomics and Environmental Monitoring

Introduction:
When analyzing a very complex mixture of chemicals, a single mode chromatographic separation will lose a substantial part of the targets. To address this problem, we use one C18 trapping column and two commercially available mixed-mode ion-exchange reversed-phase columns (WAX-1 and WCX-1) in sequence in a novel multimode separation LC-MS method.

Methods:
10 µL samples were loaded onto a C18-trap column connected to the divert valve of the MS. Two columns (100x2 mm, Acclaim MM WAX-1 and WCX-1, Thermo) were connected in sequence. An isocratic pump delivered eluent C (H2O, 5% ACN, 10 mM NH5CO3) and the sample to the C18-trap column, while the gradient pump ran eluent A (H2O, 97% ACN, 3 mM NH5CO3) and B (H2O, 10% ACN, 30 mM NH5CO3). Total runtime 50'.

Results:
The single-run multimode separation trapped hydrophobic compounds in the first part of the run, while hydrophilics were separated within 20 minutes by HILIC (10% to 90% H2O, ACN, pH 7.3) and IEX (3 to 30 mM NH5CO3, pH 7.3) gradients on the sequential multimode columns (WAX-1 and WCX-1). In the second half, hydrophobics were separated by a RP gradient (10% to 90% ACN). Flow rate was 260 µL/min and total runtime 50'. Amino acids (8), sugars (9), small fatty acids metabolites (10) and other compounds (15) representing diverse endogenous compound classes were analyzed. A good chromatographic separation was achieved for the whole set of compounds. Molecules with hydrophobic and ionic moieties, for instance aliphatic acids, were the most retained. Data gathered so far serves as a basic data set that allows to predict retention times of other metabolites and environmental contaminants.

Novel Aspect:
Four different separation mechanisms (RP, HILIC, anion and cation exchange), all in gradient mode, were combined in a single run.

Tähkä, Sari,
Division of Pharmaceutical Chemistry and Technology, Faculty of Pharmacy, University of Helsinki, Finland

Contributing Authors:
Ville Jokinen, Department of Materials Science and Engineering, Aalto University, School of Chemical Technology, Finland; Ashkan Bonabi, Division of Pharmaceutical Chemistry and Technology, Faculty of Pharmacy, University of Helsinki, Finland; Tiina Sikanen Division of Pharmaceutical Chemistry and Technology, Faculty of Pharmacy, University of Helsinki, Finland.

Title:
Non-aqueous capillary electrophoresis-electrospray ionization mass spectrometry on thiol-ene-based microchips

Introduction:
Thiol-ene (TE) polymers show good solvent compatibility and are promising material for microfabrication in standard laboratory conditions. In this work, we demonstrate the feasibility of TE polymerization for fabrication of fully integrated microchip capillary electrophoresis-electrospray chips (MCE-ESI) for mass spectrometry (MS), and their exploitation to aqueous and non-aqueous MCE.

Methods:
The MCE-ESI chips were prepared by curing a mixture of thiol and allyl monomers against a mold under UV. The microchannels were sealed with another thiol-ene layer by manually aligning the emitter tip. The materials compatibility was examined with a set of organic solvents and the feasibility for peptide separation by MCE was compared between aqueous and nonaqueous background electrolytes (BGE).

Results:
Adaptation of the microchip technology for routine use in MS-based bioanalysis is often limited by the laborious chip fabrication that requires expensive cleanroom techniques. Here, we present a simple, non-cleanroom replication approach for chip fabrication using TE polymerization. The stability of the ESI signal of the microchip emitters was typically 7.9-11.9 % RSD (TIC, n=5) and was maintained as long as 20 min at best. Unlike other common polymers used for chip fabrication (e.g., polydimethylsiloxane), the TEs were well compatible with the common organic solvents and provided better separation of the selected peptides in nonaqueous than in aqueous conditions. The migration times were repeatable in both cases, i.e., 8.8-11.5 % RSD (n=5 runs, nonaq., ethanol as BGE) and 4.6-5.0 % RSD (n=5, aq.) and the plate numbers were in the range of 13000-77000/m (nonaq.) and 7700-11000/m (aq.).

Novel Aspect:
First nonaqueous electrophoresis on a TE chip featuring fully integrated ESI emitter and fabricated in standard laboratory conditions by replication.

TAIRA, Shu,
Fukui Prefectural University

Contributing Authors:
Yasuko Kawamura-Konishi, Ishikawa Prefectural University; Hajime KATANO, Fukui Prefectural University.

Title:
Localization analysis of pesticide residue in cucumber sample by Imaging mass spectrometry

Introduction:
The localization of procymidone (proc)fungicide residue in cucumbers was investigated byIMS. Cucumbers were grown, harvested, and then divided into two groups that were either sprayed or not sprayed with proc. We quantitatively determined the content of proc in the cucumbers by GC-MS. procymidone by IMS. Proc reached the central part of the cucumbers following spraying compared with the control.

Methods:
After 3 months of cultivation, proc was applied once to the treated cucumbers on day 0 after harvest. The control group was left untreated. The harvested cucumbers were stored at 25 °C for 0, 1, 3, and 5 days. Samples for nano-PALDI IMS were cut from the fruit using a blade and the frozen samples were set axially in a cryostat at -25 °C. IMS was achieved by a FT-ICR MS.

Results:
We imaged the localization of proc in cucumbers by IMSI at a resolution of 100 um to show whether proc was localized in any of the regions. The protonated ion signal was chosen as the proc target in the cucumber section. In the pericarp, the intensity for the day 0 sample was threefold higher than that for the day 3 and 5 samples. The intensity of the samples decreased over time. However, for the flesh the intensity increased in the day 1 to 3 samples and decreased in the day 5 sample. IMS showed that pesticide residue migrated toward the center of the cucumber over 5 days.

Novel Aspect:
Visualization of pesticide gives us the time dependent permeation aspect. S. Taira et al. Journal of Agricultural and Food Chemistry 63, 6109 (2015)

TAIRA, Shu,
Fukui Prefectural University

Contributing Authors:
Hajime KATANO, Fukui Prefectural University.

Title:
Hybrid nanoparticle-assisted laser desorption/ionization (Nano-PALDI) MS for bioanalysis

Introduction:
Iron oxide-based nanoparticles (NP) were covalently modified with sinapic acid (SA) through a condensation reaction to assist the ionization of both large and small molecules. SA-NP-assisted laser desorption/ionization was carried out on small molecules, such as pesticides and plant hormone, and large molecules, such as peptides and proteins.

Methods:
SA and carbonyldiimidazole were added to a NP suspension and heated to 35 ºC to synthesized SA-NP. The usefulness of SA-NPs as an ionization-assisting material in mass spectrometry was investigated using a MALDI time-of-flight (TOF) instrument. The analyte surface was irradiated with 1,000 laser shots and TOF spectra were acquired in positive ion detection mode.

Results:
The SA-NPs were analyzed by FT-IR to confirm chemical bonding between the SA molecules and the surface of the NPs.To verify the ionization ability of SA-NP, carbendazim, ABA, angiotensin II, Aβ, insulin, and cytochrome C were analyzed using MS under MALDI conditions.The SA-NPs were used to ionize insulin, resulting in a 28-fold increase in the ionization intensity compared with that provided by the original NPs. The saturated SA had 1.6-fold greater intensity than that of the SA-NPs. However, at same density of SA condition, SA-NP showed 4.0-fold greater intensity than that of the SA, indicating that the ionization efficiency of large molecule depends the amount of SA. Thus, the use of saturated SA as an organic matrix more strongly promotes the ionization of large molecules than SA-NP.However, SA-NP could be used to detect the small molecule carbendazim and ABA without noise.

Novel Aspect:
The nano-PALDI can be used to analyze molecules of a wide range of molecular weights. S. Taira et al. Analyst 140, 8134-8137 (2015). (cover article)

Takarada, Toru,
Kobe Pharmaceutical University

Contributing Authors:

Atsuko Takeuchi, Kobe Pharmaceutical University; Toru Takarada, Kobe Pharmaceutical University; Masafumi Matsuo Kobe Gakuin University; Ken-ichi Yoshino, Kobe University.

Title:

New strategy for analysis of the splicing regulatory factors using high resolution mass spectrometry

Introduction:

Disruption of the pre-mRNA splicing causes a variety of genetic diseases. Understanding of the intricate splicing mechanism is important to perform the genetic therapy directed toward the correction of RNA mis-splicing in diseases. This paper presents the combination method of mass spectrometric and biological analysis to identify the meaningful factors involved in the splicing reaction.

Methods:

Splicing regulatory factors participate the regulation of splice site usage. To identify these factors bound to the targeted exon, we designed in vitro binding assay. The RNA probe was incubated with the cell nuclear extract. The separated band in PAGE, which contains the RNA-protein complex, was in-gel digested. This peptide mixture was applied to LC-MS/MS to identify the binding protein.

Results:

To confirm the utility of this method, we designed an RNA probe consisting of the insulin receptor exon 11. We found that hnRNPA1 (A1) forms the complex with the exon11 sequence. In general, some splicing regulatory factors directly bind to the RNA, whereas others need a protein-protein interaction to be formed the complex. To elucidate the binding model of the RNA-A1 complex, we used the technique of photo-crosslinking. UV irradiation of the complex forms a covalent bond between protein and RNA. Using the techniques indicated above, it is revealed that A1 directly bind to the RNA. Finally, we contrived the method for the binding site prediction of A1 using mass spectrometry. In case that A1 binds to RNA, the specific peptide located in RNA binding domain was hardly detected in LC-MS/MS analysis. It means that this peptide region may be related to the binding to RNA sequence.

Novel Aspect:

We established the techniques using a combination method of mass spectrometric and biological analysis to help understanding the splicing mechanism.

Takaya, Kazunari,
Rikkyo university

Contributing Authors:
Shunsuke Hoshino, Department of Physics, Rikkyo University; Yuya Hasegawa, Department of Physics, Rikkyo University; Mina Fujieda Department of Physics, Rikkyo University; Tetsuo Koizumi, Department of Physics, Rikkyo University.

Title:
Mobilities of CnH2n+1O+ ion (n = 1-3) in He gas

Introduction:
Mobilites of an ion in a buffer gas are very sensitive to the shape of the interaction potential between the ion and the gas molecule. Generally, the long-range attractive interaction decides the cross section at low energy. We can obtain the information on the interaction potential with the measured mobility of an ion in a gas.

Methods:
The apparatus consisted of an ion incident system, a drift tube and a detection system. Ions produced in the electron impact ion source are mass analyzed with a quadrupole mass spectrometer the select ions of the desired mass/charge ratio and they are injected into the drift tube.

Results:
In this work, we measured mobilities of polyatomic ions in helium gas at 83 K and room temperature to investigate the influence of the long-range interaction on the mobility. Mobilities depend on the reduced electric field E/N on the effective temperature Teff. Mobilities of polyatomic molecule ions show different Teff dependence from the typical mobility of atoms and small molesules. The mobilities of polyatomic ions fall lower than the polarization limit Kpol at all measured Teff range. We have measured mobilities of CH3O+, C2H5O+ and C3H7O+ in helium gas at 83 K and room temperature. Mobilites of C3H7O+ fall lower than the polarization limit Kpol at all measurement Teff range. On the contrary, mobilities of CH3O+and C2H5O+ have the same tendency as the typical mobility of an atoms and small molecules in He gas.

Novel Aspect:
There have been few studies of mobility measurement of polyatomic ions in helium gas at 83 K and room temperature.

Takei, Chikako,
BioChromato,Inc.

Contributing Authors:

Takehiko Sagawa, S & B Foods Inc.; Keiko Matsumoto, Shimadzu Corporation; Jun Watanabe Shimadzu Corporation; Chikako Takei, BioChromato, Inc; Motoshi Sakakura, AMR Inc.; Teruhisa Shiota, AMR Inc.; Hiroshi Matsufuji, Nihon University.

Title:

Application of a flavor release analysis method on volatile compounds of citrus fruits by DART-MS

Introduction:

Gas chromatography is typically used for analysis of the volatile compounds represented by aroma compounds. Nevertheless it's important to find relationship with the results of chemical analysis and perceptive aroma, it is hard to measurement flavor release from foods in real time. Here, we tried to measure the flavor release phenomena on volatile compounds of several kinds of citrus fruits.

Methods:

Limonene, γ-terpinene, octanal, linalool and α-terpine were used. MS spectra were obtained by triple quadrupole mass spec equipped with DART ion source and Volatimeship an optional tool for DART-MS to make highly sensitive and real time analysis of volatile compounds.MS conditions and MRM transitions were optimized using volatile compounds. Then several kinds of citrus fruits were analyzed.

Results:

Five volatile compounds were used for this study because they were known as characteristic volatile compounds in citrus fruits. Limonene has an aroma of an orange peel, octanal has a sweet aroma of citrus with low concentration, and linalool has an aroma of flower. Regarding ionization ability of these components, all compounds (their volatile) were ionized by DART. These compounds were interestingly ionized, so, they had been detected not molecular related ions (M+H) but several ions which might be fragment ions; precursor ion signals of limonene were detected by positive m/z 135, 151, 152 and terpinene, 170, 186. All components were ionized high-sensitively at positive ion mode. Whole citrus fruit were put in the closed sample cage and crushed, with analyzing the volatile compounds released from citrus fruit by this closed DART-MS system.

Novel Aspect:

Volatile compounds in citrus fruits could be detected with real time analysis by closed DART MS/MS system.

21ST INTERNATIONAL MASS SPECTROMETRY CONFERENCE
Toronto, Ontario, Canada
August 20 – 26, 2016

Takei, Chikako,
BioChromato,Inc.

Contributing Authors:

Hiroko Abe, University of Chiba; Chikako Takei, BioChromato, Inc; Yasuo Shida BioChromato,Inc; Motoshi Sakakura, AMR, Inc.; Teruhisa Shiota, AMR, Inc.; Kayako Suga, AB SCIEX; Daisuke Yajima, University of Chiba; Hirotaro Iwase, Universty of Tokyo.

Title:

Illegal Drugs Analysis by Thermal Desorption and Pyrolysis Combined with Direct Analysis in Real Time- Mass Spectrometry (TDP/DART-MS)

Introduction:

Drugs presence in biological and autopsy specimens cannot be detected without first selecting pretreatment and analytical conditions appropriate for drugs.Thus, the comprehensive analysis for drugs that require minimal selection of pretreatment and analytical conditions are desired. So, we investigate a direct analytical method for drugs in blood and urine that doesn't require any pretreatment.

Methods:

Mass spectra were obtained by using a Q-Tof mass spec equipped with a DART ion source and an ionRocket thermal desorption and pyrolysis unit. The ionRocket was mounted between the DART ion source and the mass spec. The samples were standard drugs mixture solutions, and drugs mixture loaded blood and urine. We conducted our study by assessing whether drugs could be detected.

Results:

Each drugs were separated and detected through thermal gradient heating for all samples, and thermal desorption profiles were highly reproducible for individual drugs. The detected ions were correctly identified according to their measured accurate mass and product ion spectra. Moreover, this analysis method was seemed to have the potential of quantitative analysis for drugs, as the drug's ion intensity were increased with increasing drugs concentrations in the samples. However, the drug's ion intensity was decreasing in the order standard drugs solution > urine added drugs > blood added drugs, even at the same drug concentration, with the drug's ion intensity were markedly reduced in blood in particular. Accordingly, we analyzed the blood samples that was pretreated by using SPE for removing phospholipid, whereupon the drug's ion intensity were dramatically improved.

Novel Aspect:

Each drugs were separated and detected through thermal gradient heating with avoiding each ions suppression by TDP/DART-MS.

Takeuchi, Atsuko,
Kobe Pharmaceutical University

Contributing Authors:
Toru Takarada, Kobe Pharmaceutical University; Atsuko Takeuchi, Kobe Pharmaceutical University; Yoshihiro Urade Tsukuba University; Masafumi Matsuo, Kobe Gakuin University.

Title:
Clinical evaluation of prostaglandin D2 and E2 metabolites in urine

Introduction:
The inflammatory mediator, prostaglandin D2 (PGD2) plays a role in Duchenne muscular dystrophy (DMD) pathology. PGD2 metabolite (PGDM) was shown to be a major urinary metabolite of PGD2 and to reflect its biosynthesis. The hypothesis that PGD2-mediated inflammation is involved in the pathology of DMD was examined by PGDM levels, besides PGE2 metabolite (PGEM) levels, in urine.

Methods:
The acidified urine samples (0.4ml) were added respective internal standards. The mixtures were purified by SPE. The eluted solution containing PGDM and PGEM was then applied into an API3000 LC–MS/MS system equipped with an electrospray interface. The LC–MS/MS was operated in the negative ion mode. PGDM, together with PGEM in urine was measured in the selected reaction monitoring mode.

Results:
The monitored product ion peaks of PGDM and PGEM were clearly separated in the mass chromatogram with each single peak but no interference peaks. The concentration of PGDM and PGEM was quantitated using the peak area ratios to the respective internal standard, on the basis of the calibration curve. The mean concentration was significantly higher in DMD patients than in controls. Remarkably, urinary PGDM concentrations in DMD patients showed chronological changes: it stayed slightly higher than control until 7 years old but surged at the age 8 years old. Urinary PGDM concentrations were increased and became higher with advancing age in DMD patients. It was confirmed that PGD2-mediated inflammation plays a role in the pathology of DMD and PGDM levels are influenced by physical activity, while PGEM levels are more stable.

Novel Aspect:
Clinical evaluation of prostaglandin D2 and E2 metabolites in urine was successfully performed.

Tan, Jun Hong,
Health Sciences Authority

Contributing Authors:
Dr Kee Chee Leong, Health Sciences Authority.

Title:
Identification of tadalafil-related analogue in one dietary supplement by orbitrap mass spectrometer

Introduction:
This study describes the finding of different tadalafil analogues in one dietary supplement. An unknown peak eluted very late in the high performance liquid chromatography (HPLC) alongside nortadalafil and N-ethyl nortadalafil. This unknown showed an intense peak at m/z 826 in single quadrupole liquid chromatography mass spectrometer (LCMS).

Methods:
The unknown compound was purified by semi-prep HPLC system, followed by direct infusion of the purified solution using LTQ Orbitrap XLTM hybrid FTMS system.

Results:
Further analysis by direct infusion of the purified solution using LTQ Orbitrap XLTM hybrid FTMS system discovered an intense peak at m/z 826.3015. Under high collision energy dissociation (HCD), the MS2 result showed very distinct fragments at m/z 491, 448, 336, 264 and 135. From recent literature, the accurate mass data supported the unknown to be a dimeric type of analogue, trans-bisprehomotadalafil.

Novel Aspect:
Identification of a relatively unknown tadalafil-related analogue

Tang, Qi,
University of Waterloo

Title:
Protein identification using de novo dequencing results

Introduction:
The tandem mass spectrometry (MS/MS) is widely used in proteomics nowadays to identify proteins from a sequence database. There are many de novo sequencing softwares available for sequencing peptides in relatively short running time. However, identifying proteins requires better running time for huge databases.

Methods:
Regular methods use mass spectrometry data for protein identification. We come up with a novel algorithm which use tag sequences from the result of de novo sequencing to locate proteins in sequence databases.

Results:
Our method is able to present results of same quality as regular methods within much less running time, which is very useful for protein identification from a huge sequence database. With this method, we can do sequence identification in a DNA est database, which is not avaliable for regular methods as the running time will be intolerable.

Novel Aspect:
Using string type input for protein identification, which let us be able to locate proteins from huge databases

Tanna, Sangeeta,
De Montfort University

Contributing Authors:
Graham Lawson, De Montfort University; Dennis Bernieh, De Montfort University.

Title:
Adherence to cardiovascular pharmacotherapy assessed by quantitative LC-HRMS analysis of dried blood spots

Introduction:
Up to ~50% of patients prescribed cardiovascular (CVD) drugs do not adhere to their prescribed regimen. Non-adherence impacts on patient health and healthcare system costs. Dried blood spot (DBS) sampling was used to collect micro-volume blood samples for the LC-HRMS quantitative analysis of candidate CVD drugs for objectively assessing adherence to prescribed pharmacotherapy.

Methods:
An LC-HRMS method was used for the quantification of target CVD drugs in DBS samples. The target CVD drugs were solvent extracted from disks punched from DBS and analysed using gradient chromatographic elution with a run time of 2.5 min. MS detection was carried out in electrospray positive ion mode using the precise mass for each target drug. The method was applied to volunteer DBS samples.

Results:
Drug recoveries from spiked calibration blood spots were ≥82% for all target CVD drugs except for simvastatin at ~57% and the drugs were stable in DBS for at least 12 weeks. Validation of the LC-HRMS method showed good linearity and the accuracy and precision values were within the pre-defined limits of ≤15% at all concentrations. Factors with potential to affect drug quantification measurements such as matrix effects and the volume of blood applied on the collection card were investigated. This bioanalytical method was able to identify and quantify the residual levels of target drugs up to 24 hours after the initial dose. The LC-HRMS DBS based assay successfully identified volunteers who were known to be either adherent or non-adherent.

Novel Aspect:
This LC-HRMS method is the first reported DBS assay for the simultaneous analysis and quantification of the candidate CVD drugs in human whole blood.

Tarui, Akira,
School of Materials Science, Japan Advanced Institute of Science and Technology

Contributing Authors:
Tomoyasu Fukuda, Faculty of Chemistry, Materials and Bioengineering, Kansai University; Shinya Ohki, Center for Nano Materials and Technology, Japan Advanced Institute of Science and Technology; Hideya Kawasaki Faculty of Chemistry, Materials and Bioengineering, Kansai University; Ryuichi Arakawa, Faculty of Chemistry, Materials and Bioengineering, Kansai University; Issey Osaka, Center for Nano Materials and Technology, Japan Advanced Institute of Science and Technology.

Title:
2D-patterning technique for analysis of post-translationally modified proteins using LC-MS with different chromatographic separations

Introduction:
2D LC (LCxLC) -MS has been developed for proteomic analysis. In the LCxLC-MS, small amount of eluant trapped in a sample loop during separation of 1st column has to be separated in 2nd column with gradient LC. Therefore, long-time analyses were needed in the LCxLC-MS. In this study, we have developed rapid and simple 2D patterning technique using LC-MS with different chromatographic separations.

Methods:
Two chromatograms of protein digests were obtained by LC-MS with different chromatographic separations differed in mobile phase, columns in LC, or ionization methods of MS. The obtained retention time (tR), m/z, and ion intensities were digitalized. The tR with the same m/z value were plotted along the abscissa and the ordinate for each LC-MS condition, originating a scatter diagram of 2D map.

Results:
Tryptic digests of casein were prepared with or without dephosphorylation. 2D maps of them were obtained by LC-ESI-MS or LC-MALDI-MS. 2D map was prepared to integrate two LC-MS chromatograms varied in pH of mobile phase in LC. The phosphorylated peptide could be discriminated from non-phosphorylated peptide simply to compare the 2D map of casein digests with that of dephosphorylated casein digests. When the peptide was observed near a diagonal line in the 2D map, its pI was close to 7. That indicates 2D map could describe not only the component pattern of protein digests but physicochemical property of each peptide simply. Moreover, tryptic digests of ribonuclease B were also prepared with or without deglycosylation. 2D maps of them were obtained by LC-MALDI-MS varied in polarity of column in LC. Similarly, glycosylated peptide and oligosaccharide were discriminated simply and rapidly.

Novel Aspect:
Rapid and simple 2D patterning technique using LC-MS with different chromatographic separations

Tawa, Paul,
Merck & Co

Contributing Authors:
Jane Wen, Merck & Co, Department of In-Vitro Pharmacology; Todd Mayhood, Merck & Co, Department of In-Vitro Pharmacology; Aileen Soriano Merck & Co, Department of In-Vitro Pharmacology; Deborah Zink, Merck & Co, Department of In-Vitro Pharmacology; Payal Sheth, Merck & Co, Department of In-Vitro Pharmacology.

Title:
A High-Throughput LC/MS Biochemical Assay for Monitoring Activity of the Early Wall Teichoic Acid Pathway Gene Product TarO

Introduction:
Previous assays for monitoring the activity of bactoprenol-phosphate conjugating enzymes invariably suffer from one or more limitations associated with throughput, sensitivity, or the handling of regulated material. We describe optimization of biochemical assay conditions and the development of an LC/MS method to quantify the product (C55-P-P-GlcNAc) of the TarO-catalyzed reaction.

Methods:
Using S. aureus membranes as a source of TarO activity, we optimized assay conditions in order to maximize enzyme activity, inhibitor potency, and LC/MS monitoring of the reaction product C55-P-P-GlcNac. The LC/MS method was optimized to separate the product C55-PP-GlcNAc from the substrate C55-P as much as possible. Mass spectrometric detection was carried out in negative-ion mode MRM.

Results:
Detergent type and concentration was found to be important in order to maximize the amount TarO activity observed as well as the performance of the LC/MS system. A final concentration of 6 mM CHAPS was established as ideal since it maximized TarO activity while minimizing 1) nonspecific activity, 2) shifts in the apparent potency of certain inhibitors (Tunicamycin), and 3) degradation of the LC/MS response. A membrane concentration of 0.1 mg/ml total protein was chosen because this concentration was determined to be the lowest concentration that still preserved satisfactory assay performance (Z' > 0.5). The Km's for UDP-GlcNAc (1500 uM) and C55-P (76 uM) were deduced and used for routine screening. Correlations between intrinsic compound potencies determined using this LC/MS assay and the activity of compounds in in-vitro antibacterial assays will be presented and discussed.

Novel Aspect:
Quantifying the activity of a bactoprenol-phosphate modifying enzyme using a label-free high-throughput assay.

Tellstroem, Verena,
Bruker Daltonics

Contributing Authors:

Xian Luo, Department of Chemistry, University of Alberta, Edmonton, AB; Wei Han, Department of Chemistry, University of Alberta, Edmonton, AB; Zhendong Li Department of Chemistry, University of Alberta, Edmonton, AB; Dorothea Mung, Department of Chemistry, University of Alberta, Edmonton, AB; Yunong Li, Department of Chemistry, University of Alberta, Edmonton, AB; Shuang Zhao, Department of Chemistry, University of Alberta, Edmonton, AB; Liang Li, Department of Chemistry, University of Alberta, Edmonton, AB; Verena Tellstroem, Bruker Daltonics GmbH, Bremen, Germany; Aiko Barsch, Bruker Daltonics GmbH, Bremen, Germany; Kevin Hooton, Department of Chemistry, University of Alberta, Edmonton, AB; Jaspaul Tatlay, Department of Chemistry, University of Alberta, Edmonton, AB; Adriana Buzatto, Department of Chemistry, University of Alberta, Edmonton, AB; Nan Wang, Department of Chemistry, University of Alberta, Edmonton, AB; Ulrike Schweiger-Hufnagel, Bruker Daltonics GmbH, Bremen, Germany; Yiman Wu, Department of Chemistry, University of Alberta, Edmonton, AB; Xiaohang Wang, Department of Chemistry, University of Alberta, Edmonton, AB.

Title:

Enabling High-Confidence Human Endogenous Metabolite Identification via High-Resolution MS/MS Retention-Time Library

Introduction:

Metabolite identification remains a major bottleneck in metabolomics. Our work focuses on building a human metabolite library. It contains multiple layers of information including accurate masses, high-resolution MS/MS spectra and normalized retention time in both RPLC and HILIC separations. We also illustrate the application of this library for rapid metabolite identification in human biofluids.

Methods:

831 standards were obtained from the Human Metabolite Database (HMDB). These were analyzed via a Dionex RSLC (Thermo Fisher) and detected by an impact HD Q-TOF MS (Bruker Daltonics). For retention time measurement, metabolite standards were pooled into mixtures. Each mix was spiked with a retention-time standard and separated on a RP or HILIC column.

Results:

MS/MS spectra acquired in positive and negative ion modes were manually curated before adding to the library. The addition of retention-time information significantly improved the identification confidence, especially for low intensity or isobaric compounds and isomers. 20-minute linear solvent gradients for reversed-phase (RP) and hydrophilic interaction (HILIC) separation were established. We pooled 32 compounds which were evenly spaced during the separations as a retention-time (RT) standard mixture. Reproducibility of the gradient was assessed using this mixture, and the RT standard deviations were found to be within ±1.3 seconds (n = 5). Untargeted LC-MS/MS analyses were performed on urine and serum to illustrate the performance of this library. Matching both MS/MS spectra and retention time enabled rapid and high-confidence identification of compounds in both matrices.

Novel Aspect:

High-confidence identification of human endogenous metabolites using high-resolution MS/MS retention-time library

Tellstroem, Verena,
Bruker Daltonics

Contributing Authors:

Verena Tellstroem, Bruker Daltonics GmbH, Bremen, Germany; Aiko Barsch, Bruker Daltonics GmbH, Bremen, Germany; Christian J. Wachsmuth University of Regensburg, Institute of Functional Genomics, Regensburg, Germany; Katja Dettmer, University of Regensburg, Institute of Functional Genomics, Regensburg, Germany; Peter J. Oefner, University of Regensburg, Institute of Functional Genomics, Regensburg, Germany.

Title:

A step forward in GC-HRAM-MS based metabolomics - A novel atmospheric pressure GC-APCI Source increases quantitative and qualitative performance

Introduction:

Common requests in GC-MS based metabolomics are an increased sensitivity for detecting metabolites and for identification of unknown compounds. Electron ionization (EI) is the standard ionization for GC-MS. Soft ionization by atmospheric pressure chemical ionization preserves molecular ion information enabling identification of unknowns, a reason why GC-APCI became more popular recently.

Methods:

Standards and cell culture supernatants of MiaPaCa-2 pancreatic cancer cells were derivatized and samples were separated using a Rxi-5ms (Restek) GC column which was coupled to a novel APCI source and analysed in an oTOF-MS (micrOTOF , Bruker Daltonics). Data was acquired with a rate of at least 3 spectra per second. PFTBA was injected into the APCI source as a calibrant for data recalibration.

Results:

Here we will report the application of a novel GC-APCI source design coupled with a high resolution, accurate mass oTOF to analyse derivatized metabolite extracts and reference standards. In comparison with earlier results, the analytical performance could be improved resulting in a higher number of compounds identified in human cancer cell extracts. Additionally, lower limits of quantification in the sub-micromolar range were found for twenty metabolites. At the same time, the analytical linear response range was either maintained or slightly increased. In summary, the novel GC-APCI source is expected to reveal more significantly regulated metabolites in metabolomics research and to enable their sensitive quantification by low LLOQs, broad linear ranges and decreased relative standard deviations of replicates analyses.

Novel Aspect:

Improved sensitivity and unknown ID capabilities in GC-MS based metabolomics by novel GC atmospheric pressure chemical ionization (APCI) source.

Teramoto, Kanae,
JEOL Ltd.

Contributing Authors:
Robert Cody, JEOL USA, Inc.; Hiroaki Sato, National Institute of Advanced Industrial Science and Technology, AIST; Sayaka Nakamura National Institute of Advanced Industrial Science and Technology, AIST.

Title:
A Combination of On-plate Degradation and Kendrick Mass Defect Analysis for Characterization of High Molecular-weight Polyesters by MALDI spiral-TOFMS

Introduction:
MALDI TOF-MS with high mass-resolution is used for the detailed characterization of polymers. However, accurate mass analysis is limited to low molecular-weight compounds (basically < 3000 Da). In this study, on-plate degradation was attempted to reduce the molecular weight of copolymers. Kendrick mass defect (KMD) plot analysis was applied to visualize the distribution of copolymer compositions.

Methods:
Biodegradable polyester samples were deposited on the target plate with NaOH-methanol solution applied to the surface of the sample spot for on-plate degradation. After removing NaOH with water, DHB matrix was pipetted onto the sample spot and air-dried. MALDI mass spectra were obtained with a MALDI spiral-TOFMS (JEOL JMS-S3000). KMD plots were constructed with msRepeatFinder software (JEOL).

Results:
Poly(3-hydroxybutyrate) [P(3HB)] and poly(3-hydroxybutyrate-co-3-hydroxyvalerate) [P(3HB-co-3HV)] with MW > 10,000 were used as model samples. The pristine samples gave few peaks in the lower m/z range. In contrast, many peaks were observed in the same range after the on-plate degradation procedures. Interestingly, the observed peaks were assigned as [M + Na]+ ions of the cyclic oligomers based on the accurate masses for each sample. However, peaks corresponding to the methyl-esterified molecules were essentially absent. These results suggested that decomposition of the samples into low molecular-weight fragments would be caused by intramolecular transesterification reaction. The KMD plots were constructed using high-resolution mass spectra. The distributions of copolymer compositions were effectively displayed by a combination of the KMD plot and the results of the peak assignment.

Novel Aspect:
On-plate degradation and high-resolution mass spectrometry by MALDI spiral-TOFMS allows characterization of polyesters with high molecular-weight.

Tessier, Luc,
National Research Council

Contributing Authors:
Arsalan Haqqani, National Research Council.

Title:
A Novel nanoLC-MRM Workflow For Analyzing Deoxycholic Acid (DOC)-Solubilized Protein Digests From Biological Samples

Introduction:
Absolute quantitation of therapeutic antibody at the site of its action is an essential step in antibody development, providing evidence of delivery and concentration of the therapeutic at target site. Use of DOC has been investigated to increase protein solubilisation. Developing a nanoLC-MRM workflow allows absolute quantitation and identification of these antibodies with high sensitivity.

Methods:
A Waters nano-Acquity (Binary Solvent Manager, Sample Manager with column compartment), is modified by replacing the trap valve of the Sample Manager with a second injection pod valve. The LC plumbing keeps the trapping injection cycle isolated from the nano-column avoiding possible contamination of the analytical column, and heated blanks (55 C) are used to ensure column regeneration.

Results:
The modified nano-Acquity, in combination with the use of heated blanks (55 C) and an increase in analytical nanoflow (going from 400 nL/min to 600 nL/min), greatly reduced system pressure variability, column contamination, and widening of chromatographic peaks. System pressure variability after just 5-6 deoxycholic acid containing samples could be higher than 25%, but with the modifications, that number drops below 5%. Peak retention times were also stabilized when using the modified nano-Acquity, and peak shape resolution greatly improved: from 0.8 minutes baseline width after as low as 2-3 deoxycholic acid containing sample runs, to 0.25 minutes after > 24 samples. In addition, the reduction amount of deoxycholic acid sprayed onto the mass spectrometer source has reduced background and contaminating peak in the mass spectra.

Novel Aspect:
Utilizing an extra 6-port valve injection pod valve, and making use of the column heater, we have increased the overall robustness of the nano-LC.

Theiner, Sarah,
Insitute of Analytical Chemsitry, University of Vienna

Contributing Authors:
Bernhard K. Keppler, Institute of Inorganic Chemistry, University of Vienna, Vienna, Austria; Luis Galvez Montano, Insitute of Analytical Chemistry, University of Vienna, Vienna, Austria; Marko Grabarics Institute of Analytical Chemistry, University of Vienna, Vienna, Austria; Gunda Koellenspeger, Insitute of Analytical Chemistry, University of Vienna, Vienna, Austria.

Title:
The fate of platinum(IV) anticancer prodrugs in different biological fluids by LC-ICP-MS

Introduction:
Pt(IV) complexes as potential anticancer agents offer advantages over their Pt(II)-based analogues, applied in clinics. Acting as prodrugs, Pt(IV) complexes are assumed to stay in their intact form until they reach the tumor cells, where they should be reduced to release their active species. Elemental speciation analysis offers the ideal tool-set for studying the reduction in vitro and in vivo.

Methods:
An analytical approach based on hydrophilic interaction liquid chromatography (HILIC) combined to inductively coupled plasma-tandem mass spectrometry (ICP-MS) and reversed phase (RP-)LC-ICP-MS was developed to study the stability and formation of different products of platinum(IV) complexes in different biological context.

Results:
In the current study, we aimed to investigate the kinetic inertness of platinum(IV) prodrugs with varying ligand spheres in different biological fluids such as blood and serum. RP- and HILIC-ICP-MS will be discussed regarding their analytical figures of merit. The finally applied workflow consisted in extraction by organic solvents and monitoring of platinum species by HILIC-ICP-MS being highly compatible with this sample preparation method. The procedure was cross-validated by RP-ICP-MS to study the stability of Pt(IV) compounds. Metabolisation was observed and interestingly, the kinetic inertness of Pt(IV) prodrugs strongly depends on the biological fluid used as well as on the ligand sphere of the Pt(IV) complexes.

Novel Aspect:
HILIC hyphenated to ICP-MS was applied for the first time to monitor the formation of metabolites of platinum(IV) complexes in biological fluids.

Thibault, Pierre,
Universite de Montreal/IRIC

Contributing Authors:
Peter Kubiniok, IRIC and Department of Chemistry, Universite de Montreal; Alison McCracken, Department of Developmental and Cell Biology, University of California at Irvine; Michael Perryman Department of Chemistry, Universite de Montreal; Stephen Hanessian, Department of Chemistry, Universite de Montreal; Aimee Edinger, Department of Developmental and Cell Biology, University of California at Irvine; Pierre Thibault, IRIC and Departement of Chemistry, Universite de Montreal.

Title:
Deconvolution of drug-specific phenotypes associated with nutrient transport using dynamic phosphoproteomics

Introduction:
Novel sphingolipid analogs like SH-BC-893 impair nutrient transport, and offer an interesting strategy to starve cancer cells to death. While both ceramides and SH-BC-893 disrupt transporter trafficking by activating protein phosphatase 2A (PP2A), SH-BC-893 uniquely blocks lysosomal fusion. Here, we used dynamic phosphoproteomics to understand how drugs affecting PP2A activity shape phenotypes.

Methods:
Murine hematopoietic cells (FL5.12) were cultured in the presence of SH-BC-893, LB100 (PP2A inhibitor), or C2-ceramide for up to 1h using SILAC media. Dynamic phosphoproteomic experiments were performed over a period of 1h with a time resolution of 5 min. Phosphopeptides were extracted using TiO_2 beads and fractionated on SCX stage tips prior to LC-MS/MS analysis on an Orbitrap Fusion.

Results:
Dynamic phosphoproteomics of FL5.12 cells quantified more than 14,000 phosphorylation sites, of which 647 sites were dynamically regulated. The comparison of phosphorylation sites regulated by LB-100 and SH-BC-893 identified 116 sites with opposite behaviors corresponding to known (i.e. KSR1 and PRKAR1A) and novel (e.g. ARHGEF1 and CD2AP) PP2A substrates. Dynamic phosphoproteomics data from C2-ceramide and SH-BC-893 revealed 123 regulated sites common to both stimulations such as GEFs (ARHGEF2), GTPases (AGAP2) and actin regulators (FNBP1 and FMNL1). We also identified several proteins involved in vesicular trafficking and lysosomal fusion events that are dynamically regulated only with SH-BC-893. Dynamic phosphoproteomics of FL5.12 cells treated with compounds affecting PP2A activity enabled the correlation of cell signaling with phenotypes to rationalize their mode of action.

Novel Aspect:
Dynamic phosphoproteomics enables the correlation of cell signaling with phenotypes to rationalize the mode of action of drugs affecting PP2A activity

Thomas, Gilian,
Carleton University

Contributing Authors:
Hillary Weinert, Department of Chemistry, Carleton University, Ottawa, Ontario, Canada; Samuel W.J. Shields, Department of Chemistry, Carleton University, Ottawa, Ontario, Canada; Jeffrey C. Smith Department of Chemistry, Carleton University, Ottawa, Ontario, Canada; Jeffrey M. Manthorpe, Department of Chemistry, Carleton University, Ottawa, Ontario, Canada; Carlos R. Canez, Department of Chemistry, Carleton University, Ottawa, Ontario, Canada.

Title:
Optimization of TrEnDi derivatization to enhance sphingomyelin signal in mass spectrometry-based studies

Introduction:
TrEnDi is a derivatization method that applies a permanent positive charge to lipids, thus increasing visibility and sensitivity for detection. However, sphingomyelin (SM) exhibits signal splitting due to partial methylation of the hydroxyl moiety which limits the full potential of TrEnDi. Here we investigate approaches aimed at ameliorating TrEnDi-based signal splitting of SM.

Methods:
Solvent conditions, acid strength, amount of diazomethane, and incubation times were tested during the methylation reaction on SM (16:0/18:1) to investigate their effects on peak splitting and improved sensitivity. Lipids were derivatized in other ways prior to TrEnDi in order to protect the hydroxyl group from being methylated; acid chloride, and dimethylglycine derivatizations were tested.

Results:
TrEnDi conditions different from what have been previously reported are required in order to drive the reaction to completion and optimize detection sensitivity for SM. It was found that an alcohol is required to mediate the TrEnDi reaction and prevent excess methylation. Overall, methanol resulted in less signal splitting compared to ethanol; however, in every case it was evident that the hydroxyl moiety on the sn-1 chain of SM requires protection from methylation by diazomethane. As a result the abilities of acid chloride and dimethylglycine derivatizations reactions to protect the sn-1 hydroxyl moiety were investigated. It was found that the dimethylglycine derivatization was successful, leaving only approximately 1% of the lipid unmodified. However, further studies are currently being pursued to examine the effectiveness of this derivatization in combination with TrEnDi.

Novel Aspect:
First study to optimize TrEnDi chemistry with sphingomyelin lipids by examining the effect of alcohol and complementary derivatization techniques.

Thurlow, Sophie,
LGC

Contributing Authors:
Kate Groves, LGC; Chris Hopley, LGC; Emily Schepens LGC; Bryan McCullough, LGC; Milena Quaglia, LGC.

Title:
The detection of allergens in foods using 2D-LC-MS/MS

Introduction:
Contamination of the food supply chain with allergenic species poses a risk to human health. Typically the allergenic molecules are proteins whose presence is confirmed by ELISA. Although ELISA offers excellent sensitivity to low levels of contaminant protein allergens, the selectivity of immunoassays can be compromised by issues with cross reactivity particularly in a complex matrix.

Methods:
Samples of spices and possible nut contaminants were subjected to tryptic digests facilitating a bottom-up proteomics analytical work flow. DDA experiments were performed on an Agilent 6530 Q-ToF instrument to identify peptides distinctive for the contaminants. A SRM method was applied on an ABSciex QTrap 4000 allowing for the detection of the distinctive peptides in the contaminated spice sample.

Results:
An Agilent Infinity II 2D-LC system was operated using RPxRP high-low pH separations in the first and second dimensions respectively. When operating the system in fully comprehensive mode and identifying peptides of interest using DDA experiments it was possible to characterize distinct peptides that can be used to distinguish between closely related species of nut. Targeting these distinctive peptides in a SRM method can be done in either fully comprehensive or heart cutting modes to further enhance the separation space and therefore the sensitivity of the bottom-up proteomics approach when detecting low levels of peptides of interest in a highly complex matrix. This can be achieved with a tolerable increase in the sample analysis time.

Novel Aspect:
This work describes the improved confidence in the identification of trace level contaminants of allergens in food offered through the use of 2D-LC.

Tierbach, Alena,
Eawag

Contributing Authors:

Marc J-F Suter, Eawag, Swiss Federal Institute of Aquatic Science and Technology, 8600 Dübendorf, Switzerland; Kristin Schirmer, Eawag, Swiss Federal Institute of Aquatic Science and Technology, 8600 Dübendorf, Switzerland; Ksenia J Groh Eawag, Swiss Federal Institute of Aquatic Science and Technology, 8600 Dübendorf, Switzerland.

Title:

Expression of Glutathione S-Transferases in the developing zebrafish embryo - a targeted proteomic approach

Introduction:

Zebrafish embryos are increasingly employed as alternative to the conventional acute fish toxicity test. One important requirement for the establishment of an alternative test model is that both models should be comparable in terms of uptake and biotransformation of xenobiotics. However, little is known about the expression of xenobiotic metabolizing enzymes in the developing zebrafish embryo.

Methods:

In this study, we have developed a targeted proteomic approach to characterize the expression of Glutathione S-transferases. Selected members of the family were monitored in zebrafish embryos by mass spectrometry on the basis of proteotypic peptides and peptides characteristic for the enzyme groups of interest.

Results:

The basal expression of GSTs on the protein level was investigated at 4, 8, 24, 48, 72, 96, 120 hpf and compared to adult liver samples. Samples collected at 4 and 8 hpf were used to estimate the relevance of maternal mRNA transfer in zebrafish eggs - samples from the later time points give an insight into the dynamic of GST expression throughout the development. The expression pattern of selected GST enzymes is presented including members of the alpha and pi class – candidates that have been suggested to be involved in xenobiotic metabolism.

Novel Aspect:

This study will help to fill existing knowledge gaps regarding the comparability of metabolic capacity of fish at different life stages.

21ST INTERNATIONAL MASS SPECTROMETRY CONFERENCE
Toronto, Ontario, Canada
August 20 – 26, 2016

Tillner, Jocelyn,
Imperial College

Contributing Authors:
James McKenzie, Imperial College; Anna Mroz, Imperial College; Tamas Karancsi Waters Corporation; Ian Gilmore, National Physical Laboratory; Dipa Gurung, Imperial College; Emrys Jones, Waters Corporation; Zoltan Takats, Imperial College; Steve Pringle, Waters Corporation; Joephine Bunch, National Physical Laboratory.

Title:
Back to basics - revisiting sprayer design for desorption electrospray ionisation mass spectrometry imaging (DESI-MSI) of biological tissues

Introduction:
Despite its growing application in various fields, the setup for DESI has remained unchanged since its original publication, including the sprayer, potentially the largest source of reproducibility issues. Here, we examine the impact of sprayer geometry on DESI-MSI performance and develop a new sprayer design. We also assess desorption electroflow focussing ionisation (DEFFI) as an alternative.

Methods:
The impact of sprayer geometry on tissue MS signal was tested. Based on these results, a new DESI sprayer with a centred, small I.D., tapered solvent capillary was designed and assessed for absolute intensity, spectral repeatability and imaging spatial resolution.The repeatability of DEFFI was evaluated and data were compared with DESI, both for spectral composition and imaging spatial resolution.

Results:
Absolute intensity and spectral repeatability were found to vary greatly with the geometry of the DESI solvent capillary. Orientation towards the MS inlet was ideal, while sideways rotations led to signal reduction and orientation away from the inlet was detrimental. A centering device was designed to fix the solvent capillary geometry without obstructing the nebulising gas flow, improving absolute intensity repeatability and spectral repeatability up to tenfold. For tissue imaging, spatial resolutions of 20 µm were achieved. The DEFFI sprayer, which produces a rotationally symmetrical electrospray, provided improved absolute intensity and spectral pattern repeatability compared to the conventional DESI sprayers. Spectra were found to be similar to those from DESI, suggesting a comparable mechanism of ionisation. The spatial resolution of DEFFI and DESI was similar.

Novel Aspect:
We present the first systematic investigation and optimisation of the DESI sprayer for biological tissue imaging and reliable DESI-MS.

TINTARU, Aura,
Aix-Marseille Université, CNRS, Institut de Chimie Radicalaire-UMR7273

Contributing Authors:
Jérémy DONON, Aix-Marseille Université, CNRS, Institut de Chimie Radicalaire-UMR7273; Ling PENG, Aix-Marseille Université, CNRS, Centre Interdisciplinaire de Nanoscience de Marseille-UMR7325; Mei CONG Aix-Marseille Université, CNRS, Centre Interdisciplinaire de Nanoscience de Marseille-UMR7325; Laurence CHARLES, Aix-Marseille Université, CNRS, Institut de Chimie Radicalaire-UMR7273.

Title:
ESI-MS/MS study of drug-conjugated dendrimers

Introduction:
In contrast to drug/dendrimer nanomicelles, conjugates consisting of amphiphilic dendrimers with drug-ended branches allow a better control of drug loading. In this context, novel dendrimers bearing biologically active triazole nucleosides covalently linked to terminal groups have been designed. Their highly symmetrical structure remains however a main issue for their accurate characterization.

Methods:
All experiments were performed in the positive ion mode, using electrospray ionization and high resolution mass spectrometry. Conjugated dendrimers were solubilized in acidified methanol to promote formation of protonated species. MS/MS experiments were conducted using collision-induced dissociation.

Results:
Mass spectrometry is shown to offer unique advantages for a thorough and accurate characterization of dendrimer conjugates samples. Using electrospray ionization to generate intact species in the gas phase, data obtained in the MS mode readily allowed the number of drug moieties actually attached to targeted dendrimer and indicated any defective molecules in the sample. In addition, once the dissociation behavior of these conjugates has been established, MS/MS data permitted to confirm precise location of drug moieties at the terminus part of each dendrimer branches. This analytical approach was successfully implemented to characterize G1 and G2 generation molecules. The MS/MS rules established for perfect species could also usefully be employed to determine the structure of dendrimer impurities.

Novel Aspect:
MS/MS structural characterization of drug/dendrimer covalent conjugates and associated defective analogous.

Tomczyk, Nick,
Waters Corporation

Contributing Authors:
Arthur Moseley, Duke University School of Medicine; Chris Hughes, Waters, Wilmslow; Erik J. Soderblom Duke University School of Medicine; Keith Richardson, Waters, Wilmslow; Will Thompson, Duke University School of Medicine; Jason Wildgoose, Waters, Wilmslow; James Langridge, Waters, Wilmslow.

Title:
Qualitative and Quantitative Characterization of a Novel DIA Method for Omics Analysis and its Application to Biomedical Analyses

Introduction:
In typical data independent acquisition (DIA) methods on Q-TOF instruments, the quadrupole mass filter or ion trap either operates in wide pass mode or in stepped mode with typical transmission windows in the range of 1-20Da. Here we describe a mode of DIA operation whereby a resolving quadrupole is scanned repetitively over alternating low and elevated energy scans

Methods:
The m/z range of the quadrupole was continuously and repetitively scanned with data acquired using a ToF acquisition system capable of delivering 2000 ToF spectra / s. Alternate scans contain low energy data for precursor and high energy data for fragment ions. The resulting 2D data format can be processed using both commercial and open source software for identification / quantitative results

Results:
The effect of the scanning quadrupole transmission window has been investigated to assess qualitative performance. Tryptic digest standards were injected onto a LC system and separated using a 90 minute gradient. It was found that quadrupole transmission windows of 20-30 Da provided optimum protein identifications. Over 1,000 proteins were identified from a cytosolic E.coli digest standard (4% FDR). Additional evaluations of this methodology for qualitative and quantitative proteomic analyses will be made via the analyses of two disparate sample cohorts - characterization of synaptic proteomes as part of a study of developmental brain disorders, and differential analyses of protein:protein complexes of calcineurin (Aspergillus fumigatus) as a function of mechanism-of-action of several antifungal drugs.

Novel Aspect:
DIA method incorporating scanning quadrupole provides greater specificity in 'Omics experiments compared with default DIA acquisition method

Touboul, David,
CNRS ICSN

Contributing Authors:

Roussi, CNRS ICSN UPR2301, France; Touboul, CNRS ICSN UPR2301, France; Méjean CNRS ICSN UPR2301, France; Litaudon, CNRS ICSN UPR2301, France; Laboureur, CNRS ICSN UPR2301, France; Nothias, University of California - San Diego; Brunelle, CNRS ICSN UPR2301, France; Champy, Pharmacy Faculty, Chaenay Malabry, France.

Title:

Supercritical fluid chromatography coupled to high resolution mass spectrometry for dereplication and quantification of lipids

Introduction:

Taking into account growing public concerns on environmental questions, analytical studies and the consequent use of toxic solvents have increased to such a level that it becomes untenable to continue without changing towards green methods. One promising way is the use of supercritical fluid CO2 mixed with ethanol as green eluents. SFC coupled to HRMS/MS offers a unique tool for lipid analysis.

Methods:

SFC-MS analyses were performed using a 1260 Infinity Analytical SFC system coupled to a quadrupole time-of-flight mass spectrometer 6540, equipped with ESI, APCI, and APPI sources, all from Agilent Technologies. To improve reproducibility and robustness, a make-up solvent was added between the UV detector and the back pressure regulator. Molecular networks were generated using GNPS web platform.

Results:

First, quantification of tocopherols and tocotrienols (vitamin E) in soybean oil was performed by SFC-UV-HRMS. After column screening and chromatographic optimization, APPI appeared more sensitive and robust than ESI and APCI. This method led to short analysis time (< 5 min) with limits of quantification in the tens of µg /L. Second SFC-HRMS/MS was used for the dereplication of diterpene esters from Euphorbia extracts. MS/MS data were classified using the molecular network approach. Unknown jatrophane esters and phorboids, bearing inhibition of chikungunya virus replication, were put in light, isolated and fully characterized. Finally, a unique set-up for post-column cationization by lithium salts has been optimized. Dereplication of acetogenins from Annona muricata L. was performed with similar sensitivity but better LC resolution and a reduced analysis time compared to LC-MS/MS.

Novel Aspect:

SFC-HRMS/MS is efficient for the structural characterization and quantification of various lipids sub-classes (vitamin, diterpene and polyketides).

Troiani, Anna,
Sapienza Rome University, Italy

Contributing Authors:
Marzio Rosi, Dipartimento di Ingegneria Civile e Ambientale, University of Perugia and ISTM-CNR; Chiara Salvitti, Dipartimento di Chimica e Tecnologie del Farmaco, "Sapienza" University of Rome; Giulia de Petris Dipartimento di Chimica e Tecnologie del Farmaco, "Sapienza" University of Rome; Stefania Garzoli, Dipartimento di Chimica e Tecnologie del Farmaco, "Sapienza" University of Rome.

Title:
Reaction of [(η5-C5H5)-M]+ (M = Mn, Fe, Co) with Dichloromethane in the Gas-Phase: C-C Bond Formation and Production of MCl2

Introduction:
Reactions of cyclopentadienylic transition metal ions with substrates can be used to obtain C–C bond formation or C–H bond activation. To gain insight into the intrinsic features of these ligated metal ions in the gas phase, we recently reported on the reaction of (η5-C5H5)-Fe+ with CH2Cl2 that disclosed an unusual iron transfer and C-C bond formation, giving protonated benzenium ions and FeCl2.

Methods:
The experiments were carried out using a combination of mass spectrometric techniques. An in-house modified linear quadrupole ion trap mass spectrometer was used to measure rate constants. Collisionally Activated Dissociation and Neutral Fragment Reionization experiments were performed using a modified ZABSpec oa-TOF double-focussing sector field instrument.

Results:
Here we report on the activation of dicholoromethane by (η5-C5H5)-M+ ions (M = Mn, Co) as an extension of our interest in cyclopentadienylic transition metal ions. The reactions have been studied in the gas phase at room temperature by mass spectrometry in the high pressure regime characteristic of ion trap experiments (ITMS). Results have been compared with those obtained for the (η5-C5H5)-Fe+/CH2Cl2 system. The reactions lead to the same products, namely protonated benzenium ions and the corresponding metal chloride. Main differences are found in the cobalt system, which shows a higher formation of the addition ion [(η5-C5H5)-Co-CH2Cl2]+ that is not observed with manganese and only poorly with iron. The rate constant, efficiency and the branching ratio of all products of the reactions have been measured.

Novel Aspect:
Fast and efficient reactions between CH2Cl2 and (η5-C5H5)M+ were reported. Both products, MCl2 and C6H7+, were detected by independent experiments.

Trubitsyn, Andrey,
Ryazan State Radio Engineering University

Contributing Authors:
S.B. Bimurzaev, Almaty university of power engineering and telecommunication; M.V. Dubkov, Ryazan State Radio Engineering University.

Title:
Numerical simulation of time-of-flight analyzers

Introduction:
Unlike analytical techniques, numerical ones provide an analysis of the systems which are close to real systems. The numerical technique of condition search of the high order time-of-flight focusing is proposed for electron optic systems with arbitrary electrode configuration. The technique allows us defining a value of the independent parameter providing a high order time-of-flight focusing.

Methods:
Mathematical order of the time-of-flight focusing is expressed by a number of zero derivatives in expansion of the particle flight time from a source to a collector according to initial data. Focusing plane (collector) position is found as a ratio of values of first derivatives of the flight time and final velocity at points corresponding to the found central value of the initial parameter.

Results:
Suggested method is integrated into the copyright computer application FOCUS. Testing of the method has been carried out on model tasks with known solutions, in particular, numerical analysis of time-of-flight mass-spectrometer (TOF MS) with a straight optical axis has confirmed all conclusions of analytical researches. Additionally influence of the wall thickness and output window diameter of ionization area and also impact of values of interelectrode gaps on focusing properties of concerned TOF MS has been studied.

Novel Aspect:
The central value of initial parameter (energy, angle or position) is based only on trajectory analysis.

Trubitsyn, Andrey,
Ryazan State Radio Engineering University

Contributing Authors:
V.S. Gurov, Ryazan State Radio Engineering University.

Title:
SOFTWARE FOCUS FOR SIMULATION OF STATISTICAL AND DYNAMICAL SYSTEMS OF ION OPTICS

Introduction:
Main method to optimize parameters and to improve a design of mass-spectrometers is mathematical simulation. The present paper has represented a copyright computer application for simulation of axially symmetric and planar systems of ion optics, including statistical and dynamical mass-analyzers. Theoretical base of the software is original mathematical methods.

Methods:
Method of boundary elements of the Dirichlet's outer problem solution has been developed. Method for calculation of magnetic fields of a set of arbitrarily oriented solenoids has been suggested. Solution of equations of ion motions in electromagnetic fields is carried out by the Runge-Kutta schemes. Methods for search of terms for spatial and time-of-flight focusing have been developed.

Results:
FOCUS is represented as a combination of modules – 1) formation and modification of the design, 2) calculation of the electrostatic field, 3) calculation of the magnetic field, 4) trajectory analysis of statistical and 5) dynamical systems of ion-optical systems (IOS). Cross-section of IOS electrodes can be represented by a combination of the following primitives – segment, circular arc, parabola, hyperbola, spline, rectangle and ellipse. Error of the potential calculation in areas of the practically random form by the method of boundary elements in real time does not exceed 0.0000001. Calculation of ion trajectories is carried out controlling an accuracy of calculations. Methods for search of terms for spatial and time-of-flight focusing allow evaluating an order of focusing, initial data of ion motions and collector position under which the present level of focusing is provided.

Novel Aspect:
Proposed soft allows designing ion-optical systems according to the criterion "order of the spatial and time-of-flight focusing".

Tseng, Mei-Chun,
Institute of Chemistry Academia Sinica

Contributing Authors:
Jia-Hui Wang, Department of Chemistry, National Taiwan Normal University, Taoyuan, Taiwan; Ling-Chieh Lee, Department of Chemistry, National Taiwan Normal University, Taoyuan, Taiwan; Yu-Ju Chen Institute of Chemistry, Academia Sinica, Taipei, Taiwan; Cedric Po-Wen Chung, Institute of Chemistry, Academia Sinica, Taipei, Taiwan.

Title:
Functional Carbon Surface for Structural Characterization of Disaccharides by Matrix-assisted Laser Desorption/Ionization-Mass Spectrometry

Introduction:
The functionality of saccharides depends on their degree of isomeric diversity, which results from subtle changes in their sequential arrangements. Herein, we have demonstrated that functional carbon materials can be employed as the MALDI matrix to characterize the structure differences of saccharides without any chemical derivatization or using tandem mass spectrometry.

Methods:
unctional carbon materials were developed through hard-template nano-casting, followed by post-chemical treatments. The carbon materials were mixed with different types of disaacharides. Then, the solution of the mixture was loaded on the MALDI-plate for the identification and structure analysis as described in the MALDI-I-TOF MS analysis.

Results:
We compared the ionization efficiency and extent of fragmentation by using conventional matrix and carbon materials with various concentrations. At 750 ppm concentration of carbon materials, the dissacharides sodium adducted molecular ion peaks, were predominantly present with relatively clean background. For the fingerprint fragments for the differentiation of six linkage disaccharides, the unique product ion m/z = 316.4 and m/z = 248.1 served to distinguish 1,3 linkage and 1, 4 linkage and the other diagnostic fragments. Interestingly, we observed that m/z = 201.1 fragments only occurred as a signature pattern to determine ß anomeric configuration. Based on these specific fragmentations, we can easily deduce the structure of the carbohydrates by using carbon materials without employing the MS/MS function of the instrument.

Novel Aspect:
The specific fragment pattern of disaccharides was observed by using functional carbon materials without using tandem MS/MS.

Tshuma, Nkazimulo,
University of Nottingham

Contributing Authors:
Neil Oldham, School of Chemistry, University of Nottingham, University Park, Nottingham NG7 2RD, U.K; Simon MacDonald, Fibrosis Discovery Performance Unit, GlaxoSmithKline, Medicines Research Centre, Gunnels Wood Road, Stevenage SG1 2NY, U.K; Tom McInally School of Chemistry, University of Nottingham, University Park, Nottingham NG7 2RD, U.K.

Title:
Screening of αvβ6 antagonists using ultrafiltration-liquid chromatography-mass spectrometry

Introduction:
Idiopathic pulmonary fibrosis (IPF) has mortality rates higher than most cancers. The integrin αvβ6 is a potential target for treatment of IPF. Screening the inhibitors of αvβ6 is time consuming and expensive, thus a simplified, high throughput screening method is urgently needed. Ultrafiltration (UF) was investigated for its ability to report ligand binding in an αvβ6 inhibitor screening assay.

Methods:
An ultrafiltration (UF) assay was developed that significantly improves throughput and cost of ligand screening. The assay involves the exposure of αvβ6 to a library of potential antagonists, followed by depletion of the unbound compounds through ultrafiltration. The concentration of bound ligands in the retentate was quantified using liquid chromatography-mass spectrometry (LC-MS)

Results:
The UF-LC-MS analysis allowed for the competitive screening of a library containing over 20 avb6 inhibitors, with binding affinities spanning 2 orders of magnitude, in a single experiment. The antagonists clustered into two groups: those with pIC50 values below 6.0, which possessed UF fb values that were no different from the controls, and those with pIC50 values of 6.5 and above which exhibited detectable integrin binding (fb 20-40%) in the UF assay. Moreover, this MS-based approach enabled the separation and ranking of inhibitors according to their binding affinity. A good correlation between UF-derived fb and pIC50 was observed despite the two assays utilising quite different outputs. Ten of the high binding compounds were then investigated for specificity towards avb6 over other related integrins avb3 and avb8.

Novel Aspect:
UF-LC-MS is fast, requires minimal method development and is amenable to automation. Making it ideally suited to high through put drug screening.

Turecek, Frantisek,
University of Washington

Contributing Authors:
Lubomir Rulisek, Institute of Organic Chemistry and Biochemistry, Czech Academy of Sciences; Prokopis C. Andrikopoulos, Institute of Organic Chemistry and Biochemistry, Czech Academy of Sciences; Emilie Viglino University of Washington; Chritopher J. Shaffer, University of Washington; Robert Pepin, University of Washington.

Title:
Ground and Excited-State Dissociations of Biomolecular Cation-Radicals

Introduction:
Several methods have recently become available to generate cation-radicals from biomolecules. Perturbation of the electron structure in biomolecules results in enhanced reactivity, generating unusual structures and opening new dissociation channels. Electron transfer in combination with UV and IR photodissociation represents a new approach to study peptide and DNA radicals and cation-radicals.

Methods:
UV-photodissociation and action spectroscopy are combined with the generation of cation-radicals by electron transfer into a multi-MS protocol. A linear quadrupole ion trap (ThermoElectron LTQ-XL ETD) has been modified to allow UV photodissociation and action spectroscopy of trapped ions. Ab initio, DFT and molecular dynamics (MD) calculations are used to characterize the species of interest.

Results:
Hydrogen-rich and hydrogen-deficient peptide cation-radicals are produced by different electron transfer methods. Collisional electron transfer to multiply cationized peptides and their crown-ether complexes is used to generate long-lived peptide cation radicals of the hydrogen-rich type and their fragment ions. The dissociations of long-lived cation-radicals often differ from direct conventional dissociations upon electron transfer. UV action spectroscopy has been used to characterize the excited states of biomolecular cation-radicals and provide wavelength-resolved breakdown diagrams for several dissociation channels. Time-dependent DFT calculations have been used to provide UV-VIS absorption spectra that were matched with the UV-VIS action spectra for several peptide cation-radicals. MD-aided search of transition states is a new approach to treat the radical reaction kinetics.

Novel Aspect:
Combining UV action spectroscopy with MD-aided search of transition states for structure analysis of peptide cation-radicals.

Turner, Mathew,
Centre for Analytical Science, Loughborough University

Contributing Authors:
Amy Giles, Loughborough University; Pilar Perez-Hurtado, Loughborough University; James Reynolds Loughborough University; Lourdes Arce-Jimenez, University of Cordoba; Thomas Owen, Loughborough University.

Title:
The assessment of volatile organic compounds in foods using APCI-MS and chemometric methods.

Introduction:
With concerns over food security increasing, there is a need for diagnostic techniques that enable rapid analysis and identification of potentially harmful or suspect foods. The combination of direct analysis and chemometrics offers a potential and effective tool for the detection of food spoilage, to explore flavour and aroma properties and to authenticate and assess food products.

Methods:
The headspace of olive oil and a selection of cheeses and meats were analysed directly using an Advion Expression compact mass spectrometer equipped with a Venturi pump and heated transferline. The samples were prepared by transferring 1g of sample into a headspace vial and sealing. The prepared samples were heated at 70oC for a maximum of 30 minutes and the resultant headspace was analysed.

Results:
Samples of beef, chicken, lamb, pork, and venison were analysed over 7 days. The meats were stored at ambient temperature to induce spoilage. The spectral data obtained was subjected to PLS-DA and PCA analysis. From the volatile profiles obtained it was possible to distinguish both between the different meats and to detect distinct clustering due to aging for each meat. The volatile profiles of 4 cheeses were analysed and subjected to the same statistical analysis. The 4 cheese varieties were successfully separated and this highlighted 9 organic acids as markers responsible for the separation. Finally a selection of extra virgin, virgin and lampante olive oils from different suppliers were assessed. Spectral profiles obtained from each group allowed for the successful discrimination of the extra virgin, virgin and lampante oils types that were concordant with organoleptic assignments.

Novel Aspect:
Direct analysis and statistical classification of meats, cheeses and olive oils by volatile organic profiles using APCI-MS.

Uetrecht, Jack,
University of Toronto

Contributing Authors:
Nadezhda Galeva, University of Kansas; Robert Hanzlik, University of Kansas; Yakov Koen University of Kansas; Imir Metushi, University of Toronto.

Title:
Use of Mass Spectrometry to Study Drug Covalent Binding and Idiosyncratic Drug Reactions

Introduction:
Idiosyncratic drug reactions (IDRs) increase the risk of drug development. Most are caused by reactive metabolites. Isoniazid is metabolized to a reactive metabolite and is associated with one of the highest risks of idiosyncratic liver failure. Most IDRs are immune mediated, but it appears that the dominant response is immune tolerance, and biomarkers of this response might predict IDR risk.

Methods:
Treatment of mice with isoniazid led to covalent binding to hepatic proteins and detected with an anti-isonicotinoyl antibody. These proteins were separated by SDS-PAGE and digested with trypsin followed by LC-MS/MS. Modified peptides and proteins were also concentrated with beads coated with the same antibody. LC-MS/MS can also be used to detect HMGB1, a potential biomarker of IDR risk.

Results:
2961 peptides were identified and assigned to 466 unique proteins. However, only one adducted peptide, derived from D-dopachrome decarboxylase, was detected. Peptide capture with beads coated with anti-isonicotinoyl antibodies was used to concentrate the modified peptides. Five adducted peptides were detected and assigned to ATP synthase subunit g, phosphoglycerate mutase 2, microsomal glutathione S-transferase 1, ADP/ATP translocase 2, transmembrane protein 256. When undigested protein adducts were concentrated with the beads, two different proteins were identified: fatty acid binding protein and tubulin polyglutamilase. Other proteins were isolated but not unambiguously identified. Another use of LC-MS/MS is to detect biomarkers of immune response in sera. . One candidate is HMGB1, which is released in its native form during cell necrosis, and in its acetylated by activated macrophages

Novel Aspect:
Immuno-concentration of drug-modified proteins greatly facilitates their identification. Identification of biomarkers is another use of LC-MS/MS

Uetrecht, Charlotte,
Heinrich Pette Institute/ European XFEL

Title:
Biophysics of clathrin adaptors and new opportunities for native MS in single particle imaging at XFELs

Introduction:
Viruses affect basically all organisms on earth. Some are detrimental to human development, whereas those targeting pathogenic bacteria or crop pathogens can be beneficial for us. Many viruses used clathrin coated vesicles to enter host cells. Clathrin, plasma membrane and cytoskeleton are linked via clathrin adaptor proteins.

Methods:
Native mass spectrometry was used to study lipid binding of two clathrin adaptor proteins, which were shown to interact in a lipid dependent fashion. Furthermore, the lipid mediated assembly pathway was investigated with native MS in conjunction with crystallography, DLS, ITC and SAXS. How native MS benefits single particle imaging at X-ray free-electron lasers (XFELs) will be illustrated.

Results:
Additional lipid binding sites were identified, which are required for complex assembly. Defined complex stoichiometries are observed and are conserved between species (from yeast to human). Assembly starts with lipid dependent formation of a core structure, to which the second protein attaches. The complex then gets extended reaching its final stoichiometry (8+8+lipid), which is the most stable structure as was confirmed in time-resolved native MS. Native MS delivers sufficient amount of ions to enable diffraction before destruction at XFELs. Results from feasibility studies and first tests at a soft XFEL (no diffraction but dissociation) are presented.

Novel Aspect:
Lipid dependent complex assembly New concept to derive high resolution structures of transient species

Uittenbogaard, Joost,
Intravacc

Contributing Authors:
Hugo Meiring, Intravacc; Gideon Kersten, Intravacc; Bernard Metz Intravacc.

Title:
Characterization of formaldehyde-induced modifications in diphtheria toxoid

Introduction:
Diphtheria vaccine is prepared by detoxification of the diphtheria toxin with formaldehyde and excess of glycine. This formaldehyde treatment is an important step in the vaccine production: it ensures the safety and the efficacy of the vaccine. The characterization of formaldehyde-induced modifications is important to monitor vaccine product consistency.

Methods:
Experimental diphtheria toxoids were prepared by treatment with (A) formaldehyde, (B) formaldehyde and glycine and (C) formaldehyde and glycine after reductive alkylation. Reactions were prepared with native and deuterium-labeled formaldehyde. Toxoids were mixed in a 1:1 ratio and digested. Targeted fragmentation of the mass spectral doublets was performed using a panel of methods (CID, HCD, ETD).

Results:
Toxoid (A) mostly contained cross-links, toxoid (B) contained a combination of cross-links and adducts, and toxoid (C) exclusively contained adducts. CID fragmentation of formaldehyde-modified peptides in general resulted in neutral losses of parts of the formaldehyde adducts. CID fragmentation of cross-linked peptides caused fragmentation at the cross-link. In both cases, no sequence information could be obtained. With ETD fragmentation, however, the modifications were retained and no neutral losses were observed. We compiled a mass list of all mass spectral doublets and subjected these parent masses to targeted ETD fragmentation. The ETD fragment ions were detected in the Orbitrap to get high mass accuracy and charge state information of the fragments.

Novel Aspect:
ETD fragmentation and high mass accuracy applied for characterization of formaldehyde-induced modifications in diphtheria toxoid.

21ST INTERNATIONAL MASS SPECTROMETRY CONFERENCE
Toronto, Ontario, Canada
August 20 – 26, 2016

Upton, Rosie,
University of Manchester

Contributing Authors:
David Firth, Covance Laboratories Ltd., Harrogate; Perdita E. Barran, University of Manchester.

Title:
Conformational differences of the monoclonal antibody Herceptin® demonstrated by Native TWIM-MS, DTIM-MS and HDX-MS.

Introduction:
Monoclonal antibody therapeutics currently dominate the pharmaceutical market and development of mAb biosimilars is growing. Use of heterogeneous mammalian cells during manufacture leads to the introduction of dissimilarity within and between batches making acceptance criteria for biosimilarity non-trivial. We use complementary mass spectrometry to compare three originator batches of Herceptin.

Methods:
Stability comparisons were made using Collision Induced Unfolding experiments on a Travelling Wave Ion Mobility Mass Spectrometer and freeze-thaw cycles were employed alongside Drift-Tube Ion Mobility Mass Spectrometry to observe the impact upon structural conformation. Hydrogen-Deuterium Exchange Mass Spectrometry was used to interrogate the mAb structural dynamics with and without PTMs.

Results:
Afucosylation levels can affect efficacy of mAb therapeutics and within the authentic Herceptin batches tested, significant differences were observed. CIU experiments revealed that different levels of glycosylation affect the unfolding of the mAb structure. Larger glycans preserve more of the compact structure, suggesting the presence of different glycans affects the overall stability of the mAb therapeutic. HDX-MS analyses demonstrated a faster rate of deuterium uptake by the deglycosylated Herceptin, indicative of sterical and/or conformational changes within the Fc region of the antibody. Freezing Herceptin resulted in decreased CCS values suggesting a conformational change or structural collapse as a result of freeze-thaw cycles. This observation will be coupled with SPR binding assays and ADCC potency assays to assess whether this has a knock on effect upon therapeutic functionality

Novel Aspect:
Different global conformations observed between fresh and frozen Herceptin and batch-to-batch variation in mAb stability, resultant of glycan profile.

Vahidi, Siavash,
University of Western Ontario

Contributing Authors:

Lars Konermann, University of Western Ontario.

Title:

Probing the Time Scale of FPOP (Fast Photochemical Oxidation of Proteins): Radical Reactions Extend Over Tens of Milliseconds

Introduction:

"Fast photochemical oxidation of proteins" (FPOP) involves ·OH production via H2O2 photolysis by UV laser pulses inside a flow tube. The experiments are conducted in the presence of a scavenger that is meant to shorten the ·OH lifetime. The literature claims that FPOP takes place within one microsecond. Surprisingly, the FPOP time scale has never been validated in direct kinetic measurements.

Methods:

Kinetic of processes in the FPOP solution were monitored by the bleaching of a reporter dye by flash photolysis. Time-dependent chemical changes triggered by the 248 nm pulse were probed in orthogonal geometry using a continuous-wave laser. After passing through the sample the probe laser intensity was recorded using a fast photodiode connected to a digital oscilloscope.

Results:

Bleaching of the reporter dye cyanine-5 (Cy5) served as readout of the time-dependent radical milieu. Surprisingly, Cy5 oxidation extends over tens of milliseconds. This time range is four orders of magnitude longer than expected from the FPOP literature. We demonstrate that the glutamine scavenger generates metastable secondary radicals, and that these radicals lengthen the time frame of Cy5 oxidation. Cy5 and similar dyes are widely used for monitoring the radical dose experienced by proteins in solution. The measured Cy5 kinetics thus strongly suggest that protein oxidation in FPOP extends over a much longer time window than previously thought, i.e., many milliseconds instead of one microsecond. The optical approach developed here should be suitable for assessing the performance of future FPOP-like techniques with improved temporal labeling characteristics.

Novel Aspect:

Direct kinetic measurements of the time scale of FPOP and other ·OH labeling methodologies.

Van Berkel, Gary,
Oak Ridge National Laboratory

Title:
Direct Liquid Extraction Approaches to Sample Analysis

Introduction:
Direct liquid extraction surface sampling techniques coupled with mass spectrometry are a convenient and effective means to analyze drug related materials from thin tissue sections, biopsies, skin, etc. This presentation will focus on two of the newest approaches from our group with distinct capabilities and advantages, viz., the droplet probe which couples with HPLC and the open port sampling in

Methods:
A Sciex 4000, 4000Qtrap, 5500 or 5600+ mass spectrometer was used for all experiments. The droplet probe was configured from a commercially available CTC PAL autosampler and a Agilent HPLC system. The open port sampling interface (OPSI) used a vertically aligned, co-axial tube arrangement enabling solvent delivery to the sampling end of the device through the tubing annulus and solvent aspiratio

Results:
The droplet probe samples by contacting a liquid droplet (1-2 mm dia.) to the sample surface and then injects the extract for HPLC-MS analysis. The operation of this automated 'point and click' analysis system will be overviewed. Extraction efficiency and quantitation will be addressed as well as several analysis applications, e.g., drug and biomarker detection in tissues. The OPSI is a system for 'on the spot' detection/identification of materials. The solvent delivery rate to the OPSI was set to exceed the aspiration rate, creating a continuous sampling interface along with a constant, self-cleaning spillover of solvent from the top of the probe. Analysis of samples was achieved by simple contact to the liquid interface. With the OPSI the extracted samples were transferred directly to the mass spectrometer for analysis, the chemical composition reported, and the system cleaned an

Novel Aspect:
Automated or simple manual liquid extraction probes for subsequent HPLC-MS analysis or 'on-the-spot' direct analysis of materials.

Van Berkel, Gary,
Oak Ridge National Laboratory

Title:
High Spatial Resolution Laser Ablation Sampling/Ionization under Ambient Conditions using a Hybrid Laser Microdissection/Liquid Vortex Capture/Mass Sp

Introduction:
An approach to high spatial resolution ambient sampling/ionization coupling a commercial laser microdissection (LMD) system with electrospray ionization (ESI)-MS via a liquid vortex capture (LVC) probe is presented. The ability to deliver sub-micrometer mass spectrometry imaging (MSI) spatial resolution, analyze single algae cells and perform absolute quantitation of drug in dosed thin tissues us

Methods:
A Leica LMD7000 integrated with an AB Sciex 5500 or 5600+ mass spectrometer using a LVC probe was used for all experiments. MSI was conducted on photoresist patterns, poly(2-vinylpyridine) (P2VP):poly(N-vinylcarbazol) (PVK) polymer thin films and on cocaine dosed tissue sections. Chlamydomonas reinhardtii and Allium Cepa were used for single cell analysis. The internal standard propranolol-d7 w

Results:
Mass spectral images of a P2VP:PVK phase separated polymer thin film were acquired. Features 0.62, 0.86, and 0.98 µm in size were measured in mass spectral images, and were confirmed using co-registered brightfield and fluorescence images of the sample. A pixel size of ~6 µm was achieved for MSI of cocaine in mouse brain tissue sections and features as small as 15 µm were observed in the mass spectral images. Single Chlamydomonas reinhardtii and Allium cepa cells were analyzed, for which several lipids were monitored (monogalactosyldiacylglycerols, diacylglyceryltrimethylhomo-Ser, polysaccharides, etc.). Automated software developed in house recognized the cells in the microscope images, calculated sizes and setup the system for LA sampling resulting in the ability to automatically analyze about 10 cells/min. Absolute quantitation was demonstrated for 40 µm x 40 µm and 20 µm x 20 µm

Novel Aspect:
MSI, single algae cell analysis, and absolute quantitation of drugs in tissue using a laser microdissection/liquid vortex capture probe/ mass spectrom

van Oosten, Luuk,
Heidelberg University

Contributing Authors:
Peter D.E.M. Verhaert, Delft University of Technology, Delft, The Netherlands; Mervin Pieterse, Delft University of Technology, Delft, The Netherlands; Martijn W. H. Pinkse Delft University of Technology, Delft, The Netherlands; Pierre Escoubas, Venometech, Valbonne, France.

Title:
Identification of novel cysteine-containing peptides in tropical ant venoms using 3D mass mapping

Introduction:
It has been shown that the majority of bioactive peptides isolated from animal toxins/venoms contain disulfide bridges. We developed an LC-MS based method that uses two properties of the sulfur atom as selection criterion to identify cysteine containing peptides prior to de novo sequencing efforts. In this research, we applied the method to the venom of two tropical ant species.

Methods:
Ant venom was fractionated by SCX-HPLC. Fractions were analyzed by nanoflow LC-MS on an LTQ-Orbitrap Velos. High resolution zoom scans and fragmentation spectra were recorded for the most abundant eluting peptides. The raw data files were further processed with an in-house developed MATLAB script [JASMS 2015 26(12) pp. 2039] to identify peptides containing disulfide bridges for de novo sequencing.

Results:
The 3D mass map generated from the MS data of the venom isolated from Odontomachus hastatus ants showed two distinct clusters. Investigation of spectra from one cluster revealed that 37 out of 43 peptides contained a disulfide bridge, whereas from the second cluster only 2 out of 70 proved to contain a disulfide bridge. This confirms the benefit of this method to search for cysteine containing peptides. From each cluster, 3 peptides were completely sequenced. The 3D mass map from the venom of Ectatomma tuberculatum showed to contain a group of 6 kDa and one of 8 kDa ant peptides. The 6 kDa peptides all contained one disulfide bridge and showed high similarity towards each other. The 8 kDa peptides proved to be dimeric peptides, similar to a known antimicrobial peptide, ectatomin. Analysis showed that 6 cysteines in these peptides form 2 inter- and 1 intra-molecular disulfide bridge.

Novel Aspect:
We sequenced 8 novel linear peptides which contain one disulfide bridge and we report 3 dimeric peptides with three disulfide bridges per dimer.

Vertes, Akos,
George Washington University

Contributing Authors:
Akos Vertes, Department of Chemistry, George Washington University, Washington, DC 20052, USA; Andrew R. Korte, Department of Chemistry, George Washington University, Washington, DC 20052, USA; Camille Lombard-Banek Department of Chemistry, George Washington University, Washington, DC 20052, USA; Peter Nemes, Department of Chemistry, George Washington University, Washington, DC 20052, USA; Lida Parvin, Department of Chemistry, George Washington University, Washington, DC 20052, USA; Ziad J. Sahab, Department of Chemistry, George Washington University, Washington, DC 20052, USA; Bindesh Shrestha, Department of Chemistry, George Washington University, Washington, DC 20052, USA; Sylwia A. Stopka, Department of Chemistry, George Washington University, Washington, DC 20052, USA; Wei Yuan, Department of Chemistry, George Washington University, Washington, DC 20052, USA; Deborah I. Bunin, SRI International, Menlo Park, CA 94025, USA; Merrill Knapp, SRI International, Menlo Park, CA 94025, USA; Ian Mason, SRI International, Menlo Park, CA 94025, USA; Denise M. Nishita, SRI International, Menlo Park, CA 94025, USA; Andrew Poggio, SRI International, Menlo Park, CA 94025, USA; Carolyn L. Talcott, SRI International, Menlo Park, CA 94025, USA; Maneesh Yadav, SRI International, Menlo Park, CA 94025, USA; Chrystal Chadwick, GE Global Research, Niskayuna, NY 12309, USA; Brian M. Davis, GE Global Research, Niskayuna, NY 12309, USA; Adriana I. Larriera, GE Global Research, Niskayuna, NY 12309, USA; Christine A. Morton, GE Global Research, Niskayuna, NY 12309, USA; Christopher J. Sevinsky, GE Global Research, Niskayuna, NY 12309, USA; Maria I. Zavodszky, GE Global Research, Niskayuna, NY 12309, USA; Nicholas J. Morris, Protea Biosciences Inc., Morgantown, WV 26505, USA; Heather R. Anderson, Protea Biosciences Inc., Morgantown, WV 26505, USA; Matthew J. Powell, Protea Biosciences Inc., Morgantown, WV 26505, USA; Trust T. Razunguzwa, Protea Biosciences Inc., Morgantown, WV 26505, USA.

Title:
Rapid Identification of Mechanisms of Action through Systems Biology

Introduction:
Timely response to emerging threat agents and expeditious testing of drug candidates require the identification of their mechanisms of action (MoA), a process that often takes several years. In order to develop rapid and reliable methods for MoA identification, compounds with known MoAs were interrogated using systems biology approaches, and our results were compared with the canonical MoAs.

Methods:
Hepatocytes were exposed to rotenone, forskolin, and nocodazole. Cellular responses were followed by microarray-based transcriptomics, high-throughput proteomics, and rapid untargeted metabolomics. The MoA behind the observed changes was explored by knowledge-based pathway analysis and network inference. Confirmatory assays were carried out to verify the MoA discerned from the multi-omics data.

Results:
Time dependent quantitative cellular responses to the test compounds were determined for >67,000 transcripts, >5,000 protein and phosphoprotein groups, and >400 metabolites. The latter was based on novel untargeted metabolomic technologies: LAESI-IMS-MS, and MS using laser desorption ionization from NAPA. Compounds exceeding a minimum fold change in abundance and exhibiting statistically significant responses were selected for knowledge-based pathway analysis. This approach was assessed by comparing the recovered MoA to the canonical MoA, e.g., in the case of forskolin, to the stimulation of cAMP signaling pathway. Heuristic and network inference methods discovered multiple molecular events downstream from the canonical MoA, and in one case the MoA itself. Phenotypic, qRT-PCR, immunofluorescence, and intracellular signaling assays were conducted to confirm the identified MoA.

Novel Aspect:

High-throughput transcriptomic, proteomic, and metabolomic methods were implemented to identify the MoA of three test compounds, each one in 30 days.

Voisin, Sebastien,
Association Plateforme BioPark d'Archamps

Contributing Authors:

Philippe Durand, Kallistem, SAS, ENS, Lyon F-69342, France; Sebastien Voisin, Plateforme BioPark d'Archamps, Archamps Technopole, Saint Julien en Genevois F-74166, France; Odette Prat CEA, Direction de la Recherche Fondamentale, BIAM-Marcoule, Bagnols-sur-Cèze F-30207, France; Karim Arafah, Plateforme BioPark d'Archamps, Archamps Technopole, Saint Julien en Genevois F-74166, France; Marie-Hélène Perrard, INSERM U1208, Stem Cell and Brain Research Institute, Université Lyon 1, Bron F-69500, France; Cédric Pisani, CEA, Direction de la Recherche Fondamentale, BIAM Marcoule, Bagnols-sur-Cèze F-30207, France; Philippe Bulet, Université Grenoble Alpes, Institut Albert Bonniot, Inserm 1209, CNRS UMR5309, Equipe Immunologie Analytique des Pathologies Chroniques, Grenoble F-38000, France; Marie-Roberte Guichaoua, IMBE, UMR CNRS 7263 / IRD 237, Aix-Marseille Université (AMU), Marseille F-13385, France.

Title:

An ex vivo model of rat spermatogenesis used for an integrated Omics assessment of toxicity induced by carbendazim and iprodione.

Introduction:

Close to 800 molecules are suspected to be interfering with the mammalian hormonal system; among the fungicides, benzimidazoles and dicarboximides are suspected to impair fertility. In this study, ex vivo cultures of rat seminiferous tubes were used to assess the toxicity of two fungicides, carbendazim and iprodione, alone or together, using cytogenetic, transcriptomic and proteomic methods.

Methods:

Rat seminiferous tubules were pooled and cultivated in bicameral chambers during 3 weeks, in presence of carbendazim and/or iprodione, at either 50 nM or 500 nM. Samples were collected on days 7, 14 and 21 for transcriptome and secretome analysis, and for cytological immunostaining. Pathways significantly modulated after fungicide exposure were further studied by Ingenuity Pathway Analysis.

Results:

Pursuing the Replacement, Reduction, and Refinement rule, we used an ex vivo rat model of seminiferous tubules, reducing rodents number and allowing a fast screening of suspected endocrine disruptors. Cytogenetic analyses confirmed that both fungicides altered the course of the first meiotic prophase. Whole-gene expression profiles revealed strong steroidogenesis alteration and meiosis impairment in all cases. Proteomics identified systematically differentially expressed proteins in the exposed cultures' secretome, e.g. Fuco and Pebp1, proteins linked with the motility and fertilizing ability of spermatozoa, respectively. A pathway-driven analysis showed that both substances had different initial molecular events. Finally, the mixture of both substances triggered effects greater than the sum of their cumulative effects, compromising future sperm motility after a shorter time of exposure.

Novel Aspect:

Transversal pesticide toxicity analysis on rat seminiferous tubules using an ex vivo model, reducing the number and suffering of test animals.

Voisin, Sebastien,
Association Plateforme BioPark d'Archamps

Contributing Authors:

Laurent GAUTHIER, Institut du Liebefeld (Berne, Suisse); Anne DALMON, INRA UR 406 Abeilles et Environnement (Avignon, France); Cédric ALAUX INRA UR 406 Abeilles et Environnement (Avignon, France); Katarina BILIKOVA, Institute of Forest Ecology, Slovak Academy of Sciences (Bratislava, Slovakia); Michel BOCQUET, Apimedia (Pringy, France); Jozef SIMUTH, Institute of Forest Ecology, Slovak Academy of Sciences (Bratislava, Slovakia); Philippe BULET, Université Joseph Fourier, IAB U823 (Grenoble, France); Yves LE CONTE, INRA UR 406 Abeilles et Environnement (Avignon, France); Karim ARAFAH, Plateforme BioPark d'Archamps (Archamps, France); Sébastien VOISIN, Plateforme BioPark d'Archamps (Archamps, France) .

Title:

HematoBeeTest project: Identifying patterns of honeybee immune response by Biotyper-like mass spectrometry, in the context of bee health monitoring

Introduction:

Honeybee population decline is being attributed to stressors such as parasites (viruses, Varroa mite), pesticides, and environmental changes. Most researches focused on identifying the stressors' presence instead of their impact on honeybee colonies. The HematoBeeTest project aims at profiling the infected bees' immunoproteome, in order to deliver practical applications for bee health management.

Methods:

Hemolymph samples were collected from individuals from monitored colonies with a diagnosed infection, and from individuals artificially infected with a pathogen. Virus presence was confirmed by quantitative PCR. Protein content was analyzed and compared by MALDI-MS, directly or after reduction-alkylation of the hemolymph. Top-down analysis by LC-MS/MS was conducted to confirm protein identities.

Results:

Hemolymph analyses tracked key peptides and proteins of the bee immunoproteome (apidaecin, hymenoptaecin...), and resulted in different molecular fingerprints in function of the bees' infectious conditions. These differences were confirmed by statistical comparison of MS profiles by principal component analysis. Virus-infected bee, with or without Varroa destructor co-infection, ended up in a cluster of their own inside the overall Varroa cluster. These first results strongly support the robustness of our monitoring approach in the case of co-infections, its potential as a plausible strategy to monitor honeybees' health, and a mean for a better understanding of the molecular immune response of this social insect, in both experimental and natural infections. Other infection models are currently being investigated, notably for microsporidia (Nosema) and entomobacteria.

Novel Aspect:

Diagnostic model based on the rapid comparison by MALDI-MS of honeybee hemolymph between infected and non-infected individuals.

Vojs Stanova, Andrea,
Comenius University in Bratislava, Faculty of Natural Sciences, Department of Analytical Chemistry

Contributing Authors:

Jozef Marak, Comenius University in Bratislava, Faculty of Natural Sciences, Department of Analytical Chemistry, Mlynska Dolina Ilkovicova 6, SK-842 15 Bratislava, Slovakia; Monika Radicova, Comenius University in Bratislava, Faculty of Natural Sciences, Department of Analytical Chemistry, Mlynska Dolina Ilkovicova 6, SK-842 15 Bratislava, Slovakia; Frantisek Golais Comenius University in Bratislava, Faculty of Natural Sciences, Department of Microbiology and Virology, Mlynska Dolina Ilkovicova 6, SK-842 15 Bratislava, Slovakia; Pavol Kois, Comenius University in Bratislava, Faculty of Natural Sciences, Department of Organic Chemistry, Mlynska Dolina Ilkovicova 6, SK-842 15 Bratislava, Slovakia; Miroslava Supolikova, Comenius University in Bratislava, Faculty of Natural Sciences, Department of Microbiology and Virology, Mlynska Dolina Ilkovicova 6, SK-842 15 Bratislava, Slovakia.

Title:

High-performance separation techniques in combination with high-resolution mass spectrometry in the study of MHGF-68

Introduction:

When studying the interaction of some herpes viruses with different types of host cells, a new virus-encoded substance probably resembling growth factors have been discovered. Presented work deals with the detailed study of MHGF-68 based on the combination of different sample pretreatment, high-performance separation techniques, multistage mass spectrometry and chemometric data evaluation.

Methods:

All HPLC-MS experiments were performed on instrument LCMS-IT-TOF (Shimadzu) equipped with electrospray ionization. HPLC experiments were performed on Ascentis C18 column (100/2.1 mm; 3 µm) (Sigma-Aldrich) using a gradient elution (water – acetonitrile, 10-90 %) with a 0.2 ml/min flow rate. Chemometric evaluation was realized by MSXelerator software ver. 2.4 (MSMetrix, Maarssen, Nederland).

Results:

Application of the medium of MHV infected cells on RP-HPLC C18 column resulted in appearance of 2 fractions with transforming activity and devoid of transformation repressing activity. Different results dependent on elution conditions were obtained with MHV-68-associated factor on FPLC Sephadex G-15 column. In the case of phosphate buffer highly active fractions with both activities were obtained, while water as mobile phase eluated fractions only with transformation repressing activity. The chromatogram and MS spectra obtained from HPLC-ESI-MS analyses in positive and negative ionization modes showed different composition of individual fractions. Results obtained from the chemometric data evaluation allowed characterizing and classifying all obtained fractions.
Acknowledgement: This work was financially supported by the grants APVV-0621-12 and VEGA 1/0262/13.

Novel Aspect:

Obtained complex information on MHGF-68, using powerful analytical techniques, can open new ways in the study of the properties of oncogenic MHV.

Vuckovic, Dajana,
Concordia University

Contributing Authors:
Reajean Sivakumar, Concordia University; Irina Slobodchikova, Concordia University; Samiur Rahman Concordia University; Cian Monnin, Concordia University.

Title:
Development of a highly-sensitive multi-mycotoxin LC-MS method in human plasma for exposure studies

Introduction:
Worldwide mycotoxin contamination of agricultural commodities poses a problem to human health due to their toxicity. The assessment of the human exposure to mycotoxins in Canada is needed and requires a sensitive and robust multi-analyte assay(s) allowing simultaneous detection and quantification of common toxicologically important mycotoxins and their metabolites in human plasma.

Methods:
Different columns, solvents, sample pH and gradients were investigated to achieve separation of all mycotoxins, particularly isomers. Solvent precipitation with acetonitrile, solid-phase extraction (Oasis HLB sorbent and mixed mode sorbent (Bond Elute Certify II)), and liquid-liquid extraction (ethyl acetate and methyl tert-butyl ether) ware compared to reach high analyte recovery and selectivity.

Results:
The appropriate choice of mobile phase additives and their concentrations depending on the mode of ionization contributed to further improvements of sensitivity, and can be expanded to other analyte classes. Finally, mycotoxins were successfully separated with pentafluorophenyl column and the mobile phase containing water/methanol with 0.1% acetic acid and water/methanol with 0.02% acetic acid for positive and negative electrospray ionization, respectively. The performance of two different MS platforms (Agilent 6550 QTOF and Orbitrap-LTQ Velos) was compared for this application. The optimized method was validated according to regulatory guidelines. Microsomal incubations of ochratoxin A, HT-2 toxin, aflatoxin B1, fumonisin B1, zearalenone, and deoxynivalenol were also performed to build in house library of mycotoxin metabolites to use during exposure monitoring studies.

Novel Aspect:
This is the first LC-MS assay to cover >15 mycotoxins in human plasma with required sensitivity and selectivity for exposure monitoring.

Walker, Stephen,
University Waterloo

Contributing Authors:

J. Larry Campbell, AB Sciex and University of Waterloo; W. Scott Hopkins, University of Waterloo; Brent Verbuyst Unversity of Waterloo; Alison Mark, Unversity of Waterloo; Bogdan Bogdanov, Perkin Elmer.

Title:

Investigating Tautomers of Anilinium with Differential Mobility Spectrometry

Introduction:

It is of great practical and fundamental relevance to determine the most favourable sites for protonation of molecules exhibiting multiple basic sites. The separation, and detection of gas phase tautomers has, however, been challenging. In aqueous phase aniline is a nitrogen base, in gas phase the protonation site is less clear with protonation at the nitrogen and para-carbon possible.

Methods:

Here we demonstrate that Differential Mobility Spectrometry (DMS) can be used to separate the nitrogen-protonated and para-carbon-protonated tautomers of anilinium, prior to characterization with mass spectrometry. Hydrogen-Deuterium Exchange experiments are used to help support the findings along with extensive computational modelling.

Results:

We find that the both protonation sites are observerd and clear evidence that the relative populations of the N- and para-protonated tautomers is tunable with ESI conditions.

Novel Aspect:

This experiment shows the utility of DMS with to differentiate between the binding energies of different tautomers of molecules.

Wallace, William,
National Institute of Standards and Technology

Contributing Authors:
Weihua Ji, National Institute of Standards and Technology; Stephen E. Stein, National Institute of Standards and Technology; Karen W. Phinney National Institute of Standards and Technology.

Title:
Illicit-Substance Mass Spectral Libraries: Data Quality Control Measures

Introduction:
Having high-quality illicit-substance libraries is critical to the criminal justice system. NIST has developed spectrum quality control measures suitable for forensic drug analysis. We have reviewed each SWGDRUG library spectrum and brought the level of curation up to the standards of the NIST library.

Methods:
Each spectrum in the SWGDRUG library was compared via standard library search methods to spectra of the same compound found in any of five mass spectral libraries. Confirmation of a good quality SWGDRUG spectrum was contingent primarily on matching with a match factor of 750 or better (out of 999) any and all spectra in the five reference libraries.

Results:
Spectral errors are generally of three types 1) missing peaks, 2) spurious peaks unattributable to the compound structure, and 3) incorrect mass assignments (often from incorrect rounding of mass values). In addition there are a variety of compound identification errors. These include 1) incorrect chemical structure, 2) incorrect chemical name, 3) incorrect CAS number or other identifier, and 4) incorrect isomer identification. The use of NIST's MS Interpreter program was instrumental few adjudicating cases where only a few measurements were available. MS Interpreter assigns peaks to plausible molecular substructures based on a set of rules for estimation of fragmentation energetics using a chemical structure (connection table) type analysis.

Novel Aspect:
Mass spectral library quality control by consensus comparison to extant spectral library entries and by comparison to simple thermodynamic models

Wang, Lin,
Beijing University of Technology

Contributing Authors:
Guangsheng Guo, Department of Chemistry and Chemical Engineering, Beijing University of Technology, Beijing 100124, PR China; Cuirong Sun, College of Pharmaceutical Sciences, Zhejiang University, Hangzhou 310058, Zhejiang, PR China; Xiayan Wang Department of Chemistry and Chemical Engineering, Beijing University of Technology, Beijing 100124, PR China; Yuanjiang Pan, Department of Chemistry, Zhejiang University, Hangzhou 310027, Zhejiang, PR China; Lin Wang, Department of Chemistry and Chemical Engineering, Beijing University of Technology, Beijing 100124, PR China.

Title:
Cellular Metabolomic Analysis Using Photo-controlled Probes Coupled With Ambient Mass Spectrometry

Introduction:
Metabolomics is an emerging field that provides critical insight into physiologic status of cells. One key aspect of metabolomics analysis involves the annotation of LC-MS feature. This platform advantageously enable separation analysis but with a well-known disadvantage that is LC-MS analysis generally requires effective chromatographic separation which increases analysis time.

Methods:
Ambient mass spectrometry (AMS) is an excellent choice for rapid and direct analysis of complicated samples because this technique operates under ambient and open-air condition without requiring sample pretreatment and separation.

Results:
In this work, a photo-controlled, gold-plated probe was designed and applied in rapid examination of the cellular metabolites using AMS. We proposed to modify the inner surface of capillaries with spiropyran derivative to form novel monolithic column probes, which could be used for selectively absorbing the hydrophilic compounds under UV light or hydrophobic compounds under visible light condition respectively in a single run. The initial results showed that the functional probe together with AMS has an outstanding capacity for the detection of gefitinib and its metabolites such as the N-propoxymorpholino group, demethylation of the methoxy substituent on the quinazoline and oxidative defluorination of the halogenated phenyl group in A549 cells.

Novel Aspect:
This strategy gives the opportunity for complex metabolites analysis without using sophisticated sample pretreatment and chromatographic separation.

21ST INTERNATIONAL MASS SPECTROMETRY CONFERENCE
Toronto, Ontario, Canada
August 20 – 26, 2016

Wang, Zhechen,
University of Colorado at Boulder

Contributing Authors:
Veronica M. Bierbaum, University of Colorado, Boulder.

Title:
Experimental and Theoretical Studies of Ionic Reactions of Astrochemical Relevance

Introduction:
Carbon-containing cations and anions have been observed in many important extraterrestrial environments, from the atmosphere of Titan to the interstellar medium. However, the reactions of these ions with the abundant atomic species, H, N, and O, are largely unexplored due to the complexity of the experiments. This presentation will provide an overview of our experimental and theoretical studies.

Methods:
Measurements of the reaction rate constants and product distributions were made using the tandem flowing afterglow-selected ion flow tube (FA-SIFT) at the University of Colorado, Boulder. This instrument has been described elsewhere, and only the salient details for these experiments will be discussed here. Ions are generated using electron and chemical ionization methods in the source flow tube

Results:
The rate constants and products for reactions of the azine anions with N (4S) and O (3P) atoms are summarized in Table 1. The dominant process for the reactions of azine anions with N atoms is associative electron detachment (AED) to form electrons and neutral products. There is no ionic product detected for the reactions of N atoms with C6H5–, C5H4N–, 1,2- and 1,3 C4H3N2– and 1,3,5-C3H2N3–. The only minor ionic product detected from the reaction of 1,4-C4H3N2– with N atoms is C4H3N–, corresponding to the formation of neutral N2. The reactions of deprotonated azine anions with O atoms are complex, because various ionic products are observed experimentally in addition to AED. However, the AED process is still dominant. The reaction rate constant of C6H5– with O atoms (9.0 ± 4.5 × 10-10 cm3 molecule-1 s 1) is roughly 5 times faster than that with N atoms.

Novel Aspect:
Shed light on elucidating the roles of cyclic anions and atoms as well as the identity and abundances of negative ions in the astrochemical enviroment

Wang, Yating,
NSERC CREATE York University and University of Toronto

Contributing Authors:

Alan C. Hopkinson, York University; K.W.Michael, University of Windsor.

Title:

Application of Tandem Mass Spectrometry: [b3+H]2+ and [a3+H]2+ derived from Lanthanide-Peptide Complexes

Introduction:

Protonated a ions are of draw interest because of the high charge density in small organic ions. The fragmentation of protonated a3 ions derived from lanthanide-tripeptide complexes have been investigated using tandem mass spectrometry, and the more fragile b3 ions have been observed for the first time.

Methods:

Protonated a3 ions were created by collision-induced dissociation (CID) of the metal(III)/tripeptide complex.
[M(tripeptide)(CH3CN)n]3+, where n is 1 or 2 and M is La or Ce, dissociate by loss of nCH3CN, CO and MO+ thereby creating (a3+H)2+ ions. (b3+H)2+ ions have been created by CID of [M(tripeptide)]3+.

Results:

(a3+H)2+ ions usually cleave at one of the amide bonds creating two singly charged ions, a b2 ion and an iminium ion derived from the C-terminal residue. Some (a3+H)2+ ions also lose small neutral molecules (ammonia, carbon monoxide or an imine. The relative amounts of (b3+H)2+ to (a3+H)2+ ions created by CID of [M(tripeptide)]3+ depend on the composition of the peptide. Fragmentation of the (b3+H)2+ ions that have been isolated gives (a3+H)2+ ions in low abundance; however, the preferred mode of fragmentation frequently is the loss of CO plus H2O, a pathway common in the dissociation of protonated carboxylic acids. This indicates that the (b3+H)2+ ions are best described as diprotonated tripeptides that have lost a molecule water from an amide group while retaining the terminal carboxylic acid group.

Novel Aspect:

Fragmentations of (a3+H)2+ ions are investigated with tandem mass spectrometry. (b3+H)2+ ions have been observed for the first time in the gas phase.

Warzok, Ulrike,
Freie Universität Berlin

Contributing Authors:

Kari Rissanen, Department of Chemistry, NanoScience Center, University of Jyvaskyla, P.O. Box 35, 40014 Jyvaskyla, Finnland; Christoph A. Schalley, Institut of Chemistry and Biochemistry, Free University Berlin, Takustr. 3, 14195 Berlin, Germany; Lotta Turunen Department of Chemistry, NanoScience Center, University of Jyvaskyla, P.O. Box 35, 40014 Jyvaskyla, Finnland; Ulrike Warzok, Institut of Chemistry and Biochemistry, Free University Berlin, Takustr. 3, 14195 Berlin, Germany; Rakesh Puttreddy, Department of Chemistry, NanoScience Center, University of Jyvaskyla, P.O. Box 35, 40014 Jyvaskyla, Finnland.

Title:

Investigation of halogen bonded supramolecular capsules by ESI-MS and IM-MS

Introduction:

Supramolecular chemistry focuses on the formation of complex molecular assemblies on the basis of non-covalent interactions. Within this framework, supramolecular capsules which employ diverse binding motives including hydrogen bonding, metal coordination, ion pair interaction or, very recently, also halogen bonding have been of particular interest.

Methods:

ESI-MS, CID tandem MS and travelling wave IMS experiments were performed with a Synapt G2-S HDMS.

Results:

Herein, we present the formation of supramolecular capsules based on the novel interaction of pyridyl-substituted Resorcin[4]arene cavitands with positively charged halonium ions by multiple three-center halogen bonds. The complexes were thoroughly studied by mass spectrometry including ESI-MS, CID tandem MS and travelling wave IMS experiments performed with a Synapt G2-S HDMS. Further characterization of the capsules includes NMR investigation in solution and single crystal X-ray analysis.

Novel Aspect:

The presented extensive gas phase studies are the first of their kind performed on halogen bonded supramolecular capsules.

Wei, Pu,
Purdue University

Contributing Authors:
Christopher Pulliam, Purdue University; Rafal M. Pielak, L'Oreal California Research Center, San Francisco, CA; R. Graham Cooks Purdue University.

Title:
ANALYSIS OF SQUALENE OXIDATION ON HUMAN SKIN BY PAPER SPRAY MASS SPECTROMETRY

Introduction:
Squalene (SQ) is a major component of human skin surface lipids. SQ oxidation induces various skin problems, such as comedogenesis, skin wrinkling and dermal changes. SQ and its oxidized forms have been recently reported as new biomarkers of environmental pollution. Here, paper spray and a miniature mass spectrometer were evaluated for direct analysis of SQ photooxidation on human skin in field.

Methods:
For artificial oxidation of SQ experiments, 1 µL of 100 ppm SQ solution was deposited onto a tip and placed under a 6W UV lamp at a distance of 5 cm vertically. For skin wipe experiments on human subjects, triangle tips were precut and gently wiped onto skin surface three times without any solvent. Standard PS setup was used for analysis. 10ppm $AgNO_3$ in IPA solution was prepared as spray solvent.

Results:
SQ oxidation has been studied using PS-MS. Remarkably, 10 oxygen addition steps to yield ($[Ag+SQ+nO]^+$, n = 1-10) were detected in the sample over the course of 19 hours' UVA exposure. SQ hydroperoxides were generated as the primary products induced by UVA. Off-line monitoring of SQ oxidation on human skin was performed by wiping the skin surface with paper spray triangles. Differences were observed between samples with cream application and a control group. A miniature mass spectrometer (Mini 12) was used to study SQ oxidation for comparison with data from benchtop instruments. The Mini 12 was taken to New Delhi, India and used to analyze skin wipes from 18 subjects immediately after sample collection. Same set of samples was taken back to be analyzed by benchtop instrument. The mass spectra obtained using the Mini 12 were nearly identical to those from a commercial linear ion trap.

Novel Aspect:
PS and Mini 12 displayed simplicity, robustness, portability, and excellent performance for in situ analysis of squalene photooxidation on human skin.

Weijman, Johannes,
Department of Biochemistry, University of Otago, NZ

Contributing Authors:
Peter D. Mace, Department of Biochemistry, University of Otago, NZ; Elizabeth C. Ledgerwood, Department of Biochemistry, University of Otago, NZ.

Title:
Biochemistry of redox regulation of apoptosis signal-regulating kinase 1: identification and isolation of disulfide-linked intermediates.

Introduction:
Apoptosis signal-regulating kinase 1 (ASK1) is a redox sensitive kinase, which is regulated by Thioredoxin 1 (Trx1). Previous work from our lab has shown that a major antioxidant protein, peroxiredoxin 1 (Prdx1), can transmit signals from hydrogen peroxide to substrates, such as ASK1, via transient inter-molecular disulfide bonds. However, the precise mechanism remains unknown.

Methods:
Recombinant wild type and cysteine mutant ASK1, Prdx1 and Trx1 constitute our model in vitro system. Comparative reducing and non-reducing SDS-PAGE have been used to identify disulfide-linked species. In gel digestion and LC-MS/MS is on going in order to identify disulfide-linked peptides that are relevant to signalling.

Results:
To date, we have successfully expressed and purified a variety of recombinant human ASK1 constructs. We are able to isolate disulfide linked ASK1-Trx1 and preliminary LC-MS/MS data has suggested that four cysteines within ASK1 may contribute towards this. Work is on going to isolate cross-linked, ASK1-Trx1 peptides. Various Trx1-fusion proteins have enabled purification and isolation by FPLC of intact disulfide linked ASK1-Trx protein complexes. We hope to use and expand upon these results to infer how ASK1 is redox-regulated via disulfide bonds and how Prdx1 may act on a more global regulatory level within a mammalian cell.

Novel Aspect:
All previous ASK1-Trx1-Prdx1 studies are cell based. We are uniquely placed to study this system in an in vitro setting.

Weisbrod, Chad,
Thermo Fisher Scientific

Contributing Authors:

Jae C. Schwartz, Thermo Fisher Scientific; Eugene Zhuk, Thermo Fisher Scientific; Jean-Jacques Dunyach Thermo Fisher Scientific; Mike Senko, Thermo Fisher Scientific.

Title:

Performance considerations for ultraviolet photo-dissociation using the Nd:YAG 5th harmonic (213nm)

Introduction:

We demonstrate that 213 nm light is sufficient in photon energy and near enough to the absorbance band of the amide backbone to produce high quality UVPD spectra from polypeptides. A detailed comparison between excimer and SS laser generated spectra has been established via sequence coverage, fragment ion identifications, and fragment ion partitioning.

Methods:

An ArF excimer (193nm) and a SS (213 nm) laser were each separately coupled directly to an Orbitrap Fusion Tribrid mass spectrometer. All irradiation of ions took place within the dual cell linear RF ion trap assembly. No additional light focusing optics were utilized, however, to maintain a photon beam with a diameter similar to that of the ion cloud an iris was used with the excimer laser.

Results:

Myoglobin has been utilized in a preliminary comparison between the excimer and SS lasers. In this comparison, fragmentation spectra were generated using the 15, 19, and 24 plus charge states of apomyoglobin. In this comparison, we show that sequence coverage between the two wavelengths is approximately the same (nominally >60%). The types of fragment ions observed are also very similar; predominantly a/x ions with some b/y and c/z ions with similar numbers of identified fragment ions of each type. For example, fragmentation from the [M+15H]15+ion produced 141 of 200 (70.5%) a/x type fragment ions using 213 nm , compared to 146 of 219 (66.7%) using 193nm. These percentages hold across the various charge states we studied. Moreover, the SS laser outperformed the excimer in terms of raw fragment ions identified (229@213nm vs. 212@193) at optimized fragmentation energy.

Novel Aspect:

UVPD fragmentation using 213nm light. Over fragmentation protection schema. Normalization of fragmentation energy to enable better acquisitions.

Welch, Leslie,
Genentech a Member of the Roche Group

Contributing Authors:
Jasper Lin, Genentech, A Member of the Roche Group; Madhura Deshpande, Genentech, A Member of the Roche Group; Sreedhara Alavattam Genentech, A Member of the Roche Group; Galahad Deperalta, Genentech, A Member of the Roche Group.

Title:
Investigating the Differential Photosensitivity of Tryptophan in a Model Trp-Cage Peptide

Introduction:
A Trp-cage model peptide was used to investigate mechanisms underlying tryptophan (W) photosensitivity in mAbs. Peptide analogues with W's substituted into different positions were exposed to light and characterized for oxidation by UPLC-MS and H2O2 content by an Amplex® Red kit. Unexpected observations concerning oxidation and H2O2 content suggested novel mechanisms for W photo-oxidation.

Methods:
Photo-oxidation of peptides was achieved by light-exposure to 250 W/m2, 1.32 Mlux-h. Photo-generated H2O2 content from peptides was determined by analysis with Amplex® Red kit. RP-UHPLC separation of Trp-cage peptides was performed using a Waters Acquity UPLC® instrument and CSH130™ C18 column. Oxidation levels were determined using a Thermo Scientific Q Exactive™ Hybrid-Orbitrap.

Results:
The results demonstrate that photo-degradation of the central W in the Trp-cage peptide, and its analogues, is dependent on solvent accessibility and neighboring residue effects of the hydrophobic cage. The Y-to-F substitution in the peptide, removing a hydroxyl group near the central W, showed an increase in oxidation, but an unexpected reduction in H2O2 production. Another analogue with an additional W oriented edge-to-face to the central W showed an unexpected decrease in oxidation indicating the importance of neighboring amino acids in W oxidation. Utilization of the Trp-cage peptide model, with targeted amino acid residue substitutions, allowed protection from or exacerbation of W photo-oxidation, thus supporting the use of small W-containing peptides to understand various mechanisms of light-mediated oxidation in peptides and proteins.

Novel Aspect:
Trp-cage analogues showed unexpected levels of photo-induced oxidation and H2O2 generation, suggesting potentially novel photo-degradation mechanisms.

Welch, Leslie,
Genentech a Member of the Roche Group

Contributing Authors:

Jasper Lin, Genentech, A Member of the Roche Group; Madhura Deshpande, Genentech, A Member of the Roche Group; Galahad Deperalta Genentech, A Member of the Roche Group; Sreedhara Alavattam, Genentech, A Member of the Roche Group.

Title:

Investigating the Differential Photosensitivity of Tryptophan in a Model Trp-Cage Peptide

Introduction:

A Trp-cage model peptide was used to investigate mechanisms underlying tryptophan (W) photosensitivity in mAbs. Peptide analogues with W's substituted into different positions were exposed to light and characterized for oxidation by UPLC-MS and H2O2 content by an Amplex® Red kit. Unexpected observations concerning oxidation and H2O2 content suggested novel mechanisms for W photo-oxidation.

Methods:

Photo-oxidation of peptides was achieved by light-exposure to 250 W/m2, 1.32 Mlux-h. Photo-generated H2O2 content from peptides was determined by analysis with Amplex® Red kit. RP-UHPLC separation of Trp-cage peptides was performed using a Waters Acquity UPLC® instrument and CSH130™ C18 column. Oxidation levels were determined using a Thermo Scientific Q Exactive™ Hybrid-Orbitrap.

Results:

The results demonstrate that photo-degradation of the central W in the Trp-cage peptide, and its analogues, is dependent on solvent accessibility and neighboring residue effects of the hydrophobic cage. The Y-to-F substitution in the peptide, removing a hydroxyl group near the central W, showed an increase in oxidation, but an unexpected reduction in H2O2 production. Another analogue with an additional W oriented edge-to-face to the central W showed an unexpected decrease in oxidation indicating the importance of neighboring amino acids in W oxidation. Utilization of the Trp-cage peptide model, with targeted amino acid residue substitutions, allowed protection from or exacerbation of W photo-oxidation, thus supporting the use of small W-containing peptides to understand various mechanisms of light-mediated oxidation in peptides and proteins.

Novel Aspect:

Trp-cage analogues showed unexpected levels of photo-induced oxidation and H2O2 generation, suggesting potentially novel photo-degradation mechanisms.

Wellman, Sydney,
NSERC CREATE York University and University of Toronto

Contributing Authors:
Rebecca Jockusch, University of Toronto.

Title:
Unravelling Environmental Effects on Light-harvesters: Photodissociation Action Spectroscopy and Fluorescence Emission of Gas-phase Chlorophylls

Introduction:
The design of photosynthesis-inspired solar devices would greatly benefit from an improved understanding of the photophysics of nature's most abundant light-harvester, chlorophyll a (Chla). Towards this end, here we probe isolated gaseous Chla by examining both its photodissociation (PD) action spectra and fluorescence emission profile in complex with a series of cation tags.

Methods:
A modified Bruker Esquire 3000+ quadrupole ion trap is used to monitor fluorescence and measure the PD action spectra of Chla. Protonated and metallated Chla ions are generated via nano-ESI. Action spectra are generated by irradiating mass-selected ions using laser light tuned from 350 to 455 nm. For fluorescence measurements, ions were excited at 430 or 447.5 nm.

Results:
PD action spectra of gaseous protonated, lithiated and cesiated Chla all feature a complex Soret band in the 350-445 nm region, but with notable differences in intensity of spectral features. The spectra are strikingly different from the absorption spectrum of neutral Chl a in methanol and from the PD action spectrum of the gaseous anion but are broadly similar to those measured for Chla when tagged with quaternary ammonium cations. Fluorescence emission was not detected from gaseous protonated Chla, but is "turned on" when cationized by alkali metals. Emission maxima increase in energy from 661 nm to 654 nm as the size of the metal cation increases. These emission maxima are close in energy (less than 250 cm-1 difference) to that of neutral Chla in methanol. Overall, these results highlight the degree to which the photophysics of Chl a is sensitive to its micro-environment.

Novel Aspect:
Intrinsic fluorescence of chlorophyll a is supressed when protonated, but "turns on" when bound to alkali metal cations

Wellman, Sydney,
NSERC CREATE York University and University of Toronto

Contributing Authors:
Matthew Gibb, Centre of Forensic Sciences.

Title:
Extending Quantitative Capabilities within Forensic Toxicology using Standard Addition and UPLC-QTOF-MS

Introduction:
High-resolution mass spectroscopy (HRMS) is used for drug screening within forensic toxicology. The rate at which new drugs emerge outpaces the ability of forensic laboratories to develop and validate quantitative methods. Thus, a quantitative method employing standard addition and UPLC-QTOF-MS has been developed to complement existing screening and which provides matrix-matched sample analysis.

Methods:
Six aliquots (250 µL) of authentic blood (or serum) samples were spiked with a mix of 11 deuterated internal standards (IS), followed by 25 µL of drug(s) (4 concentrations) or blank solvent, precipitated with acetonitrile, centrifuged, then extracted using Waters Oasis HLB PRiME or Phenomenex Phree cartridges. Following a timed dry down, extracts were analyzed using UPLC-QTOF (Xevo G2-S, Waters).

Results:
The evaporation step had a significant effect on chromatographic resolution, analyte recovery and quantitative accuracy, particularly when the composition of percentage of organic in the injected sample was >20%. A linear response range was identified and observed to vary depending on the drug. In general, the adjusted R2 value was more susceptible to failing (i.e., <0.985) when the response (peak area) was <5000. No significant correlations (i.e., p > 0.01) appeared between adjusted R2 and retention time, mass, drug class or drug lipophilicity. Finally, significant effort was invested to identify a reliable IS to control for analyte recovery. Of the 52 deuterated drugs assessed for inter- and intra-sample reproducibility, 11 were finally selected based on performance, cost, and RT to span the chromatogram; this led to significantly better results compared to the use of a single IS.

Novel Aspect:
A novel HRMS method enabling quantification of new or rare drugs to complement the services of a high-throughput forensic toxicology laboratory.

Wesdemiotis, Chrys,
The University of Akron

Contributing Authors:

Xinqiao Jia, University of Delaware; Selim Gerislioglu, The University of Akron; Sahar Sallam The University of Akron; Bradford A. Paik, University of Delaware.

Title:

Multidimensional Mass Spectrometry of Peptides and Proteins Conjugated with Poly(ethylene glycol)

Introduction:

Poly(ethylene glycol) (PEG) conjugation (pegylation) helps to preserve the α-helical conformation of peptides and proteins and reduces their tendency to denature and self-associate into fibrils. The performance of a conjugate depends on the number and location of its PEG chain(s). Here, we present a multidimensional mass spectrometry (MS) approach for ascertaining these important properties.

Methods:

The composition, primary structure, and architecture of the peptides/proteins and their PEG conjugates were determined by utilizing various combinations of polarity-based fractionation via ultrahigh performance liquid chromatography (LC), MALDI- and/or ESI-MS, tandem mass spectrometry (MS/MS) fragmentation, and orthogonal separation by shape and charge via ion mobility mass spectrometry (IM-MS).

Results:

The peptides K(AAAQ)4K and K(AAAQ)4K-PEG were prepared by solid-phase synthesis which yields acetylated N-termini and amidated (peptide) or pegylated (conjugate) C-termini. IM-MS detected globular and α-helical conformers. The portion of the helical species increased significantly after pegylation affirming that PEG conjugation stabilizes helical geometries. Pegylation was also performed with succinimide-functionalized PEG, which reacts with free amino groups. LC-MS and IM-MS of pegylated substance P (one Lys) showed the addition of two PEG chains, providing evidence that the N-terminus was also derivatized. With insulin, which contains one Lys and two N-termini, LC-IM-MS had to be used to resolve unreacted PEG from mono- and multi-pegylated protein. ESI-MS analysis required the use of triethylamine to lower the charge state or in-source decay to reduce the size of the bound PEG chain.

Novel Aspect:

Determination of PEG attachment sites in peptides and proteins and evaluation of the conformational stabilization provided by covalent pegylation

West, Brandi,
University of Auckland

Contributing Authors:

Angus C Grey, University of Auckland; Kathleen G Mountjoy, University of Auckland; Ashton Partridge University of Auckland; Shannon Cornett, Bruker Daltonics; Katherine A Kellersberger, Bruker Daltonics; Michael L Easterling, Bruker Daltonics; Peter J Derrick, University of Auckland.

Title:

Mapping peptide hormones in mouse brain and pituitary using MALDI Imaging Mass Spectrometry

Introduction:

Peptide hormones are molecules which are critical to most aspects of brain function and behaviour. Melanocortin peptides, are some of the most important hormones involved in the regulation of body weight, obesity and type 2 diabetes. They are enzymatically derived from the precursor protein Pro-opiomelanocortin (POMC), and have unique functions that can either increase or decrease appetite.

Methods:

Mouse pituitary and brain tissue were cryosectioned (12μm), before various tissue washes, matrices and matrix application methods were trialled to optimize peptide signals (m/z 1600-1720). Tissue from wild type mice (WT) and mice homozygous for deletion of melanocortin peptides a-MSH and desacetyl-a-MSH were imaged using MALDI-TOF/TOF and MALDI-FTICR with spatial resolutions of 50—200um.

Results:

In WT pituitary tissue, α-MSH (m/z 1664.78), desacetyl-α-MSH (m/z 1622.77) and diacetyl-α-MSH (m/z 1706.79) were located in the intermediate lobe and were not detected in the homozygous tissue. In addition, seven other peptides were also positively identified and localised to specific lobes of the pituitary gland. In the brain, several ion signals for putative neuropeptides were different between WT and homozygous tissue, and will be discussed. Accurate mass measurements have been made on these same peptides by MALDI-FTICR with the ParaCell ion cell to strengthen identifications of the peptides and as part of a programme of instrument development.

Novel Aspect:

tissue-wide distribution of melanocortin peptides directly from a single tissue section and using two different MALDI instruments for comparison.

Whitty-Léveillé, Laurence,
Université Laval

Contributing Authors:
Elisabeth Drouin, Département de chimie, Université Laval; Claude Bazin, Département de génie des mines, de la métallurgie et des matériaux, Université Laval; Dominic Larivière Département de chimie, Université Laval; Marc Constantin, Département de géologie et de génie géologique, Université Laval.

Title:
Proper quantification of scandium by ICP-MS/MS: the challenge

Introduction:
Inductively coupled plasma mass spectrometry (ICP-MS) is a very sensitive and rapid analytical technique for the quantification of elemental composition down to the ng·L-1 range. However, scandium determination by this technique is plagued by significant spectral interferences which hamper the proper quantification of Sc in complex matrices.

Methods:
This presentation emphasizes on a new method that has been developed for the accurate determination of Sc in mineral samples by inductively coupled plasma tandem mass spectrometry (ICP-MS/MS) using reaction gas.

Results:
Using helium gas was ineffective to overcome all of the spectral interferences observed at m/z = 45 and particularly for Si-based interferences. However, conversion of Sc+ ions into ScO+ ions (after bombardment with O2 in the octopole reaction system coupled with the use of the instrument in MS/MS mass-shift mode) provided interference-free conditions and sufficiently low limits of detection, down to 3 ng L−1, to accurately detect Sc. The accuracy of the proposed methodology was assessed by analyzing five different reference materials (BX-N, OKA-2, NIM-L, SY-3 and GH).

Novel Aspect:
A new and undocumented method for the accurate determination of the amount of scandium in silicon-containing minerals, based on the use of ICP-MS/MS

Wiegelmann, Marcel,
Biomedical mass spectrometry, University of Münster, Germany

Contributing Authors:
Jens Soltwisch, Institute for Hygiene, Biomedical Mass Spectrometry, University of Münster, Germany; Klaus Dreisewerd, Institute for Hygiene, Biomedical Mass Spectrometry, University of Münster, Germany.

Title:
Effect of excitation laser beam diameter and tissue type on the phospholipid ion signal intensities obtained by MALDI-MS imaging in oversampling mode

Introduction:
The lateral resolution in MALDI-MS imaging can be increased by use of a small focal laser spot or by utilizing oversampling. Here, we performed oversampling experiments with a 20 x 20 μm^2 pixel size and focal spot sizes between 30 – 200 µm in diameter to investigate the effect of these parameters on phospholipid ion signal intensities generated in MALDI-MSI of mouse brain tissue sections.

Methods:
DHB matrix was sublimated onto serial mouse brain sections. A fundamental beam shaper and telescope optics were used to adjust focal diameters of flat-top (FT) and Gaussian (Gauss) beams of a Nd:YAG laser (λ=355 nm; 300 Hz), controlled online with a beam profiler. MS data were recorded with a QStar mass spectrometer in continuous raster mode. Mostly phosphatidylcholine (PC) signals were evaluated.

Results:
Despite of full material ablation within the oversampled 20 x 20 μm^2-wide pixels for all beam diameters, strong variations in the ion signals were found. Notably, PC signals were affected by both beam diameter and tissue type. For focal diameters < 80 µm, only weak MS images were generally detected from gray matter (GM), while high PC signals were obtained from myelin-containing white matter (WM). Increasing the overall laser diameter, while keeping the freshly irradiated oversampling area constant, led to contrast-rich PC images also from GM, with marginal improvement in detection of PCs from WM. FT and Gauss profiles produced similar oversampling results. To determine if the elevated fluences needed for ion generation from small spots led to changes in the desorption/ablation mechanisms and thus ion yields, a photoacoustic study to monitor the material ablation step was performed.

Novel Aspect:
MALDI-MSI experiments of phospholipids in continuous raster mode are affected by the beam diameter of the excitation laser and the tissue type

Wielsch, Natalie,
Max Planck Institute for Chemical Ecology

Contributing Authors:

Aleš Svatoš, Max Planck Institute for Chemical Ecology, Mass Spectrometry & Proteomics Group; Natalie Wielsch, Max Planck Institute for Chemical Ecology, Mass Spectrometry & Proteomics Group; Yvonne Hupfer Max Planck Institute for Chemical Ecology, Mass Spectrometry & Proteomics Group; Ding Wang, Max Planck Institute for Chemical Ecology, Chemical Defense of Leaf Beetles Research Group; Antje Burse, Max Planck Institute for Chemical Ecology, Chemical Defense of Leaf Beetles Research Group; Lydia Schmidt, Max Planck Institute for Chemical Ecology, Chemical Defense of Leaf Beetles Research Group.

Title:

Proteomic analysis of integral membrane proteins in different larval tissues from the poplar leaf beetle Chrysomela populi

Introduction:

Larvae of the poplar leaf beetle C. populi use the sequestered glycoside salicin to produce salicylaldehyde for chemical defense against predators. Integral membrane proteins (IMPs) are mainly involved in this process as they mediate transport of sequestered glycosides through the gut membrane.

Methods:

Roughly 100-200 mg of larval tissue (gut, Malpighian tubules and defensive glands) was taken for the fractionation of proteins using differential centrifugation. The proteins were then separated by 1D SDS-PAGE, tryptically digested and analyzed by nanoUPLC-MS/MS. The acquired spectra were searched against insect (Uniprot) and internal C. populi databases.

Results:

In all 3 tissues we identified 3069 proteins with at least 1 TMD. In detail, 2383 were found in the gut, 1483 in the Malpighian tubules and 1235 in the defensive glands. IMPs were defined as proteins having at least 6 TMDs and exhibiting a positive GRAVY value. According to these criteria, we identified 334 IMPs in the gut, 248 IMPs in the Malpighian tubules and 139 IMPs in the defensive glands. Many of the identified IMPs were tissues specific proteins: 136 for the gut, 86 for the Malpighian tubules and 32 for the defensive glands. BLAST analysis revealed that gut specific IMPs include solute carriers, ABC transporters, Ion channels and Ion pumps. Within gland and Malpighian tubule specific IMPs we found different ABC transporters and solute carriers.The obtained data build a fundament for our future investigations.

Novel Aspect:

study that compares membrane proteome of the different insect larval tissues: gut, Malpighian tubules and defensive glands

Williams, Christopher,
EPSRC NMSF

Contributing Authors:
Ruth Godfrey, Swansea University; Gareth Brenton, Swansea University; Charlotte Davies Swansea University.

Title:
Development of a Flowing Atmospheric Pressure Afterglow (FAPA) Ionisation Source to Extend the Capability of Ion Trap and Orbitrap Mass Spectrometry

Introduction:
Flowing Atmospheric Pressure Afterglow (FAPA) (Hieftje et al. (2008)), is a direct atmospheric-pressure ionisation technique. This study utilises our second prototype FAPA probe, a two pin-to-capillary FAPA source design (modified from Shelley et al (2011)). This probe is compatible with the Thermo Scientific Ion Max Source housing. Results are presented from samples of varying chemistries.

Methods:
A LTQ Orbitrap XL (Thermo Fisher Scientific Inc., CA, USA) was fitted with modified Ion Max API housing. The front quartz window of the Ion Max housing was replaced with an aluminium flange which held the FAPA probe in position on-axis to the mass spectrometer (MS) inlet. The Ion Max housing was modified to facilitate sample introduction orthogonal to the FAPA discharge, using a glass tube.

Results:
A variety of analytes of differing chemistries have been successfully ionised by FAPA with strong signal intensities obtained. On-axis positioning of the FAPA probe to the MS inlet showed better signal intensities compared to the off-axis position. Increasing the discharge current and inter-electrode gap showed a brighter glow discharge with a hotter afterglow temperature. Optimal conditions found for the majority of compounds studied were: current 30 mA; helium 6 psi; inter-electrode gap 7 mm. A distance of 7 mm between the FAPA probe and MS inlet was optimal for most analyses, with the sample positioned 2 mm from the FAPA probe exit. This was found to be important to achieve good sensitivity. Comparison with ESI and APCI revealed that FAPA has the potential to provide equivalent or better sensitivity for a range of compounds. Results utilising a heated sample probe are also presented.

Novel Aspect:
Interchangeable with existing ionisation source. Rapid results. Sensitive.

Wilson, John,
Cold Spring Harbor Laboratory

Contributing Authors:
Nikita Saha Turna, Cold Spring Harbor Laboratory; Rosamonde Banks, University of Leeds; Darryl Pappin Cold Spring Harbor Laboratory; Alexandre Zougman, University of Leeds.

Title:
S-Trap sample processing for rapid (< 1 hr) and unbiased digestion of submicrogram to milligram scales

Introduction:
S-Trap technology enables rapid, reproducible and universal sample processing for bottom-up proteomics. It combines strong SDS-based protein solubilization with integrated sample cleanup and rapid reactor-type protein digestion, is simple, can be performed in any lab with standard equipment and works with multiple protease. S-Trap processing works without change on all protein sources tested.

Methods:
S-Trap micros, minis and midis were purchased from ProtiFi LLC (www.protifi.com). Tissue acetone powders and serum were bought. HeLa cells and E. coli were grown. Protein sources were solubilized in 5% SDS and subjected to SDS removal and protein capture as directed (required < 5 min). Trypsin or other protease was added and incubated for 1 hr at 47 C. Samples were analyzed by gel or Orbitrap MS.

Results:
1) 5% SDS solubilized all protein sources tested. 2) The S-Trap protocol effectively removed contaminants without change (PEG, glycerol, detergents, salts, urea, etc.) in < 5 min and captured all applied protein. 3) No bias of S-Trap sample processing was detected as a function of protein molecular weight or cellular compartment. 4) S-Trap sample processing worked without change for all protein sources tested including membrane proteins and difficult-to-digest samples such as muscle tissue. 5) Within the S-Trap, protein and protease are confined to micron-sized pores where they are forced to rapidly interact. Digestion thus occurred in as little as 3 min. 6) Digestion reproducibility was high. 7) S-Trap midis allowed processing of milligram quantities of serum in 1 hr without change. This was especially useful for enrichment studies (e.g. SISCAPA/immunoaffinity MS).

Novel Aspect:
Unbiased sample cleanup and digestion in 1 hr from submicrogram to milligram scales for all samples greatly improves throughput. Helpful to core labs.

Winkler, Robert,
CINVESTAV Unidad Irapuato

Contributing Authors:
Abigail Moreno Pedraza, CINVESTAV Irapuato, Dep. Biotechnology and Biochemistry, Irapuato, Mexico; Sebastian Gibb, Univ. Medicine Greifswald, Dep. Anesthesiology and Intensive Care, Greifswald, Germany; Sandra Martínez Jarquín CINVESTAV Irapuato, Dep. Biotechnology and Biochemistry, Irapuato, Mexico.

Title:
Imaging volatile and semi-volatile compounds by low-temperature plasma MS

Introduction:
Organic compounds with low vapor pressure have important biological functions. The release of volatiles allows communication and regulation between individuals of the same or different species (e.g. in plant defense and attraction of pollinators or prey). Analyzing the distribution of volatile organic compounds (VOCs) in tissues with mass spectrometry (MS) requires ambient ionization methods.

Methods:
We developed a low-temperature plasma (LTP) probe, which is which is based on 3D printed parts. The template files are public and free for non-commercial use. With the 3D printer company MakerMex we designed a sampling/imaging robot based on RepRap technology. For instrument control, creating and analyzing .imzML data files, we wrote a R -based software with browser interface, rmsiGUI.

Results:
Analysis of biological tissues and microbial cultures on surfaces revealed the distribution of molecules in a mass range of about 50 to 500 Da. Notably, small molecules <200 m/z could be detected and localized. Identified compounds include representatives of diverse chemical classes, such as terpenes, alkaloids and hydrocarbons. Using hue-color-luminiscence (HCL) color schemes and contour lines for visualizing MSI data improves the correct perception of signal intensities, as well as the recognition of distribution patterns and abundance gradients. This augmented representation is especially beneficial for data with limited spatial resolution such as diffuse signals of volatiles.

Novel Aspect:
Low-temperature plasma ionization enables the mapping of a wide range of biologically relevant molecules, including volatile organic compounds.

Winter, Martin,
German Cancer Research Center

Contributing Authors:
Martina Schnölzer, Functional Proteome Analysis, German Cancer Research Center (DKFZ), Heidelberg, Germany; Uwe Warnken, Functional Proteome Analysis, German Cancer Research Center (DKFZ), Heidelberg, Germany; Ramona Mayer Functional Proteome Analysis, German Cancer Research Center (DKFZ), Heidelberg, Germany; Ivana Dokic, Molecular & Translational Radiation Oncology, German Cancer Research Center (DKFZ), Heidelberg, Germany; Amir Abdollahi, Molecular & Translational Radiation Oncology, German Cancer Research Center (DKFZ), Heidelberg, Germany.

Title:
Proteomic and phosphoproteomic analysis of cell response to photon and particle irradiation

Introduction:
Radiotherapy is a major field in cancer research. Recently particle radiotherapy came into focus. Contrary to X-rays, particles deposit most of their energy to a small region within tissue, the so-called Bragg peak. This enables dose escalation in the tumor while sparing healthy tissues. To better understand the biology behind the radiation qualities we applied mass spectrometry-based proteomics.

Methods:
Human lung carcinoma cells were irradiated with 2 Gy carbon, 3.5 Gy proton and 6 Gy photon (radiobiological equivalent doses) and harvested 2 h after irradiation to uncover acute cell response. For quantification cells were metabolically labeled using SILAC. GeLC-MS/MS was used for protein expression analysis and IMAC/TiO2 phosphopeptide enrichment for analysis of the phosphorylation status.

Results:
We were able to quantify 2634 proteins and 2818 phosphorylation sites in the irradiated samples. Only negligible changes occur at the protein expression level 2h after irradiation. In contrast, phosphorylation is a much faster event than synthesis or depletion of proteins. Thus we could identify 181 radiation regulated phosphorylation sites. Overall 55 of these sites show significant differences between the radiation qualities. We selected 28 candidates to be validated using synthetic isotope labeled phosphopeptides in a targeted spike-in approach. With this experiment we were able to confirm the phosphorylation sites using fragmentation pattern as well as retention time.

Novel Aspect:
This proteomic comparison of cells irradiated with particles and X-rays shows that certain phosphosites respond differentially to radiation qualities.

Wisztorski, Maxence,
Univ. Lille, INSERM U1192, PRISM

Contributing Authors:
Annie Desmons, Univ. Lille, INSERM, U1192, PRISM, Lille, France; Jusal Quanico, Univ. Lille, INSERM, U1192, PRISM, Lille, France; Julien Franck Univ. Lille, INSERM, U1192, PRISM, Lille, France; Michel Salzet, Univ. Lille, INSERM, U1192, PRISM, Lille, France; Isabelle Fournier, Univ. Lille, INSERM, U1192, PRISM, Lille, France; Jean-Pascal Gimeno, ONCOLille, Lille, France; Benoit Fatou, Univ. Lille, INSERM, U1192, PRISM, Lille, France.

Title:
Droplet-based liquid extraction for spatially-resolved microproteomics analysis from tissue sections

Introduction:
Tissue microenvironment studies is a challenge for understanding the complexity of a pathology. Making a precise picture of the disease needs an efficient microsampling method coupled to an accurate localization for performing region-dependent proteomics. In this presentation we will present all the possibilities of the liquid-microjunction extraction (LMJ) to perform microproteomics analysis.

Methods:
A modified LESA system is used to perform automatic LMJ extrcation on specific regions of a tissue section. Two strategies are possible. In one hand, a direct on-tissue trypsin digestion is performed using a chemical inkjet printer prior to an extraction the peptides. In a second strategy, intact proteins are extracted using LMJ with conventional detergent solution for proteomics analysis.

Results:
Using the on-tissue trypsin digestion strategy followed by LMJ extraction and shotgun proteomics analysis, we showed the possibility to retrieve from a spot size of 650 µm in diameter 1500 proteins on both frozen or FFPE tissue section. Using LMJ with conventional detergent solution for proteomics analysis, we successfully performed immunoblotting experiments and showed the possibility to retrieve and identify proteins from a 1-mm diameter spot size with a high degree of reproducibility. Moreover, the small size of the extracted region allows the possibility to perform multiple extractions on different tissue section points thus permitting precise distinction of the different parts and a functional mapping. We demonstrate that with this technology, it is possible to map the tissue microenvironment and gain an understanding of the molecular mechanisms at millimeter resolution.

Novel Aspect:
Regionalized proteomics using LMJ extraction clearly open the door to the understanding of molecular cross-talk in the tissue micro-environment

Wleklinski, Michael,
Purdue University

Contributing Authors:
Bradley P. Loren, Purdue University; Andy Koswara, Purdue University; R. Graham Cooks Purdue University; Caitlin E. Falcone, Purdue University; Adam Hollerbach, Purdue University; Zinia Jaman, Purdue University; Zoltan K. Nagy, Purdue University; David H. Thompson, Purdue University.

Title:
Analytic Directed Synthesis System for the Manufacture of Pharmaceuticals

Introduction:
The use of mass spectrometry (MS) as a PAT tool for pharmaceutical manufacturing has limited testing. Yet MS can be a powerful PAT tool in both batch and flow processes given its exceptional level of sensitivity and small sample requirement. MS is ideal for flow chemistry providing real-time feedback on reaction progress and impurities in order to ensure product quality.

Methods:
Continuous flow chemistry is performed in microfluidic chips using a Labtrix S1 system (Chemtrix). Purification by extraction and crystallization are done on a homebuilt microfluidic platform. Flow samples are analyzed either on-line through air-segmented inductively charged ESI or offline through nano-electrospray ionization (nESI). All samples are analyzed on a Thermo LTQ mass spectrometer.

Results:
Herein we show that MS is a useful PAT tool in flow, and that it can be useful in the screening of potential synthetic pathways. We demonstrate 1. that accelerated reactions found in microdroplets efficiently screen synthetic pathways to common pharmaceuticals including diphenhydramine, and atropine. Novel routes include synthesizing diphenhydramine from benzhydrol through a mesylated intermediate. 2. that MS and droplet chemistry can guide chemistry in microfluidic flow reactors. It may be possible to provide initial conditions for solvent, reagents, and reaction time. By controlling the degree of product formation in droplets (through reaction distance), we can predict facile vs. arduous chemical reactions. 3. that the use of on-line MS can quickly optimize chemical reactions in flow. 4. that MS can evaluate the quality of a drug or crystal based on its purity.

Novel Aspect:
The ability of MS to both screen synthetic pathways and provide analytical feedback on the quality of a pharmaceutical flow manufacturing process.

Wood, Troy,
University at Buffalo

Contributing Authors:
Emily R. Sekera, University at Buffalo; Heather L. Rudolph, University at Buffalo; Jafor Ali University at Buffalo.

Title:
18O Labeling of Stercobilin, a Putative Autism Biomarker

Introduction:
Prior research from our group showed that the metabolite stercobilin is depleted in the urine of autistic children vs. controls. We wished to determine if this depletion could be observed in an animal model of autism. A method to produce stercobilin isotopomers for quantitation by mass spectrometry was developed, and further tested in fecal samples from transgenic mice and their littermates.

Methods:
Labeling of stercobilin with 18-O water was conducted for several hours between 50-70 C. Assessment of labeling was determined by direct infusion ESI and FT-ICR of the reaction mixtures. Fecal samples from Timothy Syndrome ("autistic model") mice and litermate controls were spiked with 18O-labeled stercobilin and subjected to solid phase extraction followed by direct infusion ESI and FT-ICR.

Results:
A method for reproducibly labeling stercobilin with 18O-labeled water was developed and exceeded 70% total labeling of the four labile sites. The isotopomeric mixture maintained very stable ratios whether frozen or stored at room temperature for at least 20 days. This isotopomer was spiked to fecal samples from the Timothy Syndrome (TS) mice and their littermate controls. After SPE of stercobilin from the samples, ESI FT-ICR revealed that stercobilin levels in the TS mice is almost half the level of their littermate controls. We found that controls excreted ca. 1200 ppb/g feces, whereas the Timothy Syndrome mice excreted only 600 ppb/g feces. This observation is consistent with the earlier observation of stercoblin depletion in autistic urine specimens. Hypotheses that might account for this observation are discussed with clinical potential of stercobilin as an autism marker.

Novel Aspect:
Development of an isotopomer standard for stercobilin, and demonstration of statistically significant depletion of stercobilin in "autistic" mice.

Wu, Bai Han,
NSERC CREATE York University and University of Toronto

Contributing Authors:
Justin Chi-Kai Lau, Department of Chemistry and Centre for Research in Mass Spectrometry, York University, 4700 Keele Street, Toronto, Ontario, Canada M3J 1P3; K. W. Michael Siu, Department of Chemistry and Biochemistry, Windsor University, 401 Sunset Avenue, Windsor Ontario, Canada N9B 3P4; Alan C. Hopkinson Department of Chemistry and Centre for Research in Mass Spectrometry, York University, 4700 Keele Street, Toronto, Ontario, Canada M3J 1P3.

Title:
Dissociation of [b5-H]•+ derived from alpha-methyltryptophan-containing hexapeptides

Introduction:
The dissociation of the [b5-H]•+ derived from tryptophan-containing hexapeptides is induced by the alpha-radical located at the tryptophan residue. The [b5-H]•+ ion exhibit sequence scrambling. Dissociation of ions with an additional methyl group located on the alpha-carbon of the tryptophan residue, are reported here.

Methods:
Peptides were synthesized using solid-phase peptide synthesis. Molecular peptide radicals were generated by collision-induced dissociation (CID) of a metal-ligand-peptide complex, using a triple-quadrupole mass spectrometer with linear ion trap setup (200 QTRAP, Sciex). [b5-H]•+ was generated by CID of the corresponding peptide radical, followed by further CID to obtain the fragmentation spectra.

Results:
The [b5-H]•+ from alpha-methyltryptophan-containing hexapeptides showed non-identical spectra, which suggested no cyclization and sequence scrambling, as opposed to the native peptides. Isotopic labeling experiments suggested loss of one residue from the C-terminus, loss of imine-amides from the N-terminus, and loss of dialanine radical from the N-terminus, from the original sequence. This suggested that the radical is located on the alpha-carbon of the alanine residue at the N-terminus, which weakens the nucleophilicity of the N-terminus nitrogen and prevents it from attacking the C-terminus to macrocyclize. The [b5-H]•+ ions also lose the side chain radicals; and lose an oxazolone ring radical and a water, when the tryptophan residue is located next to the ring. Combining the results, the [b5-H]•+ we studied showed distinctively different chemistries compared to other known systems.

Novel Aspect:
Inhibiting sequence scrambling in [b5-H]•+. Loss of oxazolone ring radical from [b5-H]•+ by homolytic cleavage.

Wu, Ronghu,
Georgia Institute of Technology

Contributing Authors:
Weixuan Chen, Georgia Institute of Technology; Haopeng Xiao, Georgia Institute of Technology; Johanna M. Smeekens Georgia Institute of Technology.

Title:
A Mass Spectrometry-Based Novel Chemical Method to Specifically and Globally Analyze Cell Surface Glycoproteins

Introduction:
Glycoproteins on the cell surface are ubiquitous and essential for cells to interact with the extracellular matrix, communicate with other cells, and respond to environmental cues. Although surface glycoproteins can dramatically impact cell properties and represent different cellular statuses, global and site-specific analysis of surface glycoproteins is extraordinarily challenging.

Methods:
An effective method integrating metabolic labeling, click chemistry and MS-based proteomics was developed to globally analyze surface glycoproteins. Surface glycoproteins labeled with a functional group were specifically tagged through click chemistry, which is ideal because it is quick, specific and occurs under mild conditions. Then tagged glycoproteins were enriched for MS analysis.

Results:
Systematic and quantitative analysis of the surface N-glycoproteome in cancer cells with distinctive invasiveness demonstrated many N-glycoproteins up-regulated in invasive cells, the majority of which contained cell adhesion-related domains. Combining with multiplexed proteomics, we systematically analyze surface glycoprotein dynamics and measure their half-lives. Glycosylation sites located outside of any domain have a notably longer median half-life than those within domains, which strongly suggests the distinct functions. Furthermore, surface glycoproteins with catalytic activities are much more stable, compared with those with binding and receptor activities. This method allows us to globally analyze surface glycoproteins, and will have extensive applications in the biological and biomedical research communities.

Novel Aspect:
The MS-based novel method specifically targets surface glycoproteins, which allows us to quantify surface glycpoproteins and study their dynamics.

Wu, Jiejun,
Janssen R&D US LLC

Contributing Authors:
Leon Chang, Janssen R&D US, LLC.; Jian Zhu, Janssen R&D US, LLC.; Mandana Tootoonchi Janssen R&D US, LLC.; Xuejun Liu, Janssen R&D US, LLC.; Jiao Song, Janssen R&D US, LLC

Title:
A novel high-throughput lipidomics platform and its applications to study drug polypharmacology

Introduction:
Small-molecule drugs, despite being selective to their primary targets, exert pharmacological effects and safety liabilities through multiple pathways. Yet, drug polypharmacology has been extremely difficult to assess experimentally. Profiling of eicosanoids that are produced in response to pharmacological stimuli ex vivo can be a unique and effective way to assess drug polypharmacology.

Methods:
Human whole blood from different donors was stimulated with several inducers of immunological responses, followed by treatments with modulators of prostaglandin or leukotriene biosynthesis including clinical and investigational drugs. An optimized LC/MS method was employed to monitor 122 eicosanoids in these samples and data were analyzed using a special software developed in-house.

Results:
A high-throughput lipidomics platform was developed to separate and quantify 122 eicosanoids in 6 min via UPLC/sMRM, together with a custom software for data analysis that is accurate, fast and fully-automated. Being able to handle hundreds of samples at a time, it has been applied to multiple drug discovery projects. For example, applying this platform to an in vivo model of endotoxemia, drugs were evaluated on their ability to silence prostaglandin biosynthesis in response to bacterial lipopolysaccharide, demonstrating their differential pharmacological activities. It also revealed variability among individual donors, indicating a potential use of this methodology to prescreen patients entering clinical studies. In another study, internally synthesized FLAP inhibitors were profiled for their eicosanoid modulations, which helped to identify potent compounds with highest selectivity.

Novel Aspect:
A first in the field eicosanoid platform (fast LCMS + fully automated data analysis) that has many important applications in biomedical research.

Wu, Jiejun,
Janssen R&D US LLC

Contributing Authors:
Holly Raymond, Janssen R&D US LLC; Justin Kanerva, Janssen R&D US LLC; Leslie Nguyen Janssen R&D US LLC; Jiao Song, Janssen R&D US LLC; Changlu Liu, Janssen R&D US LLC; Steven Nguyen, Janssen R&D US LLC; Yafei Chen, Janssen R&D US LLC; Siquan Sun, Janssen R&D US LLC; Michael C. Harris, Janssen R&D US LLC; J. Eric McDuffie, Janssen R&D US LLC.

Title:
An LC-MS based workflow to separate and quantify bioactive oxysterols from animal tissues and its application to DSS-induced IBD mouse model

Introduction:
Dihydroxycholesterols 7a25 and 7b27 are bioactive metabolites that direct B-cell migration and guide Th17 cell differentiation via their respective receptors EBI2 and RORgt. A workflow that can extract, separate and quantify the structurally very similar oxysterols from tissues is required to study their roles in biology and potentially use them as disease biomarkers.

Methods:
Mice were treated with DSS to induce IBD like symptoms, and FTY720 was used as a treatment to lessen the DSS effects. Mouse colons were extracted with acidified ethanol with antioxidant, followed by ethylacetate extraction. Samples were treated with cholesterol oxidase and derivatized with Girard Reagent P (GP). Oxysterols were separated and quantified via an optimized rpHPLC/sMRM method.

Results:
In addition to using C18 SPE to minimize cholesterol and polar components, a LLE method was carefully chosen to efficiently extract oxysterols. Built on our optimized rpHPLC separation of native oxysterols, GP-labeled oxysterols (7a25, 7b25, 7a27, 7b27, and others) were all baseline separated. The novel workflow has been applied to animal tissues such as spleen, liver, brain and colon. Compared to that of naïve mice, 7a25 was elevated in DSS-treated mouse colons; Co-dosing of FTY720 at 1 and 6 mgpk with DSS reduced this elevation. Histopathological examination showed that DSS damaged colon tissues and FTY720 reduced the damages, which correlated well with the observed 7a25 levels. Preliminary data on the corresponding mouse feces indicated the same trend as in colon tissues. This opens a door for possible noninvasive monitoring of IBD disease progression and recovery in humans.

Novel Aspect:
Optimized workflow for measuring oxysterols from tissues, and the discovery of 7a25-diOHC as a potential biomarker for IBD in DSS-induced mouse model.

Xia, Yu,
Purdue University

Contributing Authors:
Joseph S. Francisco, University of Nebraska at Lincoln; Sarju Adhikari, Purdue University

Title:
Reactions of hydroxyl radical and cysteine disulfide derivatives reveal mechanistic detail in radical induced disulfide cleavage

Introduction:
Formation and cleavage of disulfide bonds are frequently used in the synthesis and delivery of peptide mimetics and prodrugs. Disulfide bonds readily react with radicals in cells; however, reaction mechanisms remain unclear due to difficulty in monitoring transient reaction products. Herein we investigate OH• induced disulfide bond cleavage via changing the electronic property of disulfide bonds.

Methods:
Hydroxyl radicals were generated by discharge in air and their reactions with cysteine disulfide derivatives (CySSR) were effected at the interface of nano-electrospray ionization-mass spectrometer. Data collection was performed on a 4000 QTRAP mass spectrometer in linear ion trap mode. Synthesis of CySSR was achieved according to literature. Cystine ethyl ester was used as an internal standard.

Results:
CySSR was chosen as a model system with R being either electron withdrawing group (EWG) or electron donating group (EDG). Four types of reaction products were observed: protonated cysteine thiyl radical (Cy-S•), cysteine (Cy-SH), cysteine sulfinyl radical (Cy-SO•), and cysteine sulfinic acid (Cy-SOH). For R being EDGs (alkyl or aryl) CySSR had a higher reactivity toward OH• as compared to R being EWGs (acetyl group or carboxylic methyl ester). EWG directed OH• addition preferably to the electron-rich sulfur (Cy side) and promoted H transfer from Cy-SOH to RS•, leading to Cy-SO• as a dominant product. Combined with theoretical calculations, a three-step mechanism was proposed involving: 1) OH• addition to the disulfide bond, 2) disulfide bond cleavage resulting from OH attachment to one of the sulfur atoms, and 3) hydrogen transfer within [CyS-O···H···•SR] complex.

Novel Aspect:
Probing reaction mechanism of disulfide bond with OH radical via changing the electronic property of disulfide bond and monitoring reaction products

21ST INTERNATIONAL MASS SPECTROMETRY CONFERENCE
Toronto, Ontario, Canada
August 20 – 26, 2016

Xiao, Zilan,
NCSTestingTechnology CO., LTD

Contributing Authors:
Li Man, NCS Testing Technonogy Co., LTD; Liu Jia, NCS Testing Technonogy Co., LTD.

Title:
The analysis of two-dimensional distribution of trace elements in a micro area by LA-ICP-MS

Introduction:
The combination of Laser abrasive and ICP-MS, cooperated with effective algorithm is used to realize two-dimensional distribution of trace elements in solid samples. The method play a key part in element contents and spatial distribution analysis, such as detecting segregation of metal blocks and weld joints, as well as mineral elements mapping of ore, thus has excellent application prospects

Methods:
We used the combination of PlasmaMS 300 ICP-MS and LA 300 laser ablation systems.By synchronizing them, time resolving spectra of intensity vs. time are obtained, which are transformed to intensity vs. position spectra, according to correlation between time and displacemen.As a result,the 2D mapping of each element represents the distribution of the element in the area of scanning can be obtained

Results:
Nd-Fe-B alloy is an important permanent magnet material. In order to modify the characteristic of the material, a small amount of other rare earth elements, such as Terbium (Tb) and Dysprosium (Dy), are used to replace Nd by certain methods, such as high-pressure penetration, etc. Analyzing the distribution of the elements provides an effective method to evaluate the penetration process and the quality of the modified material. In this paper, the distribution of Tb and Dy in a type of Nd-Fe-B alloy was analyzed using the above method. The 2D distribution of the Tb and Dy elements in the scanning area achieved, shown that the concentration of both elements decrease from the edge to the center of the area. The work proves that the LA-ICP-MS combined with algorithm, has a great potential for analyzing 2D distribution of elements in a micro region on solid materials.

Novel Aspect:
The analysis of 2D distribution of trace elements in a micro area of solid samples is realized by combining LA-ICP-MS with effective algorithm

Xu, Guowang,
Dalian Institute of Chemical Physics, CAS

Contributing Authors:
Hailong Piao, Key Laboratory of Separation Science for Analytical Chemistry, Dalian Institute of Chemical Physics, Chinese Academy of Sciences, 116023 Dalian, China; Xinjie zhao, Key Laboratory of Separation Science for Analytical Chemistry, Dalian Institute of Chemical Physics, Chinese Academy of Sciences, 116023 Dalian, China; Min Yan Key Laboratory of Separation Science for Analytical Chemistry, Dalian Institute of Chemical Physics, Chinese Academy of Sciences, 116023 Dalian, China; Xin Lu, Key Laboratory of Separation Science for Analytical Chemistry, Dalian Institute of Chemical Physics, Chinese Academy of Sciences, 116023 Dalian, China; Yaping Shao, Key Laboratory of Separation Science for Analytical Chemistry, Dalian Institute of Chemical Physics, Chinese Academy of Sciences, 116023 Dalian, China; Peiyuan Yin, Key Laboratory of Separation Science for Analytical Chemistry, Dalian Institute of Chemical Physics, Chinese Academy of Sciences, 116023 Dalian, China; Guowang Xu, Key Laboratory of Separation Science for Analytical Chemistry, Dalian Institute of Chemical Physics, Chinese Academy of Sciences, 116023 Dalian, China.

Title:
Integrated Omics Approach Centered on MS-based Metabolomics to Decipher Mechanism Implicated in Human Diseases

Introduction:
Due to the higher sensitivity and favorable structural identification ability of mass spectrometry (MS), MS-based metabolomics has shown great potential in disease research. Yet, integration with other omics datasets is necessary for an in-depth study. In this work, we combined metabolomics and transcriptomics to unravel the potential mechanism underlying human disease.

Methods:
We performed LC-MS-based metabolomics and lipidomics analyses on matched prostate cancer tissues and adjacent noncancerous tissues, 20 of which were subjected to RNA-seq analysis. Centering on the findings at the metabolic level on differential metabolites and related metabolic pathways, we further combined the transcriptional data for the mechanism investigation.

Results:
LC-MS based metabolomics approach yielded 128 differential metabolites, including 13 carnitines and acylcarnitines, 68 lipids and fatty acids, and 47 organic acids and alkaloids. Using lipidomics approach, we defined 130 differential lipid species, most of which were dominantly elevated in tumor tissues, covering a multitude of lipid classes including PC, PE, PG, PI, Cer, DAG, CE and FFA, with 1.65–15.87 fold' increase. To trace the upstream variations of the metabolome, we analyzed the expressed gene levels of the differential metabolic species. The alteration of gene expression further confirmed the metabolic pathway dysregulation in prostate cancer. Moreover, we found impaired sphingosine-1-phosphate receptor 2 signaling, downstream of sphingosine, representing a loss of tumor suppressor gene and a potential key oncogenic pathway for therapeutic targeting.

Novel Aspect:
Integrated omics provides a broad picture of the molecular network on prostate cancer and contributes to a better understanding of the mechanism.

Yamamoto, Atsushi,
Tottori University of Environmental Studies

Contributing Authors:
Naoko Matsumoto, Kansai University; Hideya Kawasaki, Kansai University; Ryuichi Arakawa Kansai University.

Title:
Reading of accurate mass spectra to find anthropogenic substances in the environment.

Introduction:
Our daily life is surrounded by enormous substances. Although mass spectrometry (MS) has been used in various routine inspections and monitoring with superior quantitative performance, one more important advantage of MS is the qualitative performance, which is not always delivered in the analyses. We conducted nontarget analysis to identify anthropogenic substances in the environment.

Methods:
High-resolution MS with a resolution of 100 000 was applied to identification of anthropogenic substances in aquatic environment. River water samples were collected from urban rivers in Osaka, Japan. Substances extracted by solid phase extraction cartridges (PreSepC Agri) were wholly analyzed by liquid chromatograph/mass spectrometer (Ultimate 3000 and Exactive) with electrospray ionization (ESI).

Results:
Numerous substances were observed in one analysis. At the start of reading of mass spectra, obvious noise was removed from the observation. Isotopologues and relative adduct ions were consolidated as one feature. The 20 most intense features in positive- and negative-ion ESI proceeded to detailed interpretation. Collision induced dissociation mass spectra were also acquired for each feature. Elemental composition candidates from accurate masses of neutral losses and ions was retrieved by Xcalibur Software. Consistency with isotope pattern and fragmentation was examined for the candidates. Sixteen substances were confirmed by agreement with authentic reagents. They included pharmaceuticals, surfactants, and industrial materials. Some of them, for example telmisartan, stearamidopropyl dimethylamine, and 1,3-di-o-tolylguanidine, have not been considered environmental persistent substances.

Novel Aspect:
Emerging environmental persistent substances were analyzed by nontarget high-resolution MS. The majority could be identified by a manual method.

21ST INTERNATIONAL MASS SPECTROMETRY CONFERENCE
Toronto, Ontario, Canada
August 20 – 26, 2016

Yan, Rongxin,
Beijing Institute of Spacecraft Environment Engineering

Title:
The discuss of the development of gas analyzer in space exploration

Introduction:
This paper introduces the relation and application background between gas composition analysis instrument and space exploration. It summarizes the character of international space exploration for gas sample, especially the sample analysis at Mars in Curiosity rover. The Chinese gas mass spectrometer at satellite is introduced. To explore the Mars gas, we are trying to develop gas analyzer.

Methods:
The plan of unidentified gas quantitative analysis for space exploration with the combination of chromatogram and mass spectrometer (MS-TOF) is put forward. A widen range, high resolution and low power consumption TOF is developed. The component of low power field emission and adjust electron energy focus electron beam collision ionization source are manufactured.

Results:
The gas analyzer backed by the NEG pump can get 10-3Pa vacuum and measure analyzer chamber pressure from 105Pa to 10-3Pa, which can sample gas from 103Pa Mars gases in CO2 background gas. The TOF mass range can be reached to 1000amu. The resolution (28amu) is 400. The sensitivity arrives at 1ppm. The scan period of full range is less 1 µs. The gas analyzer can supply with 24 VDC and consumption is 10W. The weight is 5 kg. In the laboratory, the He、Kr、Xe、SF6、CF4、C2H6、CH3COCH3、C6H4(CH3)2 are analyzed and the scan graph are got.

Novel Aspect:
The gas analyzer of using TOF has low power consumption, less weight and high sensitivity. So it has basic ability to measure Mars gases.

21ST INTERNATIONAL MASS SPECTROMETRY CONFERENCE
Toronto, Ontario, Canada
August 20 – 26, 2016

Yan, Rongxin,
Beijing Institute of Spacecraft Environment Engineering

Title:
The Effect of Sampling Pipeline on the Measure Gas Partial Pressure with Mass Spectrometer in Molecule Flow

Introduction:
In some gas analysis or partial pressure measure, the sample gas place is far from the mass spectrometer because of mass spectrometer's weight or the gas toxicity. So the long sampling pipeline connecting the sample place and mass spectrometer has to be used. But the long sampling pipe how to affect measure results such as stable time,ingredient aberration, The problems are worthwhile to study.

Methods:
Based on the study of unsteady-state gas with the effect of pipeline adsorption, the physical mechanism of gas in pipeline end flowing into mass spectrometer chamber at rate molecules and pumped by vacuum system has been studied. Transportation equation and mass spectrometer chamber pumping equation of gas partial pressure distributing in long pipeline are set up.

Results:
The relationship equation between pipeline pressure, the chamber pressure, analysis time, pumping rate to the chamber, gas molecular mass, pipeline length has been got. (1)The pressures in sampling pipeline gas are a complex exponential equation with sampling time and sampling pipeline distance; (2) The balance pressures in mass spectrometer chamber are direct proportion to the pressure of the sampling place; (3) The sampling pipeline change the ration of gas ingredient. After sampling equilibrium,the ratio of partial pressure is inverse to the square root of the ratio of gas mass.The pressure in mass spectrometer chamber is inverse to the length of pipeline. (4) The minimum pressure stable time in chamber is proportion to the length of pipeline and inverse to gas molecule thermal motion speeds.

Novel Aspect:
Using the equation, the effects of sampling pipeline on mass spectrometer partial gas measure such as Kr、 CF_4、 SF_6、 N_2H_4、 N_2O_4 in air have been analyzed.

YANG, PENGXIANG,
THERMO SCIENTIFIC

Contributing Authors:
Kristine Van Natta, Thermo Scientific; Christopher Elicone, Thermo Scientific.

Title:
Rapid LC method development for multiple drug classes by using a new four-channel LC

Introduction:
LC/MS method development is crucial but time-consuming due to the complexity of compounds. How to develop LC methods for multiple drug classes is more challenge due to the consideration of various HPLC columns, various mobile phases and different gradient. By using a new four-channel LC system, the LC method development can be done rapidly and efficiently.

Methods:
Four different UPLC/HPLC columns were installed on Thermo Scientific Prelude LX-4 liquid chromatography system which has four independent LC channels. These four columns were tested on their performance to separate four classes of drug compounds (anticonvulsants, antidepressants, antipsychotics, and antiarrhythmics) with various LC gradients.

Results:
Four different UPLC/HPLC columns were picked to separate four classed of drug compounds (anticonvulsants, antidepressants, antipsychotics, and antiarrhythmics). Three linear gradients with different initial %MP-B were programmed as multiple test methods to check column performance in a single batch. From a single batch data, we quickly confirmed the best UPLC/HPLC column for each drug class. Also the bath data gave us a clear clue on how to optimize the LC gradient. By using the new four-channel LC system, we completed LC method development for four classes of drug compounds (24 compounds) within 24 hours. Calibration standards and quality controls (QC) of 24 compounds in plasma were analyzed with the developed methods. Accuracies were better than 84% and precisions expressed as %RSD were better than 14.7% across all compounds and all concentrations.

Novel Aspect:
A fast and high efficient workflow of LC method development was demonstrated on a new four-channel HPLC system connected to a mass spec.

Yang, Xiaoyu,
National Institute of Standards and Technology

Contributing Authors:
Pedatsur Neta, National Institute of Standards and Technology; Stephen E. Stein, National Institute of Standards and Technology.

Title:
Extending a High Quality Reference Tandem Mass Spectral Library for More Comprehensive Metabolite Identification

Introduction:
Tandem mass spectral library searching is becoming an efficient and reliable data analysis technique for identifying metabolites in complex mixtures from LC/MS/MS data. We have been extending a tandem mass spectral library with MS2 spectra of various precursor types, including ions produced in-source, and MSn spectra, for more accurate and robust metabolite identification.

Methods:
Mass spectra of each authentic metabolite were acquired with an Orbitrap Elite instrument (IT, FT-IT, and HCD). First, the precursors were identified based on the chemical structure. Then the spectra with the same precursor were clustered based on the fragmentations for MS2, MS3 and MS4, respectively. Furthermore, a consensus spectrum at the same instrument condition was generated for the library.

Results:
To date, we examined 4,480 new metabolites with multiple precursor types, MS3 and MS4 spectra, as well as MS2 spectra of ions produced in-source. A total of 208,054 new spectra of high and low resolution under different conditions were added to a tandem mass spectral library after quality control and manual validation. Using this extended library with 13,164 compounds and 485,619 spectra, some metabolites were identified with not only [M+H]+ and [M-H]-, but also [M+Na]+, [M+2H]2+, [M-2H]2-, etc. The ions produced in-source by various neutral losses (e.g. [M+H-C5H7NO3]+ from oxidized glutathione) were also identified. Isomeric metabolites such as N-methylphenylalanine and α-methylphenylalanine were distinguished by searching their MS3 spectra. This library has been applied in E. Coli metabolomic studies for identifying metabolites and ions produced in-source.

Novel Aspect:
A comprehensive high quality reference tandem mass spectral library is constructed and used for more accurate and robust metabolite identification.

21ST INTERNATIONAL MASS SPECTROMETRY CONFERENCE
Toronto, Ontario, Canada
August 20 – 26, 2016

Yao, Chenyu,
University of Waterloo

Contributing Authors:
Bin Ma, My graduate supervisor who instruct me about this project.

Title:
Waters IMS Raw Data Peptide Feature Identification

Introduction:
As the development of today's mass spectrometry, Ion-Mobility Spectrometry(IMS) has been widely used in laboratory. In order to identify peptide sequence using IMS, the primary task is to find all precursors and their corresponding spectrums from raw data. This program has two main parts: peptide feature visualization and peptide feature detection.

Methods:
Peptide feature visualization extracts raw data and presents it into a graphic user interface. A heat map represents intensities among RT and DT axis. Peptide feature detection makes use of local search algorithm to find all local maximum peaks and uses a hill-climbing algorithm to determine spectrum sets for peptides.

Results:
Peptide feature visualization allows user to select a particular mass range and visualize intensities among all RT and DT in a heat map. If certain peptides falls in selected mass range, user will be able to clearly identify those peptides and their corresponding spectrum sets from map. Peptide feature detection part automatically finds all precursors and their peptide features from raw data. The first step of detection program is to find all peaks such that their intensities are local maximum within preset mass-window, RT-window and DT-window. Then, these peaks data is processed to compose peptide features which include information such as charge, isotopes and set of corresponding spectrums.

Novel Aspect:
The structure of Waters IMS raw data is imagined as a 3D space. All handling and processing of raw data is based on this 3D space.

21ST INTERNATIONAL MASS SPECTROMETRY CONFERENCE
Toronto, Ontario, Canada
August 20 – 26, 2016

Ye, Jianhuai (Jackie),
University of Toronto

Contributing Authors:
Lauren Vysohlid, University of Toronto; Arthur Chan, University of Toronto.

Title:
Chemical Speciation of Secondary Organic Aerosol Using Electrospray Ionization-Ion Mobility Time of Flight Mass Spectrometer

Introduction:
Secondary organic aerosol (SOA) generated from oxidation of volatile organic compounds accounts for a large component of urban particulate matter. It plays an important role in global climate change and poses serious adverse effects on human health. Owing to the large complexity of atmospheric oxidation processes, the underlying chemical composition of SOA remains poorly understood.

Methods:
In this work, chemical composition of SOA was analyzed using electrospray ionization-ion mobility time of flight mass spectrometer (ESI-IMS/TOF). Coupled with high performance liquid chromatography (HPLC), SOA complex can be better separated, and important SOA composition can be resolved. In all experiments, SOA was generated from naphthalene photooxidation or limonene ozonolysis reaction.

Results:
Preliminary results on naphthalene SOA analysis show that IMS is capable of analyzing SOA mixture. SOA components detected in our analysis were consistent with the observations from other studies (Chan et al., ACP, 2009; Kautzman et al., J. Phys. Chem. A., 2010). 2D IMS-MS is used for identification of chemical families, such as quinone species. Reaction mechanisms based on the chemical speciation are further discussed.

Novel Aspect:
Atmospheric organic mixture were analyzed in a three dimensional basis. Using 2D IMS-MS, we are able to group the mixture into chemical families.

21ST INTERNATIONAL MASS SPECTROMETRY CONFERENCE
Toronto, Ontario, Canada
August 20 – 26, 2016

Yeung, Bernie,
LECO Corporation

Contributing Authors:

Lorne M. Fell, Leco Corporation, Saint Joseph, MI, USA; Christina N. Kelly, Leco Corporation, Saint Joseph, MI, USA; David E. Alonso Leco Corporation, Saint Joseph, MI, USA; Joseph E. Binkley, Leco Corporation, Saint Joseph, MI, USA; Jonathan D. Byer, Leco Corporation, Saint Joseph, MI, USA.

Title:

The Distinction of cis and trans fatty acids in food products with a novel benchtop time of flight GC-MS system

Introduction:

The distinction of cis and trans fatty acids in food is important because of the upcoming ban on trans fats for consumption. Established methods, like AOAC 996.06, accomplish this with GC-FID where complete chromatographic separation is required. This is challenging due to the structural similarity of the isomers. This poster demonstrates the analysis with a novel benchtop GC-TOFMS system.

Methods:

A fatty acid standard containing both cis and trans isomers was analyzed along with fats that were derivatized to fatty acids after extraction from food samples. GC-TOFMS with a SP-2560 (75m x 0.18mm x 0.14um) column was used for all analyses. Fatty acid profiles were determined for each sample and the presence or absence of trans fat was determined.

Results:

This novel GC-TOFMS system was used to demonstrate the distinction of cis and trans fatty acids and to screen food samples to determine their representative fat profiles. Various butter and shortening samples were analyzed and a fat profile was determined for each. The levels of cis and trans fats were also determined. This analytical approach was capable of performing the analysis and TOFMS detection offered distinct benefits over FID detection. TOFMS provided mass spectral information and the ability to identify analytes within the sample as well as deconvolution capabilities in instances of chromatographic coelution. Several examples and benefits are reported.

Novel Aspect:

A novel benchtop GC-TOFMS system was used for these analyses.

Yeung, Bernie,
LECO Corporation

Contributing Authors:
Christina N. Kelly, Leco Corporation, Saint Joseph, MI, USA; Elizabeth Humston-Fulmer, Leco Corporation, Saint Joseph, MI, USA; Lorne M. Fell Leco Corporation, Saint Joseph, MI, USA; Joseph E. Binkley, Leco Corporation, Saint Joseph, MI, USA.

Title:
A Novel Benchtop GC-TOFMS for Fast Targeted Allergen Screening and Non-targeted Characterization for Personal Care Products

Introduction:
Some fragrance compounds in personal care products are allergens or skin irritants, so it is important to be aware of their presence in a product. Manufacturers must provide this information for compliance with rules, such as the European Cosmetics Directive that regulates 26 contact allergens. GC-TOFMS is well-suited for this targeted screening and also for non-targeted characterization.

Methods:
A set of allergen calibration standards with effective on-column concentrations of 1 ppb to 1 ppm were analyzed with a novel benchtop GC-TOFMS system utilizing a short (10m) and narrow (0.18mm) Rxi-17Sil MS column and a fast temperature ramp. Calibrations were determined and applied to various fragrance samples. Non-targeted comparisons of the perfumes were also performed.

Results:
A GC-TOFMS method was developed to complete the analysis in approximately 5 minutes. Chromatographic resolution along with mathematical deconvolution of the TOFMS data separated the target allergens within the standards. Calibration equations were compiled from 1 ppb to 1 ppm levels (on-column) with excellent linearity and correlation coefficients. The calibrations were applied to various commercially-available perfume and cologne samples and quantitative information for the targeted allergens was determined. TOFMS provides simultaneous non-targeted analyses because of the full-mass range data acquisition. Thus, additional non-targeted characterization and comparison of the perfume samples were also performed. Numerous similarities and differences between the samples are highlighted for an improved understanding of the aroma profile of these products.

Novel Aspect:
The reported method reduces analysis time for allergens and increases acquired characterization information for PCPs.

Yuk, Jimmy,
Waters Corporation

Contributing Authors:
Kate Yu, Waters Corporation; Bharathi Avula, National Center for Natural Products Research, Research Institute of Pharmaceutical Sciences, University of Mississippi; Ikhlas A. Khan National Center for Natural Products Research, Research Institute of Pharmaceutical Sciences, University of Mississippi; Yan-Hong Wang, National Center for Natural Products Research, Research Institute of Pharmaceutical Sciences, University of Mississippi; Mark Wrona, Waters Corporation; Giorgis Isaac, Waters Corporation.

Title:
A Metabolomics Approach for the Authentication of Herbal Supplements Using UPLC/QTof-MS

Introduction:
Metabolomics provides an unbiased, comprehensive qualitative/quantitative overview of the metabolites present in living organisms. This approach, enables a holistic comparison of samples compared to targeted screening methodologies. In this study, the LC/MS chemical profiles of the botanicals (Hoodia and chamomile) and their commercial products were investigated using a non-targeted approach.

Methods:
The botanicals and their commercial products were injected in an ACQUITY I-Class UPLC and XEVO G2-S QTOF-MS. A streamlined workflow for data processing and analysis was performed using Progenesis QI. The LC/MS data was aligned to correct any drift and then peak detection, normalization, deconvolution and identification were performed. EZinfo in Progenesis was used for the multivariate analysis.

Results:
Multivariate statistical methods using principal component analysis and partial least squares - discriminant analysis of the datasets showed similarities and differences between the reference and commercial herbal products. Many Hoodia commercial products did not match the chemical profiles of reference Hoodia species while the commercial chamomile teas matched their respective reference materials. The loading plots identified key markers that differentiate between the botanicals and their related species. To ensure confidence in the identification of the markers, the exact mass precursor ion, theoretical isotopic distribution and theoretical fragments using in-silico fragmentation were used as the matching criteria. This study shows the advantages of using a non-targeted approach such as metabolomics to authenticate complex herbal materials and their commercial products.

Novel Aspect:
This study shows the chemical profiling of botanicals and their commercial products using a streamlined informatics platform.

Zaia, Joseph,
Boston University School of Medicine

Contributing Authors:
Chun Shao, Boston University School of Medicine; Joshua Klein, Boston University School of Medicine; Joanna Phillips University of California, San Francisco.

Title:
Glycomics, proteomics and glycoproteomics of glioblastoma

Introduction:
The extracellular matrix (ECM) constitutes 10-20% of brain volume and provides a microenvironment for maintaining the essential functions of central nervous system. Over 80% of ECM proteins are glycosylated; however, little is known about the glycosylation pattern for brain ECM proteins. Here we present a sensitive platform to analyze N-glycoproteome and O-glycoproteome for brain ECM.

Methods:
We applied enzymes to human brain cortex histological tissue sections to release glycosaminoglycans and proteins, respectively. The released peptides were either enriched with hydrophilic interaction liquid chromatography (HILIC) or strong anion exchange stage tips. Half of the HILIC-enriched peptides were analyzed with C18 LC-MS/MS at Thermo-Fisher Q Exactive Plus mass spectrometer.

Results:
We observed high levels of ECM glycoproteins from surface digestion on histological tissue sections, including CSPGs/HSPGs (brevican, versican, neurocan, CSPG4, CSPG5, and perlecan), tenascins (tenascin and tenascin R), and hyaluronan and proteoglycan link protein 1. Our data identified site specific glycosylation on ECM glycoproteins from brain. We used CAD-based tandem mass spectrometry of enriched N- and O-glycopeptides to identify peptide sequences and glycan compositions. We used the in-hous GlycoReSoft software for automated data analysis for N-glycoproteome and O-glycoproteome. Taking the versican CSPG as an example, we mapped multiple un-reported N-glycosylated, mucin type O-glycosylated, and GAG-linked sites in the core protein. We made a detailed comparison for N-glycoproteome and O-glycoproteome for human glioblastoma biospecimens and tumor-adjacent normal samples.

Novel Aspect:
A comprehensive N-glycoproteome and O-glycoproteome analysis to improve fundamental understanding of the architecture of brain ECM.

Zare, Richard,
Department of Chemistry, Stanford University

Title:
C-H Bond Silylation of Aromatic Heterocycles Catalyzed by Potassium t-Butoxide in Microdroplets

Introduction:
Much excitement greeted the news (Toutov et al. Nature 2015, 518, 80–84) that potassium t-butoxide was able to catalyze the conversion of a C–H bond to a C–Si bond in aromatic heterocycles.

Methods:
In my laboratory, Dr. Shibdas (Monon) Banerjee using compounds supplied by Dr. Anton Toutov, has investigated this reaction by means of electrospray ionization (ESI) mass spectrometry.

Results:
We propose a possible catalytic cycle for this reaction in which both potassium cation and the t-butoxide anion serve as catalytic agents. We are able to identify crucial ionic intermediates present in the C–H silylation process occurring in the microdroplet.

Novel Aspect:
Most catalysts use precious metals whereas this reaction system does not.

Zarrine-Afsar, Arash,
Department of Medical Biophysics

Contributing Authors:
Adam Gribble, University of Toronto; Alex Vitkin, University of Toronto; Michael Woolman University of Toronto; Manuela Ventura, University of Toronto; Howard Ginsberg, University of Toronto; Milan Ganguly, University of Toronto; Nicholas Bernards, University of Toronto; Jinzi Zheng, University of Toronto; Alessandra Tata, University of Toronto.

Title:
Rapid Detection of Necrosis in Breast Cancer with Desorption ElectroSpray Ionization Mass Spectrometry

Introduction:
Necrosis in tumours is often associated with highly aggressive forms of cancer, and is thought to be of prognostic value in cancer treatment planning. Rapid intraoperative detection of necrosis within a breast tumour thus allows effective treatment to begin as the tumour is encountered.

Methods:
Using murine xenograft tumours of human breast cancer subjected to DESI-MS imaging, statistical analysis and pathology, the lipid MS profile of necrosis is determined. MS experiments were performed using a XEVO-G2 Q-TOF coupled to a DESI source (Waters). The assignment of lipid biomarkers was made by DESI-MS/MS, corroborated with LIPID MAPS database search and isotopic pattern distribution.

Results:
Necrosis MS profile is characterized by (1) the presence of the ion of m/z 572.48 [Cer(d18:1)(16:0)+Cl]- which is absent from the viable cancer site; (2) the absence of the ion of m/z 391.25 which only exhibits a small abundance in the viable cancer site; and (3) by an increase in the relative intensity of known breast cancer biomarker ions of m/z 281.2 [FA(18:1)-H]- and 303.2 [FA(20:4)-H]-. We further show that necrosis is accompanied by alterations in tissue depolarization rate, allowing wide-field Mueller matrix polarimetry to guide DESI-MS to necrotic areas for rapid profiling. Necrosis-specific biomarker ions are also detected in the smear of necrotic tumours, rapidly prepared (in~30s) without the need for time-consuming tissue sectioning and slide preparation steps.

Novel Aspect:
Coupled with our discovery of necrosis MS profile, our guided approach to MS profiling allows necrotic cancer characterization in as little as 1 min.

Zeegers, Guido,
ETH Zürich

Contributing Authors:
Steffen M. Weidner, Federal Institute for Materials Research and Testing; Robert F. Steinhoff, ETH Zürich; Renato Zenobi ETH Zürich.

Title:
Evidence for laser-induced redox-reactions between added trifluoroacetate salts and substrate material during polystyrene/DCTB MALDI

Introduction:
The cation addition mechanism to polystyrene and the possible interactions between the target plate material and the cationization agent have been studied.

Methods:
A range of different trifluoroacetate salts mixed with DCTB and polystyrene, deposited on several different target plate materials, have been systematically investigated with MALDI mass spectrometry.

Results:
MALDI-MS yielded polystyrene- and DCTB-Cu adducts when sample solutions were applied on a copper substrate. For silver the same effect was observed. Control experiments showed that these species could only result from laser-induced redox-reactions between the added salt and the substrate material.

Novel Aspect:
Reversed redox-reactions between added salts and target plate material are enabled by laser photon energy input during MALDI experiments.

Zenobi, Renato,
ETH Zurich

Title:
Mass Spectrometric Profiling of Biomarkers in Exhaled Breath for Medical Diagnosis and Personalized Medicine

Introduction:
Exhaled breath contains relevant information on a person's health status. The vision is to use real-time and completely non-invasive chemical analysis of exhaled breath for applications such as medical diagnosis, monitoring progress and treatment of diseases, drug compliance, pharmacokinetics, and others.

Methods:
Our methodology to analyze breath in real time is based on secondary electrospray ionization coupled to high resolution mass spectrometry (SESI-HRMS). It affords ppb ... ppt limits of detection, and analysis of compounds with molecular weights up to 1000 Da. Extensive statistical and chemometric evaluation of the large data sets is another important element for the success of this approach.

Results:
A number of interesting questions can now be addressed via on-line mass spectrometric analysis of exhaled breath: is there a core pattern for individual phenotypes visible in mass spectrometric "breathprints"? Can diurnal changes be monitored via exhaled breath? Can diseases be diagnosed via exhaled breath, and if yes, which ones? Can proper drug use (or drug abuse) be detected via analysis of the chemical composition of exhaled breath? The presentation will focus on several examples in medical diagnosis, including the detection of novel biomarkers for diseases such as obstructive sleep apnea (OSA) and chronic obstructive pulmonary disease (COPD). Monitoring of drug compliance and pharmacokinetics via real-time SESI-MS will also be shown.

Novel Aspect:
MS-based profiling of exhaled breath is a non-invasive alternative to the analysis of blood or urine in medical diagnosis and drug monitoring.

Zenobi, Renato,
ETH Zurich

Contributing Authors:

Katharina Root, ETH Zurich; Martin Köhler, ETH Zurich.

Title:

Probing Very High Mass Noncovalent Assemblies with Mass Spectrometry

Introduction:

Many noncovalent assemblies that occur in nature, such as molecular machines, amyloids, or large capsid structures have very high and sometimes heterogeneous molecular weights that are not accessible with common mass spectrometers. Moreover, such assemblies may only be stable in media containing high salt concentrations or additives such as detergent micelles.

Methods:

MALDI is the method of choice if additives in the buffer solution compromise or prohibit ionization by ESI. MALDI spectra area easily interpreted because they are dominated by singly charged ions. However, this characteristic also leads to ions with very high m/z, which are difficult to detect. A ion-to-ion conversion high mass detector for detection of ions ≥ 1 MDa is employed in our research.

Results:

Results on MALDI-MS investigations on a number of systems will be discussed: (i) Complexes of nanobodies with membrane-bound ATP-binding cassette transporters, used for stabilization of such systems in the context of x-ray crystallography. (ii) Complexes between antibodies that specifically bind to monomeric and oligomeric forms of amyloid forming proteins such as alpha-synuclein. (iii) Studies of the assembly mechanism of carboxysomes, cellular microcompartments 80-150 nm in diameter composed of a variable number of different proteins. Monomers (e.g., CcmK1, CcmK2, CcmK3 and CcmK4 paralogs) have molecular weights of 10 to 12 kDa, and some share >80% sequence identity. (iv) Investigations of an engineered variant of Lumazine synthase from Aquifex aeolicus (AaLS-13) that self-assembles into a capsid and can be used to encapsulate a supercharged GFP (+36) molecule by Coulomb attraction.

Novel Aspect:

MALDI-MS analysis of the assembly mechanism and stoichiometry of high-mass noncovalent complexes and capsids.

Zhou, Jiang,
Peking University

Contributing Authors:
Shiwei Zhang, Peking University; Xiaodi Yang, Nanjing Normal University; Huihui Li Nanjing Normal University.

Title:
Ligand induced conformation and stability change of G-quadruplexes formed by the BRCA1 gene promoter

Introduction:
G-quadruplexes are secondary DNA structures featured of stacked planar G-quartets, which are comprised of four Hoogsteen bonded guanines. The conformation of G-quadruplexes can vary when exposed to different environments, such as the change of cations and ligands spices. Here, the BRCA1 gene was chosen as target, which was believed to be the oncogene of breast cancer.

Methods:
Here, the formation, recognition and transformation of the G-quadruplexes formed by BRCA1 gene promoter were studied by ESI-MS, CD and DMS. The target sequence from BRCA1 gene was d[AG3T2G3AG5CTCG4C] and marked as S.

Results:
The peak of [Q]5- in ESI-MS spectra implied the formation of G-quadruplexes (Q) with three tetrads, which was then confirmed by CD and DMS. Furthermore, jatrorrhizine (P), a crescent alkaloid isolated from Enantia chlorantha, was found to have high affinity to the G-quadruplexes. The complexes ions of [Q+P]5- in MS Spectra became the base peak at the molar ratio of Q/P 1:2. More interestingly, when binding to P, the conformation of the G-quadruplexes transformed from mixed motif into parallel structure, and the stability of G-quadruplexes improved significantly. In DMS foot printing assay, the intensities of cleaving bands with jatrorrhizine were lower than that without ligands, indicating that the structures of the complexes were tighter than free G-quadruplexes. Thus, this natural product may provide important information to drug design and will also help in treatment of breast cancer.

Novel Aspect:
A high affinity ligand jatrorrhizine for BRCA1 G-quadruplexes was found, which could also induce the G-quadruplex conformation and stability change.

Zhou, Jiang,
Peking University

Contributing Authors:
Gu Yuan, Peking University.

Title:
Distinguish binding sites of G-quadruplexes with ligands in competitive mode by mass spectrometry

Introduction:
Guan rich sequences can fold themselve into special secondary structures, called G-quadruplexes. In this paper, a competitive model of mass spectrometry was established to distinguish the binding sites of G-quadruplexes with different small molecules.

Methods:
The competition experiments were carried out in negative mode of ESI-MS. G-quadruplexes were mixed with 2-3 small ligands in equal concentrations, and compared by the intensities of binding peaks in one MS spectrum.

Results:
The intensities of binding peaks with G-quadruplex were directly compared to evaluate the binding affinities of small ligands in competition experiments. It was observed that a G-quadruplex binding with two different small molecules. By comparing the intensity of [Q1+B] with [Q1+A+B], whether ligand A and B are in competition can be determined. The MS/MS results also verified the different binding sites. Therefore, a new high throughput method was established to efficiently screen specific inhibitors of G-quadruplexes, which will also make a contribution to early diagnosis of human cancers.

Novel Aspect:
This study established a new high throughput method to quickly determine binding mode of G-quadruplexes by ESI-MS.

Zhou, Shouming,
University of Toronto

Contributing Authors:

Leo W.Y. Yeung, Örebro University, Sweden; Matthew W. Forbes, University of Toronto; Scott Mabury University of Toronto; Jonathan P.D. Abbatt, University of Toronto.

Title:

Applications of Ambient Ionization Mass Spectrometry and LC-MS/MS for Studying Multi-phase Chemistry in Indoor Environments

Introduction:

Given the large surface area-to-volume ratio of indoor environments compared to outdoors, it has been recognized that multi-phase chemistry occurring between organics deposited on indoor surfaces and gas-phase oxidants e.g. ozone, plays a crucial role in indoor air pollution. In this work we will present method development with DART-MS and LC-MS/MS to investigate multi-phase indoor chemistry.

Methods:

We will focus on the development of a direct analysis in real time-mass spectrometry (DART-MS) to investigate the multi-phase reaction between squalene, a chemical that accounts for ~10% by mass of skin oil lipid material, with ozone. In addition, we will present polycyclic aromatic hydrocarbons (PAH)-epxoide production from environmental degradation of PAHs using LC-MS/MS.

Results:

The squalene/ozone experiments show that this chemistry is rapid, occurring on timescales of tens of minutes. In addition, carboxylic acids are observed from the reaction implying that highly oxygenated products are formed and always present on human skin. Moreover, high molecular weight products are also observed in condesed-phase. PAH-epoxides are believed to be mutagenic and carcinogenic. We will show experimental evidence for the formation of mono- and diol-epoxide in PAHs/ozone reaction. The diol-epoxide yield is estimated to be as high as ~25%. Moreover, the mono- and diol-epoxides are also observed when PAHs are oxidized in real indoor environments.

Novel Aspect:

Organic pollutants deposited on indoor surfaces could be highly oxidized with potentially toxic products formation.

Zhou, chuanzhen,
Analytical Instrumentation Facility at NCSU

Contributing Authors:
Fred Stevie, Analytical instrumentation facility; Steven C. Smith, Centers for Disease Control and Prevention.

Title:
Time of Flight SIMS quantification of bromine based insecticide in mosquito netting

Introduction:
Insecticides are used to combat diseases such as malaria by incorporation in or topical application on mosquito netting and clothing in order to repel or kill mosquitoes before they can attack a human host. Over one billion nets have been distributed since 2000 but determination of the insecticide available on the surface of the treated material has been difficult to achieve.

Methods:
A recent study of the chlorine based insecticide permethrin in mosquito netting has shown that Time-of-Flight Secondary Ion Mass Spectrometry (ToF-SIMS) was able to detect permethrin on the fiber surface and below the surface.

Results:
A similar approach to analysis and quantification has now been achieved for the bromine based insecticide deltamethrin. ToF-SIMS analysis of this insecticide shows dominant bromine ion in the negative ion mass spectrum. Ion implantation of 79Br was made into acetone washed NetProtect mosquito netting fibers, HDPE sheet , and silicon. The silicon sample was analyzed to confirm the bromine implantation dose. The bromine implant in the HDPE showed the expected Gaussian implant shape and indicated a detection limit of 2x1017 atoms/cm3, which corresponds to a level of detection of 0.0088 wt% of deltamethrin, which is way higher than the expected dose of 0.18 wt% deltamethrin in freshly manufactured nets. Analysis of unimplanted NetProtect fibers indicated needle shaped crystals of insecticide on the surface.

Novel Aspect:
The results obtained provide quantitative analysis of the insecticide on the netting and are significant to help determine the lifetime of nets.

Zhu, Shaolong,
MS-ESE York University - NSERC CREATE

Contributing Authors:
Marcy Taylor, Treventis Corp.; Derek J. Wilson, York University; Bin Deng York University; Mark Reed, Treventis Corp.; Braden Sweeting, Treventis Corp.

Title:
Time-Resolved Hydrogen Deuterium Exchange (HDX) Reveals the Structural Basis of Amyloidogenesis Inhibition by Alzheimer's Drug Candidates

Introduction:
Formation of neurofibrillary tangles as result of misfolded Tau protein is linked to neurodegeneration in Alzheimer's disease. Microfluidics-Enabled Time-Resolved Hydrogen/Deuterium Exchange (HDX) Mass Spectrometry was implemented to elucidate the mechanism of binding and inhibition of a few potential candidates from Treventis, a biotechnology company that focuses on anti-amyloid drug designs.

Methods:
Human Tau was purified using cation exchange chromatography. Small molecule inhibitors, TRVg and TRVb were provided by Treventis. 50μM Tau and 50μM inhibitor were incubated for one hour at room temperature. For TRESI-HDX experiments, Tau with the inhibitor was reacted with deuterium at a 1:1 ratio in a rapid mixing kinetic device. HDX is then quenched in acid, followed by proteolytic digestion.

Results:
The native Tau HDX profile showed high deuterium uptake across the full-length protein and this was expected due to its disordered nature. Conversely, the HDX profile for Tau incubated with TRVb showed decreased deuterium uptake in one of the two hexapeptide regions (H1) located in the microtubule binding domain (MBD) that is known to initiate the aggregation process. However, the deuterium uptake of a second hexapeptide region (H2) was comparable to that of the native. Interestingly, when Tau was incubated with TRVg, the vast majority of the MBD showed a drastic decrease in deuterium uptake including the H1 and H2 motifs. These findings suggest that TRVg induces a major conformational change in the MDB such that aggregation-prone hexapeptides are sequestered. In the case of TRVb, MDB is only partially sequestered, rendering the H2 motif exposed.

Novel Aspect:
Elucidating the mechanism of amyloid inhibition of the disordered Tau protein when bound to novel potential inhibitors.

21ST INTERNATIONAL MASS SPECTROMETRY CONFERENCE
Toronto, Ontario, Canada
August 20 – 26, 2016

Zimmermann, Ralf,
University Rostock and Helmholtz Zentrum München

Contributing Authors:
Mohammad Saraji, Photonion GmbH, Schwerin/Germany; Hendryk Czech, Helmholtz Zentrum München and University of Rostock/Germany; Chahan Yeretzian ZHAW-University of Applied Sciences, Wädenswil/Switzerland; Sven Ehlert, Photonion GmbH, Schwerin/Germany; Thomas Denner, Netzsch-Gerätebau GmbH, Selb/Germany; Jessalin Howell, Smuckers Inc., Orrville, OH/USA; Thorsten Streibel, Helmholtz Zentrum München and University of Rostock/Germany; Micheal Fischer, Helmholtz Zentrum München and University of Rostock/Germany; Claudi Schepler, Helmholtz Zentrum München and University of Rostock/Germany.

Title:
New mass spectrometric tools to study the flavor formation during the roast-ing process of coffee beans and nuts in real time

Introduction:
The mild roasting of plant seeds such as green coffee beans, cocoa beans or nuts generate a multitude of different flavour compounds, thus representing a very important value-generating process for the food industry. New on-line mass spectrometric approaches now can give insights in the formation and release kinetics of the valuable flavour compounds.

Methods:
Soft photo ionisation time-of-flight mass spectrometry is capable of on-line real-time monitor-ing of evolved flavour compounds from complex food roasting processes gases. This is per-formed for fundamental research in food chemistry (e.g. on-line sampling inside individual coffee beans during roasting) of for rather applied research (e.g. process monitoring at large scale industrial roasters).

Results:
Coffee roasting experiments were performed on laboratory and industrial roasters using different PI approaches such as SPI and REMPI. Advanced multivariate statistical methods including non-negative matrix factorisation (NMF) were used, allowing the determination of the roasting degree and the coffee species blend composition (Robusta/Arabica). A well suited tool for fundamental investigations of food roasting processes is thermal analysis (TA, e.g. thermogravimetry) coupled to PIMS. For increased selectivity, a newly developed, optically heated ultra-fast-cycling GC was connected to the TA-oven and PIMS-system. With this system isobaric and isomeric compounds can be separated in matter of seconds and the time-resolved evolution of e.g. pyrazines, phenols and furan or carbonyl derivatives from roasting of different nuts (e.g. peanuts) or individual coffee beans can be investigated.

Novel Aspect:
1) On-line monitoring of flavor formation in food processes by photoionization-TOFMS. 2) Direct sampling, ultra-fast cycling GC for isobaric separati

Zimmermann, Ralf,
University Rostock and Helmholtz Zentrum München

Contributing Authors:
Thorsten Streibel, Helmholtz Zentrum München and University of Rostock/Germany; Jürgen Wendt, Leco GmbH/Germany; Uwe Käfer ASG GmbH, Helmholtz Zentrum München and University of Rostock/Germany; Martin Sklorz, Helmholtz Zentrum München and University of Rostock/Germany; Mohammad Saraji, Photonion GmbH/Germany; Christopher Rüger, University of Rostock/Germany; Thomas Gröger, ASG GmbH, Helmholtz Zentrum München and University of Rostock/Germany; Maximilian Jennerwein, ASG GmbH, Helmholtz Zentrum München and University of Rostock/Germany.

Title:
Characterizations of heavy petrochemical fractions: High temperature GCxGC with HR-TOFMS detection and thermal analysis-photo ionization TOFMS

Introduction:
The analysis of heavier petrochemical fractions still is very challenging. However, it is of uttermost economic importance to better understand these fractions as they play an important role in petrochemical processes (e.g. catalyst fouling, pipeline clogging) or asphalt properties. Furthermore the accessible petrochemical resources will be of increasingly heavier nature.

Methods:
To resolve the isomeric distribution underlying the HRMS data, the hyphenation of HRMS to high resolution chromatography is required. High Temperature comprehensive two-dimensional gas chromatography (GCxGC) coupled to standard and ultra-high resolution TOFMS enables isomeric separation of heavy fractions. Thermal analysis-photoionisation TOFMS can characterize non-GC transferable compounds.

Results:
A high temperature GCxGC gas chromatographic method was developed and applied with TOFMS and high-resolution TOFMS for analysis of heavy petrochemical fraction (working up to > carbon number 65). The MS-scripting approach was applied for identifying compound class members. Compound class- and isomeric-compositions depict a structured order in the GCxGC separation plane. Complementary information from high res. MS (Kendrick mass defect) and GCxGC could be combined generate chromatographically resolved mass defect plots. To detect even heavier compounds (i.e. beyond HT-GC separation capability), a direct insertion probe was coupled to a TOFMS unit. Another possibility is the hyphenation of thermal analysis (TA) to MS. Soft ionization techniques such as photo ionization (PI) are applied to allow the detection of the molecular weight distributions as a function of the TA temperature program.

Novel Aspect:
1)Analyzing isomeric composition via high temperature-GCxGC. 2)HRTOF-Kendrick mass defect information 3)Soft photoionisation-MS for heavy fractions

Zimnicka, Magdalena,
Institute of Organic Chemistry Polish Academy ofSciences

Contributing Authors:
Barbara Repeć, Institute of Organic Chemistry, Polish Academy of Sciences, Kasprzaka 44/52, 01-224, Warsaw, Poland; Magdalena Ceborska, Institute of Physical Chemistry, Polish Academy of Sciences, Kasprzaka 44/52, 01-224, Warsaw, Poland.

Title:
Insight into the Noncovalent Associations of Folinic Acid and Pemetrexed with Native Cyclodextrins Using Mass Spectrometry-Based Techniques

Introduction:
A combined IM-MS and theoretical approach supported by the gas-phase stability experiments have been used to provide insight into the mode of noncovalent molecular associations of two bioactive compounds: folinic acid (FNA) and pemetrexed (PTX) with native cyclodextrins α-CD, β-CD, and γ-CD.

Methods:
IM-MS measurements were performed on a commercially available (Waters) quadrupole traveling-wave ion mobility time-of-flight spectrometer (Synapt G2-S HDMS). The initial structures of noncovalent complexes were obtained from docking studies performed with the open source docking program AutoDock Vina. These structures were further optimized at the semiempirical PM7 level of theory.

Results:
The IM-MS experiments show differences in collision cross section values between the CDs and their corresponding complexes with FNA and PTX. These differences in the cross section values for the FNA/α-CD and PTX/α-CD complexes are considerably greater as compared to the other complexes, suggesting a different type of complex association in the case of α-CD. The semiempirical theoretical calculations yield the optimized structures of complexes that differ in type (inclusion and exclusion complexes). Theoretical collision cross-sections are calculated for the various types of complexes and are compared with the experimentally derived values. The results give the evidence that depending on the size of the cyclodextrin, either 2-rotaxane or exclusion complex are formed. The significant differences in the gas-phase stability between complexes support the IM-MS results.

Novel Aspect:
Detailed structural analysis of noncovalent complexes of Folinic Acid and Pemetrexed with Native Cyclodextrins

Zimnicka, Magdalena,
Institute of Organic Chemistry Polish Academy ofSciences

Title:
Diversity of Peptide Diastereoisomers Examined by Ion Mobility Mass Spectrometry (IM-MS)

Introduction:
The stereochemical differentiation of diastereoisomeric molecules is still a challenge for mass spectrometry and mass spectrometry-based techniques such as ion mobility mass spectrometry (IM-MS). The influence of the complexation with organic acids on the ion mobility separation of selected peptide diastereoisomers is examined.

Methods:
The chiral recognition of selected peptide diastereoisomers by various organic acids was evaluated by the ion mobility mass spectrometry (IM-MS) measurements using quadrupole traveling-wave ion mobility time-of-flight spectrometer (Synapt G2-S HDMS, Waters) supported by theoretical calculations.

Results:
The preliminary results obtained for diastereoisomeric peptides having the following sequence: Phe-Arg-Trp, show increase of the diastereoisomeric ion separation after complexation with selected acids. The structures, selectivity and specificity of formation of noncovalent associates between selected diastereoisomers and acids and the overall effect of complexation with acids on ion-mobility separation of the analyzed diastereoisomers are discussed in terms of acid structure, peptide sequence, and measurement conditions.

Novel Aspect:
The effect of complexation with acids on ion-mobility separation of diastereoisomeric peptides.

Zweigenbaum, Jerry,

Agilent Technologies, Inc.

Contributing Authors:
Jerry Zweigenbaum, Agilent Technologies, Inc.; Tarun Anumol, Agilent Technologies, Inc.; Linda Kennedy Mansfield University of Pennsylvania; Craig Marvin, Agilent Technologies, Inc.

Title:
Identification of Emerging Contaminants from a Waste Water Influenced Water Body Using High Resolution Accurate Mass LC/MS and Statistical Analysis

Introduction:
Determination of what components are significant and focusing on identifying these compounds is critical, but difficult given the variability of environmental samples from the same sampling points at different times. This work examines LC/HRMS data of the effluent from a wastewater treatment plant over 6 months using multivariate statistics to determine compounds to be identified and monitored.

Methods:
Samples from the Mansfield University water treatment plant were filtered with 0.2 um PTFE filters. Samples were analyzed by injection of 1500 uL onto on-line trapping and LC/Q-TOF MS. The MS was in positive ion mode, the quadrupole in Rf only, and the collision cell at 0, 10 and 40 eV sequentially collecting 4.5 spectra/s. Pseudo-molecular ions and fragments for each component were detected.

Results:
Final effluent, upstream, and downstream, three adjoining creek samples, and blanks were processed as described. Initial PCA of the statistically significant compounds from influent samples formed a separate group in the first principle component (67% of total variance). The secondary effluent was slightly separated from the final effluent in the PC1 and PC2 (23%) but well separated in PC3 (6%). Non-targeted analysis using the accurate mass pseudo-molecular ions and fragments of real LC peaks co-eluting identified carbamazepine, diphenhydramine, fexofenadine and others and more recently pesticides. No compound unique to effluent were identified with the database. The PCA analysis demonstrates that the statistical procedures can distinguish compounds found at the different points of collection, but data database identification was limited.

Novel Aspect:
LC/HRMS analysis of treated wastewater for statistically significant emerging contaminants in a dynamically changing environment.